Springer Texts in Statistics

Advisors:
George Casella   Stephen Fienberg   Ingram Olkin

Springer Texts in Statistics

Alfred: Elements of Statistics for the Life and Social Sciences

Berger: An Introduction to Probability and Stochastic Processes

Bilodeau and Brenner: Theory of Multivariate Statistics

Blom: Probability and Statistics: Theory and Applications

Brockwell and Davis: Introduction to Times Series and Forecasting,
    Second Edition

Carmona: Statistical Analysis of Financial Data in S-Plus

Chow and Teicher: Probability Theory: Independence, Interchangeability,
    Martingales, Third Edition

Christensen: Advanced Linear Modeling: Multivariate, Time Series, and
    Spatial Data; Nonparametric Regression and Response Surface
    Maximization, Second Edition

Christensen: Log-Linear Models and Logistic Regression, Second Edition

Christensen: Plane Answers to Complex Questions: The Theory of Linear
    Models, Third Edition

Creighton: A First Course in Probability Models and Statistical Inference

Davis: Statistical Methods for the Analysis of Repeated Measurements

Dean and Voss: Design and Analysis of Experiments

du Toit, Steyn, and Stumpf: Graphical Exploratory Data Analysis

Durrett: Essentials of Stochastic Processes

Edwards: Introduction to Graphical Modelling, Second Edition

Finkelstein and Levin: Statistics for Lawyers

Flury: A First Course in Multivariate Statistics

Heiberger and Holland: Statistical Analysis and Data Display: An Intermediate
    Course with Examples in S-PLUS, R, and SAS

Jobson: Applied Multivariate Data Analysis, Volume I: Regression and
    Experimental Design

Jobson: Applied Multivariate Data Analysis, Volume II: Categorical and
    Multivariate Methods

Kalbfleisch: Probability and Statistical Inference, Volume I: Probability,
    Second Edition

Kalbfleisch: Probability and Statistical Inference, Volume II: Statistical Inference,
    Second Edition

Karr: Probability

Keyfitz: Applied Mathematical Demography, Second Edition

Kiefer: Introduction to Statistical Inference

Kokoska and Nevison: Statistical Tables and Formulae

Kulkarni: Modeling, Analysis, Design, and Control of Stochastic Systems

Lange: Applied Probability

Lange: Optimization

Lehmann: Elements of Large-Sample Theory

(continued after index)

Christian P. Robert   George Casella

Monte Carlo Statistical Methods

Second Edition

With 132 Illustrations

 Springer

Christian P. Robert
CEREMADE
Université Paris Dauphine
75775 Paris Cedex 16
France
xian@ceremade.dauphine.fr

George Casella
Department of Statistics
University of Florida
Gainesville, FL 32611-8545
USA
casella@stat.ufl.edu

Library of Congress Cataloging-in-Publication Data
Robert, Christian P., 1961-
    Monte Carlo statistical methods / Christian P. Robert, George Casella.—2nd ed.
        p.   cm. — (Springer texts in statistics)
    Includes bibliographical references and index.
    ISBN 0-387-21239-6 (alk. paper)
    1.   Mathematical statistics. 2. Monte Carlo method. I. Casella, George. II. Title. III.
Series.
    QA276.R575   2004
    519.5—dc22   2004049157

ISBN 0-387-21239-6   Printed on acid-free paper.

Printed in the United States of America.   (MVY)

9 8 7 6 5 4 3 2 1   SPIN 10945971

Springer-Verlag is a part of *Springer Science+Business Media*

springeronline.com

In memory of our colleague and dear friend, Costas Goutis (1962-1996)

To Benjamin, Joachim, Rachel and Sarah, our favorite random generators!

Preface to the Second Edition

'What do you plan to do next?' de Wetherset asked, picking up a piece of vellum covered with minute writing and studying it. Bartholomew rose to leave. The Chancellor clearly was not interested in how they went about getting the information, only in what they discovered.
—Susanna Gregory, *An Unholy Alliance*

It is a tribute to our profession that a textbook that was current in 1999 is starting to feel old. The work for the first edition of Monte Carlo Statistical Methods (MCSM1) was finished in late 1998, and the advances made since then, as well as our level of understanding of Monte Carlo methods, have grown a great deal. Moreover, two other things have happened. Topics that just made it into MCSM1 with the briefest treatment (for example, perfect sampling) have now attained a level of importance that necessitates a much more thorough treatment. Secondly, some other methods have not withstood the test of time or, perhaps, have not yet been fully developed, and now receive a more appropriate treatment.

When we worked on MCSM1 in the mid-to-late 90s, MCMC algorithms were already heavily used, and the flow of publications on this topic was at such a high level that the picture was not only rapidly changing, but also necessarily incomplete. Thus, the process that we followed in MCSM1 was that of someone who was thrown into the ocean and was trying to grab onto the biggest and most seemingly useful objects while trying to separate the flotsam from the jetsam. Nonetheless, we also felt that the fundamentals of many of these algorithms were clear enough to be covered at the textbook level, so we swam on.

In this revision, written five years later, we have the luxury of a more relaxed perspective on the topic, given that the flurry of activity in this area has slowed somewhat into a steady state. This is not to say that there is no longer any research in this area, but simply that the tree of publications, which was growing in every direction in 1998, can now be pruned, with emphasis being put on the major branches. For this new edition, we thus spent a good bit of time attempting to arrange the material (especially in the first ten chapters) to be presented in a coherent, flowing story, with emphasis on fundamental principles. In doing so the "Fundamental Theorem of Simulation" emerged, which we now see as a basis for many Monte Carlo algorithms (as developed in Chapters 3 and 8).

As a consequence of this coming-of-age of MCMC methods, some of the original parts of MCSM1 have therefore been expanded, and others shrunken. For example, reversible jump, sequential MC methods, two-stage Gibbs and perfect sampling now have their own chapters. Also, we now put less emphasis on some of the finer details of convergence theory, because of the simultaneous publication of Roberts and Tweedie (2004), which covers the theory of MCMC, with a comprehensive treatment of convergence results, and in general provides a deeper entry to the theory of MCMC algorithms.

We also spend less time on convergence control, because some of the methods presented in MCSM1 did not stand the test of time. The methods we preserved in Chapter 12 have been sufficiently tested to be considered reliable. Finally, we no longer have a separate chapter on missing (or latent) variables. While these models are usually a case study ideal for assessing simulation methods, they do not enjoy enough of a unified structure to be kept as a separate chapter. Instead, we dispatched most of the remaining models to different chapters of this edition.

From a (more) pedagogical point of view, we revised the book towards more accessibility and readability, thus removing the most theoretical examples and discussions. We also broke up the previously long chapters on Monte Carlo integration and Gibbs sampling into more readable chapters, with increasing coverage and difficulty. For instance, Gibbs sampling is first introduced via the slice sampler, which is simpler to describe and analyze, then the two-stage Gibbs sampler (or Data Augmentation) is presented on its own and only then do we launch into processing the general Gibbs sampler. Similarly, the experience of the previous edition led us to remove several problems or examples in every chapter, to include more detailed examples and fundamental problems, and to improve the help level on others.

Throughout the preparation of this book, and of its predecessors, we were fortunate to have colleagues who provided help. George Fishman, Anne Philippe, Judith Rousseau, as well as numerous readers, pointed out typos and mistakes in the previous version, We especially grateful to Christophe Andrieu, Roberto Casarin, Nicolas Chopin, Arnaud Doucet, Jim Hobert, Merrilee Hurn, Jean-Michel Marin, and Jesper Møller, François Perron, Arafat Tayeb, and an anonymous referee, for detailed reading of parts (or the whole) of the manuscript of this second version. Obviously, we, rather than they, should be held responsible for any imperfection remaining in the current edition! We also gained very helpful feedback from the audiences of our lectures on MCMC methods, especially during the summer schools of Luminy (France) in 2001 and 2002; Les Diablerets (Switzerland) and Venezia (Italy) in 2003; Orlando in 2002 and 2003; Atlanta in 2003; and Oulanka (Finland) in 2004.

Thanks to Elias Moreno for providing a retreat in Granada for the launching of this project in November 2002 (almost) from the top of Mulhacen; to Manuella Delbois who made the move to BibTeX possible by translating the entire reference list of MCMC1 into BibTeX format; to Jeff Gill for his patient answers to our R questions, and to Olivier Cappé for never-ending Linux and LATEX support, besides his more statistical (but equally helpful!) comments and suggestions.

Christian P. Robert
George Casella

June 15, 2004

Preface to the First Edition

He sat, continuing to look down the nave, when suddenly the solution to
the problem just seemed to present itself. It was so simple, so obvious he
just started to laugh...
—P.C. Doherty, *Satan in St Mary's*

Monte Carlo statistical methods, particularly those based on Markov chains,
have now matured to be part of the standard set of techniques used by statisticians. This book is intended to bring these techniques into the classroom,
being (we hope) a self-contained logical development of the subject, with
all concepts being explained in detail, and all theorems, etc. having detailed
proofs. There is also an abundance of examples and problems, relating the
concepts with statistical practice and enhancing primarily the application of
simulation techniques to statistical problems of various difficulties.

This is a textbook intended for a second-year graduate course. We do
not assume that the reader has any familiarity with Monte Carlo techniques
(such as random variable generation) or with any Markov chain theory. We do
assume that the reader has had a first course in statistical theory at the level
of *Statistical Inference* by Casella and Berger (1990). Unfortunately, a few
times throughout the book a somewhat more advanced notion is needed. We
have kept these incidents to a minimum and have posted warnings when they
occur. While this is a book on simulation, whose actual implementation must
be processed through a computer, no requirement is made on programming
skills or computing abilities: algorithms are presented in a program-like format
but in plain text rather than in a specific programming language. (Most of
the examples in the book were actually implemented in C, with the S-Plus
graphical interface.)

Chapters 1–3 are introductory. Chapter 1 is a review of various statistical
methodologies and of corresponding computational problems. Chapters 2 and
3 contain the basics of random variable generation and Monte Carlo integration. Chapter 4, which is certainly the most theoretical in the book, is an

introduction to Markov chains, covering enough theory to allow the reader to understand the workings and evaluate the performance of Markov chain Monte Carlo (MCMC) methods. Section 4.1 is provided for the reader who already is familiar with Markov chains, but needs a refresher, especially in the application of Markov chain theory to Monte Carlo calculations. Chapter 5 covers optimization and provides the first application of Markov chains to simulation methods. Chapters 6 and 7 cover the heart of MCMC methodology, the Metropolis–Hastings algorithm and the Gibbs sampler. Finally, Chapter 8 presents the state-of-the-art methods for monitoring convergence of the MCMC methods and Chapter 9 shows how these methods apply to some statistical settings which cannot be processed otherwise, namely the missing data models.

Each chapter concludes with a section of notes that serve to enhance the discussion in the chapters, describe alternate or more advanced methods, and point the reader to further work that has been done, as well as to current research trends in the area. The level and rigor of the notes are variable, with some of the material being advanced.

The book can be used at several levels and can be presented in several ways. For example, Chapters 1–3 and most of Chapter 5 cover standard simulation theory, and hence serve as a basic introduction to this topic. Chapters 6–9 are totally concerned with MCMC methodology. A one-semester course, assuming no familiarity with random variable generation or Markov chain theory could be based on Chapters 1–7, with some illustrations from Chapters 8 and 9. For instance, after a quick introduction with examples from Chapter 1 or Section 3.1, and a description of Accept–Reject techniques of Section 2.3, the course could cover Monte Carlo integration (Section 3.2, Section 3.3 [except Section 3.3.3], Section 3.4, Section 3.7), Markov chain theory through either Section 4.1 or Section 4.2–Section 4.8 (while adapting the depth to the mathematical level of the audience), mention stochastic optimization via Section 5.3, and describe Metropolis–Hastings and Gibbs algorithms as in Chapters 6 and 7 (except Section 6.5, Section 7.1.5, and Section 7.2.4). Depending on the time left, the course could conclude with some diagnostic methods of Chapter 8 (for instance, those implemented in CODA) and/or some models of Chapter 9 (for instance, the mixture models of Section 9.3 and Section 9.4). Alternatively, a more advanced audience could cover Chapter 4 and Chapters 6–9 in one semester and have a thorough introduction to MCMC theory and methods.

Much of the material in this book had its original incarnation as the French monograph *Méthodes de Monte Carlo par Chaînes de Markov* by Christian Robert (Paris: Economica, 1996), which has been tested for several years on graduate audiences (in France, Québec, and even Norway). Nonetheless, it constitutes a major revision of the French text, with the inclusion of problems, notes, and the updating of current techniques, to keep up with the advances that took place in the past two years (like Langevin diffusions, perfect sampling, and various types of monitoring).

Throughout the preparation of this book, and its predecessor, we were fortunate to have colleagues who provided help. Sometimes this was in the form of conversations or references (thanks to Steve Brooks and Sid Chib!), and a few people actually agreed to read through the manuscript. Our colleague and friend, Costas Goutis, provided many helpful comments and criticisms, mostly on the French version, but these are still felt in this version. We are also grateful to Brad Carlin, Dan Fink, Jim Hobert, Galin Jones and Krishanu Maulik for detailed reading of parts of the manuscript, to our historian Walter Piegorsch, and to Richard Tweedie, who taught from the manuscript and provided many helpful suggestions, and to his students, Nicole Benton, Sarah Streett, Sue Taylor, Sandy Thompson, and Alex Trindade. Richard, whose influence on the field was considerable, both from a theoretical and a methodological point of view, most sadly passed away last July. His spirit, his humor and his brightness will remain with us for ever, as a recurrent process. Christophe Andrieu, Virginie Braïdo, Jean-Jacques Colleau, Randall Douc, Arnaud Doucet, George Fishman, Jean-Louis Foulley, Arthur Gretton, Ana Justel, Anne Philippe, Sandrine Micaleff, and Judith Rousseau pointed out typos and mistakes in either the French or the English versions (or both), but should not be held responsible for those remaining! Part of Chapter 8 has a lot in common with a "reviewww" written by Christian Robert with Chantal Guihenneuc–Jouyaux and Kerrie Mengersen for the Valencia Bayesian meeting (and the Internet!). The input of the French working group "MC Cube," whose focus is on convergence diagnostics, can also be felt in several places of this book. Wally Gilks and David Spiegelhalter granted us permission to use their graph (Figure 2.5) and examples as Problems 10.29–10.36, for which we are grateful. Agostino Nobile kindly provided the data on which Figures 10.4 and 10.4 are based. Finally, Arnoldo Frigessi (from Roma) made the daring move of teaching (in English) from the French version in Oslo, Norway; not only providing us with very helpful feedback but also contributing to making Europe more of a reality!

Christian P. Robert
George Casella

January 2002

Contents

Preface to the Second Edition IX

Preface to the First Edition XIII

1 **Introduction** 1
  1.1 Statistical Models 1
  1.2 Likelihood Methods 5
  1.3 Bayesian Methods 12
  1.4 Deterministic Numerical Methods 19
    1.4.1 Optimization 19
    1.4.2 Integration 21
    1.4.3 Comparison 21
  1.5 Problems 23
  1.6 Notes 30
    1.6.1 Prior Distributions 30
    1.6.2 Bootstrap Methods 32

2 **Random Variable Generation** 35
  2.1 Introduction 35
    2.1.1 Uniform Simulation 36
    2.1.2 The Inverse Transform 38
    2.1.3 Alternatives 40
    2.1.4 Optimal Algorithms 41
  2.2 General Transformation Methods 42
  2.3 Accept–Reject Methods 47
    2.3.1 The Fundamental Theorem of Simulation 47
    2.3.2 The Accept–Reject Algorithm 51
  2.4 Envelope Accept–Reject Methods 53
    2.4.1 The Squeeze Principle 53
    2.4.2 Log-Concave Densities 56
  2.5 Problems 62

2.6   Notes 72
    2.6.1   The **Kiss** Generator 72
    2.6.2   Quasi-Monte Carlo Methods 75
    2.6.3   Mixture Representations 77

3   Monte Carlo Integration 79
3.1   Introduction 79
3.2   Classical Monte Carlo Integration 83
3.3   Importance Sampling 90
    3.3.1   Principles 90
    3.3.2   Finite Variance Estimators 94
    3.3.3   Comparing Importance Sampling with Accept–Reject .. 103
3.4   Laplace Approximations 107
3.5   Problems 110
3.6   Notes 119
    3.6.1   Large Deviations Techniques 119
    3.6.2   The Saddlepoint Approximation 120

4   Controling Monte Carlo Variance 123
4.1   Monitoring Variation with the CLT 123
    4.1.1   Univariate Monitoring 124
    4.1.2   Multivariate Monitoring 128
4.2   Rao–Blackwellization 130
4.3   Riemann Approximations 134
4.4   Acceleration Methods 140
    4.4.1   Antithetic Variables 140
    4.4.2   Control Variates 145
4.5   Problems 147
4.6   Notes 153
    4.6.1   Monitoring Importance Sampling Convergence 153
    4.6.2   Accept–Reject with Loose Bounds 154
    4.6.3   Partitioning 155

5   Monte Carlo Optimization 157
5.1   Introduction 157
5.2   Stochastic Exploration 159
    5.2.1   A Basic Solution 159
    5.2.2   Gradient Methods 162
    5.2.3   Simulated Annealing 163
    5.2.4   Prior Feedback 169
5.3   Stochastic Approximation 174
    5.3.1   Missing Data Models and Demarginalization 174
    5.3.2   The EM Algorithm 176
    5.3.3   Monte Carlo EM 183
    5.3.4   EM Standard Errors 186

5.4   Problems 188
5.5   Notes 200
    5.5.1   Variations on EM 200
    5.5.2   Neural Networks 201
    5.5.3   The Robbins–Monro procedure 201
    5.5.4   Monte Carlo Approximation 203

6   Markov Chains 205
6.1   Essentials for MCMC 206
6.2   Basic Notions 208
6.3   Irreducibility, Atoms, and Small Sets 213
    6.3.1   Irreducibility 213
    6.3.2   Atoms and Small Sets 214
    6.3.3   Cycles and Aperiodicity 217
6.4   Transience and Recurrence 218
    6.4.1   Classification of Irreducible Chains 218
    6.4.2   Criteria for Recurrence 221
    6.4.3   Harris Recurrence 221
6.5   Invariant Measures 223
    6.5.1   Stationary Chains 223
    6.5.2   Kac's Theorem 224
    6.5.3   Reversibility and the Detailed Balance Condition . 229
6.6   Ergodicity and Convergence 231
    6.6.1   Ergodicity 231
    6.6.2   Geometric Convergence 236
    6.6.3   Uniform Ergodicity 237
6.7   Limit Theorems 238
    6.7.1   Ergodic Theorems 240
    6.7.2   Central Limit Theorems 242
6.8   Problems 247
6.9   Notes 258
    6.9.1   Drift Conditions 258
    6.9.2   Eaton's Admissibility Condition 262
    6.9.3   Alternative Convergence Conditions 263
    6.9.4   Mixing Conditions and Central Limit Theorems 263
    6.9.5   Covariance in Markov Chains 265

7   The Metropolis–Hastings Algorithm 267
7.1   The MCMC Principle 267
7.2   Monte Carlo Methods Based on Markov Chains 269
7.3   The Metropolis–Hastings algorithm 270
    7.3.1   Definition 270
    7.3.2   Convergence Properties 272
7.4   The Independent Metropolis–Hastings Algorithm 276
    7.4.1   Fixed Proposals 276

7.4.2   A Metropolis–Hastings Version of ARS 285
7.5   Random Walks 287
7.6   Optimization and Control 292
     7.6.1   Optimizing the Acceptance Rate 292
     7.6.2   Conditioning and Accelerations 295
     7.6.3   Adaptive Schemes 299
7.7   Problems 302
7.8   Notes 313
     7.8.1   Background of the Metropolis Algorithm 313
     7.8.2   Geometric Convergence of Metropolis–Hastings
            Algorithms 315
     7.8.3   A Reinterpretation of Simulated Annealing 315
     7.8.4   Reference Acceptance Rates 316
     7.8.5   Langevin Algorithms 318

8   **The Slice Sampler** 321
8.1   Another Look at the Fundamental Theorem 321
8.2   The General Slice Sampler 326
8.3   Convergence Properties of the Slice Sampler 329
8.4   Problems 333
8.5   Notes 335
     8.5.1   Dealing with Difficult Slices 335

9   **The Two-Stage Gibbs Sampler** 337
9.1   A General Class of Two-Stage Algorithms 337
     9.1.1   From Slice Sampling to Gibbs Sampling 337
     9.1.2   Definition 339
     9.1.3   Back to the Slice Sampler 343
     9.1.4   The Hammersley–Clifford Theorem 343
9.2   Fundamental Properties 344
     9.2.1   Probabilistic Structures 344
     9.2.2   Reversible and Interleaving Chains 349
     9.2.3   The Duality Principle 351
9.3   Monotone Covariance and Rao–Blackwellization 354
9.4   The EM–Gibbs Connection 357
9.5   Transition 360
9.6   Problems 360
9.7   Notes 366
     9.7.1   Inference for Mixtures 366
     9.7.2   ARCH Models 368

10   **The Multi-Stage Gibbs Sampler** 371
10.1   Basic Derivations 371
     10.1.1   Definition 371
     10.1.2   Completion 373

10.1.3 The General Hammersley–Clifford Theorem 376
10.2 Theoretical Justifications 378
    10.2.1 Markov Properties of the Gibbs Sampler 378
    10.2.2 Gibbs Sampling as Metropolis–Hastings 381
    10.2.3 Hierarchical Structures 383
10.3 Hybrid Gibbs Samplers................................... 387
    10.3.1 Comparison with Metropolis–Hastings Algorithms 387
    10.3.2 Mixtures and Cycles 388
    10.3.3 Metropolizing the Gibbs Sampler 392
10.4 Statistical Considerations............................... 396
    10.4.1 Reparameterization 396
    10.4.2 Rao-Blackwellization 402
    10.4.3 Improper Priors 403
10.5 Problems 407
10.6 Notes 419
    10.6.1 A Bit of Background............................... 419
    10.6.2 The BUGS Software 420
    10.6.3 Nonparametric Mixtures 420
    10.6.4 Graphical Models 422

11 Variable Dimension Models and Reversible Jump
   Algorithms 425
11.1 Variable Dimension Models 425
    11.1.1 Bayesian Model Choice............................ 426
    11.1.2 Difficulties in Model Choice........................ 427
11.2 Reversible Jump Algorithms 429
    11.2.1 Green's Algorithm................................. 429
    11.2.2 A Fixed Dimension Reassessment 432
    11.2.3 The Practice of Reversible Jump MCMC 433
11.3 Alternatives to Reversible Jump MCMC................... 444
    11.3.1 Saturation....................................... 444
    11.3.2 Continuous-Time Jump Processes 446
11.4 Problems 449
11.5 Notes 458
    11.5.1 Occam's Razor 458

12 Diagnosing Convergence 459
12.1 Stopping the Chain 459
    12.1.1 Convergence Criteria 461
    12.1.2 Multiple Chains 464
    12.1.3 Monitoring Reconsidered 465
12.2 Monitoring Convergence to the Stationary Distribution....... 465
    12.2.1 A First Illustration 465
    12.2.2 Nonparametric Tests of Stationarity 466
    12.2.3 Renewal Methods 470

12.2.4 Missing Mass 474
12.2.5 Distance Evaluations 478
12.3 Monitoring Convergence of Averages 480
12.3.1 A First Illustration 480
12.3.2 Multiple Estimates 483
12.3.3 Renewal Theory.................................. 490
12.3.4 Within and Between Variances 497
12.3.5 Effective Sample Size 499
12.4 Simultaneous Monitoring 500
12.4.1 Binary Control................................... 500
12.4.2 Valid Discretization............................... 503
12.5 Problems... 504
12.6 Notes 508
12.6.1 Spectral Analysis................................. 508
12.6.2 The CODA Software 509

13 Perfect Sampling 511
13.1 Introduction 511
13.2 Coupling from the Past 513
13.2.1 Random Mappings and Coupling 513
13.2.2 Propp and Wilson's Algorithm 516
13.2.3 Monotonicity and Envelopes 518
13.2.4 Continuous States Spaces.......................... 523
13.2.5 Perfect Slice Sampling 526
13.2.6 Perfect Sampling via Automatic Coupling 530
13.3 Forward Coupling 532
13.4 Perfect Sampling in Practice 535
13.5 Problems... 536
13.6 Notes 539
13.6.1 History 539
13.6.2 Perfect Sampling and Tempering 540

14 Iterated and Sequential Importance Sampling 545
14.1 Introduction 545
14.2 Generalized Importance Sampling 546
14.3 Particle Systems 547
14.3.1 Sequential Monte Carlo 547
14.3.2 Hidden Markov Models 549
14.3.3 Weight Degeneracy 551
14.3.4 Particle Filters 552
14.3.5 Sampling Strategies............................... 554
14.3.6 Fighting the Degeneracy 556
14.3.7 Convergence of Particle Systems.................... 558
14.4 Population Monte Carlo 559
14.4.1 Sample Simulation................................ 560

14.4.2 General Iterative Importance Sampling..............560
14.4.3 Population Monte Carlo..........................562
14.4.4 An Illustration for the Mixture Model...............563
14.4.5 Adaptativity in Sequential Algorithms 565
14.5 Problems...570
14.6 Notes..577
14.6.1 A Brief History of Particle Systems 577
14.6.2 Dynamic Importance Sampling.....................577
14.6.3 Hidden Markov Models 579

A   Probability Distributions 581

B   Notation..585
B.1 Mathematical 585
B.2 Probability 586
B.3 Distributions...586
B.4 Markov Chains.......................................587
B.5 Statistics 588
B.6 Algorithms 588

References...591

Index of Names...623

Index of Subjects 631

List of Tables

1.1   Challenger data 15
1.2   Some conjugate families 31

2.1   Horse Kick Data 61

3.1   Evaluation of some normal quantiles 85
3.2   Surgery vs. radiation data 86
3.3   Cutoff points for contingency table data 88
3.4   Comparison of instrumental distributions 102
3.5   Comparison between importance sampling estimators 108
3.6   Laplace approximation of a Gamma integral 110
3.7   Saddlepoint approximation of a noncentral χ^2 integral 122

5.1   Stochastic gradient runs 163
5.2   Simulated annealing runs 169
5.3   Sequence of Bayes estimators for the Gamma distribution 172
5.4   Average grades of first-year students 173
5.5   Maximum likelihood estimates of mean grades 173
5.6   Cellular phone data.. 180
5.7   Tree swallow movements 197
5.8   Selected batting average data 198

7.1   Monte Carlo saddlepoint approximation of a noncentral chi squared integral 284
7.2   Approximation of normal moments by a random walk Metropolis–Hastings algorithm 288
7.3   Approximation of the inverse Gaussian distribution by the Metropolis–Hastings algorithm 294
7.4   Improvement in quadratic risk from Rao–Blackwellization (1) .. 297
7.5   Improvement in quadratic risk from Rao–Blackwellization (2) .. 299
7.6   Braking distances 307

7.7   Performance of the Metropolis–Hastings algorithm [A.29] 317

9.1   Frequencies of passage for 360 consecutive observations. 346
9.2   Observations of $\mathcal{N}_2(0, \Sigma)$ with missing data 364
9.3   Estimation result for the factor ARCH model 369

10.1   Failures of pumps in a nuclear plant 386
10.2   Interquantile ranges for the Gibbs sampling and the
       modification of Liu 396
10.3   Occurrences of clinical mastitis in dairy herds 409
10.4   Calcium concentration in turnip greens..................... 414

11.1   Galaxy data 451
11.2   Yearly number of mining accidents in England 455

12.1   Estimation of the asymptotic variance for renewal control (1) .. 495
12.2   Estimation of the asymptotic variance for renewal control (2) .. 496
12.3   Gibbs approximations of expectations and variances estimated
       by renewal.. 497
12.4   Evolution of initializing and convergence times 502

List of Figures

1.1	Cauchy likelihood	11
1.2	Challenger failure probabilities	16
1.3	Newton–Raphson Algorithm	20
2.1	Plot of chaotic pairs	38
2.2	Probability of acceptance for Jöhnk's algorithm	45
2.3	Generation of Beta random variables	49
2.4	Envelope for generic Accept–Reject	50
2.5	Lower and upper envelopes of a log-concave density	57
2.6	Posterior distributions of capture log-odds ratios	59
2.7	ARS approximation to capture log-odds posterior distribution	60
2.8	ARS approximation to the intercept distribution	61
2.9	Representation of $y = 69069x$ mod 1 by uniform sampling	74
2.10	Plots of pairs from the Kiss generator	76
3.1	One-dimensional Monte Carlo integration	84
3.2	Contingency table test	88
3.3	Empirical cdf's of log-likelihoods	89
3.4	Approximated error risks	93
3.5	Range of three estimators of $\mathbb{E}_f[|X/(1-X)|^{1/2}]$	97
3.6	Range of an importance sampling estimator of $\mathbb{E}_f[|X/(1-X)|^{1/2}]$	98
3.7	Convergence of four estimators of $\mathbb{E}_f[X^5 \mathbb{I}_{X \geq 2.1}]$	99
3.8	Convergence of four estimators of $\mathbb{E}_f[h_3(X)]$	100
3.9	Convergence of four estimators of $\mathbb{E}_f[h_5(X)]$	102
3.10	Convergence of estimators of $\mathbb{E}[X/(1+X)]$	106
3.11	Convergence of the estimators of $\mathbb{E}[X/(1+X)]$	107
4.1	Range and confidence bands for a simple example	125

4.2	Range and confidence bands for the Cauchy-Normal problem (1)	127
4.3	Range and confidence bands for the Cauchy-Normal problem (2)	128
4.4	Confidence bands for running means	130
4.5	Convergence of estimators of $\mathbb{E}[\exp(-X^2)]$	132
4.6	Comparisons of an Accept–Reject and importance sampling estimator	134
4.7	Convergence of the estimators of $\mathbb{E}[X \log(X)]$	138
4.8	Convergence of estimators of $\mathbb{E}_\nu[(1 + e^X)\mathbb{I}_{X \le 0}]$	139
4.9	Convergence of estimators of $\mathbb{E}_\nu[(1 + e^X)\mathbb{I}_{X \le 0}]$	139
4.10	Approximate risks of truncated James–Stein estimators	142
4.11	Comparison of an antithetic and standard iid estimate	144
4.12	Graphs of the variance coefficients	155
5.1	Simple Monte Carlo maximization	160
5.2	Representation of the function of Example 5.3	161
5.3	Stochastic gradient paths	164
5.4	One-dimensional simulated annealing maximization	166
5.5	Simulated annealing sequence	170
5.6	Censored data likelihood	175
5.7	EM sequence for cellular phone data	181
5.8	Multiple mode EM	182
5.9	EM estimate and standard deviation	187
6.1	Trajectories of AR(1)	228
6.2	Means of AR(1)	245
6.3	Convergence of AR(1)	245
7.1	Convergence of Accept–Reject and Metropolis–Hastings (1)	280
7.2	Convergence of Accept–Reject and Metropolis–Hastings (2)	281
7.3	Estimation of logistic parameters	283
7.4	Histograms of three samples from [A.29]	289
7.5	Confidence envelopes for the random walk Metropolis–Hastings algorithm	290
7.6	Invalid adaptive MCMC (1)	301
7.7	Invalid adaptive MCMC (2)	301
7.8	Nonparametric invalid adaptive MCMC	302
7.9	Convergence of Langevin and iid simulations	320
8.1	Simple slice sampler	324
8.2	Ten steps of the slice sampler	325
8.3	Three slice samples for the truncated normal	325
8.4	A 3D slice sampler	327
8.5	A poor slice sampler	333

9.1   Gibbs output for mixture posterior 343
9.2   Evolution of the estimator δ_{rb} of [A.35] 347
9.3   Gibbs output for cellular phone data 359

10.1   Nonconnected support 380
10.2   Gibbs output for mastitis data 384
10.3   Successive moves of a Gibbs chain 389
10.4   Gibbs chain for the probit model 392
10.5   Hybrid chain for the probit model 392
10.6   Comparison of Gibbs sampling and Liu's modification 395
10.7   Evolution of the estimation of a mixture distribution 400
10.8   Convergence of parameter estimators for a mixture
        distribution 401
10.9   Iterations of a divergent Gibbs sampling algorithm 405
10.10   Evolution of $(\beta^{(t)})$ for a random effects model 407

11.1   Speed of Corona Borealis galaxies 426
11.2   Linear regression with reversible jumps 435
11.3   Reversible jump MCMC output for a mixture 440
11.4   Conditional reversible jump MCMC output for a mixture ... 441
11.5   Fit by an averaged density 442
11.6   Reversible jump output for the AR(p) model 443

12.1   Probit log-posterior 460
12.2   Probit β sequence (1) 467
12.3   Probit β sequence (2) 467
12.4   Probit β sequence (3) 467
12.5   Probit β sequence (4) 468
12.6   Probit posterior sequence 468
12.7   Probit ratio β/σ sequence 469
12.8   Plot of successive Kolmogorov–Smirnov statistics 470
12.9   Plot of renewal probabilities for the pump failure data 474
12.10   Bimodal density 475
12.11   Missing mass evaluation 476
12.12   Control curves for a mixture model 477
12.13   Convergence of empirical averages for Example 12.10 481
12.14   Evolution of the D_t^i criterion for Example 12.10 482
12.15   Evolution of CUSUMs for Example 12.1 483
12.16   Gibbs sampler stuck at a wrong mode 483
12.17   CUSUMs for the mixture posterior 484
12.18   Approximation of the density (12.17) 487
12.19   Convergence of four estimators for the density (12.17) 489
12.20   Convergence of four estimators of $\mathbb{E}[(X^{(t)})^{0.8}]$ 490
12.21   Evolutions of R_T and W_T for (12.17) 498
12.22   Convergence of the mean $\hat{\varrho}_t$ for [A.30] 503

12.23 Discretization of a continuous Markov chain 504

13.1   All possible transitions for the Beta-Binomial(2,2,4) example 515
13.2   Perfect sampling for a mixture posterior distribution 520
13.3   Nine iterations of a coupled Gibbs sampler 523
13.4   Coupling from the past on the distribution (13.7) 530
13.5   Successful CFTP for an exponential mixture 532
13.6   Forward simulation with dominating process 542

14.1   Simulated target tracking output 548
14.2   Hidden Markov model 549
14.3   Simulated sample of a stochastic volatility process 550
14.4   Importance sampling target tracking reconstruction 553
14.5   Particle filter target tracking reconstruction 555
14.6   Particle filter target tracking range 556
14.7   Mixture log-posterior distribution and PMC sample 565
14.8   Adaptivity of mixture PMC algorithm 566
14.9   MCMC sample for a stochastic volatility model 567
14.10 MCMC estimate for a stochastic volatility model 568
14.11 PMC sample for a stochastic volatility model 568
14.12 PMC estimate for a stochastic volatility model 569

1

Introduction

There must be, he thought, some key, some crack in this mystery he could
use to achieve an answer.
—P.C. Doherty, *Crown in Darkness*

Until the advent of powerful and accessible computing methods, the experimenter was often confronted with a difficult choice. Either describe an accurate model of a phenomenon, which would usually preclude the computation
of explicit answers, or choose a standard model which would allow this computation, but may not be a close representation of a realistic model. This
dilemma is present in many branches of statistical applications, for example,
in electrical engineering, aeronautics, biology, networks, and astronomy. To
use realistic models, the researchers in these disciplines have often developed
original approaches for model fitting that are customized for their own problems. (This is particularly true of physicists, the originators of Markov chain
Monte Carlo methods.) Traditional methods of analysis, such as the usual
numerical analysis techniques, are not well adapted for such settings.

In this introductory chapter, we examine some of the statistical models
and procedures that contributed to the development of simulation-based inference. The first section of this chapter looks at some statistical models,
and the remaining sections examine different statistical methods. Throughout
these sections, we describe many of the computational difficulties associated
with the methods. The final section of the chapter contains a discussion of
deterministic numerical analysis techniques.

1.1 Statistical Models

In a purely statistical setup, computational difficulties occur at both the level
of *probabilistic modeling* of the inferred phenomenon and at the level of *statistical inference* on this model (estimation, prediction, tests, variable selection,

etc.). In the first case, a detailed representation of the causes of the phenomenon, such as accounting for potential explanatory variables linked to the phenomenon, can lead to a probabilistic structure that is too complex to allow for a parametric representation of the model. Moreover, there may be no provision for getting closed-form estimates of quantities of interest. One setup with this type of complexity is *expert systems* (in medicine, physics, finance, etc.) or, more generally, *graph structures*. See Pearl (1988), Robert (1991),[1] Spiegelhalter et al. (1993), Lauritzen (1996) for examples of complex expert systems.[2]

Another situation where model complexity prohibits an explicit representation appears in econometrics (and in other areas) for structures of *latent* (or *missing*) variable models. Given a "simple" model, aggregation or removal of some components of this model may sometimes produce such involved structures that simulation is really the only way to draw an inference. In these situations, an often used method for estimation is the EM algorithm (Dempster et al. 1977), which is described in Chapter 3. In the following example, we illustrate a common missing data situation. The concept and use of missing data techniques and in particular of the two following examples will reoccur throughout the book.

Example 1.1. Censored data models. *Censored data models* are missing data models where densities are not sampled directly. To obtain estimates and make inferences in such models usually requires involved computations and precludes analytical answers.

In a typical simple statistical model, we would observe random variables[3] (rv's) $Y_1, \ldots, Y_n$, drawn independently from a population with distribution $f(y|\theta)$. The distribution of the sample would then be given by the product $\prod_{i=1}^{n} f(y_i|\theta)$. Inference about θ would be based on this distribution.

In many studies, particularly in medical statistics, we have to deal with *censored* random variables; that is, rather than observing Y_1, we may observe $\min\{Y_1, \overline{u}\}$, where $\overline{u}$ is a constant. For example, if Y_1 is the survival time of a patient receiving a particular treatment and $\overline{u}$ is the length of the study being done (say $\overline{u} = 5$ years), then if the patient survives longer than 5 years, we do not observe the survival time, but rather the censored value $\overline{u}$. This modification leads to a more difficult evaluation of the sample density.

Barring cases where the censoring phenomenon can be ignored, several types of censoring can be categorized by their relation with an underlying (unobserved) model, $Y_i \sim f(y_i|\theta)$:

---

[1] Claudine, not Christian!

[2] Section 10.6.4 also gives a brief introduction to graphical models in connection with Gibbs sampling.

[3] Throughout the book we will use uppercase Roman letters for random variables and lowercase Roman letters for their realized values. Thus, we would observe $X = x$, where the random variable X produces the observation x. (For esthetic purposes this distinction is sometimes lost with Greek letter random variables.)

(i) Given random variables Y_i, which are, for instance, times of observation or concentrations, the actual observations are $Y_i^* = \min\{Y_i, \overline{u}\}$, where $\overline{u}$ is the maximal observation duration, the smallest measurable concentration rate, or some other truncation point.

(ii) The original variables Y_i are kept in the sample with probability $\rho(y_i)$ and the number of censored variables is either known or unknown.

(iii) The variables Y_i are associated with auxiliary variables $X_i \sim g$ such that $y_i^* = h(y_i, x_i)$ is the observation. Typically, $h(y_i, x_i) = \min(y_i, x_i)$. The fact that truncation occurred, namely the variable $\mathbb{I}_{Y_i > X_i}$, may be either known or unknown.

As a particular example, if

$$X \sim \mathcal{N}(\theta, \sigma^2) \quad \text{and} \quad Y \sim \mathcal{N}(\mu, \tau^2),$$

the variable $Z = X \wedge Y = \min(X, Y)$ is distributed as

$$\left[1 - \varPhi\left(\frac{z - \theta}{\sigma}\right)\right] \times \tau^{-1}\varphi\left(\frac{z - \mu}{\tau}\right)$$

(1.1)
$$+ \left[1 - \varPhi\left(\frac{z - \mu}{\tau}\right)\right]\sigma^{-1}\varphi\left(\frac{z - \theta}{\sigma}\right),$$

where φ is the density of the normal $\mathcal{N}(0, 1)$ distribution and $\varPhi$ is the corresponding cdf, which is not easy to compute.

Similarly, if X has a Weibull distribution with two parameters, $\mathcal{W}e(\alpha, \beta)$, and density

$$f(x) = \alpha\beta x^{\alpha-1}\exp(-\beta x^\alpha)$$

on $\mathbb{R}^+$, the observation of the censored variable $Z = X \wedge \omega$, where ω is constant, has the density

(1.2)
$$f(z) = \alpha\beta z^{\alpha-1}e^{-\beta z^\alpha}\,\mathbb{I}_{z \leq \omega} + \left(\int_\omega^\infty \alpha\beta x^{\alpha-1}e^{-\beta x^\alpha}\,dx\right)\delta_\omega(z),$$

where $\delta_a(\cdot)$ is the Dirac mass at a. In this case, the weight of the Dirac mass, $P(X \geq \omega)$, can be explicitly computed (Problem 1.4).

The distributions (1.1) and (1.2) appear naturally in quality control applications. There, testing of a product may be of a duration ω, where the quantity of interest is time to failure. If the product is still functioning at the end of the experiment, the observation on failure time is censored. Similarly, in a longitudinal study of a disease, some patients may leave the study either due to other causes of death or by simply being lost to follow-up.   ∥

In some cases, the additive form of a density, while formally explicit, prohibits the computation of the density of a sample $(X_1, \ldots, X_n)$ for n large. (Here, "explicit" has the restrictive meaning that "it can be computed in a reasonable time.")

Example 1.2. Mixture models. Models of *mixtures of distributions* are based on the assumption that the observations X_i are generated from one of k elementary distributions f_j with probability p_j, the overall density being

(1.3) $$X \sim p_1 f_1(x) + \ldots + p_k f_k(x) \, .$$

If we observe a sample of independent random variables $(X_1, \ldots, X_n)$, the sample density is

$$\prod_{i=1}^{n} \{ p_1 f_1(x_i) + \cdots + p_k f_k(x_i) \} \, .$$

When $f_j(x) = f(x|\theta_j)$, the evaluation of the likelihood at a given value of $(\theta_1, \ldots, \theta_n, p_1, \ldots, p_n)$ only requires on the order[4] of $\mathcal{O}(kn)$ computations, but we will see later that likelihood and Bayesian inferences both require the expansion of the above product, which involves $\mathcal{O}(k^n)$ computations, and is thus prohibitive to compute in large samples.[5]

While the computation of standard moments like the mean or the variance of these distributions is feasible in many setups (and thus so is the derivation of moment estimators, see Problem 1.6), the representation of the likelihood function (and therefore the analytical computation of maximum likelihood or Bayes estimates) is generally impossible for mixtures.   ‖

Finally, we look at a particularly important example in the processing of temporal (or time series) data where the likelihood cannot be written explicitly.

Example 1.3. Moving average model. An MA(q) model describes variables (X_t) that can be modeled as $(t = 0, \ldots, n)$

(1.4) $$X_t = \varepsilon_t + \sum_{j=1}^{q} \beta_j \varepsilon_{t-j} \, ,$$

where for $i = -q, -(q-1), \ldots$, the ε_i's are iid random variables $\varepsilon_i \sim \mathcal{N}(0, \sigma^2)$ and for $j = 1, \ldots, q$, the β_j's are unknown parameters. If the sample consists of the observation $(X_0, \ldots, X_n)$, where $n > q$, the sample density is (Problem 1.5)

---

[4] Recall that the notation $\mathcal{O}(n)$ denotes a function that satisfies
$$0 < \limsup_{n \to \infty} \mathcal{O}(n)/n < \infty.$$
[5] This class of models will be used extensively over the book. Although the example is self-contained, detailed comments about such models are provided in Note 9.7.1 and Titterington et al. (1985).

$$\int_{\mathbb{R}^q} \sigma^{-(n+q)} \prod_{i=1}^q \varphi\left(\frac{\varepsilon_{-i}}{\sigma}\right) \varphi\left(\frac{x_0 - \sum_{i=1}^q \beta_i \varepsilon_{-i}}{\sigma}\right)$$

(1.5)
$$\times \varphi\left(\frac{x_1 - \beta_1 \hat{\varepsilon}_o - \sum_{i=2}^q \beta_i \varepsilon_{1-i}}{\sigma}\right) \cdots$$

$$\times \varphi\left(\frac{x_n - \sum_{i=1}^q \beta_i \hat{\varepsilon}_{n-i}}{\sigma}\right) d\varepsilon_{-1} \cdots d\varepsilon_{-q} ,$$

with

$$\hat{\varepsilon}_0 = x_0 - \sum_{i=1}^q \beta_i \varepsilon_{-i} ,$$

$$\hat{\varepsilon}_1 = x_1 - \sum_{i=2}^q \beta_i \varepsilon_{1-i} - \beta_1 \hat{\varepsilon}_0 ,$$

$$\vdots$$

$$\hat{\varepsilon}_n = x_n - \sum_{i=1}^q \beta_i \hat{\varepsilon}_{n-i} .$$

The iterative definition of the $\hat{\varepsilon}_i$'s is a real obstacle to an explicit integration in (1.5) and hinders statistical inference in these models. Note that for $i = -q, -(q-1), \ldots, -1$ the perturbations ε_{-i} can be interpreted as missing data (see Section 5.3.1). ‖

Before the introduction of simulation-based inference, computational difficulties encountered in the modeling of a problem often forced the use of "standard" models and "standard" distributions. One course would be to use models based on *exponential families*, defined below by (1.9) (see Brown 1986, Robert 2001, Lehmann and Casella 1998), which enjoy numerous regularity properties (see Note 1.6.1). Another course was to abandon parametric representations for nonparametric approaches which are, by definition, robust against modeling errors.

We also note that the reduction to simple, perhaps non-realistic, distributions (necessitated by computational limitations) does not necessarily eliminate the issue of nonexplicit expressions, whatever the statistical technique. Our major focus is the application of simulation-based techniques to provide solutions and inference for a more realistic set of models and, hence, circumvent the problems associated with the need for explicit or computationally simple answers.

1.2 Likelihood Methods

The statistical techniques that we will be most concerned with are *maximum likelihood* and *Bayesian* methods, and the inferences that can be drawn from their use. In their implementation, these approaches are customarily associated with specific mathematical computations, the former with maximization problems—and thus to an *implicit* definition of estimators as solutions of maximization problems— the later with integration problems—and thus to a (formally) *explicit* representation of estimators as an integral. (See Berger 1985, Casella and Berger 2001, Robert 2001, Lehmann and Casella 1998, for an introduction to these techniques.)

The method of maximum likelihood estimation is quite a popular technique for deriving estimators. Starting from an iid sample $\mathbf{x} = (x_1, \ldots, x_n)$ from a population with density $f(x|\theta_1, \ldots, \theta_k)$, the *likelihood function* is

$$
\begin{aligned}
L(\theta|\mathbf{x}) &= L(\theta_1, \ldots, \theta_k|x_1, \ldots, x_n) \\
&= \prod_{i=1}^{n} f(x_i|\theta_1, \ldots, \theta_k).
\end{aligned}
$$
(1.6)

More generally, when the X_i's are not iid, the likelihood is defined as the joint density $f(x_1, \ldots, x_n|\theta)$ taken as a function of θ. The value of θ, say $\hat{\theta}$, which is the parameter value at which $L(\theta|\mathbf{x})$ attains its maximum as a function of θ, with $\mathbf{x}$ held fixed, is known as a *maximum likelihood estimator (MLE)*. Notice that, by its construction, the range of the MLE coincides with the range of the parameter. The justifications of the maximum likelihood method are primarily asymptotic, in the sense that the MLE is converging almost surely to the true value of the parameter, under fairly general conditions (see Lehmann and Casella 1998) although it can also be interpreted as being at the fringe of the Bayesian paradigm (see, e.g., Berger and Wolpert 1988).

Example 1.4. Gamma MLE. A maximum likelihood estimator is typically calculated by maximizing the logarithm of the likelihood function (1.6). Suppose $X_1, \ldots, X_n$ are iid observations from the gamma density

$$
f(x|\alpha, \beta) = \frac{1}{\Gamma(\alpha)\beta^\alpha} x^{\alpha-1} e^{-x/\beta},
$$

where we assume that α is known. The *log likelihood* is

$$
\begin{aligned}
\log L(\alpha, \beta|x_1, \ldots, x_n) &= \log \prod_{i=1}^{n} f(x_i|\alpha, \beta) \\
&= \log \prod_{i=1}^{n} \frac{1}{\Gamma(\alpha)\beta^\alpha} x_i^{\alpha-1} e^{-x_i/\beta} \\
&= -n \log \Gamma(\alpha) - n\alpha \log \beta + (\alpha - 1) \sum_{i=1}^{n} \log x_i - \sum_{i=1}^{n} x_i/\beta,
\end{aligned}
$$

where we use the fact that the log of the product is the sum of the logs, and have done some simplifying algebra. Solving $\frac{\partial}{\partial \beta} \log L(\alpha, \beta|x_1, \ldots, x_n) = 0$ is straightforward and yields the MLE of β, $\hat{\beta} = \sum_{i=1}^{n} x_i/(n\alpha)$.

Suppose now that α was also unknown, and we additionally had to solve

$$\frac{\partial}{\partial \alpha} \log L(\alpha, \beta | x_1, \ldots, x_n) = 0.$$

This results in a particularly nasty equation, involving some difficult computations (such as the derivative of the gamma function, the *digamma function*). An explicit solution is no longer possible. ‖

Calculation of maximum likelihood estimators can sometimes be implemented through the minimization of a sum of squared residuals, which is the basis of the *method of least squares*.

Example 1.5. Least squares estimators. Estimation by *least squares* can be traced back to Legendre (1805) and Gauss (1810) (see Stigler 1986). In the particular case of linear regression, we observe (x_i, y_i), $i = 1, \ldots, n$, where

$$(1.7) \qquad Y_i = a + bx_i + \varepsilon_i, \quad i = 1, \ldots, n,$$

and the variables ε_i's represent errors. The parameter (a, b) is estimated by minimizing the distance

$$(1.8) \qquad \sum_{i=1}^{n} (y_i - ax_i - b)^2$$

in (a, b), yielding the least squares estimates. If we add more structure to the error term, in particular that $\varepsilon_i \sim \mathcal{N}(0, \sigma^2)$, independent (equivalently, $Y_i | x_i \sim \mathcal{N}(ax_i + b, \sigma^2)$), the log-likelihood function for (a, b) is proportional to

$$\log(\sigma^{-n}) - \sum_{i=1}^{n} (y_i - ax_i - b)^2 / 2\sigma^2,$$

and it follows that the maximum likelihood estimates of a and b are identical to the least squares estimates.

If, in (1.8), we assume $\mathbb{E}(\varepsilon_i) = 0$, or, equivalently, that the linear relationship $\mathbb{E}[Y | x] = ax + b$ holds, minimization of (1.8) is equivalent, from a computational point of view, to imposing a normality assumption on Y conditionally on x and applying maximum likelihood. In this latter case, the additional estimator of σ^2 is consistent if the normal approximation is asymptotically valid. (See Gouriéroux and Monfort 1996, for the related theory of *pseudo-likelihood*.) ‖

Although somewhat obvious, this formal equivalence between the optimization of a function depending on the observations and the maximization of a likelihood associated with the observations has a nontrivial outcome and applies in many other cases. For example, in the case where the parameters are constrained, Robertson et al. (1988) consider a $p \times q$ table of random variables Y_{ij} with means θ_{ij}, where the means are increasing in i and j. Estimation of the θ_{ij}'s by minimizing the sum of the $(y_{ij} - \theta_{ij})^2$'s is possible through the (numerical) algorithm called *"pool-adjacent-violators,"* developed by Robertson et al. (1988) to solve this specific problem. (See Problems 1.18 and 1.19.) An alternative is to use an algorithm based on simulation and a representation using a normal likelihood (see Section 5.2.4).

In the context of exponential families, that is, distributions with density

$$(1.9) \qquad f(x) = h(x)\, e^{\theta \cdot x - \psi(\theta)}, \qquad \theta, x \in \mathbb{R}^k,$$

the approach by maximum likelihood is (formally) straightforward. The maximum likelihood estimator of θ is the solution of

$$(1.10) \qquad x = \nabla\psi\{\hat{\theta}(x)\},$$

which also is the equation yielding a method of moments estimator, since $\mathbb{E}_\theta[X] = \nabla\psi(\theta)$. The function ψ is the log-Laplace transform, or *cumulant generating function* of h; that is, $\psi(t) = \log\mathbb{E}[\exp\{th(X)\}]$, where we recognize the right side as the log *moment generating function* of h.

Example 1.6. Normal MLE. In the setup of the normal $\mathcal{N}(\mu, \sigma^2)$ distribution, the density can be written as in (1.9), since

$$f(y|\mu, \sigma) \propto \sigma^{-1} e^{-(\mu-y)^2/2\sigma^2}$$
$$= \sigma^{-1} e^{(\mu/\sigma^2)\, y - (1/2\sigma^2)\, y^2 - \mu^2/2\sigma^2}.$$

The so-called *natural parameters* are then $\theta_1 = \mu/\sigma^2$ and $\theta_2 = -1/2\sigma^2$, with $\psi(\theta) = -\theta_1^2/4\theta_2 + \log(-\theta_2/2)/2$. While there is no MLE for a single observation from $\mathcal{N}(\mu, \sigma^2)$, equation (1.10) leads to

$$(1.11) \qquad \begin{aligned} -n\frac{\theta_1}{2\theta_2} &= \sum_{i_1}^n y_i = n\bar{y}, \\ n\frac{\theta_1^2}{4\theta_2^2} - \frac{n}{2\theta_2} &= \sum_{i_1}^n y_i^2 = n(s^2 + \bar{y}^2), \end{aligned}$$

in the case of n iid observations $y_1, \ldots, y_n$, that is, to the regular MLE, $(\hat{\mu}, \hat{\sigma}^2) = (\bar{y}, s^2)$, where $s^2 = \sum(y_i - \bar{y})^2/n$. ‖

Unfortunately, there are many settings where ψ cannot be computed explicitly. Even if it could be done, it may still be the case that the solution of (1.10) is not explicit, or there are constraints on θ such that the maximum of (1.9) is not a solution of (1.10). This last situation occurs in the estimation of the table of θ_{ij}'s in the discussion above.

Example 1.7. Beta MLE. The Beta $\mathcal{Be}(\alpha, \beta)$ distribution is a particular case of exponential family since its density,

$$f(y|\alpha, \beta) = \frac{\Gamma(\alpha + \beta)}{\Gamma(\alpha)\, \Gamma(\beta)}\, y^{\alpha-1}(1-y)^{\beta-1}\,, \qquad 0 \le y \le 1,$$

can be written as (1.9), with $\theta = (\alpha, \beta)$ and $x = (\log y, \log(1-y))$. Equation (1.10) becomes

(1.12)
$$\begin{aligned}\log y &= \Psi(\alpha) - \Psi(\alpha + \beta)\,, \\ \log(1-y) &= \Psi(\beta) - \Psi(\alpha + \beta)\,,\end{aligned}$$

where $\Psi(z) = d\log\Gamma(z)/dz$ denotes the *digamma function* (see Abramowitz and Stegun 1964). There is no explicit solution to (1.12). As in Example 1.6, although it may seem absurd to estimate both parameters of the $\mathcal{B}e(\alpha, \beta)$ distribution from a single observation, Y, the formal computing problem at the core of this example remains valid for a sample $Y_1, \ldots, Y_n$ since (1.12) is then replaced by

$$\frac{1}{n}\sum_i \log y_i = \Psi(\alpha) - \Psi(\alpha + \beta)\,,$$

$$\frac{1}{n}\sum_i \log(1 - y_i) = \Psi(\beta) - \Psi(\alpha + \beta)\,. \qquad \|$$

When the parameter of interest λ is not a one-to-one function of θ, that is, when there are *nuisance* parameters, the maximum likelihood estimator of λ is still well defined. If the parameter vector is of the form $\theta = (\lambda, \psi)$, where ψ is a nuisance parameter, a typical approach is to calculate the full MLE $\hat{\theta} = (\hat{\lambda}, \hat{\psi})$ and use the resulting $\hat{\lambda}$ to estimate λ. In principle, this does not require more complex calculations, although the distribution of the maximum likelihood estimator of λ, $\hat{\lambda}$, may be quite involved. Many other options exist, such as *conditional, marginal, or profile* likelihood (see, for example, Barndorff-Nielsen and Cox 1994).

Example 1.8. Noncentrality parameter. If $X \sim \mathcal{N}_p(\theta, I_p)$ and if $\lambda = \|\theta\|^2$ is the parameter of interest, the nuisance parameters are the angles $\mathbf{\Psi}$ in the polar representation of θ and the maximum likelihood estimator of λ is $\hat{\lambda}(x) = \|x\|^2$, which has a constant bias equal to p. Surprisingly, an observation $Y = \|X\|^2$ which has a noncentral chi squared distribution, $\chi_p^2(\lambda)$ (see Appendix A), leads to a maximum likelihood estimator of λ which differs[6] from Y, since it is the solution of the implicit equation

(1.13) $$\sqrt{\lambda}\, I_{(p-1)/2}\left(\sqrt{\lambda y}\right) = \sqrt{y}\, I_{p/2}\left(\sqrt{\lambda y}\right)\,, \qquad y > p\,,$$

where I_ν is the *modified Bessel function*

---

[6] This phenomenon is not paradoxical, as $Y = \|X\|^2$ is not a sufficient statistic in the original problem.

$$I_\nu(t) = \frac{(z/2)^\nu}{\sqrt{\pi}\Gamma(\nu + \frac{1}{2})} \int_0^\pi e^{t\cos(\theta)} \sin^{2\nu}(\theta)d\theta$$

$$= \left(\frac{t}{2}\right)^\nu \sum_{k=0}^\infty \frac{(z/2)^{2k}}{k!\Gamma(\nu + k + 1)}.$$

So even in the favorable context of exponential families, we are not necessarily free from computational problems, since the resolution of (1.13) requires us first to evaluate the special functions $I_{p/2}$ and $I_{(p-1)/2}$. Note also that the maximum likelihood estimator is not a solution of (1.13) when $y < p$ (see Problem 1.20).   ‖

When we leave the exponential family setup, we face increasingly challenging difficulties in using maximum likelihood techniques. One reason for this is the lack of a *sufficient statistic of fixed dimension* outside exponential families, barring the exception of a few families such as uniform or Pareto distributions whose support depends on θ (Robert 2001, Section 3.2). This result, known as the *Pitman–Koopman Lemma* (see Lehmann and Casella 1998, Theorem 1.6.18), implies that, outside exponential families, the complexity of the likelihood increases quite rapidly with the number of observations, n and, thus, that its maximization is delicate, even in the simplest cases.

Example 1.9. Student's t distribution. Modeling random perturbations using normally distributed errors is often (correctly) criticized as being too restrictive and a reasonable alternative is the Student's t distribution, denoted by $\mathcal{T}(p, \theta, \sigma)$, which is often more "robust" against possible modeling errors (and others). The density of $\mathcal{T}(p, \theta, \sigma)$ is proportional to

(1.14)
$$\sigma^{-1}\left(1 + \frac{(x - \theta)^2}{p\sigma^2}\right)^{-(p+1)/2}.$$

Typically, p is known and the parameters θ and σ are unknown. Based on an iid sample $(X_1, \ldots, X_n)$ from (1.14), the likelihood is proportional to a power of the product

$$\sigma^{n\frac{p+1}{2}} \prod_{i=1}^n \left(1 + \frac{(x_i - \theta)^2}{p\sigma^2}\right).$$

When σ is known, for some configurations of the sample, this polynomial of degree $2n$ may have n local minima, each of which needs to be calculated to determine the global maximum, the maximum likelihood estimator (see also Problem 1.14). Figure 1.1 illustrates this multiplicity of modes of the likelihood from a Cauchy distribution $\mathcal{C}(\theta, 1)$ ($p = 1$) when $n = 3$ and $X_1 = 0$, $X_2 = 5$, and $X_3 = 9$.   ‖

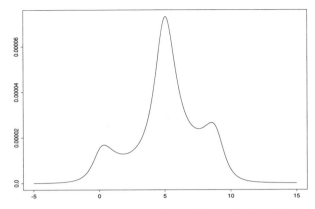

Fig. 1.1. Likelihood of the sample $(0, 5, 9)$ from the distribution $\mathcal{C}(\theta, 1)$.

Example 1.10. (Continuation of Example 1.2) In the special case of a mixture of two normal distributions,

$$p\mathcal{N}(\mu, \tau^2) + (1-p)\mathcal{N}(\theta, \sigma^2) \,,$$

an iid sample $(X_1, \ldots, X_n)$ results in a likelihood function proportional to

$$(1.15) \qquad \prod_{i=1}^{n} \left[p\tau^{-1}\varphi\left(\frac{x_i - \mu}{\tau}\right) + (1-p)\,\sigma^{-1}\,\varphi\left(\frac{x_i - \theta}{\sigma}\right) \right]$$

containing 2^n terms if expanded. Standard maximization techniques often fail to find the global maximum because of multimodality of the likelihood function, and specific algorithms must be devised (to obtain the global maximum with high probability).

The problem is actually another order of magnitude more difficult, since the likelihood is unbounded here. The expansion of the product (1.15) contains the terms

$$p^n \tau^{-n} \prod_{i=1}^{n} \varphi\left(\frac{x_i - \mu}{\tau}\right)$$

$$+ p^{n-1}(1-p)\,\tau^{-n+1}\sigma^{-1}\,\varphi\left(\frac{x_1 - \theta}{\sigma}\right) \prod_{i=2}^{n} \varphi\left(\frac{x_i - \mu}{\tau}\right) + \ldots \,.$$

This expression is unbounded in σ (let σ go to 0 when $\theta = x_1$). However, this difficulty with the likelihood function does not preclude us from using the maximum likelihood approach in this context, since Redner and Walker (1984) have shown that there exist solutions to the *likelihood equations*, that is, local maxima of (1.15), which have acceptable properties. (Similar problems occur in the context of linear regression with "errors in variables." See Casella and Berger 2001, Chapter 12, for an introduction.)   ‖

In addition to the difficulties associated with optimization problems, likelihood-related approaches may also face settings where the likelihood function is only expressible as an integral (for example, the censored data models of Example 1.1). Similar computational problems arise in the determination of the *power* of a testing procedure in the Neyman–Pearson approach (see Lehmann 1986, Casella and Berger 2001, Robert 2001)

For example, inference based on a likelihood ratio statistic requires computation of quantities such as

$$P_\theta(L(\theta|X)/L(\theta_0|X) \le k) \,,$$

with fixed θ_0 and k, where $L(\theta|x)$ represents the likelihood based on observing $X = x$. Outside of the more standard (simple) settings, this probability cannot be explicitly computed because dealing with the distribution of test statistics under the alternative hypothesis may be quite difficult. A particularly delicate case is the *Behrens–Fisher problem*, where the above probability is difficult to evaluate even under the null hypothesis (see Lehmann 1986, Lee 2004). (Note that likelihood ratio tests cannot be rigorously classified as a likelihood-related approach, since they violate the *Likelihood Principle*, see Berger and Wolpert 1988, but the latter does not provide a testing theory per se.)

1.3 Bayesian Methods

Whereas the difficulties related to maximum likelihood methods are mainly *optimization* problems (multiple modes, solution of likelihood equations, links between likelihood equations and global modes, etc.), the Bayesian approach more often results in *integration* problems. In the Bayesian paradigm, information brought by the data x, a realization of $X \sim f(x|\theta)$, is combined with prior information that is specified in a *prior distribution* with density $\pi(\theta)$ and summarized in a probability distribution, $\pi(\theta|x)$, called the *posterior distribution*. This is derived from the *joint* distribution $f(x|\theta)\pi(\theta)$, according to *Bayes formula*

$$(1.16) \qquad \pi(\theta|x) = \frac{f(x|\theta)\pi(\theta)}{\int f(x|\theta)\pi(\theta)d\theta},$$

where $m(x) = \int f(x|\theta)\pi(\theta)d\theta$ is the *marginal density* of X (see Berger 1985, Bernardo and Smith 1994, Robert 2001, for more details, in particular about the philosophical foundations of this inferential approach).

For the estimation of a particular parameter $h(\theta)$, the decision-theoretic approach to statistical inference (see, e.g. Berger 1985) requires the specification of a loss function $L(\delta, \theta)$, which represents the loss incurred by estimating $h(\theta)$ with δ. The Bayesian version of this approach leads to the minimization of the *Bayes risk*,

$$\int \int L(\delta, \theta) f(x|\theta) \pi(\theta) dx d\theta \ ,$$

that is, the loss integrated against both X and θ. A straightforward inversion of the order of integration (Fubini's theorem) leads to choosing the estimator δ that minimizes (for each x) the *posterior loss*,

(1.17) $$\mathbb{E}[L(\delta, \theta)|x] = \int L(\delta, \theta) \ \pi(\theta|x) \ d\theta \ .$$

In the particular case of the quadratic loss

$$L(\delta, \theta) = \|h(\theta) - \delta\|^2 \ ,$$

the Bayes estimator of $h(\theta)$ is $\delta^\pi(x) = \mathbb{E}^\pi[h(\theta)|x]$. (See Problem 1.22.)

Some of the difficulties related to the computation of $\delta^\pi(x)$ are, first, that $\pi(\theta|x)$ is not generally available in closed form and, second, that in many cases the integration of $h(\theta)$ according to $\pi(\theta|x)$ cannot be done analytically. Loss functions $L(\delta, \theta)$ other than the quadratic loss function are usually even more difficult to deal with.

The computational drawback of the Bayesian approach has been so great that, for a long time, the favored types of priors in a Bayesian modeling were those allowing explicit computations, namely *conjugate priors*. These are prior distributions for which the corresponding posterior distributions are themselves members of the original prior family, the Bayesian updating being accomplished through updating of parameters. (See Note 1.6.1 and Robert 2001, Chapter 3, for a discussion of the link between conjugate priors and exponential families.)

Example 1.11. Binomial Bayes estimator. For an observation X from the binomial distribution $\mathcal{B}(n, p)$, a family of conjugate priors is the family of Beta distributions $\mathcal{B}e(a, b)$. To find the Bayes estimator of p under squared error loss, we can find the minimizer of the Bayes risk, that is,

$$\min_\delta \int_0^1 \sum_{x=1}^n [p - \delta(x)]^2 \binom{n}{x} \frac{\Gamma(a+b)}{\Gamma(a)\Gamma(b)} p^{x+a-1}(1-p)^{n-x+b-1} dp.$$

Equivalently, we can work with the posterior expected loss (1.17) and find the estimator that yields

$$\min_\delta \frac{\Gamma(a+b+n)}{\Gamma(a+x)\Gamma(n-x+b)} \int_0^1 [p - \delta(x)]^2 p^{x+a-1}(1-p)^{n-x+b-1} dp \ ,$$

where we note that the posterior distribution of p (given x) is $\mathcal{B}e(x+a, n-x+b)$. The solution is easily obtained through differentiation, and the Bayes estimator δ^π is the posterior mean

$$\delta^\pi(x) = \frac{\Gamma(a+b+n)}{\Gamma(a+x)\Gamma(n-x+b)} \int_0^1 p \, p^{x+a-1}(1-p)^{n-x+b-1} dp = \frac{x+a}{a+b+n}.$$

The use of squared error loss results in the Bayes estimator being the mean of the posterior distribution, which usually simplifies calculations. If, instead, we had specified a *absolute error loss* $|p - \delta(x)|$ or had used a nonconjugate prior, the calculations could have become somewhat more involved (see Problem 1.22). ||

The use of conjugate priors for computational reasons implies a restriction on the modeling of the available prior information and may be detrimental to the usefulness of the Bayesian approach as a method of statistical inference. This is because it perpetuates an impression both of subjective "manipulation" of the background (prior information) and of formal expansions unrelated to reality. The considerable advances of Bayesian decision theory have often highlighted the negative features of modeling using only conjugate priors. For example, Bayes estimators are the optimal estimators for the three main classes of optimality (admissibility, minimaxity, invariance), but those based on conjugate priors only partially enjoy these properties (see Berger 1985, Section 4.7, or Robert 2001, Chapter 8) .

Example 1.12. (Continuation of Example 1.8). For the estimation of $\lambda = \|\theta\|^2$, a *reference prior*[7] on θ is $\pi(\theta) = \|\theta\|^{-(p-1)}$ (see Berger et al. 1998), with corresponding posterior distribution

$$(1.18) \qquad \pi(\theta|x) \propto \frac{e^{-\|x-\theta\|^2/2}}{\|\theta\|^{p-1}} .$$

The normalizing constant corresponding to $\pi(\theta|x)$ is not easily obtainable and the Bayes estimator of λ, the posterior mean

$$(1.19) \qquad \frac{\int_{\mathbb{R}^p} \|\theta\|^{2-p} e^{-\|x-\theta\|^2/2} \, d\theta}{\int_{\mathbb{R}^p} \|\theta\|^{1-p} e^{-\|x-\theta\|^2/2} \, d\theta} ,$$

cannot be explicitly computed. (See Problem 1.20.) ||

The computation of the normalizing constant of $\pi(\theta|x)$ is not just a formality. Although the derivation of a posterior distribution is generally done through proportionality relations, that is, using *Bayes Theorem* in the form

$$\pi(\theta|x) \propto \pi(\theta) \, f(x|\theta) ,$$

it is sometimes necessary to know the posterior distribution or, equivalently, the marginal distribution, exactly. For example, this is the case in the Bayesian

---

[7] A reference prior is a prior distribution which is derived from maximizing a distance measure between the prior and the posterior distributions. When there is no nuisance parameter in the model, the standard reference prior is Jeffreys (1961) prior (see Bernardo and Smith 1994, Robert 2001, and Note 1.6.1.).

comparison of (statistical) models. If $\mathcal{M}_1, \mathcal{M}_2, \ldots, \mathcal{M}_k$ are possible models for the observation X, with densities $f_j(\cdot|\theta_j)$, if the associated parameters $\theta_1, \theta_2, \ldots, \theta_k$ are *a priori* distributed from $\pi_1, \pi_2, \ldots, \pi_k$, and if these models have the prior weights $p_1, p_2, \ldots, p_k$, the posterior probability that X originates from model $\mathcal{M}_j$ is (Problem 1.21)

$$(1.20) \qquad \frac{p_j \int f_j(x|\theta_j)\,\pi_j(\theta_j)\,d\theta_j}{\sum_{i=1}^{k} p_i \int f_i(x|\theta_i)\,\pi_i(\theta_i)\,d\theta_i}.$$

In particular, the comparison of two models $\mathcal{M}_1$ and $\mathcal{M}_2$ is often implemented through the *Bayes factor*

$$B^\pi(x) = \frac{\int f_1(x|\theta_1)\,\pi_1(\theta_1)\,d\theta_1}{\int f_2(x|\theta_2)\,\pi_2(\theta_2)\,d\theta_2},$$

for which the proportionality constant is quite important (see Kass and Raftery 1995 and Goutis and Robert 1998 for different perspectives on Bayes factors). Unsurprisingly, there has been a lot of research in the computation of these normalizing constants (Gelman and Meng 1998, Kong et al. 2003).

Example 1.13. Logistic regression. A useful regression model for binary $(0-1)$ responses is the *logit model*, where the distribution of Y conditional on the explanatory (or dependent) variables $x \in \mathbb{R}^p$ is modeled by the relation

$$(1.21) \qquad P(Y=1) = p = \frac{\exp(\alpha + x\beta)}{1 + \exp(\alpha + x\beta)}.$$

Equivalently, the *logit* transform of p, $\mathrm{logit}(p) = \log[p/(1-p)]$, satisfies the linear relationship $\mathrm{logit}(p) = \alpha + x\beta$.

In 1986, the space shuttle Challenger exploded during take off, killing the seven astronauts aboard. The explosion was the result of an *O-ring* failure, a splitting of a ring of rubber that seals the parts of the ship together. The accident was believed to be caused by the unusually cold weather (31° F or 0° C) at the time of launch, as there is reason to believe that the O-ring failure probabilities increase as temperature decreases (Dalal et al. 1989).

Flight	14	9	23	10	1	5	13	15	4	3	8	17	2	11	6	7	16	21	19	22	12	20	18
Failure	1	1	1	1	0	0	0	0	0	0	0	0	1	1	0	0	0	1	0	0	0	0	0
Temp.	53	57	58	63	66	67	67	67	68	69	70	70	70	72	73	75	75	76	76	78	79	81	

Table 1.1. Temperature at flight time (degrees F) and failure of O-rings *(1 stands for failure, 0 for success)*.

Data on previous space shuttle launches, and O-ring failures, is given in Table 1.1. It is reasonable to fit a logistic regression, as in (1.21) with $p =$

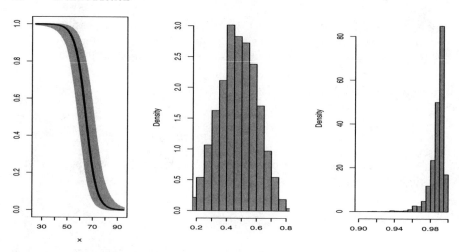

Fig. 1.2. The figure shows the result of 10,000 Monte Carlo simulations of the model (1.21). The left panel shows the average logistic function and variation, the middle panel shows predictions of failure probabilities at 65° Fahrenheit, and the right panel shows predictions of failure probabilities at 45° Fahrenheit.

probability of an O-ring failure and x = temperature. The results are shown in Figure 1.2 for an exponential prior on $\log \alpha$ and a flat prior on β.

The left panel in Figure 1.2 shows the logistic regression line, and the grey curves represent the results of a Monte Carlo simulation explained in Example 7.3 from the posterior distribution of the model showing the variability in the data. It is clear that the ends of the function have little variability, while there is some in the middle. However, the next two panels are most important, as they show the variability in failure probability predictions at two different temperatures. The middle panel, which gives the failure probability at 65° Fahrenheit, shows that, at this temperature, a failure is just about as likely as a success. However, at 45° Fahrenheit, the failures are strongly skewed toward 1. Given this trend, imagine what the failure probabilities look like at 31° Fahrenheit, the temperature at Challenger launch time: At that temperature, failure was almost a certainty.

The logistic regression Monte Carlo analysis of this data is quite straightforward, and gives easy-to-understand answers to the relevant questions. Non-Monte Carlo alternatives would typically be based on likelihood theory and asymptotics, and would be more difficult to implement and interpret.   ∥

The computational problems encountered in the Bayesian approach are not limited to the computation of integrals or normalizing constants. For instance, the determination of confidence regions (also called *credible regions*) with *highest posterior density*,

$$C^{\pi}(x) = \{\theta; \pi(\theta|x) \geq k\} \, ,$$

requires the solution of the equation $\pi(\theta|x) = k$ for the value of k that satisfies

$$P(\theta \in C^\pi(x)|x) = P(\pi(\theta|x) \geq k|x) = \gamma \,,$$

where γ is a predetermined confidence level.

Example 1.14. Bayes credible regions. For iid observations $X_1, \ldots, X_n$ from a normal distribution $\mathcal{N}(\theta, \sigma^2)$, and a prior distribution $\theta \sim \mathcal{N}(0, \tau^2)$, the posterior density of θ is normal with mean and variance

$$\delta^\pi = \frac{n\tau^2}{n\tau^2 + \sigma^2}\bar{x} \qquad \text{and} \qquad \frac{n\tau^2\sigma^2}{n\tau^2 + \sigma^2} \,,$$

respectively. If we assume that σ^2 and τ^2 are known, the highest posterior density region is

$$\left\{ \theta : \sqrt{\frac{n\tau^2 + \sigma^2}{2\pi n\tau^2}} \exp\left[-\frac{n\tau^2 + \sigma^2}{2n\tau^2}(\theta - \delta^\pi)^2 \right] \geq k \right\}.$$

Since the posterior distribution is symmetric and unimodal, this set is equivalent to (Problem 1.24)

$$\{\theta;\ \delta^\pi - k' \leq \theta \leq \delta^\pi + k'\} \,,$$

for some constant k' that is chosen to yield a specified posterior probability. Since the posterior distribution of θ is normal, this can be done by hand using a normal probability table.

For the situation of Example 1.11, the posterior distribution of p, $\pi(p|x, a, b)$, was found to be $\mathcal{B}e(x+a, n-x+b)$, which is not necessarily symmetric. To find the 90% highest posterior density region for p, we must find limits $l(x)$ and $u(x)$ that satisfy

$$\int_{l(x)}^{u(x)} \pi(p|x, a, b)dp = .9 \qquad \text{and} \qquad \pi(l(x)|x, a, b) = \pi(u(x)|x, a, b).$$

This cannot be solved analytically.   ‖

Computation of a confidence region can be quite delicate when $\pi(\theta|x)$ is not explicit. In particular, when the confidence region involves only one component of a vector parameter, calculation of $\pi(\theta|x)$ requires the integration of the joint distribution over all the other parameters. Note that knowledge of the normalizing factor is of minor importance in this setup. (See Robert 2001, Chapter 6, for other examples.)

Example 1.15. Cauchy confidence regions. Consider $X_1, \ldots, X_n$, an iid sample from the Cauchy distribution $\mathcal{C}(\theta, \sigma)$, with associated prior distribution $\pi(\theta, \sigma) = \sigma^{-1}$. The confidence region on θ is then based on

$$\pi(\theta|x_1, \ldots, x_n) \propto \int_0^\infty \sigma^{-n-1} \prod_{i=1}^n \left[1 + \left(\frac{x_i - \theta}{\sigma}\right)^2\right]^{-1} d\sigma\,,$$

an integral which cannot be evaluated explicitly. Similar computational problems occur with likelihood estimation in this model. One method for obtaining a likelihood confidence interval for θ is to use the *profile likelihood*

$$\ell^P(\theta|x_1, \ldots, x_n) = \max_\sigma \ell(\theta, \sigma|x_1, \ldots, x_n)$$

and consider the region $\{\theta : \ell^P(\theta|x_1, \ldots, x_n) \geq k\}$. Explicit computation is also difficult here.   ‖

Example 1.16. Linear calibration. In a standard regression model, $Y = \alpha + \beta x + \varepsilon$, there is interest in estimating or predicting features of Y from knowledge of x. In *linear calibration models* (see Osborne 1991, for an introduction and review of these models), the interest is in determining values of x from observed responses y. For example, in a chemical experiment, one may want to relate the precise but expensive measure y to the less precise but inexpensive measure x. A simplified version of this problem can be put into the framework of observing the independent random variables

$$Y \sim \mathcal{N}_p(\beta, \sigma^2 I_p), \; Z \sim \mathcal{N}_p(x_0\beta, \sigma^2 I_p), \; S \sim \sigma^2 \chi_q^2\,,$$

with $x_0 \in \mathbb{R}$, $\beta \in \mathbb{R}^p$. The parameter of interest is now x_0 and this problem is equivalent to Fieller (1954) problem (see, e.g Lehmann and Casella 1998).

A reference prior on (x_0, β, σ) is given in Kubokawa and Robert (1994), and yields the joint posterior distribution

(1.22)
$$\pi(x_0, \beta, \sigma^2|y, z, s) \propto \sigma^{-(3p+q)-\frac{1}{2}} \exp\{-(s + \|y - \beta\|^2 \\ + \|z - x_0\beta\|^2)/2\sigma^2\} (1 + x_0^2)^{-1/2}\,.$$

This can be analytically integrated to obtain the marginal posterior distribution of X_0 to be
(1.23)
$$\pi(x_0|y, z, s) \propto \frac{(1 + x_0^2)^{(p+q-1)/2}}{\left\{\left(x_0 - \frac{y^t z}{s + \|y\|^2}\right)^2 + \frac{\|z\|^2 + s}{\|y\|^2 + s} - \frac{(y^t z)^2}{(s + \|y\|^2)^2}\right\}^{(2p+q)/2}}\,.$$

However, the computation of the posterior mean, the Bayes estimate of x_0, is not feasible analytically; neither is the determination of the confidence region $\{\pi(x_0|\mathcal{D}) \geq k\}$. Nonetheless, it is desirable to determine this confidence region since alternative solutions, for example the *Fieller–Creasy* interval, suffer from defects such as having infinite length with positive probability (see Gleser and Hwang 1987, Casella and Berger 2001, Ghosh et al. 1995, Philippe and Robert 1998b).   ‖

1.4 Deterministic Numerical Methods

The previous examples illustrated the need for techniques, in both the construction of complex models and estimation of parameters, that go beyond the standard analytical approaches. However, before starting to describe simulation methods, which is the purpose of this book, we should recall that there exists a well-developed alternative approach for integration and optimization, based on *numerical* methods. We refer the reader to classical textbooks on numerical analysis (see, for instance, Fletcher 1980 or Evans 1993 for a description of these methods, which are generally efficient and can deal with most of the above examples (see also Lange 1999 or Gentle 2002 for presentations in statistical settings).

1.4.1 Optimization

We briefly recall here the more standard approaches to optimization and integration problems, both for comparison purposes and for future use. When the goal is to solve an equation of the form $f(x) = 0$, a common approach is to use a *Newton–Raphson* algorithm, which produces a sequence x_n such that

$$(1.24) \qquad x_{n+1} = x_n - \left(\left. \frac{\partial f}{\partial x} \right|_{x=x_n} \right)^{-1} f(x_n)$$

until it stabilizes around a solution of $f(x) = 0$. (Note that $\frac{\partial f}{\partial x}$ is a matrix in multidimensional settings.) Optimization problems associated with smooth functions F are then based on this technique, using the equation $\nabla F(x) = 0$, where ∇F denotes the *gradient* of F, that is, the vector of derivatives of F. (When the optimization involves a constraint $G(x) = 0$, F is replaced by a Lagrangian form $F(x) - \lambda G(x)$, where λ is used to satisfy the constraint.) The corresponding techniques are then the *gradient methods*, where the sequence x_n is such that

$$(1.25) \qquad x_{n+1} = x_n - \left(\nabla \nabla^t F \right)^{-1} (x_n) \nabla F(x_n) ,$$

where $\nabla \nabla^t F$ denotes the matrix of second derivatives of F.

Example 1.17. A simple Newton–Raphson Algorithm. As a simple illustration, we show how the Newton–Raphson algorithm can be used to find the square root of a number. If we are interested in the square root of b, this is equivalent to solving the equation

$$f(x) = x^2 - b = 0.$$

Applying (1.24) results in the iterations

$$x^{(j+1)} = x^{(j)} - \frac{f(x^{(j)})}{f'(x^{(j)})} = x^{(j)} - \frac{x^{(j)2} - b}{2x^{(j)}} = \frac{1}{2}\left(x^{(j)} + \frac{b}{x^{(j)}}\right).$$

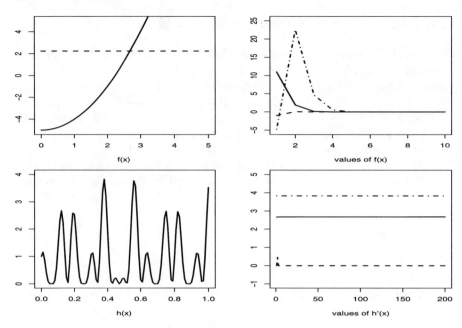

Fig. 1.3. Calculation of the root of $f(x) = 0$ and $h'(x) = 0$ for the functions defined in Example 1.17. The top left panel is $f(x)$, and the top right panel shows that from different starting points the Newton–Raphson algorithm converges rapidly to the square root. The bottom left panel is $h(x)$, and the bottom right panel shows that the Newton–Raphson algorithm cannot find the maximum of $h(x)$, but rather converges to whatever mode is closest to the starting point.

Figure 1.3 shows that the algorithm converges rapidly to the correct answer from different starting points (for $b = 2$, three runs are shown, starting at $x = .5, 2, 4$).

However, if we consider the function

$$(1.26) \qquad h(x) = [\cos(50x) + \sin(20x)]^2,$$

and try to derive its maximum, we run into problems. The "greediness" of the Newton–Raphson algorithm, that is, the fact that it always goes toward the nearest mode, does not allow it to escape from local modes. The bottom two panels in Figure 1.3 show the function, and the convergence of the algorithm from three different starting points, $x = .25, .379, .75$ (the maximum occurs at $x = .379$). From the figure we see that wherever we start the algorithm, it goes to the closest mode, which is often not the maximum. In Chapter 5, we will compare this solution to those of Examples 5.2 and 5.5 where the (global) maximum is obtained.   ‖

Numerous variants of Newton–Raphson-type techniques can be found in the literature, among which one can mention the *steepest descent* method, where each iteration results in a unidimensional optimizing problem for $F(x_n + td_n)$ $(t \in \mathbb{R})$, d_n being an acceptable direction, namely such that

$$\left. \frac{d^2 F}{dt^2}(x_n + td_n) \right|_{t=0}$$

is of the proper sign. The direction d_n is often chosen as ∇F or as the smoothed version of (1.25),

$$\left[\nabla \nabla^t F(x_n) + \lambda I \right]^{-1} \nabla F(x_n),$$

in the *Levenberg–Marquardt version*. Other versions are available that do not require differentiation of the function F.

1.4.2 Integration

Turning to integration, the numerical computation of an integral

$$\mathfrak{I} = \int_a^b h(x)dx$$

can be done by simple *Riemann integration* (see Section 4.3), or by improved techniques such as the *trapezoidal rule*

$$\hat{\mathfrak{I}} = \frac{1}{2} \sum_{i=1}^{n-1} (x_{i+1} - x_i)(h(x_i) + h(x_{i+1})) ,$$

where the x_i's constitute an ordered partition of $[a, b]$, or yet *Simpson's rule*, whose formula is

$$\tilde{\mathfrak{I}} = \frac{\delta}{3} \left\{ f(a) + 4 \sum_{i=1}^{n} h(x_{2i-1}) + 2 \sum_{i=1}^{n} h(x_{2i}) + f(b) \right\}$$

in the case of equally spaced samples with $(x_{i+1} - x_i) = \delta$. Other approaches involve orthogonal polynomials (Gram–Charlier, Legendre, etc.), as illustrated by Naylor and Smith (1982) for statistical problems, or splines (see Wahba 1981, for a statistical connection). See also Note 2.6.2 for the *quasi–Monte Carlo* methods that, despite their name, pertain more to numerical integration than to simulation, since they are totally deterministic. However, due to the *curse of dimensionality*, these methods may not work well in high dimensions, as stressed by Thisted (1988).

1.4.3 Comparison

Comparison between the approaches, simulation versus numerical analysis, is delicate because both approaches can provide well-suited tools for many problems (possibly needing a preliminary study) and the distinction between these two groups of techniques can be vague. So, rather than addressing the issue of a general comparison, we will focus on the requirements of each approach and the objective conditions for their implementation in a statistical setup.

By nature, standard numerical methods do not take into account the probabilistic aspects of the problem; that is, the fact that many of the functions involved in the computations are related to probability densities. Therefore, a numerical integration method may consider regions of a space which have zero (or low) probability under the distribution of the model, a phenomenon which usually does not appear in a simulation experiment.[8] Similarly, the occurrence of local modes of a likelihood will often cause more problems for a deterministic gradient method than for a simulation method that explores high-density regions. (But multimodality must first be identified for these efficient methods to apply, as in Oh 1989 or Oh and Berger 1993.)

On the other hand, simulation methods very rarely take into account the specific analytical form of the functions involved in the integration or optimization, while numerical methods often use higher derivatives to provide bounds on the error of approximation. For instance, because of the randomness induced by the simulation, a gradient method yields a much faster determination of the mode of a unimodal density. For small dimensions, integration by Riemann sums or by quadrature converges faster than the mean of a simulated sample. Moreover, existing scientific software (for instance, Gauss, Maple, Mathematica, Matlab, R) and scientific libraries like IMSL often provide highly efficient numerical procedures, whereas simulation is, at best, implemented through pseudo-random generators for the more common distributions. However, software like BUGS (see Note 10.6.2) are progressively bridging the gap.

Therefore, it is often reasonable to use a numerical approach when dealing with regular functions in small dimensions and in a given single problem. On the other hand, when the statistician needs to study the details of a likelihood surface or posterior distribution, or needs to simultaneously estimate several features of these functions, or when the distributions are highly multimodal (see Examples 1.2 and 1.8), it is preferable to use a simulation-based approach. Such an approach captures (if only approximately through the generated sample) the different characteristics of the density and thus allows, at little cost, extensions of the inferential scope to, perhaps, another test or estimator.

However, given the dependence on specific problem characteristics, it is fruitless to advocate the superiority of one method over the other, say of the simulation-based approach over numerical methods. Rather, it seems more reasonable to justify the use of simulation-based methods by the statistician in terms of *expertise*. The intuition acquired by a statistician in his or her

---

[8] Simulation methods using a distribution other than the distribution of interest, such as importance sampling (Section 3.3) or Metropolis–Hastings algorithms (Chapter 7), may suffer from such a drawback.

everyday processing of random models can be directly exploited in the implementation of simulation techniques (in particular in the evaluation of the variation of the proposed estimators or of the stationarity of the resulting output), while purely numerical techniques rely on less familiar branches of mathematics. Finally, note that many desirable approaches are those which efficiently combine both perspectives, as in the case of *simulated annealing* (see Section 5.2.3) or Riemann sums (see Section 4.3).

1.5 Problems

1.1 For both the censored data density (1.1) and the mixture of two normal distributions (1.15), plot the probability density function. Use various values for the parameters μ, θ, σ and τ.

1.2 In the situation of Example 1.1, establish that the densities are indeed (1.1) and (1.2).

1.3 In Example 1.1, the distribution of the random variable $Z = \min(X, Y)$ was of interest. Derive the distribution of Z in the following case of *informative censoring*, where $Y \sim \mathcal{N}(\theta, \sigma^2)$ and $X \sim \mathcal{N}(\theta, \theta^2\sigma^2)$. Pay attention to the identifiability issues.

1.4 In Example 1.1, show that the integral

$$\int_{\omega}^{\infty} \alpha\beta x^{\alpha-1} e^{-\beta x^{\alpha}} dx$$

can be explicitly calculated. (*Hint:* Use a change of variables.)

1.5 For the model (1.4), show that the density of $(X_0, \ldots, X_n)$ is given by (1.5).

1.6 In the setup of Example 1.2, derive the moment estimator of the weights $(p_1, \ldots, p_k)$ when the densities f_j are known.

1.7 In the setup of Example 1.6, show that the likelihood equations are given by (1.11) and that their solution is the standard $(\bar{y}, s^2)$ statistic.

1.8 (Titterington et al. 1985) In the case of a mixture of two exponential distributions with parameters 1 and 2,

$$\pi \mathcal{E}xp(1) + (1 - \pi)\mathcal{E}xp(2) ,$$

show that $\mathbb{E}[X^s] = \{\pi + (1 - \pi)2^{-s}\}\Gamma(s + 1)$. Deduce the best (in s) moment estimator based on $t_s(x) = x^s/\Gamma(s + 1)$.

1.9 Give the moment estimator for a mixture of k Poisson distributions, based on $t_s(x) = x(x - 1) \cdots (x - s + 1)$. (*Note:* Pearson 1915 and Gumbel 1940 proposed partial solutions in this setup. See Titterington et al. 1985, pp. 80–81, for details.)

1.10 In the setting of Example 1.9, plot the likelihood based on observing $(x_1, x_2, x_3) = (0, 5, 9)$ from the Student's t density (1.14) with $p = 1$ and $\sigma = 1$ (which is the standard Cauchy density). Observe the effect on multimodality of adding a fourth observation x_4 when x_4 varies.

1.11 The *Weibull distribution* $\mathcal{W}e(\alpha, c)$ is widely used in engineering and reliability. Its density is given by

$$f(x|\alpha, c) = c\alpha^{-1}(x/\alpha)^{c-1} e^{-(x/\alpha)^c}.$$

(a) Show that when c is known, this model is equivalent to a Gamma model.

(b) Give the likelihood equations in α and c and show that they do not allow for explicit solutions.

(c) Consider an iid sample $X_1, \ldots, X_n$ from $\mathcal{W}e(\alpha, c)$ censored from the right in y_0. Give the corresponding likelihood function when α and c are unknown and show that there is no explicit maximum likelihood estimator in this case either.

1.12 (Continuation of Problem 1.11) Show that the cdf of the Weibull distribution $\mathcal{W}e(\alpha, \beta)$ can be written explicitly, and show that the scale parameter α determines the behavior of the hazard rate $h(t) = \frac{f(t)}{1-F(t)}$, where f and F are the density and the cdf, respectively.

1.13 (Continuation of Problem 1.11) The following sample gives the times (in days) at which carcinoma was diagnosed in rats exposed to a carcinogen:

$$143, \ 164, \ 188, \ 188, \ 190, \ 192, \ 206, \ 209, \ 213, \ 216, \ 220,$$
$$227, \ 230, \ 234, \ 246, \ 265, \ 304, \ 216^*, \ 244^*,$$

where the observations with an asterisk are censored (see Pike 1966, for details). Fit a three parameter Weibull $\mathcal{W}e(\alpha, \beta, \gamma)$ distribution to this dataset, where γ is a translation parameter, for (a) $\gamma = 100$ and $\alpha = 3$, (b) $\gamma = 100$ and α unknown and (c) γ and α unknown. (*Note:* Treat the asterisked observations as ordinary data here. See Problem 5.24 for a method of dealing with the censoring.)

1.14 Let $X_1, X_2, \ldots, X_n$ be iid with density $f(x|\theta, \sigma)$, the Cauchy distribution $\mathcal{C}(\theta, \sigma)$, and let $L(\theta, \sigma|\mathbf{x}) = \prod_{i=1}^{n} f(x_i|\theta, \sigma)$ be the likelihood function.

(a) If σ is known, show that a solution to the likelihood equation $\frac{d}{d\theta} L(\theta, \sigma|\mathbf{x}) = 0$ is the root of a $2n - 1$ degree polynomial. Hence, finding *the* likelihood estimator can be challenging.

(b) For $n = 3$, if both θ and σ are unknown, find the maximum likelihood estimates and show that they are unique.

(c) For $n \geq 3$, if both θ and σ are unknown, show that the likelihood is unimodal.

(*Note:* See Copas 1975 and Ferguson 1978 for details.)

1.15 Referring to Example 1.16, show that the posterior distribution (1.22) can be written in the form (1.23).

1.16 Consider a Bernoulli random variable $Y \sim \mathcal{B}([1 + e^\theta]^{-1})$.

(a) If $y = 0$, show that the maximum likelihood estimator of θ is ∞.

(b) Show that the same problem occurs when $Y_1, Y_2 \sim \mathcal{B}([1 + e^\theta]^{-1})$ and $y_1 = y_2 = 0$ or $y_1 = y_2 = 1$. Give the maximum likelihood estimator in the other cases.

1.17 Consider n observations $x_1, \ldots, x_n$ from $\mathcal{B}(k, p)$ where both k and p are unknown.

(a) Show that the maximum likelihood estimator of k, $\hat{k}$, satisfies

$$(\hat{k}(1 - \hat{p}))^n \geq \prod_{i=1}^{n} (\hat{k} - x_i) \quad \text{and} \quad ((\hat{k} + 1)(1 - \hat{p}))^n < \prod_{i=1}^{n} (\hat{k} + 1 - x_i),$$

where $\hat{p}$ is the maximum likelihood estimator of p.

(b) If the sample is $16, 18, 22, 25, 27$, show that $\hat{k} = 99$.

(c) If the sample is $16, 18, 22, 25, 28$, show that $\hat{k} = 190$. Discuss the stability of the maximum likelihood estimator.

(*Note:* Olkin et al. 1981 were one of the first to investigate the stability of the MLE for the binomial parameter n; see also Carroll and Lombard 1985, Casella 1986, and Hall 1994.)

1.18 (Robertson et al. 1988) For a sample $X_1, \ldots, X_n$, and a function f on $\mathcal{X}$, the isotonic regression of f with weights ω_i is the solution of the minimization in g of

$$\sum_{i=1}^{n} \omega_i (g(x_i) - f(x_i))^2,$$

under the constraint $g(x_1) \leq \cdots \leq g(x_n)$.

(a) Show that a solution to this problem is obtained by the *pool-adjacent-violators* algorithm:

Algorithm A.1 –Pool-adjacent-violators–

> If f is not isotonic, find i such that $f(x_{i-1}) > f(x_i)$, replace $f(x_{i-1})$ and $f(x_i)$ by
>
> $$f^*(x_i) = f^*(x_{i-1}) = \frac{\omega_i f(x_i) + \omega_{i-1} f(x_{i-1})}{\omega_i + \omega_{i-1}},$$
>
> and repeat until the constraint is satisfied. Take $g = f^*$.

(b) Apply this algorithm to the case $n = 4$, $f(x_1) = 23$, $f(x_2) = 27$, $f(x_3) = 25$, and $f(x_4) = 28$, when the weights are all equal.

1.19 (Continuation of Problem 1.18) The simple *tree ordering* is obtained when one compares treatment effects with a control state. The isotonic regression is then obtained under the constraint $g(x_i) \geq g(x_1)$ for $i = 2, \ldots, n$.

(a) Show that the following provides the isotonic regression g^*:

Algorithm A.2 –Tree ordering–

> If f is not isotonic, assume w.l.o.g.
> that the $f(x_i)$'s are in increasing order $(i \geq 2)$.
> Find the smallest j such that
>
> $$A_j = \frac{\omega_1 f(x_1) + \cdots + \omega_j f(x_j)}{\omega_1 + \cdots \omega_j} < f(x_{j+1}),$$
>
> take $g^*(x_1) = A_j = g^*(x_2) = \cdots = g^*(x_j)$, $g^*(x_{j+1}) = f(x_{j+1})$, ...

(b) Apply this algorithm to the case where $n = 5$, $f(x_1) = 18$, $f(x_2) = 17$, $f(x_3) = 12$, $f(x_4) = 21$, $f(x_5) = 16$, with $\omega_1 = \omega_2 = \omega_5 = 1$ and $\omega_3 = \omega_4 = 3$.

1.20 For the setup of Example 1.8, where $X \sim \mathcal{N}_p(\theta, I_p)$:

(a) Show that the maximum likelihood estimator of $\lambda = \|\theta\|^2$ is $\hat{\lambda}(x) = \|x\|^2$ and that it has a constant bias equal to p.

(b) If we observe $Y = \|X\|^2$, distributed as a noncentral chi squared random variable (see Appendix A), show that the MLE of λ is the solution in (1.13). Discuss what happens if $y < p$.

(c) In part (a), if the reference prior $\pi(\theta) = \|\theta\|^{-(p-1)}$ is used, show that the posterior distribution is given by (1.18), with posterior mean (1.19).

1.21 Show that under the specification $\theta_j \sim \pi_j$ and $X \sim f_j(x|\theta_j)$ for model $\mathcal{M}_j$, the posterior probability of model $\mathcal{M}_j$ is given by (1.20).

1.22 Suppose that $X \sim f(x|\theta)$, with prior distribution $\pi(\theta)$, an interest is in the estimation of the parameter $h(\theta)$.

(a) Using the loss function $L(\delta, h(\theta))$, show that the estimator that minimizes the Bayes risk

$$\int \int L(\delta, h(\theta)) f(x|\theta) \pi(\theta) dx d\theta$$

is given by the estimator δ that minimizes (for each x)

$$\int L(\delta, h(\theta))\, \pi(\theta|x)\, d\theta\ .$$

(b) For $L(\delta, \theta) = \|h(\theta) - \delta\|^2$, show that the Bayes estimator of $h(\theta)$ is $\delta^\pi(x) = \mathbb{E}^\pi[h(\theta)|x]$.

(c) For $L(\delta, \theta) = |h(\theta) - \delta|$, show that the Bayes estimator of $h(\theta)$ is the median of the posterior distribution.

1.23 For each of the following cases, give the posterior and marginal distributions.
  (a) $X|\sigma \sim \mathcal{N}(0, \sigma^2)$, $\ 1/\sigma^2 \sim \mathcal{G}(1, 2)$;
  (b) $X|\lambda \sim \mathcal{P}(\lambda)$, $\ \lambda \sim \mathcal{G}(2, 1)$;
  (c) $X|p \sim \mathcal{N}eg(10, p)$, $\ p \sim \mathcal{B}e(1/2, 1/2)$.

1.24 Let $f(x)$ be a unimodal continuous density, and for a given value of α, let the interval $[a, b]$ satisfy $\int_a^b f = 1 - \alpha$.
  (a) Show that the shortest interval satisfying the probability constraint is given by $f(a) = f(b)$, where a and b are on each side of the mode of f.
  (b) Show that if f is symmetric, then the shortest interval satisfies $a = -b$.
  (c) Find the 90% highest posterior credible regions for the posterior distributions of Problem 1.23.

1.25 (Bauwens 1991) Consider $X_1, \ldots, X_n$ iid $\mathcal{N}(0, \sigma^2)$ with prior

$$\pi(\theta, \sigma^2) = \sigma^{-2(\alpha+1)} \exp(-s_0^2/2\sigma^2).$$

  (a) Compute the posterior distribution $\pi(\theta, \sigma^2|x_1, \ldots, x_n)$ and show that it depends only on $\bar{x}$ and $s^2 = \sum_{i=1}^n (x_i - \bar{x})^2$.
  (b) Derive the posterior expectation $\mathbb{E}^\pi[\sigma^2|x_1, \ldots, x_n]$ and show that its behavior when α and s_0 both converge to 0 depends on the limit of $(s_0^2/\alpha) - 1$.

1.26 In the setting of Section 1.3,
  (a) Show that, if the prior distribution is improper, the marginal distribution is also improper.
  (b) Show that if the prior $\pi(\theta)$ is improper and the sample space $\mathcal{X}$ is finite, the posterior distribution $\pi(\theta|x)$ is not defined for some value of x.
  (c) Consider $X_1, \ldots, X_n$ distributed according to $\mathcal{N}(\theta_j, 1)$, with $\theta_j \sim \mathcal{N}(\mu, \sigma^2)$ $(1 \le j \le n)$ and $\pi(\mu, \sigma^2) = \sigma^{-2}$. Show that the posterior distribution $\pi(\mu, \sigma^2|x_1, \ldots, x_n)$ is not defined.

1.27 Assuming that $\pi(\theta) = 1$ is an acceptable prior for real parameters, show that this generalized prior leads to $\pi(\sigma) = 1/\sigma$ if $\sigma \in \mathbb{R}^+$ and to $\pi(\varrho) = 1/\varrho(1 - \varrho)$ if $\varrho \in [0, 1]$ by considering the "natural" transformations $\theta = \log(\sigma)$ and $\theta = \log(\varrho/(1 - \varrho))$.

1.28 For each of the following situations, exhibit a conjugate family for the given distribution:
  (a) $X \sim \mathcal{G}(\theta, \beta)$; that is, $f_\beta(x|\theta) = \beta^\theta x^{\theta-1} e^{-\beta x}/\Gamma(\theta)$.
  (b) $X \sim \mathcal{B}e(1, \theta)$, $\theta \in \mathbb{N}$.

1.29 Show that, if $X \sim \mathcal{B}e(\theta_1, \theta_2)$, there exist conjugate priors on $\theta = (\theta_1, \theta_2)$ but that they do not lead to tractable posterior quantities, except for the computation of $\mathbb{E}^\pi[\theta_1/(\theta_1 + \theta_2)|x]$.

1.30 Consider $X_1, \ldots, X_n \sim \mathcal{N}(\mu + \nu, \sigma^2)$, with $\pi(\mu, \nu, \sigma) \propto 1/\sigma$.

(a) Show that the posterior distribution is not defined for every n.

(b) Extend this result to overparameterized models with improper priors.

1.31 Consider estimation in the *linear model*

$$Y = b_1 X_1 + b_2 X_2 + \epsilon,$$

under the constraint $0 \le b_1$, $b_2 \le 1$, for a sample $(Y_1, X_{11}, X_{21}), \ldots, (Y_n, X_{1n}, X_{2n})$ when the errors ϵ_i are independent and distributed according to $\mathcal{N}(0, 1)$. A noninformative prior is

$$\pi(b_1, b_2) = \mathbb{I}_{[0,1]}(b_1)\mathbb{I}_{[0,1]}(b_2) .$$

(a) Show that the posterior means are given by $(i = 1, 2)$

$$\mathbb{E}^\pi[b_i|y_1, \ldots, y_n] = \frac{\int_0^1 \int_0^1 b_i \prod_{j=1}^n \varphi(y_j - b_1 X_{1j} - b_2 X_{2j})\, db_1\, db_2}{\int_0^1 \int_0^1 \prod_{j=1}^n \varphi(y_j - b_1 X_{1j} - b_2 X_{2j})\, db_1\, db_2},$$

where φ is the density of the standard normal distribution.

(b) Show that an equivalent expression is

$$\delta_i^\pi(y_1, \ldots, y_n) = \frac{\mathbb{E}^\pi\left[b_i \mathbb{I}_{[0,1]^2}(b_1, b_2)|y_1, \ldots, y_n\right]}{P^\pi\left[(b_1, b_2) \in [0, 1]^2|y_1, \ldots, y_n\right]},$$

where the right-hand term is computed under the distribution

$$\begin{pmatrix} b_1 \\ b_2 \end{pmatrix} \sim \mathcal{N}_2 \left(\begin{pmatrix} \hat{b}_1 \\ \hat{b}_2 \end{pmatrix}, (X^t X)^{-1} \right),$$

with $(\hat{b}_1, \hat{b}_2)$ the unconstrained least squares estimator of (b_1, b_2) and

$$X = \begin{pmatrix} X_{11} & X_{21} \\ \vdots & \vdots \\ X_{1n} & X_{2n} \end{pmatrix}.$$

(c) Show that the posterior means cannot be written explicitly, except in the case where $(X^t X)$ is diagonal.

1.32 (Berger 1985) Consider the hierarchical model

$$X|\theta \sim \mathcal{N}_p(\theta, \sigma^2 I_p),$$
$$\theta|\xi \sim \mathcal{N}_p(\xi \mathbf{1}, \sigma_\pi^2 I_p),$$
$$\xi \sim \mathcal{N}(\xi_0, \tau^2) \quad , \quad \sigma_\pi^2 \sim \pi_2(\sigma_\pi^2)$$

where $\mathbf{1} = (1, \ldots, 1)^t \in \mathbb{R}^p$, and σ^2, ξ_0, and τ^2 are fixed.

(a) Show that

$$\delta(x|\xi, \sigma_\pi) = x - \frac{\sigma^2}{\sigma^2 + \sigma_\pi^2}(x - \xi\mathbf{1}),$$

$$\pi_2(\xi, \sigma_\pi^2|x) \propto (\sigma^2 + \sigma_\pi^2)^{-p/2} \exp\left\{-\frac{\|x - \xi\mathbf{1}\|^2}{2(\sigma^2 + \sigma_\pi^2)}\right\} e^{-(\xi-\xi_0)^2/2\tau^2} \pi_2(\sigma_\pi^2)$$

$$\propto \frac{\pi_2(\sigma_\pi^2)}{(\sigma^2 + \sigma_\pi^2)^{p/2}} \exp\left\{-\frac{p(\bar{x} - \xi)^2 + s^2}{2(\sigma^2 + \sigma_\pi^2)} - \frac{(\xi - \xi_0)^2}{2\tau^2}\right\}$$

with $s^2 = \sum_i (x_i - \bar{x})^2$. Deduce that $\pi_2(\xi|\sigma_\pi^2, x)$ is a normal distribution. Give the mean and variance of the distribution.

(b) Show that

$$\delta^{\pi}(x) = \mathbb{E}^{\pi_2(\sigma_{\pi}^2|x)}\left[x - \frac{\sigma^2}{\sigma^2 + \sigma_{\pi}^2}(x - \bar{x}\mathbf{1}) - \frac{\sigma^2 + \sigma_{\pi}^2}{\sigma_1^2 + \sigma_{\pi}^2 + p\tau^2}(\bar{x} - \xi_0)\mathbf{1}\right]$$

and

$$\pi_2(\sigma_{\pi}^2|x) \propto \frac{\tau \exp-\frac{1}{2}\left[\dfrac{s^2}{\sigma^2 + \sigma_{\pi}^2} + \dfrac{p(\bar{x} - \xi_0)^2}{p\tau^2 + \sigma^2 + \sigma_{\pi}^2}\right]}{(\sigma^2 + \sigma_{\pi}^2)^{(p-1)/2}(\sigma^2 + \sigma_{\pi}^2 + p\tau^2)^{1/2}}\,\pi_2(\sigma_{\pi}^2).$$

(c) Deduce the representation

$$\delta^{\pi}(x) = x - \mathbb{E}^{\pi_2(\sigma_{\pi}^2|x)}\left[\frac{\sigma^2}{\sigma^2 + \sigma_{\pi}^2}\right](x - \bar{x}\mathbf{1})$$

$$- \mathbb{E}^{\pi_2(\sigma_{\pi}^2|x)}\left[\frac{\sigma^2 + \sigma_{\pi}^2}{\sigma_1^2 + \sigma_{\pi}^2 + p\tau^2}\right](\bar{x} - \xi_0)\mathbf{1}.$$

and discuss the appeal of this expression from an integration point of view.

1.33 A classical linear regression can be written as $Y \sim \mathcal{N}_p(X\beta, \sigma^2 I_p)$ with X a $p \times q$ matrix and $\beta \in \mathbb{R}^q$.

(a) When X is known, give the natural parameterization of this exponential family and derive the conjugate priors on (β, σ^2).

(b) Generalize to $\mathcal{N}_p(X\beta, \Sigma)$.

1.34 [9] An *autoregressive model* AR(1) connects the random variables in a sample $X_1, \dots, X_n$ through the relation $X_{t+1} = \varrho X_t + \epsilon_t$, where $\epsilon_t \sim \mathcal{N}(0, \sigma^2)$ is independent of X_t.

(a) Show that the X_t's induce a Markov chain and derive a stationarity condition on ϱ. Under this condition, what is the stationary distribution of the chain?

(b) Give the covariance matrix of $(X_1, \dots, X_n)$.

(c) If x_0 is a (fixed) starting value for the chain, express the likelihood function and derive a conjugate prior on (ϱ, σ^2). (*Hint:* Note that $X_t|x_{t-1} \sim \mathcal{N}(\varrho x_{t-1}, \sigma^2)$.)

Note: The next four problems involve properties of the exponential family, conjugate prior distributions, and Jeffreys prior distributions. Brown (1986) is a book-length introduction to exponential families, and shorter introductions can be found in Casella and Berger (2001, Section 3.3), Robert (2001, Section 3.2), or Lehmann and Casella (1998, Section 1.5). For conjugate and Jeffreys priors, in addition to Note 1.6.1, see Berger (1985, Section 3.3) or Robert (2001, Sections 3.3 and 3.5).

1.35 Consider $\mathbf{x} = (x_{ij})$ and $\Sigma = (\sigma_{ij})$ symmetric positive-definite $m \times m$ matrices. The *Wishart* distribution, $\mathcal{W}_m(\alpha, \Sigma)$, is defined by the density

$$p_{\alpha, \Sigma}(\mathbf{x}) = \frac{|\mathbf{x}|^{\frac{\alpha - (m+1)}{2}}\exp(-(\mathrm{tr}(\Sigma^{-1}\mathbf{x})/2)}{\Gamma_m(\alpha)|\Sigma|^{\alpha/2}},$$

with $\mathrm{tr}(A)$ the trace of A and

---

[9] This problem requires material that will be covered in Chapter 6. It is put here for those already familiar with Markov chains.

$$\Gamma_m(\alpha) = 2^{\alpha m/2} \pi^{m(m-1)/4} \prod_{i=1}^{m} \Gamma\left(\frac{\alpha - i + 1}{2}\right).$$

(a) Show that this distribution belongs to the exponential family. Give its natural representation and derive the mean of $\mathcal{W}_m(\alpha, \Sigma)$.

(b) Show that, if $Z_1, \ldots, Z_n \sim \mathcal{N}_m(0, \Sigma)$,

$$\sum_{i=1}^{n} Z_i Z_i' \sim \mathcal{W}_m(n, \Sigma).$$

1.36 Consider $X \sim \mathcal{N}(\theta, \theta)$ with $\theta > 0$.

(a) Indicate whether the distribution of X belongs to an exponential family and derive the conjugate priors on θ.

(b) Determine the Jeffreys prior $\pi^J(\theta)$.

1.37 Show that a Student's t distribution $\mathcal{T}_p(\nu, \theta, \tau^2)$ does not allow for a conjugate family, apart from $\mathcal{F}_0$, the (trivial) family that contains all distributions.

1.38 (Robert 1991) The generalized inverse normal distribution $\mathcal{IN}(\alpha, \mu, \tau)$ has the density

$$K(\alpha, \mu, \tau)|y|^{-\alpha} \exp\left\{-\left(\frac{1}{y} - \mu\right)^2 / 2\tau^2\right\},$$

with $\alpha > 0$, $\mu \in \mathbb{R}$, and $\tau > 0$.

(a) Show that this density is well defined and that the normalizing factor is

$$K(\alpha, \mu, \tau)^{-1} = \tau^{\alpha-1} e^{-\mu^2/2\tau^2} 2^{(\alpha-1)/2} \, \Gamma\left(\frac{\alpha-1}{2}\right) \, {}_1F_1\left(\frac{\alpha-1}{2}; 1/2; \frac{\mu^2}{2\tau^2}\right),$$

where ${}_1F_1$ is the *confluent hypergeometric function*

$$\,{}_1F_1(a; b; z) = \sum_{k=0}^{\infty} \frac{\Gamma(a+k)}{\Gamma(b+k)} \frac{\Gamma(b)}{\Gamma(a)} \frac{z^k}{k!}$$

(see Abramowitz and Stegun 1964).

(b) If $X \sim \mathcal{N}(\mu, \tau^2)$, show that the distribution of $1/X$ is in the $\mathcal{IN}(\alpha, \mu, \tau)$ family.

(c) Deduce that the mean of $Y \sim \mathcal{IN}(\alpha, \mu, \tau)$ is defined for $\alpha > 2$ and is

$$\mathbb{E}_{\alpha,\mu,\tau}[Y] = \frac{\mu}{\tau^2} \frac{{}_1F_1(\frac{\alpha-1}{2}; 3/2; \mu^2/2\tau^2)}{{}_1F_1(\frac{\alpha-1}{2}; 1/2; \mu^2/2\tau^2)}.$$

(d) Show that $\theta \sim \mathcal{IN}(\alpha, \mu, \tau)$ constitutes a conjugate family for the multiplicative model $X \sim \mathcal{N}(\theta, \theta^2)$.

1.39 Recall the situation of Example 1.8 (see also Example 1.12), where $X \sim \mathcal{N}_p(\theta, I_p)$.

(a) For the prior $\pi(\lambda) = 1/\sqrt{\lambda}$, show that the Bayes estimator of $\lambda = ||\theta||^2$ under quadratic loss can be written as

$$\delta^\pi(x) = \frac{{}_1F_1(3/2; p/2; ||x||^2/2)}{{}_1F_1(1/2; p/2; ||x||^2/2)},$$

where ${}_1F_1$ is the confluent hypergeometric function.

(b) Using the series development of $_1F_1$ in Problem 1.38, derive an asymptotic expansion of δ^π (for $||x||^2 \to +\infty$) and compare it with $\delta_0(x) = ||x||^2 - p$.

(c) Compare the risk behavior of the estimators of part (b) under the weighted quadratic loss

$$L(\delta, \theta) = \frac{(||\theta||^2 - \delta)^2}{2||\theta||^2 + p}.$$

1.40 (Smith and Makov 1978) Consider the mixture density

$$X \sim f(x|p) = \sum_{i=1}^{k} p_i f_i(x),$$

where $p_i > 0$, $\sum_i p_i = 1$, and the densities f_i are known. The prior $\pi(p)$ is a Dirichlet distribution $\mathcal{D}(\alpha_1, \ldots, \alpha_k)$.

(a) Explain why the computing time could get prohibitive as the sample size increases.

(b) A sequential alternative which approximates the Bayes estimator is to replace $\pi(p|x_1, \ldots, x_n)$ by $\mathcal{D}(\alpha_1^{(n)}, \ldots, \alpha_k^{(n)})$, with

$$\alpha_1^{(n)} = \alpha_1^{(n-1)} + P(Z_{n1} = 1|x_n), \ldots, \alpha_k^{(n)} = \alpha_k^{(n-1)} + P(Z_{nk} = 1|x_n),$$

and Z_{ni} $(1 \leq i \leq k)$ is the component indicator vector of X_n. Justify this approximation and compare with the updating of $\pi(\theta|x_1, \ldots, x_{n-1})$ when x_n is observed.

(c) Examine the performances of the approximation in part (b) for a mixture of two normal distributions $\mathcal{N}(0, 1)$ and $\mathcal{N}(2, 1)$ when $p = 0.1, 0.25$, and 0.5.

(d) If $\pi_i^n = P(Z_{ni} = 1|x_n)$, show that

$$\hat{p}_i^{(n)}(x_n) = \hat{p}_i^{(n-1)}(x_{n-1}) - a_{n-1}\{\hat{p}_i^{(n-1)} - \pi_i^n\},$$

where $\hat{p}_i^{(n)}$ is the quasi-Bayesian approximation of $\mathbb{E}^\pi(p_i|x_1, \ldots, x_n)$.

1.6 Notes

1.6.1 Prior Distributions

(i) *Conjugate Priors*

When prior information about the model is quite limited, the prior distribution is often chosen from a parametric family. Families $\mathcal{F}$ that are *closed under sampling* (that is, such that, for every prior $\pi \in \mathcal{F}$, the posterior distribution $\pi(\theta|x)$ also belongs to $\mathcal{F}$) are of particular interest, for both parsimony and invariance motivations. These families are called *conjugate* families. Most often, the main motivation for using conjugate priors is their tractability; however, such choices may constrain the subjective input.

For reasons related to the *Pitman–Koopman Lemma* (see the discussion following Example 1.8), conjugate priors can only be found in exponential families. In fact, if the sampling density is of the form

(1.27) $$f(x|\theta) = C(\theta)h(x)\exp\{R(\theta) \cdot T(x)\},$$

which include many common continuous and discrete distributions (see Brown 1986), a conjugate family for $f(x|\theta)$ is given by

$$\pi(\theta|\mu, \lambda) = K(\mu, \lambda)\, e^{\theta \cdot \mu - \lambda \psi(\theta)},$$

since the posterior distribution is $\pi(\theta|\mu + x, \lambda + 1)$. Table 1.2 presents some standard conjugate families.

Distribution	Sample Density	Prior Density
Normal	$\mathcal{N}(\theta, \sigma^2)$	$\theta \sim \mathcal{N}(\mu, \tau^2)$
Normal	$\mathcal{N}(\mu, 1/\theta)$	$\theta \sim \mathcal{G}(\alpha, \beta)$
Poisson	$\mathcal{P}(\theta)$	$\theta \sim \mathcal{G}(\alpha, \beta)$
Gamma	$\mathcal{G}(\nu, \theta)$	$\theta \sim \mathcal{G}(\alpha, \beta)$
Binomial	$\mathcal{B}(n, \theta)$	$\theta \sim \mathcal{Be}(\alpha, \beta)$
Multinomial	$\mathcal{M}_k(\theta_1, \ldots, \theta_k)$	$\theta_1, \ldots, \theta_k \sim \mathcal{D}(\alpha_1, \ldots, \alpha_k)$

Table 1.2. Some conjugate families of distributions.

Extensions of (1.27) which allow for parameter-dependent support enjoy most properties of the exponential families. In particular, they extend the applicability of conjugate prior analysis to other types of distributions like the uniform or the Pareto distribution. (See Robert 1994a, Section 3.2.2.)

Another justification of conjugate priors, found in Diaconis and Ylvisaker (1979), is that some Bayes estimators are then linear. If $\xi(\theta) = \mathbb{E}_\theta[x]$, which is equal to $\nabla\psi(\theta)$, the prior mean of $\xi(\theta)$ for the prior $\pi(\theta|\mu, \lambda)$ is x_0/λ and if $x_1, \ldots, x_n$ are iid $f(x|\theta)$,

$$\mathbb{E}^\pi[\xi(\theta)|x_1, \ldots, x_n] = \frac{x_0 + n\bar{x}}{\lambda + n}.$$

For more details, see Bernardo and Smith (1994, Section 5.2).

(ii) Noninformative Priors

If there is no strong prior information, a Bayesian analysis may proceed with a "noninformative" prior; that is, a prior distribution which attempts to impart no information about the parameter of interest. A classic noninformative prior is the *Jeffreys prior* (Jeffreys 1961). For a sampling density $f(x|\theta)$, this prior has a density that is proportional to $\sqrt{|I(\theta)|}$, where $|I(\theta)|$ is the determinant of the Fisher information matrix,

$$I(\theta) = E_\theta\left[\frac{\partial}{\partial\theta}\log f(X|\theta)\right]^2 = -E_\theta\left[\frac{\partial^2}{\partial\theta^2}\log f(X|\theta)\right],$$

equal to the variance of the *score* vector.

For further details see Berger (1985), Bernardo and Smith (1994), Robert (1994a), or Lehmann and Casella (1998).

(iii) Reference Priors

An alternative approach to constructing a "noninformative" prior is that of *reference priors* (Bernardo 1979, Berger and Bernardo 1992). We start with

Kullback–Leibler information, $K[f, g]$, also known as Kullback–Leibler *information for discrimination* between two densities. For densities f and g, it is given by

$$K[f, g] = \int \log \left[\frac{f(t)}{g(t)} \right] f(t) \, dt.$$

The interpretation is that as $K[f, g]$ gets larger, it is easier to discriminate between the densities f and g.

A reference prior can be thought of as the density $\pi(\cdot)$ that maximizes (asymptotically) the expected Kullback-Leibler information (also known as Shannon information)

$$\int K[\pi(\theta|x), \pi(\theta)] m_\pi(x) \, dx,$$

where $m_\pi(x) = \int f(x|\theta)\pi(\theta) \, d\theta$ is the marginal distribution.

Further details are given in Bernardo and Smith (1994) and Robert (2001, Section 3.5) and there are approximations due to Clarke and Barron (1990) and Clarke and Wasserman (1993).

1.6.2 Bootstrap Methods

Bootstrap (or *resampling*) techniques are a collection of computationally intensive methods that are based on resampling from the observed data. They were first introduced by Efron (1979) and are described more fully in Efron (1982), Efron and Tibshirani (1994), or Hjorth (1994). (See also Hall 1992 and Barbe and Bertail 1995 for more theoretical treatments.) Although these methods do not call for, in principle, a simulation-based implementation, in many cases where their use is particularly important, intensive simulation is required. The basic idea of the bootstrap[10] is to evaluate the properties of an arbitrary estimator $\hat{\theta}(x_1, \ldots, x_n)$ through the empirical cdf of the sample $X_1, \ldots, X_n$,

$$F_n(x) = \frac{1}{n} \sum_{i=1}^{n} \mathbb{I}_{X_i \leq x},$$

instead of the theoretical cdf F. More precisely, if an estimate of $\theta(F) = \int h(x)dF(x)$ is desired, an obvious candidate is $\theta(F_n) = \int h(x)dF_n(x)$. When the X_i's are iid, the *Glivenko–Cantelli Theorem* (see Billingsley 1995) guarantees the *sup-norm* convergence of F_n to F, and hence guarantees that $\theta(F_n)$ is a consistent estimator of $\theta(F)$. The bootstrap provides a somewhat "automatic" method of computing $\theta(F_n)$, by resampling the data.

It has become common to denote a bootstrap sample with a superscript "$\star$", so we can draw bootstrap samples

$$\mathbf{X}^{*i} = (X_1^*, \ldots, X_n^*) \sim F_n, \quad \text{iid.}$$

(Note that the X_i^*'s are equal to one of the x_j's and that a same value x_j can appear several times in $\mathbf{X}^{*i}$.) Based on drawing $\mathbf{X}^{*1}, \ldots, \mathbf{X}^{*B}$, $\theta(F_n)$ can be approximated by the bootstrap estimator

---

[10] This name comes from the German novel *Adventures of Baron Munchausen* by Rudolph Raspe where the hero saves himself from drowning by pulling on his own... bootstraps!

$$(1.28) \qquad \hat{\theta}(F_n) \approx \frac{1}{B} \sum_{i=1}^{B} h(\mathbf{x}^{*i})$$

with the approximation becoming more accurate as B increases.

If $\hat{\theta}$ is an arbitrary estimator of $\theta(F)$, the bias, the variance, or even the error distribution, of $\hat{\theta}$,

$$\mathbb{E}_F[\hat{\theta} - \theta(F)], \quad \mathrm{var}_F(\hat{\theta}), \quad \text{and} \quad P_F(\hat{\theta} - \theta(F) \leq u),$$

can then be approximated by replacing F with F_n. For example, the bootstrap estimator of the bias will thus be

$$\begin{aligned} b_n &= \mathbb{E}_{F_n}[\hat{\theta}(F_n) - \theta(F_n)] \\ &\approx \frac{1}{B} \sum_{i=1}^{B} \{\hat{\theta}(F_{ni}^*) - \hat{\theta}(F_n)\}, \end{aligned}$$

where $\hat{\theta}(F_{ni}^*)$ is constructed as in (1.28), and a *confidence interval* $[\hat{\theta} - \beta, \hat{\theta} - \alpha]$ on θ can be constructed by imposing the constraint

$$P_{F_n}(\alpha \leq \hat{\theta}(F_n) - \theta(F_n) \leq \beta) = c$$

on (α, β), where c is the desired confidence level. There is a huge body of literature,[11] not directly related to the purpose of this book, in which the authors establish different optimality properties of the bootstrap estimates in terms of bias and convergence (see Hall 1992, Efron and Tibshirani 1994, Lehmann 1998).

Although direct computation of $\hat{\theta}$ is possible in some particular cases, most setups require simulation to approximate the distribution of $\hat{\theta} - \theta(F_n)$. Indeed, the distribution of $(X_1^*, \ldots, X_n^*)$ has a discrete support $\{x_1, \ldots, x_n\}^n$, but the cardinality of this support, n^n, increases much too quickly to permit an exhaustive processing of the points of the support even for samples of average size. There are some algorithms (such as those based on *Gray Codes*; see Diaconis and Holmes 1994) which may allow for exhaustive processing in larger samples.

---

[11] The May 2003 issue of *Statistical Science* is devoted to the bootstrap, providing both an introduction and overview of its properties, and an examination of its influences in various areas of Statistics.

Random Variable Generation

"Have you any thought," resumed Valentin, "of a tool with which it could be done?"
"Speaking within modern probabilities, I really haven't," said the doctor.
—G.K. Chesterton, *The Innocence of Father Brown*

The methods developed in this book mostly rely on the possibility of producing (with a computer) a supposedly endless flow of random variables (usually iid) for well-known distributions. Such a simulation is, in turn, based on the production of uniform random variables. Although we are not directly concerned with the *mechanics* of producing uniform random variables (see Note 2.6.1), we are concerned with the *statistics* of producing uniform and other random variables.

In this chapter we first consider what statistical properties we want a sequence of simulated uniform random variables to have. Then we look at some basic methodology that can, starting from these simulated uniform random variables, produce random variables from both standard and nonstandard distributions.

2.1 Introduction

Methods of simulation are based on the production of random variables, originally independent random variables, that are distributed according to a distribution f that is not necessarily explicitly known (see, for example, Examples 1.1, 1.2, and 1.3). The type of random variable production is formalized below in the definition of a *pseudo-random number generator*. We first concentrate on the generation of random variables that are uniform on the interval $[0, 1]$, because the uniform distribution $\mathcal{U}_{[0,1]}$ provides the basic probabilistic representation of randomness and also because all other distributions require a sequence of uniform variables to be simulated.

2.1.1 Uniform Simulation

The logical paradox[1] associated with the generation of "random numbers" is the problem of producing a *deterministic* sequence of values in $[0, 1]$ which imitates a sequence of *iid* uniform random variables $\mathcal{U}_{[0,1]}$. (Techniques based on the physical imitation of a "random draw" using, for example, the internal clock of the machine have been ruled out. This is because, first, there is no guarantee on the *uniform* nature of numbers thus produced and, second, there is no reproducibility of such samples.) However, we really do not want to enter here into the philosophical debate on the notion of "random," and whether it is, indeed, possible to "reproduce randomness" (see, for example, Chaitin 1982, 1988).

For our purposes, there are methods that use a fully deterministic process to produce a random sequence in the following sense: Having generated $(X_1, \ldots, X_n)$, knowledge of X_n [or of $(X_1, \ldots, X_n)$] imparts no discernible knowledge of the value of X_{n+1} if the transformation function is not available. Of course, given the initial value X_0 and the transformation function, the sample $(X_1, \ldots, X_n)$ is always the same. Thus, the "pseudo-randomness" produced by these techniques is limited since two samples $(X_1, \ldots, X_n)$ and $(Y_1, \ldots, Y_n)$ produced by the algorithm will not be independent, nor identically distributed, nor comparable in any probabilistic sense. This limitation should not be forgotten: The validity of a random number generator is based on a single sample $X_1, \ldots, X_n$ when n tends to $+\infty$ and not on replications $(X_{11}, \ldots, X_{1n}), (X_{21}, \ldots, X_{2n}), \ldots (X_{k1}, \ldots, X_{kn})$, where n is fixed and k tends to infinity. In fact, the distribution of these n-tuples depends only on the manner in which the initial values X_{r1} $(1 \leq r \leq k)$ were generated.

With these limitations in mind, we can now introduce the following operational definition, which avoids the difficulties of the philosophical distinction between a deterministic algorithm and the reproduction of a random phenomenon.

Definition 2.1. *A uniform pseudo-random number generator* is an algorithm which, starting from an initial value u_0 and a transformation D, produces a sequence $(u_i) = (D^i(u_0))$ of values in $[0, 1]$. For all n, the values $(u_1, \ldots, u_n)$ reproduce the behavior of an iid sample $(V_1, \ldots, V_n)$ of uniform random variables when compared through a usual set of tests.

This definition is clearly restricted to *testable* aspects of the random variable generation, which are connected through the deterministic transformation

---

[1] Von Neumann (1951) summarizes this problem very clearly by writing *"Any one who considers arithmetical methods of reproducing random digits is, of course, in a state of sin. As has been pointed out several times, there is no such thing as a random number—there are only methods of producing random numbers, and a strict arithmetic procedure of course is not such a method."*

$u_i = D(u_{i-1})$. Thus, the validity of the algorithm consists in the verification that the sequence $U_1, \ldots, U_n$ leads to acceptance of the hypothesis

$$H_0 : U_1, \ldots, U_n \quad \text{are iid} \quad \mathcal{U}_{[0,1]}.$$

The set of tests used is generally of some consequence. There are classical tests of uniformity, such as the Kolmogorov–Smirnov test. Many generators will be deemed adequate under such examination. In addition, and perhaps more importantly, one can use methods of *time series* to determine the degree of correlation between between U_i and $(U_{i-1}, \ldots, U_{i-k})$, by using an ARMA$(p, q)$ model, for instance. One can use nonparametric tests, like those of Lehmann (1975) or Randles and Wolfe (1979), applying them on arbitrary decimals of U_i. Marsaglia[2] has assembled a set of tests called `Die Hard`.

Definition 2.1 is therefore *functional*: An algorithm that generates uniform numbers is acceptable if it is not rejected by a set of tests. This methodology is not without problems, however. Consider, for example, particular applications that might demand a large number of iterations, as the theory of large deviations (Bucklew 1990), or particle physics, where algorithms resistant to standard tests may exhibit fatal faults. In particular, algorithms having hidden periodicities (see below) or which are not uniform for the smaller digits may be difficult to detect. Ferrenberg et al. (1992) show, for instance, that an algorithm of Wolff (1989), reputed to be "good," results in systematic biases in the processing of Ising models (see Example 5.8), due to long-term correlations in the generated sequence.

The notion that a deterministic system can imitate a random phenomenon may also suggest the use of *chaotic* models to create random number generators. These models, which result in complex deterministic structures (see Bergé et al. 1984, Gleick 1987, Ruelle 1987) are based on dynamic systems of the form $X_{n+1} = D(X_n)$ which are very sensitive to the initial condition X_0.

Example 2.2. The logistic function. The logistic function $D_\alpha(x) = \alpha x(1 - x)$ produces, for some values of $\alpha \in [3.57, 4.00]$, chaotic configurations. In particular, the value $\alpha = 4.00$ yields a sequence (X_n) in $[0, 1]$ that, theoretically, has the same behavior as a sequence of random numbers (or random variables) distributed according to the *arcsine distribution* with density $1/\pi\sqrt{x(1 - x)}$. (See Problem 2.4 for another random number generator based on the "tent" function.)

Although the limit distribution (also called the stationary distribution) associated with a dynamic system $X_{n+1} = D(X_n)$ is sometimes defined and known, the chaotic features of the system do not guarantee acceptable behavior (in the probabilistic sense) of the associated generator. Figure 2.1 illustrates the properties of the generator based on the logistic function D_α. The

---

[2] These tests are now available as a freeware on the site `http://stat.fsu.edu/~geo/diehard.html`

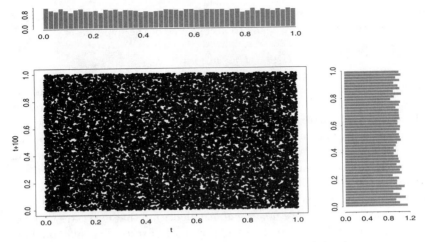

Fig. 2.1. Plot of the sample (y_n, y_{n+100}) $(n = 1, \ldots, 9899)$ for the sequence $x_{n+1} = 4x_n(1 - x_n)$ and $y_n = F(x_n)$, along with the (marginal) histograms of y_n *(on top)* and y_{n+100} *(right margin)*.

histogram of the transformed variables $Y_n = 0.5 + \arcsin(X_n)/\pi$, of a sample of successive values $X_{n+1} = D_\alpha(X_n)$ fits the uniform density extremely well. Moreover, while the plots of (Y_n, Y_{n+1}) and (Y_n, Y_{n+10}) do not display characteristics of uniformity, Figure 2.1 shows that the sample of (Y_n, Y_{n+100}) satisfactorily fills the unit square. However, even when these functions give a good approximation of randomness in the unit square $[0, 1] \times [0, 1]$, the hypothesis of randomness is rejected by many standard tests.

Classic examples from the theory of chaotic functions do not lead to acceptable pseudo-random number generators. Moreover, the 100 calls to D_α between two generations are excessive in terms of computing time.   ‖

We have presented in this introduction some necessary basic notions to now understand a very good pseudo-random number generator, the algorithm Kiss[3] of Marsaglia and Zaman (1993). However, many of the details involve notions that are a bit tangential to the main topic of this text and, in addition, most computer packages now include a well-behaved uniform random generator. Thus, we leave the details of the Kiss generator to Note 2.6.1.

2.1.2 The Inverse Transform

In describing the structure of a space of random variables, it is always possible to represent the generic probability triple $(\Omega, \mathcal{F}, P)$ (where Ω represents the whole space, $\mathcal{F}$ represents a σ-algebra on Ω, and P is a probability measure) as

---

[3] The name is an acronym of the saying *Keep it simple, stupid!*, and not reflective of more romantic notions. After all, this is a Statistics text!

$([0,1], \mathcal{B}, \mathcal{U}_{[0,1]})$ (where $\mathcal{B}$ are the Borel sets on $[0,1]$) and therefore equate the variability of $\omega \in \Omega$ with that of a uniform variable in $[0,1]$ (see, for instance, Billingsley 1995, Section 2). The random variables X are then functions from $[0,1]$ to $\mathcal{X}$, that is, functions of uniform variates transformed by the *generalized inverse* function.

Definition 2.3. For a non-decreasing function F on $\mathbb{R}$, the *generalized inverse* of F, F^-, is the function defined by

$$(2.1) \qquad F^-(u) = \inf\{x :\ F(x) \geq u\} \ .$$

We then have the following lemma, sometimes known as the *probability integral transform*, which gives us a representation of any random variable as a transform of a uniform random variable.

Lemma 2.4. *If $U \sim \mathcal{U}_{[0,1]}$, then the random variable $F^-(U)$ has the distribution F.*

Proof. For all $u \in [0,1]$ and for all $x \in F^-([0,1])$, the generalized inverse satisfies

$$F(F^-(u)) \geq u \quad \text{and} \quad F^-(F(x)) \leq x \ .$$

Therefore,

$$\{(u,x) :\ F^-(u) \leq x\} = \{(u,x) :\ F(x) \geq u\}$$

and

$$P(F^-(U) \leq x) = P(U \leq F(x)) = F(x) \ .$$

$\square$

Thus, formally, in order to generate a random variable $X \sim F$, it suffices to generate U according to $\mathcal{U}_{[0,1]}$ and then make the transformation $x = F^-(u)$.

Example 2.5. Exponential variable generation. If $X \sim \mathcal{E}xp(1)$, so $F(x) = 1 - e^{-x}$, then solving for x in $u = 1 - e^{-x}$ gives $x = -\log(1 - u)$. Therefore, if $U \sim \mathcal{U}_{[0,1]}$, the random variable $X = -\log U$ has the exponential distribution (as U and $1 - U$ are both uniform). ‖

The generation of uniform random variables is therefore a key determinant in the behavior of simulation methods for other probability distributions, since those distributions can be represented as a deterministic transformation of uniform random variables. (Although, in practice, we often use methods other than that of Lemma 2.4, this basic representation is usually a good way to think about things. Note also that Lemma 2.4 implies that a bad choice of a uniform random number generator can invalidate the resulting simulation procedure.)

As mentioned above, from a theoretical point of view an operational version of any probability space $(\Omega, \mathcal{A}, P)$ can be created from the uniform distribution $\mathcal{U}_{[0,1]}$ and Lemma 2.4. Thus, the generation of any sequence of random variables can be formally implemented through the uniform generator Kiss. In practice, however, this approach only applies when the cumulative distribution functions are "explicitly" available, in the sense that there exists an algorithm allowing the computation of $F^-(u)$ in acceptable time. In particular, for distributions with explicit forms of F^- (for instance, the exponential, double-exponential, or Weibull distributions; see Problem 2.5 for other examples), Lemma 2.4 does lead to a practical implementation. But this situation only covers a small number of cases, described in Section 2.2 and additional problems. Other methods, like the Accept–Reject method of Section 2.3, are more general and do not use any strong analytic property of the densities. Thus, they can handle more general cases as, for example, the simulation of distributions in dimensions greater than one.

2.1.3 Alternatives

Although computation by Monte Carlo methods can be thought of as an exact calculation (as the order of accuracy is only a function of computation time), it is probably more often thought of as an approximation. Thus, numerical approximation is an alternative to Monte Carlo, and should also be considered a candidate for solving any particular problem. The following example shows how numerical approximations can work in the calculation of normal probabilities (see Sections 1.4, 3.6.2 and 3.4 for other approaches).

Example 2.6. Normal probabilities. Although Φ, the cumulative distribution function of the normal distribution cannot be expressed explicitly, since

$$\Phi(x) = \frac{1}{\sqrt{2\pi}} \int_{-\infty}^{x} \exp\{-z^2/2\}dz,$$

there exist approximations of Φ and of Φ^{-1} up to an arbitrary precision. For instance, Abramowitz and Stegun (1964) give the approximation

$$\Phi(x) \simeq 1 - \varphi(x) \left[b_1 t + b_2 t^2 + b_3 t^3 + b_4 t^4 + b_5 t^5\right] \qquad (x > 0),$$

where φ denotes the normal density, $t = (1 + px)^{-1}$ and

$$p = 0.2316419, \quad b_1 = 0.31938, \quad b_2 = -0.35656,$$
$$b_3 = 1.78148, \quad b_4 = -1.82125, \quad b_5 = 1.33027.$$

Similarly, we also have the approximation

$$\Phi^{-1}(\alpha) \simeq t - \frac{a_0 + a_1 t}{1 + b_1 t + b_2 t^2},$$

where $t^2 = \log(\alpha^{-2})$ and

$$a_0 = 2.30753, \quad a_1 = 0.27061, \quad b_1 = 0.99229, \quad b_2 = 0.04481.$$

These two approximations are exact up to an error of order 10^{-8}, the error being absolute. If no other fast simulation method was available, this approximation could be used in settings which do not require much precision in the tails of $\mathcal{N}(0,1)$. (However, as shown in Example 2.8, there exists an exact and much faster algorithm.)   ‖

2.1.4 Optimal Algorithms

Devroye 1985 presents a more comprehensive (one could say almost exhaustive!) treatment of the methods of random variable generation than the one presented in this chapter, in particular looking at refinements of existing algorithms in order to achieve uniformly optimal performances. (We strongly urge the reader to consult this book[4] for a better insight on the implications of this goal in terms of probabilistic and algorithmic complexity.)

Some refinements of the simulation techniques introduced in this chapter will be explored in Chapter 4, where we consider ways to accelerate Monte Carlo methods. At this point, we note that the concepts of "optimal" and "efficient" algorithms are particularly difficult to formalize. We can naively compare two algorithms, $[B_1]$ and $[B_2]$ say, in terms of time of computation, for instance through the average generation time of one observation. However, such a comparison depends on many subjective factors like the quality of the programming, the particular programming language used to implement the method, and the particular machine on which the program runs. More importantly, it does not take into account the conception and programming (and debugging) times, nor does it incorporate the specific use of the sample produced, partly because a quantification of these factors is generally impossible. For instance, some algorithms have a decreasing efficiency when the sample size increases. The reduction of the efficiency of a given algorithm to its average computation time is therefore misleading and we only use this type of measurement in settings where $[B_1]$ and $[B_2]$ are already of the same complexity. Devroye (1985) also notes that the simplicity of algorithms should be accounted for in their evaluation, since complex algorithms facilitate programming errors and, therefore, may lead to important time losses.[5]

A last remark to bring this section to its end is that simulation of the standard distributions presented here is accomplished quite efficiently by many statistical programming packages (for instance, **Gauss**, **Mathematica**, **Matlab**, **R**, **Splus**). When the generators from these general-purpose packages are easily accessible (in terms of programming), it is probably preferable to use such

---

[4] The book is now out-of-print but available for free on the author's website, at McGill University, Montréal, Canada.

[5] In fact, in numerous settings, the time required by a simulation is overwhelmingly dedicated to programming. This is, at least, the case for the authors themselves!!!

a generator rather than to write one's own. However, if a generation technique will get extensive use or if there are particular features of a problem that can be exploited, the creation of a personal library of random variable generators can accelerate analyses and even improve results, especially if the setting involves "extreme" cases (sample size, parameter values, correlation structure, rare events) for which the usual generators are poorly adapted. The investment represented by the creation and validation of such a personal library must therefore be weighed against the potential benefits.

2.2 General Transformation Methods

When a distribution f is linked in a relatively simple way to another distribution that is easy to simulate, this relationship can often be exploited to construct an algorithm to simulate variables from f. In this section we present alternative (to Lemma 2.4) techniques for generating nonuniform random variables. Some of these methods are rather case-specific, and are difficult to generalize as they rely on properties of the distribution under consideration and its relation with other probability distributions.

We begin with an illustration of some distributions that are simple to generate.

Example 2.7. Building on exponential random variables. In Example 2.5 we saw how to generate an exponential random variable starting from a uniform. Now we illustrate some of the random variables that can be generated starting from an exponential distribution. If the X_i's are iid $\mathcal{E}xp(1)$ random variables, then

$$Y = 2\sum_{j=1}^{\nu} X_j \sim \chi_{2\nu}^2 , \qquad \nu \in \mathbb{N}^* ,$$

(2.2)
$$Y = \beta \sum_{j=1}^{a} X_j \sim \mathcal{G}a(a,\beta) , \qquad a \in \mathbb{N}^* ,$$

$$Y = \frac{\sum_{j=1}^{a} X_j}{\sum_{j=1}^{a+b} X_j} \sim \mathcal{B}e(a,b) , \qquad a,b \in \mathbb{N}^* .$$

Other derivations are possible (see Problem 2.6).   ∥

These transformations are quite simple to use and, hence, will often be a favorite. However, there are limits to their usefulness, both in scope of variables that can be generated and in efficiency of generation. For example, as we will see, there are more efficient algorithms for Gamma and Beta random variables. Also, we cannot use exponentials to generate Gamma random variables with a non-integer shape parameter. For instance, we cannot get a χ_1^2

variable, which would, in turn, get us a $\mathcal{N}(0,1)$ variable. For that, we look at the following example of the *Box–Muller* algorithm (1958) for the generation of $\mathcal{N}(0,1)$ variables.

Example 2.8. Normal variable generation. If r and θ are the polar coordinates of (X_1, X_2), then, since the distribution of (X_1, X_2) is rotation invariant (see Problem 2.7)

$$r^2 = X_1^2 + X_2^2 \sim \chi_2^2 = \mathcal{E}xp(1/2)\,,$$
$$\theta \sim \mathcal{U}_{[0,2\pi]}\,.$$

If U_1 and U_2 are iid $\mathcal{U}_{[0,1]}$, the variables X_1 and X_2 defined by

$$X_1 = \sqrt{-2\log(U_1)}\,\cos(2\pi U_2)\,,\qquad X_2 = \sqrt{-2\log(U_1)}\,\sin(2\pi U_2)\,,$$

are then iid $\mathcal{N}(0,1)$. The corresponding algorithm is

Algorithm A.3 –Box–Muller–

1 Generate U_1, U_2 iid $\mathcal{U}_{[0,1]}$;
2 Define   [A.3]

$$\begin{cases} x_1 = \sqrt{-2\log(u_1)}\cos(2\pi u_2)\,, \\ x_2 = \sqrt{-2\log(u_1)}\sin(2\pi u_2)\,; \end{cases}$$

3 Take x_1 and x_2 as two independent draws from $\mathcal{N}(0,1)$.

In comparison with algorithms based on the Central Limit Theorem, this algorithm is exact, producing two normal random variables from two uniform random variables, the only drawback (in speed) being the necessity of calculating functions such as log, cos, and sin. If this is a concern, Devroye (1985) gives faster alternatives that avoid the use of these functions (see also Problems 2.8 and 2.9). ‖

Example 2.9. Poisson generation. The Poisson distribution is connected to the exponential distribution through the Poisson process; that is, if $N \sim \mathcal{P}(\lambda)$ and $X_i \sim \mathcal{E}xp(\lambda)$, $i \in \mathbb{N}^*$, then

$$P_\lambda(N = k) = P_\lambda(X_1 + \cdots + X_k \leq 1 < X_1 + \cdots + X_{k+1})\,.$$

Thus, the Poisson distribution can be simulated by generating exponential random variables until their sum exceeds 1. This method is simple, but is really practical only for smaller values of λ. On average, the number of exponential variables required is λ, and this could be prohibitive for large values of λ. In these settings, Devroye (1981) proposed a method whose computation time is uniformly bounded (in λ) and we will see another approach, suitable for large λ's, in Example 2.23. Note also that a generator of Poisson random variables can produce negative binomial random variables since, when $Y \sim \mathcal{G}a(n, (1-p)/p)$ and $X|y \sim \mathcal{P}(y)$, $X \sim \mathcal{N}eg(n,p)$. (See Problem 2.13.)   ‖

Example 2.9 shows a specific algorithm for the generation of Poisson random variables. Based on an application of Lemma 2.4, we can also construct a generic algorithm that will work for any discrete distribution.

Example 2.10. Discrete random variables. To generate $X \sim P_\theta$, we can calculate (once for all) the probabilities

$$p_0 = P_\theta(X \le 0), \quad p_1 = P_\theta(X \le 1), \quad p_2 = P_\theta(X \le 2), \quad \ldots$$

then generate $U \sim \mathcal{U}_{[0,1]}$ and take

$$X = k \text{ if } p_{k-1} < U < p_k.$$

For example, to generate $X \sim \mathcal{B}in(10, .3)$, the first values are

$$p_0 = 0.028, \quad p_1 = 0.149, \quad p_2 = 0.382, \ldots, p_{10} = 1,$$

and to generate $X \sim \mathcal{P}(7)$, take

$$p_0 = 0.0009, \quad p_1 = 0.0073, \quad p_2 = 0.0296, \ldots$$

the sequence being stopped when it reaches 1 with a given number of decimals. (For instance, $p_{20} = 0.999985$.) Specific algorithms, such as Example 2.9, are usually more efficient but it is mostly because of the storage problem. See Problem 2.12 and Devroye (1985).   ‖

Example 2.11. Beta generation. Consider $U_1, \ldots, U_n$, an iid sample from $\mathcal{U}_{[0,1]}$. If $U_{(1)} \le \cdots \le U_{(n)}$ denotes the ordered sample, that is, the *order statistics* of the original sample, $U_{(i)}$ is distributed as $\mathcal{B}e(i, n-i+1)$ and the vector of the differences $(U_{(i_1)}, U_{(i_2)} - U_{(i_1)}, \ldots, U_{(i_k)} - U_{(i_{k-1})}, 1 - U_{(i_k)})$ has a Dirichlet distribution $\mathcal{D}(i_1, i_2 - i_1, \ldots, n - i_k + 1)$ (see Problem 2.17). However, even though these probabilistic properties allow the direct generation of Beta and Dirichlet random variables from uniform random variables, they do not yield efficient algorithms. The calculation of the order statistics can, indeed, be quite time-consuming since it requires sorting the original sample. Moreover, it only applies for integer parameters in the Beta distribution.

The following result allows for an alternative generation of Beta random variables from uniform random variables: Jöhnk's Theorem (see Jöhnk 1964 or Devroye 1985) states that if U and V are iid $\mathcal{U}_{[0,1]}$, the distribution of

$$\frac{U^{1/\alpha}}{U^{1/\alpha} + V^{1/\beta}},$$

conditional on $U^{1/\alpha} + V^{1/\beta} \leq 1$, is the $\mathcal{B}e(\alpha, \beta)$ distribution. However, given the constraint on $U^{1/\alpha} + V^{1/\beta}$, this result does not provide a good algorithm to generate $\mathcal{B}e(\alpha, \beta)$ random variables for large values of α and β, as shown by the fast decrease of the probability of accepting a pair (U, V) as a function of $\alpha = \beta$ in Figure 2.2.   ‖

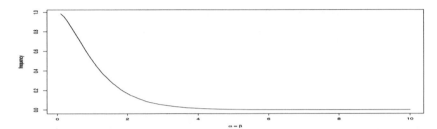

Fig. 2.2. Probability of accepting a pair (U, V) in Jöhnk (1964) algorithm as a function of α, when $\alpha = \beta$.

Example 2.12. Gamma generation. Given a generator of Beta random variables, we can derive a generator of Gamma random variables $\mathcal{G}a(\alpha, 1)$ ($\alpha < 1$) the following way: If $Y \sim \mathcal{B}e(\alpha, 1 - \alpha)$ and $Z \sim \mathcal{E}xp(1)$, then $X = YZ \sim \mathcal{G}a(\alpha, 1)$. Indeed, by making the transformation $x = yz, w = z$ and integrating the joint density, we find

$$f(x) = \frac{\Gamma(1)}{\Gamma(\alpha)\Gamma(1-\alpha)} \int_x^\infty \left(\frac{x}{w}\right)^{\alpha-1} \left(1 - \frac{x}{w}\right)^{-\alpha} w^{-1} e^{-w} dw$$

$$(2.3) \qquad = \frac{1}{\Gamma(\alpha)} x^{\alpha-1} e^{-x}.$$

Alternatively, if we can start with a Gamma random variable, a more efficient generator for $\mathcal{G}a(\alpha, 1)$ ($\alpha < 1$) can be constructed: If $Y \sim \mathcal{G}a(\alpha + 1, 1)$ and $U \sim \mathcal{U}_{[0,1]}$, independent, then $X = YU^{1/\alpha}$ is distributed according to $\mathcal{G}a(\alpha, 1)$, since

$$(2.4) \qquad f(x) \propto \int_x^\infty w^\alpha e^{-w} \left(\frac{x}{w}\right)^{\alpha-1} w^{-1} dw = x^{\alpha-1} e^{-x}.$$

(See Stuart 1962 or Problem 2.14).   ‖

The representation of a probability density as in (2.3) is a particular case of a *mixture of distributions*. Not only does such a representation induce relatively efficient simulation methods, but it is also related to methods in Chapters 9 and 10. The principle of a mixture representation is to write a density f as the marginal of another distribution, in the form

$$(2.5) \qquad f(x) = \int_{\mathcal{Y}} g(x,y)\, dy \qquad \text{or} \qquad f(x) = \sum_{i \in \mathcal{Y}} p_i\, f_i(x) \, ,$$

depending on whether $\mathcal{Y}$ is continuous or discrete. For instance, if the joint distribution $g(x,y)$ is simple to simulate, then the variable X can be obtained as a component of the generated (X,Y). Alternatively, if the component distributions $f_i(x)$ can be easily generated, X can be obtained by first choosing f_i with probability p_i and then generating an observation from f_i.

Example 2.13. Student's t generation. A useful form of (2.5) is

$$(2.6) \qquad f(x) = \int_{\mathcal{Y}} g(x,y)\, dy = \int_{\mathcal{Y}} h_1(x|y)h_2(y)\, dy \, ,$$

where h_1 and h_2 are the conditional and marginal densities of $X|Y = y$ and Y, respectively. For example, we can write Student's t density with ν degrees of freedom in this form, where

$$X|y \sim \mathcal{N}(0, \nu/y) \quad \text{and} \quad Y \sim \chi_\nu^2.$$

Such a representation is also useful for discrete distributions. In Example 2.9, we noted an alternate representation for the negative binomial distribution. If X is negative binomial, $X \sim \mathcal{N}eg(n,p)$, then $P(X = x)$ can be written as (2.6) with

$$X|y \sim \mathcal{P}(y) \quad \text{and} \quad Y \sim \mathcal{G}(n, \beta),$$

where $\beta = (1-p)/p$. Note that the discreteness of the negative binomial distribution does not result in a discrete mixture representation of the probability. The mixture is continuous, as the distribution of Y is itself continuous. ‖

Example 2.14. Noncentral chi squared generation. The noncentral chi squared distribution, $\chi_p^2(\lambda)$, also allows for a mixture representation, since it can be written as a sum of central chi squared densities. In fact, it is of the form (2.6) with h_1 the density of a χ_{p+2K}^2 distribution and h_2 the density of $\mathcal{P}(\lambda/2)$. However, this representation is not as efficient as the algorithm obtained by generating $Z \sim \chi_{p-1}^2$ and $Y \sim \mathcal{N}(\sqrt{\lambda}, 1)$, and using the fact that $Z + Y^2 \sim \chi_p^2(\lambda)$. Note that the noncentral chi squared distribution does not have an explicit form for its density function. It is either represented as an infinite mixture (see (3.31)) or by using modified Bessel functions (see Problem 1.8). ‖

In addition to the above two examples, other distributions can be represented as mixtures (see, for instance, Gleser 1989). In many cases this representation can be exploited to produce algorithms for random variable generation (see Problems 2.24–2.26, and Note 2.6.3).

2.3 Accept–Reject Methods

There are many distributions from which it is difficult, or even impossible, to directly simulate by an inverse transform. Moreover, in some cases, we are not even able to represent the distribution in a usable form, such as a transformation or a mixture. In such settings, it is impossible to exploit direct probabilistic properties to derive a simulation method. We thus turn to another class of methods that only requires us to know the functional form of the density f of interest up to a multiplicative constant; no deep analytical study of f is necessary. The key to this method is to use a simpler (simulationwise) density g from which the simulation is actually done. For a given density g— called the *instrumental density*— there are thus many densities f—called the *target densities*—which can be simulated this way. The corresponding algorithm, called *Accept–Reject*, is based on a simple connection with the uniform distribution, discussed below.

2.3.1 The Fundamental Theorem of Simulation

There exists a fundamental (simple!) idea that underlies the Accept–Reject methodology, and also plays a key role in the construction of the slice sampler (Chapter 8). If f is the density of interest, on an arbitrary space, we can write

$$(2.7) \qquad f(x) = \int_0^{f(x)} du \,.$$

Thus, f appears as the *marginal density* (in X) of the joint distribution,

$$(2.8) \qquad (X, U) \sim \mathcal{U}\{(x, u) : 0 < u < f(x)\} \,.$$

Since U is not directly related to the original problem, it is called an *auxiliary variable*, a notion to be found again in later chapters like Chapters 8–10.

Although it seems like we have not gained much, the introduction of the auxiliary uniform variable in (2.7) has brought a considerably different perspective: Since (2.8) is the joint density of X and U, we can generate from this joint distribution by just generating uniform random variables on the constrained set $\{(x, u) : 0 < u < f(x)\}$. Moreover, since the marginal distribution of X *is* the original target distribution, f, by generating a uniform variable on $\{(x, u) : 0 < u < f(x)\}$, we have generated a random variable from f. And this generation was produced without using f other than through the calculation of $f(x)$! The importance of this equivalence is stressed in the following theorem:

Theorem 2.15 (Fundamental Theorem of Simulation). *Simulating*

$$X \sim f(x)$$

is equivalent to simulating

$$(X, U) \sim \mathcal{U}\{(x, u) : 0 < u < f(x)\} \,.$$

While this theorem is fundamental in many respects, it appears mostly as a formal representation at this stage because the simulation of the uniform pair (X, U) is often not straightforward. For example, we could simulate $X \sim f(x)$ and $U|X = x \sim \mathcal{U}(0, f(x))$, but then this makes the whole representation useless. And the symmetric approach, which is to simulate U from its marginal distribution, and then X from the distribution conditional on $U = u$, does not often result in a feasible calculation. The solution is to simulate the entire pair (X, U) at once in a bigger set, where simulation is easier, and then take the pair if the constraint is satisfied.

For example, in a one-dimensional setting, suppose that

$$\int_a^b f(x)dx = 1$$

and that f is bounded by m. We can then simulate the random pair $(Y, U) \sim \mathcal{U}(0 < u < m)$ by simulating $Y \sim \mathcal{U}(a, b)$ and $U|Y = y \sim \mathcal{U}(0, m)$, and take the pair only if the further constraint $0 < u < f(y)$ is satisfied. This results in the correct distribution of the accepted value of Y, call it X, because

$$P(X \leq x) = P(Y \leq x | U < f(Y))$$

$$(2.9) \qquad = \frac{\int_a^x \int_0^{f(y)} du \, dy}{\int_a^b \int_0^{f(y)} du \, dy} = \int_a^x f(y) \, dy.$$

This amounts to saying that, if $A \subset B$ and if we generate a uniform sample on B, keeping only the terms of this sample that are in A will result in a uniform sample on A (with a random size that is independent of the values of the sample).

Example 2.16. Beta simulation. We have seen (Example 2.11) that direct simulation of Beta random variables can be difficult. However, we can easily use Theorem 2.15 for this simulation when $\alpha \geq 1$ and $\beta \geq 1$. Indeed, to generate $X \sim \mathcal{B}e(\alpha, \beta)$, we take $Y \sim \mathcal{U}_{[0,1]}$ and $U \sim \mathcal{U}_{[0,m]}$, where m is the maximum of the Beta density (Problem 2.15). For $\alpha = 2.7$ and $\beta = 6.3$ Figure 2.3 shows the results of generating 1000 pairs (Y, U). The pairs that fall under the density function are those for which we accept $X = Y$, and we reject those pairs that fall outside. ‖

In addition, it is easy to see that the probability of acceptance of a given simulation in the box $[a, b] \times [0, m]$ is given by

$$P(\text{Accept}) = P(U < f(Y)) = \frac{1}{m} \int_0^1 \int_0^{f(y)} du \, dy = \frac{1}{m}.$$

For Example 2.16, $m = 2.67$, so we accept approximately $1/2.67 = 37\%$ of the values.

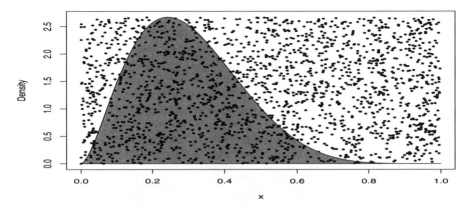

Fig. 2.3. Generation of Beta random variables: Using Theorem 2.15, 1000 (Y, U) pairs were generated, and 365 were accepted (the black circles under the Beta $\mathcal{B}e(2.7, 6.3)$ density function).

The argument leading to (2.9) can easily be generalized to the situation where the larger set is not a box any longer, as long as simulating uniformly over this larger set is feasible. This generalization may then allow for cases where either or both of the support of f and the maximum of f are unbounded. If the larger set is of the form

$$\mathscr{L} = \{(y, u) : 0 < u < m(y)\} ,$$

the constraints are thus that $m(x) \geq f(x)$ and that simulation of a uniform on $\mathscr{L}$ is feasible. Obviously, efficiency dictates that m be as close as possible to f in order to avoid wasting simulations. A remark of importance is that, because of the constraint $m(x) \geq f(x)$, m cannot be a probability density. We then write

$$m(x) = Mg(x) \text{ where } \int_{\mathcal{X}} m(x)\, dx = \int_{\mathcal{X}} Mg(x)\, dx = M ,$$

since m is necessarily integrable (otherwise, $\mathscr{L}$ would not have finite mass and a uniform distribution would not exist on $\mathscr{L}$). As mentioned above, a natural way of simulating the uniform on $\mathscr{L}$ is then to use (2.7) backwards, that is, to simulate $Y \sim g$ and then $U | Y = y \sim \mathcal{U}(0, Mg(y))$. If we only accept the y's such that the constraint $u < f(y)$ is satisfied, we have

$$P(X \in \mathcal{A}) = P(Y \in \mathcal{A} | U < f(Y))$$
$$= \frac{\int_{\mathcal{A}} \int_0^{f(y)} \frac{du}{Mg(y)}\, g(y) dy}{\int_{\mathcal{A}} \int_0^{f(y)} \frac{du}{Mg(y)}\, g(y) dy} = \int_{\mathcal{A}} f(y)\, dy$$

for every measurable set $\mathcal{A}$ and the accepted X's are indeed distributed from f. We have thus derived a more general implementation of the fundamental theorem, as follows:

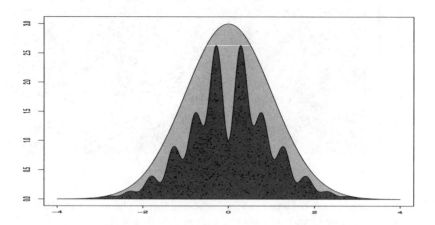

Fig. 2.4. Plot of a uniform sample over the set $\{(x, u) : 0 < u < f(x)\}$ for $f(x) \propto \exp(-x^2/2)(\sin(6x)^2 + 3\cos(x)^2 \sin(4x)^2 + 1)$ and of the envelope function $g(x) = 5\exp(-x^2/2)$.

Corollary 2.17. *Let $X \sim f(x)$ and let $g(x)$ be a density function that satisfies $f(x) \leq Mg(x)$ for some constant $M \geq 1$. Then, to simulate $X \sim f$, it is sufficient to generate*

$$Y \sim g \qquad and \qquad U|Y = y \sim \mathcal{U}(0, Mg(y)),$$

until $0 < u < f(y)$.

Figure 2.4 illustrates Corollary 2.17 for the target density

$$f(x) \propto \exp(-x^2/2)(\sin(6x)^2 + 3\cos(x)^2 \sin(4x)^2 + 1)$$

with upper bound (or, rather, dominating density) the normal density

$$g(x) = \exp(-x^2/2)/\sqrt{2\pi},$$

which is obviously straightforward to generate.

Corollary 2.17 has two consequences. First, it provides a generic method to simulate from any density f that is known *up to a multiplicative factor*; that is, the normalizing constant of f need not be known, since the method only requires input of the ratio f/M, which does not depend on the normalizing constant. This is for instance the case of Figure 2.4, where the normalizing constant of f is unknown. This property is particularly important in Bayesian calculations. There, a quantity of interest is the posterior distribution, defined according to Bayes Theorem by

$$(2.10) \qquad \qquad \pi(\theta|x) \propto \pi(\theta) \, f(x|\theta) \, .$$

Thus, the posterior density $\pi(\theta|x)$ is easily specified up to a normalizing constant and, to use Corollary 2.17, this constant need not be calculated. (See Problem 2.29.)

Of course, there remains the task of finding a density g satisfying $f \leq Mg$, a bound that need not be tight, in the sense that Corollary 2.17 remains valid when M is replaced with any larger constant. (See Problem 2.30.)

A second consequence of Corollary 2.17 is that the probability of acceptance is exactly $1/M$ (a geometric waiting time), when evaluated for the properly normalized densities, and the expected number of trials until a variable is accepted is M (see Problem 2.30). Thus, a comparison between different simulations based on different instrumental densities $g_1, g_2, \ldots$ can be undertaken through the comparison of the respective bounds $M_1, M_2, \ldots$ (as long as the corresponding densities $g_1, g_2, \ldots$ are correctly normalized). In particular, a first method of optimizing the choice of g in $g_1, g_2, \ldots$ is to find the smallest bound M_i. However, this first and rudimentary comparison technique has some limitations, which we will see later in this chapter.

2.3.2 The Accept–Reject Algorithm

The implementation of Corollary 2.17 is known as the *Accept–Reject method*, which is usually stated in the slightly modified, but equivalent form. (See Problem 2.28 for extensions.)

Algorithm A.4 −Accept–Reject Method−

1. Generate $X \sim g$, $U \sim \mathcal{U}_{[0,1]}$;
2. Accept $Y = X$ if $U \leq f(X)/Mg(X)$;   [A.4]
3. Return to 1. otherwise.

In cases where f and g are normalized so they are both probability densities, the constant M is necessarily larger than 1. Therefore, the size of M, and thus the efficiency of [A.4], becomes a function of how closely g can imitate f, especially in the tails of the distribution. Note that for f/g to remain bounded, it is necessary for g to have tails thicker than those of f. It is therefore impossible for instance to use [A.4] to simulate a Cauchy distribution f using a normal distribution g; however, the reverse works quite well. (See Problem 2.34.) Interestingly enough, the opposite case when g/f is bounded can also be processed by a tailored Markov chain Monte Carlo algorithm derived from Doukhan et al. (1994) (see Problems 7.5 and 7.6).

A limited optimization of the Accept–Reject algorithm is possible by choosing the instrumental density g in a parametric family, and then determining the value of the parameter which minimizes the bound M. A similar comparison between two parametric families is much more delicate since it is then necessary to take into account the computation time of one generation from g in [A.4]. In fact, pushing the reasoning to the limit, if $g = f$

and if we simulate $X \sim f$ by numerical inversion of the distribution function, we formally achieve the minimal bound $M = 1$, but this does not guarantee that we have an efficient algorithm, as can be seen in the case of the normal distribution.

Example 2.18. Normals from double exponentials. Consider generating a $\mathcal{N}(0,1)$ by [A.4] using a double-exponential distribution $\mathcal{L}(\alpha)$, with density $g(x|\alpha) = (\alpha/2)\exp(-\alpha|x|)$. It is then straightforward to show that

$$\frac{f(x)}{g(x|\alpha)} \leq \sqrt{2/\pi}\,\alpha^{-1}e^{\alpha^2/2}$$

and that the minimum of this bound (in α) is attained for $\alpha = 1$. The probability of acceptance is then $\sqrt{\pi/2e} = .76$, which shows that to produce one normal random variable, this Accept–Reject algorithm requires on the average $1/.76 \approx 1.3$ uniform variables, to be compared with the fixed single uniform required by the Box–Muller algorithm. ‖

A real advantage of the Accept–Reject algorithm is illustrated in the following example.

Example 2.19. Gamma Accept–Reject We saw in Example 2.7 that if $\alpha \in \mathbb{N}$, the Gamma distribution $\mathcal{G}a(\alpha, \beta)$ can be represented as the sum of α exponential random variables $\epsilon_i \sim \mathcal{E}xp(\beta)$, which are very easy to simulate, since $\epsilon_i = -\log(U_i)/\beta$, with $U_i \sim \mathcal{U}([0,1])$. In more general cases (for example when $\alpha \notin \mathbb{N}$), this representation does not hold.

A possible approach is to use the Accept–Reject algorithm with instrumental distribution $\mathcal{G}a(a, b)$, with $a = [\alpha]$ ($\alpha \geq 1$). (Without loss of generality, suppose $\beta = 1$.) The ratio f/g is $b^{-a}x^{\alpha-a}\exp\{-(1-b)x\}$, up to a normalizing constant, yielding the bound

$$M = b^{-a}\left(\frac{\alpha - a}{(1 - b)e}\right)^{\alpha-a}$$

for $b < 1$. Since the maximum of $b^{-a}(1 - b)^{\alpha-a}$ is attained at $b = a/\alpha$, the optimal choice of b for simulating $\mathcal{G}a(\alpha, 1)$ is $b = a/\alpha$, which gives the same mean for $\mathcal{G}a(\alpha, 1)$ and $\mathcal{G}a(a, b)$. (See Problem 2.31.) ‖

It may also happen that the complexity of the optimization is very expensive in terms of analysis or of computing time. In the first case, the construction of the optimal algorithm should still be undertaken when the algorithm is to be subjected to intensive use. In the second case, it is most often preferable to explore the use of another family of instrumental distributions g.

Example 2.20. Truncated normal distributions. *Truncated normal distributions* appear in many contexts, such as in the discussion after Example 1.5. When constraints $x \geq \underline{\mu}$ produce densities proportional to

$$e^{-(x-\mu)^2/2\sigma^2}\ \mathbb{I}_{x \geq \underline{\mu}}$$

for a bound $\underline{\mu}$ large compared with μ, there are alternatives which are far superior to the naïve method in which a $\mathcal{N}(\mu,\sigma^2)$ distribution is simulated until the generated value is larger than $\underline{\mu}$. (This approach requires an average number of $1/\varPhi((\mu - \underline{\mu})/\sigma)$ simulations from $\mathcal{N}(\mu,\sigma^2)$ for one acceptance.) Consider, without loss of generality, the case $\mu = 0$ and $\sigma = 1$. A potential instrumental distribution is the translated exponential distribution, $\mathcal{E}xp(\alpha,\underline{\mu})$, with density

$$g_\alpha(z) = \alpha e^{-\alpha(z-\underline{\mu})}\ \mathbb{I}_{z \geq \underline{\mu}}\ .$$

The ratio $f/g_\alpha(z) = e^{-\alpha(z-\underline{\mu})}\ e^{-z^2/2}$ is then bounded by $\exp(\alpha^2/2 - \alpha\underline{\mu})$ if $\alpha > \underline{\mu}$ and by $\exp(-\underline{\mu}^2/2)$ otherwise. The corresponding (upper) bound is

$$\begin{cases} 1/\alpha\ \exp(\alpha^2/2 - \alpha\underline{\mu}) & \text{if } \alpha > \underline{\mu}, \\ 1/\alpha\ \exp(-\underline{\mu}^2/2) & \text{otherwise.} \end{cases}$$

The first expression is minimized by

(2.11)
$$\alpha^* = \underline{\mu} + \frac{1}{2}\sqrt{\underline{\mu}^2 + 4}\ ,$$

whereas $\tilde{\alpha} = \underline{\mu}$ minimizes the second bound. The optimal choice of α is therefore (2.11), which requires the computation of the square root of $\underline{\mu}^2 + 4$. Robert (1995b) proposes a similar algorithm for the case where the normal distribution is restricted to the interval $[\underline{\mu}, \overline{\mu}]$. For some values of $[\underline{\mu}, \overline{\mu}]$, the optimal algorithm is associated with a value of α, which is a solution to an implicit equation. (See also Geweke 1991 for a similar resolution of this simulation problem and Marsaglia 1964 for an earlier solution.)   ‖

One criticism of the Accept–Reject algorithm is that it generates "useless" simulations when rejecting. We will see in Chapter 3 how the method of importance sampling (Section 3.3) can be used to bypass this problem and also how both methods can be compared.

2.4 Envelope Accept–Reject Methods

2.4.1 The Squeeze Principle

In numerous settings, the distribution associated with the density f is difficult to simulate because of the complexity of the function f itself, which may require substantial computing time at each evaluation. In the setup of Example 1.9 for instance, if a Bayesian approach is taken with θ distributed (a posteriori) as

$$(2.12) \qquad \prod_{i=1}^{n} \left[1 + \frac{(x_i - \theta)^2}{p\sigma^2} \right]^{-\frac{p+1}{2}},$$

where σ is known, each single evaluation of $\pi(\theta|x)$ involves the computation of n terms in the product. It turns out that an acceleration of the simulation of densities such as (2.12) can be accomplished by an algorithm that is "one step beyond" the Accept–Reject algorithm. This algorithm is an *envelope* algorithm and relies on the evaluation of a simpler function g_l which bounds the target density f from below. The algorithm is based on the following extension of Corollary 2.17 (see Problems 2.35 and 2.36).

Lemma 2.21. *If there exist a density g_m, a function g_l and a constant M such that*

$$g_l(x) \leq f(x) \leq M g_m(x),$$

then the algorithm

Algorithm A.5 –Envelope Accept–Reject–

> 1. Generate $X \sim g_m(x)$, $U \sim \mathcal{U}_{[0,1]}$;
> 2. Accept X if $U \leq g_l(X)/M g_m(X)$;   [A.5]
> 3. otherwise, accept X if $U \leq f(X)/M g_m(X)$

produces random variables that are distributed according to f.

By the construction of a lower envelope on f, based on the function g_l, the number of evaluations of f is potentially decreased by a factor

$$\frac{1}{M} \int g_l(x)\, dx,$$

which is the probability that f is not evaluated. This method is called the *squeeze principle* by Marsaglia (1977) and the ARS algorithm [A.7] in Section 2.4 is based on it. A possible way of deriving the bounds g_l and $M g_m$ is to use a Taylor expansion of $f(x)$.

Example 2.22. Lower bound for normal generation. It follows from the Taylor series expansion of $\exp(-x^2/2)$ that $\exp(-x^2/2) \geq 1 - (x^2/2)$, and hence

$$\left(1 - \frac{x^2}{2} \right) \leq f(x),$$

which can be interpreted as a lower bound for the simulation of $\mathcal{N}(0,1)$. This bound is obviously useless when $|X| < \sqrt{2}$, an event which occurs with probability 0.61 for $X \sim \mathcal{C}(0,1)$. ‖

Example 2.23. Poisson variables from logistic variables. As indicated in Example 2.9, the simulation of the Poisson distribution $\mathcal{P}(\lambda)$ using a Poisson process and exponential variables can be rather inefficient. Here, we describe a simpler alternative of Atkinson (1979), who uses the relationship between the Poisson $\mathcal{P}(\lambda)$ distribution and the *logistic distribution*. The logistic distribution has density and distribution function

$$f(x) = \frac{1}{\beta} \frac{\exp\{-(x-\alpha)/\beta\}}{[1 + \exp\{-(x-\alpha)/\beta\}]^2} \quad \text{and} \quad F(x) = \frac{1}{1 + \exp\{-(x-\alpha)/\beta\}}$$

and is therefore analytically invertible.

To better relate the continuous and discrete distributions, we consider $N = \lfloor x + 0.5 \rfloor$, the integer part of $x + 0.5$. Also, the range of the logistic distribution is $(-\infty, \infty)$, but to better match it with the Poisson, we restrict the range to $[-1/2, \infty)$. Thus, the random variable N has distribution function

$$P(N = n) = \frac{1}{1 + e^{-(n+0.5-\alpha)/\beta}} - \frac{1}{1 + e^{-(n-0.5-\alpha)/\beta}}$$

if $x > 1/2$ and

$$P(N = n) = \left(\frac{1}{1 + e^{-(n+0.5-\alpha)/\beta}} - \frac{1}{1 + e^{-(n-0.5-\alpha)/\beta}} \right) \frac{1 + e^{-(0.5+\alpha)/\beta}}{e^{-(0.5+\alpha)/\beta}}$$

if $-1/2 < x \leq 1/2$ and the ratio of the densities is

$$\lambda^n / P(N = n) \, e^\lambda \, n! \, . \tag{2.13}$$

Although it is difficult to compute a bound on (2.13) and, hence, to optimize it in (α, β), Atkinson (1979) proposed the choice $\alpha = \lambda$ and $\beta = \pi/\sqrt{3\lambda}$. This identifies the two first moments of X with those of $\mathcal{P}(\lambda)$. For this choice of α and β, analytic optimization of the bound on (2.13) remains impossible, but numerical maximization and interpolation yields the bound $c = 0.767 - 3.36/\lambda$. The resulting algorithm is then

Algorithm A.6 –Atkinson's Poisson Simulation–

0. Define $\beta = \pi/\sqrt{3\lambda}$, $\quad \alpha = \lambda\beta \quad$ and $\quad k = \log c - \lambda - \log \beta$;
1. Generate $U_1 \sim \mathcal{U}_{[0,1]}$ and calculate

$$x = \{\alpha - \log\{(1 - u_1)/u_1\}\}/\beta$$

   until $X > -0.5$;
2. Define $N = \lfloor X + 0.5 \rfloor$ and generate $U_2 \sim \mathcal{U}_{[0,1]}$;
3. Accept $N \sim \mathcal{P}(\lambda)$ if   [A.6]

$$\alpha - \beta x + \log\left(u_2/\{1 + \exp(\alpha - \beta x)\}^2\right) \leq k + N \log \lambda - \log N! \, .$$

Although the resulting simulation is exact, this algorithm is based on a number of approximations, both through the choice of (α, β) and in the computation of the majorization bounds and the density ratios. Moreover, note that it requires the computation of factorials, $N!$, which may be quite time-consuming. Therefore, although [A.6] usually has a reasonable efficiency, more complex algorithms such as those of Devroye (1985) may be preferable.   ∥

2.4.2 Log-Concave Densities

The particular case of *log-concave densities* (that is, densities whose logarithm is concave) allows the construction of a generic algorithm that can be quite efficient.

Example 2.24. Log-concave densities.   Recall the exponential family (1.9)

$$f(x) = h(x) \, e^{\theta \cdot x - \psi(\theta)}, \qquad \theta, x \in \mathbb{R}^k.$$

This density is log-concave if

$$\frac{\partial^2}{\partial x^2} \log f(x) = \frac{\partial^2}{\partial x^2} \log h(x) = \frac{h(x)h''(x) - [h'(x)]^2}{h^2(x)} < 0,$$

which will often be the case for the exponential family. For example, if $X \sim \mathcal{N}(\theta, 1)$, then $h(x) \propto \exp\{-x^2/2\}$ and $\partial^2 \log h(x)/\partial x^2 = -1$. See Problems 2.40–2.42 for properties and examples of log-concave densities.   ∥

Devroye (1985) describes some algorithms that take advantage of the log-concavity of the density, but here we present a universal method. The algorithm, which was proposed by Gilks (1992) and Gilks and Wild (1992), is based on the construction of an envelope and the derivation of a corresponding Accept–Reject algorithm. The method is called *adaptive rejection sampling* (ARS) and it provides a sequential evaluation of lower and upper envelopes of the density f when $h = \log f$ is concave.

Let $\mathcal{S}_n$ be a set of points $x_i, i = 0, 1, \ldots, n+1$, in the support of f such that $h(x_i) = \log f(x_i)$ is known up to the same constant. Given the concavity of h, the line $L_{i,i+1}$ through $(x_i, h(x_i))$ and $(x_{i+1}, h(x_{i+1}))$ is below the graph of h in $[x_i, x_{i+1}]$ and is above this graph outside this interval (see Figure 2.5). For $x \in [x_i, x_{i+1}]$, if we define

$$\overline{h}_n(x) = \min\{L_{i-1,i}(x), L_{i+1,i+2}(x)\} \quad \text{and} \quad \underline{h}_n(x) = L_{i,i+1}(x),$$

the envelopes are

(2.14) $$\underline{h}_n(x) \le h(x) \le \overline{h}_n(x)$$

uniformly on the support of f. (We define

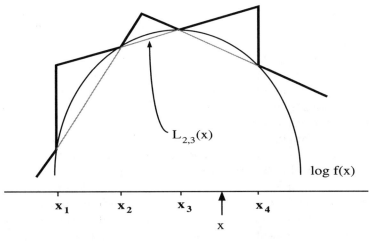

Fig. 2.5. Lower and upper envelopes of $h(x) = \log f(x)$, f a log-concave density (*Source:* Gilks et al. 1995).

$$\underline{h}_n(x) = -\infty \quad \text{and} \quad \overline{h}_n(x) = \min(L_{0,1}(x), L_{n,n+1}(x))$$

on $[x_0, x_{n+1}]^c$.) Therefore, for $\underline{f}_n(x) = \exp \underline{h}_n(x)$ and $\overline{f}_n(x) = \exp \overline{h}_n(x)$, (2.14) implies that

$$\underline{f}_n(x) \le f(x) \le \overline{f}_n(x) = \varpi_n \, g_n(x) \,,$$

where ϖ_n is the normalized constant of f_n; that is, g_n is a density. The ARS algorithm to generate an observation from f is thus

Algorithm A.7 –ARS Algorithm–

```
0. Initialize n and S_n.
1. Generate X ~ g_n(x), U ~ U_[0,1].
2. If U ≤ f_n(X)/ϖ_n g_n(X), accept X;            [A.7]
   otherwise, if U ≤ f(X)/ϖ_n g_n(X), accept X
   and update S_n to S_{n+1} = S_n ∪ {X}.
```

An interesting feature of this algorithm is that the set S_n is only updated when $f(x)$ has been previously computed. As the algorithm produces variables $X \sim f(x)$, the two envelopes $\underline{f}_n$ and $\overline{f}_n$ become increasingly accurate and, therefore, we progressively reduce the number of evaluations of f. Note that in the initialization of S_n, a necessary condition is that $\varpi_n < +\infty$ (i.e., that g_n is actually a probability density). To achieve this requirement, $L_{0,1}$ needs to have positive slope if the support of f is not bounded on the left and $L_{n,n+1}$ needs to have a negative slope if the support of f is not bounded on the right. (See Problem 2.39 for more on simulation from g_n.)

The ARS algorithm is not optimal in the sense that it is often possible to devise a better specialized algorithm for a given log-concave density. However, although Gilks and Wild (1992) do not provide theoretical evaluations of simulation speeds, they mention reasonable performances in the cases they consider. Note that, in contrast to the previous algorithms, the function g_n is updated during the iterations and, therefore, the average computation time for one generation from f decreases with n. This feature makes the comparison with other approaches quite delicate.

The major advantage of [A.7] compared with alternatives is its *universality*. For densities f that are only known through their functional form, the ARS algorithm yields an automatic Accept–Reject algorithm that only requires checking f for log-concavity. Moreover, the set of log-concave densities is wide; see Problems 2.40 and 2.41. The ARS algorithm thus allows for the generation of samples from distributions that are rarely simulated, without requiring the development of case-specific Accept–Reject algorithms.

Example 2.25. Capture–recapture models. In a heterogeneous *capture–recapture model* (see Seber 1983, 1992 or Borchers et al. 2002), animals are captured at time i with probability p_i, the size N of the population being unknown. The corresponding likelihood is therefore

$$L(p_1, \ldots, p_I | N, n_1, \ldots, n_I) = \frac{N!}{(N-r)!} \prod_{i=1}^{I} p_i^{n_i} (1 - p_i)^{N-n_i},$$

where I is the number of captures, n_i is the number of captured animals during the ith capture, and r is the total number of different captured animals. If N is a priori distributed as a $\mathcal{P}(\lambda)$ variable and the p_i's are from a *normal logistic* model,

$$\alpha_i = \log\left(\frac{p_i}{1 - p_i}\right) \sim \mathcal{N}(\mu_i, \sigma^2),$$

as in George and Robert (1992), the posterior distribution satisfies

$$\pi(\alpha_i | N, n_1, \ldots, n_I) \propto \exp\left\{\alpha_i n_i - \frac{1}{2\sigma^2}(\alpha_i - \mu_i)^2\right\} \Big/ (1 + e^{\alpha_i})^N .$$

If this conditional distribution must be simulated (for reasons which will be made clearer in Chapters 9 and 10), the ARS algorithm can be implemented. In fact, the log of the posterior distribution

$$(2.15) \qquad \alpha_i n_i - \frac{1}{2\sigma^2}(\alpha_i - \mu_i)^2 - N \log(1 + e^{\alpha_i})$$

is concave in α_i, as can be shown by computing the second derivative (see also Problem 2.42).

As an illustration, consider the dataset $(n_1, \ldots, n_{11}) = (32, 20, 8, 5, 1, 2, 0, 2, 1, 1, 0)$ which describes the number of recoveries over the years 1957–1968 of

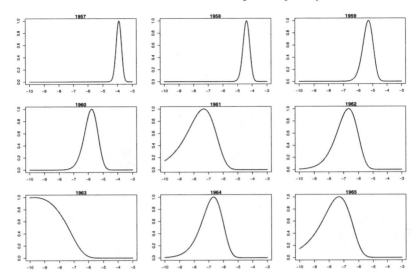

Fig. 2.6. Posterior distributions of the capture log-odds ratios for the Northern Pintail duck dataset of Johnson and Hoeting (2003) for the years 1957–1965.

$N = 1612$ Northern Pintail ducks banded in 1956, as reported in Johnson and Hoeting (2003). Figure 2.6 provides the corresponding posterior distributions for the first 9 α_i's. The ARS algorithm can then be used independently for each of these distributions. For instance, if we take the year 1960, the starting points in $\mathcal{S}$ can be $-10, -6$ and -3. The set $\mathcal{S}$ then gets updated along iterations as in Algorithm [A.7], which provides a correct simulation from the posterior distributions of the α_i's, as illustrated in Figure 2.7 for the year 1960. ‖

The above example also illustrates that checking for log-concavity of a Bayesian posterior distribution is straightforward, as log $\pi(\theta|x) = $ log $\pi(\theta)$ +log $f(x|\theta)+c$, where c is a constant (in θ). This implies that the log-concavity of $\pi(\theta)$ and of $f(x|\theta)$ (in θ) are sufficient to conclude the log-concavity of $\pi(\theta|x)$.

Example 2.26. Poisson regression. Consider a sample $(Y_1, x_1), \ldots, (Y_n, x_n)$ of integer-valued data Y_i with explanatory variable x_i, where Y_i and x_i are connected via a Poisson distribution,

$$Y_i|x_i \sim \mathcal{P}(\exp\{a + bx_i\}) \, .$$

If the prior distribution of (a, b) is a normal distribution $\mathcal{N}(0, \sigma^2) \times \mathcal{N}(0, \tau^2)$, the posterior distribution of (a, b) is given by

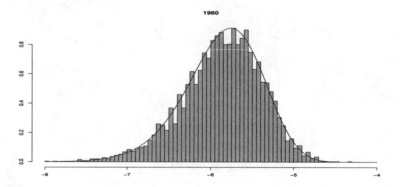

Fig. 2.7. Histogram of an ARS sample of 5000 points and corresponding posterior distribution of the log-odds ratio α_{1960}.

$$\pi(a,b|\mathbf{x},\mathbf{y}) \propto \exp\left\{ a\sum_i y_i + b\sum_i y_i x_i - e^a \sum_i e^{x_i b} \right\} e^{-a^2/2\sigma^2} e^{-b^2/2\tau^2}.$$

We will see in Chapter 9 that it is often of interest to simulate successively the (full) conditional distributions $\pi(a,|\mathbf{x},\mathbf{y},b)$ and $\pi(b|\mathbf{x},\mathbf{y},a)$. Since

$$\log \pi(a|\mathbf{x},\mathbf{y},b) = a\sum_i y_i - e^a \sum_i e^{x_i b} - a^2/2\sigma^2 \,,$$

$$\log \pi(b|\mathbf{x},\mathbf{y},a) = b\sum_i y_i x_i - e^a \sum_i e^{x_i b} - b^2/2\tau^2 \,,$$

and

$$\frac{\partial^2}{\partial a^2} \log \pi(a|\mathbf{x},\mathbf{y},b) = -\sum_i e^{x_i b} \, e^a - \sigma^{-2} < 0 \,,$$

$$\frac{\partial^2}{\partial b^2} \log \pi(b|\mathbf{x},\mathbf{y},a) = -e^a \sum_i x_i^2 \, e^{x_i b} - \tau^{-2} < 0 \,,$$

the ARS algorithm directly applies for both conditional distributions.

As an illustration, consider the data in Table 2.1. This rather famous data set gives the deaths in the Prussian Army due to kicks from horses, gathered by von Bortkiewicz (1898). A question of interest is whether there is a trend in the deaths over time. For illustration here, we show how to generate the conditional distribution of the intercept, $\pi(a|\mathbf{x},\mathbf{y},b)$, since the generation of the other conditional is quite similar.

Before implementing the ARS algorithm, we note two simplifying things. One, if $f(x)$ is easy to compute (as in this example), there is really no need to construct $\underline{f}_n(x)$, and we just skip that step in Algorithm [A.7]. Second, we do not need to construct the function g_n, we only need to know how to simulate

Table 2.1. Data from the 19th century study by Bortkiewicz (1898) of deaths in the Prussian army due to horse kicks. The data are the number of deaths in fourteen army corps from 1875 to 1894.

Year	75	76	77	78	79	80	81	82	83	84
Deaths	3	5	7	9	10	18	6	14	11	9
Year	85	86	87	88	89	90	91	92	93	94
Deaths	5	11	15	6	11	17	12	15	8	4

from it. To do this, we only need to compute the area of each segment above the intervals $[x_i, x_{i+1}]$.

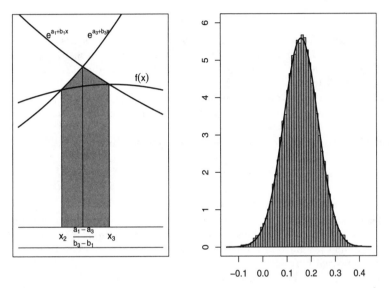

Fig. 2.8. Left panel is the area of integration for the weight of the interval $[x_2, x_3]$. The right panel is the histogram and density of the sample from g_n, with $b = .025$ and $\sigma^2 = 5$.

The left panel of Figure 2.8 shows the region of the support of $f(x)$ between x_2 and x_3, with the grey shaded area proportional to the probability of selecting the region $[x_2, x_3]$. If we denote by $a_i + b_i x$ the line through $(x_i, h(x_i))$ and $(x_{i+1}, h(x_{i+1}))$, then the area of the grey region of Figure 2.8 is

$$
\omega_2 = \int_{x_2}^{\frac{a_1 - a_3}{b_3 - b_1}} e^{a_1 + b_1 x} \, dx + \int_{\frac{a_1 - a_3}{b_3 - b_1}}^{x_3} e^{a_3 + b_3 x} \, dx
$$

(2.16)
$$
= \frac{e^{a_1}}{b_1} \left[e^{\frac{a_1 - a_3}{b_3 - b_1} b_1} - e^{b_1 x_2} \right] + \frac{e^{a_3}}{b_3} \left[e^{b_3 x_3} - e^{\frac{a_1 - a_3}{b_3 - b_1} b_3} \right].
$$

Thus, to sample from g_n we choose a region $[x_i, x_{i+1}]$ proportional to ω_i, generate $U \sim \mathcal{U}_{[0,1]}$ and then take

$$X = x_i + U(x_{i+1} - x_i.)$$

The right panel of Figure 2.8 shows the good agreement between the histogram and density of g_n. (See Problems 2.37 and 2.38 for generating the g_n corresponding to $\pi(b|\mathbf{x}, \mathbf{y}, a)$, and Problems 9.7 and 9.8 for full Gibbs samplers.)   ‖

2.5 Problems

2.1 Check the uniform random number generator on your computer:
   (a) Generate 1,000 uniform random variables and make a histogram
   (b) Generate uniform random variables $(X_1, \ldots, X_n)$ and plot the pairs (X_i, X_{i+1}) to check for autocorrelation.
2.2 (a) Generate a binomial $Bin(n, p)$ random variable with $n = 25$ and $p = .2$. Make a histogram and compare it to the binomial mass function, and to the R binomial generator.
   (b) Generate $5,000$ *logarithmic series* random variables with mass function

$$P(X = x) = \frac{-(1-p)^x}{x \log p}, \quad x = 1, 2, \ldots \quad 0 < p < 1.$$

Make a histogram and plot the mass function.
2.3 In each case generate the random variables and compare to the density function
   (a) Normal random variables using a Cauchy candidate in Accept–Reject;
   (b) Gamma $\mathcal{G}a(4.3, 6.2)$ random variables using a Gamma Ga$(4, 7)$;
   (c) Truncated normal: Standard normal truncated to $(2, \infty)$.
2.4 The arcsine distribution was discussed in Example 2.2.
   (a) Show that the *arcsine distribution*, with density $f(x) = 1/\pi\sqrt{x(1-x)}$, is invariant under the transform $y = 1 - x$, that is, $f(x) = f(y)$.
   (b) Show that the uniform distribution $\mathcal{U}_{[0,1]}$ is invariant under the "tent" transform,

$$D(x) = \begin{cases} 2x & \text{if } x \le 1/2 \\ 2(1-x) & \text{if } x > 1/2. \end{cases}$$

   (c) As in Example 2.2, use both the arcsine and "tent" distributions in the dynamic system $X_{n+1} = D(X_n)$ to generate 100 uniform random variables. Check the properties with marginal histograms, and plots of the successive iterates.
   (d) The tent distribution can have disastrous behavior. Given the finite representation of real numbers in the computer, show that the sequence (X_n) will converge to a fixed value, as the tent function progressively eliminates the last decimals of X_n. (For example, examine what happens when the sequence starts at a value of the form $1/2^n$.)

2.5 For each of the following distributions, calculate the explicit form of the distribution function and show how to implement its generation starting from a uniform random variable: (a) exponential; (b) double exponential; (c) Weibull; (d) Pareto; (e) Cauchy; (f) extreme value; (g) arcsine.

2.6 Referring to Example 2.7:
  (a) Show that if $U \sim \mathcal{U}_{[0,1]}$, then $X = -\log U/\lambda \sim \mathcal{E}xp(\lambda)$.
  (b) Verify the distributions in (2.2).
  (c) Show how to generate an $\mathcal{F}_{m,n}$ random variable, where both m and n are even integers.
  (d) Show that if $U \sim \mathcal{U}_{[0,1]}$, then $X = \log \frac{u}{1-u}$ is a Logistic$(0,1)$ random variable. Show also how to generate a Logistic(μ, β) random variable.

2.7 Establish the properties of the *Box–Muller algorithm* of Example 2.8. If U_1 and U_2 are iid $\mathcal{U}_{[0,1]}$, show that:
  (a) The transforms

$$X_1 = \sqrt{-2\log(U_1)} \, \cos(2\pi U_2), \qquad X_2 = \sqrt{-2\log(U_1)} \, \sin(2\pi U_2),$$

  are iid $\mathcal{N}(0,1)$.
  (b) The polar coordinates are distributed as

$$r^2 = X_1^2 + X_2^2 \sim \chi_2^2,$$

$$\theta = \arctan\frac{X_1}{X_2} \sim \mathcal{U}[0, 2\pi].$$

  (c) Establish that $\exp(-r^2/2) \sim \mathcal{U}[0,1]$, and so r^2 and θ can be simulated directly.

2.8 (Continuation of Problem 2.7)
  (a) Show that an alternate version of the Box–Muller algorithm is
  Algorithm A.8 –Box–Muller (2)–

  1. Generate   [A.8]

$$U_1, U_2 \sim \mathcal{U}([-1, 1])$$

  until $S = U_1^2 + U_2^2 \leq 1.$
  2. Define $Z = \sqrt{-2\log(S)/S}$ and take

$$X_1 = Z\, U_1, \qquad X_2 = Z\, U_2.$$

  (*Hint:* Show that (U_1, U_2) is uniform on the unit sphere and that X_1 and X_2 are independent.)
  (b) Give the average number of generations in **1.** and compare with the original Box–Muller algorithm [A.3] on a small experiment.
  (c) Examine the effect of not constraining (U_1, U_2) to the unit circle.

2.9 Show that the following version of the Box–Muller algorithm produces one normal variable and compare the execution time with both versions [A.3] and [A.8]:

Algorithm A.9 –Box–Muller (3)–

1. Generate
$$Y_1, Y_2 \sim \mathcal{E}xp(1)$$
   until $Y_2 > (1 - Y_1)^2/2$.   [A.9]
2. Generate $U \sim \mathcal{U}([0, 1])$ and take

$$X = \begin{cases} Y_1 & \text{if } U < 0.5 \\ -Y_1 & \text{if } U > 0.5. \end{cases}$$

2.10 Examine the properties of an algorithm to simulate $\mathcal{N}(0, 1)$ random variables based on the Central Limit Theorem, which takes the appropriately adjusted mean of a sequence of uniforms $U_1, \ldots, U_n$ for $n = 12$, $n = 48$ and $n = 96$. Consider, in particular, the moments, ranges, and tail probability calculations based on the generated variables.

2.11 For the generation of a Cauchy random variable, compare the inversion method with one based on the generation of the normal pair of the polar Box–Muller method (Problem 2.8).

(a) Show that, if X_1 and X_2 are iid normal, $Y = X_1/X_2$ is distributed as a Cauchy random variable.

(b) Show that the Cauchy distribution function is $F(x) = \tan^{-1}(x)/\pi$, so the inversion method is easily implemented.

(c) Is one of the two algorithms superior?

2.12 Use the algorithm of Example 2.10 to generate the following random variables. In each case make a histogram and compare it to the mass function, and to the generator in your computer.

(a) Binomials and Poisson distributions;

(b) The hypergeometric distribution;

(c) The *logarithmic series* distribution. A random variable X has a logarithmic series distribution with parameter p if

$$P(X = x) = \frac{-(1 - p)^x}{x \log p}, \quad x = 1, 2, \ldots, \quad 0 < p < 1.$$

(d) Referring to part (a), for different parameter values, compare the algorithms there with those of Problems 2.13 and 2.16.

2.13 Referring to Example 2.9.

(a) Show that if $N \sim \mathcal{P}(\lambda)$ and $X_i \sim \mathcal{E}xp(\lambda)$, $i \in \mathbb{N}^*$, independent, then

$$P_\lambda(N = k) = P_\lambda(X_1 + \cdots + X_k \leq 1 < X_1 + \cdots + X_{k+1}).$$

(b) Use the results of part (a) to justify that the following algorithm simulates a Poisson $\mathcal{P}(\lambda)$ random variable:

Algorithm A.10 –Poisson simulation–

$p = 1$, $N = 0$, $c = e^{-\lambda}$.

1. Repeat   [A.10]
      $N = N + 1$
      generate U_i
      update $p = pU_i$
   until $p < c$.
2. Take $X = N - 1$.

(*Hint:* For part (*a*), integrate the Gamma density by parts.)

2.14 There are (at least) two ways to establish (2.4) of Example 2.12:
  (a) Make the transformation $x = yu^{1/\alpha}$, $w = y$, and integrate out w.
  (b) Make the transformation $x = yz$, $w = z$, and integrate out w.

2.15 In connection with Example 2.16, for a Beta distribution $\mathcal{B}e(\alpha, \beta)$, find the maximum of the $\mathcal{B}e(\alpha, \beta)$ density.

2.16 Establish the validity of Knuth (1981) $\mathcal{B}(n, p)$ generator:

Algorithm A.11 –Binomial–

> Define $k = n$, $\theta = p$ and $x = 0$.
> 1. Repeat $i = [1 + k\theta]$
> $V \sim \mathcal{B}e(i, k + 1 - i)$ [A.11]
>   if $\theta > V$, $\theta = \theta/V$ and $k = i - 1$;
>   otherwise, $x = x + i$, $\theta = (\theta - V)/(1 - V)$ and $k = k - i$
>   until $k \leq K$.
> 2. For $i = 1, 2, \dots, k$,
>   generate U_i
>   if $U_i < p$, $x = x + 1$.
> 3. Take x.

2.17 Establish the claims of Example 2.11: If $U_1, \dots, U_n$ is an iid sample from $\mathcal{U}_{[0,1]}$ and $U_{(1)} \leq \cdots \leq U_{(n)}$ are the corresponding order statistics, show that
  (a) $U_{(i)} \sim \mathcal{B}e(i, n - i + 1)$;
  (b) $(U_{(i_1)}, U_{(i_2)} - U_{(i_1)}, \dots, U_{(i_k)} - U_{(i_{k-1})}, 1 - U_{(i_k)}) \sim \mathcal{D}(i_1, i_2 - i_1, \dots, n - i_k + 1)$;
  (c) If U and V are iid $\mathcal{U}_{[0,1]}$, the distribution of

  $$\frac{U^{1/\alpha}}{U^{1/\alpha} + V^{1/\beta}},$$

  conditional on $U^{1/\alpha} + V^{1/\beta} \leq 1$, is the $\mathcal{B}e(\alpha, \beta)$ distribution.
  (d) Show that the order statistics can be directly generated via the *Renyi representation* $u_{(i)} = \sum_{j=1}^{i} \nu_j / \sum_{j=1}^{n} \nu_j$, where the ν_j's are iid $\mathcal{E}xp(1)$.

2.18 For the generation of a Student's t distribution, $\mathcal{T}(\nu, 0, 1)$, Kinderman et al. (1977) provide an alternative to the generation of a normal random variable and a chi squared random variable.

Algorithm A.12 –Student's t–

> 1. Generate $U_1, U_2 \sim \mathcal{U}([0, 1])$.
> 2. If $U_1 < 0.5$, $X = 1/(4U_1 - 1)$ and $V = X^{-2}U_2$;
>   otherwise, $X = 4U_1 - 3$ and $V = U_2$.   [A.12]
> 3. If $V < 1 - (|X|/2)$ or $V < (1 + (X^2/\nu))^{-(\nu+1)/2}$, take X;
>   otherwise, repeat.

Validate this algorithm and compare it with the algorithm of Example 2.13.

2.19 For $\alpha \in [0, 1]$, show that the algorithm

Algorithm A.13

> Generate
>
> $$U \sim \mathcal{U}([0, 1])$$ [A.13]
>
> until $U < \alpha$.

produces a simulation from $\mathcal{U}([0, \alpha])$. Compare it with the transform αU, $U \sim \mathcal{U}(0, 1)$ for values of α close to 0 and close to 1.

2.20 In each case generate the random variables and compare the histogram to the density function

(a) Normal random variables using a Cauchy candidate in Accept–Reject
(b) Gamma$(4.3, 6.2)$ random variables using a Gamma$(4, 7)$.
(c) Truncated normal: Standard normal truncated to $(2, \infty)$

2.21 An efficient algorithm for the simulation of Gamma $\mathcal{G}a(\alpha, 1)$ distributions is based on *Burr's distribution,* a distribution with density

$$g(x) = \lambda \mu \, \frac{x^{\lambda - 1}}{(\mu + x^\lambda)^2} , \qquad x > 0 .$$

It has been developed by Cheng (1977) and Cheng and Feast (1979). (See Devroye 1985.) For $\alpha > 1$, it is

Algorithm A.14 –Cheng and Feast's Gamma–

> Define $c_1 = \alpha - 1$, $c_2 = (\alpha - (1/6\alpha))/c_1$, $c_3 = 2/c_1$, $c_4 = 1 + c_3$,
> and $c_5 = 1/\sqrt{\alpha}$.
> 1. Repeat
>   generate U_1, U_2
>   take $U_1 = U_2 + c_5(1 - 1.86U_1)$ if $\alpha > 2.5$
>   until $0 < U_1 < 1$.   [A.14]
> 2. $W = c_2 U_2 / U_1$.
> 3. If $c_3 U_1 + W + W^{-1} \leq c_4$ or $c_3 \log U_1 - \log W + W \leq 1$,
>   take $c_1 W$;
>   otherwise, repeat.

(a) Show g is a density.
(b) Show that this algorithm produces variables generated from $\mathcal{G}a(\alpha, 1)$.

2.22 Ahrens and Dieter (1974) propose the following algorithm to generate a Gamma $\mathcal{G}a(\alpha, 1)$ distribution:

Algorithm A.15 –Ahrens and Dieter's Gamma–

> 1. Generate U_0, U_1.
> 2. If $U_0 > e/(e + \alpha)$, $x = -\log\{(\alpha + e)(1 - U_0)/\alpha e\}$ and $y = x^{\alpha - 1}$;
>   otherwise, $x = \{(\alpha + e)U_0/e\}^{1/\alpha}$ and $y = e^{-x}$.   [A.15]
> 3. If $U_1 < y$, take x;
>   otherwise, repeat.

Show that this algorithm produces variables generated from $\mathcal{G}a(\alpha, 1)$. Compare with Problem 2.21.

2.23 To generate the Beta distribution $\mathcal{B}e(\alpha, \beta)$ we can use the following representation:

(a) Show that, if $Y_1 \sim \mathcal{G}a(\alpha, 1)$, $Y_2 \sim \mathcal{G}a(\beta, 1)$, then

$$X = \frac{Y_1}{Y_1 + Y_2} \sim \mathcal{B}e(\alpha, \beta) .$$

(b) Use part (a) to construct an algorithm to generate a Beta random variable.

(c) Compare this algorithm with the method given in Problem 2.17 for different values of (α, β).

(d) Compare this algorithm with an Accept–Reject algorithm based on (i) the uniform distribution; (ii) the truncated normal distribution (when $\alpha \geq 1$ and $\beta \geq 1$).

(*Note:* See Schmeiser and Shalaby 1980 for an alternative Accept–Reject algorithm to generate Beta rv's.)

2.24 (a) Show that Student's t density can be written in the form (2.6), where $h_1(x|y)$ is the density of $\mathcal{N}(0, \nu/y)$ and $h_2(y)$ is the density of χ^2_ν.

(b) Show that Fisher's $\mathcal{F}_{m,\nu}$ density can be written in the form (2.6), with $h_1(x|y)$ the density of $\mathcal{G}a(m/2, \nu/m)$ and $h_2(y)$ the density of χ^2_ν.

2.25 The noncentral chi squared distribution, $\chi^2_p(\lambda)$, can be defined by a mixture representation (2.6), where $h_1(x|K)$ is the density of χ^2_{p+2K} and $h_2(k)$ is the density of $\mathcal{P}(\lambda/2)$.

(a) Show that it can also be expressed as the sum of a χ^2_{p-1} random variable and of the square of a standard normal variable.

(b) Compare the two algorithms which can be derived from these representations.

(c) Discuss whether a direct approach via an Accept–Reject algorithm is at all feasible.

2.26 (Walker 1997) Show that the Weibull distribution, $\mathcal{W}e(\alpha, \beta)$, with density

$$f(x|\alpha, \beta) = \beta \alpha x^{\alpha-1} \exp\left(-\beta x^\alpha\right),$$

can be represented as a mixture of $X \sim \mathcal{B}e(\alpha, \omega^{1/\alpha})$ by $\omega \sim \mathcal{G}a(2, \beta)$. Examine whether this representation is helpful from a simulation point of view.

2.27 An application of the mixture representation can be used to establish the following result (see Note 2.6.3):

Lemma 2.27. *If*

$$f(x) = \frac{f_1(x) - \varepsilon f_2(x)}{1 - \varepsilon},$$

where f_1 and f_2 are probability densities such that $f_1(x) \geq \varepsilon f_2(x)$, the algorithm

Generate

$$(X, U) \sim f_1(x)\, \mathbb{I}_{[0,1]}(u)$$

until $U \geq \varepsilon f_2(X)/f_1(X)$.

produces a variable X distributed according to f.

(a) Show that the distribution of X satisfies

$$P(X \leq x_0) = \int_{-\infty}^{x_0} \left(1 - \frac{\varepsilon f_2(x)}{f_1(x)}\right) f_1(x)\, dx \sum_{i=0}^{\infty} \varepsilon^i.$$

(b) Evaluate the integral in (a) to complete the proof.

2.28 (a) Demonstrate the equivalence of Corollary 2.17 and the Accept–Reject algorithm.

(b) Generalize Corollary 2.17 to the multivariate case. That is, for $\mathbf{X} = (X_1, X_2, \ldots, X_p) \sim f(x_1, x_2, \ldots, x_p) = f(\mathbf{x})$, formulate a joint distribution $(\mathbf{X}, U) \sim \mathbb{I}(0 < u < f(\mathbf{x}))$, and show how to generate a sample from $f(\mathbf{x})$ based on uniform random variables.

2.29 (a) Referring to (2.10), if $\pi(\theta|x)$ is the target density in an Accept–Reject algorithm, and $\pi(\theta)$ is the candidate density, show that the bound M can be taken to be the likelihood function evaluated at the MLE.

(b) For estimating a normal mean, a robust prior is the Cauchy. For $X \sim N(\theta, 1)$, $\theta \sim \text{Cauchy}(0, 1)$, the posterior distribution is

$$\pi(\theta|x) \propto \frac{1}{\pi(1 + \theta^2)} \frac{1}{2\pi} e^{-(x-\theta)^2/2}.$$

Use the Accept–Reject algorithm, with a Cauchy candidate, to generate a sample from the posterior distribution.
(*Note:* See Problem 3.19 and Smith and Gelfand 1992.)

2.30 For the Accept–Reject algorithm [A.4], with f and g properly normalized,

(a) Show that the probability of accepting a random variable is

$$P\left(U < \frac{f(X)}{Mg(X)}\right) = \frac{1}{M}.$$

(b) Show that $M \geq 1$.

(c) Let N be the number of trials until the kth random variable is accepted. Show that, for the normalized densities, N has the negative binomial distribution $\mathcal{N}eg(k, p)$, where $p = 1/M$. Deduce that the expected number of trials until k random variables are obtained is kM.

(d) Show that the bound M does not have to be tight; that is, there may be $M' < M$ such that $f(x) \leq M'g(x)$. Give an example where it makes sense to use M instead of M'.

(e) When the bound M is too tight (i.e., when $f(x) > Mg(x)$ on a non-negligible part of the support of f), show that the algorithm [A.4] does not produce a generation from f. Give the resulting distribution.

(f) When the bound is not tight, show that there is a way, using Lemma 2.27, to recycle part of the rejected random variables. (*Note:* See Casella and Robert 1998 for details.)

2.31 For the Accept–Reject algorithm of the $\mathcal{G}a(n, 1)$ distribution, based on the $\mathcal{E}xp(\lambda)$ distribution, determine the optimal value of λ.

2.32 This problem looks at a generalization of Example 2.19.

(a) If the target distribution of an Accept–Reject algorithm is the Gamma distribution $\mathcal{G}a(\alpha, \beta)$, where $\alpha \geq 1$ is not necessarily an integer, show that the instrumental distribution $\mathcal{G}a(a, b)$ is associated with the ratio

$$\frac{f(x)}{g(x)} = \frac{\Gamma(a)}{\Gamma(\alpha)} \frac{\beta^\alpha}{b^a} x^{\alpha-a} e^{-(\beta-b)x}.$$

(b) Why do we need $a < \alpha$ and $b < \beta$?

(c) For $a = \lfloor \alpha \rfloor$, show that the bound is maximized (in x) at $x = (\alpha - a)/(\beta - b)$.

(d) For $a = \lfloor \alpha \rfloor$, find the optimal choice of b.

(e) Compare with $a' = \lfloor \alpha \rfloor - 1$, when $\alpha > 2$.

2.33 The right-truncated Gamma distribution $\mathcal{TG}(a, b, t)$ is defined as the restriction of the Gamma distribution $\mathcal{G}a(a, b)$ to the interval $(0, t)$.

(a) Show that we can consider $t = 1$ without loss of generality.

(b) Give the density f of $\mathcal{TG}(a, b, 1)$ and show that it can be expressed as the following mixture of Beta $\mathcal{B}e(a, k + 1)$ densities:

$$f(x) = \sum_{k=0}^{\infty} \frac{b^a e^{-b}}{\gamma(a, b)} \frac{b^k}{k!} x^{a-1}(1 - x)^k,$$

where $\gamma(a, b) = \int_0^1 x^{a-1} e^{-x}$.

(c) If f is replaced with g_n which is the series truncated at term $k = n$, show that the acceptance probability of the Accept–Reject algorithm based on (g_n, f) is

$$\frac{1 - \dfrac{\gamma(n + 1, b)}{n!}}{1 - \dfrac{\gamma(a + n + 1, b)\Gamma(a)}{\Gamma(a + n + 1)\gamma(a, b)}}.$$

(d) Evaluate this probability for different values of (a, b).

(e) Give an Accept–Reject algorithm based on the pair (g_n, f) and a computable bound. (*Note:* See Philippe 1997c for a complete resolution of the problem.)

2.34 Let $f(x) = \exp(-x^2/2)$ and $g(x) = 1/(1 + x^2)$, densities of the normal and Cauchy distributions, respectively (ignoring the normalization constants).

(a) Show that the ratio

$$\frac{f(x)}{g(x)} = (1 + x^2)\, e^{-x^2/2} \leq 2/\sqrt{e},$$

which is attained at $x = \pm 1$.

(b) Show that for the normalized densities, the probability of acceptance is $\sqrt{e/2\pi} = 0.66.$, which implies that, on the average, one out of every three simulated Cauchy variables is rejected. Show that the mean number of trials to success is $1/.66 = 1.52$.

(c) Replacing g by a Cauchy density with scale parameter σ,

$$g_\sigma(x) = 1/\{\pi\sigma(1 + x^2/\sigma^2)\},$$

show that the bound on f/g_σ is $2\sigma^{-1}\exp\{\sigma^2/2 - 1\}$ and is minimized by $\sigma^2 = 1$. (This shows that $\mathcal{C}(0, 1)$ is the best choice among the Cauchy distributions for simulating a $\mathcal{N}(0, 1)$ distribution.)

2.35 There is a direct generalization of Corollary 2.17 that allows the proposal density to change at each iteration.

Algorithm A.16 –Generalized Accept–Reject–

At iteration i $(i \geq 1)$
1. Generate $X_i \sim g_i$ and $U_i \sim \mathcal{U}([0, 1])$, independently.
2. If $U_i \leq \epsilon_i f(X_i)/g_i(X_i)$, accept $X_i \sim f$;
3. otherwise, move to iteration $i + 1$.

(a) Let Z denote the random variable that is output by this algorithm. Show that Z has the cdf

$$P(Z \le z) = \sum_{i=1}^{\infty} \epsilon_i \prod_{j=1}^{i-1}(1 - \epsilon_j) \int_{-\infty}^{z} f(x)dx .$$

(b) Show that

$$\sum_{i=1}^{\infty} \epsilon_i \prod_{j=1}^{i-1}(1 - \epsilon_j) = 1 \text{ if and only if } \sum_{i=1}^{\infty} \log(1 - \epsilon_i) \text{ diverges}.$$

Deduce that we have a valid algorithm if the second condition is satisfied.

(c) Give examples of sequences ϵ_i that satisfy, and do not satisfy, the requirement of part (b).

2.36 (a) Prove the validity of the ARS Algorithm [A.7], without the envelope step, by applying Algorithm [A.16].

(b) Prove the validity of the ARS Algorithm [A.7], with the envelope step, directly. Note that

$$P(X \le x | \text{Accept}) = P\left(X \le x \Big| \left\{ U < \frac{g_\ell}{M g_m} \text{ or } U < \frac{f}{M g_m} \right\} \right)$$

and

$$\left\{ U < \frac{g_\ell}{M g_m} \text{ or } U < \frac{f}{M g_m} \right\} = \left\{ U < \frac{g_\ell}{M g_m} \right\} \cup \left\{ \frac{g_\ell}{M g_m} < U < \frac{f}{M g_m} \right\},$$

which are disjoint.

2.37 Based on the discussion in Example 2.26, write an alternate algorithm to Algorithm [A.17] that does not require the calculation of the density g_n.

2.38 The histogram and density of Figure 2.8 give the candidate g_n for $\pi(a | \mathbf{x}, \mathbf{y}, b)$, the conditional density of the intercept a in $\log \lambda = a + bt$, where we set $b = .025$ and $\sigma^2 = 5$. Produce the same picture for the slope, b, when we set the intercept $a = .15$ and $\tau^2 = 5$.

2.39 Step 1 of [A.7] relies on simulations from g_n. Show that we can write

$$g_n = \varpi_n^{-1} \left\{ \sum_{i=0}^{r_n} e^{\alpha_i x + \beta_i} \, \mathbb{I}_{[x_i, x_{i+1}]}(x) + e^{\alpha_{-1} x + \beta_{-1}} \mathbb{I}_{[-\infty, x_0]}(x) \right.$$
$$\left. e^{\alpha_{r_n+1} x + \beta_{r_n+1}} \, \mathbb{I}_{[x_{n+1}, +\infty]}(x) \right\},$$

where $y = \alpha_i x + \beta_i$ is the equation of the segment of line corresponding to g_n on $[x_i, x_{i+1}]$, r_n denotes the number of segments, and

$$\varpi_n = \int_{-\infty}^{x_0} e^{\alpha_{-1} x + \beta_{-1}} dx + \sum_{i=0}^{n} \int_{x_i}^{x_{i+1}} e^{\alpha_i x + \beta_i} dx + \int_{x_{n+1}}^{+\infty} e^{\alpha_{r_n+1} x + \beta_{r_n+1}} dx$$
$$= \frac{e^{\alpha_{-1} x_0 + \beta_{-1}}}{\alpha_{-1}} + \sum_{i=0}^{n} e^{\beta_i} \frac{e^{\alpha_i x_{i+1}} - e^{\alpha_i x_i}}{\alpha_i} - \frac{e^{\alpha_{r_n+1} x_{n+1}}}{\alpha_{r_n+1}},$$

when $\text{supp } f = \mathbb{R}$.

Verify that this representation as a sequence validates the following algorithm for simulation from g_n:

Algorithm A.17 –Supplemental ARS Algorithm–

1. Select the interval $[x_i, x_{i+1}]$ with probability

$$e^{\beta_i} \frac{e^{\alpha_i x_{i+1}} - e^{\alpha_i x_i}}{\varpi_n \alpha_i} . \qquad [A.17]$$

2. Generate $U \sim \mathcal{U}_{[0,1]}$ and take

$$X = \alpha_i^{-1} \log[e^{\alpha_i x_i} + U(e^{\alpha_i x_{i+1}} - e^{\alpha_i x_i})].$$

Note that the segment $\alpha_i + \beta_i x$ is not the same as the line $a_i + b_i x$ used in (2.16)

2.40 As mentioned in Section 2.4, many densities are log-concave.

(a) Show that the so-called *natural* exponential family,

$$dP_\theta(x) = \exp\{x \cdot \theta - \psi(\theta)\}d\nu(x)$$

is log-concave.

(b) Show that the logistic distribution of (2.23) is log-concave.

(c) Show that the *Gumbel* distribution

$$f(x) = \frac{k^k}{(k-1)!} \exp\left\{-kx - ke^{-x}\right\} , \quad k \in \mathbb{N}^* ,$$

is log-concave (Gumbel 1958).

(d) Show that the *generalized inverse Gaussian* distribution,

$$f(x) \propto x^\alpha e^{-\beta x - \alpha/x} , \qquad x > 0, \ \alpha > 0, \ \beta > 0 ,$$

is log-concave.

2.41 (George et al. 1993) For the natural exponential family, the conjugate prior measure is defined as

$$d\pi(\theta|x_0, n_0) \propto \exp\{x_0 \cdot \theta - n_0\psi(\theta)\}d\theta,$$

with $n_0 > 0$. (See Brown 1986, Chapter 1, for properties of exponential families.)

(a) Show that

$$\varphi(x_0, n_0) = \log \int_\Theta \exp\{x_0 \cdot \theta - n_0\psi(\theta)\}d\theta$$

is convex.

(b) Show that the so-called *conjugate likelihood distribution*

$$L(x_0, n_0|\theta_1, \ldots, \theta_p) \propto \exp\left\{x_0 \cdot \sum_{i=1}^p \theta_i - n_0 \sum_{i=1}^p \psi(\theta) - p\varphi(x_0, n_0)\right\}$$

is log-concave in (x_0, n_0).

(c) Deduce that the ARS algorithm applies in hierarchical Bayesian models with conjugate priors on the natural parameters and log-concave hyperpriors on (x_0, n_0).

(d) Apply the ARS algorithm to the case

$$X_i|\theta_i \sim \mathcal{P}(\theta_i t_i), \qquad \theta_i \sim \mathcal{G}a(\alpha, \beta), \qquad i = 1, \ldots, n,$$

with fixed α and $\beta \sim \mathcal{G}a(0.1, 1)$.

2.42 In connection with Example 2.25,

    (a) Show that a sum of log-concave functions is a log-concave function.

    (b) Deduce that (2.15) is log-concave.

2.43 (Casella and Berger 2001, Section 8.3) This problem examines the relationship of the property of log-concavity with other desirable properties of density functions.

    (a) The property of *monotone likelihood ratio* is very important in the construction of hypothesis tests, and in many other theoretical investigations. A family of pdfs or pmfs $\{g(t|\theta): \theta \in \Theta\}$ for a univariate random variable T with real-valued parameter θ has a *monotone likelihood ratio* (MLR) if, for every $\theta_2 > \theta_1$, $g(t|\theta_2)/g(t|\theta_1)$ is a monotone (nonincreasing or nondecreasing) function of t on $\{t: g(t|\theta_1) > 0 \text{ or } g(t|\theta_2) > 0\}$. Note that $c/0$ is defined as ∞ if $0 < c$.

    Show that if a density is log-concave, it has a monotone likelihood ratio.

    (b) Let $f(x)$ be a pdf and let a be a number such that, if $a \geq x \geq y$ then $f(a) \geq f(x) \geq f(y)$ and, if $a \leq x \leq y$ then $f(a) \geq f(x) \geq f(y)$. Such a pdf is called *unimodal* with a *mode* equal to a.

    Show that if a density is log-concave, it is unimodal.

2.44 This problem will look into one of the failings of congruential generators, the production of parallel lines of output. Consider a congruential generator $D(x) = ax \mod 1$, that is, the output is the fractional part of ax.

    (a) For $k = 1, 2, \ldots, 333$, plot the pairs $(k*0.003, D(k*0.003))$ for $a = 5, 20, 50$. What can you conclude about the parallel lines?

    (b) Show that each line has slope a and the lines repeat at intervals of $1/a$ (hence, larger values of a will increase the number of lines). (*Hint:* Let $x = \frac{i}{a} + \delta$, for $i = 1, \ldots, a$ and $0 < \delta < \frac{1}{a}$. For this x, show that $D(x) = a\delta$, regardless of the value of i.)

2.6 Notes

2.6.1 The Kiss Generator

Although this book is not formally concerned with the generation of uniform random variables (as we start from the assumption that we have an endless supply of such variables), it is good to understand the basic workings and algorithms that are used to generate these variables. In this note we describe the way in which uniform pseudo-random numbers are generated, and give a particularly good algorithm.

    To keep our presentation simple, rather than give a catalog of random number generators, we only give details for a single generator, the Kiss algorithm of Marsaglia and Zaman (1993). For details on other random number generators, the books of Knuth (1981), Rubinstein (1981), Ripley (1987), and Fishman (1996) are excellent sources.

    As we have remarked before, the finite representation of real numbers in a computer can radically modify the behavior of a dynamic system. Preferred generators are those that take into account the specifics of this representation and provide a uniform sequence. It is important to note that such a sequence does not really take values in the interval $[0, 1]$ but rather on the integers $\{0, 1, \ldots, M\}$, where M is the largest integer accepted by the computer. One manner of characterizing the performance of these integer generators is through the notion of *period*.

Definition 2.28. The *period*, T_0, of a generator is the smallest integer T such that $u_{i+T} = u_i$ for every i; that is, such that D^T is equal to the identity function.

The period is a very important parameter, having direct impact on the usefulness of a random number generator. If the number of needed generations exceeds the period of a generator, there may be noncontrollable artifacts in the sequence (cyclic phenomena, false orderings, etc.). Unfortunately, a generator of the form $X_{n+1} = f(X_n)$ has a period no greater than $M+1$, for obvious reasons. In order to overcome this bound, a generator must utilize several sequences X_n^i simultaneously (which is a characteristic of Kiss) or must involve $X_{n-1}, X_{n-2}, \ldots$ in addition to X_n, or must use other methods such as *start-up tables*, that is, using an auxiliary table of random digits to restart the generator.

Kiss simultaneously uses two generation techniques, namely *congruential* generation and *shift register* generation.

Definition 2.29. A *congruential generator* on $\{0, 1, \ldots, M\}$ is defined by the function

$$D(x) = (ax + b) \bmod (M + 1).$$

The period and, more generally, the performance of congruential generators depend heavily on the choice of (a, b) (see Ripley 1987). When transforming the above generator into a generator on $[0, 1]$, with $\tilde{D}(x) = (ax+b)/(M+1) \bmod 1$, the graph of D should range throughout $[0, 1]^2$, and a choice of the constant $a \notin \mathbb{Q}$ would yield a "recovery " of $[0, 1]^2$; that is, an infinite sequence of points should fill the space.

Although ideal, the choice of an irrational a is impossible (since a needs be specified with a finite number of digits). With a rational, a congruential generator will produce pairs $(x_n, D(x_n))$ that lie on parallel lines. Figure 2.9 illustrates this phenomenon for $a = 69069$, representing the sequence $(3k\ 10^{-4}, D(3k))$ for $k = 1, 2, \ldots, 333$. It is thus important to select a in such a way as to maximize the number of parallel segments in $[0, 1]^2$ (see Problem 2.44).

Most commercial generators use congruential methods, with perhaps the most disastrous choice of (a, b) being that of the old (and notorious) procedure *RANDU* (see Ripley 1987). Even when the choice of (a, b) assures the acceptance of the generator by standard tests, nonuniform behavior will be observed in the last digits of the real numbers produced by this method, due to round-up errors.

The second technique employed by Kiss is based on the (theoretical) independence between the k binary components of $X_n \sim \mathcal{U}_{\{0,1,\ldots,M\}}$ (where $M = 2^k - 1$) and is called a *shift register* generator.

Definition 2.30. For a given $k \times k$ matrix T, whose entries are either 0 or 1, the associated *shift register generator* is given by the transformation

$$x_{n+1} = Tx_n,$$

where x_n is represented as a vector of binary coordinates e_{ni}, that is to say,

$$x_n = \sum_{i=0}^{k-1} e_{ni} 2^i,$$

with e_{ni} equal to 0 or 1.

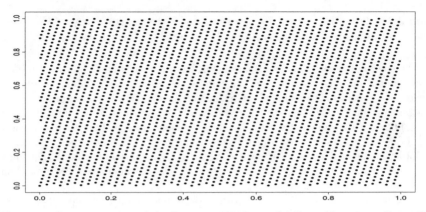

Fig. 2.9. Representation of the line $y = 69069x \mod 1$ by uniform sampling with sampling step $3 \; 10^{-4}$.

This second class of generators is motivated by both the internal (computer-dependent) representation of numbers as sequences of *bits* and the speed of manipulation of these elementary algebraic operations. Since the computation of Tx_n is done modulo 2, each addition is then the equivalent of a logical OR. Moreover, as the matrix T only contains 0 and 1 entries, multiplication by T amounts to shifting the content of coordinates, which gives the technique its name.

For instance, if the ith line of T contains a 1 in the ith and jth positions uniquely, the ith coordinate of x_{n+1}, will be obtained by

$$e_{(n+1)i} = (e_{ni} + e_{nj}) \mod 2$$
$$= e_{ni} \vee e_{nj} - e_{ni} \wedge e_{nj} ,$$

where $a \wedge b = \min(a, b)$ and $a \vee b = \max(a, b)$. This is a comparison of the ith coordinate of x_n and the coordinate corresponding to a *shift* of $(j - i)$. There also exist sufficient conditions on T for the associated generator to have period 2^k (see Ripley 1987).

The generators used by Kiss are based on the matrices

$$T_L = \begin{pmatrix} 1 & 1 & & \\ & \ddots & & 0 \\ & & \ddots & 1 \\ 0 & & & 1 \end{pmatrix} \quad \text{and} \quad T_R = \begin{pmatrix} 1 & & & \\ 1 & \ddots & 0 & \\ & & \ddots & \\ 0 & & 1 & 1 \end{pmatrix},$$

whose entries are 1 on the main diagonal and on the first upper diagonal and first lower diagonal, respectively, the other elements being 0. They are related to the right and left shift matrices,

$$R(e_1, \ldots, e_k)^t = (0, e_1, \ldots, e_{k-1})^t,$$
$$L(e_1, \ldots, e_k)^t = (e_2, e_3, \ldots, e_k, 0)^t,$$

since $T_R = (I + R)$ and $T_L = (I + L)$.

To generate a sequence of integers $X_1, X_2, \ldots$, the **Kiss** algorithm generates three sequences of integers. First, the algorithm uses a congruential generator to obtain

$$I_{n+1} = (69069 \times I_n + 23606797) \ (\mathrm{mod} \ 2^{32}) \,,$$

and then two shift register generators of the form,

$$J_{n+1} = (I + L^{15})(I + R^{17}) \, J_n \ (\mathrm{mod} \ 2^{32}) \,,$$
$$K_{n+1} = (I + L^{13})(I + R^{18}) \, K_n \ (\mathrm{mod} \ 2^{31}) \,.$$

These are then combined to produce

$$X_{n+1} = (I_{n+1} + J_{n+1} + K_{n+1}) \quad \mathrm{mod} \ 2^{32} \,.$$

Formally, this algorithm is not of the type specified in Definition 2.1, since it uses three parallel chains of integers. However, this feature yields advantages over algorithms based on a single dynamic system $X_{n+1} = f(X_n)$ since the period of **Kiss** is of order 2^{95}, which is almost $(2^{32})^3$. In fact, the (usual) congruential generator I_n has a maximal period of 2^{32}, the generator of K_n has a period of $2^{31} - 1$ and that of J_n a period of $2^{32} - 2^{21} - 2^{11} + 1$ for almost all initial values J_0 (see Marsaglia and Zaman 1993 for more details). The **Kiss** generator has been successfully tested on the different criteria of **Die Hard**, including tests on random subsets of *bits*. Figure 2.10 presents plots of (X_n, X_{n+1}), (X_n, X_{n+2}), (X_n, X_{n+5}) and (X_n, X_{n+10}) for $n = 1, \ldots, 5000$, where the sequence (X_n) has been generated by **Kiss**, without exhibiting any nonuniform feature on the square $[0, 1]^2$. A version of this algorithm in the programming language C is given below.

Algorithm A.18 –The Kiss Algorithm–

```
long int kiss (i,j,k)
unsigned long *i,*j,*k
{
*j = *j ∧ (*j << 17);                          [A.18]
*k = (*k ∧ (*k << 18)) & 0X7FFFFFFF ;
return  ((*i = 69069 * (*i) + 23606797) +
(*j ∧ = (*j >> 15)) + (*k ∧ = (*k >> 13)) ;
}
```

(See Marsaglia and Zaman 1993 for a **Fortran** version of **Kiss**). Note that some care must be exercised in the use of this program as a generator on $[0, 1]$, since it implies dividing by the largest integer available on the computer and may sometimes result in uniform generation on $[-1, 1]$.

2.6.2 Quasi-Monte Carlo Methods

Quasi-Monte Carlo methods were proposed in the 1950s to overcome some drawbacks of regular Monte Carlo methods by replacing probabilistic bounds on the errors with deterministic bounds. The idea at the core of quasi-Monte Carlo methods is to substitute the randomly (or pseudo-randomly) generated (uniform $[0, 1]$) sequences

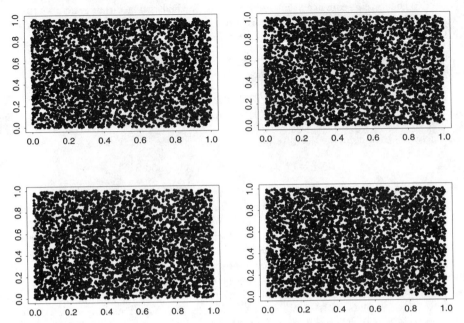

Fig. 2.10. Plots of pairs (X_t, X_{t+1}), (X_t, X_{t+2}), (X_t, X_{t+5}) and (X_t, X_{t+10}) for a sample of 5000 generations from Kiss.

used in regular Monte Carlo methods with a deterministic sequence (x_n) on $[0, 1]$ in order to minimize the so-called *divergence*,

$$D(x_1, \ldots, x_n) = \sup_{0 \le u \le 1} \left| \frac{1}{n} \sum_{i=1}^{n} \mathbb{I}_{[0,u]}(x_i) - u \right|.$$

This is also the *Kolmogorov–Smirnov distance* between the empirical cdf and that of the uniform distribution, used in nonparametric tests. For fixed n, the solution is obviously $x_i = (2i-1)/2n$ in dimension 1, but the goal here is to get a *low-discrepancy* sequence (x_n) which provides small values of $D(x_1, \ldots, x_n)$ for all n's (i.e., such that $x_1, \ldots, x_{n-1}$ do not depend on n) and can thus be updated sequentially.

As shown in Niederreiter (1992), there exist such sequences, which ensure a divergence rate of order $\mathcal{O}(n^{-1} \log(n)^{d-1})$, where d is the dimension of the integration space.[6] Since, for any function h defined on $[0, 1]$, it can be shown that the divergence is related to the overall approximation error by

$$(2.17) \qquad \left| \frac{1}{n} \sum_{i=1}^{n} h(x_i) - \int_0^1 h(x)dx \right| \le V(h)D(x_1, \ldots, x_n)$$

(see Niederreiter 1992), where $V(h)$ is the total variation of h,

---

[6] The notation $\mathcal{O}(1/n)$ denotes a function that satisfies $0 < \lim_{n \to \infty} n\mathcal{O}(1/n) < \infty$.

$$V(h) = \lim_{N \to \infty} \sup_{0 \leq x_1 \cdots x_{N-1} \leq 1} \sum_{j=1}^{N} |h(x_j) - h(x_{j-1})|,$$

with $x_0 = 0$ and $x_N = 1$, the gain over standard Monte Carlo methods can be substantial since standard methods lead to order $\mathcal{O}(n^{-1/2})$ errors (see Section 4.3). The advantage over standard integration techniques such as Riemann sums is also important when the dimension d increases since the latter are of order $n^{4/d}$ (see Yakowitz et al. 1978).

The true comparison with regular Monte Carlo methods is, however, more delicate than a simple assessment of the order of convergence. Construction of these sequences, although independent from h, can be quite involved (see Fang and Wang 1994 for examples), even though they only need to be computed *once*. More importantly, the construction requires that the functions to be integrated have bounded support, which can be a hindrance in practice because the choice of the transformation to $[0, 1]^d$ is crucial for the efficiency of the method. See Niederreiter (1992) for extensions in optimization setups.

2.6.3 Mixture Representations

Mixture representations, such as those used in Examples 2.13 and 2.14, can be extended (theoretically, at least) to many other distributions. For instance, a random variable X (and its associated distribution) is called *infinitely divisible* if for every n there exist iid random variables $X_1', \cdots, X_n'$ such that $X \sim X_1' + \cdots + X_n'$ (see Feller 1971, Section XVII.3 or Billingsley 1995, Section 28). It turns out that most distributions that are infinitely divisible can be represented as mixtures of Poisson distributions, the noncentral $\chi_p^2(\lambda)$ distribution being a particular case of this phenomenon. (However, this theoretical representation does not necessarily guarantee that infinitely divisible distributions are always easy to simulate.)

We also note that if the finite mixture

$$\sum_{i=1}^{k} p_i \, f_i(x)$$

can result in a decomposition of $f(x)$ into simple components (for instance, uniform distributions on intervals) and a last residual term with a small weight, the following approximation applies: We can use a trapezoidal approximation of f on intervals $[a_i, b_i]$, the weight p_i being of the order of $\int_{a_i}^{b_i} f(x)dx$. Devroye (1985) details the applications of this method in the case where f is a polynomial on $[0, 1]$.

3

Monte Carlo Integration

> Cadfael had heard the words without hearing them and enlightenment fell
> on him so dazzlingly that he stumbled on the threshold.
> —Ellis Peter, *The Heretic's Apprentice*

While Chapter 2 focussed on developing techniques to produce random variables by computer, this chapter introduces the central concept of Monte Carlo methods, that is, taking advantage of the availability of computer generated random variables to approximate univariate and multidimensional integrals. In Section 3.2, we introduce the basic notion of Monte Carlo approximations as a byproduct of the Law of Large Numbers, while Section 3.3 highlights the universality of the approach by stressing the versatility of the representation of an integral as an expectation.

3.1 Introduction

Two major classes of numerical problems that arise in statistical inference are *optimization* problems and *integration* problems. (An associated problem, that of *implicit equations*, can often be reformulated as an optimization problem.) Although optimization is generally associated with the likelihood approach, and integration with the Bayesian approach, these are not strict classifications, as shown by Examples 1.5 and 1.15, and Examples 3.1, 3.2 and 3.3, respectively.

Examples 1.1–1.15 have also shown that it is not always possible to derive explicit probabilistic models and that it is even less possible to analytically compute the estimators associated with a given paradigm (maximum likelihood, Bayes, method of moments, etc.). Moreover, other statistical methods, such as *bootstrap* methods (see Note 1.6.2), although unrelated to the Bayesian approach, may involve the integration of the empirical cdf. Similarly, alternatives to standard likelihood, such as *marginal* likelihood, may require the integration of the nuisance parameters (Barndorff-Nielsen and Cox 1994).

Although many calculations in Bayesian inference require integration, this is not always the case. Integration is clearly needed when the Bayes estimators are posterior expectations (see Section 1.3 and Problem 1.22), however Bayes estimators are not always posterior expectations. In general, the Bayes estimate under the loss function $L(\theta, \delta)$ and the prior π is the solution of the minimization program

$$(3.1) \qquad \min_{\delta} \int_{\Theta} L(\theta, \delta)\, \pi(\theta)\, f(x|\theta)\, d\theta\ .$$

Only when the loss function is the quadratic function $\|\theta - \delta\|^2$ will the Bayes estimator be a posterior expectation. While some other loss functions lead to general solutions $\delta^{\pi}(x)$ of (3.1) in terms of $\pi(\theta|x)$ (see, for instance, Robert 1996b, 2001 for the case of *intrinsic losses*), a specific setup where the loss function is constructed by the decision-maker almost always precludes analytical integration of (3.1). This necessitates an approximate solution of (3.1) either by numerical methods or by simulation.

Thus, whatever the type of statistical inference, we are led to consider numerical solutions. The previous chapter has illustrated a number of methods for the generation of random variables with any given distribution and, hence, provides a basis for the construction of solutions to our statistical problems. Thus, just as the search for a stationary state in a dynamical system in physics or in economics can require one or several simulations of successive states of the system, statistical inference on complex models will often require the use of simulation techniques. (See, for instance, Bauwens 1984, Bauwens and Richard 1985 and Gouriéroux and Monfort 1996 for illustrations in econometrics.) We now look at a number of examples illustrating these situations before embarking on a description of simulation-based integration methods.

Example 3.1. L_1 loss. For $\theta \in \mathbb{R}$ and $L(\theta, \delta) = |\theta - \delta|$, the Bayes estimator associated with π is the posterior median of $\pi(\theta|x)$, $\delta^{\pi}(x)$, which is the solution to the equation

$$(3.2) \qquad \int_{\theta \leq \delta^{\pi}(x)} \pi(\theta)\, f(x|\theta)\, d\theta = \int_{\theta \geq \delta^{\pi}(x)} \pi(\theta)\, f(x|\theta)\, d\theta\ .$$

In the setup of Example 1.7, that is, when $\lambda = \|\theta\|^2$ and $X \sim \mathcal{N}_p(\theta, I_p)$, this equation is quite complex, since, when using the reference prior of Example 1.12,

$$\pi(\lambda|x) \propto \lambda^{p-1/2} \int e^{-\|x - \theta\|^2/2} \prod_{i=1}^{p-2} \sin(\varphi_i)^{p-i-1}\, d\varphi_1 \ldots d\varphi_{p-1}\ ,$$

where $\lambda, \varphi_1, \ldots, \varphi_{p-1}$ are the polar coordinates of θ, that is, $\theta_1 = \lambda \cos(\varphi_1)$, $\theta_2 = \lambda \sin(\varphi_1) \cos(\varphi_2)$, ... $\qquad\qquad\qquad\qquad\qquad\qquad\qquad \|$

Example 3.2. Piecewise linear and quadratic loss functions. Consider a loss function which is piecewise quadratic,

$$(3.3) \qquad L(\theta, \delta) = w_i(\theta - \delta)^2 \quad \text{when} \quad \theta - \delta \in [a_i, a_{i+1}), \quad w_i > 0.$$

Differentiating the posterior expectation (3.3) shows that the associated Bayes estimator satisfies

$$\sum_i w_i \int_{a_i}^{a_{i+1}} (\theta - \delta^\pi(x)) \, \pi(\theta|x) \, d\theta = 0 \, ,$$

that is,

$$\delta^\pi(x) = \frac{\sum_i w_i \int_{a_i}^{a_{i+1}} \theta \, \pi(\theta) \, f(x|\theta) \, d\theta}{\sum_i w_i \int_{a_i}^{a_{i+1}} \pi(\theta) \, f(x|\theta) \, d\theta} \, .$$

Although formally explicit, the computation of $\delta^\pi(x)$ requires the computation of the posterior means restricted to the intervals $[a_i, a_{i+1})$ and of the posterior probabilities of these intervals.

Similarly, consider a piecewise linear loss function,

$$L(\theta, \delta) = w_i|\theta - \delta| \quad \text{if} \quad \theta - \delta \in [a_i, a_{i+1}),$$

or Huber's (1972) loss function,

$$L(\theta, \delta) = \begin{cases} \rho(\theta - \delta)^2 & \text{if } |\theta - \delta| < c, \\ 2\rho c\{|\theta - \delta| - c/2\} & \text{otherwise,} \end{cases}$$

where ρ and c are specified constants. Although a specific type of prior distribution leads to explicit formulas, most priors result only in integral forms of δ^π. Some of these may be quite complex. ‖

Inference based on *classical decision theory* evaluates the performance of estimators (maximum likelihood estimator, best unbiased estimator, moment estimator, etc.) through the loss imposed by the decision-maker or by the setting. Estimators are then compared through their expected losses, also called risks. In most cases, it is impossible to obtain an analytical evaluation of the risk of a given estimator, or even to establish that a new estimator (uniformly) dominates a standard estimator.

It may seem that the topic of *James–Stein* estimation is an exception to this observation, given the abundant literature on the topic. In fact, for some families of distributions (such as exponential or spherically symmetric) and some types of loss functions (such as quadratic or concave), it is possible to analytically establish domination results over the maximum likelihood estimator or unbiased estimators (see Lehmann and Casella 1998, Chapter 5 or Robert 2001, Chapter 2). Nonetheless, in these situations, estimators such as *empirical Bayes estimators*, which are quite attractive in practice, will rarely allow for analytic expressions. This makes their evaluation under a given loss problematic.

Given a sampling distribution $f(x|\theta)$ and a conjugate prior distribution $\pi(\theta|\lambda, \mu)$, the empirical Bayes method estimates the *hyperparameters* λ and μ from the *marginal distribution*

$$m(x|\lambda, \mu) = \int f(x|\theta)\, \pi(\theta|\lambda, \mu)\, d\theta$$

by maximum likelihood. The estimated distribution $\pi(\theta|\hat\lambda, \hat\mu)$ is often used as in a standard Bayesian approach (that is, without taking into account the effect of the substitution) to derive a point estimator. See Searle et al. (1992, Chapter 9) or Carlin and Louis (1996) for a more detailed discussion on this approach. (We note that this approach is sometimes called *parametric* empirical Bayes, as opposed to the *nonparametric* empirical Bayes approach developed by Herbert Robbins. See Robbins 1964, 1983 or Maritz and Lwin 1989 for details.) The following example illustrates some difficulties encountered in evaluating empirical Bayes estimators (see also Example 4.12).

Example 3.3. Empirical Bayes estimator. Let X have the distribution $X \sim \mathcal{N}_p(\theta, I_p)$ and let $\theta \sim \mathcal{N}_p(\mu, \lambda I_p)$, the corresponding conjugate prior. The hyperparameter μ is often specified, and here we take $\mu = 0$. In the empirical Bayes approach, the scale hyperparameter λ is replaced by the maximum likelihood estimator, $\hat\lambda$, based on the marginal distribution $X \sim \mathcal{N}_p(0, (\lambda + 1)I_p)$. This leads to the maximum likelihood estimator $\hat\lambda = (\|x\|^2/p - 1)^+$. Since the posterior distribution of θ given λ is $\mathcal{N}_p(\lambda x/(\lambda + 1), \lambda I_p/(\lambda + 1))$, empirical Bayes inference may be based on the pseudo-posterior $\mathcal{N}_p(\hat\lambda x/(\hat\lambda + 1), \hat\lambda I_p/(\hat\lambda + 1))$. If, for instance, $\|\theta\|^2$ is the quantity of interest, and if it is evaluated under a quadratic loss, the empirical Bayes estimator is

$$
\begin{aligned}
\delta^{eb}(x) = \mathbb{E}(\|\theta\|^2|x) &= \left(\frac{\hat\lambda}{\hat\lambda + 1}\right)^2 \|x\|^2 + \frac{\hat\lambda p}{\hat\lambda + 1} \\
&= \left[\left(1 - \frac{p}{\|x\|^2}\right)^+\right]^2 \|x\|^2 + p\left(1 - \frac{p}{\|x\|^2}\right)^+ \\
&= (\|x\|^2 - p)^+ .
\end{aligned}
$$

This estimator dominates both the best unbiased estimator, $\|x\|^2 - p$, and the maximum likelihood estimator based on $\|x\|^2 \sim \chi_p^2(\|\theta\|^2)$ (see Saxena and Alam 1982 and Example 1.8). However, since the proof of this second domination result is quite involved, one might first check for domination through a simulation experiment that evaluates the risk function,

$$R(\theta, \delta) = \mathbb{E}_\theta[(\|\theta\|^2 - \delta)^2],$$

for the three estimators. This quadratic risk is often normalized by $1/(2\|\theta\|^2 + p)$ (which does not affect domination results but ensures the existence of a minimax estimator; see Robert 2001). Problem 3.8 contains a complete solution to the evaluation of risk.   ‖

A general solution to the different computational problems contained in the previous examples and in those of Section 1.1 is to use simulation, of either the true or approximate distributions to calculate the quantities of interest. In the setup of Decision Theory, whether it is classical or Bayesian, this solution is natural, since risks and Bayes estimators involve integrals with respect to probability distributions. We will see in Chapter 5 why this solution also applies in the case of maximum likelihood estimation. Note that the possibility of producing an almost infinite number of random variables distributed according to a given distribution gives us access to the use of *frequentist* and *asymptotic* results much more easily than in usual inferential settings (see Serfling 1980 or Lehmann and Casella 1998, Chapter 6) where the sample size is most often fixed. One can, therefore, apply probabilistic results such as the Law of Large Numbers or the Central Limit Theorem, since they allow for an assessment of the convergence of simulation methods (which is equivalent to the deterministic bounds used by numerical approaches.)

3.2 Classical Monte Carlo Integration

Before applying our simulation techniques to more practical problems, we first need to develop their properties in some detail. This is more easily accomplished by looking at the generic problem of evaluating the integral

$$(3.4) \qquad \mathbb{E}_f[h(X)] = \int_{\mathcal{X}} h(x)\, f(x)\, dx \;.$$

Based on previous developments, it is natural to propose using a sample $(X_1, \ldots, X_m)$ generated from the density f to approximate (3.4) by the empirical average[1]

$$\overline{h}_m = \frac{1}{m} \sum_{j=1}^{m} h(x_j)\;,$$

since $\overline{h}_m$ converges almost surely to $\mathbb{E}_f[h(X)]$ by the Strong Law of Large Numbers. Moreover, when h^2 has a finite expectation under f, the speed of convergence of $\overline{h}_m$ can be assessed since the variance

$$\mathrm{var}(\overline{h}_m) = \frac{1}{m} \int_{\mathcal{X}} (h(x) - \mathbb{E}_f[h(X)])^2 \, f(x)\, dx$$

---

[1] This approach is often referred to as the *Monte Carlo method*, following Metropolis and Ulam (1949). We will meet Nicolas Metropolis (1915–1999) again in Chapters 5 and 7, with the simulated annealing and MCMC methods.

can also be estimated from the sample $(X_1, \ldots, X_m)$ through

$$v_m = \frac{1}{m^2} \sum_{j=1}^{m} [h(x_j) - \overline{h}_m]^2 .$$

For m large,

$$\frac{\overline{h}_m - \mathbb{E}_f[h(X)]}{\sqrt{v_m}}$$

is therefore approximately distributed as a $\mathcal{N}(0,1)$ variable, and this leads to the construction of a convergence test and of confidence bounds on the approximation of $\mathbb{E}_f[h(X)]$.

Example 3.4. A first Monte Carlo integration. Recall the function (1.26) that we saw in Example 1.17, $h(x) = [\cos(50x) + \sin(20x)]^2$. As a first example, we look at integrating this function, which is shown in Figure 3.1 *(left)*. Although it is possible to integrate this function analytically, it is a good first test case. To calculate the integral, we generate $U_1, U_2, \ldots, U_n$ iid $\mathcal{U}(0,1)$ random variables, and approximate $\int h(x)dx$ with $\sum h(U_i)/n$. The center panel in Figure 3.1 shows a histogram of the values of $h(U_i)$, and the last panel shows the running means and standard errors. It is clear that the Monte Carlo average is converging, with value of 0.963 after 10,000 iterations. This compares favorably with the exact value of 0.965. (See Example 4.1 for a more formal monitoring of convergence.) ‖

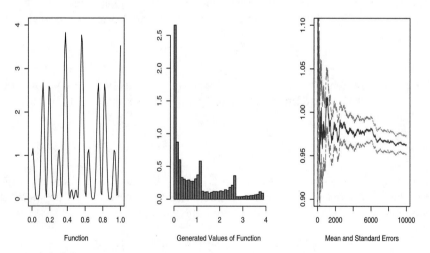

Fig. 3.1. Calculation of the integral of the function (1.26): *(left)* function (1.26), *(center)* histogram of 10,000 values $h(U_i)$, simulated using a uniform generation, and *(right)* mean $\pm$ one standard error.

n	0.0	0.67	0.84	1.28	1.65	2.32	2.58	3.09	3.72
10^2	0.485	0.74	0.77	0.9	0.945	0.985	0.995	1	1
10^3	0.4925	0.7455	0.801	0.902	0.9425	0.9885	0.9955	0.9985	1
10^4	0.4962	0.7425	0.7941	0.9	0.9498	0.9896	0.995	0.999	0.9999
10^5	0.4995	0.7489	0.7993	0.9003	0.9498	0.9898	0.995	0.9989	0.9999
10^6	0.5001	0.7497	0.8	0.9002	0.9502	0.99	0.995	0.999	0.9999
10^7	0.5002	0.7499	0.8	0.9001	0.9501	0.99	0.995	0.999	0.9999
10^8	0.5	0.75	0.8	0.9	0.95	0.99	0.995	0.999	0.9999

Table 3.1. Evaluation of some normal quantiles by a regular Monte Carlo experiment based on n replications of a normal generation. The last line gives the exact values.

The approach followed in the above example can be successfully utilized in many cases, even though it is often possible to achieve greater efficiency through numerical methods (Riemann quadrature, Simpson method, etc.) in dimension 1 or 2. The scope of application of this Monte Carlo integration method is obviously not limited to the Bayesian paradigm since, similar to Example 3.3, the performances of complex procedures can be measured in any setting where the distributions involved in the model can be simulated. For instance, we can use Monte Carlo sums to calculate a normal cumulative distribution function (even though the normal cdf can now be found in all software and many pocket calculators).

Example 3.5. Normal cdf. Since the normal cdf cannot be written in an explicit form, a possible way to construct normal distribution tables is to use simulation. Consider the generation of a sample of size n, $(x_1, \ldots, x_n)$, based on the Box–Muller algorithm $[A_4]$ of Example 2.2.2.

The approximation of

$$\Phi(t) = \int_{-\infty}^{t} \frac{1}{\sqrt{2\pi}} e^{-y^2/2} dy$$

by the Monte Carlo method is thus

$$\hat{\Phi}(t) = \frac{1}{n} \sum_{i=1}^{n} \mathbb{I}_{x_i \leq t},$$

with (exact) variance $\Phi(t)(1 - \Phi(t))/n$ (as the variables $\mathbb{I}_{x_i \leq t}$ are independent Bernoulli with success probability $\Phi(t)$). For values of t around $t = 0$, the variance is thus approximately $1/4n$, and to achieve a precision of four decimals, the approximation requires on average $n = (\sqrt{2}\, 10^4)^2$ simulations, that is, 200 million iterations. Table 3.1 gives the evolution of this approximation for several values of t and shows an accurate evaluation for 100 million iterations. Note that greater (absolute) accuracy is achieved in the tails and that more efficient simulations methods could be used, as in Example 3.8 below. ∥

We mentioned in Section 3.1 the potential of this approach in evaluating estimators based on a decision-theoretic formulation. The same applies for testing, when the level of significance of a test, and its power function, cannot be easily computed, and simulation thus can provide a useful improvement over asymptotic approximations when explicit computations are impossible. The following example illustrates this somewhat different application of Monte Carlo integration.

Many tests are based on an asymptotic normality assumption as, for instance, the *likelihood ratio test*. Given H_0, a null hypothesis corresponding to r independent constraints on the parameter $\theta \in \mathbb{R}^k$, denote by $\hat{\theta}$ and $\hat{\theta}^0$ the unconstrained and constrained (under H_0) maximum likelihood estimators of θ, respectively. The likelihood ratio $\ell(\hat{\theta}|x)/\ell(\hat{\theta}^0|x)$ then satisfies

$$(3.5) \qquad \log[\ell(\hat{\theta}|x)/\ell(\hat{\theta}^0|x)] = 2 \left\{\log \ell(\hat{\theta}|x) - \log \ell(\hat{\theta}^0|x)\right\} \xrightarrow{\mathcal{L}} \chi_r^2 ,$$

when the number of observations goes to infinity (see Lehmann 1986, Section 8.8, or Gouriéroux and Monfort 1996). However, the χ_r^2 approximation only holds asymptotically and, further, this convergence only holds under regularity constraints on the likelihood function (see Lehmann and Casella 1998, Chapter 6, for a full development); hence, the asymptotics may even not apply.

Example 3.6. Contingency Tables. Table 3.2 gives the results of a study comparing radiation therapy with surgery in treating cancer of the larynx.

	Cancer Controlled	Cancer not Controlled	
Surgery	21	2	23
Radiation	15	3	18
	36	5	41

Table 3.2. Comparison of cancer treatment success from surgery or radiation only (*Source:* Agresti 1996, p.50).

Typical sampling models for contingency tables may condition on both margins, one margin, or only the table total, and often the choice is based on philosophical reasons (see, for example, Agresti 1992). In this case we may argue for conditioning on the number of patients in each group, or we may just condition on the table total (there is little argument for conditioning on both margins). Happily, in many cases the resulting statistical conclusion is not dependent on this choice but, for definiteness, we will choose to condition only on the table total, $n = 41$.

Under this model, each observation X_i comes from a multinomial distribution with four cells and cell probabilities $\mathbf{p} = (p_{11}, p_{12}, p_{21}, p_{22})$, with $\sum_{ij} p_{ij} = 1$, that is,

$$X_i \sim \mathcal{M}_4(1, \mathbf{p}), \quad i = 1, \ldots, n.$$

If we denote by y_{ij} the number of x_i that are in cell ij, the likelihood function can be written

$$\ell(\mathbf{p}|\mathbf{y}) \propto \prod_{ij} p_{ij}^{y_{ij}}.$$

The null hypothesis to be tested is one of independence, which is to say that the treatment has no bearing on the control of cancer. To translate this into a parameter statement, we note that the full parameter space corresponding to Table 3.2 is

$$
\begin{array}{cc|c}
p_{11} & p_{12} & p_1 \\
p_{21} & p_{22} & 1 - p_1 \\
\hline
p_2 & 1 - p_2 & 1
\end{array}
$$

and the null hypothesis of independence is $H_0 : p_{11} = p_1 p_2$. The likelihood ratio statistic for testing this hypothesis is

$$\lambda(\mathbf{y}) = \frac{\max_{\mathbf{p}:p_{11}=p_1 p_2} \ell(\mathbf{p}|\mathbf{y})}{\max_{\mathbf{p}} \ell(\mathbf{p}|\mathbf{y})}.$$

It is straightforward to show that the numerator maximum is attained at $\hat{p}_1 = (y_{11} + y_{12})/n$ and the denominator maximum at $\hat{p}_{ij} = y_{ij}/n$.

As mentioned above, under H_0, $-2 \log \lambda$ is asymptotically distributed as χ_1^2. However, with only 41 observations, the asymptotics do not necessarily apply. One alternative is to use an exact permutation test (Mehta et al. 2000), and another alternative is to devise a Monte Carlo experiment to simulate the null distribution of $-2 \log \lambda$ or equivalently of λ in order to obtain a cutoff point for a hypothesis test. If we denote this null distribution by $f_0(\lambda)$, and we are interested in an α level test, we specify α and solve for λ_α the integral equation

$$(3.6) \qquad \int_0^{\lambda_\alpha} f_0(\lambda) d\lambda = 1 - \alpha.$$

The standard Monte Carlo approach to this problem is to generate random variables $\lambda^t \sim f_0(\lambda)$, $t = 1, \ldots, M$, then order the sample $\lambda^{(1)} \leq \lambda^{(2)} \leq \cdots \lambda^{(M)}$ and finally calculate the empirical $1 - \alpha$ percentile $\lambda^{(\lfloor (1-\alpha)M \rfloor)}$. We then have

$$\lim_{M \to \infty} \lambda^{(\lfloor (1-\alpha)M \rfloor)} \to \lambda_\alpha.$$

(Note that this is a slightly unusual Monte Carlo experiment in that α is known and λ_α is not, but it is nonetheless based on the same convergence of empirical measures.)

Percentile	Monte Carlo	χ_1^2
.10	2.84	3.87
.05	3.93	4.68
.01	6.72	6.36

Table 3.3. Cutoff points for the null distribution f_0 compared to χ_1^2.

To run the Monte Carlo experiment, we need to generate values from $f_0(\lambda)$. Since this distribution is not completely specified (the parameters p_1 and p_2 can be any value in $(0, 1)$), to generate a value from $f_0(\lambda)$ we generate

$$(3.7) \qquad p_i \sim \mathcal{U}(0, 1), \quad i = 1, 2,$$
$$\mathbf{X} \sim \mathcal{M}_4(p_1 p_2, p_1(1 - p_2), (1 - p_1)p_2, (1 - p_1)(1 - p_2)),$$

and calculate $\lambda(\mathbf{x})$. The results, given in Table 3.3 and Figure 3.2, show that the Monte Carlo null distribution has a slightly different shape than the χ_1^2 distribution, being slightly more concentrated around 0 but with longer tails.

The analysis of the given data is somewhat anticlimactic, as the observed value of $\lambda(\mathbf{y})$ is .594, which according to any calibration gives overwhelming support to H_0.   ‖

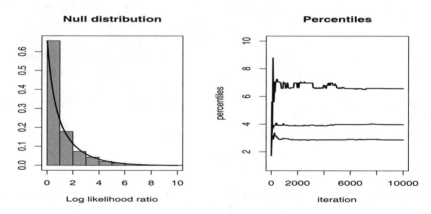

Fig. 3.2. For Example 3.6, histogram of null distribution and approximating χ_1^2 density(left panel). The right panel gives the running empirical percentiles $(.90, .95, .99)$, from bottom to top. Notice the higher variability in the higher percentiles (10,000 simulations).

Example 3.7. Testing the number of components. A situation where the standard χ_r^2 regularity conditions do not apply for the likelihood ratio test is that of the normal mixture (see Example 1.10)

$$p\,\mathcal{N}(\mu, 1) + (1 - p)\,\mathcal{N}(\mu + \theta, 1)\,,$$

where the constraint $\theta > 0$ ensures *identifiability*. A test on the existence of a mixture cannot be easily represented in a hypothesis test since $H_0 : p = 0$ effectively eliminates the mixture and results in the identifiability problem related with $\mathcal{N}(\mu + \theta, 1)$. (The inability to estimate the nuisance parameter p under H_0 results in the likelihood not satisfying the necessary regularity conditions; see Davies 1977. However, see Lehmann and Casella 1998, Section 6.6 for mixtures where it is possible to construct efficient estimators.)

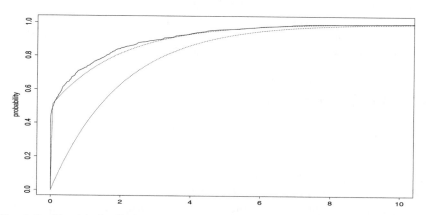

Fig. 3.3. Empirical cdf of a sample of log-likelihood ratios for the test of presence of a Gaussian mixture (solid lines) and comparison with the cdf of a χ_2^2 distribution (dotted lines, below) and with the cdf of a $.5 - .5$ mixture of a χ_2^2 distribution and of a Dirac mass at 0 (dotted lines, above) (based on 1000 simulations of a normal $\mathcal{N}(0, 1)$ sample of size 100).

A slightly different formulation of the problem will allow a solution, however. If the identifiability constraint is taken to be $p \geq 1/2$ instead of $\theta > 0$, then H_0 can be represented as

$$H_0 : \quad p = 1 \quad \text{or} \quad \theta = 0\,.$$

We therefore want to determine the limiting distribution of (3.5) under this hypothesis and under a local alternative. Figure 3.3 represents the empirical cdf of $2\,\{\log\,\ell(\hat{p}, \hat{\mu}, \hat{\theta}|x) - \log\,\ell(\hat{\mu}^0|x)\}$ and compares it with the χ_2^2 cdf, where $\hat{p}, \hat{\mu}, \hat{\theta}$, and $\hat{\mu}^0$ are the respective MLEs for 1000 simulations of a normal $\mathcal{N}(0, 1)$ sample of size 100. The poor agreement between the asymptotic approximation and the empirical cdf is quite obvious. Figure 3.3 also shows how the χ_2^2 approximation is improved if the limit (3.5) is replaced by an equally weighted mixture of a Dirac mass at 0 and a χ_2^2 distribution.

Note that the resulting sample of the log-likelihood ratios can also be used for inferential purposes, for instance to derive an exact test via the estimation of the quantiles of the distribution of (3.5) under H_0 or to evaluate the power of a standard test.   ‖

It may seem that the method proposed above is sufficient to approximate integrals like (3.4) in a controlled way. However, while the straightforward Monte Carlo method indeed provides good approximations of (3.4) in most regular cases, there exist more efficient alternatives which not only avoid a direct simulation from f but also can be used repeatedly for several integrals of the form (3.4). The repeated use can be for either a family of functions h or a family of densities f. In particular, the usefulness of this flexibility is quite evident in Bayesian analyses of *robustness*, of *sensitivity* (see Berger 1990, 1994), or for the computation of power functions of specific tests (see Lehmann 1986, or Gouriéroux and Monfort 1996).

3.3 Importance Sampling

3.3.1 Principles

The method we now study is called *importance sampling* because it is based on so-called *importance functions*, and although it would be more accurate to call it "weighted sampling," we will follow common usage. We start this section with a somewhat unusual example, borrowed from Ripley (1987), which shows that it may actually pay to generate from a distribution other than the distribution f of interest or, in other words, to modify the representation of an integral as an expectation against a given density. (See Note 3.6.1 for a global approach to the approximation of tail probabilities by *large deviation* techniques.)

Example 3.8. Cauchy tail probability. Suppose that the quantity of interest is the probability, p, that a Cauchy $\mathcal{C}(0,1)$ variable is larger than 2, that is,

$$p = \int_2^{+\infty} \frac{1}{\pi(1+x^2)} \, dx .$$

When p is evaluated through the empirical average

$$\hat{p}_1 = \frac{1}{m} \sum_{j=1}^{m} \mathbb{I}_{X_j > 2}$$

of an iid sample $X_1, \ldots, X_m \sim \mathcal{C}(0,1)$, the variance of this estimator is $p(1-p)/m$ (equal to $0.127/m$ since $p = 0.15$). This variance can be reduced by taking into account the symmetric nature of $\mathcal{C}(0,1)$, since the average

$$\hat{p}_2 = \frac{1}{2m} \sum_{j=1}^{m} \mathbb{I}_{|x_j| > 2}$$

has variance $p(1-2p)/2m$ equal to $0.052/m$.

The (relative) inefficiency of these methods is due to the generation of values outside the domain of interest, $[2,+\infty)$, which are, in some sense, irrelevant for the approximation of p. If p is written as

$$p = \frac{1}{2} - \int_0^2 \frac{1}{\pi(1+x^2)}\, dx \, ,$$

the integral above can be considered to be the expectation of $h(X) = 2/\pi(1+X^2)$, where $X \sim \mathcal{U}_{[0,2]}$. An alternative method of evaluation for p is therefore

$$\hat{p}_3 = \frac{1}{2} - \frac{1}{m} \sum_{j=1}^{m} h(U_j)$$

for $U_j \sim \mathcal{U}_{[0,2]}$. The variance of $\hat{p}_3$ is $(\mathbb{E}[h^2] - \mathbb{E}[h]^2)/m$ and an integration by parts shows that it is equal to $0.0285/m$. Moreover, since p can be written as

$$p = \int_0^{1/2} \frac{y^{-2}}{\pi(1+y^{-2})}\, dy \, ,$$

this integral can also be seen as the expectation of $\frac{1}{4} h(Y) = 1/2\pi(1+Y^2)$ against the uniform distribution on $[0,1/2]$ and another evaluation of p is

$$\hat{p}_4 = \frac{1}{4m} \sum_{j=1}^{m} h(Y_j)$$

when $Y_j \sim \mathcal{U}_{[0,1/2]}$. The same integration by parts shows that the variance of $\hat{p}_4$ is then $0.95 \; 10^{-4}/m$.

Compared with $\hat{p}_1$, the reduction in variance brought by $\hat{p}_4$ is of order 10^{-3}, which implies, in particular, that this evaluation requires $\sqrt{1000} \approx 32$ times fewer simulations than $\hat{p}_1$ to achieve the same precision.   ‖

The evaluation of (3.4) based on simulation from f is therefore not necessarily optimal and Theorem 3.12 shows that this choice is, in fact, always suboptimal. Note also that the integral (3.4) can be represented in an infinite number of ways by triplets $(\mathcal{X}, h, f)$. Therefore, the search for an optimal estimator should encompass all these possible representations (as in Example 3.8). As a side remark, we should stress that the very notion of "optimality" of a representation is quite difficult to define precisely. Indeed, as already noted in Chapter 2, the comparison of simulation methods cannot be equated with the comparison of the variances of the resulting estimators. Conception and computation times should also be taken into account. At another level, note that the optimal method proposed in Theorem 3.12 depends on the function h involved in (3.4). Therefore, it cannot be considered as optimal when several integrals related to f are simultaneously evaluated. In such cases, which often occur in Bayesian analysis, only generic methods can be compared (that is to say, those which are independent of h).

The principal alternative to direct sampling from f for the evaluation of (3.4) is to use importance sampling, defined as follows:

Definition 3.9. The method of *importance sampling* is an evaluation of (3.4) based on generating a sample $X_1, \ldots, X_n$ from a given distribution g and approximating

$$(3.8) \qquad \mathbb{E}_f[h(X)] \approx \frac{1}{m} \sum_{j=1}^{m} \frac{f(X_j)}{g(X_j)} h(X_j) .$$

This method is based on the alternative representation of (3.4):

$$(3.9) \qquad \mathbb{E}_f[h(X)] = \int_{\mathcal{X}} h(x) \frac{f(x)}{g(x)} g(x) \, dx ,$$

which is called the *importance sampling fundamental identity*, and the estimator (3.8) converges to (3.4) for the same reason the regular Monte Carlo estimator $\bar{h}_m$ converges, whatever the choice of the distribution g (as long as $\mathrm{supp}(g) \supset \mathrm{supp}(f)$).

Note that (3.9) is a very general representation that expresses the fact that a given integral is not intrinsically associated with a given distribution. Example 3.8 shows how much of an effect this choice of representation can have. Importance sampling is therefore of considerable interest since it puts very little restriction on the choice of the instrumental distribution g, which can be chosen from distributions that are easy to simulate. Moreover, the same sample (generated from g) can be used repeatedly, not only for different functions h but also for different densities f, a feature which is quite attractive for robustness and Bayesian sensitivity analyses.

Example 3.10. Exponential and log-normal comparison. Consider X as an estimator of λ, when $X \sim \mathcal{E}xp(1/\lambda)$ or when $X \sim \mathcal{LN}(0, \sigma^2)$ (with $e^{\sigma^2/2} = \lambda$, see Problem 3.11). If the goal is to compare the performances of this estimator under both distributions for the scaled squared error loss

$$L(\lambda, \delta) = (\delta - \lambda)^2 / \lambda^2,$$

a *single* sample from $\mathcal{LN}(0, \sigma^2)$, $X_1, \ldots, X_T$, can be used for both purposes, the risks being evaluated by

$$\hat{R}_1 = \frac{1}{T\lambda^2} \sum_{t=1}^{T} X_t e^{-X_t/\lambda} \lambda^{-1} e^{\log(X_t)^2/2\sigma^2} \sqrt{2\pi\sigma} (X_t - \lambda)^2$$

in the exponential case and by

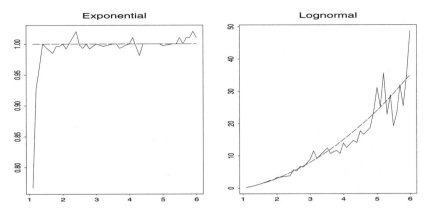

Fig. 3.4. Graph of approximate scaled squared error risks of X vs. λ for an exponential and a log-normal observation, compared with the theoretical values (dashes) for $\lambda \in [1,6]$ (10,000 simulations).

$$\hat{R}_2 = \frac{1}{T\lambda^2} \sum_{t=1}^{T} (X_t - \lambda)^2$$

in the log-normal case. In addition, the scale nature of the parameterization allows a single sample $(Y_1^0, \ldots, Y_T^0)$ from $\mathcal{N}(0,1)$ to be used for all σ's, with $X_t = \exp(\sigma Y_t^0)$.

The comparison of these evaluations is given in Figure 3.4 for $T = 10,000$, each point corresponding to a sample of size T simulated from $\mathcal{LN}(0,\sigma^2)$ by the above transformation. The exact values are given by 1 and $(\lambda + 1)(\lambda - 1)$, respectively. Note that implementing importance sampling in the opposite way offers little appeal since the weights $\exp\{-\log(X_t)^2/2\sigma^2\} \times \exp(\lambda X_t)/X_t$ have infinite variance (see below). The graph of the risk in the exponential case is then more stable than for the original sample from the log-normal distribution. ‖

We close this section by revisiting a previous example with a new twist.

Example 3.11. Small tail probabilities. In Example 3.5 we calculated normal tail probabilities with Monte Carlo sums, and found the method to work well. However, the method breaks down if we need to go too far into the tail. For example, if $Z \sim \mathcal{N}(0,1)$, and we are interested in the probability $P(Z > 4.5)$ (which we know is very small), we could simulate $Z^{(i)} \sim \mathcal{N}(0,1)$ for $i = 1, \ldots M$ and calculate

$$P(Z > 4.5) \approx \frac{1}{M} \sum_{i=1}^{M} \mathbb{I}(Z^{(i)} > 4.5).$$

If we do this, a value of $M = 10,000$ usually produces all zeros of the indicator function.

Of course, the problem is that we are calculating the probability of a very rare event, and naïve simulation will need a lot of iterations to get a reasonable answer. However, with importance sampling we can greatly improve our accuracy.

Let $Y \sim \mathcal{TE}(4.5, 1)$, an exponential distribution (left) truncated at 4.5 with scale 1, with density

$$f_Y(y) = e^{-(x-4.5)} / \int_{4.5}^{\infty} e^{-x} dx .$$

If we now simulate from f_Y and use importance sampling, we obtain (see Problem 3.16)

$$P(Z > 4.5) \approx \frac{1}{M} \sum_{i=1}^{M} \frac{\varphi(Y^{(i)})}{f_Y(Y^{(i)})} \mathbb{I}(Y^{(i)} > 4.5) = .000003377 .$$

$\parallel$

3.3.2 Finite Variance Estimators

Although the distribution g can be almost any density for the estimator (3.8) to converge, there are obviously some choices that are better than others, and it is natural to try to compare different distributions g for the evaluation of (3.4). First, note that, while (3.8) does converge almost surely to (3.4), its variance is finite only when the expectation

$$\mathbb{E}_g \left[h^2(X) \frac{f^2(X)}{g^2(X)} \right] = \mathbb{E}_f \left[h^2(X) \frac{f(X)}{g(X)} \right] = \int_{\mathcal{X}} h^2(x) \frac{f^2(x)}{g(x)} dx < \infty .$$

Thus, instrumental distributions with tails lighter than those of f (that is, those with unbounded ratios f/g) are not appropriate for importance sampling. In fact, in these cases, the variances of the corresponding estimators (3.8) will be infinite for many functions h. More generally, if the ratio f/g is unbounded, the weights $f(x_j)/g(x_j)$ will vary widely, giving too much importance to a few values x_j. This means that the estimator (3.8) may change abruptly from one iteration to the next one, even after many iterations. Conversely, distributions g with thicker tails than f ensure that the ratio f/g does not cause the divergence of $\mathbb{E}_f[h^2 f/g]$. In particular, Geweke (1989) mentions two types of sufficient conditions:

(a) $f(x)/g(x) < M \; \forall x \in \mathcal{X}$ and $\mathrm{var}_f(h) < \infty$;
(b) $\mathcal{X}$ is compact, $f(x) < F$ and $g(x) > \varepsilon \; \forall x \in \mathcal{X}$.

These conditions are quite restrictive. In particular, $f/g < M$ implies that the Accept–Reject algorithm [A.4] also applies. (A comparison between the two approaches is given in Section 3.3.3.)

An alternative to (3.8) which addresses the finite variance issue, and generally yields a more stable estimator, is to use

$$(3.10) \qquad \frac{\sum_{j=1}^{m} h(x_j) \, f(x_j)/g(x_j)}{\sum_{j=1}^{m} f(x_j)/g(x_j)} \, ,$$

where we have replaced m with the sum of the weights. Since $(1/m) \sum_{j=1}^{m}$ $f(x_j)/g(x_j)$ converges to 1 as $m \to \infty$, this estimator also converges to $\mathbb{E}_f h(X)$ by the Strong Law of Large Numbers.. Although this estimator is biased, the bias is small, and the improvement in variance makes it a preferred alternative to (3.8) (see also Lemma 4.3). In fact, Casella and Robert (1998) have shown that the weighted estimator (3.10) may perform better (when evaluated under squared error loss) in some settings. (See also Van Dijk and Kloeck 1984.) For instance, when h is nearly constant, (3.10) is close to this value, while (3.8) has a higher variation since the sum of the weights is different from one.

Among the distributions g leading to finite variances for the estimator (3.8), it is, in fact, possible to exhibit the optimal distribution corresponding to a given function h and a fixed distribution f, as stated by the following result of Rubinstein (1981); see also Geweke (1989).

Theorem 3.12. *The choice of g that minimizes the variance of the estimator (3.8) is*

$$g^*(x) = \frac{|h(x)| \, f(x)}{\int_{\mathcal{X}} |h(z)| \, f(z) \, dz} \, .$$

Proof. First note that

$$\mathrm{var}\left[\frac{h(X)f(X)}{g(X)}\right] = \mathbb{E}_g\left[\frac{h^2(X)f^2(X)}{g^2(X)}\right] - \left(\mathbb{E}_g\left[\frac{h(X)f(X)}{g(X)}\right]\right)^2,$$

and the second term does not depend on g. So, to minimize variance, we only need minimize the first term. From Jensen's inequality it follows that

$$\mathbb{E}_g\left[\frac{h^2(X)f^2(X)}{g^2(X)}\right] \geq \left(\mathbb{E}_g\left[\frac{|h(X)|f(X)}{g(X)}\right]\right)^2 = \left(\int |h(x)|f(x)dx\right)^2,$$

which provides a lower bound that is independent of the choice of g. It is straightforward to verify that this lower bound is attained by choosing $g = g^*$. □

This optimality result is rather formal since, when $h(x) > 0$, the optimal choice $g^*(x)$ requires the knowledge of $\int h(x)f(x)dx$, the integral of interest! A practical alternative taking advantage of Theorem 3.12 is to use the estimator (3.10) as

$$(3.11) \qquad \frac{\sum_{j=1}^{m} h(x_j) \, f(x_j)/g(x_j)}{\sum_{j=1}^{m} f(x_j)/g(x_j)} = \frac{\sum_{j=1}^{m} h(x_j)|h(x_j)|^{-1}}{\sum_{j=1}^{m} |h(x_j)|^{-1}},$$

where $x_j \sim g \propto |h|f$. Note that the numerator is the number of times $h(x_j)$ is positive minus the number of times it is negative. In particular, when h is positive, (3.11) is the *harmonic mean*. Unfortunately, the optimality of Theorem 3.12 does not transfer to (3.11), which is biased and may exhibit severe instability.[2]

From a practical point of view, Theorem 3.12 suggests looking for distributions g for which $|h|f/g$ is almost constant with finite variance. It is important to note that although the finite variance constraint is not necessary for the convergence of (3.8) and of (3.11), importance sampling performs quite poorly when

$$(3.12) \qquad \int \frac{f^2(x)}{g(x)}\, dx = +\infty ,$$

whether in terms of behavior of the estimator (high-amplitude jumps, instability of the path of the average, slow convergence) or of comparison with direct Monte Carlo methods. Distributions g such that (3.12) occurs are therefore not recommended.

The next two examples show that importance sampling methods can bring considerable improvement over naïve Monte Carlo estimates when implemented with care. However, they can encounter disastrous performances and produce extremely poor estimates when the variance conditions are not met.

Example 3.13. Student's t distribution. Consider $X \sim T(\nu, \theta, \sigma^2)$, with density

$$f(x) = \frac{\Gamma((\nu+1)/2)}{\sigma\sqrt{\nu\pi}\,\Gamma(\nu/2)} \left(1 + \frac{(x-\theta)^2}{\nu\sigma^2}\right)^{-(\nu+1)/2}$$

Without loss of generality, we take $\theta = 0$ and $\sigma = 1$. We choose the quantities of interest to be $\mathbb{E}_f[h_i(X)]$ $(i = 1, 2, 3)$, with

$$h_1(x) = \sqrt{\left|\frac{x}{1-x}\right|}, \quad h_2(x) = x^5 \mathbb{I}_{[2.1,\infty[}(x), \quad h_3(x) = \frac{x^5}{1+(x-3)^2}\, \mathbb{I}_{x \geq 0} \ .$$

Obviously, it is possible to generate directly from f. Importance sampling alternatives are associated here with a Cauchy $C(0, 1)$ distribution and a normal $\mathcal{N}(0, \nu/(\nu-2))$ distribution (scaled so that the variance is the same as $T(\nu, \theta, \sigma^2)$). The choice of the normal distribution is not expected to be efficient, as the ratio

$$\frac{f^2(x)}{g(x)} \propto \frac{e^{x^2(\nu-2)/2\nu}}{[1+x^2/\nu]^{(\nu+1)}}$$

does not have a finite integral. However, this will give us an opportunity to study the performance of importance sampling in such a situation. On the

---

[2] In fact, the optimality only applies to the numerator, while another sequence should be used to better approximate the denominator.

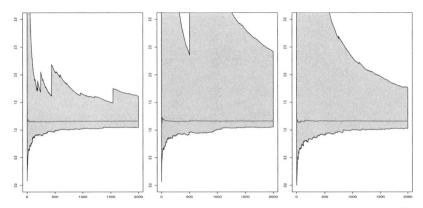

Fig. 3.5. Empirical range of three series of estimators of $\mathbb{E}_f[|X/(1-X)|^{1/2}]$ for $\nu = 12$ and 500 replications: sampling from f *(left)*, importance sampling with a Cauchy instrumental distribution *(center)* and importance sampling with normal importance distribution *(right)*. Average of the 500 series in overlay.

other hand, the $\mathcal{C}(0,1)$ distribution has larger tails than f and ensures that the variance of f/g is finite.

Figure 3.5 illustrates the performances of the three corresponding estimators for the function h_1 when $\nu = 12$ by representing the range of 500 series over 2000 iterations. The average of these series is quite stable over iterations and does not depend on the choice of the importance function, while the range exhibits wide jumps for all three. This phenomenon is due to the fact that the function h_1 has a singularity at $x = 1$ such that h_1^2 is not integrable under f but also such that none of the two other importance sampling estimators has a finite variance (Problem 3.20)! Were we to repeat this experiment with 5000 series rather than 500 series, we would then see larger ranges. There is thus no possible comparison between the three proposals in this case, since they all are inefficient. An alternative choice devised purposely for this function h_1 is to choose g such that $(1-x)g(x)$ is better behaved in $x = 1$. If we take for instance the double Gamma distribution folded at 1, that is, the distribution of X symmetric around 1 such that

$$|X - 1| \sim \mathcal{G}a(\alpha, 1),$$

the ratio

$$h_1(x)\frac{f^2(x)}{g(x)} \propto \sqrt{x}\, f^2(x)\, |1 - x|^{1-\alpha-1}\, exp|1 - x|$$

is integrable around $x = 1$ when $\alpha < 1$. Obviously, the exponential part creates problems at ∞ and leads once more to an infinite variance, but it has much less influence on the stability of the estimator, as shown in Figure 3.6.

Since both h_2 and h_3 have restricted supports, we could benefit by having the instrumental distributions take this information into account. In the case

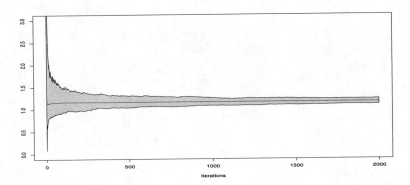

Fig. 3.6. Empirical range of the importance sampling estimator of $\mathbb{E}_f[|X/(1-X)|^{1/2}]$ for $\nu = 12$ and 500 replications based on the double Gamma $\mathcal{G}a(\alpha, 1)$ distribution folded at 1 when $\alpha = .5$. Average of the 500 series in overlay.

of h_2, a uniform distribution on $[0, 1/2.1]$ is reasonable, since the expectation $\mathbb{E}_f[h_2(X)]$ can be written as

$$\int_0^{1/2.1} u^{-7} f(1/u) \, du = \frac{1}{2.1} \int_0^{1/2.1} 2.1 \, u^{-7} f(1/u) \, du \,,$$

as in Example 3.8. The corresponding importance sampling estimator is then

$$\delta_2 = \frac{1}{2.1m} \sum_{j=1}^m U_j^{-7} f(1/U_j) \,,$$

where the U_j's are iid $\mathcal{U}([0, 1/2.1])$. Figure 3.7 shows the improvement brought by this choice, with the estimator δ_2 converging to the true value after only a few hundred iterations. The importance sampling estimator associated with the Cauchy distribution is also quite stable, but it requires more iterations to achieve the same precision. Both of the other estimators (which are based on the true distribution and the normal distribution, respectively) fluctuate around the exact value with high-amplitude jumps, because their variance is infinite.

In the case of h_3, a reasonable candidate for the instrumental distribution is $g(x) = \exp(-x)\mathbb{I}_x \geq 0$, leading to the estimation of

$$\mathbb{E}_f[h_3(X)] = \int_0^\infty \frac{x^5}{1 + (x-3)^2} f(x) \, dx$$
$$= \int_0^\infty \frac{x^5 e^x}{1 + (x-3)^2} f(x) \, e^{-x} \, dx$$

by

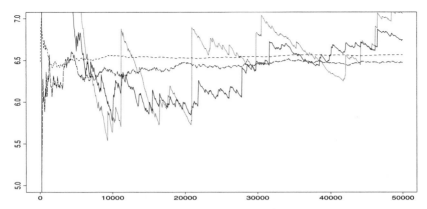

Fig. 3.7. Convergence of four estimators of $\mathbb{E}_f[X^5\mathbb{I}_{X\geq 2.1}]$ for $\nu = 12$: Sampling from f (solid lines), importance sampling with Cauchy instrumental distribution (short dashes), importance sampling with uniform $\mathcal{U}([0, 1/2.1])$ instrumental distribution (long dashes) and importance sampling with normal instrumental distribution (dots). The final values are respectively 6.75, 6.48, 6.57, and 7.06, for an exact value of 6.54.

$$(3.13) \qquad\qquad \frac{1}{m}\sum_{j=1}^{m} h_3(X_j)\, w(X_j)\,,$$

where the X_j's are iid $\mathcal{E}xp(1)$ and $w(x) = f(x)\exp(x)$. Figure 3.8 shows that, although this weight does not have a finite expectation under $\mathcal{T}(\nu,0,1)$, meaning that the variance is infinite, the estimator (3.13) provides a good approximation of $\mathbb{E}_f[h_3(X)]$, having the same order of precision as the estimation provided by the exact simulation, and greater stability. The estimator based on the Cauchy distribution is, as in the other case, stable, but its bias is, again, slow to vanish, and the estimator associated with the normal distribution once more displays large fluctuations which considerably hinder its convergence. ‖

Example 3.14. Transition matrix estimation. Consider a Markov chain with two states, 1 and 2, whose transition matrix is

$$T = \begin{pmatrix} p_1 & 1 - p_1 \\ 1 - p_2 & p_2 \end{pmatrix},$$

that is,

$$P(X_{t+1} = 1|X_t = 1) = 1 - P(X_{t+1} = 2|X_t = 1) = p_1,$$
$$P(X_{t+1} = 2|X_t = 2) = 1 - P(X_{t+1} = 1|X_t = 2) = p_2\,.$$

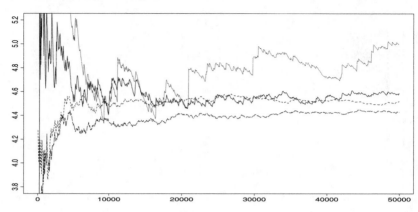

Fig. 3.8. Convergence of four estimators of $\mathbb{E}_f[h_3(X)]$: Sampling from f (solid lines), importance sampling with Cauchy instrumental distribution (short dashes), with normal instrumental distribution (dots), and with exponential instrumental distribution (long dashes). The final values after $50,000$ iterations are respectively 4.58, 4.42, 4.99, and 4.52, for a true value of 4.64.

Assume, in addition, that the constraint $p_1 + p_2 < 1$ holds (see Geweke 1989 for a motivation related to continuous time processes). If the sample is $X_1, \ldots, X_m$ and the prior distribution is

$$\pi(p_1, p_2) = 2 \, \mathbb{I}_{p_1 + p_2 < 1} \, ,$$

the posterior distribution of (p_1, p_2) is

$$\pi(p_1, p_2 | m_{11}, m_{12}, m_{21}, m_{22}) \propto p_1^{m_{11}} (1 - p_1)^{m_{12}} (1 - p_2)^{m_{21}} p_2^{m_{22}} \, \mathbb{I}_{p_1 + p_2 < 1} \, ,$$

where m_{ij} is the number of passages from i to j, that is,

$$m_{ij} = \sum_{t=2}^{m} \mathbb{I}_{x_t = i} \mathbb{I}_{x_{t+1} = j},$$

and it follows that $\mathcal{D} = (m_{11}, \ldots, m_{22})$ is a sufficient statistic.

Suppose now that the quantities of interest are the posterior expectations of the probabilities and the associated odds:

$$h_1(p_1, p_2) = p_1, \ h_2(p_1, p_2) = p_2, \ h_3(p_1, p_2) = \frac{p_1}{1 - p_1}$$

and

$$h_4(p_1, p_2) = \frac{p_2}{1 - p_2}, \quad h_5(p_1, p_2) = \log\left(\frac{p_1(1 - p_2)}{p_2(1 - p_1)}\right) \, ,$$

respectively.

We now look at a number of ways in which to calculate these posterior expectations.

(i) The distribution $\pi(p_1, p_2 | \mathcal{D})$ is the restriction of the product of two distributions $\mathcal{B}e(m_{11} + 1, m_{12} + 1)$ and $\mathcal{B}e(m_{22} + 1, m_{21} + 1)$ to the simplex $\{(p_1, p_2) : p_1 + p_2 < 1\}$. So a reasonable first approach is to simulate these two distributions until the sum of two realizations is less than 1. Unfortunately, this naïve strategy is rather inefficient since, for the given data $(m_{11}, m_{12}, m_{21}, m_{22}) = (68, 28, 17, 4)$ we have $P^{\pi}(p_1 + p_2 < 1 | \mathcal{D}) = 0.21$ (Geweke 1989). The importance sampling alternatives are to simulate distributions which are restricted to the simplex.

(ii) A solution inspired from the shape of $\pi(p_1, p_2 | \mathcal{D})$ is a Dirichlet distribution $\mathcal{D}(m_{11} + 1, m_{22} + 1, m_{12} + m_{21} + 1)$, with density

$$\pi_1(p_1, p_2 | \mathcal{D}) \propto p_1^{m_{11}} p_2^{m_{22}} (1 - p_1 - p_2)^{m_{12} + m_{21}} .$$

However, the ratio $\pi(p_1, p_2 | \mathcal{D}) / \pi_1(p_1, p_2 | \mathcal{D})$ is not bounded and the corresponding variance is infinite.

(iii) Geweke's (1989) proposal is to use the normal approximation to the binomial distribution, that is,

$$\pi_2(p_1, p_2 | \mathcal{D}) \propto \exp\{-(m_{11} + m_{12})(p_1 - \hat{p}_1)^2 / 2\, \hat{p}_1(1 - \hat{p}_1)\}$$
$$\times \exp\{-(m_{21} + m_{22})(p_2 - \hat{p}_2)^2 / 2\, \hat{p}_2(1 - \hat{p}_2)\}\, \mathbb{I}_{p_1 + p_2 \leq 1} ,$$

where $\hat{p}_i$ is the maximum likelihood estimator of p_i, that is, $m_{ii} / (m_{ii} + m_{i(3-i)})$. An efficient way to simulate π_2 is then to simulate p_1 from the normal distribution $\mathcal{N}(\hat{p}_1, \hat{p}_1(1 - \hat{p}_1) / (m_{12} + m_{11}))$ restricted to $[0, 1]$, then p_2 from the normal distribution $\mathcal{N}(\hat{p}_2, \hat{p}_2(1 - \hat{p}_2) / (m_{21} + m_{22}))$ restricted to $[0, 1 - p_1]$, using the method proposed by Geweke (1991) and Robert (1995b). The ratio π / π_2 then has a finite expectation under π, since (p_1, p_2) is restricted to $\{(p_1, p_2) : p_1 + p_2 < 1\}$.

(iv) Another possibility is to keep the distribution $\mathcal{B}(m_{11} + 1, m_{12} + 1)$ as the marginal distribution on p_1 and to modify the conditional distribution $p_2^{m_{22}} (1 - p_2)^{m_{21}} \mathbb{I}_{p_2 < 1 - p_1}$ into

$$\pi_3(p_2 | p_1, \mathcal{D}) = \frac{2}{(1 - p_1)^2}\, p_2\, \mathbb{I}_{p_2 < 1 - p_1} .$$

The ratio $w(p_1, p_2) \propto p_2^{m_{22} - 1} (1 - p_2)^{m_{21}} (1 - p_1)^2$ is then bounded in (p_1, p_2).

Table 3.4 provides the estimators of the posterior expectations of the functions h_j evaluated for the true distribution π (simulated the naïve way, that is, until $p_1 + p_2 < 1$) and for the three instrumental distributions π_1, π_2 and π_3. The distribution π_3 is clearly preferable to the two other instrumental distributions since it provides the same estimation as the true distribution, at a lower computational cost. Note that π_1 does worse in all cases.

Figure 3.9 describes the evolution of the estimators (3.10) of $\mathbb{E}[h_5]$ as m increases for the three instrumental distributions considered. Similarly to

Distribution	h_1	h_2	h_3	h_4	h_5
π_1	0.748	0.139	3.184	0.163	2.957
π_2	0.689	0.210	2.319	0.283	2.211
π_3	0.697	0.189	2.379	0.241	2.358
π	0.697	0.189	2.373	0.240	2.358

Table 3.4. Comparison of the evaluations of $\mathbb{E}_f[h_j]$ for the estimators (3.10) corresponding to three instrumental distributions π_i and to the true distribution π (10,000 simulations).

Table 3.4, it shows the improvement brought by the distribution π_3 upon the alternative distributions, since the precision is of the same order as the true distribution, for a significantly lower simulation cost. The jumps in the graphs of the estimators associated with π_2 and, especially, with π_1 are characteristic of importance sampling estimators with infinite variance. ‖

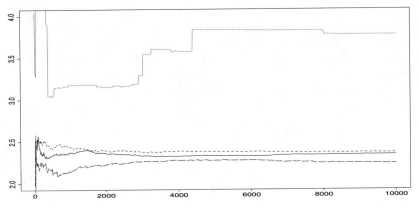

Fig. 3.9. Convergence of four estimators of $\mathbb{E}_f[h_5(X)]$ for the true distribution π (solid lines) and for the instrumental distributions π_1 (dots), π_2 (long dashes), and π_3 (short dashes). The final values after 10,000 iterations are 2.373, 3.184, 2.319, and 2.379, respectively.

We therefore see that importance sampling cannot be applied blindly. Rather, care must be taken in choosing an instrumental density as the almost sure convergence of (3.8) is only formal (in the sense that it may require an enormous number of simulations to produce an accurate approximation of the quantity of interest). These words of caution are meant to make the user aware of the problems that might be encountered if importance sampling is used when $\mathbb{E}_f[\|f(X)/g(X)\|]$ is infinite. (When $\mathbb{E}_f[f(X)/g(X)]$ is finite, the stakes are not so high, as convergence is more easily attained.) If the issue of finiteness of the variance is ignored, and not detected, it may result in strong biases. For example, it can happen that the obvious divergence behavior of the previous examples does not occur. Thus, other measures, such as monitoring of the range of the weights $f(X_i)/g(X_i)$ (which are of mean 1 in all cases), can help to detect convergence problems. (See also Note 4.6.1.)

The finiteness of the ratio $\mathbb{E}_f[f(X)/g(X)]$ can be achieved by substituting a mixture distribution for the density g,

$$(3.14) \qquad\qquad \rho g(x) + (1 - \rho)\ell(x),$$

where ρ is close to 1 and ℓ is chosen for its heavy tails (for instance, a Cauchy or a Pareto distribution). From an operational point of view, this means that the observations are generated with probability ρ from g and with probability $1 - \rho$ from ℓ. However, the mixture (g versus ℓ) does not play a role in the computation of the importance weights; that is, by construction, the estimator integrates out the uniform variable used to decide between g and ℓ. (We discuss in detail such a marginalization perspective in Section 4.2, where uniform variables involved in the simulation are integrated out in the estimator.) Obviously, (3.14) replaces $g(x)$ in the weights of (3.8) or (3.11), which can then ensure a finite variance for integrable functions h^2. Hesterberg (1998) studies the performances of this approach, called a *defensive mixture*.

3.3.3 Comparing Importance Sampling with Accept–Reject

Theorem[3] 3.12 formally solves the problem of comparing Accept–Reject and importance sampling methods, since with the exception of the constant functions $h(x) = h_0$, the optimal density g^* is always different from f. However, a more realistic comparison should also take account of the fact that Theorem 3.12 is of limited applicability in a practical setup, as it prescribes an instrumental density that depends on the function h of interest. This may not only result in a considerable increase of the computation time for every new function h (especially if the resulting instrumental density is not easy to generate from), but it also eliminates the possibility of reusing the generated sample to estimate a number of different quantities, as in Example 3.14. Now, when the Accept–Reject method is implemented with a density g satisfying $f(x) \leq Mg(x)$ for a constant $1 < M < \infty$, the density g can serve as the instrumental density for importance sampling. A positive feature is that f/g is bounded, thus ensuring finiteness of the variance for the corresponding importance sampling estimators. Bear in mind, though, that in the Accept–Reject method the resulting sample, $X_1, \ldots, X_n$, is a subsample of $Y_1, \ldots, Y_t$, where the Y_i's are simulated from g and where t is the (random) number of simulations from g required for produce the n variables from f.

---

[3] This section contains more specialized material and may be omitted on a first reading.

To undertake a comparison of estimation using Accept–Reject and estimation using importance sampling, it is reasonable to start with the two traditional estimators

$$(3.15) \qquad \delta_1 = \frac{1}{n} \sum_{i=1}^{n} h(X_i) \qquad \text{and} \qquad \delta_2 = \frac{1}{t} \sum_{j=1}^{t} h(Y_j) \frac{f(Y_j)}{g(Y_j)} \,.$$

These estimators correspond to the straightforward utilization of the sample produced by Accept–Reject and to an importance sampling estimation derived from the overall sample, that is, to a recycling of the variables rejected by algorithm [A.4].[4] If the ratio f/g is only known up to a constant, δ_2 can be replaced by

$$\delta_3 = \sum_{j=1}^{t} h(Y_j) \frac{f(Y_j)}{g(Y_j)} \left/ \sum_{j=1}^{t} \frac{f(Y_j)}{g(Y_j)} \right. \,.$$

If we write δ_2 in the more explicit form

$$\delta_2 = \frac{n}{t} \left\{ \frac{1}{n} \sum_{i=1}^{n} h(X_i) \frac{f(X_i)}{g(X_i)} + \frac{t-n}{n} \frac{1}{t-n} \sum_{i=1}^{t-n} h(Z_i) \frac{f(Z_i)}{g(Z_i)} \right\} \,,$$

where $\{Y_1, \ldots, Y_t\} = \{X_1, \ldots, X_n\} \cup \{Z_1, \ldots, Z_{t-n}\}$ (the Z_i's being the variables rejected by the Accept–Reject algorithm [A.4]), one might argue that, based on sample size, the variance of δ_2 is smaller than that of the estimator

$$\frac{1}{n} \sum_{i=1}^{n} h(X_i) \frac{f(X_i)}{g(X_i)} \,.$$

If we could apply Theorem 3.12, we could then conclude that this latter estimator dominates δ_1 (for an appropriate choice of g) and, hence, that it is better to recycle the Z_i's than to discard them. Unfortunately, this reasoning is flawed since t is a random variable, being the *stopping rule* of the Accept–Reject algorithm. The distribution of t is therefore a negative binomial distribution, $\mathcal{N}eg(n, 1/M)$ (see Problem 2.30) so $(Y_1, \ldots, Y_t)$ is not an iid sample from g. (Note that the Y_j's corresponding to the X_i's, including Y_t, have distribution f, whereas the others do not.)

The comparison between δ_1 and δ_2 can be reduced to comparing $\delta_1 = f(y_t)$ and δ_2 for $t \sim \mathcal{G}eo(1/M)$ and $n = 1$. However, even with this simplification, the comparison is quite involved (see Problem 3.34 for details), so a general comparison of the bias and variance of δ_2 with $\text{var}_f(h(X))$ is difficult (Casella and Robert 1998).

While the estimator δ_2 is based on an incorrect representation of the distribution of $(Y_1, \ldots, Y_t)$, a reasonable alternative based on the correct distribution of the sample is

---

[4] This obviously assumes a relatively tight control on the simulation methods rather than the use of a (black box) pseudo-random generation software, which only delivers the accepted variables.

$$(3.16) \qquad \delta_4 = \frac{n}{t}\, \delta_1 + \frac{1}{t} \sum_{j=1}^{t-n} h(Z_j)\, \frac{(M-1)f(Z_j)}{Mg(Z_j) - f(Z_j)} \,,$$

where the Z_j's are the elements of $(Y_1, \ldots, Y_t)$ that have been rejected. This estimator is also unbiased and the comparison with δ_1 can also be studied in the case $n = 1$; that is, through the comparison of the variances of $h(X_1)$ and of δ_4, which now can be written in the form

$$\delta_4 = \frac{1}{t}\, h(X_1) + (1 - \rho) \frac{1}{t} \sum_{j=1}^{t-1} h(Z_j) \left(\frac{g(Z_j)}{f(Z_j)} - \rho \right)^{-1}.$$

Assuming again that $\mathbb{E}_f[h(X)] = 0$, the variance of δ_4 is

$$\mathrm{var}(\delta_4) = \mathbb{E}\left[\frac{t-1}{t^2} \int h^2(x) \frac{f^2(x)(M-1)}{Mg(x) - f(x)}\, dx + \frac{1}{t^2}\, \mathbb{E}_f[h^2(X)] \right],$$

which is again too case-specific (that is, too dependent on f, g, and h) to allow for a general comparison.

The marginal distribution of the Z_i's from the Accept–Reject algorithm is $(Mg - f)/(M - 1)$, and the importance sampling estimator δ_5 associated with this instrumental distribution is

$$\delta_5 = \frac{1}{t-n} \sum_{j=1}^{t-n} \frac{(M-1)f(Z_j)}{Mg(Z_j) - f(Z_j)} h(Z_j),$$

which allows us to write δ_4 as

$$\delta_4 = \frac{n}{t} \delta_1 + \frac{t-n}{t} \delta_5,$$

a weighted average of the usual Monte Carlo estimator and of δ_5.

According to Theorem 3.12, the instrumental distribution can be chosen such that the variance of δ_5 is lower than the variance of δ_1. Since this estimator is unbiased, δ_4 will dominate δ_1 for an appropriate choice of g. This domination result is of course as formal as Theorem 3.12, but it indicates that, for a fixed g, there exist functions h such that δ_4 improves on δ_1.

If f is only known up to the constant of integration (hence, f and M are not properly scaled), δ_4 can replaced by

$$\delta_6 = \frac{n}{t} \delta_1 + \frac{t-n}{t} \sum_{j=1}^{t-n} \frac{h(Z_j)f(Z_j)}{Mg(Z_j) - f(z_j)}$$

$$(3.17) \qquad \Big/ \sum_{j=1}^{t-n} \frac{f(Z_j)}{Mg(Z_j) - f(Z_j)} \,.$$

Although the above domination of δ_1 by δ_4 does not extend to δ_6, nonetheless, δ_6 correctly estimates constant functions while being asymptotically equivalent to δ_4. See Casella and Robert (1998) for additional domination results of δ_1 by weighted estimators.

Example 3.15. Gamma simulation. For illustrative purposes, consider the simulation of $\mathcal{G}a(\alpha, \beta)$ from the instrumental distribution $\mathcal{G}a(a, b)$, with $a = [\alpha]$ and $b = a\beta/\alpha$. (This choice of b is justified in Example 2.19 as maximizing the acceptance probability in an Accept–Reject scheme.) The ratio f/g is therefore

$$w(x) = \frac{\Gamma(a)}{\Gamma(\alpha)} \frac{\beta^{\alpha}}{b^{a}} \, x^{\alpha-a} e^{(b-\beta)x} \,,$$

which is bounded by

$$M = \frac{\Gamma(a)}{\Gamma(\alpha)} \frac{\beta^{\alpha}}{b^{a}} \left(\frac{\alpha - a}{\beta - b} \right)^{\alpha - a} e^{-(\alpha-a)}$$

(3.18)
$$= \frac{\Gamma(a)}{\Gamma(\alpha)} \exp\{\alpha(\log(\alpha) - 1) - a(\log(a) - 1)\} \,.$$

Since the ratio $\Gamma(a)/\Gamma(\alpha)$ is bounded from above by 1, an approximate bound that can be used in the simulation is

$$M' = \exp\{a(\log(a) - 1) - \alpha(\log(\alpha) - 1)\} \,,$$

with $M'/M = 1 + \varepsilon = \Gamma(\alpha)/\Gamma([\alpha])$. In this particular setup, the estimator δ_4 is available since f/g and M are explicitly known. In order to assess the effect of the approximation (3.17), we also compute the estimator δ_6 for the following functions of interest:

$$h_1(x) = x^3, \quad h_2(x) = x \log x, \quad \text{and} \quad h_3(x) = \frac{x}{1 + x} \,.$$

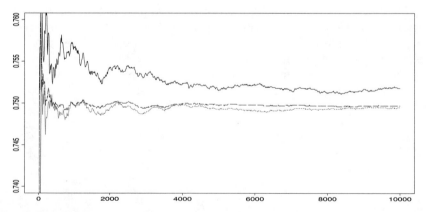

Fig. 3.10. Convergence of the estimators of $\mathbb{E}[X/(1+X)]$, δ_1 (solid lines), δ_4 (dots) and δ_6 (dashes), for $\alpha = 3.7$ and $\beta = 1$. The final values are respectively 0.7518, 0.7495, and 0.7497, for a true value of the expectation equal to 0.7497.

Figure 3.10 describes the convergence of the three estimators of h_3 in m for $\alpha = 3.7$ and $\beta = 1$ (which yields an Accept–Reject acceptance probability of $1/M = .10$). Both estimators δ_4 and δ_6 have more stable graphs than the empirical average δ_1 and they converge much faster to the theoretical expectation 0.7497, δ_6 then being equal to this value after 6,000 iterations. For $\alpha = 3.08$ and $\beta = 1$ (which yields an Accept–Reject acceptance probability of $1/M = .78$), Figure 3.11 illustrates the change of behavior of the three estimators of h_3 since they now converge at similar speeds. Note the proximity of δ_4 and δ_1, δ_6 again being the estimator closest to the theoretical expectation 0.7081 after 10,000 iterations.

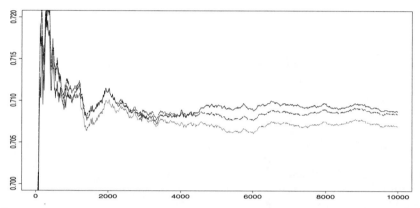

Fig. 3.11. Convergence of estimators of $\mathbb{E}[X/(1 + X)]$, δ_1 (solid lines), δ_4 (dots) and δ_6 (dashes) for $\alpha = 3.08$ and $\beta = 1$. The final values are respectively 0.7087, 0.7069, and 0.7084, for a true value of the expectation equal to 0.7081.

Table 3.5 provides another evaluation of the three estimators in a case which is a priori very favorable to importance sampling, namely for $\alpha = 3.7$. The table exhibits, in most cases, a strong domination of δ_4 and δ_6 over δ_1 and a moderate domination of δ_4 over δ_6.   ‖

In contrast to the general setup of Section 3.3, δ_4 (or its approximation δ_6) can always be used in an Accept–Reject sampling setup since this estimator does not require additional simulations. It provides a second evaluation of $\mathbb{E}_f[h]$, which can be compared with the Monte Carlo estimator for the purpose of convergence assessment.

3.4 Laplace Approximations

As an alternative to simulation of integrals, we can also attempt analytic approximations. One of the oldest and most useful approximations is the integral

m	100			1000			5 000		
	δ_1	δ_4	δ_6	δ_1	δ_4	δ_6	δ_1	δ_4	δ_6
h_1	87.3	55.9	64.2	36.5	0.044	0.047	2.02	0.54	0.64
h_2	1.6	3.3	4.4	4.0	0.00	0.00	0.17	0.00	0.00
h_3	6.84	0.11	0.76	4.73	0.00	0.00	0.38	0.02	0.00

Table 3.5. Comparison of the performances of the Monte Carlo estimator (δ_1) with two importance sampling estimators (δ_4 and δ_6) under squared error loss after m iterations for $\alpha = 3.7$ and $\beta = 1$. The squared error loss is multiplied by 10^2 for the estimation of $\mathbb{E}[h_2(X)]$ and by 10^5 for the estimation of $\mathbb{E}[h_3(X)]$. The squared errors are actually the difference from the theoretical values (99.123, 5.3185, and 0.7497, respectively) and the three estimators are based on the same unique sample, which explains the lack of monotonicity (in m) of the errors. (*Source:* Casella and Robert 1998.)

Laplace approximation. It is based on the following argument: Suppose that we are interested in evaluating the integral

$$(3.19) \qquad \int_A f(x|\theta)dx$$

for a fixed value of θ. (The function f needs to be non-negative and integrable; see Tierney and Kadane 1986 and Tierney et al. 1989 for extensions.). Write $f(x|\theta) = \exp\{nh(x|\theta)\}$, where n is the sample size or another parameter which can go to infinity, and use a Taylor series expansion of $h(x|\theta)$ about a point x_0 to obtain

$$h(x|\theta) \approx h(x_0|\theta) + (x - x_0)h'(x_0|\theta) + \frac{(x - x_0)^2}{2!}h''(x_0|\theta)$$

$$(3.20) \qquad + \frac{(x - x_0)^3}{3!}h'''(x_0|\theta) + R_n(x) \,,$$

where we write

$$h'(x_0|\theta) = \left. \frac{\partial h(x|\theta)}{\partial x} \right|_{x=x_0} \,,$$

and similarly for the other terms, while the remainder $R_n(x)$ satisfies

$$\lim_{x \to x_0} R_n(x)/(x - x_0)^3 = 0.$$

Now choose $x_0 = \hat{x}_\theta$, the value that satisfies $h'(\hat{x}_\theta|\theta) = 0$ and maximizes $h(x|\theta)$ for the given value of θ. Then, the linear term in (3.20) is zero and we have the approximation

$$\int_A e^{nh(x|\theta)}dx \simeq e^{nh(\hat{x}_\theta|\theta)} \int_A e^{n\frac{(x-\hat{x}_\theta)^2}{2}h''(\hat{x}_\theta|\theta)}e^{n\frac{(x-\hat{x}_\theta)^3}{3!}h'''(\hat{x}_\theta|\theta)}dx,$$

which is valid within a neighborhood of $\hat{x}_\theta$. (See Schervish 1995, Section 7.4.3, for detailed conditions.) Note the importance of choosing the point x_0 to be a maximum.

The cubic term in the exponent is now expanded in a series around $\hat{x}_\theta$. Recall that the second order Taylor expansion of e^y around 0 is $e^y \approx 1 + y + y^2/2!$, and hence expanding $\exp\{n(x - \hat{x}_\theta)^3 h'''(\hat{x}_\theta|\theta)/3!\}$ around $\hat{x}_\theta$, we obtain the approximation

$$1 + n\frac{(x - \hat{x}_\theta)^3}{3!}h'''(\hat{x}_\theta|\theta) + n^2\frac{(x - \hat{x}_\theta)^6}{2!(3!)^2}[h'''(\hat{x}_\theta|\theta)]^2$$

and thus

$$\int_A e^{nh(x|\theta)}dx \simeq e^{nh(\hat{x}_\theta|\theta)}\int_A e^{n\frac{(x - \hat{x}_\theta)^2}{2}h''(\hat{x}_\theta|\theta)}$$

$$(3.21) \quad \times \left[1 + n\frac{(x - \hat{x}_\theta)^3}{3!}h'''(\hat{x}_\theta|\theta) + n^2\frac{(x - \hat{x}_\theta)^6}{2!(3!)^2}[h'''(\hat{x}_\theta|\theta)]^2 + R_n\right]dx,$$

where R_n again denotes a remainder term.

Excluding R_n, we call the integral approximations in (3.21) a *first-order* approximation if it includes only the first term in the right-hand side, a *second-order* approximation if it includes the first two terms; and a *third-order* approximation if it includes all three terms.

Since the above integrand is the kernel of a normal density with mean $\hat{x}_\theta$ and variance $-1/n\,h''(\hat{x}_\theta|\theta)$, we can evaluate these expressions further. More precisely, letting $\Phi(\cdot)$ denote the standard normal cdf, and taking $A = [a, b]$, we can evaluate the integral in the first-order approximation to obtain (see Problem 3.25)

$$\int_a^b e^{nh(x|\theta)}dx \simeq e^{nh(\hat{x}_\theta|\theta)}\sqrt{\frac{2\pi}{-nh''(\hat{x}_\theta|\theta)}}$$

$$(3.22) \quad \times \left\{\Phi[\sqrt{-nh''(\hat{x}_\theta|\theta)}(b - \hat{x}_\theta)] - \Phi[\sqrt{-nh''(\hat{x}_\theta|\theta)}(a - \hat{x}_\theta)]\right\}.$$

Example 3.16. Gamma approximation. As a simple illustration of the Laplace approximation, consider estimating a Gamma $\mathcal{G}a(\alpha, 1/\beta)$ integral, say

$$(3.23) \qquad \int_a^b \frac{x^{\alpha-1}}{\Gamma(\alpha)\beta^\alpha}\,e^{-x/\beta}dx.$$

Here we have $h(x) = -\frac{x}{\beta} + (\alpha - 1)\log(x)$ with second order Taylor expansion (around a point x_0)

$$h(x) \approx h(x_0) + h'(x_0)(x - x_0) + h''(x_0)\frac{(x - x_0)^2}{2!}$$

$$= -\frac{x_0}{\beta} + (\alpha - 1)\log(x_0) + \left(\frac{\alpha - 1}{x_0} - \frac{1}{\beta}\right)(x - x_0) - \frac{\alpha - 1}{2x_0^2}(x - x_0)^2.$$

Choosing $x_0 = \hat{x}_\theta = (\alpha - 1)\beta$ (the mode of the density and maximizer of h) yields

$$h(x) \approx \frac{\hat{x}_\theta}{\beta} + (\alpha - 1)\log(\hat{x}_\theta) + \frac{\alpha - 1}{2\hat{x}_\theta^2}(x - \hat{x}_\theta)^2$$

Now substituting into (3.22) yields the Laplace approximation

$$\int_a^b \frac{x^{\alpha-1}}{\Gamma(\alpha)\beta^\alpha} e^{-x/\beta} dx = \hat{x}_\theta^{\alpha-1} e^{\hat{x}_\theta/\beta} \sqrt{\frac{2\pi\hat{x}_\theta^2}{\alpha - 1}}$$

$$\times \left\{ \Phi\left(\sqrt{\frac{\alpha - 1}{\hat{x}_\theta^2}}(b - \hat{x}_\theta)\right) - \Phi\left(\sqrt{\frac{\alpha - 1}{\hat{x}_\theta^2}}(a - \hat{x}_\theta)\right) \right\}$$

For $\alpha = 5$ and $\beta = 2$, $\hat{x}_\theta = 8$, and the approximation will be best in that area. In Table 3.6 we see that although the approximation is reasonable in the central region of the density, it becomes quite unacceptable in the tails. ‖

Interval	Approximation	Exact
$(7, 9)$	0.193351	0.193341
$(6, 10)$	0.375046	0.37477
$(2, 14)$	0.848559	0.823349
$(15.987, \infty)$	0.0224544	0.100005

Table 3.6. Laplace approximation of a Gamma integral for $\alpha = 5$ and $\beta = 2$.

Thus, we see both the usefulness and the limits of the Laplace approximation. In problems where Monte Carlo calculations are prohibitive because of computing time, the Laplace approximation can be useful as a guide to the solution of the problem. Also, the corresponding Taylor series can be used as a proposal density, which is particularly useful in problems where no obvious proposal exists. (See Example 7.12 for a similar situation.)

3.5 Problems

3.1 For the normal-Cauchy Bayes estimator

$$\delta(x) = \frac{\int_{-\infty}^\infty \frac{\theta}{1+\theta^2} e^{-(x-\theta)^2/2} d\theta}{\int_{-\infty}^\infty \frac{1}{1+\theta^2} e^{-(x-\theta)^2/2} d\theta}$$

(a) Plot the integrand and use Monte Carlo integration to calculate the integral.
(b) Monitor the convergence with the standard error of the estimate. Obtain three digits of accuracy with probability .95.

3.2 (Continuation of Problem 3.1)
(a) Use the Accept–Reject algorithm, with a Cauchy candidate, to generate a sample from the posterior distribution and calculate the estimator.

(b) Design a computer experiment to compare Monte Carlo error when using
   (i) the same random variables θ_i in numerator and denominator, or (ii)
   different random variables.

3.3 (a) For a standard normal random variable Z, calculate $P(Z > 2.5)$ using
   Monte Carlo sums based on indicator functions. How many simulated random variables are needed to obtain three digits of accuracy?

(b) Using Monte Carlo sums verify that if $X \sim \mathcal{G}(1,1)$, $P(X > 5.3) \approx .005$.
   Find the exact .995 cutoff to three digits of accuracy.

3.4 (a) If $X \sim \mathcal{N}(0, \sigma^2)$, show that

$$\mathbb{E}[e^{-X^2}] = \frac{1}{\sqrt{2\sigma^2 + 1}} .$$

(b) Generalize to the case $X \sim \mathcal{N}(\mu, \sigma^2)$.

3.5 Referring to Example 3.6:

(a) Verify the maximum of the likelihood ratio statistic.

(b) Generate 5000 random variables according to (3.7), recreating the left panel
   of Figure 3.2. Compare this distribution to a null distribution where we *fix*
   null values of p_1 and p_2, for example, $(p_1, p_2) = (.25, .75)$. For a range of
   values of (p_1, p_2), compare the histograms both with the one from (3.7) and
   the χ_1^2 density. What can you conclude?

3.6 An alternate analysis to that of Example 3.6 is to treat the contingency table
as two binomial distributions, one for the patients receiving surgery and one for
those receiving radiation. Then the test of hypothesis becomes a test of equality
of the two binomial parameters. Repeat the analysis of the data in Table 3.2
under the assumption of two binomials. Compare the results to those of Example
3.6.

3.7 A famous medical experiment was conducted by Joseph Lister in the late 1800s
to examine the relationship between the use of a disinfectant, carbolic acid, and
surgical success rates. The data are

	Disinfectant	
	Yes	No
Success	34	19
Failure	6	16

Using the techniques of Example 3.6, analyze these data to examine the associ-
ation between disinfectant and surgical success rates. Use both the multinomial
model and the two-binomial model.

3.8 Referring to Example 3.3, we calculate the expected value of $\delta^\pi(x)$ from the pos-
terior distribution $\pi(\theta|x) \propto \|\theta\|^{-2} \exp\{-\|x - \theta\|^2/2\}$, arising from a normal
likelihood and noninformative prior $\|\theta\|^{-2}$ (see Example 1.12).

(a) Show that if the quadratic loss of Example 3.3 is normalized by $1/(2\|\theta\|^2 + p)$, the resulting Bayes estimator is

$$\delta^\pi(x) = \mathbb{E}^\pi \left[\frac{\|\theta\|^2}{2\|\theta\|^2 + p} \,\bigg|\, x, \lambda \right] \Big/ \mathbb{E}^\pi \left[\frac{1}{2\|\theta\|^2 + p} \,\bigg|\, x, \lambda \right] .$$

(b) Simulation of the posterior can be done by representing θ in polar co-
   ordinates $(\rho, \varphi_1, \varphi_2)$ $(\rho > 0, \varphi_1 \in [-\pi/2, \pi/2], \varphi_2 \in [-\pi/2, \pi/2])$, with
   $\theta = (\rho \cos \varphi_1, \ \rho \sin \varphi_1 \cos \varphi_2, \ \rho \sin \varphi_1 \sin \varphi_2)$. If we denote $\xi = \theta/\rho$, which

depends only on (φ_1, φ_2), show that $\rho|\varphi_1, \varphi_2 \sim \mathcal{N}(x \cdot \xi, 1)$ and then integration of ρ then leads to

$$\pi(\varphi_1, \varphi_2|x) \propto \exp\{(x \cdot \xi)^2/2\} \sin(\varphi_1),$$

where $x \cdot \xi = x_1 \cos(\varphi_1) + x_2 \sin(\varphi_1) \cos(\varphi_2) + x_3 \sin(\varphi_1) \sin(\varphi_2)$.

(c) Show how to simulate from $\pi(\varphi_1, \varphi_2|x)$ using an Accept–Reject algorithm with instrumental function $\sin(\varphi_1) \exp\{||x||^2/2\}$.

(d) For $p = 3$ and $x = (0.1, 1.2, -0.7)$, demonstrate the convergence of the algorithm. Make plots of the iterations of the integral and its standard error.

3.9 For the situation of Example 3.10, recreate Figure 3.4 using the following simulation strategies with a sample size of $10,000$ points:

(a) For each value of λ, simulate a sample from the $\mathcal{E}xp(1/\lambda)$ distribution and a separate sample from the log-normal $\mathcal{LN}(0, 2\log \lambda)$ distribution. Plot the resulting risk functions.

(b) For each value of λ, simulate a sample from the $\mathcal{E}xp(1/\lambda)$ distribution and then transform it into a sample from the $\mathcal{LN}(0, 2\log \lambda)$ distribution. Plot the resulting risk functions.

(c) Simulate a sample from the $\mathcal{E}xp(1)$ distribution. For each value of λ, transform it into a sample from $\mathcal{E}xp(1/\lambda)$, and then transform it into a sample from the $\mathcal{LN}(0, 2\log \lambda)$ distribution. Plot the resulting risk functions.

(d) Compare and comment on the accuracy of the plots.

3.10 Compare (in a simulation experiment) the performances of the regular Monte Carlo estimator of

$$\int_1^2 \frac{e^{-x^2/2}}{\sqrt{2\pi}} \, dx = \Phi(2) - \Phi(1)$$

with those of an estimator based on an optimal choice of instrumental distribution (see (3.11)).

3.11 In the setup of Example 3.10, give the two first moments of the log-normal distribution $\mathcal{LN}(\mu, \sigma^2)$.

3.12 In the setup of Example 3.13, examine whether or not the different estimators of the expectations $\mathbb{E}_f[h_i(X)]$ have finite variances.

3.13 Establish the equality (3.18) using the representation $b = \beta a/\alpha$.

3.14 (Ó Ruanaidh and Fitzgerald 1996) For simulating random variables from the density $f(x) = \exp\{-\sqrt{x}\}[\sin(x)]^2$, $0 < x < \infty$, compare the following choices of instrumental densities:

$$g_1(x) = \tfrac{1}{2} e^{-|x|}, \qquad g_2(x) = \tfrac{1}{2\sqrt{2}} \operatorname{sech}^2(x/\sqrt{2}),$$
$$g_3(x) = \tfrac{1}{2\pi} \tfrac{1}{1+x^2/4}, \qquad g_4(x) = \tfrac{1}{\sqrt{2\pi}} e^{-x/2}.$$

(a) For $M = 100, 1000$, and $10,000$, compare the standard deviations of the estimates based on simulating M random variables.

(b) For each of the instrumental densities, estimate the size of M needed to obtain three digits of accuracy in estimating $\mathbb{E}_f X$.

3.15 Use the techniques of Example 3.11 to redo Problem 3.3. Compare the number of variables needed to obtain three digits of accuracy with importance sampling to the answers obtained from Problem 3.3.

3.16 Referring to Example 3.11:

(a) Show that to simulate $Y \sim \mathcal{TE}(a,1)$, an exponential distribution left truncated at a, we can simulate $X \sim \mathcal{E}(1)$ and take $Y = a + X$.

(b) Use this method to calculate the probability that a χ_3^2 random variable is greater that 25, and that a t_5 random variable is greater than 50.

(c) Explore the gain in efficiency from this method. Take $a = 4.5$ in part (a) and run an experiment to determine how many random variables would be needed to calculate $P(Z > 4.5)$ to the same accuracy obtained from using 100 random variables in an importance sampler.

3.17 In this chapter, the importance sampling method is developed for an iid sample $(Y_1, \ldots, Y_n)$ from g.

(a) Show that the importance sampling estimator is still unbiased if the Y_i's are correlated while being marginally distributed from g.

(b) Show that the importance sampling estimator can be extended to the case when Y_i is generated from a conditional distribution $q(y_i|Y_{i-1})$.

(c) Implement a scheme based on an iid sample $(Y_1, Y_3, \ldots, Y_{2n-1})$ and a secondary sample $(Y_2, Y_4, \ldots, Y_{2n})$ such that $Y_{2i} \sim q(y_{2i}|Y_{2i-1})$. Show that the covariance

$$\mathrm{cov}\left(h(Y_{2i-1})\frac{f(Y_{2i-1})}{g(Y_{2i-1})}, h(Y_{2i})\frac{f(Y_{2i})}{q(Y_{2i}|Y_{2i-1})}\right)$$

is null. Generalize.

3.18 For a sample $(Y_1, \ldots, Y_h)$ from g, the weights ω_i are defined as

$$\omega_i = \frac{f(Y_i)/g(Y_i)}{\sum_{j=1}^h f(Y_j)/g(Y_j)}.$$

Show that the following algorithm (Rubin 1987) produces a sample from f such that the empirical average

$$\frac{1}{M}\sum_{m=1}^M h(X_m)$$

is asymptotically equivalent to the importance sampling estimator based on $(Y_1, \ldots, Y_N)$:

For $m = 1, \ldots, M$,

take $X_m = Y_i$ with probability ω_i

(*Note:* This is the SIR algorithm.)

3.19 (Smith and Gelfand 1992) Show that, when evaluating an integral based on a posterior distribution

$$\pi(\theta|x) \propto \pi(\theta)\ell(\theta|x),$$

where π is the prior distribution and ℓ the likelihood function, the prior distribution can always be used as instrumental distribution (see Problem 2.29).

(a) Show that the variance is finite when the likelihood is bounded.

(b) Compare with choosing $\ell(\theta|x)$ as instrumental distribution when the likelihood is proportional to a density. (*Hint:* Consider the case of exponential families.)

(c) Discuss the drawbacks of this (these) choice(s) in specific settings.

(d) Show that a mixture between both instrumental distributions can ease some of the drawbacks.

3.20 In the setting of Example 3.13, show that the variance of the importance sampling estimator associated with an importance function g and the integrand $h(x) = \sqrt{x/(1-x)}$ is infinite for all g's such that $g(1) < \infty$.

3.21 Monte Carlo marginalization is a technique for calculating a marginal density when simulating from a joint density. Let $(X_i, Y_i) \sim f_{XY}(x, y)$, independent, and the corresponding marginal distribution $f_X(x) = \int f_{XY}(x, y) dy$.
   (a) Let $w(x)$ be an arbitrary density. Show that

$$\lim_n \frac{1}{n} \sum_{i=1}^{n} \frac{f_{XY}(x^*, y_i) w(x_i)}{f_{XY}(x_i, y_i)} = \int \int \frac{f_{XY}(x^*, y) w(x)}{f_{XY}(x, y)} f_{XY}(x, y) dx dy = f_X(x^*)$$

and so we have a Monte Carlo estimate of f_X, the marginal distribution of X, from only knowing the form of the joint distribution.
   (b) Let $X|Y = y \sim \mathcal{G}a(y, 1)$ and $Y \sim \mathcal{E}xp(1)$. Use the technique of part (a) to plot the marginal density of X. Compare it to the exact marginal.
   (c) Choosing $w(x) = f_{X|Y}(x|y)$ works to produce the marginal distribution, and it is optimal. In the spirit of Theorem 3.12, can you prove this?

3.22 Given a real importance sample $X_1, \ldots, X_n$ with importance function g and target density f,
   (a) show that the sum of the weights $\omega_i = f(X_i)/g(X_i)$ is only equal to 1 in expectation and deduce that the weights need to be renormalized even when both densities have know normalizing constants.
   (b) Assuming that the weights ω_i have been renormalized to sum to one, we sample, with replacement, n points $\tilde{X}_j$ from the X_i's using those weights. Show that the $\tilde{X}_j$'s satisfy

$$\mathbb{E}\left[\frac{1}{n} \sum_{j=1}^{n} h(\tilde{X}_j)\right] = \mathbb{E}\left[\sum_{i=1}^{n} \omega_i h(X_i)\right].$$

   (c) Deduce that, if the above formula is satisfied for $\omega_i = f(X_i)/g(X_i)$ instead, the empirical distribution associated with the $\tilde{X}_j$'s is unbiased.

3.23 (Evans and Swartz 1995) Devise and implement a simulation experiment to approximate the probability $P(Z \in (0, \infty)^6)$ when $Z \sim \mathcal{N}_6(0, \Sigma)$ and

$$\Sigma^{-1/2} = \text{diag}(0, 1, 2, 3, 4, 5) + e \cdot e^t,$$

with $e^t = (1, 1, 1, 1, 1, 1)$:
   (a) when using the $\Sigma^{-1/2}$ transform of a $\mathcal{N}_6(0, I_6)$ random variables;
   (b) when using the Choleski decomposition of Σ;
   (c) when using a distribution restricted to $(0, \infty)^6$ and importance sampling.

3.24 Using the facts

$$\int y^3 e^{-cy^2/2} dy = \frac{-1}{2c}\left[y^2 + \frac{1}{c}\right] e^{-cy^2/2},$$

$$\int y^6 e^{-cy^2/2} dy = \frac{-1}{2c}\left[y^5 + \frac{5y^3}{2c} + \frac{15y}{4c}\right] e^{-cy^2/2} + 30\sqrt{\frac{\pi}{c^7}} \Phi(\sqrt{2c}y),$$

derive expressions similar to (3.22) for the second- and third-order approximations (see also Problem 5.6).

3.25 By evaluating the normal integral for the first order approximation from (3.21), establish (3.22).

3.26 Referring to Example 3.16, derive the Laplace approximation for the Gamma density and reproduce Table 3.6.

3.27 (Gelfand and Dey 1994) Consider a density function $f(x|\theta)$ and a prior distribution $\pi(\theta)$ such that the marginal $m(x) = \int_\Theta f(x|\theta)\pi(\theta)d\theta$ is finite a.e. The marginal density is of use in the comparison of models since it appears in the Bayes factor (see Section 1.3).

(a) Give a Laplace approximation of m and derive the corresponding approximation of the Bayes factor. (See Tierney et al. 1989 for details.)

(b) Give the general shape of an importance sampling approximation of m.

(c) Detail this approximation when the importance function is the posterior distribution and when the normalizing constant is unknown.

(d) Show that for a proper density τ,

$$m(x)^{-1} = \int_\Theta \frac{\tau(\theta)}{f(x|\theta)\pi(\theta)}\pi(\theta|x)d\theta \;,$$

and deduce that when the θ_i^*'s are generated from the posterior,

$$\hat{m}(x) = \left\{ \frac{1}{T}\sum_{t=1}^T \frac{\tau(\theta_i^*)}{f(x|\theta_i^*)\pi(\theta_i^*)} \right\}^{-1}$$

is another importance sampling estimator of m.

3.28 (Berger et al. 1998) For Σ a $p \times p$ positive-definite symmetric matrix, consider the distribution

$$\pi(\theta) \propto \frac{\exp\left(-(\theta-\mu)^t \Sigma^{-1}(\theta-\mu)/2\right)}{||\theta||^{p-1}} \;.$$

(a) Show that the distribution is well defined; that is, that

$$\int_{\mathbb{R}^p} \frac{\exp\left(-(\theta-\mu)^t \Sigma^{-1}(\theta-\mu)/2\right)}{||\theta||^{p-1}} d\theta < \infty.$$

(b) Show that an importance sampling implementation based on the normal instrumental distribution $\mathcal{N}_p(\mu, \Sigma)$ is not satisfactory from both theoretical and practical points of view.

(c) Examine the alternative based on a Gamma distribution $\mathcal{G}a(\alpha,\beta)$ on $\eta = ||\theta||^2$ and a uniform distribution on the angles.

Note: Priors such as these have been used to derive *Bayes minimax estimators* of a multivariate normal mean. See Lehmann and Casella (1998).

3.29 From the Accept–Reject Algorithm we get a sequence $Y_1, Y_2, \ldots$ of independent random variables generated from g along with a corresponding sequence $U_1, U_2, \ldots$ of uniform random variables. For a fixed sample size t (i.e. for a fixed number of accepted random variables), the number of generated Y_i's is a random integer N.

(a) Show that the joint distribution of $(N, Y_1, \ldots, Y_N, U_1, \ldots, U_N)$ is given by

$$P(N = n, Y_1 \leq y_1, \ldots, Y_n \leq y_n, U_1 \leq u_1, \ldots, U_n \leq u_n)$$
$$= \int_{-\infty}^{y_n} g(t_n)(u_n \wedge w_n)dt_n \int_{-\infty}^{y_1} \cdots \int_{-\infty}^{y_{n-1}} g(t_1)\ldots g(t_{n-1})$$
$$\times \sum_{(i_1,\cdots,i_{t-1})} \prod_{j=1}^{t-1}(w_{i_j} \wedge u_{i_j}) \prod_{j=t}^{n-1}(u_{i_j} - w_{i_j})^+ dt_1 \cdots dt_{n-1},$$

where $w_i = f(y_i)/Mg(y_i)$ and the sum is over all subsets of $\{1, \ldots, n-1\}$ of size $t-1$.

(b) There is also interest in the joint distribution of $(Y_i, U_i)|N = n$, for any $i = 1, \ldots, n-1$, as we will see in Problem 4.17. Since this distribution is the same for each i, we can just derive it for (Y_1, U_1). (Recall that $Y_n \sim f$.) Show that

$$P(N = n, Y_1 \le y, U_1 \le u_1)$$
$$= \binom{n-1}{t-1}\left(\frac{1}{M}\right)^{t-1}\left(1 - \frac{1}{M}\right)^{n-t-1}$$
$$\times \left[\frac{t-1}{n-1}(w_1 \wedge u_1)\left(1 - \frac{1}{M}\right) + \frac{n-t}{n-1}(u_1 - w_1)^+\left(\frac{1}{M}\right)\right]\int_{-\infty}^{y} g(t_1)dt_1.$$

(c) Show that part (b) yields the negative binomial marginal distribution of N,

$$P(N = n) = \binom{n-1}{t-1}\left(\frac{1}{M}\right)^t\left(1 - \frac{1}{M}\right)^{n-t},$$

the marginal distribution of Y_1, $m(y)$,

$$m(y) = \frac{t-1}{n-1}f(y) + \frac{n-t}{n-1}\frac{g(y) - \rho f(y)}{1 - \rho},$$

and

$$P(U_1 \le w(y)|Y_1 = y, N = n) = \frac{g(y)w(y)M^{\frac{t-1}{n-1}}}{m(y)}.$$

3.30 If $(Y_1, \ldots, Y_N)$ is the sample produced by an Accept–Reject method based on (f, g), where $M = \sup(f/g)$, $(X_1, \ldots, X_t)$ denotes the accepted subsample and $(Z_1, \ldots, Z_{N-t})$ the rejected subsample.

(a) Show that both

$$\delta_2 = \frac{1}{N-t}\sum_{i=1}^{N-t} h(Z_i)\frac{(M-1)f(Z_i)}{Mg(Z_i) - f(Z_i)}$$

and

$$\delta_1 = \frac{1}{t}\sum_{i=1}^{t} h(X_i)$$

are unbiased estimators of $I = \mathbb{E}_f[h(X)]$ (when $N > t$).

(b) Show that δ_1 and δ_2 are independent.

(c) Determine the optimal weight β^* in $\delta_3 = \beta\delta_1 + (1-\beta)\delta_2$ in terms of variance. (*Note*: β may depend on N but not on $(Y_1, \ldots, Y_N)$.)

3.31 Given a sample $Z_1, \ldots, Z_{n+t}$ produced by an Accept–Reject algorithm to accept n values, based on (f, g, M), show that the distribution of a rejected variable is

$$\left(1 - \frac{f(z)}{Mg(z)}\right)g(z) = \frac{g(z) - \rho f(z)}{1 - \rho},$$

where $\rho = 1/M$, that the marginal distribution of Z_i ($i < n + t$) is

$$f_m(z) = \frac{n-1}{n+t-1}f(z) + \frac{t}{n+t-1}\frac{g(z) - \rho f(z)}{1 - \rho},$$

and that the joint distribution of (Z_i, Z_j) $(1 \le i \ne j < n + t)$ is

$$\frac{(n-1)(n-2)}{(n+t-1)(n+t-2)} f(z_i) f(z_j)$$

$$+ \frac{(n-1)t}{(n+t-1)(n+t-2)} \left\{ f(z_i) \frac{g(z_j) - \rho f(z_j)}{1 - \rho} \frac{g(z_i) - \rho f(z_i)}{1 - \rho} f(z_j) \right\}$$

$$+ \frac{n(n-1)}{(n+t-1)(n+t-2)} \frac{g(z_i) - \rho f(z_i)}{1 - \rho} \frac{g(z_j) - \rho f(z_j)}{1 - \rho}.$$

3.32 (Continuation of Problem 3.31) If $Z_1, \ldots, Z_{n+t}$ is the sample produced by an Accept–Reject algorithm to generate n values, based on (f, g, M), show that the Z_i's are negatively correlated in the sense that for every square integrable function h,

$$\text{cov}(h(Y_i), h(Y_j)) = -\mathbb{E}_g[h]^2 \mathbb{E}_N \left[\frac{(t-1)(n-t)}{(n-1)^2(n-2)} \right]$$
$$= -\mathbb{E}_g[h]^2 \{ \rho^t {}_2F_1(t-1, t-1; t-1; 1-\rho) - \rho^2 \},$$

where ${}_2F_1(a, b; c; z)$ is the confluent hypergeometric function (see Abramowitz and Stegun 1964 or Problem 1.38).

3.33 Given an Accept–Reject algorithm based on (f, g, ρ), we denote by

$$b(y_j) = \frac{(1 - \rho)f(y_j)}{g(y_j) - \rho f(y_j)}$$

the importance sampling weight of the rejected variables $(Y_1, \ldots, Y_t)$, and by $(X_1, \ldots, X_n)$ the accepted variables.

(a) Show that the estimator

$$\delta_1 = \frac{n}{n+t} \delta^{AR} + \frac{t}{n+t} \delta_0,$$

with

$$\delta_0 = \frac{1}{t} \sum_{j=1}^{t} b(Y_j) h(Y_j)$$

and

$$\delta^{AR} = \frac{1}{n} \sum_{i=1}^{n} h(X_i),$$

does not uniformly dominate δ^{AR}. (*Hint:* Consider the constant functions.)

(b) Show that

$$\delta_{2w} = \frac{n}{n+t} \delta^{AR} + \frac{t}{n+t} \sum_{j=1}^{t} b(Y_j) h(Y_j) / \sum_{j=1}^{t} b(Y_j)$$

is asymptotically equivalent to δ_1 in terms of bias and variance.

(c) Deduce that δ_{2w} asymptotically dominates δ^{AR} if (4.20) holds.

3.34 For the Accept–Reject algorithm of Section 3.3.3:

(a) Show that conditionally on t, the joint density of $(Y_1, \ldots, Y_t)$ is indeed

$$\prod_{j=1}^{t-1} \left(\frac{Mg(y_j) - f(y_j)}{M-1} \right) f(y_t)$$

and the expectation of δ_2 of (3.15) is given by

$$\mathbb{E}\left[\frac{t-1}{t} \left\{ \frac{M}{M-1} \mathbb{E}_f[h(X)] - \frac{1}{M-1} \mathbb{E}_f\left[h(X) \frac{f(X)}{g(X)} \right] \right\} \right. $$
$$\left. + \frac{1}{t} \mathbb{E}_f\left[h(X) \frac{f(X)}{g(X)} \right] \right].$$

(b) If we denote the acceptance probability of the Accept–Reject algorithm by $\rho = 1/M$ and assume $\mathbb{E}_f[h(X)] = 0$, show that the *bias* of δ_2 is

$$\left(\frac{1}{1-\rho} \mathbb{E}[t^{-1}] - \frac{\rho}{1-\rho} \right) \mathbb{E}_f\left[h(X) \frac{f(X)}{g(X)} \right].$$

(c) Establish that for $t \sim \mathcal{G}eo(\rho)$, $\mathbb{E}[t^{-1}] = -\rho \log(\rho)/(1-\rho)$, and that the bias of δ_2 can be written as

$$-\frac{\rho}{1-\rho} (1 + \log(\rho)) \mathbb{E}_f\left[h(X) \frac{f(X)}{g(X)} \right].$$

(d) Assuming that $\mathbb{E}_f[h(X)] = 0$, show that the variance of δ_2 is

$$\mathbb{E}\left[\frac{t-1}{t^2} \right] \frac{1}{1-\rho} \mathbb{E}_f\left[h^2(X) \frac{f(X)}{g(X)} \right]$$
$$+ \mathbb{E}\left[\frac{1}{t^2} \left\{ 1 - \frac{\rho(t-1)}{1-\rho} \right\} \right] \operatorname{var}_f(h(X)f(X)/g(X)).$$

3.35 Using the information from Note 3.6.1, for a binomial experiment $X_n \sim \mathcal{B}(n, p)$ with $p = 10^{-6}$, determine the minimum sample size n so that

$$P\left(\left| \frac{X_n}{n} - p \right| \le \epsilon p \right) > .95$$

when $\epsilon = 10^{-1}, 10^{-2}$, and 10^{-3}.

3.36 When random variables Y_i are generated from (3.25), show that $J^{(m)}$ is distributed as $\lambda(\theta_0)^{-n} \exp(-n\theta J)$. Deduce that (3.26) is unbiased.

3.37 Starting with a density f of interest, we create the exponential family

$$\mathcal{F} = \{f(\cdot|\tau); f(x|\tau) = \exp[\tau x - K(\tau)]f(x)\},$$

where $K(\tau)$ is the cumulant generating function of f given in Section 3.6.2. It immediately follows that if $X_1, X_2, \ldots, X_n$ are iid from $f(x|\tau)$, the density of $\bar{X}$ is

(3.24) $$f_{\bar{X}}(x|\tau) = \exp\{n[\tau x - K(\tau)]\} f_{\bar{X}}(x),$$

where $f_{\bar{X}}(x)$ is the density of the average of an iid sample from f

(a) Show that $f(x|\tau)$ is a density.

(b) If $X_1, X_2, \ldots, X_n$ is an iid sample from $f(x)$, and $f_{\bar{X}}(x)$ is the density of the sample mean, show that $\int e^{n[\tau x - K(\tau)]} f_{\bar{X}}(x) dx = 1$ and hence $f_{\bar{X}}(x|\tau)$ of (3.24) is a density.

(c) Show that the mgf of $f_{\bar{X}}(x|\tau)$ is $e^{n[K(\tau + t/n)x - K(\tau)]}$

(d) In (3.29), for each x we choose τ so that μ_τ, the mean of $f_{\bar{X}}(x|\tau)$, satisfies $\mu_\tau = x$. Show that this value of τ is the solution to the equation $K'(\tau) = x$.

3.38 For the situation of Example 3.18:

(a) Verify the mgf in (3.32).

(b) Show that the solution to the saddlepoint equation is given by (3.33).

(c) Plot the saddlepoint density for $p = 7$ and $n = 1, 5, 20$. Compare your results to the exact density.

3.6 Notes

3.6.1 Large Deviations Techniques

When we introduced importance sampling methods in Section 3.3, we showed in Example 3.8 that alternatives to direct sampling were preferable when sampling from the tails of a distribution f. When the event A is particularly rare, say $P(A) \leq 10^{-6}$, methods like importance sampling are needed to get an acceptable approximation (see Problem 3.35). Since the optimal choice given in Theorem 3.12 is formal, in the sense that it involves the unknown constant I, more practical choices have been proposed in the literature. In particular, Bucklew (1990) indicates how the *theory of large deviations* may help in devising proposal distributions in this purpose.

Briefly, the theory of large deviations is concerned with the approximation of tail probabilities $P(|\bar{X}_n - \mu| > \varepsilon)$ when $\bar{X}_n = (X_1 + \cdots + X_n)/n$ is a mean of iid random variables, n goes to infinity, and ε is large. (When ε is small, the normal approximation based on the Central Limit Theorem works well enough.)

If $M(\theta) = \mathbb{E}[\exp(\theta X_1)]$ is the moment generating function of X_1 and we define $I(x) = \sup_\theta \{\theta x - \log M(\theta)\}$, the large deviation approximation is

$$\frac{1}{n} \log P(S_n \in F) \approx -\inf_F I(x).$$

This result is sometimes called *Cramér's Theorem* and a simulation device based on this result and called *twisted simulation* is as follows.

To evaluate

$$I = P\left(\frac{1}{n} \sum_{i=1}^{n} h(x_i) \geq 0\right),$$

when $\mathbb{E}[h(X_1)] < 0$, we use the proposal density

(3.25) $$t(x) \propto f(x) \exp\{\theta_0 h(x)\},$$

where the parameter θ_0 is chosen such that $\int h(x) f(x) e^{\theta_0 h(x)} dx = 0$. (Note the similarity with exponential tilting in saddlepoint approximations in Section 3.6.2.) The corresponding estimate of I is then based on blocks $(m = 1, \ldots, M)$

$$J^{(m)} = \frac{1}{n} \sum_{i=1}^{n} h(Y_i^{(m)}),$$

where the $Y_i^{(m)}$ are iid from t, as follows:

$$(3.26) \qquad \hat{I} = \frac{1}{M} \sum_{m=1}^{M} \mathbb{I}_{[0,\infty[}(J^{(m)}) e^{-n\theta_0 J^{(m)}} \lambda(\theta_0)^n,$$

with $\lambda(\theta) = \int f(x) e^{\theta_0 f(x)} dx$. The fact that (3.26) is unbounded follows from a regular importance sampling argument (Problem 3.36). Bucklew (1990) provides arguments about the fact that the variance of $\hat{I}$ goes to 0 exponentially twice as fast as the regular (direct sampling) estimate.

Example 3.17. Laplace distribution. Consider $h(x) = x$ and the sampling distribution $f(x) = \frac{1}{2a} \exp\{-|x - \mu|/a\}$, $\mu < 0$. We then have

$$t(x) \propto \exp\{-|x - \mu|/a + \theta_0\},$$
$$\theta_0 = \sqrt{\mu^{-2} + a^{-2}} - \mu^{-1},$$
$$\lambda(\theta_0) = \frac{\mu^2}{2} \exp(-C) a^2 C,$$

with $C = \sqrt{1 + \frac{\mu^2}{a^2}} - 1$. A large deviation computation then shows that (Bucklew 1990, p. 139)

$$\lim_n \frac{1}{n} \log(M \operatorname{var} \hat{I}) = 2 \log \lambda(\theta_0),$$

while the standard average $\bar{I}$ satisfies

$$\lim_n \frac{1}{n} \log(M \operatorname{var} \bar{I}) = \log \lambda(\theta_0) .$$

‖

Obviously, this is not the entire story, Further improvements can be found in the theory, while the computation of θ_0 and $\lambda(\theta_0)$ and simulation from $t(x)$ may become quite intricate in realistic setups.

3.6.2 The Saddlepoint Approximation

The saddlepoint approximation, in contrast to the Laplace approximation, is mainly a technique for approximating a function rather than an integral, although it naturally leads to an integral approximation. (For introductions to the topic see Goutis and Casella 1999, the review papers of Reid 1988, 1991, or the books by Field and Ronchetti 1990, Kolassa 1994, or Jensen 1995.)

Suppose we would like to evaluate

$$(3.27) \qquad g(\theta) = \int_A f(x|\theta) dx$$

for a range of values of θ. One interpretation of a *saddlepoint approximation* is that for each value of θ, we do a Laplace approximation centered at $\hat{x}_\theta$ (the *saddlepoint*).[5]

---

[5] The saddlepoint approximation got its name because its original derivation (Daniels 1954) used a complex analysis argument, and the point $\hat{x}_\theta$ is a saddlepoint in the complex plane.

One way to derive the saddlepoint approximation is to use an *Edgeworth expansion* (see Hall 1992 or Reid 1988 for details). As a result of a quite detailed derivation, we obtain the approximation to the density of $\bar{X}$ to be

$$f_{\bar{X}}(x) = \frac{\sqrt{n}}{\sigma}\varphi\left(\frac{x-\mu}{\sigma/\sqrt{n}}\right)$$

(3.28)
$$\times \left[1 + \frac{\kappa}{6\sqrt{n}}\left\{\left(\frac{x-\mu}{\sigma/\sqrt{n}}\right)^3 - 3\left(\frac{x-\mu}{\sigma/\sqrt{n}}\right)\right\} + \mathcal{O}(1/n)\right].$$

Ignoring the term within braces produces the usual normal approximation, which is accurate to $\mathcal{O}(1/\sqrt{n})$. If we are using (3.28) for values of x near μ, then the value of the expression in braces is close to zero, and the approximation will then be accurate to $\mathcal{O}(1/n)$. The trick of the saddlepoint approximation is to make this always be the case.

To do so, we use a family of densities such that, for each x, we can choose a density from the family to cancel the term in braces in (3.28). One method of creating such a family is through a technique known as *exponential tilting* (see Efron 1981, Stuart and Ord 1987, Section 11.13, Reid 1988, or Problem 3.37). The result of the exponential tilt is a family of Edgeworth expansions for $f_{\bar{X}}(x)$ indexed by a parameter τ, that is,

$$f_{\bar{X}}(x) = \exp\{-n[\tau x - K(\tau)]\}\frac{\sqrt{n}}{\sigma_\tau}\varphi\left(\frac{x-\mu_\tau}{\sigma_\tau/\sqrt{n}}\right)$$

(3.29)
$$\times \left[1 + \frac{\kappa_\tau}{6\sqrt{n}}\left\{\left(\frac{x-\mu_\tau}{\sigma_\tau/\sqrt{n}}\right)^3 - 3\left(\frac{x-\mu_\tau}{\sigma_\tau/\sqrt{n}}\right)\right\} + \mathcal{O}(1/n)\right].$$

As the parameter τ is free for us to choose in (3.29), for each x we choose $\tau = \tau(x)$ so that the mean satisfies $\mu_\tau = x$. This choice cancels the middle term in the square brackets in (3.29), thereby improving the order of the approximation. If $K(\tau) = \log(\mathbb{E}\exp(\tau X))$ is the cumulant generating function, we can choose τ so that $K'(\tau) = x$, which is the saddlepoint equation. Denoting this value by $\hat{\tau} = \hat{\tau}(x)$ and noting that $\sigma_\tau = K''(\tau)$, we get the saddlepoint approximation

$$f_{\bar{X}}(x) = \frac{\sqrt{n}}{\sigma_{\hat{\tau}}}\varphi(0)\exp\{n[K(\hat{\tau}) - \hat{\tau}x]\}[1 + \mathcal{O}(1/n)]$$

(3.30)
$$\approx \left(\frac{n}{2\pi\,K''(\hat{\tau}(x))}\right)^{1/2}\exp\{n\,[K\,(\hat{\tau}(x)) - \hat{\tau}(x)x]\}.$$

Example 3.18. Saddlepoint tail area approximation. The noncentral chi squared density has the rather complex form

(3.31)
$$f(x|\lambda) = \sum_{k=0}^{\infty}\frac{x^{p/2+k-1}e^{-x/2}}{\Gamma(p/2+k)2^{p/2+k}}\frac{\lambda^k e^{-\lambda}}{k!},$$

where p is the number of degrees of freedom and λ is the noncentrality parameter. It turns out that calculation of the moment generating function is simple, and it can be expressed in closed form as

(3.32)
$$\phi_X(t) = \frac{e^{2\lambda t/(1-2t)}}{(1-2t)^{p/2}}.$$

Solving the saddlepoint equation $\partial \log \phi_X(t)/\partial t = x$ yields the saddlepoint

$$(3.33) \qquad \hat{t}(x) = \frac{-p + 2x - \sqrt{p^2 + 8\lambda x}}{4x}$$

and applying (3.30) yields the approximate density (see Hougaard 1988 or Problem 3.38).   ‖

The saddlepoint can also be used to approximate the tail area of a distribution. From (3.30), we have the approximation

$$P(\bar{X} > a) = \int_a^\infty \left(\frac{n}{2\pi K_X''(\hat{\tau}(x))} \right)^{1/2} \exp\left\{ n\left[K_X\left(\hat{\tau}(x)\right) - \hat{\tau}(x)x \right] \right\} dx$$

$$= \int_{\hat{\tau}(a)}^{1/2} \left(\frac{n}{2\pi} \right)^{1/2} \left[K_X''(t) \right]^{1/2} \exp\left\{ n\left[K_X(t) - tK_X'(t) \right] \right\} dt,$$

where we make the transformation $K_X'(t) = x$ and $\hat{\tau}(a)$ satisfies $K_X'(\hat{\tau}(a)) = a$. This transformation was noted by Daniels (1983, 1987) and allows the evaluation of the integral with only one saddlepoint evaluation.

Interval	Approximation	Renormalized approximation	Exact
$(36.225, \infty)$	0.1012	0.0996	0.10
$(40.542, \infty)$	0.0505	0.0497	0.05
$(49.333, \infty)$	0.0101	0.0099	0.01

Table 3.7. Saddlepoint approximation of a noncentral chi squared tail probability for $p = 6$ and $\lambda = 9$.

To examine the accuracy of the saddlepoint tail area, we return to the noncentral chi squared distribution of Example 3.18. Table 3.7 compares the tail areas calculated by integrating the exact density and using the regular and renormalized saddlepoints. As can be seen, the accuracy is quite impressive.

The discussion above shows only that the order of the approximation is $\mathcal{O}(1/n)$, not the $\mathcal{O}(n^{-3/2})$ that is often claimed. This better error rate is obtained by renormalizing (3.30) so that it integrates to 1.

Saddlepoint approximations for tail areas have seen much more development than given here. For example, the work of Lugannani and Rice (1980) produced a very accurate approximation that only requires the evaluation of one saddlepoint and *no* integration. There are other approaches to tail area approximations; for example, the work of Barndorff-Nielsen (1991) using ancillary statistics or the Bayes-based approximation of DiCiccio and Martin (1993). Wood et al. (1993) give generalizations of the Lugannani and Rice formula.

4

Controling Monte Carlo Variance

> The others regarded him uncertainly, none of them sure how he had arrived
> at such a conclusion or on how to refute it.
> —Susanna Gregory, *A Deadly Brew*

In Chapter 3, the Monte Carlo method was introduced (and discussed) as a
simulation-based approach to the approximation of complex integrals. There
has been a considerable body of work in this area and, while not all of it is
completely relevant for this book, in this chapter we discuss the specifics of
variance estimation and control. These are fundamental concepts, and we will
see connections with similar developments in the realm of MCMC algorithms
that are discussed in Chapters 7–12.

4.1 Monitoring Variation with the CLT

In Chapter 3, we mentioned the use of the Central Limit Theorem for assessing
the convergence of a Monte Carlo estimate,

$$\overline{h}_m = \frac{1}{m} \sum_{j=1}^{m} h(X_j) \qquad X_j \sim f(x) \, ,$$

to the integral of interest

(4.1) $$\mathfrak{I} = \int h(x) f(x) \, dx \, .$$

Figure 3.1 *(right)* was, for example, an illustration of the use of a normal confidence interval for this assessment. It also shows the limitation of a straightforward application of the CLT to a sequence $(\overline{h}_m)$ of estimates that are *not*
independent. Thus, while a given slice (that is, for a given m) in Figure 3.1
(right) indeed provides an asymptotically valid confidence interval, the envelope built over iterations and represented in this figure has no overall validity,

that is, another Monte Carlo sequence $(\overline{h}_m)$ will not stay in this envelope with probability 0.95. To gather a valid assessment of convergence of Monte Carlo estimators, we need to either derive the joint distribution of the sequence $(\overline{h}_m)$ or recover independence by running several sequences in parallel. The former is somewhat involved, but we will look at it in Section 4.1.2. The latter is easier to derive and more widely applicable, but greedy in computing time. However, this last "property" is a feature we will meet repeatedly in the book, namely that validation of the assessment of variation is of an higher order than convergence of the estimator itself. Namely, this requires much more computing time than validation of the pointwise convergence (except in very special cases like regeneration).

4.1.1 Univariate Monitoring

In this section we look at monitoring methods that are univariate in nature. That is, the bounds placed on the estimate at iteration k depend on the values at time k and essentially ignore any correlation structure in the iterates. We begin with an example.

Example 4.1. Monitoring with the CLT. When considering the evaluation of the integral of $h(x) = [\cos(50x) + \sin(20x)]^2$ over $[0, 1]$, Figure 3.1 *(right)* provides one convergence path with a standard error evaluation. As can be seen there, the resulting confidence band is moving over iterations in a rather noncoherent fashion, that is, the band exhibits the same "wiggles" as the point estimate.[1]

If, instead, we produce parallel sequences of estimates, we get the output summarized in Figure 4.1. The main point of this illustration is that the range *and* the empirical 90% band (derived from the set of estimates at each iteration by taking the empirical 5% and 95% quantiles) are much wider than the 95% confidence interval predicted by the CLT, where the variance was computed by averaging the empirical variances over the parallel sequences.   ‖

This simple example thus warns even further against the blind use of a normal approximation when repeatedly invoked over iterations with dependent estimators, simply because the normal confidence approximation only has a marginal and static validation. Using a band of estimators in parallel is obviously more costly but it provides the correct assessment on the variation of these estimators.

Example 4.2. Cauchy prior. For the problem of estimating a normal mean, it is sometimes the case that a *robust* prior is desired (see, for example, Berger 1985, Section 4.7). A degree of robustness can be achieved with a Cauchy prior, so we have the model

---

[1] This behavior is fairly natural when considering that, for each iteration, the confidence band is centered at the point estimate.

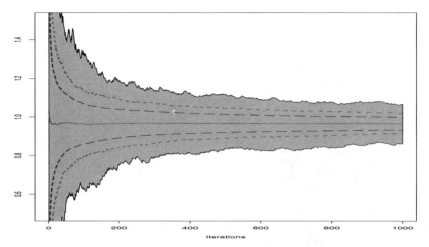

Fig. 4.1. Convergence of 1000 parallel sequences of Monte Carlo estimators of the integral of $h(x) = [\cos(50x) + \sin(20x)]^2$: The straight line is the running average of 1000 estimates, the dotted line is the empirical 90% band and the dashed line is the normal approximation 95% confidence interval. The grey shading represents the range of the entire set of estimates.

$$(4.2) \qquad\qquad X \sim \mathcal{N}(\theta, 1), \quad \theta \sim \mathcal{C}(0, 1).$$

Under squared error loss, the posterior mean is

$$(4.3) \qquad \delta^\pi(x) = \int_{-\infty}^{\infty} \frac{\theta}{1 + \theta^2} e^{-(x-\theta)^2/2} d\theta \bigg/ \int_{-\infty}^{\infty} \frac{1}{1 + \theta^2} e^{-(x-\theta)^2/2} d\theta\,.$$

From the form of $\delta^\pi(x)$ we see that we can simulate iid variables $\theta_1, \cdots, \theta_m \sim \mathcal{N}(x, 1)$ and calculate

$$(4.4) \qquad \hat{\delta}_m^\pi(x) = \sum_{i=1}^{m} \frac{\theta_i}{1 + \theta_i^2} \bigg/ \sum_{i=1}^{m} \frac{1}{1 + \theta_i^2}\,.$$

The Law of Large Numbers implies that $\hat{\delta}_m^\pi(x)$ goes to $\delta^\pi(x)$ as m goes to ∞, since both the numerator and the denominator are convergent (in m).   ‖

Note the "little problem" associated with this example: computing the variance of the estimator $\hat{\delta}_m^\pi(x)$ directly "on the run" gets fairly complicated because the estimator is a ratio of estimators and the variance of a ratio is not the ratio of the variances! This is actually a problem of some importance in that ratios of estimators are very common, from importance sampling (where the weights are most often unnormalized) to the computation of Bayes factors in model choice (see Problems 4.1 and 4.2).

Consider, thus, a ratio estimator, represented under the importance sampling form

$$\delta_h^n = \sum_{i=1}^{n} \omega_i\, h(x_i) \Big/ \sum_{i=1}^{n} \omega_i$$

without loss of generality. We assume that the x_i's are realizations of random variables $X_i \sim g(y)$, where g is a candidate distribution for target f. The ω_i's are realizations of random variables W_i such that $\mathbb{E}[W_i|X_i = x] = \kappa f(x)/g(x)$, κ being an arbitrary constant (that corresponds to the lack of normalizing constants in f and g). We denote

$$S_h^n = \sum_{i=1}^{n} W_i h(X_i), \quad S_1^n = \sum_{i=1}^{n} W_i.$$

(Note that we do not assume independence between the X_i's as in regular importance sampling.) Then, as shown in Liu (1996a) and Gåsemyr (2002), the asymptotic variance of δ_h^n can be derived in general:

Lemma 4.3. *The asymptotic variance of δ_h^n is*

$$\mathrm{var}(\delta_h^n) = \frac{1}{n^2 \kappa^2}\left(\mathrm{var}(S_h^n) - 2\mathbb{E}^\pi[h]\,\mathrm{cov}(S_h^n, S_1^n) + \mathbb{E}^\pi[h]^2\,\mathrm{var}(S_1^n)\right).$$

Proof. As shown in Casella and Berger (2001), the variance of a ratio X/Y of random variables can be approximated by the *delta method* (also called *Cramér-Wold's theorem*) as

$$(4.5) \qquad \mathrm{var}\left(\frac{X}{Y}\right) \approx \frac{\mathrm{var}(X)}{\mathbb{E}[Y^2]} - 2\frac{\mathbb{E}[X]}{\mathbb{E}[Y^3]}\,\mathrm{cov}(X, Y) + \frac{\mathbb{E}[X^2]}{\mathbb{E}[Y^4]}\,\mathrm{var}(Y).$$

The result then follows from straightforward computation (Problem 4.7).   □

As in Liu (1996a), we can then deduce that, for the regular importance sampling estimator and the right degree of approximation, we have

$$\mathrm{var}\delta_h^n \approx \frac{1}{n}\,\mathrm{var}^\pi(h(X))\left\{1 + \mathrm{var}_g(W)\right\},$$

which evaluates the additional degree of variability due to the denominator in the importance sampling ratio. (Of course, this is a rather crude approximation, as can be seen through the fact that this variance is always higher than $\mathrm{var}_f(h(X))$, which is the variance for an iid sample with the same size, and there exist choices of g and h where this does not hold!)

Example 4.4 (Continuation of Example 4.2). If we generate normal $\mathcal{N}(x, 1)$ samples for the importance sampling approximation[2] of δ^π in (4.3),

---

[2] The estimator (4.4) is, formally, an importance sampler because the target distribution is the Cauchy. However, the calculation is just a straightforward Monte Carlo sum, generating the variables from $\mathcal{N}(x, 1)$. This illustrates that we can consider any Monte Carlo sum an importance sampling calculation.

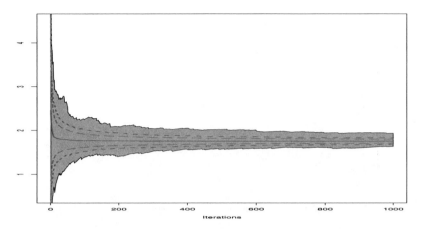

Fig. 4.2. Convergence of 1000 parallel sequences of Monte Carlo estimators for the posterior mean in the Cauchy-Normal problem when $x = 2.5$: The straight line is the running average of 1000 estimates, the dotted line is the empirical 90% band and the dashed line is the normal approximation 95% confidence interval, using the variance approximation of Lemma 4.3. The grey shading represents the range of the entire set of estimates at each iteration.

the variance approximation of Lemma 4.3 can be used to assess the variability of these estimates, but, again, the asymptotic nature of the approximation must be taken into account. Figure 4.2 compares the asymptotic variance (computed over $1,000$ parallel sequences of estimators of $\delta^\pi(x)$ for $x = 2.5$) with the actual variation of the estimates, evaluated over the $1,000$ parallel sequences. Even though the scales are not very different, there is once more a larger variability than predicted.

Figure 4.3 reproduces this evaluation in the case where the θ_i's are simulated from the prior $\mathcal{C}(0,1)$ distribution and associated with the importance sampling estimate[3]

$$\tilde{\delta}_h^n = \sum_{i=1}^n \theta_i \exp\left\{-(x - \theta_i)^2/2\right\} \Big/ \sum_{i=1}^n \exp\left\{-(x - \theta_i)^2/2\right\}.$$

In this case, since the corresponding functions h are bounded for both choices, the variabilities of the estimates are quite similar, with a slight advantage to the normal sampling. ‖

---

[3] The inversion of the roles of the $\mathcal{N}(x, 1)$ and $\mathcal{C}(0, 1)$ distributions illustrates once more both the ambiguity of the integral representation and the opportunities open by importance sampling.

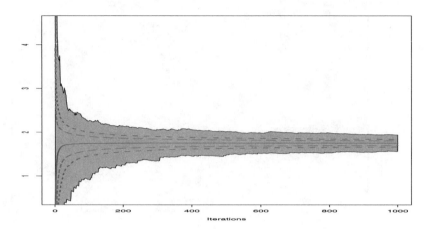

Fig. 4.3. Same plot as Figure 4.2 when the θ_i's are simulated from the prior $\mathcal{C}(0,1)$ distribution.

4.1.2 Multivariate Monitoring

As mentioned in the introduction to this chapter, one valid method for attaching variances to a mean plot, and of having a valid Central Limit Theorem, is to derive the bounds using a multivariate approach. Although the entire calculation is not difficult, and is even enlightening, it does get a bit notation-intensive, like many multivariate calculations.

Suppose that $X_1, X_2, \ldots$ is a sequence of independent (iid) random variables that are simulated with the goal of estimating $\mu = \mathbb{E}_f(X_1)$. (Without loss of generality we work with the X_i, when we are typically interested in $h(X_i)$ for some function h.) Define $\bar{X}_m = (1/m) \sum_{i=1}^{m} X_i$, for $m = 1, 2, \ldots, n$, where n is the number of random variables that we will simulate (typically a large number). A *running mean plot*, something that we have already seen, is a plot of $\bar{X}_m$ against m, and our goal is to put valid error bars on this plot.

For simplicity's sake, we assume that $X_i \sim \mathcal{N}(0, \sigma^2)$, independent, and want the distribution of the vector $\bar{\mathbf{X}} = (\bar{X}_1, \bar{X}_2, \ldots, \bar{X}_n)$. Since each element of this random variable has mean μ, a simultaneous confidence interval, based on the multivariate normal distribution, will be a valid assessment of variance.

Let $\mathbf{1}$ denote the $n \times 1$ vector of ones. Then $\mathbb{E}[\bar{\mathbf{X}}] = \mathbf{1}\mu$. Moreover, it is straightforward to calculate

$$\mathrm{cov}(\bar{\mathbf{X}}_k, \bar{\mathbf{X}}_{k'}) = \frac{\sigma^2}{\max\{k, k'\}}.$$

It then follows that $\bar{\mathbf{X}} \sim \mathcal{N}_n(\mathbf{1}\mu, \Sigma)$, where

$$\Sigma = \sigma^2 \begin{pmatrix} 1 & \frac{1}{2} & \frac{1}{3} & \frac{1}{4} & \frac{1}{5} & \cdots & \frac{1}{n} \\ \frac{1}{2} & \frac{1}{2} & \frac{1}{3} & \frac{1}{4} & \frac{1}{5} & \cdots & \frac{1}{n} \\ \frac{1}{3} & \frac{1}{3} & \frac{1}{3} & \frac{1}{4} & \frac{1}{5} & \cdots & \frac{1}{n} \\ \vdots & \vdots & \vdots & \vdots & \vdots & \cdots & \vdots \\ \frac{1}{n} & \frac{1}{n} & \frac{1}{n} & \frac{1}{n} & \frac{1}{n} & \cdots & \frac{1}{n} \end{pmatrix}$$

and

(4.6) $\qquad (\bar{\mathbf{X}} - \mathbf{1}\mu)'\Sigma^{-1}(\bar{\mathbf{X}} - \mathbf{1}\mu) \sim \begin{cases} \chi_n^2 & \text{if } \sigma^2 \text{ is known} \\ nF_{n,\nu} & \text{if } \sigma^2 \text{ is unknown,} \end{cases}$

when we have an estimate $\hat{\sigma}^2 \sim \chi_\nu^2$, independent of $\bar{\mathbf{X}}$.

Our goal now is to make a running mean plot of $(\bar{X}_1, \bar{X}_2, \ldots, \bar{X}_n)$ and, at each value of $\bar{X}_k$, attach error bars according to (4.6). At first, this seems like a calculational nightmare: Since n will typically be 5–20 *thousand* or more, we are in the position of inverting a huge matrix Σ a huge number of times. This could easily get prohibitive.

However, it turns out that the inverse of Σ is not only computable in closed form, the elements can be computed with a recursion relation *and* the matrix Σ^{-1} is tridiagonal (Problem 4.9) and is given by

(4.7) $\qquad \Sigma^{-1} = \frac{1}{\sigma^2} \begin{pmatrix} 2 & -2 & 0 & 0 & 0 & \cdots & 0 \\ -2 & 8 & -6 & 0 & 0 & \cdots & 0 \\ 0 & -6 & 18 & -12 & 0 & \cdots & 0 \\ 0 & 0 & -12 & 32 & -20 & \cdots & 0 \\ \vdots & \vdots & \vdots & \vdots & \vdots & \cdots & -n(n-1) \\ 0 & 0 & 0 & 0 & \cdots & -n(n-1) & n^2 \end{pmatrix}.$

Finally, if we choose d_n to be the appropriate χ_n^2 or $F_{n,\nu}$ cutoff point, the confidence limits on μ are given by

$$\left\{ \mu : (\bar{\mathbf{x}} - \mathbf{1}\mu)'\Sigma^{-1}(\bar{\mathbf{x}} - \mathbf{1}\mu) \leq d \right\},$$

and a bit of algebra will show that this is equivalent to
(4.8)

$$\left\{ \mu : n\mu^2 - 2n\bar{x}_n\mu + \bar{\mathbf{x}}'\Sigma^{-1}\bar{\mathbf{x}} \leq d \right\} = \left\{ \mu : \mu \in \bar{x}_n \pm \sqrt{\bar{x}_n^2 - \frac{\mathbf{x}'\Sigma^{-1}\bar{\mathbf{x}} - d_n}{n}} \right\}.$$

To implement this procedure, we can plot the estimate of μ,

(4.9) $\qquad \bar{x}_k \pm \sqrt{\bar{x}_k^2 - \frac{\mathbf{x}'\Sigma^{-1}\bar{\mathbf{x}} - d_k}{k}} \quad \text{for } k = 1, 2, \ldots, n.$

Figure 4.4 shows this plot, along with the univariate normal bands of Section 4.1.1. The difference is striking, and shows that the univariate bands are extremely overoptimistic. The more conservative multivariate bands give a much more accurate picture of the confidence in the convergence.

Note that if when we implement (4.8) in a plot such as Figure 4.4, the width of the band is not dependent on $k = 1, 2, \ldots, n$, so will produce horizontal line with width equal to the final value. Instead, we suggest using (4.9), which gives the appropriate band for each value of k, showing the evolution of the band and equaling (4.8) at $k = n$.

This procedure, although dependent on the normality assumption, can be used as an approximation even when normality does not hold, with the gain from the more conservative procedure outweighing any loss due to the violation of normality.

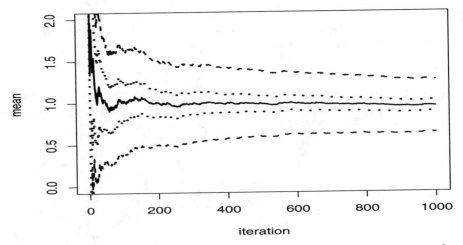

Fig. 4.4. Monte Carlo estimator of the integral of $h(x) = [\cos(50x) + \sin(20x)]^2$ (solid line). The narrow bands (grey shading) are the univariate normal approximate 95% confidence interval, and the wider bands (lighter grey) are the multivariate bands of (4.9).

4.2 Rao–Blackwellization

An approach to reduce the variance of an estimator is to use the conditioning inequality

$$\text{var}(\mathbb{E}[\delta(X)|Y]) \leq \text{var}(\delta(X)) \,,$$

sometimes called *Rao–Blackwellization* (Gelfand and Smith 1990; Liu et al. 1994; Casella and Robert 1996) because the inequality is associated with the Rao–Blackwell Theorem (Lehmann and Casella 1998), although the conditioning is not always in terms of sufficient statistics.

In a simulation context, if $\delta(X)$ is an estimator of $\mathfrak{J} = \mathbb{E}_f[h(X)]$ and if X can be simulated from the joint distribution $f(x, y)$ satisfying

$$\int f(x, y)dy = f(x),$$

the estimator $\delta^*(Y) = \mathbb{E}_f[\delta(X)|Y]$ dominates δ in terms of variance (and in squared error loss, since the bias is the same). Obviously, this result only applies in settings where $\delta^*(Y)$ can be explicitly computed.

Example 4.5. Student's t expectation. Consider the expectation of $h(x) = \exp(-x^2)$ when $X \sim T(\nu, \mu, \sigma^2)$. The Student's t distribution can be simulated as a mixture of a normal distribution and of a gamma distribution by Dickey's decomposition (1968),

$$X|y \sim \mathcal{N}(\mu, \sigma^2 y) \text{ and } Y^{-1} \sim \mathcal{G}a(\nu/2, \nu/2) \Rightarrow X \sim T(\nu, \mu, \sigma^2).$$

Therefore, the empirical average

$$\delta_m = \frac{1}{m} \sum_{j=1}^{m} \exp(-X_j^2)$$

can be improved upon when the X_j are parts of the sample $((X_1, Y_1), \ldots, (X_m, Y_m))$, since

$$(4.10) \qquad \delta_m^* = \frac{1}{m} \sum_{j=1}^{m} \mathbb{E}[\exp(-X^2)|Y_j] = \frac{1}{m} \sum_{j=1}^{m} \frac{1}{\sqrt{2\sigma^2 Y_j + 1}}$$

is the conditional expectation when $\mu = 0$ (see Problem 4.4). Figure 4.5 provides an illustration of the difference of the convergences of δ_m and δ_m^* to $\mathbb{E}_g[\exp(-X^2)]$ for $(\nu, \mu, \sigma) = (4.6, 0, 1)$. For δ_m to have the same precision as δ_m^* requires 10 times as many simulations.   ‖

Unfortunately, this conditioning method seems to enjoy a limited applicability since it involves a particular type of simulation (joint variables) and requires functions that are sufficiently regular for the conditional expectations to be explicit.

There exists, however, a specific situation where Rao–Blackwellization is always possible.[4] This is in the general setup of Accept–Reject methods, which are not always amenable to the other acceleration techniques mentioned later in Section 4.4.1 and Section 4.4.2. (Casella and Robert 1996 distinguish between *parametric* Rao–Blackwellization and *nonparametric* Rao–Blackwellization, the parametric version being more restrictive and only used in specific setups such as Gibbs sampling. See Section 7.6.2.)

---

[4] This part contains rather specialized material, which will not be used again in the book. It can be omitted at first reading.

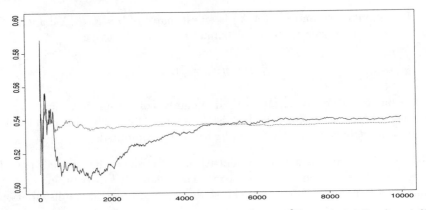

Fig. 4.5. Convergence of the estimators of $\mathbb{E}[\exp(-X^2)]$, δ_m (solid lines) and δ_m^* (dots) for $(\nu, \mu, \sigma) = (4.6, 0, 1)$. The final values are 0.5405 and 0.5369, respectively, for a true value equal to 0.5373.

Consider an Accept–Reject method based on the instrumental distribution g. If the original sample produced by the algorithm is $(X_1, \ldots, X_m)$, it can be associated with two iid samples, $(U_1, \ldots, U_N)$ and $(Y_1, \ldots, Y_N)$, with corresponding distributions $\mathcal{U}_{[0,1]}$ and g; N is then the stopping time associated with the acceptance of m variables Y_j. An estimator of $\mathbb{E}_f[h]$ based on $(X_1, \ldots, X_m)$ can therefore be written

$$\delta_1 = \frac{1}{m} \sum_{i=1}^{m} h(X_i) = \frac{1}{m} \sum_{j=1}^{N} h(Y_j) \, \mathbb{I}_{U_j \leq w_j} \,,$$

with $w_j = f(Y_j)/Mg(Y_j)$. A reduction of the variance of δ_1 can be obtained by integrating out the U_i's, which leads to the estimator

$$\delta_2 = \frac{1}{m} \sum_{j=1}^{N} \mathbb{E}[\mathbb{I}_{U_j \leq w_j} | N, Y_1, \ldots, Y_N] \, h(Y_j) = \frac{1}{m} \sum_{i=1}^{N} \rho_i h(Y_i),$$

where, for $i = 1, \ldots, n - 1$, ρ_i satisfies

$$\rho_i = P(U_i \leq w_i | N = n, Y_1, \ldots, Y_n)$$

(4.11)
$$= w_i \frac{\sum_{(i_1, \ldots, i_{m-2})} \prod_{j=1}^{m-2} w_{i_j} \prod_{j=m-1}^{n-2} (1 - w_{i_j})}{\sum_{(i_1, \ldots, i_{m-1})} \prod_{j=1}^{m-1} w_{i_j} \prod_{j=m}^{n-1} (1 - w_{i_j})},$$

while $\rho_n = 1$. The numerator sum is over all subsets of $\{1, \ldots, i - 1, i + 1, \ldots, n - 1\}$ of size $m - 2$, and the denominator sum is over all subsets of size $m - 1$. The resulting estimator δ_2 is an average over all the possible permutations of the realized sample, the permutations being weighted by their probabilities. The Rao–Blackwellized estimator is then a function only of $(N, Y_{(1)}, \ldots, Y_{(N-1)}, Y_N)$, where $Y_{(1)}, \ldots, Y_{(N-1)}$ are the order statistics.

Although the computation of the ρ_i's may appear formidable, a recurrence relation of order n^2 can be used to calculate the estimator. Define, for $k \leq m < n$,

$$S_k(m) = \sum_{(i_1, \ldots, i_k)} \prod_{j=1}^{k} w_{i_j} \prod_{j=k+1}^{m} (1 - w_{i_j}),$$

with $\{i_1, \ldots, i_m\} = \{1, \ldots, m\}$, $S_k(m) = 0$ for $k > m$, and $S_k^i(i) = S_k(i-1)$. Then we can recursively calculate

(4.12)
$$S_k(m) = w_m S_{k-1}(m-1) + (1 - w_m)S_k(m-1),$$
$$S_k^i(m) = w_m S_{k-1}^i(m-1) + (1 - w_m)S_k^i(m-1)$$

and note that weight ρ_i of (4.11) is given by

$$\rho_i = w_i\, S_{t-2}^i(n-1)/S_{t-1}(n-1) \qquad (i < n).$$

Note that, if the random nature of N and its dependence on the sample are ignored when taking the conditional expectation, this leads to the importance sampling estimator,

$$\delta_3 = \sum_{j=1}^{N} w_j\, h(Y_j) \Big/ \sum_{j=1}^{N} w_j,$$

which does not necessarily improve upon δ_1 (see Section 3.3.3).

Casella and Robert (1996) establish the following proposition, showing that δ_2 can be computed and dominates δ_1. (The proof is left to Problem 4.6.)

Proposition 4.6. *The estimator $\delta_2 = \frac{1}{m} \sum_{j=1}^{N} \rho_j\, h(y_j)$ dominates the estimator δ_1 under quadratic loss.*

The computation of the weights ρ_i is obviously more costly than the derivation of the weights of the importance sampling estimator δ_3 or of the corrected estimator of Section 3.3.3. However, the recursive formula (4.12) leads to an overall simplification of the computation of the coefficients ρ_i. Nonetheless, the increase in computing time can go as high as seven times (Casella and Robert 1996), but the corresponding variance decrease is even greater (80%).

Example 4.7. (Continuation of Example 3.15) If we repeat the simulation of $\mathcal{G}a(\alpha, \beta)$ from the $\mathcal{G}a(a, b)$ distribution, with $a \in \mathbb{N}$, it is of interest to compare the estimator obtained by Rao–Blackwellization, δ_2, with the standard estimator based on the Accept–Reject sample and with the biased importance sampling estimator δ_3. Figure 4.6 illustrates the substantial improvement brought by conditioning, since δ_2 (uniformly) dominates both

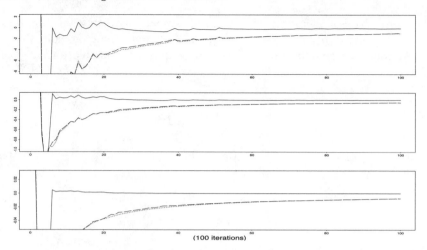

Fig. 4.6. Comparisons of the errors $\delta - \mathbb{E}[h_i(X)]$ of the Accept–Reject estimator δ_1 (long dashes), of the importance sampling estimator δ_3 (dots), and of the conditional version of δ_1, δ_2 (solid lines), for $h_1(x) = x^3$ (top), $h_2(x) = x \log(x)$ (middle), and $h_3(x) = x/(1+x)$ (bottom) and $\alpha = 3.7$, $\beta = 1$. The final errors are respectively -0.998, -0.982, and -0.077 (top), -0.053, -0.053, and -0.001 (middle), and -0.0075, -0.0074, and -0.00003 (bottom).

alternatives for the observed simulations. Note also the strong similarity between δ_1 and the importance sampling estimator, the latter failing to bring any noticeable improvement in this setup.   ∥

For further discussion, estimators and examples see Casella (1996), Casella and Robert (1996). See also Perron (1999), who does a slightly different calculation, conditioning on N and the order statistics $Y_{(1)}, \ldots, Y_{(n)}$ in (4.11).

4.3 Riemann Approximations

In approximating an integral $\mathfrak{J}$ like (4.1), the simulation-based approach is justified by *probabilistic* convergence result for the empirical average

$$\frac{1}{m} \sum_{i=1}^{m} h(X_i) \,,$$

when the X_i's are simulated according to f. As briefly mentioned in Section 1.4, numerical integration (for one-dimensional integrals) is based on the *analytical* definition of the integral, namely as a limit of Riemann sums. In fact, for every sequence $(a_{i,n})_i$ ($0 \le i \le n$) such that $a_{0,n} = a$, $a_{n,n} = b$, and $a_{i,n} - a_{i,n-1}$ converges to 0 (in n), the (Riemann) sum

$$\sum_{i=0}^{n-1} h(a_{i,n}) \, f(a_{i,n})(a_{i+1,n} - a_{i,n}) \,,$$

converges to (4.1) as n goes to infinity. When $\mathcal{X}$ has dimension greater than 1, the same approximation applies with a grid on the domain $\mathcal{X}$ (see Rudin 1976 for more details.)

When the two approaches are put together, the result is a Riemann sum with random steps, with the $a_{i,n}$'s simulated from f (or from an instrumental distribution g). This method was first introduced by Yakowitz et al. (1978) as *weighted Monte Carlo integration* for uniform distributions on $[0,1]^d$. In a more general setup, we call this approach *simulation by Riemann sums* or *Riemannian simulation*, following Philippe (1997a,b), although it is truly an integration method.

Definition 4.8. The method of *simulation by Riemann sums* approximates the integral $\mathfrak{I}$ by

(4.13)
$$\sum_{i=0}^{m-1} h(X_{(i)}) \, f(X_{(i)})(X_{(i+1)} - X_{(i)}) \,,$$

where $X_0, \ldots, X_m$ are iid random variables from f and $X_{(0)} \leq \cdots \leq X_{(m)}$ are the order statistics associated with this sample.

Suppose first that the integral $\mathfrak{I}$ can be written

$$\mathfrak{I} = \int_0^1 h(x) \, dx,$$

and that h is a differentiable function. We can then establish the following result about the validity of the Riemannian approximation. (See Problem 4.14 for the proof.)

Proposition 4.9. Let $U = (U_0, U_1, \ldots, U_m)$ be an ordered sample from $\mathcal{U}_{[0,1]}$. If the derivative h' is bounded on $[0,1]$, the estimator

$$\delta(U) = \sum_{i=0}^{m-1} h(U_i)(U_{i+1} - U_i) + h(0)U_0 + h(U_m)(1 - U_m)$$

has a variance of order $\mathcal{O}(m^{-2})$.

Yakowitz et al. (1978) improve on the order of the variance by symmetrizing δ into

$$\tilde{\delta} = \sum_{i=-1}^{m} (U_{i+1} - U_i) \frac{h(U_i) + h(U_{i+1})}{2} \,.$$

When the second derivative of h is bounded, the error of $\tilde{\delta}$ is then of order $\mathcal{O}(m^{-4})$.

Even when the additional assumption on the second derivative is not satisfied, the practical improvement brought by Riemann sums (when compared with regular Monte Carlo integration) is substantial since the magnitude of the variance decreases from m^{-1} to m^{-2}. Unfortunately, this dominance fails to extend to the case of multidimensional integrals, a phenomenon that is related to the so-called "curse of dimensionality"; that is, the subefficiency of numerical methods compared with simulation algorithms for dimensions d larger than 4 since the error is then of order $\mathcal{O}(m^{-4/d})$ (see Yakowitz et al. 1978). The intuitive reason behind this phenomenon is that a numerical approach like the Riemann sum method basically covers the entire space with a grid. When the dimension of the space increases, the number of points on the grid necessary to obtain a given precision increases too, which means, in practice, a much larger number of iterations for the same precision.

The result of Proposition 4.9 holds for an arbitrary density, due to the property that the integral $\mathfrak{I}$ can also be written as

$$(4.14) \qquad \int_0^1 H(x)\,dx\,,$$

where $H(x) = h(F^-(x))$, and F^- is the generalized inverse of F, cdf of f (see Lemma 2.4). Although this is a formal representation when F^- is not available in closed form and cannot be used for simulation purposes in most cases (see Section 2.1.2), (4.14) is central to this extension of Proposition 4.9. Indeed, since

$$X_{(i+1)} - X_{(i)} = F^-(U_{i+1}) - F^-(U_i),$$

where the $X_{(i)}$'s are the order statistics of a sample from F and the U_i's are the order statistics of a sample from $\mathcal{U}_{[0,1]}$,

$$\sum_{i=0}^{m-1} h(X_{(i)})f(X_{(i)})(X_{(i+1)} - X_{(i)})$$

$$= \sum_{i=0}^{m-1} H(U_i)f(F^-(U_i))(F^-(U_{i+1}) - F^-(U_i))$$

$$\simeq \sum_{i=0}^{m-1} H(U_i)(U_{i+1} - U_i)\,,$$

given that $(F^-)'(u) = 1/f(F^-(u))$. Since the remainder is negligible in the first-order expansion of $F^-(U_{i+1}) - F^-(U_i)$, the above Riemann sum can then be expressed in terms of uniform variables and Proposition 4.9 does apply in this setup, since the extreme terms $h(0)U_0$ and $h(U_m)(1 - U_m)$ are again of order m^{-2} (variancewise). (See Philippe 1997a,b, for more details on the convergence of (4.13) to $\mathfrak{I}$ under some conditions on the function h.)

The above results imply that the Riemann sums integration method will perform well in unidimensional setups when the density f is known. It thus provides an efficient alternative to standard Monte Carlo integration in this setting, since it does not require additional computations (although it requires keeping track of and storing all the $X_{(i)}$'s). Also, as the convergence is of a higher order, there is no difficulty in implementing the method. When f is known only up to a constant (that is, $f_0(x) \propto f(x)$), (4.13) can be replaced by

$$(4.15) \qquad \frac{\sum_{i=0}^{m-1} h(X_{(i)}) \, f_0(X_{(i)})(X_{(i+1)} - X_{(i)})}{\sum_{i=0}^{m-1} f_0(X_{(i)})(X_{(i+1)} - X_{(i)})} \,,$$

since both the numerator and the denominator almost surely converge. This approach thus provides, in addition, an efficient estimation method for the normalizing constant via the denominator in (4.15). Note also that when f is entirely known, the denominator converges to 1, which can be used as a convergence assessment device (see Philippe and Robert 2001, and Section 12.2.4).

Example 4.10. (Continuation of Example 3.15) When $X \sim \mathcal{G}a(3.7, 1)$, assume that $h_2(x) = x \log(x)$ is the function of interest. A sample $X_1, \ldots, X_m$ from $\mathcal{G}a(3.7, 1)$ can easily be produced by the algorithms $[A.14]$ or $[A.15]$ of Chapter 2 and we compare the empirical mean, δ_{1m}, with the Riemann sum estimator

$$\delta_{2m} = \frac{1}{\Gamma(3.7)} \sum_{i=1}^{m-1} h_2(X_{(i)}) \, X_{(i)}^{2.7} e^{-X_{(i)}} \, (X_{(i+1)} - X_{(i)}) \,,$$

which uses the known normalizing constant. Figure 4.7 clearly illustrates the difference in convergence speed between the two estimators and the much greater stability of δ_{2m}, which is close to the theoretical value 5.3185 after 3000 iterations. ‖

If the original simulation is done by importance sampling (that is, if the sample $X_1, \ldots, X_m$ is generated from an instrumental distribution g), since the integral $\mathfrak{I}$ can also be written

$$\mathfrak{I} = \int h(x) \frac{f(x)}{g(x)} g(x) dx,$$

the Riemann sum estimator (4.13) remains unchanged. Although it has similar convergence properties, the boundedness conditions on h are less explicit and, thus, more difficult to check. As in the original case, it is possible to establish an equivalent to Theorem 3.12, namely to show that $g(x) \propto |h(x)| f(x)$ is optimal (in terms of variance) (see Philippe 1997c), with the additional advantage that the normalizing constant does not need to be known, since g does not appear in (4.13).

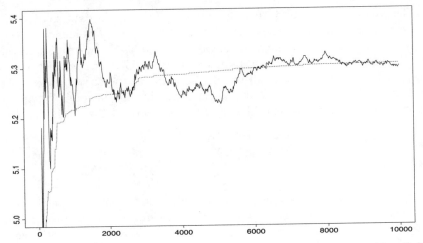

Fig. 4.7. Convergence of estimators of $\mathbb{E}[X \log(X)]$, the Riemann sum δ_{1m} (solid lines and smooth curve) and the empirical average δ_{2m} (dots and wiggly curve) for $\alpha = 3.7$ and $\beta = 1$. The final values are 5.3007 and 5.3057, respectively, for a true value of 5.31847.

Example 4.11. (Continuation of Example 3.13) If $\mathcal{T}(\nu, 0, 1)$ is simulated by importance sampling from the normal instrumental distribution $\mathcal{N}(0, \nu/(\nu - 2))$, the difference between the two distributions is mainly visible in the tails. This makes the importance sampling estimator δ_{1m} very unstable (see Figures 3.5, 3.7, and 3.8). Figure 4.8 compares this estimator to the Riemann sum estimator

$$\delta_{2m} = \sum_{i=1}^{m-1} h_4(X_{(i)}) \frac{\Gamma((\nu + 1)/2)}{\sqrt{\nu \pi}\, \Gamma(\nu/2)} \left[1 + X_{(i)}^2/\nu\right]^{-\frac{\nu+1}{2}} (X_{(i+1)} - X_{(i)})$$

and its normalized version

$$\delta_{3m} = \frac{\sum_{i=0}^{m-1} h_4(X_{(i)}) \left[1 + X_{(i)}^2/\nu\right]^{-\frac{\nu+1}{2}} (X_{(i+1)} - X_{(i)})}{\sum_{i=0}^{m-1} \left[1 + X_{(i)}^2/\nu\right]^{-\frac{\nu+1}{2}} (X_{(i+1)} - X_{(i)})},$$

for $h_4(X) = (1 + e^X)\, \mathbb{I}_{X \le 0}$ and $\nu = 2.3$.

We can again note the stability of the approximations by Riemann sums, the difference between δ_{2m} and δ_{3m} mainly due to the bias introduced by the approximation of the normalizing constant in δ_{3m}. For the given sample, note that δ_{2m} dominates the other estimators.

If, instead, the instrumental distribution is chosen to be the Cauchy distribution $\mathcal{C}(0, 1)$, the importance sampling estimator is much better behaved. Figure 4.9 shows that the speed of convergence of the associated estimator

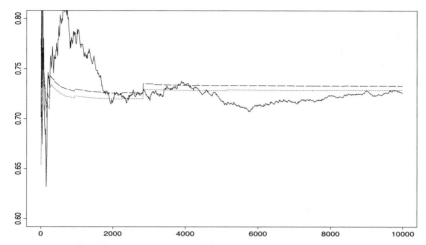

Fig. 4.8. Convergence of estimators of $\mathbb{E}_\nu[(1 + e^X)\mathbb{I}_{X \leq 0}]$, δ_{1m} (solid lines), δ_{2m} (dots), and δ_{3m} (dashes), for a normal instrumental distribution and $\nu = 2.3$. The final values are respectively 0.7262, 0.7287, and 0.7329, for a true value of 0.7307.

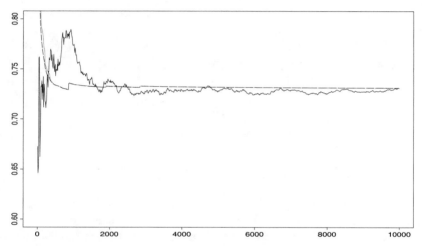

Fig. 4.9. Convergence of estimators of $\mathbb{E}_\nu[(1 + e^X)\mathbb{I}_{X \leq 0}]$, δ_{1m} (solid lines), δ_{2m} (dots), and δ_{3m} (dashes) for a Cauchy instrumental distribution and $\nu = 2.3$. The two Riemann sum approximations are virtually equal except for the beginning simulations. The final values are respectively 0.7325, 0.7314, and 0.7314, and the true value is 0.7307.

is much faster than with the normal instrumental distribution. Although δ_{2m} and δ_{3m} have the same representation for $\mathcal{N}(0, \nu/(\nu-2))$ and $\mathcal{C}(0,1)$, the corresponding samples differ, and these estimators exhibit an even higher stability in this case, giving a good approximation of $\mathbb{E}[h_4(X)]$ after only a few hundred iterations. The two estimators are actually identical almost from the start, a fact which indicates how fast the denominator of δ_{3m} converges to the normalizing constant.   ‖

4.4 Acceleration Methods

While the different methods proposed in Chapter 3 and in this chapter seem to require a comparison, we do not expect there to be any clear-cut domination (as was the case with the comparison between Accept–Reject and importance sampling in Section 3.3.3). Instead, we look at more global *acceleration* strategies, which are more or less independent of the simulation setup but try to exploit the output of the simulation in more efficient ways.

The acceleration methods described below can be used not only in a single implementation but also as a control device to assess the convergence of a simulation algorithm, following the argument of *parallel estimators*. For example, if $\delta_{1m}, \ldots, \delta_{pm}$ are p convergent estimators of the same quantity $\mathfrak{I}$, a stopping rule for convergence is that $\delta_{1m}, \ldots, \delta_{pm}$ are identical or, given a minimum precision requirement ε, that

$$\max_{1 \le i < j \le p} |\delta_{im} - \delta_{jm}| < \varepsilon,$$

as in Section 4.1.

4.4.1 Antithetic Variables

Although the usual simulation methods lead to iid samples (or quasi-iid, see Section 2.6.2), it may actually be preferable to generate samples of correlated variables when estimating an integral $\mathfrak{I}$, as they may reduce the variance of the corresponding estimator.

A first setting where the generation of independent samples is less desirable corresponds to the comparison of two quantities which are close in value. If

$$(4.16) \qquad \mathfrak{I}_1 = \int g_1(x) f_1(x) dx \quad \text{and} \quad \mathfrak{I}_2 = \int g_2(x) f_2(x) dx$$

are two such quantities, where δ_1 estimates $\mathfrak{I}_1$ and δ_2 estimates $\mathfrak{I}_2$, independently of δ_1, the variance of $(\delta_1 - \delta_2)$, is then $\text{var}(\delta_1) + \text{var}(\delta_2)$, which may be too large to support a fine enough analysis on the difference $\mathfrak{I}_1 - \mathfrak{I}_2$. However, if δ_1 and δ_2 are positively correlated, the variance is reduced by a factor $-2 \, \text{cov}(\delta_1, \delta_2)$, which may greatly improve the analysis of the difference.

A convincing illustration of the improvement brought by correlated samples is the comparison of (regular) statistical estimators via simulation. Given a density $f(x|\theta)$ and a loss function $L(\delta, \theta)$, two estimators δ_1 and δ_2 are evaluated through their risk functions, $R(\delta_1, \theta) = \mathbb{E}[L(\delta_1, \theta)]$ and $R(\delta_2, \theta)$. In general, these risk functions are not available analytically, but they may be approximated, for instance, by a regular Monte Carlo method,

$$\hat{R}(\delta_1, \theta) = \frac{1}{m} \sum_{i=1}^{m} L(\delta_1(X_i), \theta), \qquad \hat{R}(\delta_2, \theta) = \frac{1}{m} \sum_{i=1}^{m} L(\delta_2(Y_i), \theta),$$

the X_i's and Y_i's being simulated from $f(\cdot|\theta)$. Positive correlation between $L(\delta_1(X_i), \theta)$ and $L(\delta_2(Y_i), \theta)$ then reduces the variability of the approximation of $R(\delta_1, \theta) - R(\delta_2, \theta)$.

Before we continue with the development in this section, we pause to make two elementary remarks that should be observed in any simulation comparison.

(i) First, the *same sample* $(X_1, \ldots, X_m)$ should be used in the evaluation of $R(\delta_1, \theta)$ and of $R(\delta_2, \theta)$. This repeated use of a single sample greatly improves the precision of the estimated difference $R(\delta_1, \theta) - R(\delta_2, \theta)$, as shown by the comparison of the variances of $\hat{R}(\delta_1, \theta) - \hat{R}(\delta_2, \theta)$ and of

$$\frac{1}{m} \sum_{i=1}^{m} \{L(\delta_1(X_i), \theta) - L(\delta_2(X_i), \theta)\} .$$

(ii) Second, *the same sample* should be used for the comparison of risks *for every value of θ*. Although this sounds like an absurd recommendation since the sample $(X_1, \ldots, X_m)$ is usually generated from a distribution depending on θ, it is often the case that the same uniform sample can be used for the generation of the X_i's for every value of θ. Also, in many cases, there exists a transformation M_θ on $\mathcal{X}$ such that if $X^0 \sim f(X|\theta_0)$, $M_\theta X^0 \sim f(X|\theta)$. A single sample $(X_1^0, \ldots, X_m^0)$ from $f(X|\theta_0)$ is then sufficient to produce a sample from $f(X|\theta)$ by the transform M_θ. (This second remark is somewhat tangential for the theme of this section; however, it brings significant improvement in the practical implementation of Monte Carlo methods.)

The variance reduction associated with the conservation of the underlying uniform sample is obvious in the graphs of the resulting risk functions, which then miss the irregular peaks of graphs obtained with independent samples and allow for an easier comparison of estimators. See, for instance, the graphs in Figure 3.4, which are based on samples generated independently for each value of λ. By comparison, an evaluation based on a single sample corresponding to $\lambda = 1$ would give a constant risk in the exponential case.

Example 4.12. James–Stein estimation. In the case $X \sim \mathcal{N}_p(\theta, I_p)$, the transform is the location shift $M_\theta X = X + \theta - \theta_0$. When studying *positive-part James–Stein estimators*

$$\delta_a(x) = \left(1 - \frac{a}{\|x\|^2}\right)^+ x, \qquad 0 < a < 2(p-2)$$

(see Robert 2001, Chapter 2, for a motivation), the squared error risk of δ_a can be computed "explicitly," but the resulting expression involves several special functions (Robert 1988) and the approximation of the risks by simulation is much more helpful in comparing these estimators. Figure 4.10 illustrates this comparison in the case $p = 5$ and exhibits a crossing phenomenon for the risk functions in the same region; however, as shown by the inset, the crossing point for the risks of δ_a and δ_c depends on (a, c). $\qquad\qquad\|$

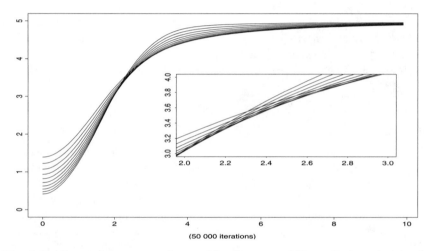

Fig. 4.10. Approximate squared error risks of truncated James–Stein estimators for a normal distribution $\mathcal{N}_5(\theta, I_5)$, as a function of $\|\theta\|$. The inset gives a magnification of the intersection zone for the risk functions.

In a more general setup, creating a strong enough correlation between δ_1 and δ_2 is rarely so simple, and the quest for correlation can result in an increase in the conception and simulation burdens, which may even have a negative overall effect on the efficiency of the analysis. Indeed, to use the same uniform sample for the generation of variables distributed from f_1 and f_2 in (4.16) is only possible when there exists a simple transformation from f_1 to f_2. For instance, if f_1 or f_2 must be simulated by Accept–Reject methods, the use of a random number of uniform variables prevents the use of a common sample.

The method of *antithetic variables* is based on the same idea that higher efficiency can be brought about by correlation. Given two samples $(X_1, \ldots, X_m)$ and $(Y_1, \ldots, Y_m)$ from f used for the estimation of

$$\mathfrak{I} = \int_{\mathbb{R}} h(x)f(x)dx \ ,$$

the estimator

(4.17)
$$\frac{1}{2m} \sum_{i=1}^{m} [h(X_i) + h(Y_i)]$$

is more efficient than an estimator based on an iid sample of size $2m$ if the variables $h(X_i)$ and $h(Y_i)$ are *negatively correlated*. In this setting, the Y_i's are the *antithetic variables*, and it remains to develop a method for generating these variables in an optimal (or, at least, useful) way. However, the correlation between $h(X_i)$ and $h(Y_i)$ depends both on the pair (X_i, Y_i) and on the function h. (For instance, if h is even, X_i has mean 0, and $X_i = -Y_i$, X_i and Y_i are negatively correlated, but $h(X_i) = h(Y_i)$.) A solution proposed in Rubinstein (1981) is to use the uniform variables U_i to generate the X_i's and the variables $1 - U_i$ to generate the Y_i's. The argument goes as follows: If $H = h \circ F^-$, $X_i = F^-(U_i)$, and $Y_i = F^-(1 - U_i)$, then $h(X_i)$ and $h(Y_i)$ are negatively correlated *when H is a monotone function*. Again, such a constraint is often difficult to verify and, moreover, the technique only applies for direct transforms of uniform variables, thus excluding the Accept–Reject methods.

Geweke (1988) proposed the implementation of an inversion at the level of the X_i's by taking $Y_i = 2\mu - X_i$ when f *is symmetric around μ*. With some additional conditions on the function h, the improvement brought by

$$\frac{1}{2m} \sum_{i=1}^{m} [h(X_i) + h(2\mu - X_i)]$$

upon

$$\frac{1}{2m} \sum_{i=1}^{2m} h(X_i)$$

is quite substantial for large sample sizes m. Empirical extensions of this approach can then be used in cases where f is not symmetric, by replacing μ with the mode of f or the median of the associated distribution. Moreover, if f is unknown or, more importantly, if μ is unknown, μ can be estimated from a first sample (but caution is advised!). More general group actions can also be considered, as in Kong et al. (2003), where the authors replace the standard average by an average (over i) of the average of the $h(gx_i)$ (over the transformations g).

Example 4.13. (Continuation of Example 3.3) Assume, for the sake of illustration, that the noncentral chi squared variables $\|X_i\|^2$ are simulated from normal random variables $X_i \sim \mathcal{N}_p(\theta, I_p)$. We can create negative correlation by using $Y_i = 2\theta - X_i$, which has a correlation of -1 with X_i, to produce a second sample, $||Y_i||^2$. However, the negative correlation does not necessarily transfer to the pairs $(h(||X_i||^2), h(||Y_i||^2))$. Figure 4.11 illustrates the behavior of (4.17) for

$$h_1(||x||^2) = ||x||^2 \qquad \text{and} \qquad h_2(||x||^2) = \mathbb{I}_{||x||^2 < ||\theta||^2 + p},$$

when $m = 500$ and $p = 4$, compared to an estimator based on an iid sample of size $2m$. As shown by the graphs in Figure 4.11, although the correlation between $h(||X_i||^2)$ and $h(||Y_i||^2 i)$ is actually positive for small values of $||\theta||^2$, the improvement brought by (4.17) over the standard average is quite impressive in the case of h_1. The setting is less clear for h_2, but the variance of the terms of (4.17) is much smaller than its independent counterpart. $\parallel$

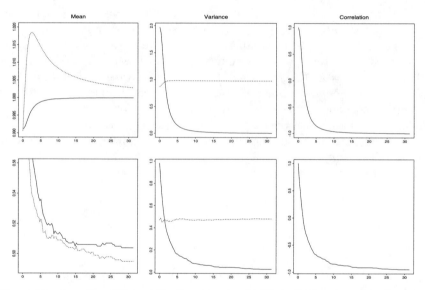

Fig. 4.11. Average of the antithetic estimator (4.17) (solid lines) against the average of an standard iid estimate (dots) for the estimation of $\mathbb{E}[h_1(||X||^2)]$ *(upper left)* and $\mathbb{E}[h_2(||X||^2)]$ *(lower left)*, along with the empirical variance of $h_1(X_i) + h_1(Y_i)$ *(upper center)* and $h_2(X_i) + h_2(Y_i)$ *(lower center)*, and the correlation between $h_1(||X_i||^2)$ and $h_1(||Y_i||^2)$ *(upper right)* and between $h_2(||X_i||^2)$ and $h_2(||Y_i||^2)$ *(lower right)*, for $m = 500$ and $p = 4$. The horizontal axis is scaled in terms of $||\theta||$ and the values in the upper left graph are divided by the true expectation, $||\theta||^2 + p$, and the values in the upper central graph are divided by $8||\theta||^2 + 4p$.

4.4.2 Control Variates

In some settings, there exist functions h_0 whose mean under f is known. For instance, if f is symmetric around μ, the mean of $h_0(X) = \mathbb{I}_{X \geq \mu}$ is $1/2$. We also saw a more general example in the case of Riemann sums with known density f, with a convergent estimator of 1. This additional information can reduce the variance of an estimator of $\mathfrak{I} = \int h(x) f(x) dx$ in the following way. If δ_1 is an estimator of $\mathfrak{I}$ and δ_3 an unbiased estimator of $\mathbb{E}_f[h_0(X)]$, consider the weighted estimator $\delta_2 = \delta_1 + \beta(\delta_3 - \mathbb{E}_f[h_0(X)])$. The estimators δ_1 and δ_2 have the same mean and

$$\text{var}(\delta_2) = \text{var}(\delta_1) + \beta^2 \, \text{var}(\delta_3) + 2\beta \, \text{cov}(\delta_1, \delta_3) \; .$$

For the optimal choice

$$\beta^* = -\frac{\text{cov}(\delta_1, \delta_3)}{\text{var}(\delta_3)} \; ,$$

we have

$$\text{var}(\delta_2) = (1 - \rho_{13}^2) \, \text{var}(\delta_1) \; ,$$

ρ_{13}^2 being the correlation coefficient between δ_1 and δ_3, so the control variate strategy will result in decreased variance. In particular, if

$$\delta_1 = \frac{1}{m} \sum_{i=1}^{m} h(X_i) \qquad \text{and} \qquad \delta_3 = \frac{1}{m} \sum_{i=1}^{m} h_0(X_i) \; ,$$

the control variate estimator is

$$\delta_2 = \frac{1}{m} \sum_{i=1}^{m} (h(X_i) + \beta^* h_0(X_i)) - \beta^* \mathbb{E}_f[h_0(X)] \; ,$$

with $\beta^* = -\text{cov}(h(X), h_0(X))/\text{var}(h_0(X))$. Note that this construction is only formal since it requires the computation of β^*. An incorrect choice of β may lead to an increased variance; that is, $\text{var}(\delta_2) > \text{var}(\delta_1)$. (However, in practice, the sign of β^* can be evaluated by a regression of the $h(x_i)$'s over the $h_0(x_i)$'s. More generally, functions with known expectations can be used as side controls in convergence diagnoses.)

Example 4.14. Control variate integration. Let $X \sim f$, and suppose that we want to evaluate

$$P(X > a) = \int_a^{\infty} f(x) dx.$$

The natural place to start is with $\delta_1 = \frac{1}{n} \sum_{i=1}^{n} \mathbb{I}(X_i > a)$, where the X_i's are iid from f.

Suppose now that f is symmetric or, more, generally, that for some parameter μ we know the value of $P(X > \mu)$ (where we assume that $a > \mu$). We can then take $\delta_3 = \frac{1}{n} \sum_{i=1}^{n} \mathbb{I}(X_i > \mu)$ and form the control variate estimator

$$\delta_2 = \frac{1}{n}\sum_{i=1}^{n}\mathbb{I}(X_i > a) + \beta\left(\frac{1}{n}\sum_{i=1}^{n}\mathbb{I}(X_i > \mu) - P(X > \mu)\right).$$

Since $\text{var}(\delta_2) = \text{var}(\delta_1) + \beta^2\text{var}(\delta_3) + 2\beta\text{cov}(\delta_1, \delta_3)$ and

$$\text{cov}(\delta_1, \delta_3) = \frac{1}{n}P(X > a)[1 - P(X > \mu)],$$

(4.18) $$\text{var}(\delta_3) = \frac{1}{n}P(X > \mu)[1 - P(X > \mu)],$$

it follows that δ_2 will be an improvement over δ_1 if

$$\beta < 0 \quad \text{and} \quad |\beta| < 2\frac{\text{cov}(\delta_1, \delta_3)}{\text{var}(\delta_3)} = 2\frac{P(X > a)}{P(X > \mu)}.$$

If $P(X > \mu) = 1/2$ and we have some idea of the value of $P(X > a)$, we can choose an appropriate value for β (see Problems 4.15 and 4.18).   ‖

Example 4.15. Logistic regression. Consider the logistic regression model introduced in Example 1.13,

$$P(Y_i = 1) = \exp(x_i^t\theta)/\{1 + \exp(x_i^t\theta)\}.$$

The likelihood associated with a sample $((x_1, Y_1), \ldots, (x_n, Y_n))$ can be written

$$\exp\left(\theta^t\sum_i Y_i x_i\right)\prod_{i=1}^{n}\{1 + \exp(x_i^t\theta)\}^{-1}.$$

When $\pi(\theta)$ is a *conjugate prior* (see Note 1.6.1),

(4.19) $$\pi(\theta|\zeta, \lambda) \propto \exp(\theta^t\zeta)\prod_{i=1}^{n}\{1 + \exp(x_i^t\theta)\}^{-\lambda}, \quad \lambda > 0,$$

the posterior distribution of θ is of the same form, with $(\sum_i Y_i x_i + \zeta, \lambda + 1)$ replacing (ζ, λ).

The expectation $\mathbb{E}^\pi[\theta|\sum_i Y_i x_i + \zeta, \lambda + 1]$ is derived from variables θ_j $(1 \leq j \leq m)$ generated from (4.19). Since the logistic distribution is in an exponential family, the following holds (see Brown 1986 or Robert 2001, Section 3.3):

$$\mathbb{E}_\theta\left[\sum_i Y_i x_i\right] = n\nabla\psi(\theta)$$

and

$$\mathbb{E}^\pi\left[\nabla\psi(\theta)\Big|\sum_i Y_i x_i + \zeta, \lambda + 1\right] = \frac{\sum_i Y_i x_i + \zeta}{n(\lambda + 1)}.$$

Therefore, the posterior expectation of the function

$$n \nabla \psi(\theta) = \sum_{i=1}^{n} \frac{\exp(x_i^t \theta)}{1 + \exp(x_i^t \theta)} \, x_i$$

is known and equal to $(\sum_i Y_i x_i + \zeta)/(\lambda + 1)$ under the prior distribution $\pi(\theta|\zeta, \lambda)$. Unfortunately, a control variate version of

$$\delta_1 = \frac{1}{m} \sum_{j=1}^{m} \theta_j$$

is not available since the optimal constant β^* (or even its sign) cannot be evaluated, except by the regression of the θ_j's upon the

$$\sum_i \frac{\exp x_i^t \theta_j}{1 + \exp x_i^t \theta_j} \, x_i \, .$$

Thus the fact that the posterior mean of $\nabla \psi(\theta)$ is known does not help us to establish a control variate estimator. This information can be used in a more informal way to study convergence of δ_1 (see, for instance, Robert 1993).   ‖

In conclusion, the technique of control variates is manageable only in very specific cases: the control function h must be available, as well as the optimal weight β^*. See, however, Brooks and Gelman (1998b) for a general approach based on the score function (whose expectation is null under general regularity conditions).

4.5 Problems

4.1 (Chen and Shao 1997) As mentioned, normalizing constants are superfluous in Bayesian inference except in the case when several models are considered at once (as in the computation of Bayes factors). In such cases, where $\pi_1(\theta) = \tilde{\pi}_1(\theta)/c_1$ and $\pi_2(\theta) = \tilde{\pi}_2(\theta)/c_2$, and only $\tilde{\pi}_1$ and $\tilde{\pi}_2$ are known, the quantity to approximate is $\varrho = c_1/c_2$ or $\xi = \log(c_1/c_2)$.
(a) Show that the ratio ϱ can be approximated by

$$\frac{1}{n} \sum_{i=1}^{n} \frac{\tilde{\pi}_1(\theta_i)}{\tilde{\pi}_2(\theta_i)}, \qquad \theta_1, \ldots, \theta_n \sim \pi_2.$$

(*Hint:* Use an importance sampling argument.)
(b) Show that

$$\frac{\int \tilde{\pi}_1(\theta)\alpha(\theta)\pi_2(\theta)d\theta}{\int \tilde{\pi}_2(\theta)\alpha(\theta)\pi_1(\theta)d\theta} = \frac{c_1}{c_2}$$

holds for every function $\alpha(\theta)$ such that both integrals are finite.

(c) Deduce that

$$\frac{\frac{1}{n_2}\sum_{i=1}^{n_2}\tilde{\pi}_1(\theta_{2i})\alpha(\theta_{2i})}{\frac{1}{n_1}\sum_{i=1}^{n_1}\tilde{\pi}_2(\theta_{1i})\alpha(\theta_{1i})},$$

with $\theta_{1i}\sim\pi_1$ and $\theta_{2i}\sim\pi_2$, is a convergent estimator of ϱ.

(d) Show that part (b) covers the case of the Newton and Raftery (1994) representation

$$\frac{c_1}{c_2}=\frac{\mathbb{E}^{\pi_2}[\tilde{\pi}_2(\theta)^{-1}]}{\mathbb{E}^{\pi_1}[\tilde{\pi}_1(\theta)^{-1}]}.$$

(e) Show that the optimal choice (in terms of mean square error) of α in part (c) is

$$\alpha(\theta)=c\frac{n_1+n_2}{n_1\pi_1(\theta)+n_2\pi_2(\theta)},$$

where c is a constant. (*Note:* See Meng and Wong 1996.)

4.2 (Continuation of Problem 4.1) When the priors π_1 and π_2 belong to a parameterized family (that is, $\pi_i(\theta)=\pi(\theta|\lambda_i)$), the corresponding constants are denoted by $c(\lambda_i)$.

(a) Verify the identity

$$-\log\left(\frac{c(\lambda_1)}{c(\lambda_2)}\right)=\mathbb{E}\left[\frac{U(\theta,\lambda)}{\pi(\lambda)}\right],$$

where

$$U(\theta,\lambda)=\frac{d}{d\lambda}\log(\tilde{\pi}(\theta|\lambda))$$

and $\pi(\lambda)$ is an arbitrary distribution on λ.

(b) Show that ξ can be estimated with the *bridge estimator* of Gelman and Meng (1998),

$$\hat{\xi}=\frac{1}{n}\sum_{i=1}^{n}\frac{U(\theta_i,\lambda_i)}{\pi(\lambda_i)},$$

when the (θ_i,λ_i)'s are simulated from the joint distribution induced by $\pi(\lambda)$ and $\pi(\theta|\lambda_i)$.

(c) Show that the minimum variance estimator of ξ is based on

$$\pi(\lambda)\propto\sqrt{\mathbb{E}_\lambda[U^2(\theta,\lambda)]}$$

and examine whether this solution gives the Jeffreys prior.

4.3 For the situation of Example 4.2:

(a) Show that $\hat{\delta}_m^\pi(x)\to\delta^\pi(x)$ as $m\to\infty$.

(b) Show that the Central Limit Theorem can be applied to $\hat{\delta}_m^\pi(x)$.

(c) Generate random variables $\theta_1,\ldots,\theta_m\sim\mathcal{N}(x,1)$ and calculate $\hat{\delta}_m^\pi(x)$ for $x=0,1,4$. Use the Central Limit Theorem to construct a measure of accuracy of your calculation.

4.4 Verify equation (4.10), that is, show that

$$\mathbb{E}\left[e^{-X^2}|y\right]=\frac{1}{\sqrt{2\pi\sigma^2 y}}\int_{-\infty}^{\infty}e^{-x^2}e^{-(x-\mu)^2/2\sigma^2 y}dx$$

$$=\frac{1}{\sqrt{2\sigma^2 y+1}}\exp-\frac{\mu^2}{1+2\sigma^2 y},$$

by completing the square in the exponent to evaluate the integral.

4.5 In simulation from mixture densities, it is always possible to set up a Rao–Blackwellized alternative to the empirical average.

(a) Show that, if $f(x) = \int g(x|y)h(y)dy$, then

$$\frac{1}{M}\sum_{i=1}^{M}h(X_i), \quad X_i \sim f \text{ and } \frac{1}{M}\sum_{i=1}^{M}\mathbb{E}(h(X)|Y_i), \quad Y_i \sim g,$$

each converge to $\mathbb{E}_f(X)$.

(b) For each of the following cases, generate random variables X_i and Y_i, and compare the empirical average and Rao–Blackwellized estimator of $\mathbb{E}_f(X)$ and $\text{var}_f(X)$:

a) $X|y \sim \mathcal{P}(y)$, $Y \sim \mathcal{G}a(a,b)$ (X is negative binomial);

b) $X|y \sim \mathcal{N}(0,y)$, $Y \sim \mathcal{G}a(a,b)$ (X is a generalized t);

c) $X|y \sim \mathcal{B}in(y)$, $Y \sim \mathcal{B}e(a,b)$ (X is beta-binomial).

4.6 For the estimator δ_2 of Section 4.2:

(a) Verify the expression (4.11) for ρ_i.

(b) Verify the recursion (4.12).

(c) Prove Proposition 4.6. (*Hint:* Show that $\mathbb{E}[\delta_2] = \mathbb{E}[\delta_1]$ and apply the Rao–Blackwell Theorem.)

4.7 Referring to Lemma 4.3, show how to use (4.5) to derive the expression for the asymptotic variance of δ_h^n.

4.8 Given an Accept–Reject algorithm based on (f,g,ρ), we denote by

$$b(y_j) = \frac{(1-\rho)f(y_j)}{g(y_j) - \rho f(y_j)}$$

the importance sampling weight of the rejected variables $(Y_1,\ldots,Y_t)$, and by $(X_1,\ldots,X_n)$ the accepted variables.

(a) Show that the estimator

$$\delta_1 = \frac{n}{n+t}\,\delta^{AR} + \frac{t}{n+t}\,\delta_0,$$

with

$$\delta_0 = \frac{1}{t}\sum_{j=1}^{t}b(Y_j)h(Y_j)$$

and

$$\delta^{AR} = \frac{1}{n}\sum_{i=1}^{n}h(X_i),$$

does not uniformly dominate δ^{AR}. (*Hint:* Consider the constant functions.)

(b) Show that

$$\delta_{2w} = \frac{n}{n+t}\,\delta^{AR} + \frac{t}{n+t}\sum_{j=1}^{t}\frac{b(Y_j)}{S_t}h(Y_j)$$

is asymptotically equivalent to δ_1 in terms of bias and variance.

(c) Deduce that δ_{2w} asymptotically dominates δ^{AR} if (4.20) holds.

4.9 Referring to Section 4.1.2:

(a) Show that $\text{cov}(\bar{\mathbf{X}}_k, \bar{\mathbf{X}}_{k'}) = \sigma^2/\max\{k,k'\}$, regardless of the distribution of X_i.

(b) Verify that Σ^{-1} is given by (4.7) for $n = 3, 10, 25$.

(c) Establish a recursion relation for calculating the elements a_{ij} of Σ^{-1}:

$$a_{ij} = \begin{cases} 2i^2 & \text{if } i = j < n \\ n^2 & \text{if } i = j = n \\ -ij & \text{if } |i - j| = 1 \\ 0 & \text{otherwise.} \end{cases}$$

(d) If we denote by Σ_k^{-1} the inverse matrix corresponding to $(\bar{X}_1, \bar{X}_2, \ldots, \bar{X}_k)$, show that to get Σ_{k+1}^{-1} you only have to change one element, then add one row and one column to Σ_k^{-1}.

4.10 (a) Establish (4.8) by showing that (i) $\mathbf{1}'\Sigma^{-1}\bar{\mathbf{x}} = n$ and (ii) $\mathbf{1}'\Sigma^{-1}\mathbf{1} = n$.

(b) Show that a reasonable approximation for d_n is $\hat{\sigma}^2(n + 2\sqrt{2n})$. (*Hint:* Consider the mean and variance of the χ_n^2 distribution.)

4.11 Referring to Example 4.2:

(a) Compare a running mean plot with ordinary univariate normal error bars to the variance assessment of (4.9). Discuss advantages and disadvantages.

(b) Compare a running mean plot with empirical error bars to the variance assessment of (4.9). Use 500 estimates to calculate 95% error bars. Discuss advantages and disadvantages.

(c) Repeat parts (*a*) and (*b*) for $h(x) = x^2$ and thus assess the estimate of the posterior variance of θ.

4.12 In the setting of Section 4.3, examine whether the substitution of

$$\sum_{i=1}^{n-1}(X_{(i+1)} - X_{(i)}) \frac{h(X_{(i)}) + h(X_{(i+1)})}{2}$$

into

$$\sum_{i=1}^{n-1}(X_{(i+1)} - X_{(i)}) h(X_{(i)})$$

improves the speed of convergence. (*Hint:* Examine the influence of the remainder terms

$$\int_{-\infty}^{X_{(1)}} h(x)f(x)dx \quad \text{and} \quad \int_{X_{(n)}}^{+\infty} h(x)f(x)dx.)$$

4.13 Show that it is always possible to express an integral $\Im$ as $\Im = \int_0^1 \tilde{h}(y)\tilde{f}(y)dy$, where the integration is over $(0, 1)$ and $\tilde{h}$ and $\tilde{f}$ are transforms of the original functions.

4.14 In this problem we will prove Proposition 4.9

(a) Define $U_{-1} = 0$ and $U_{m+1} = 1$, and show that δ can be written

$$\delta = \sum_{i=-1}^{m} h(U_i)(U_{i+1} - U_i) = \sum_{i=-1}^{m} \int_{U_i}^{U_{i+1}} h(U_i)du,$$

and thus the difference $(\Im - \delta)$ can be written as $\sum_{i=-1}^{m} \int_{U_i}^{U_{i+1}} (h(u) - h(U_i))\, du$.

(b) Show that the first-order expansion $h(u) = h(U_i) + h'(\zeta)(u - U_i)$, $\zeta \in [U_i, u]$, implies that $|h(u) - h(U_i)| < c(u - U_i)$, with $c = \sup_{[0,1]} |h'(X)|$.

(c) Show that

$$\text{var}(\delta) = \mathbb{E}[(\Im - \delta)^2] < c^2 \left\{ (m+2)\,\mathbb{E}[Z_i^4] + (m+1)(m+2)\,\mathbb{E}[Z_i^2 Z_j^2] \right\},$$

where $Z_i = U_{i+1} - U_i$.

(d) The U_i's are the order statistics of a uniform distribution. Show that (i) the variables Z_i are jointly distributed according to a Dirichlet distribution $\mathcal{D}_m(1, \ldots, 1)$, with $Z_i \sim \mathcal{B}e(1, m)$ and $(Z_i, Z_j, 1 - Z_i - Z_j) \sim \mathcal{D}_3(1, 1, m-1)$, (ii) $\mathbb{E}[Z_i^4] = m$, $\int_0^1 z^4 (1-z)^{m-1} dz = \frac{24\,m!}{(m+4)!}$, and $\mathbb{E}[Z_i^2 Z_j^2] = \frac{4\,m!}{(m+4)!}$.

(e) Finally, establish that

$$\text{var}(\delta) \leq c \left\{ \frac{24}{(m+4)(m+3)(m+1)} + \frac{4}{(m+4)(m+3)} \right\} = \mathcal{O}(m^{-2}),$$

proving the proposition.

4.15 For the situation of Example 4.14:

(a) Verify (4.18).

(b) Verify the conditions on β in order for δ_2 to improve on δ_1.

(c) For f the density of $\mathcal{N}(0, 1)$, find $P(X > a)$ for $a = 3, 5, 7$.

(d) For f the density of $\mathcal{T}_5$, find $P(X > a)$ for $a = 3, 5, 7$.

(e) For f the density of $\mathcal{T}_5$, find a such that $P(X > a) = .01, .001, .0001$.

4.16 A naïve way to implement the antithetic variable scheme is to use both U and $(1 - U)$ in an inversion simulation. Examine empirically whether this method leads to variance reduction for the distributions (i) $f_1(x) = 1/\pi(1 + x^2)$, (ii) $f_2(x) = \frac{1}{2} e^{-|x|}$, (iii) $f_3(x) = e^{-x} \mathbb{I}_{x>0}$, (iv) $f_4(x) = \frac{2}{\pi\sqrt{3}} \left(1 + x^2/3\right)^{-2}$, and (v) $f_5(x) = 2x^{-3} \mathbb{I}_{x>1}$. Examine variance reductions of the mean, second moment, median, and 75th percentile.

To calculate the weights for the Rao–Blackwellized estimator of Section 4.2, it is necessary to derive properties of the distribution of the random variables in the Accept–Reject algorithm [A.4]. The following problem is a rather straightforward exercise in distribution theory and is only made complicated by the stopping rule of the Accept–Reject algorithm.

4.17 This problem looks at the performance of a *termwise* Rao–Blackwellized estimator. Casella and Robert (1998) established that such an estimator does not sacrifice much performance over the full Rao–Blackwellized estimator of Proposition 4.6. Given a sample $(Y_1, \ldots, Y_N)$ produced by an Accept–Reject algorithm to accept m values, based on (f, g, M):

(a) Show that

$$\frac{1}{n} \sum_{i=1}^n \mathbb{E}[\mathbb{I}_{U_i \leq w_i} | Y_i] h(Y_i)$$

$$= \frac{1}{n-m} \left(h(Y_n) + \sum_{i=1}^{n-1} b(Y_i) h(Y_i) \right),$$

with

$$b(Y_i) = \left(1 + \frac{m(g(Y_i) - \rho f(Y_i))}{(n-m-1)(1-\rho) f(Y_i)} \right)^{-1}.$$

(b) If $S_n = \sum_1^{n-1} b(y_i)$, show that

$$\delta = \frac{1}{n-m}\left(h(Y_n) + \frac{n-m-1}{S_n}\sum_{i=1}^{nt-1} b(Y_i)h(Y_i)\right)$$

asymptotically dominates the usual Monte Carlo approximation, conditional on the number of rejected variables m under quadratic loss. (*Hint:* Show that the sum of the weights S_n can be replaced by $(n-m-1)$ in δ and assume $\mathbb{E}_f[h(X)] = 0$.)

4.18 Strawderman (1996) adapted the control variate scheme to the Accept–Reject algorithm. When $Y_1, \ldots, Y_N$ is the sample produced by an Accept–Reject algorithm based on g, let m denote the density

$$m(y) = \frac{t-1}{n-1}f(y) + \frac{n-t}{n-1}\frac{g(y) - \rho f(y)}{1-\rho} ,$$

when $N = n$ and $\rho = \dfrac{1}{M}$.

(a) Show that

$$\Im = \int h(x)f(x)dx = \mathbb{E}_N\left[\mathbb{E}\left[\frac{h(Y)f(Y)}{m(Y)}\Big| N\right]\right] ,$$

where m is the marginal density of Y_i (see Problem 3.29).

(b) Show that for any function $c(\cdot)$ and some constant β,

$$\Im = \beta\mathbb{E}[c(Y)] + \mathbb{E}\left[\frac{h(Y)f(Y)}{m(Y)} - \beta c(Y)\right] .$$

(c) Setting $d(y) = h(y)f(y)/m(y)$, show that the optimal choice of β is

$$\beta^* = \mathrm{cov}[d(Y), c(Y)]/\mathrm{var}[c(Y)].$$

(d) Examine choices of c for which the optimal β can be constructed and, thus, where the control variate method applies.

(*Note:* Strawderman 1996 suggests estimating β with $\hat{\beta}$, the estimated slope of the regression of $d(y_i)$ on $c(y_i)$, $i = 1, 2, \ldots, n-1$.)

4.19 For $t \sim \mathcal{G}eo(\rho)$, show that

$$\mathbb{E}[t^{-1}] = -\frac{\rho\log\rho}{1-\rho} , \qquad \mathbb{E}[t^{-2}] = \frac{\rho\mathrm{Li}(1-\rho)}{1-\rho} ,$$

where $\mathrm{Li}(x)$ is the dilog function (see Note 4.6.2).

4.20 (Continuation of Problem 4.19) If $N \sim \mathcal{N}eg(n, \rho)$, show that

$$\mathbb{E}[(N-1)^{-1}] = \frac{\rho}{n-1} , \qquad \mathbb{E}[(N-2)^{-1}] = \frac{\rho(n+\rho-2)}{(n-1)(n-2)} ,$$

$$\mathbb{E}[((N-1)(N-2))^{-1}] = \frac{\rho^2}{(n-1)(n-2)} ,$$

$$\mathbb{E}[(N-1)^{-2}] = \frac{\rho^t {}_2F_1(n-1, n-1; n; 1-\rho)}{(n-1)^2} .$$

4.21 (Continuation of Problem 4.19) If Li is the dilog function, show that

$$\lim_{\rho \to 1} \frac{\rho \text{Li}(1 - \rho)}{1 - \rho} = 1 \qquad \text{and} \qquad \lim_{\rho \to 1} \log(\rho) \text{Li}(1 - \rho) = 0 .$$

Deduce that the domination corresponding to (4.20) occurs on an interval of the form $[\rho_0, 1]$.

4.22 Given an integral

$$\Im = \int_{\mathcal{X}} h(x) f(x) dx$$

to be evaluated by simulation, compare the usual Monte Carlo estimator

$$\delta_a = \frac{1}{np} \sum_{i=1}^{np} h(x_i)$$

based on an iid sample $(X_1, \ldots, X_{np})$ with a stratified estimator (see Note 4.6.3)

$$\delta_a = \sum_{j=1}^{p} \frac{p_j}{n} \sum_{i=1}^{n} h(Y_i^p) ,$$

where $p_j = \int_{\mathcal{X}_j} f(x) dx$, $\mathcal{X} = \bigcup_{j=1}^{p} \mathcal{X}_j$ and $(Y_1^j, \ldots, Y_n^j)$ is a sample from $f \mathbb{I}_{\mathcal{X}_j}$.

Show that δ_2 does not bring any improvement if the p_j's are unknown and must be estimated.

4.6 Notes

4.6.1 Monitoring Importance Sampling Convergence

With reference to convergence control for simulation methods, importance sampling methods can be implemented in a *monitored* way; that is, in parallel with other evaluation methods. These can be based on alternative instrumental distributions or other techniques (standard Monte Carlo, Markov chain Monte Carlo, Riemann sums, Rao–Blackwellization, etc.). The respective samples then provide separate evaluations of $\mathbb{E}[h(X)]$, through (3.8), (3.11), or yet another estimator (as in Section 3.3.3), and the convergence criterion is to stop when most estimators are close enough. Obviously, this empirical method is not completely foolproof, but it generally prevents pseudo-convergences when the instrumental distributions are sufficiently different. On the other hand, this approach is rather *conservative*, as it is only as fast as the slowest estimator. However, it may also point out instrumental distributions with variance problems. From a computational point of view, an efficient implementation of this control method relies on the use of *parallel programming* in order to weight each distribution more equitably, so that a distribution f of larger variance, compared with another distribution g, may compensate this drawback by a lower computation time, thus producing a larger sample in the same time.[5]

---

[5] This feature does not necessarily require a truly parallel implementation, since it can be reproduced by the cyclic allocation of uniform random variables to each of the distributions involved.

4.6.2 Accept–Reject with Loose Bounds

For the Accept–Reject algorithm [A.4], some interesting results emerge if we assume that

$$f(x) \leq M \, g(x)/(1 + \varepsilon) \, ;$$

that is, the bound M is not tight. If we assume $\mathbb{E}_f[h(X)] = 0$, the estimator δ_4 of (3.16) then satisfies

$$\mathrm{var}(\delta_4) \leq \left(\mathbb{E}\left[\frac{t-1}{t^2} \right] \frac{M-1}{\varepsilon} + \mathbb{E}[t^{-2}] \right) \mathbb{E}_f[h^2(X)] \, ,$$

where

$$\mathbb{E}[t^{-2}] = \frac{\rho}{1 - \rho} \mathrm{Li}(1 - \rho) \, ,$$

and $\mathrm{Li}(x)$ denotes the *dilog* function,

$$\mathrm{Li}(x) = \sum_{k=1}^{\infty} \frac{x^k}{k^2}$$

(see Abramowitz and Stegun 1964, formula 27.7, who also tabulated the function), which can also be written as

$$\mathrm{Li}(x) = \int_x^1 \frac{\log(u)}{1 - u} du$$

(see Problem 4.19). The bound on the variance of δ_4 is thus

$$\mathrm{var}(\delta_4) \leq \left\{ \left(\frac{\rho}{1 - \rho} - \epsilon^{-1} \right) \mathrm{Li}(1 - \rho) - \epsilon^{-1} \log(\rho) \right\} \mathbb{E}_f[h^2(X)]$$

and δ_4 uniformly dominates the usual Accept–Reject estimator δ_1 of (3.15) as long as

$$(4.20) \qquad \varepsilon^{-1}\{-\log(\rho) - \mathrm{Li}(1 - \rho)\} + \frac{\rho \mathrm{Li}(1 - \rho)}{1 - \rho} < 1 \, .$$

This result rigorously establishes the advantage of recycling the rejected variables for the computation of integrals, since (4.20) does not depend on the function h. Note, however, that the assumption $\mathbb{E}_f[h(X)] = 0$ is quite restrictive, because the sum of the weights of δ_4 does not equal 1 and, therefore, δ_4 does not correctly estimate constant functions (except for the constant function $h = 0$). Therefore, δ_1 will dominate δ_4 for constant functions, and a uniform comparison between the two estimators is impossible.

Figure 4.12 gives the graphs of the left-hand side of (4.20) for $\varepsilon = 0.1, 0.2, \ldots,$ 0.9. A surprising aspect of these graphs is that domination (that is, where the curve is less than 1) occurs for larger values of ρ, which is somewhat counterintuitive, since smaller values of ρ lead to higher rejection rates, therefore to larger rejected subsamples and to a smaller variance of δ_4 for an adequate choice of density functions g. On the other hand, the curves are correctly ordered in ϵ since larger values of ϵ lead to wider domination zones.

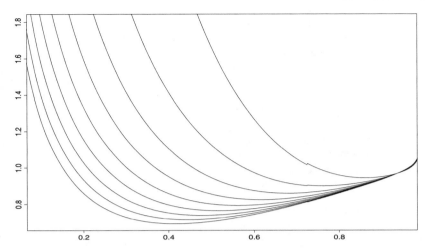

Fig. 4.12. Graphs of the variance coefficients of δ_4 for $\epsilon = 0.1, \ldots, 0.9$, the curves decreasing in ϵ. The domination of δ_1 by δ_4 occurs when the variance coefficient is less than 1.

4.6.3 Partitioning

When enough information is available on the function f, *stratified sampling* may be used. This technique (see Hammersley and Handscomb 1964 or Rubinstein 1981 for references) decomposes the integration domain $\mathcal{X}$ in a partition $\mathcal{X}_1, \ldots, \mathcal{X}_p$, with separate evaluations of the integrals on each region $\mathcal{X}_i$; that is, the integral $\int_{\mathcal{X}} h(x)f(x)dx$ is expressed as

$$\int_{\mathcal{X}} h(x)f(x)dx = \sum_{i=1}^{p} \int_{\mathcal{X}_i} h(x)f(x)dx = \sum_{i=1}^{p} \varrho_i \int_{\mathcal{X}_i} h(x)f_i(x)dx = \sum_{i=1}^{p} \varrho_i \mathfrak{I}_i \;,$$

where the weights ϱ_i are the probabilities of the regions $\mathcal{X}_i$ and the f_i's are the restrictions of f to these regions.

Then samples of size n_i are generated from the f_i's to evaluate each integral $\mathfrak{I}_i$ separately by a regular estimator $\hat{\mathfrak{I}}_i$.

The motivation for this approach is that the variance of the resulting estimator, $\varrho_1 \hat{\mathfrak{I}}_1 + \cdots + \varrho_p \hat{\mathfrak{I}}_p$, that is,

$$\sum_{i=1}^{p} \varrho_i^2 \frac{1}{n_i} \int_{\mathcal{X}_i} (h(x) - \mathfrak{I}_i)^2 f_i(x)dx,$$

may be much smaller than the variance of the standard Monte Carlo estimator based on a sample of size $n = n_1 + \cdots + n_p$. The optimal choice of the n_i's in this respect is such that

$$(n_i^*)^2 \propto \varrho_i^2 \int_{\mathcal{X}_i} (h(x) - \mathfrak{I}_i)^2 f_i(x)dx.$$

Thus, if the regions $\mathcal{X}_i$ can be chosen, the variance of the stratified estimator can be reduced by selecting $\mathcal{X}_i$'s with similar variance factors $\int_{\mathcal{X}_i} (h(x) - \mathfrak{I}_i)^2 f_i(x)dx$.

While this approach may seem far-fetched, because of its requirements on the distribution f, note that it can be combined with importance sampling, where the importance function g may be chosen in such a way that its quantiles are well known. Also, it can be iterated, with each step producing an evaluation of the ϱ_i's and of the corresponding variance factors, which may help in selecting the partition and the n_i's in the next step (see, however, Problem 4.22).

An extension proposed by McKay et al. (1979) introduces stratification on all input dimensions. More precisely, if the domain is represented as the unit hypercube $\mathcal{H}$ in $\mathbb{R}^d$ and the integral (3.5) is written as $\int_{\mathcal{H}} h(x)dx$, *Latin hypercube sampling* relies on the simulation of (a) d random permutations π_j of $\{1,\ldots,n\}$ and (b) uniform $\mathcal{U}([0,1])$ rv's $U_{i_1,\ldots,i_d,j}$ $(j = 1,\ldots,d,\ 1 \leq i_1,\ldots,i_d \leq n)$. A sample of n vectors $\mathbf{X}_i$ in the unit hypercube is then produced as $\mathbf{X}(\pi_1(i),\ldots,\pi_d(i))$ with

$$\mathbf{X}(i_1,\ldots,i_d) = (X_1(i_1,\ldots,i_d),\ldots,X_d(i_1,\ldots,i_d)),$$
$$X_j(i_1,\ldots,i_d) = \frac{i_j - U_{i_1,\ldots,i_d,j}}{n},\qquad 1 \leq j \leq d.$$

The component $X_j(i_1,\ldots,i_d)$ is therefore a point taken at random on the interval $[i_{j-1}/n, i_j/n]$ and the permutations ensure that no uniform variable is taken twice from the same interval, for every dimension. Note that we also need only generate $n \times d$ uniform random variables. (Latin hypercubes are also used in agricultural experiments to ensure that all parcels and all varieties are used, at a minimal cost. See Mead 1988 or Kuehl 1994.) McKay et al. (1979) show that when h is a real-valued function, the variance of the resulting estimator is substantially reduced, compared with the regular Monte Carlo estimator based on the same number of samples. Asymptotic results about this technique can be found in Stein (1987) and Loh (1996). (It is, however, quite likely that the curse of dimensionality, see Section 4.3, occurs for this technique.)

5

Monte Carlo Optimization

"Remember, boy," Sam Nakai would sometimes tell Chee, "when you're tired of walking up a long hill you think about how easy it's going to be walking down."
—Tony Hillerman, *A Thief of Time*

This chapter is the equivalent for optimization problems of what Chapter 3 is for integration problems. Here we distinguish between two separate uses of computer generated random variables. The first use, as seen in Section 5.2, is to produce stochastic techniques to reach the maximum (or minimum) of a function, devising random explorations techniques on the surface of this function that avoid being trapped in a local maximum (or minimum) but also that are sufficiently attracted by the global maximum (or minimum). The second use, described in Section 5.3, is closer to Chapter 3 in that it approximates the function to be optimized. The most popular algorithm in this perspective is the EM (Expectation–Maximization) algorithm.

5.1 Introduction

Similar to the problem of integration, differences between the numerical approach and the simulation approach to the problem

$$(5.1) \qquad \max_{\theta \in \Theta} h(\theta)$$

lie in the treatment of the function[1] h. (Note that (5.1) also covers minimization problems by considering $-h$.) In approaching an optimization problem

---

[1] Although we use θ as the running parameter and h typically corresponds to a (possibly penalized) transform of the likelihood function, this setup applies to inferential problems other than likelihood or posterior maximization. As noted in the introduction to Chapter 3, problems concerned with complex loss functions or confidence regions also require optimization procedures.

using deterministic numerical methods, the analytical properties of the target function (convexity, boundedness, smoothness) are often paramount. For the simulation approach, we are more concerned with h from a probabilistic (rather than analytical) point of view. Obviously, this dichotomy is somewhat artificial, as there exist simulation approaches where the probabilistic interpretation of h is not used. Nonetheless, the use of the analytical properties of h plays a lesser role in the simulation approach.

Numerical methods enjoy a longer history than simulation methods (see, for instance, Kennedy and Gentle 1980 or Thisted 1988), but simulation methods have gained in appeal due to the relaxation of constraints both on the regularity of the domain Θ and on the function h. Of course, there may exist an alternative numerical approach which provides an exact solution to (5.1), a property rarely achieved by a stochastic algorithm, but simulation has the advantage of bypassing the preliminary steps of devising an algorithm and studying whether some regularity conditions on h hold. This is particularly true when the function h is very costly to compute.

Example 5.1. Signal processing. Ó Ruanaidh and Fitzgerald (1996) study signal processing data, of which a simple model is $(i = 1, \ldots, N)$

$$x_i = \alpha_1 \cos(\omega t_i) + \alpha_2 \sin(\omega t_i) + \epsilon_i, \qquad \epsilon_i \sim \mathcal{N}(0, \sigma^2),$$

with unknown parameters $\alpha = (\alpha_1, \alpha_2), \omega$, and σ and observation times $t_1, \ldots, t_N$. The likelihood function is then of the form

$$\sigma^{-N} \exp\left(-\frac{(\mathbf{x} - G\alpha^t)^t(\mathbf{x} - G\alpha^t)}{2\sigma^2}\right),$$

with $\mathbf{x} = (x_1, \ldots, x_N)$ and

$$G = \begin{pmatrix} \cos(\omega t_1) & \sin(\omega t_1) \\ \vdots & \vdots \\ \cos(\omega t_N) & \sin(\omega t_N) \end{pmatrix}.$$

The prior $\pi(\alpha, \omega, \sigma) = \sigma^{-1}$ leads to the marginal distribution

$$(5.2) \qquad \pi(\omega|\mathbf{x}) \propto \left(\mathbf{x}^t\mathbf{x} - \mathbf{x}^t G(G^t G)^{-1} G^t \mathbf{x}\right)^{(2-N)/2} (\det G^t G)^{-1/2},$$

which, although explicit in ω, is not particularly simple to compute. This setup is also illustrative of functions with many modes, as shown by Ó Ruanaidh and Fitzgerald (1996). ‖

Following Geyer (1996), we want to distinguish between two approaches to Monte Carlo optimization. The first is an *exploratory* approach, in which the goal is to optimize the function h by describing its entire range. The actual properties of the function play a lesser role here, with the Monte Carlo aspect more closely tied to the exploration of the entire space Θ, even though, for instance, the slope of h can be used to speed up the exploration. (Such a technique can be useful in describing functions with multiple modes, for example.) The second approach is based on a probabilistic *approximation* of the objective function h and is somewhat of a preliminary step to the actual optimization. Here, the Monte Carlo aspect exploits the probabilistic properties of the function h to come up with an acceptable approximation and is less concerned with exploring Θ. We will see that this approach can be tied to *missing data methods*, such as the EM algorithm. We note also that Geyer (1996) only considers the second approach to be "Monte Carlo optimization." Obviously, even though we are considering these two different approaches separately, they might be combined in a given problem. In fact, methods like the EM algorithm (Section 5.3.2) or the Robbins–Monro algorithm (Section 5.5.3) take advantage of the Monte Carlo approximation to enhance their particular optimization technique.

5.2 Stochastic Exploration

5.2.1 A Basic Solution

There are a number of cases where the exploration method is particularly well suited. First, if Θ is bounded, which may sometimes be achieved by a reparameterization, a first approach to the resolution of (5.1) is to simulate from a uniform distribution on Θ, $u_1, \ldots, u_m \sim \mathcal{U}_\Theta$, and to use the approximation $h_m^* = \max(h(u_1), \ldots, h(u_m))$. This method converges (as m goes to ∞), but it may be very slow since it does not take into account any specific feature of h. Distributions other than the uniform, which can possibly be related to h, may then do better. In particular, in setups where the likelihood function is extremely costly to compute, the number of evaluations of the function h is best kept to a minimum.

Example 5.2. A first Monte Carlo maximization. Recall the function that we looked at in Example 3.4, $h(x) = [\cos(50x) + \sin(20x)]^2$. Since the function is defined on a bounded interval, we try our naïve strategy and simulate $u_1, \ldots, u_m \sim \mathcal{U}(0,1)$, and use the approximation $h_m^* = \max(h(u_1), \ldots, h(u_m))$. The results are shown in Figure 5.1. There we see that the random search has done a fine job of mimicking the function. The Monte Carlo maximum is 3.832, which agrees perfectly with the "true" maximum, obtained by an exhaustive evaluation.

Of course, this is a small example, and as mentioned above, this naïve method can be costly in many situations. However, the example illustrates the fact that in low-dimensional problems, if function evaluation is rapid, this method is a reasonable choice. ‖

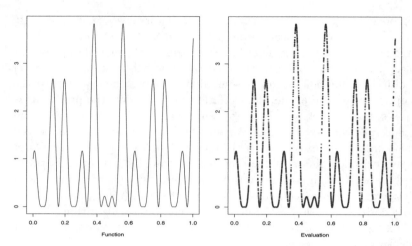

Fig. 5.1. Calculation of the maximum of the function (1.26), given in Example 3.4. The left panel is a graph of the function, and the right panel is a scatterplot of 5000 random $\mathcal{U}(0, 1)$ and the function evaluated at those points.

This leads to a second, and often more fruitful, direction, which relates h to a probability distribution. For instance, if h is positive and if

$$\int_{\Theta} h(\theta) \, d\theta < +\infty \, ,$$

the resolution of (5.1) amounts to finding the *modes* of the density h. More generally, if these conditions are not satisfied, then we may be able to transform the function $h(\theta)$ into another function $H(\theta)$ that satisfies the following:

(i) The function H is non-negative and satisfies $\int H < \infty$.
(ii) The solutions to (5.1) are those which maximize $H(\theta)$ on Θ.

For example, we can take

$$H(\theta) = \exp(h(\theta)/T) \quad \text{or} \quad H(\theta) = \exp\{h(\theta)/T\}/(1 + \exp\{h(\theta)/T\})$$

and choose T to accelerate convergence or to avoid local maxima (as in simulated annealing; see Section 5.2.3). When the problem is expressed in statistical terms, it becomes natural to then generate a sample $(\theta_1, \ldots, \theta_m)$ from h (or H) and to apply a standard mode estimation method (or to simply compare the $h(\theta_i)$'s). (In some cases, it may be more useful to decompose $h(\theta)$ into $h(\theta) = h_1(\theta)h_2(\theta)$ and to simulate from h_1.)

Example 5.3. Minimization of a complex function. Consider minimizing the (artificially constructed) function in $\mathbb{R}^2$

$$h(x, y) = (x \sin(20y) + y \sin(20x))^2 \cosh(\sin(10x)x)$$
$$+ (x \cos(10y) - y \sin(10x))^2 \cosh(\cos(20y)y) ,$$

whose global minimum is 0, attained at $(x, y) = (0, 0)$. (This is the big brother of the function in Example 5.2.) Since this function has many local minima, as shown by Figure 5.2, it does not satisfy the conditions under which standard minimization methods are guaranteed to provide the global minimum. On the other hand, the distribution on $\mathbb{R}^2$ with density proportional to $\exp(-h(x, y))$ can be simulated, even though this is not a standard distribution, by using, for instance, Markov chain Monte Carlo techniques (introduced in Chapters 7 and 8), and a convergent approximation of the minimum of $h(x, y)$ can be derived from the minimum of the resulting $h(x_i, y_i)$'s. An alternative is to simulate from the density proportional to

$$h_1(x, y) = \exp\{-(x \sin(20y) + y \sin(20x))^2 - (x \cos(10y) - y \sin(10x))^2\},$$

which eliminates the computation of both cosh and sinh in the simulation step.   ∥

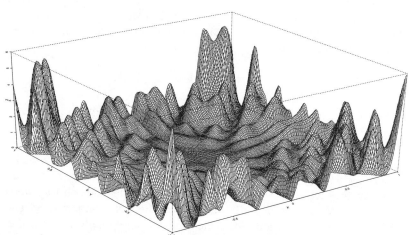

Fig. 5.2. Grid representation of the function $h(x, y)$ of Example 5.3 on $[-1, 1]^2$.

Exploration may be particularly difficult when the space Θ is not convex (or perhaps not even connected). In such cases, the simulation of a sample $(\theta_1, \ldots, \theta_m)$ may be much faster than a numerical method applied to (5.1). The appeal of simulation is even clearer in the case when h can be represented as

$$h(\theta) = \int H(x, \theta) dx .$$

In particular, if $H(x, \theta)$ is a density and if it is possible to simulate from this density, the solution of (5.1) is the mode of the marginal distribution of θ. (Although this setting may appear contrived or even artificial, we will see in Section 5.3.1 that it includes the case of missing data models.)

We now look at several methods to find maxima that can be classified as *exploratory methods*.

5.2.2 Gradient Methods

As mentioned in Section 1.4, the *gradient method* is a deterministic numerical approach to the problem (5.1). It produces a sequence (θ_j) that converges to the exact solution of (5.1), θ^*, when the domain $\Theta \subset \mathbb{R}^d$ and the function $(-h)$ are both convex. The sequence (θ_j) is constructed in a recursive manner through

$$(5.3) \qquad \theta_{j+1} = \theta_j + \alpha_j \nabla h(\theta_j) , \qquad \alpha_j > 0 ,$$

where ∇h is the gradient of h. For various choices of the sequence (α_j) (see Thisted 1988), the algorithm converges to the (unique) maximum.

In more general setups (that is, when the function or the space is less regular), equation (5.3) can be modified by stochastic perturbations to again achieve convergence, as described in detail in Rubinstein (1981) or Duflo (1996, pp. 61–63). One of these stochastic modifications is to choose a second sequence (β_j) to define the chain (θ_j) by

$$(5.4) \qquad \theta_{j+1} = \theta_j + \frac{\alpha_j}{2\beta_j} \, \Delta h(\theta_j, \beta_j \zeta_j) \, \zeta_j .$$

The variables ζ_j are uniformly distributed on the unit sphere $||\zeta|| = 1$ and $\Delta h(x, y) = h(x + y) - h(x - y)$ approximates $2||y||\nabla h(x)$. In contrast to the deterministic approach, this method does not necessarily proceed along the steepest slope in θ_j, but this property is sometimes a *plus* in the sense that it may avoid being trapped in local maxima or in saddlepoints of h.

The convergence of (θ_j) to the solution θ^* again depends on the choice of (α_j) and (β_j). We note in passing that (θ_j) can be seen as a *nonhomogeneous Markov chain* (see Definition 6.4) which almost surely converges to a given value. The study of these chains is quite complicated given their ever-changing transition kernel (see Winkler 1995 for some results in this direction). However, sufficiently strong conditions such as the decrease of α_j toward 0 and of α_j/β_j to a nonzero constant are enough to guarantee the convergence of the sequence (θ_j).

Example 5.4. (Continuation of Example 5.3) We can apply the iterative construction (5.4) to the multimodal function $h(x, y)$ with different sequences of α_j's and β_j's. Figure 5.3 and Table 5.1 illustrate that, depending on the starting value, the algorithm converges to different local minima of the function h. Although there are occurrences when the sequence $h(\theta_j)$ increases and avoids some local minima, the solutions are quite distinct for the three different sequences, both in location and values.

As shown by Table 5.1, the number of iterations needed to achieve stability of θ_T also varies with the choice of (α_j, β_j). Note that Case 1 results in a very poor evaluation of the minimum, as the fast decrease of (α_j) is associated with big jumps in the first iterations. Case 2 converges to the closest local minima, and Case 3 illustrates a general feature of the stochastic gradient method, namely that slower decrease rates of the sequence (α_j) tend to achieve better minima. The final convergence along a valley of h after some initial big jumps is also noteworthy.   ‖

α_j	β_j	θ_T	$h(\theta_T)$	$\min_t h(\theta_t)$	Iteration T
$1/10j$	$1/10j$	$(-0.166, 1.02)$	1.287	0.115	50
$1/100j$	$1/100j$	$(0.629, 0.786)$	0.00013	0.00013	93
$1/10\log(1+j)$	$1/j$	$(0.0004, 0.245)$	4.24×10^{-6}	2.163×10^{-7}	58

Table 5.1. Results of three stochastic gradient runs for the minimization of the function h in Example 5.3 with different values of (α_j, β_j) and starting point $(0.65, 0.8)$. The iteration T is obtained by the stopping rule $\|\theta_T - \theta_{T-1}\| < 10^{-5}$.

This approach is still quite close to numerical methods in that it requires a precise knowledge on the function h, which may not necessarily be available.

5.2.3 Simulated Annealing

The simulated annealing algorithm[2] was introduced by Metropolis et al. (1953) to minimize a criterion function on a finite set with very large size[3], but it also applies to optimization on a continuous set and to simulation (see Kirkpatrick et al. 1983, Ackley et al. 1985, and Neal 1993, 1995).

The fundamental idea of simulated annealing methods is that a change of scale, called *temperature*, allows for faster moves on the surface of the function h to maximize, whose negative is called *energy*. Therefore, rescaling partially avoids the trapping attraction of local maxima. Given a temperature parameter $T > 0$, a sample $\theta_1^T, \theta_2^T, \ldots$ is generated from the distribution

---

[2] This name is borrowed from wetallurgy: a metal manufactured by a slow decrease of temperature (*annealing*) is stronger than a metal manufactured by a fast decrease of temperature. There is also input from physics, as the function to be minimized is called *energy* and the variance factor T, which controls convergence, is called *temperature*. We will try to keep these idiosyncrasies to a minimal level, but they are quite common in the literature.

[3] This paper is also the originator of the *Markov chain Monte Carlo methods* developed in the following chapters.

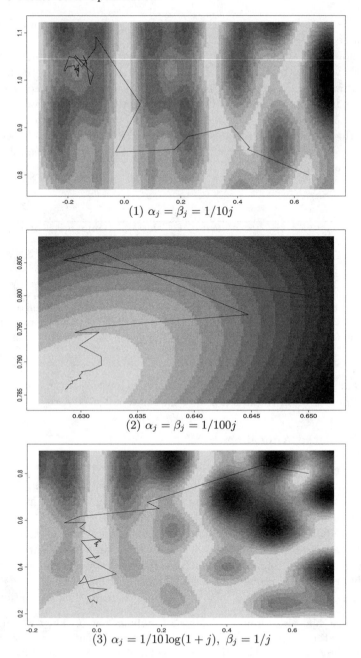

Fig. 5.3. Stochastic gradient paths for three different choices of the sequences (α_j) and (β_j) and starting point $(0.65, 0.8)$ for the same sequence (ζ_j) in (5.4). The gray levels are such that darker shades mean higher elevations. The function h to be minimized is defined in Example 5.3.

$$\pi(\theta) \propto \exp(h(\theta)/T)$$

and can be used as in Section 5.2.1 to come up with an approximate maximum of h. As T decreases toward 0, the values simulated from this distribution become concentrated in a narrower and narrower neighborhood of the local maxima of h (see Theorem 5.10 and Winkler 1995).

The fact that this approach has a moderating effect on the attraction of the local maxima of h becomes more apparent when we consider the simulation method proposed by Metropolis et al. (1953). Starting from θ_0, ζ is generated from a uniform distribution on a neighborhood $\mathcal{V}(\theta_0)$ of θ_0 or, more generally, from a distribution with density $g(|\zeta-\theta_0|)$, and the new value of θ is generated as follows:

$$\theta_1 = \begin{cases} \zeta & \text{with probability } \rho = \exp(\Delta h/T) \wedge 1 \\ \theta_0 & \text{with probability } 1 - \rho, \end{cases}$$

where $\Delta h = h(\zeta) - h(\theta_0)$. Therefore, if $h(\zeta) \geq h(\theta_0)$, ζ is accepted with probability 1; that is, θ_0 is always changed into ζ. On the other hand, if $h(\zeta) < h(\theta_0)$, ζ may still be accepted with probability $\rho \neq 0$ and θ_0 is then changed into ζ. This property allows the algorithm to escape the attraction of θ_0 if θ_0 is a local maximum of h, with a probability which depends on the choice of the scale T, compared with the range of the density g. (This method is in fact the *Metropolis algorithm*, which simulates the density proportional to $\exp\{h(\theta)/T\}$, as the limiting distribution of the chain $\theta_0, \theta_1, \ldots$, as described and justified in Chapter 7.)

In its most usual implementation, the simulated annealing algorithm modifies the temperature T at each iteration; it is then of the form

Algorithm A.19 –Simulated Annealing–

1. Simulate ζ from an instrumental distribution
   with density $g(|\zeta - \theta_i|)$;
2. Accept $\theta_{i+1} = \zeta$ with probability
   $\rho_i = \exp\{\Delta h_i/T_i\} \wedge 1$;   [A.19]
   take $\theta_{i+1} = \theta_i$ otherwise.
3. Update T_i to T_{i+1}.

Example 5.5. A first simulated annealing maximization. We again look at the function from Example 5.2,

$$h(x) = [\cos(50x) + \sin(20x)]^2,$$

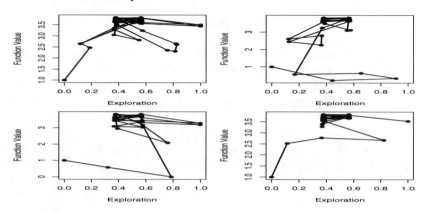

Fig. 5.4. Calculation of the maximum of the function (1.26), given in Example 5.5. The four different panels show the trajectory of 2500 pairs $(x^{(t)}, h(x^{(t)}))$ for each of four different runs.

and apply a simulated annealing algorithm to find the maximum. The specific algorithm we use is

> At iteration t the algorithm is at $(x^{(t)}, h(x^{(t)}))$:
>
> 1. Simulate $u \sim \mathcal{U}(a_t, b_t)$ where $a_t = \max(x^{(t)} - r, 0)$ and $b_t = \min(x^{(t)} + r, 1)$
> 2. Accept $x^{(t+1)} = u$ with probability
>
> $$\rho^{(t)} = \min\left\{ \exp\left(\frac{h(u) - h(x^{(t)})}{T_t}\right), 1 \right\},$$
>
> take $x^{(t+1)} = x^{(t)}$ otherwise.
> 3. Update T_t to T_{t+1}.

For $r = .5$ and $T_t = 1/\log(t)$, the results of the algorithm are shown in Figure 5.4. The four panels show different trajectories of the points $(x^{(t)}, h(x^{(t)}))$. It is interesting to see how the path moves toward the maximum fairly rapidly, and then remains there, oscillating between the two maxima (remember that h is symmetric around $1/2$.

The value of r controls the size of the interval around the current point (we truncate to stay in $(0, 1)$ and the value of T_t controls the cooling. For different values of r and T the path will display different properties. See Problem 5.4 and the more complex Example 5.9.   ‖

An important feature of the simulated annealing algorithm is that there exist convergence results in the case of finite spaces, as Theorem 5.7 below, which was proposed by Hàjek (1988). (See Winkler 1995, for extensions.)

Consider the following notions, which are used to impose restrictions on the decrease rate of the temperature:

Definition 5.6. Given a finite state-space $\mathcal{E}$ and a function h to be maximized:

(i) a state $e_j \in \mathcal{E}$ can be *reached at altitude* $\underline{h}$ from state $e_i \in \mathcal{E}$ if there exists a sequence of states $e_1, \ldots, e_n$ linking e_i and e_j, such that $h(e_k) \geq \underline{h}$ for $k = 1, \ldots, n$;

(ii) the *height of a maximum* e_i is the largest value d_i such that there exists a state e_j such that $h(e_j) > h(e_i)$ which can be reached at altitude $h(e_i) + d_i$ from e_i.

Thus, $h(e_i) + d_i$ is the altitude of the highest pass linking e_i and e_j through an optimal sequence. (In particular, $h(e_i) + d_i$ can be larger than the altitude of the closest pass relating e_i and e_j.) By convention, we take $d_i = -\infty$ if e_i is a global maximum. If $\mathcal{O}$ denotes the set of local maxima of E and $\underline{\mathcal{O}}$ is the subset of $\mathcal{O}$ of global maxima, Hàjek (1988) establishes the following result:

Theorem 5.7. *Consider a system in which it is possible to link two arbitrary states by a finite sequence of states. If, for every $\underline{h} > 0$ and every pair (e_i, e_j), e_i can be reached at altitude $\underline{h}$ from e_j if and only if e_j can be reached at altitude $\underline{h}$ from e_i, and if (T_i) decreases toward 0, the sequence (θ_i) defined by Algorithm $[A.19]$ satisfies*

$$\lim_{i \to \infty} P(\theta_i \in \underline{\mathcal{O}}) = 1$$

if and only if

$$\sum_{i=1}^{\infty} \exp(-D/T_i) = +\infty ,$$

with $D = \min\{d_i : e_i \in \mathcal{O} - \underline{\mathcal{O}}\}$.

This theorem therefore gives a necessary and sufficient condition, on the rate of decrease of the temperature, so that the simulated annealing algorithm converges to the set of global maxima. This remains a relatively formal result since D is, in practice, unknown. For example, if $T_i = \Gamma/\log i$, there is convergence to a global maximum if and only if $\Gamma \geq D$. Numerous papers and books have considered the practical determination of the sequence (T_n) (see Geman and Geman 1984, Mitra et al. 1986, Van Laarhoven and Aarts 1987, Aarts and Kors 1989, Winkler 1995, and references therein). Instead of the above logarithmic rate, a geometric rate, $T_i = \alpha^i T_0$ $(0 < \alpha < 1)$, is also often adopted in practice, with the constant α calibrated at the beginning of the algorithm so that the acceptance rate is high enough in the Metropolis algorithm (see Section 7.6).

The fact that approximate methods are necessary for optimization problems in finite state-spaces may sound rather artificial and unnecessary, but the spaces involved in some modeling can be huge. For instance, a black-and-white TV image with 256×256 pixels corresponds to a state-space with cardinality $2^{256 \times 256} \simeq 10^{20,000}$. Similarly, the analysis of DNA sequences may involve 600 thousand bases (A, C, G, or T), which corresponds to state-spaces of size $4^{600,000}$ (see Churchill 1989, 1995).

Example 5.8. Ising model. The *Ising model* can be applied in electromagnetism (Cipra 1987) and in image processing (Geman and Geman 1984). It models two-dimensional tables s, of size $D \times D$, where each term of s takes the value $+1$ or -1. The distribution of the entire table is related to the (so-called energy) function

$$(5.5) \qquad h(s) = -J \sum_{(i,j) \in \mathcal{N}} s_i s_j - H \sum_i s_i \,,$$

where i denotes the index of a term of the table and $\mathcal{N}$ is an equivalence neighborhood relation, for instance, when i and j are neighbors either vertically or horizontally. (The scale factors J and H are supposedly known.) The model (5.5) is a particular case of models used in spatial statistics (Cressie 1993) to describe multidimensional correlated structures.

Note that the conditional representation of (5.5) is equivalent to a *logit* model on $\tilde{s}_i = (s_i + 1)/2$,

$$(5.6) \qquad P(\tilde{S}_i = 1 | s_j, j \neq i) = \frac{e^g}{1 + e^g} \,,$$

with $g = g(s_j) = 2(H + J \sum_j s_j)$, the sum being taken on the neighbors of i. For known parameters H and J, the inferential question may be to obtain the most likely configuration of the system; that is, the minimum of $h(s)$. The implementation of the Metropolis et al. (1953) approach in this setup, starting from an initial value $s^{(0)}$, is to modify the sites of the table s one at a time using the conditional distributions (5.6), with probability $\exp(-\Delta h / T)$, ending up with a modified table $s^{(1)}$, and to iterate this method by decreasing the temperature T at each step. The reader can consult Swendson and Wang (1987) and Swendson et al. (1992) for their derivation of efficient simulation algorithms in these models and accelerating methods for the Gibbs sampler (see Problem 7.43). ‖

Duflo (1996, pp. 264–271) also proposed an extension of these simulated annealing methods to the general (continuous) case. Andrieu and Doucet (2000) give a detailed proof of convergence of the simulated annealing algorithm, as well as sufficient conditions on the cooling schedule, in the setup of hidden Markov models (see Section 14.6.3). Their proof, which is beyond our scope, is based on the developments of Haario and Sacksman (1991).

Case	T_i	θ_T	$h(\theta_T)$	$\min_t h(\theta_t)$	Accept. rate
1	$1/10i$	$(-1.94, -0.480)$	0.198	$4.02\,10^{-7}$	0.9998
2	$1/\log(1+i)$	$(-1.99, -0.133)$	3.408	3.823×10^{-7}	0.96
3	$100/\log(1+i)$	$(-0.575, 0.430)$	0.0017	4.708×10^{-9}	0.6888
4	$1/10\log(1+i)$	$(0.121, -0.150)$	0.0359	2.382×10^{-7}	0.71

Table 5.2. Results of simulated annealing runs for different values of T_i and starting point $(0.5, 0.4)$.

Example 5.9. (Continuation of Example 5.3) We can apply the algorithm $[A.19]$ to find a local minimum of the function h of Example 5.3, or equivalently a maximum of the function $\exp(-h(x,y)/T_i)$. We choose a uniform distribution on $[-0.1, 0.1]$ for g, and different rates of decrease of the temperature sequence (T_i). As illustrated by Figure 5.5 and Table 5.2, the results change with the rate of decrease of the temperature T_i. Case 3 leads to a very interesting exploration of the valleys of h on both sides of the central zone. Since the theory (Duflo 1996) states that rates of the form $\Gamma/\log(i+1)$ are satisfactory for Γ large enough, this shows that $\Gamma = 100$ should be acceptable. Note also the behavior of the acceptance rate in Table 5.2 for Step 2 in algorithm $[A.19]$. This is indicative of a rule we will discuss further in Chapter 7 with Metropolis–Hastings algorithms, namely that superior performances are not always associated with higher acceptance rates.   ‖

5.2.4 Prior Feedback

Another approach to the maximization problem (5.1) is based on the result of Hwang (1980) of convergence (in T) of the so-called *Gibbs measure* $\exp(h(\theta)/T)$ (see Section 5.5.3) to the uniform distribution on the set of global maxima of h. This approach, called *recursive integration* or *prior feedback* in Robert (1993) (see also Robert and Soubiran 1993), is based on the following convergence result.

Theorem 5.10. *Consider h a real-valued function defined on a closed and bounded set, Θ, of $\mathbb{R}^p$. If there exists a unique solution θ^* satisfying*

$$\theta^* = \arg\max_{\theta \in \Theta} h(\theta),$$

then

$$\lim_{\lambda \to \infty} \frac{\int_\Theta \theta\, e^{\lambda h(\theta)}\, d\theta}{\int_\Theta e^{\lambda h(\theta)}\, d\theta} = \theta^*,$$

provided h is continuous at θ^*.

See Problem 5.6 for a proof. More details can be found in Pincus (1968) (see also Robert 1993 for the case of exponential families and Duflo 1996, pp.

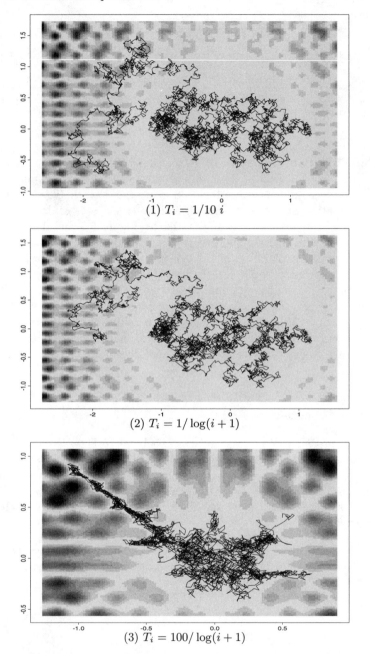

$$(1)\ T_i = 1/10\,i$$

$$(2)\ T_i = 1/\log(i+1)$$

$$(3)\ T_i = 100/\log(i+1)$$

Fig. 5.5. Simulated annealing sequence of 5000 points for three different choices of the temperature T_i in [A.19] and starting point $(0.5, 0.4)$, aimed at minimizing the function h of Example 5.3.

244–245, for a sketch of a proof). A related result can be found in D'Epifanio (1989, 1996). A direct corollary to Theorem 5.10 then justifies the recursive integration method which results in a Bayesian approach to maximizing the log-likelihood, $\ell(\theta|x)$.

Corollary 5.11. *Let π be a positive density on Θ. If there exists a unique maximum likelihood estimator θ^*, it satisfies*

$$\lim_{\lambda \to \infty} \frac{\int \theta e^{\lambda \ell(\theta|x)} \pi(\theta)d\theta}{\int e^{\lambda \ell(\theta|x)} \pi(\theta)d\theta} = \theta^* \ .$$

This result uses the same technique as in Theorem 5.10, namely the Laplace approximation of the numerator and denominator integrals (see also Tierney et al. 1989). It mainly expresses the fact that the maximum likelihood estimator can be written as a limit of Bayes estimators associated with an *arbitrary* distribution π and with *virtual* observations corresponding to the λth power of the likelihood, $\exp\{\lambda \ell(\theta|x)\}$. When $\lambda \in \mathbb{N}$,

$$\delta_\lambda^\pi(x) = \frac{\int \theta e^{\lambda \ell(\theta|x)} \pi(\theta)d\theta}{\int e^{\lambda \ell(\theta|x)} \pi(\theta)d\theta}$$

is simply the Bayes estimator associated with the prior distribution π and a corresponding sample which consists of λ replications of the initial sample x. The intuition behind these results is that as the size of the sample goes to infinity, the influence of the prior distribution vanishes and the distribution associated with $\exp(\lambda \ell(\theta|x))\pi(\theta)$ gets more and more concentrated around the global maxima of $\ell(\theta|x)$ when λ increases (see, e.g., Schervish 1995).

From a practical point of view, the recursive integration method can be implemented by computing the Bayes estimators $\delta_{\lambda_i}^\pi(x)$ for $i = 1, 2, \ldots$ until they stabilize.

Obviously, it is only interesting to maximize the likelihood by this method when more standard methods like the ones above are difficult or impossible to implement and the computation of Bayes estimators is straightforward. (Chapters 7 and 9 show that this second condition is actually very mild.) It is, indeed, necessary to compute the Bayes estimators $\delta_\lambda^\pi(x)$ corresponding to a sequence of λ's until they stabilize. Note that when iterative algorithms are used to compute $\delta_\lambda^\pi(x)$, the previous solution (in λ) of $\delta_\lambda^\pi(x)$ can serve as the new initial value for the computation of $\delta_\lambda^\pi(x)$ for a larger value of λ. This feature increases the analogy with simulated annealing. The differences with simulated annealing are:

(i) for a fixed temperature $(1/\lambda)$, the algorithm converges to a fixed value, δ_λ^π;

(ii) a continuous decrease of $1/\lambda$ is statistically meaningless;

λ	5	10	100	1000	5000	10^4
δ_λ^π	2.02	2.04	1.89	1.98	1.94	2.00

Table 5.3. Sequence of Bayes estimators of δ_λ^π for the estimation of α when $X \sim \mathcal{G}(\alpha, 1)$ and $x = 1.5$.

(iii) the speed of convergence of λ to $+\infty$ does not formally matter[4] for the convergence of $\delta_\lambda^\pi(x)$ to θ^*;

(iv) the statistical motivation of this method is obviously stronger, in particular because of the meaning of the parameter λ;

(v) the only analytical constraint on $\ell(\theta|x)$ is the existence of a global maximum, θ^* (see Robert and Titterington 1998 for extensions).

Example 5.12. Gamma shape estimation. Consider the estimation of the shape parameter, α, of a $\mathcal{G}(\alpha, \beta)$ distribution with β known. Without loss of generality, take $\beta = 1$. For a constant (improper) prior distribution on α, the posterior distribution satisfies

$$\pi_\lambda(\alpha|x) \propto x^{\lambda(\alpha-1)} e^{-\lambda x} \Gamma(\alpha)^{-\lambda} .$$

For a fixed λ, the computation of $\mathbb{E}[\alpha|x, \lambda]$ can be obtained by simulation with the Metropolis–Hastings algorithm (see Chapter 7 for details) based on the instrumental distribution $\mathcal{E}xp(1/\alpha^{(n-1)})$, where $\alpha^{(n-1)}$ denotes the previous value of the associated Markov chain. Table 5.3 presents the evolution of $\delta_\lambda^\pi(x) = \mathbb{E}[\alpha|x, \lambda]$ against λ, for $x = 1.5$.

An analytical verification (using a numerical package like Mathematica) shows that the maximum of $x^\alpha/\Gamma(\alpha)$ is, indeed, close to 2.0 for $x = 1.5$.   ‖

The appeal of recursive integration is also clear in the case of constrained parameter estimation.

Example 5.13. Isotonic regression. Consider a table of normal observations $X_{i,j} \sim \mathcal{N}(\theta_{i,j}, 1)$ with means that satisfy

$$\theta_{i-1,j} \vee \theta_{i,j-1} \leq \theta_{i,j} \leq \theta_{i+1,j} \wedge \theta_{i,j+1}.$$

Dykstra and Robertson (1982) have developed an efficient *deterministic* algorithm which maximizes the likelihood under these restrictions (see Problems 1.18 and 1.19.) However, a direct application of recursive integration also provides the maximum of likelihood estimator of $\theta = (\theta_{ij})$, requiring neither an extensive theoretical study nor high programming skills.

---

[4] However, we note that if λ increases too quickly, the performance is affected in that there may be convergence to a local mode (see Robert and Titterington 1998 for an illustration).

Table 5.4 presents the data of Robertson et al. (1988), which relates the notes at the end of first year with two entrance exams at the University of Iowa. Although these values are bounded, it is possible to use a normal model if the function to minimize is the least squares criterion (1.8), as already pointed out in Example 1.5. Table 5.5 provides the solution obtained by recursive integration in Robert and Hwang (1996); it coincides with the result of Robertson et al. (1988).   ‖

ACT	1–12	13–15	16–18	19–21	22–24	25–27	28–30	31–33	34–36
91- 99	1.57 (4)	2.11 (5)	2.73 (18)	2.96 (39)	2.97 (126)	3.13 (219)	3.41 (232)	3.45 (47)	3.51 (4)
81- 90	1.80 (6)	1.94 (15)	2.52 (30)	2.68 (65)	2.69 (117)	2.82 (143)	2.75 (70)	2.74 (8)	– (0)
71- 80	1.88 (10)	2.32 (13)	2.32 (51)	2.53 (83)	2.58 (115)	2.55 (107)	2.72 (24)	2.76 (4)	– (0)
61- 70	2.11 (6)	2.23 (32)	2.29 (59)	2.29 (84)	2.50 (75)	2.42 (44)	2.41 (19)	– (0)	– (0)
51- 60	1.60 (11)	2.06 (16)	2.12 (49)	2.11 (63)	2.31 (57)	2.10 (40)	1.58 (4)	2.13 (1)	– (0)
41- 50	1.75 (6)	1.98 (12)	2.05 (31)	2.16 (42)	2.35 (34)	2.48 (21)	1.36 (4)	– (0)	– (0)
31- 40	1.92 (7)	1.84 (6)	2.15 (5)	1.95 (27)	2.02 (13)	2.10 (13)	1.49 (2)	– (0)	– (0)
21- 30	1.62 (1)	2.26 (2)	1.91 (5)	1.86 (14)	1.88 (11)	3.78 (1)	1.40 (2)	– (0)	– (0)
00- 20	1.38 (1)	1.57 (2)	2.49 (5)	2.01 (7)	2.07 (7)	– (0)	0.75 (1)	– (0)	– (0)

Table 5.4. Average grades of first-year students at the University of Iowa given their rank at the end of high school (HSR) and at the ACT exam. Numbers in parentheses indicate the number of students in each category. (*Source:* Robertson et al. 1988.)

ACT	1 − 12	13 − 15	16 − 18	19 − 21	22 − 24	25 − 27	28 − 39	31 − 32	34 − 36
91 − 99	1.87	2.18	2.73	2.96	2.97	3.13	3.41	3.45	3.51
81 − 89	1.87	2.17	2.52	2.68	2.69	2.79	2.79	2.80	−
71 − 79	1.86	2.17	2.32	2.53	2.56	2.57	2.72	2.76	−
61 − 69	1.86	2.17	2.29	2.29	2.46	2.46	2.47	−	−
51 − 59	1.74	2.06	2.12	2.13	2.24	2.24	2.24	2.27	−
41 − 49	1.74	1.98	2.05	2.13	2.24	2.24	2.24	−	−
31 − 39	1.74	1.94	1.99	1.99	2.02	2.06	2.06	−	−
21 − 29	1.62	1.93	1.97	1.97	1.98	2.05	2.06	−	−
00 − 20	1.38	1.57	1.97	1.97	1.97	−	1.97	−	−

Table 5.5. Maximum likelihood estimates of the mean grades under lexicographical constraint. (*Source:* Robert and Hwang 1996).

5.3 Stochastic Approximation

We next turn to methods that work more directly with the objective function rather than being concerned with fast explorations of the space. Informally speaking, these methods are somewhat preliminary to the true optimization step, in the sense that they utilize approximations of the objective function h. We note that these approximations have a different purpose than those we have previously encountered (for example, Laplace and saddlepoint approximations in Section 3.4 and Section 3.6.2). In particular, the methods described here may sometimes result in an additional level of error by looking at the maximum of an approximation to h.

Since most of these approximation methods only work in so-called *missing data models*, we start this section with a brief introduction to these models. We return to the assumption that the objective function h satisfies $h(x) = \mathbb{E}[H(x, Z)]$ and (as promised) show that this assumption arises in many realistic setups. Moreover, note that artificial extensions (or *demarginalization*), which use this representation, are only computational devices and do not invalidate the overall inference.

5.3.1 Missing Data Models and Demarginalization

In the previous chapters, we have already met structures where some missing (or latent) element greatly complicates the observed model. Examples include the obvious censored data models (Example 1.1), mixture models (Example 1.2), where we do not observe the indicator of the component generating the observation, or logistic regression (Example 1.13), where the observation Y_i can be interpreted as an indicator that a continuous variable with logistic distribution is less than $X_i^t \beta$.

Missing data models are best thought of as models where the likelihood can be expressed as

$$(5.7) \qquad g(x|\theta) = \int_{\mathcal{Z}} f(x, z|\theta) \, dz$$

or, more generally, where the function $h(x)$ to be optimized can be expressed as the expectation

$$(5.8) \qquad h(x) = \mathbb{E}[H(x, Z)] \,.$$

This assumption is relevant, and useful, in the setup of censoring models:

Example 5.14. Censored data likelihood. Suppose that we observe Y_1, ..., Y_n, iid, from $f(y - \theta)$ and we have ordered the observations so that $\mathbf{y} = (y_1, \cdots, y_m)$ are uncensored and $(y_{m+1}, \ldots, y_n)$ are censored (and equal to a). The likelihood function is

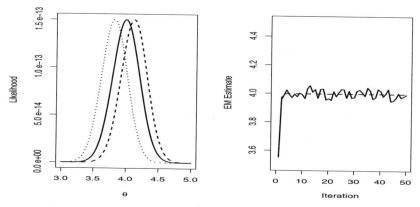

Fig. 5.6. The left panel shows three "likelihoods" of a sample of size 25 from a $\mathcal{N}(4, 1)$. The leftmost is the likelihood of the sample where values greater than 4.5 are replaced by the value 4.5 (dotted), the center (solid) is the observed-data likelihood (5.9), and the rightmost (dashed) is the likelihood using the actual data. The right panel shows EM (dashed) and MCEM (solid) estimates; see Examples 5.17 and 5.20.

$$(5.9) \qquad L(\theta|\mathbf{y}) = [1 - F(a - \theta)]^{n-m} \prod_{i=1}^{m} f(y_i - \theta) \, ,$$

where F is the cdf associated with f. If we had observed the last $n - m$ values, say $\mathbf{z} = (z_{m+1}, \ldots, z_n)$, with $z_i > a$ $(i = m + 1, \ldots, n)$, we could have constructed the (complete data) likelihood

$$L^c(\theta|\mathbf{y}, \mathbf{z}) = \prod_{i=1}^{m} f(y_i - \theta) \prod_{i=m+1}^{n} f(z_i - \theta),$$

with which it often is easier to work. Note that

$$L(\theta|\mathbf{y}) = \mathbb{E}[L^c(\theta|\mathbf{y}, \mathbf{Z})] = \int_{\mathcal{Z}} L^c(\theta|\mathbf{y}, \mathbf{z}) f(\mathbf{z}|\mathbf{y}, \theta) \, d\mathbf{z},$$

where $f(\mathbf{z}|\mathbf{y}, \theta)$ is the density of the missing data conditional on the observed data. For $f(y - \theta) = \mathcal{N}(\theta, 1)$ three likelihoods are shown in Figure 5.6. Note how the observed-data likelihood is biased down from the true value of θ. ‖

When (5.7) holds, the Z vector merely serves to simplify calculations, and the way Z is selected to satisfy (5.8) should not affect the value of the estimator. This is a *missing data model*, and we refer to the function $L^c(\theta|x, z)) = f(x, z|\theta)$ as the "complete-model" or "complete-data" likelihood, which corresponds to the observation of the *complete data* (x, z). This complete model is often within the exponential family framework, making it much easier to work with (see Problem 5.14).

More generally, we refer to the representation (5.7) as *demarginalization*, a setting where a function (or a density) of interest can be expressed as an integral of a more manageable quantity. We will meet such setups again in Chapters 8–9. They cover models such as missing data models (censoring, grouping, mixing, etc.), latent variable models (tobit, probit, arch, stochastic volatility, etc.) and also artificial embedding, where the variable Z in (5.8) has no meaning for the inferential or optimization problem, as illustrated by *slice sampling* (Chapter 8).

5.3.2 The EM Algorithm

The EM (*Expectation–Maximization*) algorithm was originally introduced by Dempster et al. (1977) to overcome the difficulties in maximizing likelihoods by taking advantage of the representation (5.7) and solving a sequence of easier maximization problems whose limit is the answer to the original problem. It thus fits naturally in this demarginalization section, even though it is not a stochastic algorithm in its original version. Monte Carlo versions are examined in Section 5.3.3 and in Note 5.5.1. Moreover, the EM algorithm relates to MCMC algorithms in the sense that it can be seen as a forerunner of the Gibbs sampler in its Data Augmentation version (Section 10.1.2), replacing simulation by maximization.

Suppose that we observe $X_1, \ldots, X_n$, iid from $g(x|\theta)$ and want to compute $\hat{\theta} = \arg\max L(\theta|\mathbf{x}) = \prod_{i=1}^{n} g(x_i|\theta)$. We augment the data with $\mathbf{z}$, where $\mathbf{X}, \mathbf{Z} \sim f(\mathbf{x}, \mathbf{z}|\theta)$ and note the identity (which is a basic identity for the EM algorithm)

$$(5.10) \qquad k(\mathbf{z}|\theta, \mathbf{x}) = \frac{f(\mathbf{x}, \mathbf{z}|\theta)}{g(\mathbf{x}|\theta)},$$

where $k(\mathbf{z}|\theta, \mathbf{x})$ is the conditional distribution of the missing data $\mathbf{Z}$ given the observed data $\mathbf{x}$. The identity (5.10) leads to the following relationship between the complete-data likelihood $L^c(\theta|\mathbf{x}, \mathbf{z})$ and the observed-data likelihood $L(\theta|\mathbf{x})$. For any value θ_0,

$$(5.11) \qquad \log L(\theta|\mathbf{x}) = \mathbb{E}_{\theta_0}[\log L^c(\theta|\mathbf{x}, \mathbf{z})] - \mathbb{E}_{\theta_0}[\log k(\mathbf{z}|\theta, \mathbf{x})],$$

where the expectation is with respect to $k(\mathbf{z}|\theta_0, \mathbf{x})$. We now see the EM algorithm as a demarginalization model. However, the strength of the EM algorithm is that it can go further. In particular, to maximize $\log L(\theta|\mathbf{x})$, we only have to deal with the first term on the right side of (5.11), as the other term can be ignored.

Common EM notation is to denote the expected log-likelihood by

$$(5.12) \qquad Q(\theta|\theta_0, \mathbf{x}) = \mathbb{E}_{\theta_0}[\log L^c(\theta|\mathbf{x}, \mathbf{z})].$$

We then maximize $Q(\theta|\theta_0, \mathbf{x})$, and if $\hat{\theta}_{(1)}$ is the value of θ maximizing $Q(\theta|\theta_0, \mathbf{x})$, the process can then be repeated with θ_0 replaced by the updated value $\hat{\theta}_{(1)}$. In this manner, a sequence of estimators $\hat{\theta}_{(j)}$, $j = 1, 2, \ldots$, is obtained iteratively where $\hat{\theta}_{(j)}$ is defined as the value of θ maximizing $Q(\theta|\hat{\theta}_{(j-1)}, \mathbf{x})$; that is,

$$(5.13) \qquad Q(\hat{\theta}_{(j)}|\hat{\theta}_{(j-1)}, \mathbf{x}) = \max_\theta Q(\theta|\hat{\theta}_{(j-1)}, \mathbf{x}).$$

The iteration described above contains both an expectation step and a maximization step, giving the algorithm its name. At the jth step of the iteration, we calculate the expectation (5.12), with θ_0 replaced by $\hat{\theta}_{(j-1)}$ (the *E-step*), and then maximize it (*the M-step*).

Algorithm A.20 –The EM Algorithm–

1. Compute

$$Q(\theta|\hat{\theta}_{(m)}, \mathbf{x}) = \mathbb{E}_{\hat{\theta}_{(m)}}[\log L^c(\theta|\mathbf{x}, \mathbf{z})],$$

where the expectation is with respect to $k(\mathbf{z}|\hat{\theta}_m, \mathbf{x})$ (*the E-step*) .

2. Maximize $Q(\theta|\hat{\theta}_{(m)}, \mathbf{x})$ in θ and take (*the M-step*) [A.20]

$$\theta_{(m+1)} = \arg\max_\theta Q(\theta|\hat{\theta}_{(m)}, \mathbf{x}).$$

The iterations are conducted until a fixed point of Q is obtained.

The theoretical core of the EM Algorithm is based on the fact that by maximizing $Q(\theta|\hat{\theta}_{(m)}, \mathbf{x})$ at each step, the likelihood on the left side of (5.11) is increased at each step. The following theorem was established by Dempster et al. (1977).

Theorem 5.15. *The sequence $(\hat{\theta}_{(j)})$ defined by (5.13) satisfies*

$$L(\hat{\theta}_{(j+1)}|\mathbf{x}) \geq L(\hat{\theta}_{(j)}|\mathbf{x}),$$

with equality holding if and only if $Q(\hat{\theta}_{(j+1)}|\hat{\theta}_{(j)}, \mathbf{x}) = Q(\hat{\theta}_{(j)}|\hat{\theta}_{(j)}, \mathbf{x})$.

Proof. On successive iterations, it follows from the definition of $\hat{\theta}_{(j+1)}$ that

$$Q(\hat{\theta}_{(j+1)}|\hat{\theta}_{(j)}, \mathbf{x}) \geq Q(\hat{\theta}_{(j)}|\hat{\theta}_{(j)}, \mathbf{x}).$$

Thus, if we can show that

$$(5.14) \qquad \mathbb{E}_{\hat{\theta}_{(j)}}[\log k(\mathbf{Z}|\hat{\theta}_{(j+1)}, \mathbf{x})] \leq \mathbb{E}_{\hat{\theta}_{(j)}}[\log k(\mathbf{Z}|\hat{\theta}_{(j)}, \mathbf{x})],$$

it will follow from (5.11) that the value of the likelihood is increased at each iteration.

Since the difference of the logarithms is the logarithm of the ratio, (5.14) can be written as

$$(5.15) \qquad \mathbb{E}_{\hat{\theta}_{(j)}} \left[\log \left(\frac{k(\mathbf{Z}|\hat{\theta}_{(j+1)}, \mathbf{x})}{k(\mathbf{Z}|\hat{\theta}_{(j)}, \mathbf{x})} \right) \right] \leq \log \mathbb{E}_{\hat{\theta}_{(j)}} \left[\frac{k(\mathbf{Z}|\hat{\theta}_{(j+1)}, \mathbf{x})}{k(\mathbf{Z}|\hat{\theta}_{(j)}, \mathbf{x})} \right] = 0 \,,$$

where the inequality follows from Jensen's inequality (see Problem 5.15). The theorem is therefore established.   □

Although Theorem 5.15 guarantees that the likelihood will increase at each iteration, we still may not be able to conclude that the sequence $(\hat{\theta}_{(j)})$ converges to a maximum likelihood estimator. To ensure convergence we require further conditions on the mapping $\hat{\theta}_{(j)} \to \hat{\theta}_{(j+1)}$. These conditions are investigated by Boyles (1983) and Wu (1983). The following theorem is, perhaps, the most easily applicable condition to guarantee convergence to a *stationary point*, a zero of the first derivative that may be a local maximum or saddle-point.

Theorem 5.16. *If the expected complete-data likelihood $Q(\theta|\theta_0, \mathbf{x})$ is continuous in both θ and θ_0, then every limit point of an EM sequence $(\hat{\theta}_{(j)})$ is a stationary point of $L(\theta|\mathbf{x})$, and $L(\hat{\theta}_{(j)}|\mathbf{x})$ converges monotonically to $L(\hat{\theta}|\mathbf{x})$ for some stationary point $\hat{\theta}$.*

Note that convergence is only guaranteed to a stationary point. Techniques such as running the EM algorithm a number of times with different, random starting points, or algorithms such as *simulated annealing* (see, for example, Finch et al. 1989) attempt to give some assurance that the global maximum is found. Wu (1983) states another theorem that guarantees convergence to a local maximum, but its assumptions are difficult to check. It is usually better, in practice, to use empirical methods (graphical or multiple starting values) to check that a maximum has been reached.

As a first example, we look at the censored data likelihood of Example 5.14.

Example 5.17. EM for censored data. For $Y_i \sim \mathcal{N}(\theta, 1)$, with censoring at a, the complete-data likelihood is

$$L^c(\theta|\mathbf{y}, \mathbf{z}) \propto \prod_{i=1}^{m} \exp\{-(y_i - \theta)^2/2)\} \prod_{i=m+1}^{n} \exp\{(z_i - \theta)^2/2\}.$$

The density of of the missing data $\mathbf{z} = (z_{n-m+1}, \ldots, z_n)$ is a truncated normal

$$(5.16) \qquad \mathbf{Z} \sim k(\mathbf{z}|\theta, by) = \frac{1}{(2\pi)^{(n-m)/2}} \exp \left\{ \sum_{i=m+1}^{n} (z_i - \theta)^2/2 \right\},$$

resulting in the expected complete-data log likekihood

$$-\frac{1}{2} \sum_{i=1}^{m} (y_i - \theta)^2 - \frac{1}{2} \sum_{i=n-m+1}^{n} \mathbb{E}_{\theta'}[(Z_i - \theta)^2].$$

Before evaluating the expectation we differentiate and set equal to zero, solving for the EM estimate

$$\hat{\theta} = \frac{m\bar{y} + (n-m)\mathbb{E}_{\theta'}(Z_1)}{n}.$$

This leads to the EM sequence

(5.17)
$$\hat{\theta}^{(j+1)} = \frac{m}{n}\bar{y} + \frac{n-m}{n}\hat{\theta}^{(j)} + \frac{1}{n}\frac{\phi(a - \hat{\theta}^{(j)})}{1 - \Phi(a - \hat{\theta}^{(j)})},$$

where ϕ and Φ are the normal pdf and cdf, respectively. (See Problem 5.16 for details of the calculations.)

The EM sequence is shown in the right panel of Figure 5.6. The convergence in this problem is quite rapid, giving an MLE of 3.99, in contrast to the observed data mean of 3.55. ‖

The following example has a somewhat more complex model.

Example 5.18. Cellular phone plans. A clear case of missing data occurs in the following estimation problem. It is typical for cellular phone companies to offer "plans" of options, bundling together four or five options (such as messaging, caller id, etc.) for one price, or, alternatively, selling them separately. One cellular company had offered a four-option plan in some areas, and a five-option plan (which included the four, plus one more) in another area.

In each area, customers were ask to choose their favorite plan, and the results were tabulated. In some areas they choose their favorite from four plans, and in some areas from five plans. The phone company is interested in knowing which are the popular plans, to help them set future prices. A portion of the data are given in Table 5.6. We can model the complete data as follows. In area i, there are n_i customers, each of whom choses their favorite plan from Plans $1-5$. The observation for customer i is $Z_i = (Z_{i1}, \ldots, Z_{i5})$, where Z_i is $\mathcal{M}(1, (p_1, p_2, \ldots, p_5))$. If we assume the customers are independent, in area i the data are $T_i = (T_{i1}, \ldots, T_{i5}) = \sum_{j=1}^{n_i} Z_i \sim \mathcal{M}(n_i, (p_1, p_2, \ldots, p_5))$ (see Problem 5.23). If the first m observations have the Z_{i5} missing, denote the missing data by x_i and then we have the complete-data likelihood
(5.18)
$$L(\mathbf{p}|\mathbf{T}, \mathbf{x}) = \prod_{i=1}^{m} \binom{n_i + x_i}{T_{i1}, \ldots, T_{i4}, x_i} p_1^{T_{i1}} \cdots p_4^{T_{i4}} p_5^{x_i} \times \prod_{i=m+1}^{n} \binom{n_i}{T_{i1}, \ldots, T_{i5}} \prod_{j=1}^{5} p_j^{T_{ij}},$$

where $\mathbf{p} = (p_1, p_2, \ldots, p_5)$, $\mathbf{T} = (T_1, T_2, \ldots, T_5)$, $\mathbf{x} = (x_1, x_2, \ldots, x_m)$, and $\binom{n}{n_1, n_2, \ldots, n_k}$ is the multinomial coefficient $\frac{n!}{n_1! n_2! \cdots n_k!}$. The observed-data likelihood can be calculated as $L(\mathbf{p}|\mathbf{T}) = \sum_{\mathbf{x}} L(\mathbf{p}|\mathbf{T}, \mathbf{x})$ leading to the missing data distribution

	Plan						Plan				
	1	2	3	4	5		1	2	3	4	5
1	26	63	20	0	–	20	56	18	29	5	–
2	31	14	16	51	–	21	27	53	10	0	–
3	41	28	34	10	–	22	47	29	4	11	–
4	27	29	19	25	–	23	43	66	6	1	–
5	26	48	41	0	–	24	14	30	23	23	6
6	30	45	12	14	–	25	4	24	24	32	7
7	53	39	12	11	–	26	11	30	22	23	8
8	40	25	26	6	–	27	0	55	0	28	10
9	37	26	15	0	–	28	1	20	25	33	12
10	27	59	17	0	–	29	29	25	5	34	13
11	32	39	12	0	–	30	23	71	10	0	13
12	28	46	17	12	–	31	53	29	1	5	18
13	27	5	44	3	–	32	9	31	22	20	19
14	43	51	11	5	–	33	0	49	6	32	20
15	52	30	17	0	–	34	21	25	12	37	20
16	31	55	21	0	–	35	35	26	18	0	25
17	54	51	0	0	–	36	8	15	4	54	32
18	34	29	32	10	–	37	39	19	3	8	38
19	42	33	7	13	–						

Table 5.6. Cellular phone plan preferences in 37 areas: Data are number of customers who choose the particular plan as their favorite. Some areas ranked 4 plans (with the 5th plan denoted by –) and some ranked 5 plans.

$$(5.19) \qquad k(\mathbf{x}|\mathbf{T}, \mathbf{p}) = \prod_{i=1}^{m} \binom{n_i + x_i}{x_i} p_5^{x_i} (1 - p_5)^{n_i + 1},$$

a product of negative binomial distributions.

The rest of the EM analysis follows. Define $W_j = \sum_{i=1}^{n} T_{ij}$ for $j = 1, \ldots, 4$, and $W_5 = \sum_{i=1}^{m} T_{i5}$ for $j = 5$. The expected complete-data log likelihood is

$$\sum_{j=1}^{4} W_j \log p_j + \left[W_5 + \sum_{i=1}^{m} \mathbb{E}(X_i | \mathbf{p}') \right] \log(1 - p_1 - p_2 - p_3 - p_4),$$

leading to the EM iterations

$$\mathbb{E}(X_i | \mathbf{p}^{(t)}) = (n_i + 1) \frac{\hat{p}_5^{(t)}}{1 - \hat{p}_5^{(t)}}, \qquad \hat{p}_j^{(t+1)} = \frac{W_j}{\sum_{i=1}^{m} \mathbb{E}(X_i | \mathbf{p}^{(t)}) + \sum_{j'=1}^{5} W_{j'}}$$

for $j = 1, \ldots, 4$. The MLE of $\mathbf{p}$ is $(0.273, 0.329, 0.148, 0.125, 0.125)$, with convergence being very rapid. Convergence of the estimators is shown in Figure 5.7, and further details are given in Problem 5.23. (See also Example 9.22 for a Gibbs sampling treatment of this problem.)   ‖

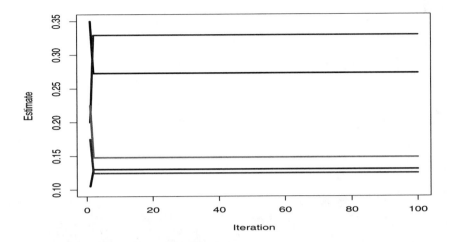

Fig. 5.7. EM sequence for cellular phone data, 25 iterations

One weakness of EM is that it is "greedy"; it always moves upward to a higher value of the likelihood. This means that it cannot escape from a local model. The following example illustrates the possible convergence of EM to the "wrong" mode, even in a well-behaved case.

Example 5.19. EM for mean mixtures of normal distributions Consider the mixture of two normal distributions already introduced in Example 1.10,

$$p\,\mathcal{N}(\mu_1, \sigma^2) + (1 - p)\,\mathcal{N}(\mu_2, \sigma^2)\,,$$

in the special case where all parameters but (μ_1, μ_2) are known. Figure 5.8 *(bottom)* shows the log-likelihood surface associated with this model and 500 observations simulated with $p = 0.7$, $\sigma = 1$ and $(\mu_1, \mu_2) = (0, 3.1)$. As easily seen from this surface, the likelihood is bimodal,[5] with one mode located near the true value of the parameters, and another located at $(2, -.5)$. Running EM five times with various starting points chosen at random, we represent the corresponding occurences: three out of five sequences are attracted by the higher mode, while the other two go to the lower mode (even though the likelihood is considerably smaller). This is because the starting points happened to be in the domain of attraction of the lower mode. We also represented in Figure 5.8 *(top)* the corresponding (increasing) sequence of log-likelihood values taken by the sequence. Note that in a very few iterations the value

---

[5] Note that this is not a special occurrence associated with a particular sample: there are two modes of the likelihood for every simulation of a sample from this mixture, even though the model is completely identifiable.

is close to the modal value, with improvement brought by further iterations being incremental. ‖

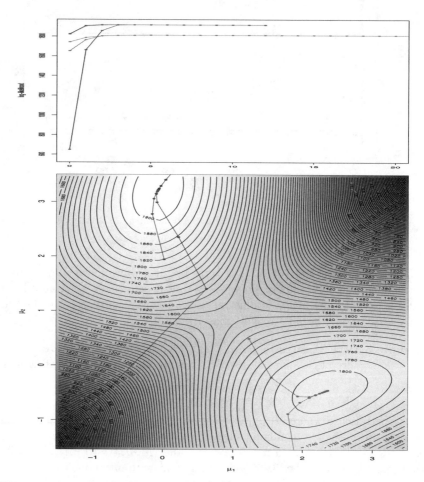

Fig. 5.8. Trajectories of five runs of the EM algorithm for Example 5.19 with their likelihood values *(top)* and their position on the likelihood surface *(bottom)*.

This example reinforces the case of the need for rerunning the algorithm a number of times, each time starting from a different initial value.

The books by Little and Rubin (1987) and Tanner (1996) provide good overviews of the EM literature. Other references include Louis (1982), Little and Rubin (1983), Laird et al. (1987), Meng and Rubin (1991), Qian and Titterington (1992), Liu and Rubin (1994), MacLachlan and Krishnan (1997), and Meng and van Dyk (1997).

5.3.3 Monte Carlo EM

A difficulty with the implementation of the EM algorithm is that each "E-step" requires the computation of the expected log likelihood $Q(\theta|\theta_0, \mathbf{x})$. Wei and Tanner (1990a,b) propose to use a Monte Carlo approach (MCEM) to overcome this difficulty, by simulating $Z_1, \ldots, Z_m$ from the conditional distribution $k(\mathbf{z}|\mathbf{x}, \theta)$ and then maximizing the approximate complete-data log-likelihood

$$(5.20) \qquad \hat{Q}(\theta|\theta_0, \mathbf{x}) = \frac{1}{m} \sum_{i=1}^{m} \log L^c(\theta|\mathbf{x}, \mathbf{z}) .$$

When m goes to infinity, this quantity indeed converges to $Q(\theta|\theta_0, \mathbf{x})$, and the limiting form of the *Monte Carlo EM* algorithm is thus the regular EM algorithm. The authors suggest that m should be increased along with the iterations. Although the maximization of a sum like (5.20) is, in general, rather involved, exponential family settings often allow for closed-form solutions.

Example 5.20. MCEM for censored data. The EM solution of Example 5.17 can easily become an MCEM solution. For the EM sequence

$$\hat{\theta}^{(j+1)} = \frac{m\bar{y} + (n - m)\mathbb{E}_{\hat{\theta}^{(j)}}(Z_1)}{n},$$

the MCEM solution replaces $\mathbb{E}_{\theta^{(j)}}(Z_1)$ with

$$\frac{1}{M} \sum_{i=1}^{M} Z_i, \quad Z_i \sim k(z|\hat{\theta}^{(j)}, \mathbf{y}).$$

The MCEM sequence is shown in the right panel of Figure 5.6. The convergence is not quite so rapid as EM. The variability is controlled by the choice of M, and a larger value would bring the sequences closer together.   ‖

Example 5.21. Genetic linkage. A classic (perhaps overused) example of the EM algorithm is the genetics problem (see Rao 1973, Dempster et al. 1977, or Tanner 1996), where observations (x_1, x_2, x_3, x_4) are gathered from the multinomial distribution

$$\mathcal{M}\left(n; \frac{1}{2} + \frac{\theta}{4}, \frac{1}{4}(1 - \theta), \frac{1}{4}(1 - \theta), \frac{\theta}{4}\right).$$

Estimation is easier if the x_1 cell is split into two cells, so we create the augmented model

$$(z_1, z_2, x_2, x_3, x_4) \sim \mathcal{M}\left(n; \frac{1}{2}, \frac{\theta}{4}, \frac{1}{4}(1 - \theta), \frac{1}{4}(1 - \theta), \frac{\theta}{4}\right),$$

with $x_1 = z_1 + z_2$. The complete-data likelihood function is then simply $\theta^{z_2+x_4}(1-\theta)^{x_2+x_3}$, as opposed to the observed-data likelihood function $(2+\theta)^{x_1}\theta^{x_4}(1-\theta)^{x_2+x_3}$. The expected complete log-likelihood function is

$$\mathbb{E}_{\theta_0}[(Z_2 + x_4)\log\theta + (x_2 + x_3)\log(1-\theta)]$$
$$= \left(\frac{\theta_0}{2+\theta_0}x_1 + x_4\right)\log\theta + (x_2 + x_3)\log(1-\theta) ,$$

which can easily be maximized in θ, leading to

$$\hat{\theta}_1 = \frac{\dfrac{\theta_0 x_1}{2+\theta_0} + x_4}{\dfrac{\theta_0 x_1}{2+\theta_0} + x_2 + x_3 + x_4} .$$

If we instead use the Monte Carlo EM algorithm, $\theta_0 x_1/(2+\theta_0)$ is replaced with the average

$$\bar{z}_m = \frac{1}{m}\sum_{i=1}^{m} z_i,$$

where the z_i's are simulated from a binomial distribution $\mathcal{B}(x_1, \theta_0/(2+\theta_0))$. The maximum in θ is then

$$\theta_1 = \frac{\bar{z}_m + x_4}{\bar{z}_m + x_2 + x_3 + x_4} . \qquad \|$$

This example is merely an illustration of the Monte Carlo EM algorithm since EM also applies. The next example, however, details a situation in which the expectation is quite complicated and the Monte Carlo EM algorithm works quite nicely.

Example 5.22. Capture–recapture models revisited. A generalization of a capture–recapture model (see Example 2.25) is to assume that an animal i, $i = 1, 2, \ldots, n$ can be captured at time j, $j = 1, 2, \ldots, t$, in one of m locations, where the location is a multinomial random variable

$$H \sim \mathcal{M}_m(\theta_1, \ldots, \theta_m).$$

Of course, the animal may not be captured (it may not be seen, or it may have died). As we track each animal through time, we can model this process with two random variables. The random variable H can take values in the set $\{1, 2, \ldots, m\}$ with probabilities $\{\theta_1, \ldots, \theta_m\}$. Given $H = k$, the random variable is $X \sim \mathcal{B}(p_k)$, where p_k is the probability of capturing the animal in location k. (See Dupuis 1995 for details.)

As an example, for $t = 6$, a typical realization for an animal might be

$$\mathbf{h} = (4, 1, -, 8, 3, -), \quad \mathbf{x} = (1, 1, 0, 1, 1, 0),$$

where **h** denotes the sequence of observed h_j's and of non-captures. Thus, we have a missing data problem. If we had observed all of **h**, the maximum likelihood estimation would be trivial, as the MLEs would simply be the cell means. For animal i, we define the random variables $X_{ijk} = 1$ if animal i is captured at time j in location k, and 0 otherwise,

$$Y_{ijk} = \mathbb{I}(H_{ij} = k)\mathbb{I}(X_{ijk} = 1)$$

(which is the observed data), and

$$Z_{ijk} = \mathbb{I}(H_{ij} = k)\mathbb{I}(X_{ijk} = 0)$$

(which is the missing data). The likelihood function is

$$
\begin{aligned}
L(\theta_1, &\ldots, \theta_m, p_1, \ldots, p_m | \mathbf{y}, \mathbf{x}) \\
&= \sum_{\mathbf{z}} L(\theta_1, \ldots, \theta_m, p_1, \ldots, p_m | \mathbf{y}, \mathbf{x}, \mathbf{z}) \\
&= \sum_{\mathbf{z}} \prod_{k=1}^{m} p_k^{\sum_{i=1}^{n} \sum_{j=1}^{t} x_{ijk}} \\
&\quad \times (1 - p_k)^{nt - \sum_{i=1}^{n} \sum_{j=1}^{t} x_{ijk}} \theta_k^{\sum_{i=1}^{n} \sum_{j=1}^{t} (y_{ijk} + z_{ijk})},
\end{aligned}
$$

where the sum over **z** represents the expectation over all the states that could have been visited. This can be a complicated expectation, but the likelihood can be calculated by first using an EM strategy and working with the complete data likelihood $L(\theta_1, \ldots, \theta_k, p_1, \ldots, p_k | \mathbf{y}, \mathbf{x}, \mathbf{z})$, then using MCEM for the calculation of the expectation. Note that calculation of the MLEs of $p_1, \ldots, p_k$ is straightforward, and for $\theta_1, \ldots, \theta_k$, we use

Algorithm A.21 –Capture–recapture MCEM Algorithm–

1. *(M-step)* Take $\hat{\theta}_k = \frac{1}{nt} \sum_{i=1}^{n} \sum_{j=1}^{t} y_{ijk} + \hat{z}_{ijk}$.
2. *(Monte Carlo E-step)* If $x_{ijk} = 0$, for $\ell = 1, \ldots, L$, generate

$$\hat{z}_{ijk\ell} \sim \mathcal{M}_k(\hat{\theta}_1, \ldots, \hat{\theta}_m)$$

and calculate

$$\hat{z}_{ijk} = \sum_{\ell} \hat{z}_{ijk\ell} / L.$$

Scherrer (1997) examines the performance of more general versions of this algorithm and shows, in particular, that they outperform the conditional likelihood approach of Brownie et al. (1993). ‖

Note that the MCEM approach does not enjoy the EM monotonicity any longer and may even face some smoothness difficulties when the sample used in

(5.20) is different for each new value θ_j. In more involved likelihoods, Markov chain Monte Carlo methods can also be used to generate the missing data sample, usually creating additional dependence structures between the successive values produced by the algorithm.

5.3.4 EM Standard Errors

There are many algorithms and formulas available for obtaining standard errors from the EM algorithm (see Tanner 1996 for a selection). However, the formulation by Oakes (1999), and its Monte Carlo version, seem both simple and useful.

Recall that the variance of the MLE, is approximated by

$$\mathrm{Var}\,\hat{\theta} \approx \left[\frac{\partial^2}{\partial \theta^2} \log L(\theta|\mathbf{x}) \right]^{-1}.$$

Oakes (1999) shows that this second derivative can be expressed in terms of the complete-data likelihood

(5.21) $\dfrac{\partial^2}{\partial \theta^2} \log L(\theta|\mathbf{x})$

$$= \left\{ \frac{\partial^2}{\partial \theta'^2} \mathbb{E}[\log L(\theta'|\mathbf{x}, \mathbf{z})] + \frac{\partial^2}{\partial \theta' \partial \theta} \mathbb{E}[\log L(\theta'|\mathbf{x}, \mathbf{z})] \right\} \Bigg|_{\theta'=\theta},$$

where the expectation is taken with respect to the distribution of the missing data. Thus, for the EM algorithm, we have a formula for the variance of the MLE.

The advantage of this expression is that it only involves the distribution of the complete data, which is often a reasonable distribution to work with. The disadvantage is that the mixed derivative may be difficult to compute. However, in complexity of implementation it compares favorably with other methods.

For the Monte Carlo EM algorithm, (5.21) cannot be used in its current form, as we would want all expectations to be on the outside. Then we could calculate the expression using a Monte Carlo sum. However, if we now take the derivative inside the expectation, we can rewrite Oakes' identity as

$$\frac{\partial^2}{\partial \theta^2} \log L(\theta|\mathbf{x})$$

(5.22) $= \mathbb{E}\left(\dfrac{\partial^2}{\partial \theta^2} \log L(\theta|\mathbf{x}, \mathbf{z}) \right)$

$$+ \mathbb{E}\left[\left(\frac{\partial}{\partial \theta} \log L(\theta|\mathbf{x}, \mathbf{z}) \right)^2 \right] - \left[\mathbb{E}\left(\frac{\partial}{\partial \theta} \log L(\theta|\mathbf{x}, \mathbf{z}) \right) \right]^2,$$

which is better suited for simulation, as all expectations are on the outside. Equation (5.22) can be expressed in the rather pleasing form

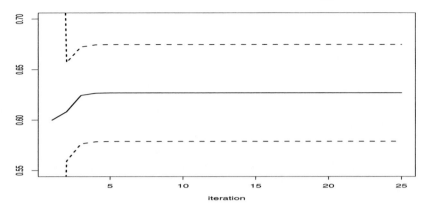

Fig. 5.9. EM sequence ± one standard deviation and for genetic linkage data, 25 iterations

$$\frac{\partial^2}{\partial\theta^2}\log L(\theta|\mathbf{x}) = \mathbb{E}\left(\frac{\partial^2}{\partial\theta^2}\log L(\theta|\mathbf{x},\mathbf{z})\right) + \mathrm{var}\left(\frac{\partial}{\partial\theta}\log L(\theta|\mathbf{x},\mathbf{z})\right),$$

which allows the Monte Carlo evaluation

$$\frac{\partial^2}{\partial\theta^2}\log L(\theta|\mathbf{x})$$

$$= \frac{1}{M}\sum_{j=1}^{M}\frac{\partial^2}{\partial\theta^2}\log L(\theta|\mathbf{x},\mathbf{z}^{(j)})$$

$$+ \frac{1}{M}\sum_{j=1}^{M}\left(\frac{\partial}{\partial\theta}\log L(\theta|\mathbf{x},\mathbf{z}^{(j)}) - \frac{1}{M}\sum_{j'=1}^{M}\frac{\partial}{\partial\theta}\log L(\theta|\mathbf{x},\mathbf{z}^{(j')})\right)^2,$$

where $(\mathbf{z}^{(j)})$, $j = 1,\ldots,M$ are generated from the missing data distribution (and have already been generated to do MCEM).

Example 5.23. Genetic linkage standard errors. For Example 5.21, the complete-data likelihood is $\theta^{z_2+x_4}(1-\theta)^{x_2+x_3}$, and applying (5.22) yields

$$\frac{\partial^2}{\partial\theta^2}\log L(\theta|\mathbf{x}) = \frac{\mathbb{E}Z_2(1-\theta)^2 + x_4(1-\theta)^2 + (x_2+x_3)\theta^2}{\theta^2(1-\theta)^2},$$

where $\mathbb{E}Z_2 = x_1\theta/(2+\theta)$. In practice, we evaluate the expectation at the converged value of the likelihood estimator. For these data we obtain $\hat{\theta} = .627$ with standard deviation .048. Figure 5.9 shows the evolution of the EM estimate and the standard error. ‖

5.4 Problems

5.1 Use a numerical maximizer to find the maximum of

$$f(x) = [\cos(50x) + \sin(20x)]^2 .$$

Compare to the results of a stochastic exploration. (*Note:* If you use R the functions are optim and optimize.)

5.2 For each of the likelihood functions in Exercise 1.1, find the maximum with both R function optimize and stochastic exploration.

5.3 (Ó Ruanaidh and Fitzgerald 1996) Consider the setup of Example 5.1.
(a) Show that (5.2) holds.
(b) Discuss the validity of the approximation

$$\mathbf{x}^t\mathbf{x} - \mathbf{x}^tG(G^tG)^{-1}G^t\mathbf{x} \simeq \sum_{i=1}^{N} x_i^2 - \frac{2}{N}\left(\sum_{i=1}^{N} x_i\cos(\omega t_i) + \sum_{i=1}^{N} x_i\sin(\omega t_i)\right)^2$$

$$= \sum_{i=1}^{N} x_i^2 - 2S_N .$$

(c) Show that $\pi(\omega|\mathbf{x})$ can be approximated by

$$\pi(\omega|\mathbf{x}) \propto \left[1 - \frac{2S_N}{\sum_{i=1}^{N} x_i^2}\right]^{(2-N)/2} .$$

5.4 For the situation of Example 5.5:
(a) Reproduce Figure 5.4.
(b) For a range of values of r, and a range of values of c, where $T_t = c/\log(t)$, examine the trajectories of $(x^{(t)}, h(x^{(t)}))$. Comment on the behaviors and recommend an optimal choice of r and c.

5.5 Given a simple Ising model with energy function

$$h(s) = \beta \sum_{(u,v)\in\mathcal{N}} s_u s_v ,$$

where $\mathcal{N}$ is the neighborhood relation that u and v are neighbors either horizontally or vertically, apply the algorithm [A.19] to the cases $\beta = 0.4$ and $\beta = 4.5$.

5.6 Here we will outline a proof of Theorem 5.10. It is based on *the Laplace approximation* (see Section 3.4), and is rather painful. The treatment of the error terms here is a little cavalier; see Tierney and Kadane (1986) or Schervish (1995) for complete details.
(a) Expand $h(\theta)$ around θ^* up to third order terms to establish

$$\int \theta e^{\lambda h(\theta)} d(\theta) = e^{\lambda h(\theta^*)} \int \theta e^{\lambda h''(\theta^*)(\theta-\theta^*)^2/2} e^{\lambda h'''(\theta^*)(\theta-\theta^*)^3/6} e^{\lambda \mathcal{O}(|\theta-\theta^*|^4)},$$

where $\mathcal{O}(|\theta-\theta^*|^4) \le C(|\theta-\theta^*|^4)$ for some constant C and $|\theta-\theta^*|$ small enough.

(b) Use Taylor series expansions to show that

$$e^{\lambda \mathcal{O}(|\theta - \theta^*|^4)} = 1 + \lambda \mathcal{O}(|\theta - \theta^*|^4)$$

and

$$e^{\lambda h'''(\theta^*)(\theta - \theta^*)^3/6} = 1 + \lambda h'''(\theta^*)(\theta - \theta^*)^3/6 + \lambda \mathcal{O}(|\theta - \theta^*|^6).$$

(c) Substitute the part (b) expressions into the integral of part (a), and write $\theta = t + \theta^*$, where $t = \theta - \theta^*$, to obtain

$$\int \theta e^{\lambda h(\theta)} d(\theta) = e^{\lambda h(\theta^*)} \theta^* \int e^{\lambda h''(\theta^*)t^2/2}[1 + \lambda h'''(\theta^*)t^3/6 + \lambda \mathcal{O}(|t|)^4] dt$$

$$+ e^{\lambda h(\theta^*)} \int t e^{\lambda h''(\theta^*)t^2/2}[1 + \lambda h'''(\theta^*)t^3/6 + \lambda \mathcal{O}(|t|)^4] dt.$$

(d) All of the integrals in part (c) can be evaluated. Show that the ones involving odd powers of t in the integrand are zero by symmetry, and

$$\left| \int \lambda |t|^4 e^{\lambda h''(\theta^*)t^2/2} dt \right| \leq \frac{M}{\lambda^{3/2}},$$

which bounds the error terms.

(e) Show that we now can write

$$\int \theta e^{\lambda h(\theta)} d(\theta) = \theta^* e^{\lambda h(\theta^*)} \left[\sqrt{-2\pi} h''(\theta^*) + \frac{M}{\lambda^{3/2}} \right],$$

$$\int e^{\lambda h(\theta)} d(\theta) = e^{\lambda h(\theta^*)} \left[\sqrt{-2\pi} h''(\theta^*) + \frac{M}{\lambda^{3/2}} \right],$$

and hence

$$\frac{\int \theta e^{\lambda h(\theta)} d(\theta)}{\int e^{\lambda h(\theta)} d(\theta)} = \theta^* \left[1 + \frac{M}{\lambda^{3/2}} \right],$$

completing the proof.

(f) Show how Corollary 5.11 follows by starting with $\int b(\theta) e^{\lambda h(\theta)} d(\theta)$ instead of $\int \theta e^{\lambda h(\theta)} d(\theta)$. Use a Taylor series on $b(\theta)$.

5.7 For the Student's t distribution $\mathcal{T}_\nu$:

(a) Show that the distribution $\mathcal{T}_\nu$ can be expressed as the (continuous) mixture of a normal and of a chi squared distribution. (*Hint:* See Section 2.2.)

(b) When $X \sim \mathcal{T}_\nu$, given a function $h(x)$ derive a representation of $h(x) = \mathbb{E}[H(x, Z)|x]$, where $Z \sim \mathcal{G}a((\alpha - 1)/2, \alpha/2)$.

5.8 For the normal mixture of Example 5.19, assume that $\sigma_1 = \sigma_2 = 1$ for simplicity's sake.

(a) Show that the model does belong to the missing data category by considering the vector $\mathbf{z} = (z_1, \ldots, z_n)$ of allocations of the observations x_i to the first and second components of the mixture.

(b) Show that the mth iteration of the EM algorithm consists in replacing the allocations z_i by their expectation

$$\mathbb{E}[Z_i|x_i, \mu_1^{(m-1)}, \mu_2^{(m-1)}] = (1 - p)\varphi(x_i; \mu_2^{(m-1)}, 1) \Big/$$

$$\{p\varphi(x_i; \mu_1^{(m-1)}, 1) + (1 - p)\varphi(x_i; \mu_2^{(m-1)}, 1)\}$$

and then setting the new values of the means as

$$\mu_1^{(m)} = \sum \mathbb{E}[1 - Z_i | x_i, \mu_1^{(m-1)}, \mu_2^{(m-1)}] x_i / \sum \mathbb{E}[1 - Z_i | x_i, \mu_1^{(m-1)}, \mu_2^{(m-1)}]$$

and

$$\mu_2^{(m)} = \sum \mathbb{E}[Z_i | x_i, \mu_1^{(m-1)}, \mu_2^{(m-1)}] x_i / \sum \mathbb{E}[Z_i | x_i, \mu_1^{(m-1)}, \mu_2^{(m-1)}].$$

5.9 Suppose that the random variable X has a mixture distribution (1.3); that is, the X_i are independently distributed as

$$X_i \sim \theta g(x) + (1 - \theta) h(x), \quad i = 1, \ldots, n,$$

where $g(\cdot)$ and $h(\cdot)$ are known. An EM algorithm can be used to find the ML estimator of θ. Introduce $Z_1, \ldots, Z_n$, where Z_i indicates from which distribution X_i has been drawn, so

$$X_i | Z_i = 1 \sim g(x),$$
$$X_i | Z_i = 0 \sim h(x).$$

(a) Show that the complete data likelihood can be written

$$L^c(\theta | \mathbf{x}, \mathbf{z}) = \prod_{i=1}^{n} [z_i g(x_i) + (1 - z_i) h(x_i)] \, \theta^{z_i} (1 - \theta)^{1 - z_i}.$$

(b) Show that $\mathbb{E}[Z_i | \theta, x_i] = \theta g(x_i) / [\theta g(x_i) + (1 - \theta) h(x_i)]$ and, hence, that the EM sequence is given by

$$\hat{\theta}_{(j+1)} = \frac{1}{n} \sum_{i=1}^{n} \frac{\hat{\theta}_{(j)} g(x_i)}{\hat{\theta}_{(j)} g(x_i) + (1 - \hat{\theta}_{(j)}) h(x_i)}.$$

(c) Show that $\hat{\theta}_{(j)}$ converges to $\hat{\theta}$, a maximum likelihood estimator of θ.

5.10 Consider the sample $\mathbf{x} = (0.12, 0.17, 0.32, 0.56, 0.98, 1.03, 1.10, 1.18, 1.23, 1.67, 1.68, 2.33)$, generated from an exponential mixture

$$p \mathcal{E}xp(\lambda) + (1 - p) \mathcal{E}xp(\mu).$$

(a) Show that the likelihood $h(p, \lambda, \mu)$ can be expressed as $\mathbb{E}[H(x, Z)]$, where $z = (z_1, \ldots, z_{12})$ corresponds to the vector of allocations of the observations x_i to the first and second components of the mixture; that is, for $i = 1, \ldots, 12$,

$$P(z_i = 1) = 1 - P(z_i = 2) = \frac{p \lambda \exp(-\lambda x_i)}{p \lambda \exp(-\lambda x_i) + (1 - p) \mu \exp(-\mu x_i)}.$$

(b) Compare the performances of [A.22] with those of the EM algorithm in this setup.

5.11 This problem refers to Section 5.5.4.

(a) Show that $\int \tilde{h}(x | \theta) dx = \frac{1}{c(\theta)}$.

(b) Show that for any two values θ and η,

$$\log \frac{h(x|\theta)}{h(x|\eta)} = \log \frac{\tilde{h}(x|\theta)}{\tilde{h}(x|\eta)} + \frac{c(\theta)}{c(\eta)} = \log \frac{\tilde{h}(x|\theta)}{\tilde{h}(x|\eta)} - \log \mathbb{E}\left(\frac{\tilde{h}(X|\theta)}{\tilde{h}(X|\eta)}\right),$$

where $X \sim h(x|\eta)$. (*Hint:* The last equality follows from part (a) and an importance sampling argument.)

(c) Thus, establish the validity of the approximation

$$\max_x h(x|\theta) \approx \max_x \left\{ \log \frac{\tilde{h}(x|\theta)}{\tilde{h}(x|\eta)} - \log \frac{1}{m} \sum_{i=1}^{m} \frac{\tilde{h}(x_i|\theta)}{\tilde{h}(x_i|\eta)} \right\},$$

where the X_i's are generated from $h(x|\eta)$

5.12 Consider $h(\alpha)$, the likelihood of the beta $\mathcal{B}(\alpha, \alpha)$ distribution associated with the observation $x = 0.345$.

(a) Express the normalizing constant, $c(\alpha)$, in $h(\alpha)$ and show that it cannot be easily computed when α is not an integer.

(b) Examine the approximation of the ratio $c(\alpha)/c(\alpha_0)$, for $\alpha_0 = 1/2$ by the method of Geyer and Thompson (1992) (see Example 5.25).

(c) Compare this approach with the alternatives of Chen and Shao (1997), detailed in Problems 4.1 and 4.2.

5.13 Consider the function

$$h(\theta) = \frac{||\theta||^2 (p + ||\theta||^2)(2p - 2 + ||\theta||^2)}{(1 + ||\theta||^2)(p + 1 + ||\theta||^2)(p + 3 + ||\theta||^2)},$$

when $\theta \in \mathbb{R}^p$ and $p = 10$.

(a) Show that the function $h(\theta)$ has a unique maximum.

(b) Show that $h(\theta)$ can be expressed as $\mathbb{E}[H(\theta, Z)]$, where $z = (z_1, z_2, z_3)$ and $Z_i \sim \mathcal{E}xp(1/2)$ ($i = 1, 2, 3$). Deduce that $f(z|x)$ does not depend on x in (5.26).

(c) When $g(z) = \exp(-\alpha\{z_1 + z_2 + z_3\})$, show that the variance of (5.26) is infinite for some values of $t = ||\theta||^2$ when $\alpha > 1/2$. Identify A_2, the set of values of t for which the variance of (5.26) is infinite when $\alpha = 2$.

(d) Study the behavior of the estimate (5.26) when t goes from A_2 to its complement A_2^c to see if the infinite variance can be detected in the evaluation of $h(t)$.

5.14 In the exponential family, EM computations are somewhat simplified. Show that if the complete data density f is of the form

$$f(y, z|\theta) = h(y, z) \exp\{\eta(\theta)T(y, z) - B(\theta)\},$$

then we can write

$$Q(\theta|\theta^*, \mathbf{y}) = \mathbb{E}_{\theta^*}\left[\log h(\mathbf{y}, \mathbf{Z})\right] + \sum \eta_i(\theta)\mathbb{E}_{\theta^*}\left[T_i|\mathbf{y}\right] - B(\theta).$$

Deduce that calculating the complete-data MLE only involves the simpler expectation $\mathbb{E}_{\theta^*}\left[T_i|\mathbf{y}\right]$.

5.15 For density functions f and g, we define the *entropy distance* between f and g, with respect to f (also known as *Kullback–Leibler information of g at f* or *Kullback–Leibler distance between g and f*) as

$$\mathbb{E}_f[\log(f(X)/g(X))] = \int \log\left[\frac{f(x)}{g(x)}\right] f(x)\, dx.$$

(a) Use Jensen's inequality to show that $\mathbb{E}_f[\log(f(X)/g(X))] \geq 0$ and, hence, that the entropy distance is always non-negative, and equals zero if $f = g$.

(b) The inequality in part (a) implies that $\mathbb{E}_f \log[g(X)] \leq \mathbb{E}_f \log[f(X)]$. Show that this yields (5.15).

(*Note:* Entropy distance was explored by Kullback 1968; for an exposition of its properties, see, for example, Brown 1986. Entropy distance has, more recently, found many uses in Bayesian analysis see, for example, Berger 1985, Bernardo and Smith 1994, or Robert 1996b.)

5.16 Refer to Example 5.17

(a) Verify that the missing data density is given by (5.16).

(b) Show that $\mathbb{E}_{\theta'} Z_1 = \frac{\phi(a-\theta')}{1-\Phi(a-\theta')}$, and that the EM sequence is given by (5.17).

(c) Reproduce Figure 5.6. In particular, examine choices of M that will bring the MCEM sequence closed to the EM sequence.

5.17 The *probit model* is a model with a covariate $X \in \mathbb{R}^p$ such that

$$Y|X = x \sim \mathcal{B}(\Phi(x^T \beta)),$$

where $\beta \in \mathbb{R}^p$ and Φ denotes the $\mathcal{N}(0, 1)$ cdf.

(a) Give the likelihood associated with a sample $((x_1, y_1), \ldots, (x_n, y_n))$.

(b) Show that, if we associate with each observation (x_i, y_i) a missing variable Z_i such that

$$Z_i|X_i = x \sim \mathcal{N}(x^T \beta, 1) \qquad Y_i = \mathbb{I}_{Z_i > 0},$$

iteration m of the associated EM algorithm is the expected least squares estimator

$$\beta_{(m)} = (X^T X)^{-1} X^T \, \mathbb{E}_{\beta_{(m-1)}}[\mathbf{Z}|\mathbf{x}, \mathbf{y}],$$

where $\mathbf{x} = (x_1, \ldots, x_n)$, $\mathbf{y} = (y_1, \ldots, y_n)$, and $\mathbf{Z} = (Z_1, \ldots, Z_n)^T$, and X is the matrix with columns made of the x_i's.

(c) Give the value of $\mathbb{E}_\beta[z_i|x_i, y_i]$.

5.18 The following are genotype data on blood type.

Genotype	Probability	Observed	Probability	Frequency
AA	p_A^2	A	$p_A^2 + 2p_A p_O$	$n_A = 186$
AO	$2p_A p_O$			
BB	p_B^2	B	$p_B^2 + 2p_B p_O$	$n_B = 38$
BO	$2p_B p_O$			
AB	$2p_A p_B$	AB	$p_A p_B$	$n_{AB} = 13$
OO	p_O^2	O	p_O^2	$n_O = 284$

Because of dominance, we can only observe the genotype in the third column, with probabilities given by the fourth column. The interest is in estimating the allele frequencies p_A, p_B, and p_O (which sum to 1).

(a) Under a multinomial model, verify that the observed data likelihood is proportional to

$$(p_A^2 + 2p_A p_O)^{n_A} (p_B^2 + 2p_B p_O)^{n_B} (p_A p_B)^{n_{AB}} (p_O^2)^{n_O}.$$

(b) With missing data Z_A and Z_B, verify the complete data likelihood

$$(p_A^2)^{Z_A} (2p_A p_O)^{n_A - Z_A} (p_B^2)^{Z_B} (2p_B p_O)^{n_B - Z_B} (p_A p_B)^{n_{AB}} (p_O^2)^{n_O}.$$

(c) Verify that the missing data distribution is

$$Z_A \sim \text{binomial}\left(n_A, \frac{p_A^2}{p_A^2 + 2p_A p_O}\right) \quad \text{and} \quad Z_B \sim \text{binomial}\left(n_B, \frac{p_B^2}{p_B^2 + 2p_B p_O}\right),$$

and write an EM algorithm to estimate p_A, p_B, and p_O.

5.19 Cox and Snell (1981) report data on survival time Y in weeks and $\log_{10}$ (initial white blood count), x, for 17 patients suffering from leukemia as follows.

x	3.36	2.88	3.63	3.41	3.78	4.02	4.0	4.23	3.73	3.85	3.97	4.51	4.54	5.0	5.0	4.72	5.0
Y	65	156	100	134	16	108	121	4	39	143	56	26	22	1	1	5	65

They suggest that an appropriate model for these data is

$$Y_i = b_0 \exp b_1 (x_x - \bar{x})\varepsilon_i, \quad \varepsilon_i \sim \mathcal{E}xp(1),$$

which leads to the likelihood function

$$L(b_0, b_1) = \prod_i b_0 \exp b_1 (x_x - \bar{x}) \exp\left\{-y_i / (b_0 \exp b_1 (x_x - \bar{x}))\right\}.$$

(a) Differentiation of the log likelihood function leads to the following equations for the MLEs:

$$\hat{b}_0 = \frac{1}{n} \sum_i y_i \exp(-\hat{b}_1 (x_i - \bar{x}))$$

$$0 = \sum_i y_i (x_i - \bar{x}) \exp(-\hat{b}_1 (x_i - \bar{x})).$$

Solve these equations for the MLEs ($\hat{b}_0 = 51.1$, $\hat{b}_1 = -1.1$).

(b) It is unusual for survival times to be uncensored. Suppose there is censoring in that the experiment is stopped after 90 weeks (hence any Y_i greater than 90 is replaced by a 90, resulting in 6 censored observations.) Order the data so that the first m observations are uncensored, and the last $n - m$ are censored. Show that the observed-data likelihood is

$$L(b_0, b_1) = \prod_{i=1}^{m} b_0 \exp b_1 (x_x - \bar{x}) \exp\left\{-y_i / (b_0 \exp(b_1 (x_x - \bar{x})))\right\}$$

$$\times \prod_{i=m+1}^{n} (1 - F(a|x_i)),$$

where

$$F(a|x_i) = \int_0^a b_0 \exp b_1 (x_x - \bar{x}) \exp\left\{-t / (b_0 \exp(b_1 (x_x - \bar{x})))\right\} dt$$

$$= 1 - \exp\left\{-a / (b_0 \exp(b_1 (x_x - \bar{x})))\right\}.$$

(c) If we let Z_i denote the missing data, show that the Z_i are independent with distribution

$$Z_i \sim \frac{b_0 \exp b_1 (x_x - \bar{x}) \exp\left\{-z / (b_0 \exp(b_1 (x_x - \bar{x})))\right\}}{1 - F(a|x_i)}, \quad 0 < z < \infty.$$

(d) Verify that the complete data likelihood is

$$L(b_0, b_1) = \prod_{i=1}^{m} b_0 \exp b_1 (x_x - \bar{x}) \exp\{-y_i/(b_0 \exp(b_1(x_x - \bar{x})))\}$$

$$\times \prod_{i=m+1}^{n} b_0 \exp b_1 (x_x - \bar{x}) \exp\{-z_i/(b_0 \exp(b_1(x_x - \bar{x})))\},$$

and the expected complete-data log-likelihood is obtained by replacing the Z_i by their expected value

$$\mathbb{E}[Z_i] = (a + b_0 \exp(b_1(x_x - \bar{x})) \frac{1 - F(a|x_i)}{F(a|x_i)}.$$

(c) Implement an EM algorithm to obtain the MLEs of b_0 and b_1.

5.20 Consider the following 12 observations from $\mathcal{N}_2(0, \Sigma)$, with σ_1^2, σ_2^2, and ρ unknown:

x_1	1	1	−1	−1	2	2	−2	−2	−	−	−	−
x_2	1	−1	1	−1	−	−	−	−	2	2	−2	−2

where "−" represents a missing value.

(a) Show that the likelihood function has global maxima at $\rho = \pm 1/2$, $\sigma_1^2 = \sigma_2^2 = 8/3$, and a saddlepoint at $\rho = 0$, $\sigma_1^2 = \sigma_2^2 = 5/2$.

(b) Show that if an EM sequence starts with $\rho = 0$, then it remains at $\rho = 0$ for all subsequent iterations.

(c) Show that if an EM sequence starts with ρ bounded away from zero, it will converge to a maximum.

(d) Take into account roundoff errors; that is, the fact that $\lfloor x_i \rfloor$ is observed instead of x_i.

(*Note:* This problem is due to Murray 1977 and is discussed by Wu 1983.)

5.21 (Ó Ruanaidh and Fitzgerald 1996) Consider an AR(p) model

$$X_t = \sum_{j=1}^{p} \theta_j X_{t-j} + \epsilon_t,$$

with $\epsilon_t \sim \mathcal{N}(0, \sigma^2)$, observed for $t = p+1, \ldots, m$. The future values $X_{m+1}, \ldots,$ X_n are considered to be missing data. The initial values $x_1, \ldots, x_p$ are taken to be zero.

(a) Give the expression of the observed and complete-data likelihoods.

(b) Give the conditional maximum likelihood estimators of θ, σ and $\mathbf{z} = (X_{m+1}, \ldots, X_n)$; that is, the maximum likelihood estimators when the two other parameters are fixed.

(c) Detail the E- and M-steps of the EM algorithm in this setup, when applied to the future values $\mathbf{z}$ and when σ is fixed.

5.22 We observe independent Bernoulli variables $X_1, \ldots, X_n$, which depend on unobservable variables Z_i distributed independently as $\mathcal{N}(\zeta, \sigma^2)$, where

$$X_i = \begin{cases} 0 & \text{if } Z_i \leq u \\ 1 & \text{if } Z_i > u. \end{cases}$$

Assuming that u is known, we are interested in obtaining MLEs of ζ and σ^2.

(a) Show that the likelihood function is

$$p^S(1-p)^{n-S},$$

where $S = \sum x_i$ and

$$p = P(Z_i > u) = \Phi\left(\frac{\zeta - u}{\sigma}\right).$$

(b) If we consider $z_1, \ldots, z_n$ to be the complete data, show that the complete data likelihood is

$$\prod_{i=1}^{n} \frac{1}{\sqrt{2\pi}\sigma} \exp\left\{-\frac{1}{2\sigma^2}(z_i - \zeta)^2\right\}$$

and the expected complete-data log-likelihood is

$$-\frac{n}{2}\log(2\pi\sigma^2) - \frac{1}{2\sigma^2}\sum_{i=1}^{n}\left(\mathbb{E}[Z_i^2|x_i] - 2\zeta\mathbb{E}[Z_i|x_i] + \zeta^2\right).$$

(c) Show that the EM sequence is given by

$$\hat{\zeta}_{(j+1)} = \frac{1}{n}\sum_{i=1}^{n} t_i(\hat{\zeta}_{(j)}, \hat{\sigma}_{(j)}^2)$$

$$\hat{\sigma}_{(j+1)}^2 = \frac{1}{n}\left[\sum_{i=1}^{n} v_i(\hat{\zeta}_{(j)}, \hat{\sigma}_{(j)}^2) - \frac{1}{n}\left(\sum_{i=1}^{n} t_i(\hat{\zeta}_{(j)}, \hat{\sigma}_{(j)}^2)\right)^2\right],$$

where

$$t_i(\zeta, \sigma^2) = \mathbb{E}[Z_i|x_i, \zeta, \sigma^2] \quad \text{and} \quad v_i(\zeta, \sigma^2) = \mathbb{E}[Z_i^2|x_i, \zeta, \sigma^2].$$

(d) Show that

$$\mathbb{E}[Z_i|x_i, \zeta, \sigma^2] = \zeta + \sigma H_i\left(\frac{u - \zeta}{\sigma}\right),$$

$$\mathbb{E}[Z_i^2|x_i, \zeta, \sigma^2] = \zeta^2 + \sigma^2 + \sigma(u + \zeta)H_i\left(\frac{u - \zeta}{\sigma}\right),$$

where

$$H_i(t) = \begin{cases} \frac{\varphi(t)}{1 - \Phi(t)} & \text{if } X_i = 1 \\ -\frac{\varphi(t)}{\Phi(t)} & \text{if } X_i = 0. \end{cases}$$

(e) Show that $\hat{\zeta}_{(j)}$ converges to $\hat{\zeta}$ and that $\hat{\sigma}_{(j)}^2$ converges to $\hat{\sigma}^2$, the MLEs of ζ and σ^2, respectively.

5.23 Referring to Example 5.18

(a) Show that if $Z_i \sim \mathcal{M}(m_i; p_1, \ldots, p_k)$, $i = 1, \ldots, n$, then $\sum_i Z_i \sim \mathcal{M}(\sum_i m_i; p_1, \ldots, p_k)$.

(b) Show that the incomplete data likelihood is given by

$$\sum_{\mathbf{x}} L(\mathbf{p}|\mathbf{z}, \mathbf{x}) = \left[\prod_{i=1}^{m} \sum_{x_i=0}^{\infty} \binom{n_i + x_i}{T_{i1}, \ldots, T_{i4}, x_i} p_1^{T_{i1}} \cdots p_4^{T_{i4}} p_5^{x_i}\right]$$

$$\times \left[\prod_{i=m+1}^{n} \binom{n_i}{T_{i1}, \ldots, T_{i4}} \prod_{j=1}^{5} p_j^{T_{ij}}\right]$$

$$= \left[\prod_{i=1}^{m} \frac{n_i!}{T_{i1}! \cdots T_{i4}!} \frac{p_1^{T_{i1}} \cdots p_4^{T_{i4}}}{(1 - p_5)^{n_i+1}}\right] \left[\prod_{i=m+1}^{n} \binom{n_i}{T_{i1}, \ldots, T_{i4}} \prod_{j=1}^{4} p_j^{T_{ij}}\right].$$

(c) Verify equation (5.19).

(d) Verify the equations of the expected log-likelihood and the EM sequence.

5.24 Recall the censored Weibull model of Problems 1.11–1.13. By extending the likelihood technique of Example 5.14, use the EM algorithm to fit the Weibull model, accounting for the censoring. Use the data of Problem 1.13 and fit the three cases outlined there.

5.25 An alternate implementation of the Monte Carlo EM might be, for $Z_1, \ldots, Z_m \sim k(\mathbf{z}|\mathbf{x}, \theta)$, to iteratively maximize

$$\log \hat{L}(\theta|\mathbf{x}) = \frac{1}{m} \sum_{i=1}^{m} \{\log L^c(\theta|\mathbf{x}, \mathbf{z}_i) - \log k(\mathbf{z}_i|\theta, x)\}$$

(which might more accurately be called Monte Carlo maximum likelihood).

(a) Show that $\hat{L}(\theta|\mathbf{x}) \to L(\theta|\mathbf{x})$ as $m \to \infty$.

(b) Show how to use $\hat{L}(\theta|\mathbf{x})$ to obtain the MLE in Example 5.22. (*Warning:* This is difficult.)

5.26 For the situation of Example 5.21, data $(x_1, x_2, x_3, x_4) = (125, 18, 20, 34)$ are collected.

(a) Use the EM algorithm to find the MLE of θ.

(b) Use the Monte Carlo EM algorithm to find the MLE of θ. Compare your results to those of part (a).

5.27 For the situation of Example 5.22:

(a) Verify the formula for the likelihood function.

(b) Show that the complete-data MLEs are given by $\hat{\theta}_k = \frac{1}{nt} \sum_{i=1}^{n} \sum_{j=1}^{t} y_{ijk} + \hat{z}_{ijk}$.

5.28 For the model of Example 5.22, Table 5.7 contains data on the movement between 5 zones of 18 tree swallows with $m = t = 5$, where a 0 denotes that the bird was not captured.

(a) Using the MCEM algorithm of Example 5.22, calculate the MLEs for $\theta_1, \ldots, \theta_5$ and $p_1, \ldots, p_5$.

(b) Assume now that state 5 represents the death of the animal. Rewrite the MCEM algorithm to reflect this, and recalculate the MLEs. Compare them to the answer in part (a).

5.29 Referring to Example 5.23

(a) Verify the expression for the second derivative of the log-likelihood.

(b) Reproduce the EM estimator and its standard error.

(c) Estimate θ, and its standard error, using the Monte Carlo EM algorithm. Compare the results to those of the EM algorithm.

Time (1 2 3 4 5)	Time (1 2 3 4 5)	Time (1 2 3 4 5)
a 2 2 0 0 0	g 1 1 1 5 0	m 1 1 1 1 1
b 2 2 0 0 0	h 4 2 0 0	n 2 2 1 0 0
c 4 1 1 2 0	i 5 5 5 5 0	o 4 2 2 0 0
d 4 2 0 0 0	j 2 2 0 0 0	p 1 1 1 1 0
e 1 1 0 0 0	k 2 5 0 0 0	q 1 0 0 4 0
f 1 1 0 0 0	1 1 1 0 0 0	s 2 2 0 0 0

Table 5.7. Movement histories of 18 tree swallows over 5 time periods (*Source: Scherrer 1997.*)

5.30 Referring to Section 5.3.4

(a) Show how (5.22) can be derived from (5.21).

(b) The derivation in Section 5.3.4 is valid for vector parameters $\theta = (\theta_1, \ldots, \theta_p)$. For a function $h(\theta)$, define

$$\frac{\partial}{\partial\theta}h(\theta) = \left\{\frac{\partial}{\partial\theta_i}h(\theta))\right\}_{ij} = h^{(1)}$$

$$\frac{\partial^2}{\partial\theta^2}h(\theta) = \left\{\frac{\partial^2}{\partial\theta_i\partial\theta_j}h(\theta))\right\}_{ij} = h^{(2)},$$

where $h^{(1)}$ is a $p \times 1$ vector and $h^{(2)}$ is a $p \times p$ matrix. Using this notation, show that (5.22) becomes

$$\log L(\theta|\mathbf{x})^{(2)} = \mathbb{E}\left(\log L(\theta|\mathbf{x},\mathbf{z})^{(2)}\right) + \mathbb{E}\left[\left(\log L(\theta|\mathbf{x},\mathbf{z})^{(1)}\right)\left(\log L(\theta|\mathbf{x},\mathbf{z})^{(1)}\right)'\right]$$
$$- \left[\mathbb{E}\left(\log L(\theta|\mathbf{x},\mathbf{z})^{(1)}\right)\right]\left[\mathbb{E}\left(\log L(\theta|\mathbf{x},\mathbf{z})^{(1)}\right)\right]'.$$

(c) Use the equation in part(b) to attach standard errors to the EM estimates in Example 5.18

5.31 (For baseball fans only) It is typical for baseball announcers to report biased information, intended to overstate a player's ability. If we consider a sequence of at-bats as Bernoulli trials, we are likely to hear the report of a maximum (the player is 8-out-of-his-last-17) rather than an ordinary average. Assuming that $X_1, X_2, \ldots, X_n$ are the Bernoulli random variables representing a player's sequence of at-bats (1=hit, 0=no hit), a biased report is the observance of k^*, m^*, and r^*, where

$$r^* = \frac{k^*}{m^*} \geq \max_{m^*\leq i<n}\frac{X_n + X_{n-1} + \cdots + X_{n-i}}{i+1}.$$

If we assume that $\mathbb{E}[X_i] = \theta$, then θ is the player's true batting ability and the parameter of interest. Estimation of θ is difficult using only k^*, m^*, and r^*, but it can be accomplished with an EM algorithm. With observed data (k^*, m^*, r^*), let $\mathbf{z} = (z_1, \ldots, z_{n-m^*-1})$ be the augmented data. (This is a sequence of 0's and 1's that are commensurate with the observed data. Note that X_{n-m^*} is certain to equal 0.)

(a) Show that the EM sequence is given by

$$\hat{\theta}_{j+1} = \frac{k^* + \mathbb{E}[S_Z|\hat{\theta}_j]}{n} ,$$

where $\mathbb{E}[S_Z|\hat{\theta}_j]$ is the expected number of successes in the missing data, assuming that $\hat{\theta}_j$ is the true value of θ.

(b) Give an algorithm for computing the sequence $(\hat{\theta}_j)$. Use a Monte Carlo approximation to evaluate the expectation.

At-Bat	k^*	m^*	$\hat{\theta}$	EM MLE
339	12	39	0.298	0.240
340	47	155	0.297	0.273
341	13	41	0.299	0.251
342	13	42	0.298	0.245
343	14	43	0.300	0.260
344	14	44	0.299	0.254
345	4	11	0.301	0.241
346	5	11	0.303	0.321

Table 5.8. A portion of the 1992 batting record of major-league baseball player Dave Winfield.

(c) For the data given in Table 5.8, implement the Monte Carlo EM algorithm and calculate the EM estimates.

(*Note:* The "true batting average" $\hat{\theta}$ cannot be computed from the given data and is only included for comparison. The selected data EM MLEs are usually biased downward, but also show a large amount of variability. See Casella and Berger 1994 for details.)

5.32 The following dataset gives independent observations of $Z = (X, Y) \sim \mathcal{N}_2(0, \Sigma)$ with missing data ∗.

x	1.17	-0.98	0.18	0.57	0.21	∗	∗	∗
y	0.34	-1.24	-0.13	∗	∗	-0.12	-0.83	1.64

(a) Show that the observed likelihood is

$$\prod_{i=1}^{3} \left\{ |\Sigma|^{-1/2} e^{-z_i^t \Sigma^{-1} z_i/2} \right\} \sigma_1^{-2} e^{-(x_4^2 + x_5^2)/2\sigma_1^2} \sigma_2^{-3} e^{-(y_6^2 + y_7^2 + y_8^2)/2\sigma_2^2} .$$

(b) Examine the consequence of the choice of $\pi(\Sigma) \propto |\Sigma|^{-1}$ on the posterior distribution of Σ.

(c) Show that the missing data can be simulated from

$$X_i^* \sim \mathcal{N}\left(\rho \frac{\sigma_1}{\sigma_2} y_i, \sigma_1^2(1 - \rho^2)\right) \qquad (i = 6, 7, 8),$$

$$Y_i^* \sim \mathcal{N}\left(\rho \frac{\sigma_2}{\sigma_1} x_i, \sigma_2^2(1 - \rho^2)\right) \qquad (i = 4, 5),$$

to derive a stochastic EM algorithm.

(d) Derive an efficient simulation method to obtain the MLE of the covariance matrix Σ.

5.33 The EM algorithm can also be implemented in a Bayesian hierarchical model to find a posterior mode. Suppose that we have the hierarchical model

$$X|\theta \sim f(x|\theta) ,$$
$$\theta|\lambda \sim \pi(\theta|\lambda) ,$$
$$\lambda \sim \gamma(\lambda) ,$$

where interest would be in estimating quantities from $\pi(\theta|x)$. Since

$$\pi(\theta|x) = \int \pi(\theta, \lambda|x)d\lambda,$$

where $\pi(\theta, \lambda|x) = \pi(\theta|\lambda, x)\pi(\lambda|x)$, the EM algorithm is a candidate method for finding the mode of $\pi(\theta|x)$, where λ would be used as the augmented data.

(a) Define $k(\lambda|\theta, x) = \pi(\theta, \lambda|x)/\pi(\theta|x)$ and show that

$$\log \pi(\theta|x) = \int \log \pi(\theta, \lambda|x)k(\lambda|\theta^*, x)d\lambda - \int \log k(\lambda|\theta, x)k(\lambda|\theta^*, x)d\lambda.$$

(b) If the sequence $(\hat{\theta}_{(j)})$ satisfies

$$\max_\theta \int \log \pi(\theta, \lambda|x)k(\lambda|\theta_{(j)}, x)d\lambda = \int \log \pi(\theta_{(j+1)}, \lambda|x)k(\lambda|\theta_{(j)}, x)d\lambda,$$

show that $\log \pi(\theta_{(j+1)}|x) \geq \log \pi(\theta_{(j)}|x)$. Under what conditions will the sequence $(\hat{\theta}_{(j)})$ converge to the mode of $\pi(\theta|x)$?

(c) For the hierarchy

$$X|\theta \sim \mathcal{N}(\theta, 1) ,$$
$$\theta|\lambda \sim \mathcal{N}(\lambda, 1) ,$$

with $\pi(\lambda) = 1$, show how to use the EM algorithm to calculate the posterior mode of $\pi(\theta|x)$.

5.34 Let $X \sim \mathcal{T}_\nu$ and consider the function

$$h(x) = \frac{\exp(-x^2/2)}{[1 + (x - \mu)^2]^\nu} .$$

(a) Show that $h(x)$ can be expressed as the conditional expectation $\mathbb{E}[H(x, Z)|x]$, when $Z \sim \mathcal{G}a(\nu, 1)$.

(b) Apply the direct Monte Carlo method of Section 5.5.4 to maximize (5.5.4) and determine whether or not the resulting sequence converges to the true maximum of h.

(c) Compare the implementation of (b) with an approach based on (5.26) for (i) $g = \mathcal{E}xp(\lambda)$ and (ii) $g = f(z|\mu)$, the conditional distribution of Z given $X = \mu$. For each choice, examine if the approximation (5.26) has a finite variance.

(d) Run [A.22] to see if the recursive scheme of Geyer (1996) improves the convergence speed.

5.5 Notes

5.5.1 Variations on EM

Besides a possible difficult computation in the E-step (see Section 5.3.3), problems with the EM algorithm can occur in the case of multimodal likelihoods. The increase of the likelihood function at each step of the algorithm ensures its convergence to the maximum likelihood estimator in the case of unimodal likelihoods but implies a dependence on initial conditions for multimodal likelihoods. Several proposals can be found in the literature to overcome this problem, one of which we now describe.

Broniatowski et al. (1984) and Celeux and Diebolt (1985, 1992) have tried to overcome the dependence of EM methods on the starting value by replacing step 1 in [A.20] with a *simulation* step, the missing data z being generated conditionally on the observation x and on the current value of the parameter θ_m. The maximization in step 2 is then done on the (simulated) complete-data log-likelihood, $\tilde{H}(x, z_m|\theta)$. The appeal of this approach is that it allows for a more systematic exploration of the likelihood surface by partially avoiding the fatal attraction of the closest mode. Unfortunately, the theoretical convergence results for these methods are limited: The Markov chain (θ_m) produced by this variant of EM called SEM (for *Stochastic EM*) is often ergodic, but the relation of the stationary distribution with the maxima of the observed likelihood is rarely known (see Diebolt and Ip 1996). Moreover, the authors mainly study the behavior of the "ergodic" average,

$$\frac{1}{M} \sum_{m=1}^{M} \theta_m,$$

instead of the "global" mode,

$$\hat{\theta}_{(M)} = \arg \max_{1 \leq m \leq M} \ell(\theta_m|x),$$

which is more natural in this setup. Celeux and Diebolt (1990) have, however, solved the convergence problem of SEM by devising a hybrid version called SAEM (for *Simulated Annealing EM*), where the amount of randomness in the simulations decreases with the iterations, ending up with an EM algorithm. This version actually relates to the simulated annealing methods, described in Section 5.2.3. Celeux et al. (1996) also propose a hybrid version, where SEM produces $\hat{\theta}_{(M)}$, the starting point of EM, when the later applies. See Lavielle and Moulines (1997) for an approach similar to Celeux and Diebolt (1990), where the authors obtain convergence conditions which are equivalent to those of EM. In the same setting, Doucet et al. (2002) develop an algorithm called the *SAME* algorithm (where SAME stands for *state augmentation for marginal estimation*), which integrates the ideas of Sections 5.2.4 and 5.3.1 within an MCMC algorithm (see Chapters 7 and 10).

Meng and Rubin (1991, 1992), Liu and Rubin (1994) , and Meng and van Dyk (1997) have also developed versions of EM called ECM which take advantage of Gibbs sampling advances by maximizing the complete-data likelihood along successive given directions (that is, through conditional likelihoods).

5.5.2 Neural Networks

Neural networks provide another type of missing data model where simulation methods are almost always necessary. These models are frequently used in classification and pattern recognition, as well as in robotics and computer vision (see Cheng and Titterington 1994 for a review on these models). Barring the biological vocabulary and the idealistic connection with actual neurons, the theory of neural networks covers

(i) modeling nonlinear relations between explanatory and dependent (explained) variables,
(ii) estimation of the parameters of these models based on a (training) sample

Although the neural network literature usually avoids probabilistic modeling, these models can be analyzed and estimated from a statistical point of view (see Neal 1999 or Ripley 1994, 1996). They can also be seen as a particular type of nonparametric estimation problem, where a major issue is then identifiability.

A simple classical example of a neural network is the *multilayer* model (also called the *backpropagation model*) which relates explanatory variables $x = (x_1, \ldots, x_n)$ with dependent variables $y = (y_1, \ldots, y_n)$ through a hidden "layer", $h = (h_1, \ldots, h_p)$, where $(k = 1, \ldots, p; \ell = 1, \ldots, n)$

$$h_k = f\left(\alpha_{k0} + \sum_j \alpha_{kj} x_j\right),$$

$$\mathbb{E}[Y_\ell|h] = g\left(\beta_{\ell 0} + \sum_{k=1}^p \beta_{\ell k} h_k\right),$$

and $\text{var}(Y_\ell) = \sigma^2$. The functions f and g are known (or arbitrarily chosen) from categories such as *threshold*, $f(t) = \mathbb{I}_{t>0}$, *hyperbolic*, $f(t) = \tanh(t)$, or *sigmoid*, $f(t) = 1/(1 + e^{-t})$.

As an example, consider the problem of character recognition, where handwritten manuscripts are automatically deciphered. The x's may correspond to geometric characteristics of a digitized character, or to pixel gray levels, and the y's are the 26 letters of the alphabet, plus side symbols. (See Le Cun et al. 1989 for an actual modeling based on a sample of 7291, 16×16 pixel images, for 9760 parameters.) The likelihood of the multilayer model then includes the parameters $\alpha = (\alpha_{kj})$ and $\beta = (\beta_{lk})$ in a nonlinear structure. Assuming normality, for observations (y_t, x_t), $t = 1, 2, \ldots, T$, the log-likelihood can be written

$$\ell(\alpha, \beta|x, y) = -\sum_{t=1}^T \sum_{l=1}^n (y_{tl} - \mathbb{E}[y_{tl}|x_{tl}])^2 / 2\sigma^2.$$

A similar objective function can be derived using a least squares criterion. The maximization of $\ell(\alpha, \beta|x, y)$ involves the detection and the elimination of numerous local modes.

5.5.3 The Robbins–Monro procedure

The Robbins–Monro algorithm (Robbins and Monro 1951) is a technique of stochastic approximation to solve for x in equations of the form

(5.23)
$$h(x) = \beta ,$$

when $h(x)$ can be written as in (5.8). It was also extended by Kiefer and Wolfowitz (1952) to the more general setup of (5.1). In the case of (5.23), the Robbins–Monro method proceeds by generating a Markov chain of the form

(5.24)
$$X_{j+1} = X_j + \gamma_j (\beta - H(Z_j, X_j)) ,$$

where Z_j is *simulated* from the conditional distribution defining (5.8). The following result then describes sufficient conditions on the γ_j's for the algorithm to be convergent (see Bouleau and Lépingle 1994 for a proof).

Theorem 5.24. *If (γ_n) is a sequence of positive numbers such that*

$$\sum_{n=1}^{\infty} \gamma_n = +\infty \quad and \quad \sum_{n=1}^{\infty} \gamma_n^2 < +\infty,$$

if the x_j's are simulated from H conditionally on θ_j such that

$$\mathbb{E}[H(X_j, \theta_j)|\theta_j] = h(\theta_j)$$

and $|x_j| \leq B$ for a fixed bound B, and if there exists $\theta^* \in \Theta$ such that

$$\inf_{\delta \leq |\theta - \theta^*| \leq 1/\delta} (\theta - \theta^*) \cdot (h(\theta) - \beta) > 0$$

for every $0 < \delta < 1$, the sequence (θ_j) converges to θ^* almost surely.

The solution of the maximization problem (5.1) can be expressed in terms of the solution of the equation $\nabla h(\theta) = 0$ if the problem is sufficiently regular; that is, if the maximum is not achieved on the boundary of the domain Θ. Note then the similarity between (5.3) and (5.24). Since its proposal, this method has seen numerous variations. Besides Benveniste et al. (1990) and Bouleau and Lépingle (1994), see Wasan (1969), Kersting (1987), Winkler (1995), or Duflo (1996) for more detailed references.

When the function h has several local maxima, the Robbins–Monro procedure converges to one of these maxima. In the particular case $H(x, \theta) = h(\theta) + x/\sqrt{a}$ $(a > 0)$, Pflug (1994) examines the relation

(5.25)
$$\theta_{j+1} = \theta_j + a\, h(\theta_j) + \sqrt{a/2}\, x_j ,$$

when the X_j are iid with $\mathbb{E}[X_j] = 0$ and $\text{var}(X_j) = T$. The relevance of this particular case is that, under the conditions

(i) $\exists k > 0$, $k_1 > 0$ such that $\theta \cdot h(\theta) \leq -k_1|\theta|$ for $|\theta| > k$,
(ii) $|h(\theta)| \leq k_2|h(\theta_0)|$ for $K < |\theta| < |\theta_0|$,
(iii) $|h'(\theta)| \leq k_3$,

the stationary measure ν_a associated with the Markov chain (5.25) weakly converges to the distribution with density

$$c(T) \exp\left(E(\theta)/T\right) ,$$

when a goes to 0, T remaining fixed and E being the primitive of h (see Duflo 1996). The hypotheses (i)–(iii) ensure, in particular, that $\exp\{E(\theta)/T\}$ is integrable on $\mathbb{R}$. Therefore, the limiting distribution of θ_j (when a goes to 0) is the so-called *Gibbs measure* with *energy function* E, which is a pivotal quantity for the *simulated annealing* method introduced in Section 5.2.3. In particular, when T goes to 0, the Gibbs measure converges to the uniform distribution on the set of (global) maxima of E (see Hwang 1980). This convergence is interesting more because of the connections it exhibits with the notions of Gibbs measure and of simulated annealing rather than for its practical consequences. The assumptions (i)–(iii) are rather restrictive and difficult to check for implicit h's, and the representation (5.25) is rather specialized! (Note, however, that the completion of $h(\theta)$ in $H(x,\theta)$ is free, since the conditions (i)–(iii) relate only to h.)

5.5.4 Monte Carlo Approximation

In cases where a function $h(x)$ can be written as $\mathbb{E}[H(x, Z)]$ but is not directly computable, it can be approximated by the empirical (Monte Carlo) average

$$\hat{h}(x) = \frac{1}{m} \sum_{i=1}^{m} H(x, z_i) \, ,$$

where the Z_i's are generated from the conditional distribution $f(z|x)$. This approximation yields a convergent estimator of $h(x)$ for every value of x, but its use in optimization setups is not recommended for at least two related reasons:

(i) Presumably $h(x)$ needs to be evaluated at many points, which will involve the generation of many samples of Z_i's of size m.
(ii) Since the sample changes with every value of x, the resulting sequence of evaluations of h will usually not be smooth.

These difficulties prompted Geyer (1996) to suggest, instead, an *importance sampling* approach to this problem, using a single sample of Z_i's simulated from $g(z)$ and estimate $h(x)$ with

(5.26) $$\hat{h}_m(x) = \frac{1}{m} \sum_{i=1}^{m} \frac{f(z_i|x)}{g(z_i)} H(x, z_i) \, ,$$

where the Z_i's are simulated from $g(z)$. Since this evaluation of h does not depend on x, points (i) and (ii) above are answered.

The problem then shifts from (5.1) to

(5.27) $$\max_{x} \hat{h}_m(x) \, ,$$

which leads to a convergent solution in most cases and also allows for the use of regular optimization techniques, since the function $\hat{h}_m$ does not vary with each iteration. However, three remaining drawbacks of this approach are as follows:

(i) As $\hat{h}_m$ is expressed as a sum, it most often enjoys fewer analytical properties than the original function h.
(ii) The choice of the importance function g can be very influential in obtaining a good approximation of the *function* $h(x)$.

(iii) The number of points z_i used in the approximation should vary with x to achieve the same precision in the approximation of $h(x)$, but this is usually impossible to assess in advance.

In the case $g(z) = f(z|x_0)$, Geyer's (1996) solution to (ii) is to use a recursive process in which x_0 is updated by the solution of the last optimization at each step. The Monte Carlo maximization algorithm then looks like the following:

Algorithm A.22 –Monte Carlo Maximization–

At step i

1. Generate $z_1, \ldots, z_m \sim f(z|x_i)$
   and compute $\hat{h}_{g_i}$ with $g_i = f(\cdot|x_i)$.
2. Find $x^* = \arg\max \hat{h}_{g_i}(x)$.
3. Update x_i to $x_{i+1} = x^*$.   [A.22]

Repeat until $x_i = x_{i+1}$.

Example 5.25. Maximum likelihood estimation for exponential families.
Geyer and Thompson (1992) take advantage of this technique to derive maximum likelihood estimators in exponential families; that is, for functions

$$h(x|\theta) = c(\theta)e^{\theta x - \nu(x)} = c(\theta)\tilde{h}(x|\theta) \ .$$

However, $c(\theta)$ may be unknown or difficult to compute. Geyer and Thompson (1992, 1995) establish that maximization of $h(x|\theta)$ is equivalent to maximizing

$$\log \frac{\tilde{h}(x|\theta)}{\tilde{h}(x|\eta)} - \log \frac{1}{m} \sum_{i=1}^{m} \frac{\tilde{h}(x_i|\theta)}{\tilde{h}(x_i|\eta)} \ ,$$

where the X_i's are generated from $h(x|\eta)$ (see Problem 5.11, and Geyer 1993, 1994).

This representation also extends to setups where the likelihood $L(\theta|x)$ is known up to a multiplicative constant (that is, where $L(\theta|x) \propto h(\theta|x)$).   ‖

6

Markov Chains

Leaphorn never counted on luck. Instead, he expected order—the natural sequence of behavior, the cause producing the natural effect, the human behaving in the way it was natural for him to behave. He counted on that and on his own ability to sort out the chaos of observed facts and find in them this natural order.

—Tony Hillerman, *The Blessing Way*

In this chapter we introduce fundamental notions of Markov chains and state the results that are needed to establish the convergence of various MCMC algorithms and, more generally, to understand the literature on this topic. Thus, this chapter, along with basic notions of probability theory, will provide enough foundation for the understanding of the following chapters. It is, unfortunately, a necessarily brief and, therefore, incomplete introduction to Markov chains, and we refer the reader to Meyn and Tweedie (1993), on which this chapter is based, for a thorough introduction to Markov chains. Other perspectives can be found in Doob (1953), Chung (1960), Feller (1970, 1971), and Billingsley (1995) for general treatments, and Norris (1997), Nummelin (1984), Revuz (1984), and Resnick (1994) for books entirely dedicated to Markov chains. Given the purely utilitarian goal of this chapter, its style and presentation differ from those of other chapters, especially with regard to the plethora of definitions and theorems and to the rarity of examples and proofs. In order to make the book accessible to those who are more interested in the implementation aspects of MCMC algorithms than in their theoretical foundations, we include a preliminary section that contains the essential facts about Markov chains.

Before formally introducing the notion of a Markov chain, note that we do not deal in this chapter with Markov models in *continuous time* (also called *Markov processes*) since the very nature of simulation leads[1] us to consider only discrete-time stochastic processes, $(X_n)_{n \in \mathbb{N}}$. Indeed, Hastings (1970) notes that the use of pseudo-random generators and the representation of numbers in a computer imply that the Markov chains related with Markov chain Monte Carlo methods are, in fact, finite state-space Markov chains. However, we also consider arbitrary state-space Markov chains to allow for continuous support distributions and to avoid addressing the problem of approximation of these distributions with discrete support distributions, since such an approximation depends on both material and algorithmic specifics of a given technique (see Roberts et al. 1995, for a study of the influence of discretization on the convergence of Markov chains associated with Markov chain Monte Carlo algorithms).

6.1 Essentials for MCMC

For those familiar with the properties of Markov chains, this first section provides a brief survey of the properties of Markov chains that are contained in the chapter and are essential for the study of MCMC methods. Starting with Section 6.2, the theory of Markov chains is developed from first principles.

In the setup of MCMC algorithms, Markov chains are constructed from a *transition kernel* K (Definition 6.2), a conditional probability density such that $X_{n+1} \sim K(X_n, X_{n+1})$. A typical example is provided by the *random walk* process, formally defined as follows.

Definition 6.1. A sequence of random variables (X_n) is a *random walk* if it satisfies

$$X_{n+1} = X_n + \epsilon_n \,,$$

where ϵ_n is generated independently of $X_n, X_{n-1}, \ldots$. If the distribution of the ϵ_n is symmetric about zero, the sequence is called a *symmetric random walk*.

There are many examples of random walks (see Examples 6.39, 6.40, and 6.73), and random walks play a key role in many MCMC algorithms, particularly those based on the Metropolis–Hastings algorithm (see Chapter 7).

The chains encountered in MCMC settings enjoy a very strong stability property, namely a *stationary probability distribution* exists by construction (Definition 6.35); that is, a distribution π such that if $X_n \sim \pi$, then $X_{n+1} \sim \pi$, if the kernel K allows for free moves all over the state space. (This freedom is called *irreducibility* in the theory of Markov chains and is formalized in

---

[1] Some Markov chain Monte Carlo algorithms still employ a diffusion representation to speed up convergence to the stationary distribution (see, for instance, Section 7.8.5, Roberts and Tweedie 1995, or Phillips and Smith 1996).

Definition 6.13 as the existence of $n \in \mathbb{N}$ such that $P(X_n \in A|X_0) > 0$ for every A such that $\pi(A) > 0$.) This property also ensures that most of the chains involved in MCMC algorithms are *recurrent* (that is, that the average number of visits to an arbitrary set A is infinite (Definition 6.29)), or even Harris recurrent (that is, such that the probability of an infinite number of returns to A is 1 (Definition 6.32)). Harris recurrence ensures that the chain has the same limiting behavior for *every* starting value instead of *almost* every starting value. (Therefore, this is the Markov chain equivalent of the notion of continuity for functions.)

This latter point is quite important in the context of MCMC algorithms. Since most algorithms are started from some arbitrary point x_0, we are in effect starting the algorithm from a set of measure zero (under a continuous dominating measure). Thus, insuring that the chain converges for almost every starting point is not enough, and we need Harris recurrence to guarantee convergence from every starting point.

The *stationary distribution* is also a *limiting distribution* in the sense that the limiting distribution of X_{n+1} is π under the total variation norm (see Proposition 6.48), notwithstanding the initial value of X_0. Stronger forms of convergence are also encountered in MCMC settings, like *geometric* and *uniform* convergences (see Definitions 6.54 and 6.58). In a simulation setup, a most interesting consequence of this convergence property is that the average

$$(6.1) \qquad \frac{1}{N} \sum_{n=1}^{N} h(X_n)$$

converges to the expectation $\mathbb{E}_{\pi}[h(X)]$ almost surely. When the chain is *reversible* (Definition 6.44) (that is, when the transition kernel is symmetric), a Central Limit Theorem also holds for this average.

In Chapter 12, diagnostics will be based on a minorization condition; that is, the existence of a set C such that there also exists $m \in \mathbb{N}$, $\epsilon_m > 0$, and a probability measure ν_m such that

$$P(X_m \in A|X_0) \geq \epsilon_m \nu_m(A)$$

when $X_0 \in C$. The set C is then called a *small set* (Definition 6.19) and visits of the chain to this set can be exploited to create independent batches in the sum (6.1), since, with probability ϵ_m, the next value of the *m-skeleton Markov chain* $(X_{mn})_n$ is generated from the minorizing measure ν_m, which is independent of X_0.

As a final essential, it is sometimes helpful to associate the probabilistic language of Markov chains with the statistical language of data analysis.

Statistics	Markov Chain
marginal distribution	$\Leftrightarrow$ invariant distribution
proper marginals	$\Leftrightarrow$ positive recurrent

Thus, if the marginals are proper, for convergence we only need our chain to be aperiodic. This is easy to satisfy; a sufficient condition is that $K(x_n, \cdot) > 0$ (or, equivalently, $f(\cdot|x_n) > 0$) in a neighborhood of x_n. If the marginals are not proper, or if they do not exist, then the chain is not positive recurrent. It is either null recurrent or transient, and both cases are bad.

6.2 Basic Notions

A Markov chain is a sequence of random variables that can be thought of as evolving over time, with probability of a transition depending on the particular set in which the chain is. It therefore seems natural and, in fact, is mathematically somewhat cleaner to define the chain in terms of its *transition kernel*, the function that determines these transitions.

Definition 6.2. A *transition kernel* is a function K defined on $\mathcal{X} \times \mathcal{B}(\mathcal{X})$ such that

(i) $\forall x \in \mathcal{X}$, $K(x, \cdot)$ is a probability measure;
(ii) $\forall A \in \mathcal{B}(\mathcal{X})$, $K(\cdot, A)$ is measurable.

When $\mathcal{X}$ is *discrete*, the transition kernel simply is a (transition) matrix K with elements

$$P_{xy} = P(X_n = y | X_{n-1} = x), \qquad x, y \in \mathcal{X}.$$

In the continuous case, the *kernel* also denotes the conditional density $K(x, x')$ of the transition $K(x, \cdot)$; that is, $P(X \in A|x) = \int_A K(x, x')dx'$.

Example 6.3. Bernoulli–Laplace Model. Consider $\mathcal{X} = \{0, 1, \dots, M\}$ and a chain (X_n) such that X_n represents the state, at time n, of a tank which contains exactly M particles and is connected to another identical tank. Two types of particles are introduced in the system, and there are M of each type. If X_n denotes the number of particles of the first kind in the first tank at time n and the moves are restricted to a single exchange of particles between the two tanks at each instant, the transition matrix is given by (for $0 < x, y < M$) $P_{xy} = 0$ if $|x - y| > 1$,

$$P_{xx} = 2\,\frac{x(M-x)}{M^2}\,, \quad P_{x(x-1)} = \left(\frac{x}{M}\right)^2, \quad P_{x(x+1)} = \left(\frac{M-x}{M}\right)^2$$

and $P_{01} = P_{M(M-1)} = 1$. (This model is the *Bernoulli–Laplace* model; see Feller 1970, Chapter XV.)　‖

The chain (X_n) is usually defined for $n \in \mathbb{N}$ rather than for $n \in \mathbb{Z}$. Therefore, the distribution of X_0, the initial state of the chain, plays an important role. In the discrete case, where the kernel K is a transition matrix, given an initial distribution $\mu = (\omega_1, \omega_2, \ldots)$, the marginal probability distribution of X_1 is obtained from the matrix multiplication

$$(6.2) \qquad\qquad \mu_1 = \mu K$$

and, by repeated multiplication, $X_n \sim \mu_n = \mu K^n$. Similarly, in the continuous case, if μ denotes the initial distribution of the chain, namely if

$$(6.3) \qquad\qquad X_0 \sim \mu ,$$

then we let P_μ denote the probability distribution of (X_n) under condition (6.3). When X_0 is fixed, in particular for μ equal to the Dirac mass δ_{x_0}, we use the alternative notation P_{x_0}.

Definition 6.4. Given a transition kernel K, a sequence $X_0, X_1, \ldots, X_n, \ldots$ of random variables is a *Markov chain*, denoted by (X_n), if, for any t, the conditional distribution of X_t given $x_{t-1}, x_{t-2}, \ldots, x_0$ is the same as the distribution of X_t given x_{t-1}; that is,

$$P(X_{k+1} \in A | x_0, x_1, x_2, \ldots, x_k) = P(X_{k+1} \in A | x_k)$$
$$(6.4) \qquad\qquad\qquad = \int_A K(x_k, dx) .$$

The chain is *time homogeneous*, or simply *homogeneous*, if the distribution of $(X_{t_1}, \ldots, X_{t_k})$ given x_{t_0} is the same as the distribution of $(X_{t_1 - t_0}, X_{t_2 - t_0}, \ldots, X_{t_k - t_0})$ given x_0 for every k and every $(k+1)$-uplet $t_0 \le t_1 \le \cdots \le t_k$.

So, in the case of a Markov chain, if the initial distribution or the initial state is known, the construction of the Markov chain (X_n) is entirely determined by its *transition*, namely by the distribution of X_n conditionally on x_{n-1}.

The study of Markov chains is almost always restricted to the time-homogeneous case and we omit this designation in the following. It is, however, important to note here that an incorrect implementation of Markov chain Monte Carlo algorithms can easily produce nonhomogeneous Markov chains for which the standard convergence properties do not apply. (See also the case of the ARMS algorithm in Section 7.4.2.)

Example 6.5. Simulated Annealing. The *simulated annealing* algorithm (see Section 5.2.3 for details) is often implemented in a nonhomogeneous form and studied in time-homogeneous form. Given a finite state-space with size K, $\Omega = \{1, 2, \ldots, K\}$, an energy function $E(\cdot)$, and a temperature T, the simulated annealing Markov chain $X_0, X_1, \ldots$ is represented by the following transition operator: Conditionally on X_n, Y is generated from a fixed probability distribution $(\pi_1, \ldots, \pi_K)$ on Ω and the new value of the chain is given by

$$X_{n+1} = \begin{cases} Y & \text{with probability} \quad \exp\{(E(Y) - E(X_n))/T\} \wedge 1 \\ X_n & \text{otherwise.} \end{cases}$$

If the temperature T depends on n, the chain is time heterogeneous. ||

Example 6.6. AR(1) Models. AR(1) models provide a simple illustration of Markov chains on continuous state-space. If

$$X_n = \theta X_{n-1} + \varepsilon_n , \qquad \theta \in \mathbb{R},$$

with $\varepsilon_n \sim N(0, \sigma^2)$, and if the ε_n's are independent, X_n is indeed independent from $X_{n-2}, X_{n-3}, \ldots$ conditionally on X_{n-1}. The Markovian properties of an AR(q) process can be derived by considering the vector $(X_n, \ldots, X_{n-q+1})$. On the other hand, ARMA(p, q) models do not fit in the Markovian framework (see Problem 6.3). ||

In the general case, the fact that the kernel K determines the properties of the chain (X_n) can be inferred from the relations

$$P_x(X_1 \in A_1) = K(x, A_1) ,$$

$$P_x((X_1, X_2) \in A_1 \times A_2) = \int_{A_1} K(y_1, A_2) \, K(x, dy_1)$$

$$\cdots$$

$$P_x((X_1, \ldots, X_n) \in A_1 \times \cdots \times A_n) = \int_{A_1} \cdots \int_{A_{n-1}} K(y_{n-1}, A_n)$$
$$\times K(x, dy_1) \cdots K(y_{n-2}, dy_{n-1}) .$$

In particular, the relation $P_x(X_1 \in A_1) = K(x, A_1)$ indicates that $K(x_n, dx_{n+1})$ is a *version* of the conditional distribution of X_{n+1} given X_n. However, as we have defined a Markov chain by first specifying this kernel, we do not need to be concerned with different versions of the conditional probabilities. This is why we noted that constructing the Markov chain through the transition kernel was mathematically "cleaner." (Moreover, in the following chapters, we will see that the objects of interest are often these conditional distributions, and it is important that we need not worry about different versions. Nonetheless, the properties of a Markov chain considered in this chapter are independent of the version of the conditional probability chosen.)

If we denote $K^1(x, A) = K(x, A)$, the kernel for n transitions is given by $(n > 1)$

$$(6.5) \qquad K^n(x, A) = \int_{\mathcal{X}} K^{n-1}(y, A) \, K(x, dy).$$

The following result provides convolution formulas of the type $K^{m+n} = K^m \star K^n$, which are called *Chapman–Kolmogorov equations*.

Lemma 6.7. Chapman–Kolmogorov equations *For every* $(m, n) \in \mathbb{N}^2$, $x \in \mathcal{X}$, $A \in \mathcal{B}(\mathcal{X})$,

$$K^{m+n}(x, A) = \int_{\mathcal{X}} K^n(y, A) \, K^m(x, dy) \ .$$

(In a very informal sense, the Chapman–Kolmogorov equations state that to get from x to A in $m + n$ steps, you must pass through some y on the nth step.) In the discrete case, Lemma 6.7 is simply interpreted as a matrix product and follows directly from (6.2). In the general case, we need to consider K as an operator on the space of integrable functions; that is, we define

$$Kh(x) = \int h(y) \, K(x, dy) \ , \qquad h \in \mathcal{L}_1(\lambda) \ ,$$

λ being the dominating measure of the model. K^n is then the nth composition of P, namely $K^n = K \circ K^{n-1}$.

Definition 6.8. A *resolvant* associated with the kernel P is a kernel of the form

$$K_\varepsilon(x, A) = (1 - \varepsilon) \sum_{i=0}^{\infty} \varepsilon^i K^i(x, A), \qquad 0 < \epsilon < 1,$$

and the chain with kernel K_ε is a K_ε-*chain*.

Given an initial distribution μ, we can associate with the kernel K_ε a chain (X_n^ε) which formally corresponds to a subchain of the original chain (X_n), where the indices in the subchain are generated from a geometric distribution with parameter $1 - \varepsilon$. Thus, K_ϵ is indeed a kernel, and we will see that the resulting Markov chain (X_n^ϵ) enjoys much stronger regularity. This will be used later to establish many properties of the original chain.

If $\mathbb{E}_\mu[\,\cdot\,]$ denotes the expectation associated with the distribution P_μ, the *(weak) Markov property* can be written as the following result, which just rephrases the limited memory properties of a Markov chain:

Proposition 6.9. Weak Markov property *For every initial distribution* μ *and every* $(n + 1)$ *sample* $(X_0, \ldots, X_n)$,

(6.6) $\mathbb{E}_\mu[h(X_{n+1}, X_{n+2}, \ldots)|x_0, \ldots, x_n] = \mathbb{E}_{x_n}[h(X_1, X_2, \ldots)]$,

provided that the expectations exist.

Note that if h is the indicator function, then this definition is exactly the same as Definition 6.4. However, (6.6) can be generalized to other classes of functions—hence the terminology "weak"—and it becomes particularly useful with the notion of *stopping time* in the convergence assessment of Markov chain Monte Carlo algorithms in Chapter 12.

Definition 6.10. Consider $A \in \mathcal{B}(\mathcal{X})$. The first n for which the chain enters the set A is denoted by

$$(6.7) \qquad \tau_A = \inf\{n \geq 1; X_n \in A\}$$

and is called the *stopping time* at A with, by convention, $\tau_A = +\infty$ if $X_n \notin A$ for every n. More generally, a function $\zeta(x_1, x_2, \ldots)$ is called a *stopping rule* if the set $\{\zeta = n\}$ is measurable for the σ-algebra induced by $(X_0, \ldots, X_n)$. Associated with the set A, we also define

$$(6.8) \qquad \eta_A = \sum_{n=1}^{\infty} \mathbb{I}_A(X_n),$$

the *number of passages* of (X_n) in A.

Of particular importance are the related quantities $\mathbb{E}_x[\eta_A]$ and $P_x(\tau_A < \infty)$, which are the *average number of passages in A* and the *probability of return to A in a finite number of steps*.

We will be mostly concerned with stopping rules of the form given in (6.7), which express the fact that τ_A takes the value n when none of the values of $X_0, X_1, \ldots, X_{n-1}$ are in the given state (or set) A, but the nth value is. The *strong Markov property* corresponds to the following result, whose proof follows from the weak Markov property and conditioning on $\{\zeta = n\}$:

Proposition 6.11. Strong Markov property *For every initial distribution μ and every stopping time ζ which is almost surely finite,*

$$\mathbb{E}_\mu[h(X_{\zeta+1}, X_{\zeta+2}, \cdots)|x_1, \ldots, x_\zeta] = \mathbb{E}_{x_\zeta}[h(X_1, X_2, \cdots)] \,,$$

provided the expectations exist.

We can thus condition on a random number of instants while keeping the fundamental properties of a Markov chain.

Example 6.12. Coin tossing. In a coin tossing game, player b has a gain of $+1$ if a head appears and player c has a gain of $+1$ if a tail appears (so player b has a "gain" of -1 (a loss) if a tail appears). If X_n is the sum of the gains of player b after n rounds of this coin tossing game, the transition matrix P is an infinite dimensional matrix with upper and lower subdiagonals equal to $1/2$. Assume that player b has B dollars and player c has C dollars, and consider the following return times:

$$\tau_1 = \inf\{n; X_n = 0\}, \quad \tau_2 = \inf\{n; X_n < -B\}, \quad \tau_3 = \inf\{n; X_n > C\},$$

which represent respectively the return to null and the ruins of the first and second players, that is to say, the first times the fortunes of both players, respectively B and C, are spent. The probability of ruin (bankruptcy) for the first player is then $P_0(\tau_2 > \tau_3)$. (Feller 1970, Chapter III, has a detailed analysis of this coin tossing game.) $\qquad \|$

6.3 Irreducibility, Atoms, and Small Sets

6.3.1 Irreducibility

The property of irreducibility is a first measure of the sensitivity of the Markov chain to the initial conditions, x_0 or μ. It is crucial in the setup of Markov chain Monte Carlo algorithms, because it leads to a guarantee of convergence, thus avoiding a detailed study of the transition operator, which would otherwise be necessary to specify "acceptable" initial conditions.

In the discrete case, the chain is *irreducible* if all states communicate, namely if

$$P_x(\tau_y < \infty) > 0 , \qquad \forall x, y \in \mathcal{X} ,$$

τ_y being the first time y is visited, defined in (6.7). In many cases, $P_x(\tau_y < \infty)$ is uniformly equal to zero, and it is necessary to introduce an auxiliary measure φ on $\mathcal{B}(\mathcal{X})$ to correctly define the notion of irreducibility.

Definition 6.13. Given a measure φ, the Markov chain (X_n) with transition kernel $K(x, y)$ is φ-*irreducible* if, for every $A \in \mathcal{B}(\mathcal{X})$ with $\varphi(A) > 0$, there exists n such that $K^n(x, A) > 0$ for all $x \in \mathcal{X}$ (equivalently, $P_x(\tau_A < \infty) > 0$). The chain is *strongly* φ-*irreducible* if $n = 1$ for all measurable A.

Example 6.14. (Continuation of Example 6.3) In the case of the Bernoulli–Laplace model, the (finite) chain is indeed irreducible since it is possible to connect the states x and y in $|x - y|$ steps with probability

$$\prod_{i=x\wedge y}^{x\vee y-1} \left(\frac{K-i}{K}\right)^2 .$$

$\|$

The following result provides equivalent definitions of irreducibility. The proof is left to Problem 6.13, and follows from (6.9) and the Chapman–Kolmogorov equations.

Theorem 6.15. *The chain* (X_n) *is* φ-*irreducible if and only if for every* $x \in \mathcal{X}$ *and every* $A \in \mathcal{B}(\mathcal{X})$ *such that* $\varphi(A) > 0$, *one of the following properties holds:*

(i) there exists $n \in \mathbb{N}^*$ *such that* $K^n(x, A) > 0$;
(ii) $\mathbb{E}_x[\eta_A] > 0$;
(iii) $K_\varepsilon(x, A) > 0$ *for an* $0 < \epsilon < 1$.

The introduction of the K_ε-chain then allows for the creation of a strictly positive kernel in the case of a φ-irreducible chain and this property is used in the following to simplify the proofs. Moreover, the measure φ in Definition 6.13 plays no crucial role in the sense that irreducibility is an intrinsic property of (X_n) and does not rely on φ.

The following theorem details the properties of the *maximal irreducibility measure* ψ.

Theorem 6.16. *If (X_n) is φ-irreducible, there exists a probability measure ψ such that:*

(i) the Markov chain (X_n) is ψ-irreducible;
(ii) if there exists a measure ξ such that (X_n) is ξ-irreducible, then ξ is dominated by ψ; that is, $\xi \ll \psi$;
(iii) if $\psi(A) = 0$, then $\psi(\{y; P_y(\tau_A < \infty) > 0\}) = 0$;
(iv) the measure ψ is equivalent to

$$(6.9) \qquad \psi_0(A) = \int_{\mathcal{X}} K_{1/2}(x, A)\, \varphi(dx), \qquad \forall A \in \mathcal{B}(\mathcal{X}) ;$$

that is, $\psi \ll \psi_0$ and $\psi_0 \ll \psi$.

This result provides a *constructive* method of determining the maximal irreducibility measure ψ through a candidate measure φ, which still needs to be defined.

Example 6.17. (Continuation of Example 6.6) When $X_{n+1} = \theta X_n + \varepsilon_{n+1}$ and ε_n are independent normal variables, the chain is irreducible, the reference measure being the *Lebesgue measure*, λ. (In fact, $K(x, A) > 0$ for every $x \in \mathbb{R}$ and every A such that $\lambda(A) > 0$.) On the other hand, if ε_n is uniform on $[-1, 1]$ and $|\theta| > 1$, the chain is not irreducible anymore. For instance, if $\theta > 1$, then

$$X_{n+1} - X_n \geq (\theta - 1)X_n - 1 \geq 0$$

for $X_n \geq 1/(\theta - 1)$. The chain is thus monotonically increasing and obviously cannot visit previous values. ‖

6.3.2 Atoms and Small Sets

In the discrete case, the transition kernel is necessarily atomic in the usual sense; that is, there exist points in the state-space with positive mass. The extension of this notion to the general case by Nummelin (1978) is powerful enough to allow for a control of the chain which is almost as "precise" as in the discrete case.

Definition 6.18. The Markov chain (X_n) has an *atom* $\alpha \in \mathcal{B}(\mathcal{X})$ if there exists an associated nonzero measure ν such that

$$K(x, A) = \nu(A), \qquad \forall x \in \alpha, \ \forall A \in \mathcal{B}(\mathcal{X}) .$$

If (X_n) is ψ-irreducible, the atom is *accessible* when $\psi(\alpha) > 0$.

While it trivially applies to every possible value of X_n in the discrete case, this notion is often too strong to be of use in the continuous case since it implies that the transition kernel is *constant* on a set of positive measure. A more powerful generalization is the so-called *minorizing condition*, namely that there exists a set $C \in \mathcal{B}(\mathcal{X})$, a constant $\varepsilon > 0$, and a probability measure ν such that

$$(6.10) \qquad K(x, A) \geq \varepsilon \nu(A), \qquad \forall x \in C, \, \forall A \in \mathcal{B}(\mathcal{X}) \,.$$

The probability measure ν thus appears as a constant component of the transition kernel on C. The minorizing condition (6.10) leads to the following notion, which is essential in this chapter and in Chapters 7 and 12 as a technique of proof and as the basis of *renewal theory*.

Definition 6.19. A set C is *small* if there exist $m \in \mathbb{N}^*$ and a nonzero measure ν_m such that

$$(6.11) \qquad K^m(x, A) \geq \nu_m(A), \qquad \forall x \in C, \, \forall A \in \mathcal{B}(\mathcal{X}) \,.$$

Example 6.20. (Continuation of Example 6.17) Since $X_n | x_{n-1} \sim \mathcal{N}(\theta x_{n-1}, \sigma^2)$, the transition kernel is bounded from below by

$$\frac{1}{\sigma\sqrt{2\pi}} \exp\left\{ (-x_n^2 + 2\theta x_n \underline{w} - \theta^2 \underline{w}^2 \wedge \overline{w}^2)/2\sigma^2 \right\} \quad \text{if } x_n > 0,$$

$$\frac{1}{\sigma\sqrt{2\pi}} \exp\left\{ (-x_n^2 + \theta x_n \overline{w}\sigma^{-2} - \theta^2 \underline{w}^2 \wedge \overline{w}^2)/2\sigma^2 \right\} \quad \text{if } x_n < 0,$$

when $x_{n-1} \in [\underline{w}, \overline{w}]$. The set $C = [\underline{w}, \overline{w}]$ is indeed a small set, as the measure ν_1, with density

$$\frac{\exp\{(-x^2 + 2\theta x \underline{w})/2\sigma^2\}\, \mathbb{I}_{x>0} + \exp\{(-x^2 + 2\theta x \overline{w})/2\sigma^2\}\, \mathbb{I}_{x<0}}{\sqrt{2\pi}\, \sigma[\Phi(-\theta\underline{w}/\sigma^2)\, \exp\{\theta^2\underline{w}^2/2\sigma^2\} + [1 - \Phi(-\theta\underline{w}/\sigma^2)]\, \exp\{\theta^2\overline{w}^2/2\sigma^2\}]} \,,$$

and

$$\varepsilon = \exp\{-\theta^2\underline{w}^2/2\sigma^2\} \left[\Phi(-\theta\underline{w}/\sigma^2)\, \exp\{\theta^2\underline{w}^2/2\sigma^2\} \right.$$
$$\left. + [1 - \Phi(-\theta\underline{w}/\sigma^2)]\, \exp\{\theta^2\overline{w}^2/2\sigma^2\} \right] \,,$$

satisfy (6.11) with $m = 1$. $\qquad\qquad \|$

A sufficient condition for C to be small is that (6.11) is satisfied by the K_ε-chain in the special case $m = 1$. The following result indicates the connection between small sets and irreducibility.

Theorem 6.21. Let (X_n) be a ψ-irreducible chain. For every set $A \in \mathcal{B}(\mathcal{X})$ such that $\psi(A) > 0$, there exist $m \in \mathbb{N}^*$ and a small set $C \subset A$ such that the associated minorizing measure satisfies $\nu_m(C) > 0$. Moreover, $\mathcal{X}$ can be decomposed in a denumerable partition of small sets.

The proof of this characterization result is rather involved (see Meyn and Tweedie 1993, pp. 107–109). The decomposition of $\mathcal{X}$ as a denumerable union of small sets is based on an arbitrary small set C and the sequence

$$C_{nm} = \{y; K^n(y, C) > 1/m\}$$

(see Problem 6.19).

Small sets are obviously easier to exhibit than atoms, given the freedom allowed by the minorizing condition (6.11). Moreover, they are, in fact, very common since, in addition to Theorem 6.21, Meyn and Tweedie (1993, p. 134) show that for sufficiently regular (in a topological sense) Markov chains, every *compact* set is small. Atoms, although a special case of small sets, enjoy stronger stability properties since the transition probability is invariant on α. However, *splitting methods* (see below) offer the possibility of extending most of these properties to the general case and it will be used as a technique of proof in the remainder of the chapter.

If the minorizing condition holds for (X_n), there are two ways of deriving a companion Markov chain $(\check{X}_n)$ sharing many properties with (X_n) and possessing an atom α. The first method is called *Nummelin's splitting* and constructs a chain made of two copies of (X_n) (see Nummelin 1978 and Meyn and Tweedie 1993, Section 5.1).

A second method, discovered at approximately the same time, is due to Athreya and Ney (1978) and uses a stopping time to create an atom. We prefer to focus on this latter method because it is related to notions of *renewal time*, which are also useful in the control of Markov chain Monte Carlo algorithms (see Section 12.2.3).

Definition 6.22. A *renewal time* (or *regeneration time*) is a stopping rule τ with the property that $(X_\tau, X_{\tau+1}, \ldots)$ is independent of $(X_{\tau-1}, X_{\tau-2}, \ldots)$.

For instance, in Example 6.12, the returns to zero gain are renewal times. The excursions between two returns to zero are independent and identically distributed (see Feller 1970, Chapter III). More generally, visits to atoms are renewal times, whose features are quite appealing in convergence control for Markov chain Monte Carlo algorithms (see Chapter 12).

If (6.10) holds and if the probability $P_x(\tau_C < \infty)$ of a return to C in a finite time is identically equal to 1 on $\mathcal{X}$, Athreya and Ney (1978) modify the transition kernel when $X_n \in C$, by simulating X_{n+1} as

$$(6.12) \qquad X_{n+1} \sim \begin{cases} \nu & \text{with probability } \varepsilon \\ \dfrac{K(X_n, \cdot) - \varepsilon\nu(\cdot)}{1 - \varepsilon} & \text{with probability } 1 - \varepsilon; \end{cases}$$

that is, by simulating X_{n+1} from ν with probability ε every time X_n is in C. This modification does not change the marginal distribution of X_{n+1} conditionally on x_n, since

$$\epsilon\nu(A) + (1-\epsilon)\frac{K(x_n, A) - \epsilon\nu(A)}{1-\epsilon} = K(x_n, A), \qquad \forall A \in \mathcal{B}(\mathcal{X}),$$

but it produces *renewal times* for each time j such that $X_j \in C$ and $X_{j+1} \sim \nu$.

Now we clearly see how the renewal times result in independent chains. When $X_{j+1} \sim \nu$, this event is totally independent of any past history, as the current state of the chain has no effect on the measure ν. Note also the key role that is played by the minorization condition. It allows us to create the split chain with the same marginal distribution as the original chain. We denote by $(j > 0)$

$$\tau_j = \inf\{n > \tau_{j-1}; X_n \in C \text{ and } X_{n+1} \sim \nu\}$$

the sequence of *renewal times* with $\tau_0 = 0$. Athreya and Ney (1978) introduce the *augmented chain*, also called the *split chain* $\check{X}_n = (X_n, \check{\omega}_n)$, with $\check{\omega}_n = 1$ when $X_n \in C$ and X_{n+1} is generated from ν. It is then easy to show that the set $\check{\alpha} = C \times \{1\}$ *is an atom of the chain* $(\check{X}_n)$, the resulting subchain (X_n) being still a Markov chain with transition kernel $K(x_n, \cdot)$ (see Problem 6.17).

The notion of small set is useful only in finite and discrete settings when the individual probabilities of states are too small to allow for a reasonable rate of renewal. In these cases, small sets are made of collections of states with ν defined as a minimum. Otherwise, small sets reduced to a single value are also atoms.

6.3.3 Cycles and Aperiodicity

The behavior of (X_n) may sometimes be restricted by deterministic constraints on the moves from X_n to X_{n+1}. We formalize these constraints here and show in the following chapters that the chains produced by Markov chain Monte Carlo algorithms do not display this behavior and, hence, do not suffer from the associated drawbacks.

In the discrete case, the *period* of a state $\omega \in \mathcal{X}$ is defined as

$$d(\omega) = \text{g.c.d.} \, \{m \geq 1; K^m(\omega, \omega) > 0\},$$

where we recall that g.c.d. is the greatest common denominator. The value of the period is constant on all states that communicate with ω. In the case of an irreducible chain on a finite space $\mathcal{X}$, the transition matrix can be written (with a possible reordering of the states) as a block matrix

(6.13)
$$P = \begin{pmatrix} 0 & D_1 & 0 & \cdots & 0 \\ 0 & 0 & D_2 & & 0 \\ & & & \ddots & \\ D_d & 0 & 0 & & 0 \end{pmatrix},$$

where the blocks D_i are stochastic matrices. This representation clearly illustrates the forced passage from one group of states to another, with a return to the initial group occurring every dth step. If the chain is irreducible (so all states communicate), there is only one value for the period. An irreducible chain is *aperiodic* if it has period 1. The extension to the general case requires the existence of a small set.

Definition 6.23. A ψ-irreducible chain (X_n) has a *cycle of length d* if there exists a small set C, an associated integer M, and a probability distribution ν_M such that d is the g.c.d. of

$$\{m \geq 1; \ \exists \ \delta_m > 0 \text{ such that } C \text{ is small for } \nu_m \geq \delta_m \nu_M\}.$$

A decomposition like (6.13) can be established in general. It is easily shown that the number d is independent of the small set C and that this number intrinsically characterizes the chain (X_n). The *period* of (X_n) is then defined as the largest integer d satisfying Definition 6.23 and (X_n) is *aperiodic* if $d = 1$. If there exists a small set A and a minorizing measure ν_1 such that $\nu_1(A) > 0$ (so it is possible to go from A to A in a single step), the chain is said to be *strongly aperiodic*. Note that the K_ε-chain can be used to transform an aperiodic chain into a strongly aperiodic chain.

In discrete setups, if one state $x \in \mathcal{X}$ satisfies $P_{xx} > 0$, the chain (X_n) is aperiodic, although this is not a necessary condition (see Problem 6.35).

Example 6.24. (Continuation of Example 6.14) The Bernoulli-Laplace chain is aperiodic and even strongly aperiodic since the diagonal terms satisfy $P_{xx} > 0$ for every $x \in \{0, \dots, K\}$.   ‖

When the chain is continuous and the transition kernel has a component which is absolutely continuous with respect to the Lebesgue measure, with density $f(\cdot|x_n)$, a sufficient condition for aperiodicity is that $f(\cdot|x_n)$ is positive in a neighborhood of x_n. The chain can then remain in this neighborhood for an arbitrary number of instants before visiting any set A. For instance, in Example 6.3, (X_n) is strongly aperiodic when ε_n is distributed according to $\mathcal{U}_{[-1,1]}$ and $|\theta| < 1$ (in order to guarantee irreducibility). The next chapters will demonstrate that Markov chain Monte Carlo algorithms lead to aperiodic chains, possibly via the introduction of additional steps.

6.4 Transience and Recurrence

6.4.1 Classification of Irreducible Chains

From an algorithmic point of view, a Markov chain must enjoy good *stability* properties to guarantee an acceptable approximation of the simulated model. Indeed, *irreducibility* ensures that every set A will be visited by the Markov chain (X_n), but this property is too weak to ensure that the trajectory of (X_n) will enter A often enough. Consider, for instance, a maximization problem using a random walk on the surface of the function to maximize (see Chapter 5). The convergence to the global maximum cannot be guaranteed without a systematic sweep of this surface. Formalizing this stability of the Markov chain leads to different notions of *recurrence*. In a discrete setup, the *recurrence of a state* is equivalent to a guarantee of a sure return. This notion is thus necessarily satisfied for irreducible chains on a finite space.

Definition 6.25. In a finite state-space $\mathcal{X}$, a state $\omega \in \mathcal{X}$ is *transient* if the average number of visits to ω, $\mathbb{E}_\omega[\eta_\omega]$, is finite, and *recurrent* if $\mathbb{E}_\omega[\eta_\omega] = \infty$.

For irreducible chains, the properties of recurrence and transience are properties of the chain, not of a particular state. This fact is easily deduced from the Chapman–Kolmogorov equations. Therefore, if η_A denotes the number of visits defined in (6.8), for every $(x, y) \in \mathcal{X}^2$ either $\mathbb{E}_x[\eta_y] < \infty$ in the transient case or $\mathbb{E}_x[\eta_y] = \infty$ in the recurrent case. The chain is then said to be *transient* or *recurrent*, one of the two properties being necessarily satisfied in the irreducible case.

Example 6.26. Branching process. Consider a population whose individuals reproduce independently of one another. Each individual has X sibling(s), $X \in \mathbb{N}$, distributed according to the distribution with generating function $\phi(s) = \mathbb{E}[s^X]$. If individuals reproduce at fixed instants (thus defining *generations*), the size of the tth generation S_t $(t > 1)$ is given by

$$S_t = X_1 + \cdots + X_{S_{t-1}},$$

where the $X_i \sim \phi$ are independent. Starting with a single individual at time 0, $S_1 = X_1$, the generating function of S_t is $g_t(s) = \phi^t(s)$, with $\phi^t = \phi \circ \phi^{t-1}$ $(t > 1)$. The chain (S_t) is an example of a *branching process* (see Feller 1971, Chapter XII).

If ϕ does not have a constant term (i.e., if $P(X_1 = 0) = 0$), the chain (S_t) is necessarily transient since it is increasing. If $P(X_1 = 0) > 0$, the probability of a return to 0 at time t is $\rho_t = P(S_t = 0) = g_t(0)$, which thus satisfies the recurrence equation $\rho_t = \phi(\rho_{t-1})$. Therefore, there exists a limit ρ different from 1, such that $\rho = \phi(\rho)$, if and only if $\phi'(1) > 1$; namely if $\mathbb{E}[X] > 1$. The chain is thus transient when the average number of siblings per individual is larger than 1. If there exists a restarting mechanism in 0, $S_{t+1}|S_t = 0 \sim \phi$, it is easily shown that when $\phi'(1) > 1$, the number of returns to 0 follows a geometric distribution with parameter ρ. If $\phi'(1) \leq 1$, one can show that the chain is recurrent (see Example 6.42). ‖

The treatment of the general (that is to say, non-discrete) case is based on chains with atoms, the extension to general chains (with small sets) following from Athreya and Ney's (1978) splitting. We begin by extending the notions of recurrence and transience.

Definition 6.27. A set A is called *recurrent* if $\mathbb{E}_x[\eta_A] = +\infty$ for every $x \in A$. The set A is *uniformly transient* if there exists a constant M such that $\mathbb{E}_x[\eta_A] < M$ for every $x \in A$. It is *transient* if there exists a covering of $\mathcal{X}$ by uniformly transient sets; that is, a countable collection of uniformly transient sets B_i such that

$$A = \bigcup_i B_i.$$

Theorem 6.28. Let (X_n) be ψ-irreducible Markov chain with an accessible atom α.

(i) If α is recurrent, every set A of $\mathcal{B}(\mathcal{X})$ such that $\psi(A) > 0$ is recurrent.
(ii) If α is transient, $\mathcal{X}$ is transient.

Property (i) is the most relevant in the Markov chain Monte Carlo setup and can be derived from the Chapman–Kolmogorov equations. Property (ii) is more difficult to establish and uses the fact that $P_\alpha(\tau_\alpha < \infty) < 1$ for a transient set when $\mathbb{E}_x[\eta_A]$ is decomposed conditionally on the last visit to α (see Meyn and Tweedie 1993, p. 181, and Problem 6.29).

Definition 6.29. A Markov chain (X_n) is *recurrent* if

(i) there exists a measure ψ such that (X_n) is ψ-irreducible, and
(ii) for every $A \in \mathcal{B}(\mathcal{X})$ such that $\psi(A) > 0$, $\mathbb{E}_x[\eta_A] = \infty$ for every $x \in A$.

The chain is *transient* if it is ψ-irreducible and if $\mathcal{X}$ is transient.

The classification result of Theorem 6.28 can be easily extended to *strongly aperiodic* chains since they satisfy a minorizing condition (6.11), thus can be split as in (6.3.2), while the chain (X_n) and its split version $(\check{X}_n)$ (see Problem 6.17) are either both recurrent or both transient. The generalization to an arbitrary irreducible chain follows from the properties of the corresponding K_ε-chain which is strongly aperiodic, through the relation

$$(6.14) \qquad \sum_{n=0}^{\infty} K_\varepsilon^n = \frac{1-\varepsilon}{\varepsilon} \sum_{n=0}^{\infty} K^n,$$

since

$$\mathbb{E}_x[\eta_A] = \sum_{n=0}^{\infty} K^n(x, A) = \frac{\epsilon}{1-\epsilon} \sum_{n=0}^{\infty} K_\epsilon^n(x, A).$$

This provides us with the following classification result:

Theorem 6.30. A ψ-irreducible chain is either recurrent or transient.

6.4.2 Criteria for Recurrence

The previous results establish a clear dichotomy between transience and recurrence for irreducible Markov chains. Nevertheless, given the requirement of Definition 6.29, it is useful to examine simpler criteria for recurrence. By analogy with discrete state-space Markov chains, a first approach is based on small sets.

Proposition 6.31. *A ψ-irreducible chain (X_n) is recurrent if there exists a small set C with $\psi(C) > 0$ such that $P_x(\tau_C < \infty) = 1$ for every $x \in C$.*

Proof. First, we show that the set C is recurrent. Given $x \in C$, consider $u_n = K^n(x, C)$ and $f_n = P_x(X_n \in C, X_{n-1} \notin C, \ldots, X_1 \notin C)$, which is the probability of first visit to C at the nth instant, and define

$$\tilde{U}(s) = 1 + \sum_{n=1}^{\infty} u_n s^n \qquad \text{and} \qquad Q(s) = \sum_{n=1}^{\infty} f_n s^n.$$

The equation

(6.15) $$u_n = f_n + f_{n-1} u_1 + \cdots + f_1 u_{n-1}$$

describes the relation between the probability of a visit of C at time n and the probabilities of first visit of C. This implies

$$\tilde{U}(s) = \frac{1}{1 - Q(s)},$$

which connects $\tilde{U}(1) = \mathbb{E}_x[\eta_C] = \infty$ with $Q(1) = P_x(\tau_C < \infty) = 1$. Equation (6.15) is, in fact, valid for the split chain $(\check{X}_n)$ (see Problem 6.17), since a visit to $C \times \{0\}$ ensures independence by renewal. Since $\mathbb{E}_x[\eta_C]$, associated with (X_n), is larger than $\mathbb{E}_{\check{x}}[\eta_{C \times \{0\}}]$, associated with $(\check{x}_n)$, and $P_x(\tau_C < \infty)$ for (X_n) is equal to $P_{\check{X}}(\tau_{C \times \{0\}} < \infty)$ for $(\check{X}_n)$, the recurrence can be extended from $(\check{x}_n)$ to (X_n). The recurrence of $(\check{X}_n)$ follows from Theorem 6.28, since $C \times \{0\}$ is a recurrent atom for $(\check{X}_n)$. $\square\square$

A second method of checking recurrence is based on a generalization of the notions of small sets and minorizing conditions. This generalization involves a *potential function V* and a *drift condition* like (6.38) and uses the transition kernel $K(\cdot, \cdot)$ rather than the sequence K^n. Note 6.9.1 details this approach, as well as its bearing on the following stability and convergence results.

6.4.3 Harris Recurrence

It is actually possible to strengthen the stability properties of a chain (X_n) by requiring not only an infinite average number of visits to every small set but also an infinite number of visits for every path of the Markov chain. Recall that η_A is the number of passages of (X_n) in A, and we consider $P_x(\eta_A = \infty)$, the probability of visiting A an infinite number of times. The following notion of recurrence was introduced by Harris (1956).

Definition 6.32. A set A is *Harris recurrent* if $P_x(\eta_A = \infty) = 1$ for all $x \in A$. The chain (X_n) is *Harris recurrent* if there exists a measure ψ such that (X_n) is ψ-irreducible and for every set A with $\psi(A) > 0$, A is Harris recurrent.

Recall that recurrence corresponds to $\mathbb{E}_x[\eta_\alpha] = \infty$, a weaker condition than $P_x(\eta_A = \infty) = 1$ (see Problem 6.30). The following proposition expresses Harris recurrence as a condition on $P_x(\tau_A < \infty)$ defined in (6.8).

Proposition 6.33. *If for every $A \in \mathcal{B}(\mathcal{X})$, $P_x(\tau_A < \infty) = 1$ for every $x \in A$, then $P_x(\eta_A = \infty) = 1$, for all $x \in \mathcal{X}$, and (X_n) is Harris recurrent.*

Proof. The average number of visits to B before a first visit to A is

$$(6.16) \qquad U_A(x, B) = \sum_{n=1}^{\infty} P_x(X_n \in B, \tau_A \geq n) \, .$$

Then, $U_A(x, A) = P_x(\tau_A < \infty)$, since, if $B \subset A$, $P_x(X_n \in B, \tau_A \geq n) = P_x(X_n \in B, \tau = n) = P_x(\tau_B = n)$. Similarly, if $\tau_A(k), k > 1$, denotes the time of the kth visit to A, $\tau_A(k)$ satisfies

$$P_x(\tau_A(2) < \infty) = \int_A P_y(\tau_A < \infty) \, U_A(x, dy) = 1$$

for every $x \in A$ and, by induction,

$$P_x(\tau_A(k+1) < \infty) = \int_A P_x(\tau_A(k) < \infty) \, U_A(x, dy) = 1.$$

Since $P_x(\eta_A \geq k) = P_x(\tau_A(k) < \infty)$ and

$$P_x(\eta_A = \infty) = \lim_{k \to \infty} P_x(\eta_A \geq k),$$

we deduce that $P_x(\eta_A = \infty) = 1$ for $x \in A$. □

Note that the property of Harris recurrence is needed only when $\mathcal{X}$ is not denumerable. If $\mathcal{X}$ is finite or denumerable, we can indeed show that $\mathbb{E}_x[\eta_x] = \infty$ if and only if $P_x(\tau_x < \infty) = 1$ for every $x \in \mathcal{X}$, through an argument similar to the proof of Proposition 6.31. In the general case, it is possible to prove that if (X_n) is Harris recurrent, then $P_x(\eta_B = \infty) = 1$ for every $x \in \mathcal{X}$ and $B \in \mathcal{B}(\mathcal{X})$ such that $\psi(B) > 0$. This property then provides a sufficient condition for Harris recurrence which generalizes Proposition 6.31.

Theorem 6.34. *If (X_n) is a ψ-irreducible Markov chain with a small set C such that $P_x(\tau_C < \infty) = 1$ for all $x \in \mathcal{X}$, then (X_n) is Harris recurrent.*

Contrast this theorem with Proposition 6.31, where $P_x(\tau_C < \infty) = 1$ only for $x \in C$. This theorem also allows us to replace recurrence by Harris recurrence in Theorem 6.72. (See Meyn and Tweedie 1993, pp. 204–205 for a discussion of the "almost" Harris recurrence of recurrent chains.) Tierney (1994) and Chan and Geyer (1994) analyze the role of Harris recurrence in the setup of Markov chain Monte Carlo algorithms and note that Harris recurrence holds for most of these algorithms (see Chapters 7 and 10).[2]

6.5 Invariant Measures

6.5.1 Stationary Chains

An increased level of stability for the chain (X_n) is attained if the marginal distribution of X_n is independent of n. More formally, this is a requirement for the existence of a probability distribution π such that $X_{n+1} \sim \pi$ if $X_n \sim \pi$, and Markov chain Monte Carlo methods are based on the fact that this requirement, which defines a particular kind of recurrence called *positive recurrence*, can be met. The Markov chains constructed from Markov chain Monte Carlo algorithms enjoy this greater stability property (except in very pathological cases; see Section 10.4.3). We therefore provide an abridged description of invariant measures and positive recurrence.

Definition 6.35. A σ-finite measure π is *invariant* for the transition kernel $K(\cdot, \cdot)$ (and for the associated chain) if

$$\pi(B) = \int_{\mathcal{X}} K(x, B)\, \pi(dx)\,, \qquad \forall B \in \mathcal{B}(\mathcal{X})\,.$$

When there exists an *invariant probability measure* for a ψ-irreducible (hence recurrent by Theorem 6.30) chain, the chain is *positive*. Recurrent chains that do not allow for a finite invariant measure are called *null recurrent*.

The invariant distribution is also referred to as *stationary* if π is a probability measure, since $X_0 \sim \pi$ implies that $X_n \sim \pi$ for every n; thus, the chain is stationary in distribution. (Note that the alternative case when π is not finite is more difficult to interpret in terms of behavior of the chain.) It is easy to show that if the chain is irreducible and allows for an σ-finite invariant measure, this measure is unique, up to a multiplicative factor (see Problem 6.60). The link between positivity and recurrence is given by the following result, which formalizes the intuition that the existence of a invariant measure prevents the probability mass from "escaping to infinity."

Proposition 6.36. *If the chain (X_n) is positive, it is recurrent.*

---

[2] Chan and Geyer (1994) particularly stress that *"Harris recurrence essentially says that there is no measure-theoretic pathology (...) The main point about Harris recurrence is that asymptotics do not depend on the starting distribution because of the 'split' chain construction."*

Proof. If (X_n) is transient, there exists a covering of $\mathcal{X}$ by uniformly transient sets, A_j, with corresponding bounds

$$\mathbb{E}_x[\eta_{A_j}] \leq M_j, \qquad \forall x \in A_j, \ \forall j \in \mathbb{N}.$$

Therefore, by the invariance of π,

$$\pi(A_j) = \int K(x, A_j)\, \pi(dx) = \int K^n(x, A_j)\, \pi(dx).$$

Therefore, for every $k \in \mathbb{N}$,

$$k\,\pi(A_j) = \sum_{n=0}^{k} \int K^n(x, A_j)\, \pi(dx) \leq \int \mathbb{E}_x[\eta_{A_j}]\, \pi(dx) \leq M_j,$$

since, from (6.8) it follows that $\sum_{n=0}^{k} K^n(x, A_j) \leq \mathbb{E}_x[\eta_{A_j}]$. Letting k go to ∞ shows that $\pi(A_j) = 0$, for every $j \in \mathbb{N}$, and hence the impossibility of obtaining an invariant probability measure. $\qquad\square$

We may, therefore, talk of *positive chains* and of *Harris positive chains*, without the superfluous denomination *recurrent* and *Harris recurrent*. Proposition 6.36 is useful only when the positivity of (X_n) can be proved, but, again, the chains produced by Markov chain Monte Carlo methods are, by nature, guaranteed to possess an invariant distribution.

6.5.2 Kac's Theorem

A classical result (see Feller 1970) on irreducible Markov chains with discrete state-space is that the stationary distribution, when it exists, is given by

$$\pi_x = (\mathbb{E}_x[\tau_x])^{-1}, \qquad x \in \mathcal{X},$$

where, from (6.7), we can interpret $\mathbb{E}_x[\tau_x]$ as the average number of excursions between two passages in x. (It is sometimes called *Kac's Theorem*.) It also follows that $(\mathbb{E}_x[\tau_x]^{-1})$ is the eigenvector associated with the eigenvalue 1 for the transition matrix $\mathbb{P}$ (see Problems 6.10 and 6.61). We now establish this result in the more general case when (X_n) has an atom, α.

Theorem 6.37. *Let (X_n) be ψ-irreducible with an atom α. The chain is positive if and only if $\mathbb{E}_\alpha[\tau_\alpha] < \infty$. In this case, the invariant distribution π for (X_n) satisfies*

$$\pi(\alpha) = (\mathbb{E}_\alpha[\tau_\alpha])^{-1}.$$

The notation $\mathbb{E}_\alpha[\,\cdot\,]$ is legitimate in this case since the transition kernel is the same for every $x \in \alpha$ (see Definition 6.18). Moreover, Theorem 6.37

indicates how positivity is a stability property stronger than recurrence. In fact, the latter corresponds to

$$P_\alpha(\tau_\alpha = \infty) = 0,$$

which is a necessary condition for $\mathbb{E}_\alpha[\tau_\alpha] < \infty$.

Proof. If $\mathbb{E}_\alpha[\tau_\alpha] < \infty$, $P_\alpha(\tau_\alpha < \infty) = 1$; thus, (X_n) is recurrent from Proposition 6.31. Consider a measure π given by

$$(6.17) \qquad \pi(A) = \sum_{n=1}^{\infty} P_\alpha(X_n \in A, \tau_\alpha \geq n)$$

as in (6.16). This measure is invariant since $\pi(\alpha) = P_\alpha(\tau_\alpha < \infty) = 1$ and

$$\int K(x, A)\pi(dx) = \pi(\alpha)K(\alpha, A) + \int_{\alpha^c} \sum_{n=1}^{\infty} K(x_n, A) \, P_\alpha(\tau_\alpha \geq n, dx_n)$$

$$= K(\alpha, A) + \sum_{n=2}^{\infty} P_\alpha(X_n \in A, \tau_\alpha \geq n) = \pi(A).$$

It is also finite as

$$\pi(\mathcal{X}) = \sum_{n=1}^{\infty} P_\alpha(\tau_\alpha \geq n) = \sum_{n=1}^{\infty} \sum_{m=n}^{\infty} P_\alpha(\tau_\alpha = m)$$

$$= \sum_{m=1}^{\infty} m \, P_\alpha(\tau_\alpha = m) = \mathbb{E}_\alpha[\tau_\alpha] < \infty \, .$$

Since π is invariant when (X_n) is positive, the uniqueness of the invariant distribution implies finiteness of $\pi(\mathcal{X})$, thus of $\mathbb{E}_\alpha[\tau_\alpha]$. Renormalizing π to $\pi/\pi(\mathcal{X})$ implies $\pi(\alpha) = (\mathbb{E}_\alpha[\tau_\alpha])^{-1}$. □

Following a now "classical" approach, the general case can be treated by splitting (X_n) to $(\check{X}_n)$ (which has an atom) and the invariant measure of $(\check{X}_n)$ induces an invariant measure for (X_n) by marginalization. A converse of Proposition 6.31 establishes the generality of invariance for Markov chains (see Meyn and Tweedie 1993, pp. 240–245, for a proof).

Theorem 6.38. *If (X_n) is a recurrent chain, there exists an invariant σ-finite measure which is unique up to a multiplicative factor.*

Example 6.39. Random walk on $\mathbb{R}$. Consider the random walk on $\mathbb{R}$, $X_{n+1} = X_n + W_n$, where W_n has a cdf Γ. Since $K(x, \cdot)$ is the distribution with cdf $\Gamma(y - x)$, the distribution of X_{n+1} is invariant by translation, and this implies that the Lebesgue measure is an invariant measure:

$$\int K(x, A)\lambda(dx) = \int \int_{A-x} \Gamma(dy)\lambda(dx)$$

$$= \int \Gamma(dy) \int \mathbb{I}_{A-y}(x)\lambda(dx) = \lambda(A) \, .$$

Moreover, the invariance of λ and the uniqueness of the invariant measure imply that the chain (X_n) cannot be positive recurrent (in fact, it can be shown that it is null recurrent). ∥

Example 6.40. Random walk on $\mathbb{Z}$. A *random walk* on $\mathbb{Z}$ is defined by

$$X_{n+1} = X_n + W_n,$$

the perturbations W_n being iid with distribution $\gamma_k = P(W_n = k)$, $k \in \mathbb{Z}$. With the same kind of argument as in Example 6.39, since the counting measure on $\mathbb{Z}$ is invariant for (X_n), (X_n) cannot be positive. If the distribution of W_n is symmetric, straightforward arguments lead to the conclusion that

$$\sum_{n=1}^{\infty} P_0(X_n = 0) = \infty \, ,$$

from which we derive the (null) recurrence of (X_n) (see Feller 1970, Durrett 1991, or Problem 6.25). ∥

Example 6.41. (Continuation of Example 6.24) Given the quasi-diagonal shape of the transition matrix, it is possible to directly determine the invariant distribution, $\pi = (\pi_0, \ldots, \pi_K)$. In fact, it follows from the equation $P^t\pi = \pi$ that

$$\pi_0 = P_{00}\pi_0 + P_{10}\pi_1,$$
$$\pi_1 = P_{01}\pi_0 + P_{11}\pi_1 + P_{21}\pi_2,$$
$$\vdots$$
$$\pi_K = P_{(K-1)K}\pi_{K-1} + P_{KK}\pi_K \, .$$

Therefore,

$$\pi_1 = \frac{P_{01}}{P_{10}}\pi_0,$$
$$\pi_2 = \frac{P_{01}P_{12}}{P_{21}P_{10}}\pi_0,$$
$$\vdots$$
$$\pi_k = \frac{P_{01}\cdots P_{(k-1)k}}{P_{k(k-1)}\cdots P_{10}}\pi_0 \, .$$

Hence,

$$\pi_k = \binom{K}{k}^2 \pi_0, \qquad k = 0, \ldots, K,$$

and through normalization,

$$\pi_K = \frac{\binom{K}{k}^2}{\binom{2K}{k}},$$

which implies that the hypergeometric distribution $\mathcal{H}(2K, K, 1/2)$ is the invariant distribution for the Bernoulli–Laplace model. Therefore, the chain is positive.   ‖

Example 6.42. (Continuation of Example 6.26) Assume $f'(1) \leq 1$. If there exists an invariant distribution for (S_t), its characteristic function g satisfies

$$(6.18) \qquad g(s) = f(s)g(0) + g[f(s)] - g(0) .$$

In the simplest case, that is to say, when the number of siblings of a given individual is distributed according to a Bernoulli distribution $\mathcal{B}(p)$, $f(s) = q + ps$, where $q = 1 - p$, and $g(s)$ is solution of

$$(6.19) \qquad g(s) = g(q + ps) + p(s - 1)g(0) .$$

Iteratively substituting (6.18) into (6.19), we obtain

$$\begin{aligned} g(s) &= g[q + p(q + ps)] + p(q + ps - 1)g(0) + p(s - 1)g(0) \\ &= g(q + pq + \cdots + p^{k-1}q + p^k s) + (p + \cdots + p^k)(s - 1)g(0) , \end{aligned}$$

for every $k \in \mathbb{N}$. Letting k go to infinity, we have

$$\begin{aligned} g(s) &= g[q/(1 - p)] + [p/(1 - p)](s - 1)g(0) \\ &= 1 + \frac{p}{q}(s - 1)g(0) , \end{aligned}$$

since $q/(1 - p) = 1$ and $g(1) = 1$. Substituting $s = 0$ implies $g(0) = q$ and, hence, $g(s) = 1 + p(s - 1) = q + ps$. The Bernoulli distribution is thus the invariant distribution and the chain is positive.   ‖

Example 6.43. (Continuation of Example 6.20) Given that the transition kernel corresponds to the $\mathcal{N}(\theta x_{n-1}, \sigma^2)$ distribution, a normal distribution $\mathcal{N}(\mu, \tau^2)$ is stationary for the AR(1) chain only if

$$\mu = \theta\mu \qquad \text{and} \qquad \tau^2 = \tau^2\theta^2 + \sigma^2 .$$

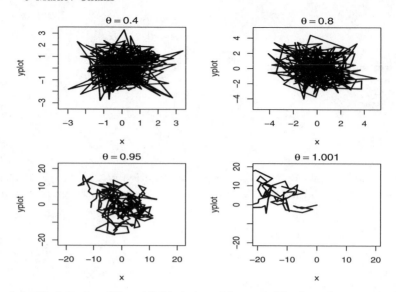

Fig. 6.1. Trajectories of four AR(1) chains with $\sigma = 1$. The first three panels show positive recurrent chains, and a θ increases the chain gets closer to transience. The fourth chain with $\theta = 1.0001$ is transient, and the trajectory never returns.

These conditions imply that $\mu = 0$ and that $\tau^2 = \sigma^2/(1 - \theta^2)$, which can only occur for $|\theta| < 1$. In this case, $\mathcal{N}(0, \sigma^2/(1 - \theta^2))$ is indeed the unique stationary distribution of the AR(1) chain.

So if $|\theta| < 1$, the marginal distribution of the chain is a proper density independent of n, and the chain is positive (hence recurrent). Figure 6.1 shows the two-dimensional trajectories of an AR(1) chain, where each coordinate is a univariate AR(1) chain. (We use two dimensions to better graphically illustrate the behavior of the chain.)

In the first three panels of Figure 6.1 we see increasing θ, but all three are positive recurrent. This results in the chain "filling" the space; and we can see as θ increases there is less dense filling in. Finally, the fourth chain, with $\theta = 1.001$, is transient, and not only it does not fill the space, but it escapes and never returns. Note the scale on that panel.

When we use a Markov chain to explore a space, we want it to fill the space. Thus, we want our MCMC chains to be positive recurrent.   ‖

Note that the converse to Proposition 6.38 does not hold: there exist transient Markov chains with stationary measures. For instance, the random walks in $\mathbb{R}^3$ and $\mathbb{Z}^3$, corresponding to Examples 6.39 and 6.40, respectively, are both transient *and* have the Lebesgue and the counting measures as stationary measures (see Problem 6.25).

In the case of a general Harris recurrent irreducible and aperiodic Markov chain (X_n) with stationary distribution π, Hobert and Robert (2004) propose a representation of π that is associated with Kac's Theorem, under the form

$$(6.20) \qquad \pi(A) = \sum_{t=1}^{\infty} \Pr(N_t \in A) \Pr(T^* = t),$$

where T^* is an integer valued random variable and, for each $t \in \mathbb{N}$, N_t is a random variable whose distribution depends on t. This representation is very closely related to Kac's (6.17), but offers a wider range of applicability since it does not require the chain to have an atom. We simply assume that the minorizing condition (6.10) is satisfied (and for simplicity's sake we take $m = 1$ in (6.10)). The random variables N_t and T^* can then be defined in terms of the split chain of Section 6.3.2. If $\tau_C \geq 1$ denotes the renewal time associated with the small set C in (6.10), then $\mathbb{E}_\nu[\tau_C] < \infty$ by recurrence (Problem 6.32), and T^* is given by the tail probabilities of τ_C as

$$(6.21) \qquad P(T^* = t) = \frac{\Pr_\nu(\tau_C \geq t)}{\mathbb{E}_\nu(\tau_C)}.$$

The random variable N_t is then logically distributed from ν if $t = 1$ and as X_t conditional on no renewal before time t otherwise, following from (6.10). Breyer and Roberts (2000b) derive the representation (6.20) by the mean of a functional equation (see Problem 6.33).

Simulating from π thus amounts to simulating T^* according to (6.21) and then, for $T^* = t$, to simulating N_t. Simulating the latter starts from the minorizing measure ν and then runs $t - 1$ steps of the residual distribution

$$\tilde{K}(x, \cdot) = \frac{K(x, \cdot) - \epsilon \mathbb{I}_C(x) \nu(\cdot)}{1 - \epsilon \mathbb{I}_C(x)}.$$

In cases when simulating from the residual is too complex, a brute force Accept–Reject implementation is to run the split chain t iterations until $\tau_C \geq t$, but this may be too time-consuming in many situations. Hobert and Robert (2004) also propose more advanced approaches to the simulation of T^*.

Note that, when the state space $\mathcal{X}$ is small, the chain is said to be *uniformly ergodic* (see Definition 6.58 below), $\tilde{K}(x, y) = (K(x, y) - \epsilon \nu(y))/(1 - \epsilon)$ and the mixture representation (6.20) translates into the following algorithm.

Algorithm A.23 –Kac's Mixture Implementation–

1. Simulate $X_0 \sim \nu$, $\omega \sim \mathcal{G}eo(\epsilon)$.
2. Run the transition $X_{t+1} \sim \tilde{K}(x_t, \cdot)$ $t = 0, \cdots, \omega - 1$, and take X_ω.

6.5.3 Reversibility and the Detailed Balance Condition

The stability property inherent to stationary chains can be related to another stability property called *reversibility*, which states that the direction of time does not matter in the dynamics of the chain (see also Problems 6.53 and 6.54).

Definition 6.44. A stationary Markov chain (X_n) is *reversible* if the distribution of X_{n+1} conditionally on $X_{n+2} = x$ is the same as the distribution of X_{n+1} conditionally on $X_n = x$.

In fact, reversibility can be linked with the existence of a stationary measure π if a condition stronger than in Definition 6.35 holds.

Definition 6.45. A Markov chain with transition kernel K satisfies the *detailed balance condition* if there exists a function f satisfying

$$(6.22) \qquad\qquad K(y,x)f(y) = K(x,y)f(x)$$

for every (x,y).

While this condition is not *necessary* for f to be a stationary measure associated with the transition kernel K, it provides a sufficient condition that is often easy to check and that can be used for most MCMC algorithms. The balance condition (6.22) express an equilibrium in the flow of the Markov chain, namely that the probability of being in x and moving to y is the same as the probability of being in y and moving back to x. When f is a density, it also implies that the chain is reversible.[3] More generally,

Theorem 6.46. *Suppose that a Markov chain with transition function K satisfies the detailed balance condition with π a probability density function. Then:*

(i) The density π is the invariant density of the chain.
(ii) The chain is reversible.

Proof. Part (i) follows by noting that, by the detailed balance condition, for any measurable set B,

$$\int_y K(y,B)\pi(y)dy = \int_y \int_B K(y,x)\pi(y)dxdy$$
$$= \int_y \int_B K(x,y)\pi(x)dxdy = \int_B \pi(x)dx,$$

since $\int K(x,y)dy = 1$. With the existence of the kernel K and invariant density π, it is clear that detailed balance and reversibility are the same property. □

If $f(x,y)$ is a joint density, then we can write (with obvious notation)

$$f(x,y) = f_{X|Y}(x|y)f_Y(y)$$
$$f(x,y) = f_{Y|X}(y|x)f_X(x),$$

and thus detailed balance requires that $f_X = f_Y$ and $f_{X|Y} = f_{Y|X}$, that is, there is symmetry in the conditionals and the marginals are the same.

---

[3] If there are no measure-theoretic difficulties with the definition of the kernel K, both notions are equivalent.

6.6 Ergodicity and Convergence

6.6.1 Ergodicity

Considering the Markov chain (X_n) from a temporal perspective, it is natural (and important) to establish the limiting behavior of X_n; that is, *To what is the chain converging?* The existence (and uniqueness) of an invariant distribution π makes that distribution a natural candidate for the limiting distribution, and we now turn to finding sufficient conditions on (X_n) for X_n to be asymptotically distributed according to π. The following theorems are fundamental convergence results for Markov chains and they are at the core of the motivation for Markov chain Monte Carlo algorithms. They are, unfortunately, if not surprisingly, quite difficult to establish and we restrict the proof to the countable case, the extension to the general case being detailed in Meyn and Tweedie (1993, pp. 322–323).

There are many conditions that can be placed on the convergence of P^n, the distribution of X_n, to π. Perhaps, the most fundamental and important is that of *ergodicity*, that is, independence of initial conditions.

Definition 6.47. For a Harris positive chain (X_n), with invariant distribution π, an atom α is *ergodic* if

$$\lim_{n\to\infty} |K^n(\alpha, \alpha) - \pi(\alpha)| = 0 .$$

In the countable case, the existence of an ergodic atom is, in fact, sufficient to establish convergence according to the *total variation norm*,

$$\|\mu_1 - \mu_2\|_{TV} = \sup_A |\mu_1(A) - \mu_2(A)|.$$

Proposition 6.48. *If (X_n) is Harris positive on $\mathcal{X}$ and denumerable, and if there exists an ergodic atom $\alpha \subset \mathcal{X}$, then, for every $x \in \mathcal{X}$,*

$$\lim_{n\to\infty} \|K^n(x, \cdot) - \pi\|_{TV} = 0 .$$

Proof. The first step follows from a decomposition formula called *"first entrance and last exit"*:

$$K^n(x, y) = P_x(X_n = y, \tau_\alpha \geq n)$$
$$+ \sum_{j=1}^{n-1} \left[\sum_{k=1}^{j} P_x(X_k \in \alpha, \tau_\alpha \geq k) K^{j-k}(\alpha, \alpha) \right]$$
(6.23)
$$\times P_\alpha(X_{n-j} = y, \tau_\alpha \geq n - j),$$

which relates $K^n(x, y)$ to the last visit to α. (See Problem 6.37.) This shows the reduced influence of the initial value x, since $P_x(X_n = y, \tau_\alpha \geq n)$ converges to 0 with n. The expression (6.17) of the invariant measure implies, in addition, that

$$\pi(y) = \pi(\alpha) \sum_{j=1}^{\infty} P_\alpha(X_j = y, \tau_\alpha \geq j).$$

These two expressions then lead to

$$||K^n(x, \cdot) - \pi||_{TV} = \sum_y |K^n(x, y) - \pi(y)|$$

$$\leq \sum_y P_x(X_n = y, \tau_\alpha \geq n)$$

$$+ \sum_y \sum_{j=1}^{n-1} \left| \sum_{k=1}^{j} P_x(X_k \in \alpha, \tau_\alpha = k) K^{j-k}(\alpha, \alpha) - \pi(\alpha) \right|$$

$$\times P_\alpha(X_{n-j} = y, \tau_\alpha \geq n - j)$$

$$+ \sum_y \pi(\alpha) \sum_{j=n-1}^{\infty} P_\alpha(X_j = y, \tau_\alpha \geq j) .$$

The second step in the proof is to show that each term in the above decomposition goes to 0 as n goes to infinity. The first term is actually $P_x(\tau_\alpha \geq n)$ and goes to 0 since the chain is Harris recurrent. The third term is the remainder of the convergent series

(6.24)
$$\sum_y \pi(\alpha) \sum_{j=1}^{\infty} P_\alpha(X_j = y, \tau_\alpha \geq j) = \sum_y \pi(y) .$$

The middle term is the sum over the y's of the convolution of the two sequences $a_n = |\sum_{k=1}^{n} P_x(X_k \in \alpha, \tau_\alpha = k) K^{n-k}(\alpha, \alpha) - \pi(\alpha)|$ and $b_n = P_\alpha(X_n = y, \tau_\alpha \geq n)$. The sequence (a_n) is converging to 0 since the atom α is ergodic and the series of the b_n's is convergent, as mentioned. An algebraic argument (see Problem 6.39) then implies that (6.24) goes to 0 as n goes to ∞. □

The decomposition (6.23) is quite revealing in that it shows the role of the atom α as the generator of a renewal process. Below, we develop an extension which allows us to deal with the general case using *coupling* techniques. (These techniques are also useful in the assessment of convergence for Markov chain Monte Carlo algorithms.) Lindvall (1992) provides an introduction to coupling.

The *coupling* principle uses two chains (X_n) and (X'_n) associated with the same kernel, the "coupling" event taking place when they meet in α; that is, at the first time n_0 such that $X_{n_0} \in \alpha$ and $X'_{n_0} \in \alpha$. After this instant, the probabilistic properties of (X_n) and (X'_n) are identical and if one of the two chains is stationary, there is no longer any dependence on initial conditions for either chain. Therefore, if we can show that the *coupling time* (that is, the time it takes for the two chains to meet), is finite for almost every starting point, the ergodicity of the chain follows.

For a recurrent atom α on a denumerable space $\mathcal{X}$, let $\tau_\alpha(k)$ denote the kth visit to α ($k = 1, 2, \ldots$), and let $p = (p(1), p(2), \cdots)$ be the distribution of the *excursion time*,

$$S_k = \tau_\alpha(k+1) - \tau_\alpha(k),$$

between two visits to α. If $q = (q(0), q(1), \ldots)$ represents the distribution of $\tau_\alpha(1)$ (which depends on the initial condition, x_0 or μ), then the distribution of $\tau_\alpha(n+1)$ is given by the convolution product $q \star p^{n\star}$ (that is, the distribution of the sum of n iid rv's distributed from p and of a variable distributed from q), since

$$\tau_\alpha(n+1) = S_n + \cdots + S_1 + \tau_\alpha(1).$$

Thus, consider two sequences (S_i) and (S_i') such that $S_1, S_2, \ldots$ and $S_1', S_2', \ldots$ are iid from p with $S_0 \sim q$ and $S_0' \sim r$. We introduce the indicator functions

$$Z_q(n) = \sum_{j=0}^{n} \mathbb{I}_{S_1 + \cdots + S_j = n} \quad \text{and} \quad Z_r(n) = \sum_{j=0}^{n} \mathbb{I}_{S_1' + \cdots + S_j' = n},$$

which correspond to the events that the chains (X_n) and (X_n') visit α at time n. The *coupling time* is then given by

$$T_{qr} = \min\left\{ j; Z_q(j) = Z_r(j) = 1 \right\},$$

which satisfies the following lemma, whose proof can be found in Problem 6.45.

Lemma 6.49. *If the mean excursion time satisfies*

$$m_p = \sum_{n=0}^{\infty} np(n) < \infty$$

and if p is aperiodic (the g.c.d. of the support of p is 1), then the coupling time T_{pq} is almost surely finite, that is,

$$P(T_{pq} < \infty) = 1,$$

for every q.

If p is aperiodic with finite mean m_p, this implies that Z_p satisfies

(6.25) $$\lim_{n \to \infty} |P(Z_q(n) = 1) - m_q^{-1}| = 0,$$

as shown in Problem 6.42. The probability of visiting α at time n is thus asymptotically independent of the initial distribution and this result implies that Proposition 6.48 holds without imposing constraints in the discrete case.

Theorem 6.50. *For a positive recurrent aperiodic Markov chain on a countable space, for every initial state x,*

$$\lim_{n\to\infty} \|K^n(x,\cdot) - \pi\|_{TV} = 0 \;.$$

Proof. Since (X_n) is positive recurrent, $\mathbb{E}_\alpha[\tau_\alpha]$ is finite by Theorem 6.37. Therefore, m_p is finite, (6.25) holds, and every atom is ergodic. The result follows from Proposition 6.48.   □

For general state-spaces $\mathcal{X}$, Harris recurrence is nonetheless necessary in the derivation of the convergence of K^n to π. (Note that another characterization of Harris recurrence is the convergence of $\|K_x^n - \pi\|_{TV}$ to 0 for *every* value x, instead of almost every value.)

Theorem 6.51. *If (X_n) is Harris positive and aperiodic, then*

$$\lim_{n\to\infty} \left\| \int K^n(x,\cdot)\mu(dx) - \pi \right\|_{TV} = 0$$

for every initial distribution μ.

This result follows from an extension of the denumerable case to strongly aperiodic Harris positive chains by splitting, since these chains always allow for small sets (see Section 6.3.3), based on an equivalent to the "first entrance and last exit" formula (6.23). It is then possible to move to arbitrary chains by the following result.

Proposition 6.52. *If π is an invariant distribution for P, then*

$$\left\| \int K^n(x,\cdot)\mu(dx) - \pi \right\|_{TV}$$

is decreasing in n.

Proof. First, note the equivalent definition of the norm (Problem 6.40)

$$(6.26) \qquad \|\mu\|_{TV} = \frac{1}{2} \sup_{|h|\le 1} \left| \int h(x)\mu(dx) \right| .$$

We then have

$$2\left\| \int K^{n+1}(x,\cdot)\mu(dx) - \pi \right\|_{TV}$$

$$= \sup_{|h|\le 1} \left| \int h(y)K^{n+1}(x,dy)\mu(dx) - \int h(y)\pi(dy) \right|$$

$$= \sup_{|h|\le 1} \left| \int h(y) \int K^n(x,dw)K(w,dy)\mu(dx) \right.$$

$$\left. - \int h(y) \int K(w,dy)\pi(dw) \right| ,$$

since, by definition, $K^{n+1}(x, dy) = \int K^n(x, dw)K(w, dy)$ and by the invariance of π, $\pi(dy) = \int K(w, dy)\pi(dw)$. Regrouping terms, we can write,

$$2 \left\| \int K^{n+1}(x, \cdot)\mu(dx) - \pi \right\|_{TV}$$

$$= \sup_{|h| \leq 1} \left| \int \left[\int h(y)K(w, dy) \right] K^n(x, dw)\mu(dx) \right.$$

$$\left. - \int \left[\int h(y)K(w, dy) \right] \pi(dw) \right|$$

$$\leq \sup_{|h| \leq 1} \left| \int h(w)K^n(x, dw)\mu(dx) - \int h(w)\pi(dw) \right|,$$

where the inequality follows from the fact that the quantity in square brackets is a function with norm less than 1. Hence, monotonicity of the total variation norm is established.   □

Note that the equivalence (6.26) also implies the convergence

$$(6.27) \qquad \lim_{n \to \infty} |\mathbb{E}_\mu[h(X_n)] - \mathbb{E}^\pi[h(X)]| = 0$$

for every *bounded function* h. This equivalence is, in fact, often taken as the defining condition for convergence of distributions (see, for example, Billingsley 1995, Theorem 25.8). We can, however, conclude (6.27) from a slightly weaker set of assumptions, where we do not need the full force of Harris recurrence (see Theorem 6.80 for an example).

The extension of (6.27) to more general functions h is called *h-ergodicity* by Meyn and Tweedie (1993, pp. 342–344).

Theorem 6.53. *Let* (X_n) *be positive, recurrent, and aperiodic.*

(a) If $\mathbb{E}^\pi[|h(X)|] = \infty$, $\mathbb{E}_x[|h(X_n)|] \to \infty$ *for every* x.
(b) If $\int |h(x)|\pi(dx) < \infty$, *then*

$$(6.28) \qquad \lim_{n \to \infty} \sup_{|m(x)| \leq |h(x)|} |\mathbb{E}_y[m(X_n)] - \mathbb{E}^\pi[m(X)]| = 0$$

on all small sets C *such that*

$$(6.29) \qquad \sup_{y \in C} \mathbb{E}_y \left[\sum_{t=0}^{\tau_C - 1} h(X_t) \right] < \infty .$$

Similar conditions appear as necessary conditions for the Central Limit Theorem (see (6.31) in Theorem 6.64). Condition (6.29) relates to a coupling argument, in the sense that the influence of the initial condition vanishes "fast enough," as in the proof of Theorem 6.63.

6.6.2 Geometric Convergence

The convergence (6.28) of the expectation of $h(x)$ at time n to the expectation of $h(x)$ under the stationary distribution π somehow ensures the proper behavior of the chain (X_n) whatever the initial value X_0 (or its distribution). A more precise description of convergence properties involves the study of *the speed of convergence* of K^n to π. An evaluation of this speed is important for Markov chain Monte Carlo algorithms in the sense that it relates to stopping rules for these algorithms; minimal convergence speed is also a requirement for the application of the Central Limit Theorem.

To study the speed of convergence more closely, we first introduce an extension of the total variation norm, denoted by $\|\cdot\|_h$, which allows for an upper bound other than 1 on the functions. The generalization is defined by

$$\|\mu\|_h = \sup_{|g| \leq h} \left| \int g(x)\mu(dx) \right| .$$

Definition 6.54. A chain (X_n) is *geometrically h-ergodic*, with $h \geq 1$ on $\mathcal{X}$, if (X_n) is Harris positive, with stationary distribution π, if (X_n) satisfies $\mathbb{E}^\pi[h] < \infty$, and if there exists $r_h > 1$ such that

$$(6.30) \qquad \sum_{n=1}^\infty r_h^n \|K^n(x, \cdot) - \pi\|_h < \infty$$

for every $x \in \mathcal{X}$. The case $h = 1$ corresponds to the *geometric ergodicity* of (X_n).

Geometric h-ergodicity means that $\|K^n(x, \cdot) - \pi\|_h$ is decreasing at least at a geometric speed, since (6.30) implies

$$\|K^n(x, \cdot) - \pi\|_h \leq M r_h^{-n}$$

with

$$M = \sum_{n=1}^\infty r_h^n \|K^n(x, \cdot) - \pi\|_h .$$

If (X_n) has an atom α, (6.30) implies that for a real number $r > 1$,

$$\mathbb{E}_x \left[\sum_{n=1}^{\tau_\alpha} h(X_n) r^n \right] < \infty \quad \text{and} \quad \sum_{n=1}^\infty |P_\alpha(X_n \in \alpha) - \pi(\alpha)| r^n < \infty.$$

The series associated with $|P_\alpha(X_n \in \alpha) - \pi(\alpha)|\, r^n$ converges outside of the unit circle if the power series associated with $P_\alpha(\tau_\alpha = n)$ converges for values of $|r|$ strictly larger than 1. (The proof of this result, called *Kendall's Theorem*, is based on the renewal equations established in the proof of Proposition 6.31.) This equivalence justifies the following definition.

Definition 6.55. An accessible atom α is *geometrically ergodic* if there exists $r > 1$ such that

$$\sum_{n=1}^{\infty} |K^n(\alpha, \alpha) - \pi(\alpha)| \, r^n < \infty$$

and α is a *Kendall atom* if there exists $\kappa > 1$ such that

$$\mathbb{E}_\alpha[\kappa^{\tau_\alpha}] < \infty.$$

If α is a Kendall atom, it is thus geometrically ergodic and ensures geometric ergodicity for (X_n):

Theorem 6.56. *If (X_n) is ψ-irreducible, with invariant distribution π, and if there exists a geometrically ergodic atom α, then there exist $r > 1$, $\kappa > 1$, and $R < \infty$ such that, for almost every $x \in \mathcal{X}$,*

$$\sum_{n=1}^{\infty} r^n \|K^n(x, \cdot) - \pi\|_{TV} < R \, \mathbb{E}_x[\kappa^{\tau_\alpha}] < \infty.$$

Example 6.57. Nongeometric returns to 0. For a chain on $\mathbb{Z}_+$ with transition matrix $\mathbb{P} = (p_{ij})$ such that

$$p_{0j} = \gamma_j, \quad p_{jj} = \beta_j, \quad p_{j0} = 1 - \beta_j, \quad \sum_j \gamma_j = 1,$$

Meyn and Tweedie (1993, p. 361) consider the return time to 0, τ_0, with mean

$$\mathbb{E}_0[\tau_0] = \sum_j \gamma_j \left\{ (1 - \beta_j) + 2\beta_j(1 - \beta_j) + \cdots \right\}$$

$$= \sum_j \gamma_j \left\{ 1 + (1 - \beta_j)^{-1} \right\}.$$

The state 0 is thus an ergodic atom when all the γ_j's are positive (yielding irreducibility) and $\sum_j \gamma_j (1 - \beta_j)^{-1} < \infty$. Now, for $r > 0$,

$$\mathbb{E}_0[r^{\tau_0}] = r \sum_j \gamma_j \mathbb{E}_j[r^{\tau_0 - 1}] = r \sum_j \gamma_j \sum_{k=0}^{\infty} r^k \beta_j^k (1 - \beta_j).$$

For $r > 1$, if $\beta_j \to 1$ as $j \to \infty$, the series in the above expectation always diverges for j large enough. Thus, the chain is not geometrically ergodic. ‖

6.6.3 Uniform Ergodicity

The property of uniform ergodicity is stronger than geometric ergodicity in the sense that the rate of geometric convergence must be uniform over the whole space. It is used in the Central Limit Theorem given in Section 6.7.

Definition 6.58. The chain (X_n) is *uniformly ergodic* if

$$\lim_{n \to \infty} \sup_{x \in \mathcal{X}} \|K^n(x, \cdot) - \pi\|_{TV} = 0 .$$

Uniform ergodicity can be established through one of the following equivalent properties:

Theorem 6.59. *The following conditions are equivalent:*

(a) (X_n) is uniformly ergodic;
(b) there exist $R < \infty$ and $r > 1$ such that

$$\|K^n(x, \cdot) - \pi\|_{TV} < Rr^{-n} , \qquad \text{for all } x \in \mathcal{X} ;$$

(c) (X_n) is aperiodic and $\mathcal{X}$ is a small set;
(d) (X_n) is aperiodic and there exist a small set C and a real $\kappa > 1$ such that

$$\sup_{x \in \mathcal{X}} \mathbb{E}_x[\kappa^{\tau_C}] < \infty .$$

If the whole space $\mathcal{X}$ is small, there exist a probability distribution, φ, on $\mathcal{X}$, and constants $\varepsilon < 1$, $\delta > 0$, and n such that, if $\varphi(A) > \varepsilon$ then

$$\inf_{x \in \mathcal{X}} K^n(x, A) > \delta .$$

This property is sometimes called *Doeblin's condition*. This requirement shows the strength of the uniform ergodicity and suggests difficulties about the verification. We will still see examples of Markov chain Monte Carlo algorithms which achieve this superior form of ergodicity (see Example 10.17). Note, moreover, that in the finite case, uniform ergodicity can be derived from the smallness of $\mathcal{X}$ since the condition

$$P(X_{n+1} = y | X_n = x) \geq \inf_z p_{zy} = \rho_y \quad \text{for every } x, y \in \mathcal{X},$$

leads to the choice of the minorizing measure ν as

$$\nu(y) = \frac{\rho_y}{\sum_{z \in \mathcal{X}} \rho_z}$$

as long as $\rho_y > 0$ for some $y \in \mathcal{X}$. (If (X_n) is recurrent and aperiodic, this positivity condition can be attained by a subchain $(Y_m) = (X_{nd})$ for d large enough. See Meyn and Tweedie 1993, Chapter 16, for more details.)

6.7 Limit Theorems

Although the notions and results introduced in the previous sections are important in justifying Markov chain Monte Carlo algorithms, in the following chapters we will see that this last section is essential to the processing of these algorithms. In fact, the different convergence results (ergodicity) obtained in Section 6.6 deal only with the probability measure P_x^n (through different norms), which is somewhat of a "snapshot" of the chain (X_n) at time n. So, it determines the probabilistic properties of *average* behavior of the chain at a fixed instant. Such properties, even though they provide justification for the simulation methods, are of lesser importance for the control of convergence of a given simulation, where the properties of the *realization* (x_n) of the chain are the only characteristics that truly matter. (Meyn and Tweedie 1993 call this type of properties "sample path" properties.)

We are thus led back to some basic ideas, previously discussed in a statistical setup by Robert (2001, Chapters 1 and 11); that is, we must consider the difference between *probabilistic analysis*, which describes the average behavior of samples, and *statistical inference*, which must reason by induction from the observed sample. While probabilistic properties can justify or refute some statistical approaches, this does not contradict the fact that statistical analysis must be done *conditional on the observed sample*. Such a consideration can lead to the Bayesian approach in a statistical setup (or at least to consideration of the *Likelihood Principle*; see, e.g., Berger and Wolpert 1988, or Robert 2001, Section 1.3). In the setup of Markov chains, a conditional analysis can take advantage of convergence properties of P_x^n to π only to verify the convergence, to a quantity of interest, of functions of the observed path of the chain. Indeed, the fact that $\|P_x^n - \pi\|$ is close to 0, or even converges geometrically fast to 0 with speed ρ^n $(0 < \rho < 1)$, does not bring direct information about the unique available observation from P_x^n, namely X_n.

The problems in directly applying the classical convergence theorems (Law of Large Numbers, Law of the Iterated Logarithm, Central Limit Theorem, etc.) to the sample $(X_1, \ldots, X_n)$ are due both to the Markovian dependence structure between the observations X_i and to the non-stationarity of the sequence. (Only if $X_0 \sim \pi$, the stationary distribution of the chain, will the chain be stationary. Since this is equivalent to integrating over the initial conditions, it eliminates the need for a conditional analysis. Such an occurrence, especially in Markov chain Monte Carlo, is somewhat rare.[4])

We therefore assume that the chain is started from a point X_0 whose distribution is not the stationary distribution of the chain, and thus we deal with non-stationary chains directly. We begin with a detailed presentation of convergence results equivalent to the Law of Large Numbers, which are often called *ergodic theorems*. We then mention in Section 6.7.2 various versions of the Central Limit Theorem whose assumptions are usually (and unfortunately) difficult to check.

---

[4] Nonetheless, there is considerable research in MCMC theory about *perfect simulation*; that is, ways of starting the algorithm with $X_0 \sim \pi$. See Chapter 13.

6.7.1 Ergodic Theorems

Given observations $X_1, \ldots, X_n$ of a Markov chain, we now examine the limiting behavior of the partial sums

$$S_n(h) = \frac{1}{n} \sum_{i=1}^{n} h(X_i)$$

when n goes to infinity, getting back to the iid case through renewal when (X_n) has an atom. Consider first the notion of *harmonic functions*, which is related to ergodicity for Harris recurrent Markov chains.

Definition 6.60. A measurable function h is *harmonic* for the chain (X_n) if

$$\mathbb{E}[h(X_{n+1})|x_n] = h(x_n).$$

These functions are *invariant* for the transition kernel (in the functional sense) and they characterize Harris recurrence as follows.

Proposition 6.61. *For a positive Markov chain, if the only bounded harmonic functions are the constant functions, the chain is Harris recurrent.*

Proof. First, the probability of an infinite number of returns, $Q(x, A) = P_x(\eta_A = \infty)$, as a function of x, $h(x)$, is clearly a harmonic function. This is because

$$\mathbb{E}_y[h(X_1)] = \mathbb{E}_y[P_{X_1}(\eta_A = \infty)] = P_y(\eta_A = \infty),$$

and thus, $Q(x, A)$ is constant (in x).

The function $Q(x, A)$ describes a *tail event*, an event whose occurrence does not depend on $X_1, X_2, \ldots, X_m$, for any finite m. Such events generally obey a $0 - 1$ law, that is, their probabilities of occurrence are either 0 or 1. However, $0 - 1$ laws are typically established in the independence case, and, unfortunately, extensions to cover Markov chains are beyond our scope. (For example, see the *Hewitt–Savage* $0 - 1$ *Law*, in Billingsley 1995, Section 36.) For the sake of our proof, we will just state that $Q(x, A)$ obeys a $0 - 1$ Law and proceed.

If π is the invariant measure and $\pi(A) > 0$, the case $Q(x, A) = 0$ is impossible. To see this, suppose that $Q(x, A) = 0$. It then follows that the chain almost surely visits A only a finite number of times and the average

$$\frac{1}{N} \sum_{i=1}^{N} \mathbb{I}_A(X_i)$$

will not converge to $\pi(A)$, contradicting the Law of Large Numbers (see Theorem 6.63). Thus, for any x, $Q(x, A) = 1$, establishing that the chain is a Harris chain. $\qquad\square$

Proposition 6.61 can be interpreted as a continuity property of the transition functional $Kh(x) = \mathbb{E}_x[h(X_1)]$ in the following sense. By induction, a harmonic function h satisfies $h(x) = \mathbb{E}_x[h(X_n)]$ and by virtue of Theorem 6.53, $h(x)$ is almost surely equal to $\mathbb{E}^\pi[h(X)]$; that is, it is constant *almost everywhere*. For Harris recurrent chains, Proposition 6.61 states that this implies $h(x)$ is constant *everywhere*. (Feller 1971, pp. 265–267, develops a related approach to ergodicity, where Harris recurrence is replaced by a regularity constraint on the kernel.)

Proposition 6.61 will be most useful in establishing Harris recurrence of some Markov chain Monte Carlo algorithms. Interestingly, the behavior of bounded harmonic functions characterizes Harris recurrence, as the converse of Proposition 6.61 is true. We state it without its rather difficult proof (see Meyn and Tweedie 1993, p. 415).

Lemma 6.62. *For Harris recurrent Markov chains, the constants are the only bounded harmonic functions.*

A consequence of Lemma 6.62 is that if (X_n) is Harris positive with stationary distribution π and if $S_n(h)$ converges μ_0-almost surely (μ_0 a.s.) to

$$\int_{\mathcal{X}} h(x)\,\pi(dx) \ ,$$

for an initial distribution μ_0, this convergence occurs for every initial distribution μ. Indeed, the convergence probability

$$P_x\left(S_N(h) \to \mathbb{E}^\pi[h]\right)$$

is then harmonic. Once again, this shows that Harris recurrence is a superior type of stability in the sense that *almost sure convergence* is replaced by convergence at every point.

Of course, we now know that if functions other than bounded functions are harmonic, the chain is not Harris recurrent. This is looked at in detail in Problem 6.59.

The main result of this section, namely the Law of Large Numbers for Markov chains (which is customarily called the *Ergodic Theorem*), guarantees the convergence of $S_n(h)$.

Theorem 6.63. Ergodic Theorem *If (X_n) has a σ-finite invariant measure π, the following two statements are equivalent:*

(i) If $f, g \in L^1(\pi)$ with $\int g(x)d\pi(x) \neq 0$, then

$$\lim_{n\to\infty} \frac{S_n(f)}{S_n(g)} = \frac{\int f(x)d\pi(x)}{\int g(x)d\pi(x)} \ .$$

(ii) The Markov chain (X_n) is Harris recurrent.

Proof. If (i) holds, take f to be the indicator function of a set A with finite measure and g an arbitrary function with finite and positive integral. If $\pi(A) > 0$,

$$P_x(X \in A \text{ infinitely often}) = 1$$

for every $x \in \mathcal{X}$, which establishes Harris recurrence.

If (ii) holds, we need only to consider the atomic case by a splitting argument. Let α be an atom and $\tau_\alpha(k)$ be the time of the $(k+1)$th visit to α. If ℓ_N is the number of visits to α at time N, we get the bounds

$$\sum_{j=0}^{\ell_N-1} \sum_{n=\tau_\alpha(j)+1}^{\tau_\alpha(j+1)} f(x_n) \le \sum_{k=1}^{N} f(x_k)$$

$$\le \sum_{j=0}^{\ell_N} \sum_{n=\tau_\alpha(j)+1}^{\tau_\alpha(j+1)} f(x_n) + \sum_{k=1}^{\tau_\alpha(0)} f(x_k) \, .$$

The blocks

$$S_j(f) = \sum_{n=\tau_\alpha(j)+1}^{\tau_\alpha(j+1)} f(x_n)$$

are independent and identically distributed. Therefore,

$$\frac{\sum_{i=1}^{n} f(x_i)}{\sum_{i=1}^{n} g(x_i)} \le \frac{\ell_N}{\ell_N - 1} \frac{\left(\sum_{j=0}^{\ell_N} S_j(f) + \sum_{k=1}^{\tau_\alpha} f(x_k)\right)/\ell_N}{\sum_{j=0}^{\ell_N-1} s_j(g)/(\ell_N - 1)} \, .$$

The theorem then follows by an application of the strong Law of Large Numbers for iid rv's.   □

An important aspect of Theorem 6.63 is that π does not need to be a probability measure and, therefore, that there can be some type of strong stability even if the chain is null recurrent. In the setup of a Markov chain Monte Carlo algorithm, this result is sometimes invoked to justify the use of improper posterior measures, although we fail to see the relevance of this kind of argument (see Section 10.4.3).

6.7.2 Central Limit Theorems

There is a natural progression from the Law of Large Numbers to the Central Limit Theorem. Moreover, the proof of Theorem 6.63 suggests that there is a direct extension of the Central Limit Theorem for iid variables. Unfortunately this is not the case, as conditions on the finiteness of the variance explicitly involve the atom α of the split chain. Therefore, we provide alternative conditions for the Central Limit Theorem to apply in different settings.

6.7.2.1 The Discrete Case

The discrete case can be solved directly, as shown by Problems 6.50 and 6.51.

Theorem 6.64. *If (X_n) is Harris positive with an atom α such that*

$$(6.31) \qquad \mathbb{E}_\alpha[\tau_\alpha^2] < \infty, \qquad \mathbb{E}_\alpha\left[\left(\sum_{n=1}^{\tau_\alpha} |h(X_n)|\right)^2\right] < \infty$$

and

$$\gamma_h^2 = \pi(\alpha)\, \mathbb{E}_\alpha\left[\left(\sum_{n=1}^{\tau_\alpha} \{h(X_n) - \mathbb{E}^\pi[h]\}\right)^2\right] > 0,$$

the Central Limit Theorem applies; that is,

$$\frac{1}{\sqrt{N}}\left(\sum_{n=1}^{N} (h(X_n) - \mathbb{E}^\pi[h])\right) \overset{\mathcal{L}}{\rightsquigarrow} \mathcal{N}(0, \gamma_h^2).$$

Proof. Using the same notation as in the proof of Theorem 6.63, if $\overline{h}$ denotes $h - \mathbb{E}^\pi[h]$, we get

$$\frac{1}{\sqrt{\ell_N}} \sum_{i=1}^{\ell_N} S_i(\overline{h}) \overset{\mathcal{L}}{\rightsquigarrow} \mathcal{N}\left(0, \mathbb{E}_\alpha\left[\sum_{n=1}^{\tau_\alpha} \overline{h}(X_n)\right]^2\right),$$

following from the Central Limit Theorem for the independent variables $S_i(\overline{f})$, while N/ℓ_N converges a.s. to $\mathbb{E}_\alpha[S_0(1)] = 1/\pi(\alpha)$. Since

$$\left|\sum_{i=1}^{\ell_{N-1}} S_i(\overline{h}) - \sum_{k=1}^{N} \overline{h}(X_k)\right| \le S_{\ell_N}(|\overline{h}|)$$

and

$$\lim_{n\to\infty} \frac{1}{n} \sum_{j=1}^{n} S_i(|\overline{h}|)^2 = \mathbb{E}_\alpha[S_0(|\overline{h}|)^2],$$

we get

$$\limsup_{N\to\infty} \frac{S_{\ell_N}(|\overline{h}|)}{\sqrt{N}} = 0,$$

and the remainder goes to 0 almost surely. $\qquad\square$

This result indicates that an extension of the Central Limit Theorem to the nonatomic case will be more delicate than for the Ergodic Theorem: Conditions (6.31) are indeed expressed in terms of the split chain $(\check{X}_n)$. (See Section 12.2.3 for an extension to cases when there exists a small set.) In Note 6.9.1, we present some alternative versions of the Central Limit Theorem involving a drift condition.

6.7.2.2 Reversibility

The following theorem avoids the verification of a drift condition, but rather requires the Markov chain to be reversible (see Definition 6.44).

With the assumption of reversibility, this Central Limit Theorem directly follows from the strict positivity of γ_g. This was established by Kipnis and Varadhan (1986) using a proof that is beyond our reach.

Theorem 6.65. *If (X_n) is aperiodic, irreducible, and reversible with invariant distribution π, the Central Limit Theorem applies when*

$$0 < \gamma_g^2 = \mathbb{E}_\pi[\bar{g}^2(X_0)] + 2 \sum_{k=1}^{\infty} \mathbb{E}_\pi[\bar{g}(X_0)\bar{g}(X_k)] < +\infty.$$

The main point here is that even though reversibility is a very restrictive assumption in general, it is often easy to impose in Markov chain Monte Carlo algorithms by introducing additional simulation steps (see Geyer 1992, Tierney 1994, Green 1995). See also Theorem 6.77 for another version of the Central Limit Theorem, which relies on a "drift condition" (see Note 6.9.1) similar to geometric ergodicity.

Example 6.66 (Continuation of Example 6.43). For the AR(1) chain, the transition kernel corresponds to the $\mathcal{N}(\theta x_{n-1}, \sigma^2)$ distribution, and the stationary distribution is $\mathcal{N}(0, \sigma^2/(1-\theta^2))$. It is straightforward to verify that the chain is reversible by showing that (Problem 6.65)

$$X_{n+1}|X_n \sim \mathcal{N}(\theta x_n, \sigma^2) \text{ and } X_n|X_{n+1} \sim \mathcal{N}(\theta x_{n+1}, \sigma^2).$$

Thus the chain satisfies the conditions for the CLT.

Figure 6.2 shows histograms of means for the cases of $\theta = .5$ and $\theta = 2$. In the first case (left panel) we have a positive recurrent chain that satisfies the conditions of the CLT. The right panel is most interesting, however, because $\theta = 2$ and the chain is transient. However, the histogram of the means "looks" quite well behaved, giving no sign that the chain is not converging.

It can happen that null recurrent and transient chains can often look well behaved when examined graphically through some output. However, another picture shows a different story. In Figure 6.3 we look at the trajectories of the cumulative mean and standard deviation from one chain of length 1000. There, the left panel corresponds to the ergodic case with $\theta = .5$, and the right panel corresponds to the (barely) transient case of $\theta = 1.0001$. However, it is clear that there is no convergence. See Section 10.4.3 for the manifestation of this in MCMC algorithms.   ‖

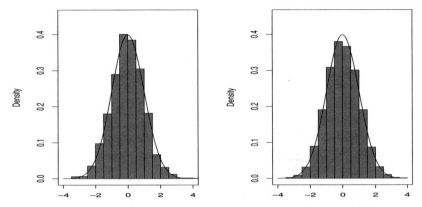

Fig. 6.2. Histogram of 2500 means (each based on 50 observations) from an AR(1) chain. The left panel corresponds to $\theta = .5$, which results in an ergodic chain. The right panel corresponds to $\theta = 2$, which corresponds to a transient chain.

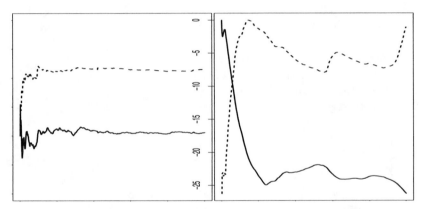

Fig. 6.3. Trajectories of mean (solid line) and standard deviation (dashed line) from the AR(1) process of Example 6.66. The left panel has $\theta = .5$, resulting in an ergodic Markov chain, and displays convergence of the mean and standard deviation. The right panel has $\theta = 1.0001$, resulting in a transient Markov chain and no convergence.

6.7.2.3 Geometric Ergodicity and Regeneration

There is yet another approach to the Central Limit Theorem for Markov chains. It relies on geometric ergodicity, a Liapounov-type moment condition on the function h, and a regeneration argument. Hobert et al. (2002), extending work of Chan and Geyer (1994) (see Problem 6.66), give specific conditions for Theorem 6.67 to apply, namely for Liapounov condition to apply and a consistent estimate of γ_h^2 to be found.

Theorem 6.67. *If (X_n) is aperiodic, irreducible, positive Harris recurrent with invariant distribution π and geometrically ergodic, and if, in addition,*

(6.32) $$\mathbb{E}^{\pi}[|h(X)|^{2+\varepsilon}] < \infty$$

for some $\varepsilon > 0$, then

(6.33) $$\sqrt{n}\,(S_n(h)/n - \mathbb{E}^{\pi}[h(X)]) \overset{\mathcal{L}}{\rightsquigarrow} \mathcal{N}(0,\gamma_h^2)\,,$$

where γ_h^2 is defined as in Theorem 6.65.

They first discuss the difficulty in finding such estimates, as fixed batch mean approximations are *not* consistent when the batch size is fixed. We can, however, use regeneration (Mykland et al. 1995) when available; that is, when a minorization condition as in Sections 6.3.2 and 6.5.2 holds: there exists a function $0 < s(x) < 1$ and a probability measure Q such that, for all $x \in \mathcal{X}$ and all measurable sets A,

(6.34) $$P(x, A) \geq s(x)\,Q(A)\,.$$

Following an idea first developed in Robert (1995a) for MCMC algorithms, Hobert et al. (2002) then construct legitimate asymptotic standard errors bypassing the estimation of γ_g^2.

The approach is to introduce the regeneration times $0 = \tau_0 < \tau_1 < \tau_2 < \cdots$ associated with the Markov chain (X_t) and to write $S_n(h)$ in terms of the regeneration times, namely, if the chain is started as $X_0 \sim Q$ and stopped after the T-th regeneration,

$$S_{\tau_T}(h) = \sum_{t=1}^{T} \sum_{j=\tau_{t-1}}^{\tau_t - 1} h(X_j) = \sum_{t=1}^{T} \tilde{S}_t\,,$$

where the $\tilde{S}_t$'s are the partial sums appearing in Theorem 6.63, which are iid. If we define the inter-regeneration lengths $N_t = \tau_t - \tau_{t-1}$, then

(6.35) $$\bar{h}_{\tau_T} = \frac{\sum_{t=1}^{T} \tilde{S}_t}{\sum_{t=1}^{T} N_t} = \frac{\bar{S}_T}{\bar{N}_T} = \frac{1}{\tau_T} \sum_{j=0}^{\tau_T - 1} g(X_j)$$

converges almost surely to $\mathbb{E}^{\pi}[h(X)]$ when T goes to infinity, by virtue of the Ergodic Theorem (Theorem 6.63), since τ_T converges almost surely to ∞.

By Theorem 6.37, $\mathbb{E}^{Q}[N_1] = 1/\mathbb{E}^{\pi}[s(X)]$ (which is assumed to be finite). It follows from the Strong Law of Large Numbers that $\bar{N}$ converges almost surely to $\mathbb{E}^{Q}[N_1]$, which together with (6.35) implies that $\bar{S}_T$ converges almost surely to $\mathbb{E}^{Q}[N_1]\,\mathbb{E}^{\pi}[h(X)]$. This implies in particular that $\mathbb{E}^{Q}[|\tilde{S}_1|] < \infty$ and $\mathbb{E}^{Q}[\tilde{S}_1] = \mathbb{E}^{Q}[N_1]\mathbb{E}^{\pi}[h(X)]$. Hence, the random variables $\tilde{S}_t - N_t\,\mathbb{E}^{\pi}[h(X)]$, are iid and centered. Thus

Theorem 6.68. *If $\mathbb{E}^{Q}[\tilde{S}_1^2]$ and $\mathbb{E}^{Q}[N_1^2]$ are both finite, the Central Limit Theorem applies:*

(6.36) $$\sqrt{R}\left(\bar{h}_{\tau_R} - \mathbb{E}^\pi[h(X)]\right) \overset{\mathcal{L}}{\leadsto} \mathcal{N}(0, \sigma_h^2),$$

where

$$\sigma_h^2 = \frac{\mathbb{E}^Q\left[(\tilde{S}_1 - N_1\mathbb{E}^\pi[h(X)])^2\right]}{\{\mathbb{E}^Q[N_1]\}^2}.$$

While it seems that (6.33) and (6.36) are very similar, the advantage in using this approach is that σ_h^2 can be estimated much more easily due to the underlying independent structure. For instance,

$$\hat{\sigma}_h^2 = \frac{\sum_{t=1}^R (\tilde{S}_t - \bar{h}_{\tau_R} N_t)^2}{R\bar{N}^2}$$

is a consistent estimator of σ_h^2.

In addition, the conditions on $\mathbb{E}^Q[\tilde{S}_1^2]$ and $\mathbb{E}^Q[N_1^2]$ appearing in Theorem 6.68 are minimal in that they hold when the conditions of Theorem 6.67 hold (see Hobert et al. 2002, for a proof).

6.8 Problems

6.1 Examine whether a Markov chain (X_t) may always be represented by the deterministic transform $X_{t+1} = \psi(X_t, \epsilon_t)$, where (ϵ_t) is a sequence of iid rv's. (*Hint:* Consider that ϵ_t can be of infinite dimension.)

6.2 Show that if (X_n) is a time-homogeneous Markov chain, the transition kernel does not depend on n. In particular, if the Markov chain has a finite state-space, the transition matrix is constant.

6.3 Show that an ARMA(p, q) model, defined by

$$X_n = \sum_{i=1}^p \alpha_i X_{n-i} + \sum_{j=1}^q \beta_j \varepsilon_{n-j} + \varepsilon_n,$$

does not produce a Markov chain. (*Hint:* Examine the relation with an AR(q) process through the decomposition

$$Z_n = \sum_{i=1}^p \alpha_i Z_{n-i} + \varepsilon_n, \qquad Y_n = \sum_{j=1}^q \beta_j Z_{n-j} + Z_n,$$

since (Y_n) and (X_n) are then identically distributed.)

6.4 Show that the resolvent kernel of Definition 6.8 is truly a kernel.

6.5 Show that the properties of the resolvent kernel are preserved if the geometric distribution $\mathcal{G}eo(\epsilon)$ is replaced by a Poisson distribution $\mathcal{P}(\lambda)$ with arbitrary parameter λ.

6.6 Derive the strong Markov property from the decomposition

$$\mathbb{E}_\mu[h(X_{\zeta+1}, X_{\zeta+2}, \ldots)|x_\zeta, x_{\zeta-1}, \ldots]$$
$$= \sum_{n=1}^\infty \mathbb{E}_\mu[h(X_{n+1}, X_{n+2}, \ldots)|x_n, x_{n-1}, \ldots, \zeta = n]P(\zeta = n|x_n, x_{n-1}, \ldots)$$

and from the weak Markov property.

6.7 Given the transition matrix

$$\mathbb{P} = \begin{pmatrix} 0.0 & 0.4 & 0.6 & 0.0 & 0.0 \\ 0.65 & 0.0 & 0.35 & 0.0 & 0.0 \\ 0.32 & 0.68 & 0.0 & 0.0 & 0.0 \\ 0.0 & 0.0 & 0.0 & 0.12 & 0.88 \\ 0.0 & 0.0 & 0.0 & 0.56 & 0.44 \end{pmatrix},$$

examine whether the corresponding chain is irreducible and aperiodic.

6.8 Show that irreducibility in the sense of Definition 6.13 coincides with the more intuitive notion that two arbitrary states are connected when the Markov chain has a discrete support.

6.9 Show that an aperiodic Markov chain on a finite state-space with transition matrix $\mathbb{P}$ is irreducible if and only if there exists $N \in \mathbb{N}$ such that $\mathbb{P}^N$ has no zero entries. (The matrix is then called *regular*.)

6.10 (Kemeny and Snell 1960) Show that for a regular matrix $\mathbb{P}$:
   (a) The sequence $(\mathbb{P}^n)$ converges to a stochastic matrix A.
   (b) Each row of A is the same probability vector π.
   (c) All components of π are positive.
   (d) For every probability vector μ, $\mu \mathbb{P}^n$ converges to π.
   (e) π satisfies $\pi = \pi \mathbb{P}$.
   (*Note:* See Kemeny and Snell 1960, p. 71 for a full proof.)

6.11 Show that for the measure ψ given by (6.9), the chain (X_n) is irreducible in the sense of Definition 6.13. Show that for two measures φ_1 and φ_2, such that (X_n) is φ_i-irreducible, the corresponding ψ_i's given by (6.9) are equivalent.

6.12 Let $Y_1, Y_2, \dots$ be iid rv's concentrated on $\mathbb{N}_+$ and Y_0 be another rv also concentrated on $\mathbb{N}_+$. Define

$$Z_n = \sum_{i=0}^{n} Y_i.$$

   (a) Show that (Z_n) is a Markov chain. Is it irreducible?
   (b) Define the forward recurrence time as

$$V_n^+ = \inf\{Z_m - n; Z_m > n\}.$$

   Show that (V_n^+) is also a Markov chain.
   (c) If $V_n^+ = k > 1$, show that $V_{n+1}^+ = k - 1$. If $V_n^+ = 1$, show that a renewal occurs at $n + 1$. (*Hint:* Show that $V_{n+1}^+ \sim Y_i$ in the latter case.)

6.13 Detail the proof of Theorem 6.15. In particular, show that the fact that K_ϵ includes a Dirac mass does not invalidate the irreducibility. (*Hint:* Establish that

$$\mathbb{E}_x[\eta_A] = \sum_n P_x^n(A) > P_x(\tau_A < \infty),$$

$$\lim_{\epsilon \to 1} K_\epsilon(x, A) > P_x(\tau_A < \infty),$$

$$K_\epsilon(x, A) = (1 - \epsilon) \sum_{i=1}^{\infty} \epsilon^i P^i(x, A) > 0$$

imply that there exists n such that $K^n(x, A) > 0$. See Meyn and Tweedie 1993, p. 87.)

6.14 Show that the multiplicative random walk

$$X_{t+1} = X_t \epsilon_t$$

is not irreducible when $\epsilon_t \sim \mathcal{E}xp(1)$ and $x_0 \in \mathbb{R}$. (*Hint:* Show that it produces two irreducible components.)

6.15 Show that in the setup of Example 6.17, the chain is not irreducible when ϵ_n is uniform on $[-1, 1]$ and $|\theta| > 1$.

6.16 In the spirit of Definition 6.25, we can define a *uniformly transient set* as a set A for which there exists $M < \infty$ with

$$\mathbb{E}_x[\eta_A] \leq M, \qquad \forall x \in A .$$

Show that *transient* sets are denumerable unions of *uniformly transient* sets.

6.17 Show that the split chain defined on $\mathcal{X} \times \{0, 1\}$ by the following transition kernel:

$$P(\check{X}_{n+1} \in A \times \{0\} | (x_n, 0))$$
$$= \mathbb{I}_C(x_n) \left\{ \frac{P(X_n, A \cap C) - \epsilon \nu(A \cap C)}{1 - \epsilon} (1 - \epsilon) \right.$$
$$\left. + \frac{P(X_n, A \cap C^c) - \epsilon \nu(A \cap C^c)}{1 - \epsilon} \right\}$$
$$+ \mathbb{I}_{C^c}(X_n) \left\{ P(X_n, A \cap C)(1 - \epsilon) + P(X_n, A \cap C^c) \right\} ,$$
$$P(\check{X}_{n+1} \in A \times \{1\} | (x_n, 0))$$
$$= \mathbb{I}_C(X_n) \frac{P(X_n, A \cap C) - \epsilon \nu(A \cap C)}{1 - \epsilon} \epsilon + \mathbb{I}_{C^c}(X_n) P(X_n, A \cap C)\epsilon,$$

$$P(\check{X}_{n+1} \in A \times \{0\} | (x_n, 1)) = \nu(A \cap C)(1 - \epsilon) + \nu(A \cap C^c),$$
$$P(\check{X}_{n+1} \in A \times \{1\} | (x_n, 1)) = \nu(A \cap C)\epsilon ,$$

satisfies

$$P(\check{X}_{n+1} \in A \times \{1\} | \check{x}_n) = \epsilon \nu(A \cap C),$$
$$P(\check{X}_{n+1} \in A \times \{0\} | \check{x}_n) = \nu(A \cap C^c) + (1 - \epsilon)\nu(A \cap C)$$

for every $\check{x}_n \in C \times \{1\}$. Deduce that $C \times \{1\}$ is an atom of the split chain $(\check{X}_n)$.

6.18 If C is a small set and $B \subset C$, under which conditions on B is B a small set?

6.19 If C is a small set and $D = \{x; P^m(x, D) > \delta\}$, show that D is a small set for δ small enough. (*Hint:* Use the Chapman–Kolmogorov equations.)

6.20 Show that the period d given in Definition 6.23 is independent of the selected small set C and that this number characterizes the chain (X_n).

6.21 Given the transition matrix

$$\mathbb{P} = \begin{pmatrix} 0.0 & 0.4 & 0.6 & 0.0 & 0.0 \\ 0.6 & 0.0 & .35 & 0.0 & 0.05 \\ 0.32 & .68 & 0.0 & 0.0 & 0.0 \\ 0.0 & 0.0 & 0.12 & 0.0 & 0.88 \\ 0.14 & 0.3 & 0.0 & 0.56 & 0.0 \end{pmatrix} ,$$

show that the corresponding chain is aperiodic, despite the null diagonal.

The *random walk* (Examples 6.40 and 6.39) is a useful probability model and has been given many colorful interpretations. (A popular one is the description of an inebriated individual whose progress along a street is composed of independent steps in random directions, and a question of interest is to describe where the individual will end up.) Here, we look at a simple version to illustrate a number of the Markov chain concepts.

6.22 A random walk on the non-negative integers $I = \{0, 1, 2, \ldots\}$ can be constructed in the following way. For $0 < p < 1$, let $Y_0, Y_1, \ldots$ be iid random variables with $P(Y_i = 1) = p$ and $P(Y_i = -1) = 1 - p$, and $X_k = \sum_{i=0}^{k} Y_i$. Then, (X_n) is a Markov chain with transition probabilities

$$P(X_{i+1} = j + 1 | X_i = j) = p, \qquad P(X_{i+1} = j - 1 | X_i = j) = 1 - p,$$

but we make the exception that $P(X_{i+1} = 1 | X_i = 0) = p$ and $P(X_{i+1} = 0 | X_i = 0) = 1 - p$.

(a) Show that (X_n) is a Markov chain.

(b) Show that (X_n) is also irreducible.

(c) Show that the invariant distribution of the chain is given by

$$a_k = \left(\frac{p}{1-p}\right)^k a_0, \quad k = 1, 2, \ldots,$$

where a_k is the probability that the chain is at k and a_0 is arbitrary. For what values of p and a_0 is this a probability distribution?

(d) If $\sum a_k < \infty$, show that the invariant distribution is also the stationary distribution of the chain; that is, the chain is ergodic.

6.23 If (X_t) is a random walk, $X_{t+1} = X_t + \epsilon_t$, such that ϵ_t has a moment generating function f, defined in a neighborhood of 0, give the moment generating function of X_{t+1}, g_{t+1} in terms of g_t and f, when $X_0 = 0$. Deduce that there is no invariant distribution with a moment generating function in this case.

Although the property of aperiodicity is important, it is probably less important than properties such as recurrence and irreducibility. It is interesting that Feller (1971, Section XV.5) notes that the classification into periodic and aperiodic states "represents a nuisance." However, this is less true when the random variables are continuous.

6.24 (Continuation of Problem 6.22)

(a) Using the definition of periodic given here, show that the random walk of Problem 6.22 is periodic with period 2.

(b) Suppose that we modify the random walk of Problem 6.22 by letting $0 < p + q < 1$ and redefining

$$Y_i = \begin{cases} 1 & \text{with probability } p \\ 0 & \text{with probability } 1 - p - q \\ -1 & \text{with probability } q. \end{cases}$$

Show that this random walk is irreducible and aperiodic. Find the invariant distribution, and the conditions on p and q for which the Markov chain is positive recurrent.

6.25 (Continuation of Problem 6.22) A Markov chain that is not positive recurrent may be either null recurrent or *transient*. In either of these latter two cases, the invariant distribution, if it exists, is not a probability distribution (it does not have a finite integral), and the difference is one of expected return times. For any integer j, the probability of returning to j in k steps is $p_{jj}^{(k)} = P(X_{i+k} = j|X_i = j)$, and the expected return time is thus $m_{jj} = \sum_{k=1}^{\infty} k p_{jj}^{(k)}$.

(a) Show that since the Markov chain is irreducible, $m_{jj} = \infty$ either for all j or for no j; that is, for any two states x and y, x is transient if and only if y is transient.

(b) An irreducible Markov chain is *transient* if $m_{jj} = \infty$; otherwise it is recurrent. Show that the random walk is positive recurrent if $p < 1/2$ and transient if $p > 1/2$.

(c) Show that the random walk is null recurrent if $p = 1/2$. This is the interesting case where each state will be visited infinitely often, but the expected return time is infinite.

6.26 Explain why the resolvent chain is necessarily strongly irreducible.

6.27 Consider a random walk on $\mathbb{R}_+$, defined as

$$X_{n+1} = (X_n + \epsilon)^+.$$

Show that the sets $(0, c)$ are small, provided $P(\epsilon < 0) > 0$.

6.28 Consider a random walk on $\mathbb{Z}$ with transition probabilities

$$P(Z_t = n + 1|Z_{t-1} = n) = 1 - P(Z_t = n - 1|Z_{t-1} = n) \propto n^{-\alpha}$$

and

$$P(Z_t = 1|Z_{t-1} = 0) = 1 - P(Z_t = -1|Z_{t-1} = 0) = 1/2 \ .$$

Study the recurrence properties of the chain in terms of α.

6.29 Establish (i) and (ii) of Theorem 6.28.

(a) Use

$$K^n(x, A) \geq K^r(x, \alpha)K^s(\alpha, \alpha)K^t(\alpha, A)$$

for $r + s + t = n$ and r and s such that

$$K^r(x, \alpha) > 0 \qquad \text{and} \qquad K^s(\alpha, A) > 0$$

to derive from the Chapman–Kolmogorov equations that $\mathbb{E}_x[\eta_A] = \infty$ when $\mathbb{E}_\alpha[\eta_\alpha] = \infty$.

(b) To show (ii):
   a) Establish that transience is equivalent to $P_\alpha(\tau_\alpha < \infty) < 1$.
   b) Deduce that $\mathbb{E}_x[\eta_\alpha] < \infty$ by using a generating function as in the proof of Proposition 6.31.
   c) Show that the covering of $\mathcal{X}$ is made of the

$$\bar{\alpha}_j = \{y; \sum_{n=1}^{j} K^n(y, \alpha) > j^{-1}\}.$$

6.30 Referring to Definition 6.32, show that if $P(\eta_A = \infty) \neq 0$ then $\mathbb{E}_x[\eta_A] = \infty$, but that $P(\eta_A = \infty) = 0$ does not imply $\mathbb{E}_x[\eta_A] < \infty$.

6.31 In connection with Example 6.42, show that the chain is null recurrent when $f'(1) = 1$.

6.32 Referring to (6.21):
  (a) Show that $\mathbb{E}_\nu[\tau_C] < \infty$;
  (b) show that $\sum_\nu P(\tau_C \geq t) = \mathbb{E}_\nu[\tau_C]$.

6.33 Let $\Gamma = \{Z_n : n = 0, 1, \ldots\}$ be a discrete time homogeneous Markov chain with state space $\mathcal{Z}$ and Markov transition kernel

$$(6.37) \qquad M(z, \cdot) = \omega\nu(\cdot) + (1 - \omega)K(z, \cdot) ,$$

where $\omega \in (0, 1)$ and ν is a probability measure.
  (a) Show that the measure

$$\phi(\cdot) = \sum_{i=1}^{\infty} \omega(1 - \omega)^{i-1} K^{i-1}(\nu, \cdot)$$

  is an invariant probability measure for Γ.
  (b) Deduce that Γ is positive recurrent.
  (c) Show that, when Φ satisfies a minorization condition with $C = \mathcal{X}$, (6.10) holds for all $x \in \mathcal{X}$ and is thus a mixture of the form (6.37).
  (*Note:* Even if the Markov chain associated with K is badly behaved, e.g., transient, Γ is still positive recurrent. Breyer and Roberts (2000b) propose another derivation of this result, through the functional identity

$$\int \phi(x)M(x, z)dx = \pi(z).)$$

6.34 Establish the equality (6.14).

6.35 Consider the simple Markov chain (X_n), where each X_i takes on the values -1 and 1 with $P(X_{i+1} = 1|X_i = -1) = 1$, $P(X_{i+1} = -1|X_i = 1) = 1$, and $P(X_0 = 1) = 1/2$.
  (a) Show that this is a stationary Markov chain.
  (b) Show that $\text{cov}(X_0, X_k)$ does not go to zero.
  (c) The Markov chain is not strictly positive. Verify this by exhibiting a set that has positive unconditional probability but zero conditional probability.
  (*Note:* The phenomenon seen here is similar to what Seidenfeld and Wasserman 1993 call a *dilation*.)

6.36 In the setup of Example 6.5, find the stationary distribution associated with the proposed transition when $\pi_i = \pi_j$ and in general.

6.37 Show the decomposition of the "first entrance and last exit" equation (6.23).

6.38 If (a_n) is a sequence of real numbers converging to a, and if $b_n = (a_1 + \cdots + a_n)/n$, then show that

$$\lim_n b_n = a .$$

  (*Note:* The sum $(1/n)\sum_{i=1}^n a_i$ is called a *Cesàro average*; see Billingsley 1995, Section A30.)

6.39 Consider a sequence (a_n) of positive numbers which is converging to a^* and a convergent series with running term b_n. Show that the convolution

$$\sum_{j=1}^{n-1} a_j b_{n-j} \overset{n \to \infty}{\longrightarrow} a^* \sum_{j=1}^{\infty} b_j.$$

  (*Hint:* Use the Dominated Convergence Theorem.)

6.40 (a) Verify (6.26), namely, that $\|\mu\|_{TV} = (1/2)\sup_{|h|\leq 1} |\int h(x)\mu(dx)|$.

(b) Show that (6.26) is compatible with the definition of the total variation norm. Establish the relation with the alternative definition

$$\|\mu\|_{TV} = \sup_A \mu(A) - \inf_A \mu(A).$$

6.41 Show that if (X_n) and (X'_n) are coupled at time N_0 and if $X_0 \sim \pi$, then $X'_n \sim \pi$ for $n > N_0$ for any initial distribution of X'_0.

6.42 Using the notation of Section 6.6.1, set

$$u(n) = \sum_{j=0}^{\infty} p^{j\star}(n)$$

with $p^{j\star}$ the distribution of the sum $S_1 + \cdots + S_j$, $p^{0\star}$ the Dirac mass at 0, and

$$Z(n) = \mathbb{I}_{\exists j; S_j = n}.$$

(a) Show that $P_q(Z(n) = 1) = q \star u(n)$.

(b) Show that

$$|q \star u(n) - p \star u(n)| \leq 2\, P(T_{pq} > n).$$

(This bound is often called *Orey's inequality*, from Orey 1971. See Problem 7.10 for a slightly different formulation.)

(c) Show that if m_p is finite,

$$e(n) = \frac{\sum_{n+1}^{\infty} p(j)}{m_p}$$

is the invariant distribution of the renewal process in the sense that $P_e(Z(n) = 1) = 1/m_p$ for every n.

(d) Deduce from Lemma 6.49 that

$$\lim_n \left| q \star u(n) - \frac{1}{m_p} \right| = 0$$

when the mean renewal time is finite.

6.43 Consider the so-called "forward recurrence time" process V_n^+, which is a Markov chain on $\mathbb{N}_+$ with transition probabilities

$$P(1, j) = p(j), \qquad j \geq 1,$$
$$P(j, j-1) = 1, \qquad j > 1,$$

where p is an arbitrary probability distribution on $\mathbb{N}_+$. (See Problem 6.12.)

(a) Show that (V_n^+) is recurrent.

(b) Show that

$$P(V_n^+ = j) = p(j + n - 1).$$

(c) Deduce that the invariant measure satisfies

$$\pi(j) = \sum_{n \geq j} p(n)$$

and show it is finite if and only if

$$m_p = \sum_n np(n) < \infty.$$

6.44 (Continuation of Problem 6.43) Consider two independent forward recurrence time processes (V_n^+) and (W_n^+) with the same generating probability distribution p.

(a) Give the transition probabilities of the joint process $V_n^* = (V_n^+, W_n^+)$.

(b) Show that (V_n^*) is irreducible when p is aperiodic. (*Hint:* Consider r and s such that g.c.d.$(r, s) = 1$ with $p(r) > 0$, $p(q) > 0$, and show that if $nr - ms = 1$ and $i \geq j$, then

$$P_{(i,j)}(V_{j+(i-j)nr}^* = (1,1)) > 0 \ .)$$

(c) Show that $\pi^* = \pi \times \pi$, with π defined in Problem 6.43 is invariant and, therefore, that (V_n^*) is positive Harris recurrent when $m_p < \infty$.

6.45 (Continuation of Problem 6.44) Consider V_n^* defined in Problem 6.44 associated with (S_n, S_n') and define $\tau_{1,1} = \min\{n; V_n^* = (1,1)\}$.

(a) Show that $T_{pq} = \tau_{1,1} + 1$.

(b) Use (c) in Problem 6.44 to show Lemma 6.49.

6.46 (Kemeny and Snell 1960) Establish (directly) the Law of Large Numbers for a finite irreducible state-space chain (X_n) and for $h(x_n) = \mathbb{I}_j(x_n)$, if j is a possible state of the chain; that is,

$$\frac{1}{N} \sum_{n=1}^{N} \mathbb{I}_j(x_n) \longrightarrow \pi_j \ ,$$

where $\pi = (\pi_1, \dots, \pi_j, \dots)$ is the stationary distribution.

6.47 (Kemeny and Snell 1960) Let $\mathbb{P}$ be a regular transition matrix, that is, $\mathbb{P}A = A\mathbb{P}$ (see Problem 6.9), with limiting (stationary) matrix A; that is, each column of A is equal to the stationary distribution.

(a) Show that the so-called *fundamental matrix* $\mathbb{Z} = (I - (\mathbb{P} - A))^{-1}$ exists.

(b) Show that $\mathbb{Z} = I + \sum_{n=1}^{\infty}(\mathbb{P}^n - A)$.

(c) Show that $\mathbb{Z}$ satisfies $\pi\mathbb{Z} = \pi$ and $\mathbb{P}\mathbb{Z} = \mathbb{Z}\mathbb{P}$, where π denotes a row of A (this is the stationary distribution).

6.48 (Continuation of Problem 6.47) Let $N_j(n)$ be the number of times the chain is in state j in the first n instants.

(a) Show that for every initial distribution μ,

$$\lim_{n \to \infty} \mathbb{E}_\mu[N_j(n)] - n\pi_j = \mu(\mathbb{Z} - A).$$

(*Note:* This convergence shows the strong stability of a recurrent chain since each term in the difference goes to infinity.)

(b) Show that for every pair of initial distributions, (μ, ν),

$$\lim_{n \to \infty} \mathbb{E}_\mu[N_j(n)] - \mathbb{E}_\nu[N_j(n)] = (\mu - \nu)\mathbb{Z}.$$

(c) Deduce that for every pair of states, (u, v),

$$\lim_{n \to \infty} \mathbb{E}_u[N_j(n)] - \mathbb{E}_v[N_j(n)] = z_{uj} - z_{vj},$$

which is called the *divergence* div$_j(u, v)$.

6.49 (Continuation of Problem 6.47) Let f_j denote the number of steps before entering state j.

(a) Show that for every state i, $\mathbb{E}_i[f_j]$ is finite.

(b) Show that the matrix M with entries $m_{ij} = \mathbb{E}_i[f_j]$ can be written $M = \mathbb{P}(M - M_d) + E$, where M_d is the diagonal matrix with same diagonal as M and E is the matrix made of 1's.

(c) Deduce that $m_{ii} = 1/\pi_i$.

(d) Show that πM is the vector of the z_{ii}/π_i's.

(e) Show that for every pair of initial distributions, (μ, ν),

$$\mathbb{E}_\mu[f_i] - \mathbb{E}_\nu[f_i] = (\mu - \nu)(I - Z)D,$$

where D is the diagonal matrix $\text{diag}(1/\pi_i)$.

6.50 If h is a function taking values on a finite state-space $\{1, \ldots, r\}$, with $h(i) = h_i$, and if (X_n) is an irreducible Markov chain, show that

$$\lim_{n \to \infty} \frac{1}{n} \text{var} \left(\sum_{t=1}^n h(x_t) \right) = \sum_{i,j} h_i c_{ij} h_j,$$

where $c_{ij} = \pi_i z_{ij} + \pi_j z_{ji} - \pi_i \delta_{ij} - \pi_i \pi_j$ and δ_{ij} is Kroenecker's 0–1 function.

6.51 For the two-state transition matrix $\mathbb{P} = \begin{pmatrix} 1 - \alpha & \alpha \\ \beta & 1 - \beta \end{pmatrix}$, show that

(a) the stationary distribution is $\pi = (\beta/(\alpha + \beta), \alpha/(\alpha + \beta))$;

(b) the mean first passage matrix is

$$M = \begin{pmatrix} (\alpha + \beta)/\beta & 1/\alpha \\ 1/\beta & (\alpha + \beta)/\alpha \end{pmatrix};$$

(c) and the limiting variance for the number of times in state j is $\alpha\beta(2 - \alpha - \beta)/(\alpha + \beta)^3$, for $j = 1, 2$.

6.52 Show that a finite state-space chain is always geometrically ergodic.

6.53 (Kemeny and Snell 1960) Given a finite state-space Markov chain, with transition matrix $\mathbb{P}$, define a second transition matrix by

$$p_{ij}(n) = \frac{P_\mu(X_{n-1} = j)P(X_n = i|X_{n-1} = j)}{P_\mu(X_n = j)}.$$

(a) Show that $p_{ij}(n)$ does not depend on n if the chain is stationary (i.e., if $\mu = \pi$).

(b) Explain why, in this case, the chain with transition matrix $\tilde{\mathbb{P}}$ made of the probabilities

$$\tilde{p}_{ij} = \frac{\pi_j p_{ji}}{\pi_i}$$

is called the *reverse* Markov chain.

(c) Show that the limiting variance C is the same for both chains.

6.54 (Continuation of Problem 6.53) A Markov chain is *reversible* if $\tilde{\mathbb{P}} = \mathbb{P}$. Show that every two-state ergodic chain is reversible and that an ergodic chain with symmetric transition matrix is reversible. Examine whether the matrix

$$\mathbb{P} = \begin{pmatrix} 0 & 0 & 1 & 0 & 0 \\ 0.5 & 0 & 0.5 & 0 & 0 \\ 0 & 0.5 & 0 & 0.5 & 0 \\ 0 & 0 & 0.5 & 0 & 0.5 \\ 0 & 0 & 1 & 0 & 0 \end{pmatrix}$$

is reversible. (*Hint:* Show that $\pi = (0.1, 0.2, 0.4, 0.2, 0.1)$.)

6.55 (Continuation of Problem 6.54) Show that an ergodic random walk on a finite state-space is reversible.

6.56 (Kemeny and Snell 1960) A Markov chain (X_n) is *lumpable* with respect to a nontrivial partition of the state-space, $(A_1, \ldots, A_k)$, if, for every initial distribution μ, the process

$$Z_n = \sum_{i=1}^{k} i \mathbb{I}_{A_i}(X_n)$$

is a Markov chain with transition probabilities independent of μ.

(a) Show that a necessary and sufficient condition for lumpability is that

$$p_{uA_j} = \sum_{v \in A_j} p_{uv}$$

is constant (in n) on A_i for every i.

(b) Examine whether

$$\mathbb{P} = \begin{pmatrix} 1 & 0 & 0 & 0 & 0 \\ 0 & 1 & 0 & 0 & 0 \\ 0.5 & 0 & 0 & 0.5 & 0 \\ 0 & 0 & 0.5 & 0 & 0.5 \\ 0 & 0.5 & 0 & 0.5 & 0 \end{pmatrix}$$

is lumpable for $A_1 = \{1, 2\}$, $A_2 = \{3, 4\}$, and $A_3 = \{5\}$.

6.57 Consider the random walk on $\mathbb{R}^+$, $X_{n+1} = (X_n + \epsilon_n)^+$, with $\mathbb{E}[\epsilon_n] = \beta$.

(a) Establish Lemma 6.70. (*Hint:* Consider an alternative V to V^* and show by recurrence that

$$V(x) \geq \int_C K(x, y)V(y)dy + \int_{C^c} K(x, y)V(y)dy$$
$$\geq \cdots \geq V^*(x) .)$$

(b) Establish Theorem 6.72 by assuming that there exists x^* such that $P_{x^*}(\tau_C < \infty) < 1$, choosing M such that $M \geq V(x^*)/[1 - P_{x^*}(\tau_C < \infty)]$ and establishing that $V(x^*) \geq M[1 - P_{x^*}(\tau_C < \infty)]$.

6.58 Show that

(a) a time-homogeneous Markov chain (X_n) is stationary if the initial distribution is the invariant distribution;

(b) the invariant distribution of a stationary Markov chain is also the marginal distribution of any X_n.

6.59 Referring to Section 6.7.1, let X_n be a Markov chain and $h(\cdot)$ a function with $\mathbb{E}h(X_n) = 0$, $\mathrm{Var}h(X_n) = \sigma^2 > 0$, and $\mathbb{E}h(X_{n+1}|x_n) = h(x_n)$, so $h(\cdot)$ is a nonconstant harmonic function.

(a) Show that $\mathbb{E}h(X_{n+1}|x_0) = h(x_0)$.

(b) Show that $\mathrm{Cov}(h(x_0), h(X_n)) = \sigma^2$.

(c) Use (6.52) to establish that $\mathrm{Var}\left(\frac{1}{n+1}\sum_{i=0}^{n} h(X_i)\right) \to \infty$ as $n \to \infty$, showing that the chain is not ergodic.

6.60 Show that if an irreducible Markov chain has a σ-finite invariant measure, this measure is unique up to a multiplicative factor. (*Hint:* Use Theorem 6.63.)

6.61 (Kemeny and Snell 1960) Show that for an aperiodic irreducible Markov chain with finite state-space and with transition matrix $\mathbb{P}$, there always exists a stationary probability distribution which satisfies

$$\pi = \pi\mathbb{P}.$$

(a) Show that if $\beta < 0$, the random walk is recurrent. (*Hint:* Use the drift function $V(x) = x$ as in Theorem 6.71.)
(b) Show that if $\beta = 0$ and $\text{var}(\epsilon_n) < \infty$, (X_n) is recurrent. (*Hint:* Use $V(x) = \log(1+x)$ for $x > R$ and $V(x) = 0$, otherwise, for an adequate bound R.)
(c) Show that if $\beta > 0$, the random walk is transient.

6.62 Show that if there exist a finite potential function V and a small set C such that V is bounded on C and satisfies (6.40), the corresponding chain is Harris positive.

6.63 Show that the random walk on $\mathbb{Z}$ is transient when $\mathbb{E}[W_n] \neq 0$.

6.64 Show that the chains defined by the kernels (6.46) and (6.48) are either both recurrent or both transient.

6.65 Referring to Example 6.66, show that the AR(1) chain is reversible.

6.66 We saw in Section 6.6.2 that a stationary Markov chain is *geometrically ergodic* if there is a non-negative real-valued function M and a constant $r < 1$ such that for any $A \in \mathcal{X}$,

$$|P(X_n \in A|X_0 \in B) - P(X_n \in A)| \leq M(x)r^n.$$

Prove that the following Central Limit Theorem (due to Chan and Geyer 1994) can be considered a corollary to Theorem 6.82 (see Note 6.9.4):

Corollary 6.69. *Suppose that the stationary Markov chain $X_0, X_1, X_2, \ldots$ is geometrically ergodic with $M^* = \int |M(x)|f(x)dx < \infty$ and satisfies the moment conditions of Theorem 6.82. Then*

$$\sigma^2 = \lim_{n\to\infty} n\,\text{var}\bar{X}_n < \infty$$

and if $\sigma^2 > 0$, $\sqrt{n}\bar{X}_n/\sigma$ tends in law to $\mathcal{N}(0, \sigma^2)$.

(*Hint:* Integrate (with respect to f) both sides of the definition of geometric ergodicity to conclude that the chain has exponentially fast α-mixing, and apply Theorem 6.82.)

6.67 Suppose that $X_0, X_1, \ldots, X_n$ have a common mean ξ and variance σ^2 and that $\text{cov}(X_i, X_j) = \rho_{j-i}$. For estimating ξ, show that

(a) $\bar{X}$ may not be consistent if $\rho_{j-i} = \rho \neq 0$ for all $i \neq j$. (*Hint:* Note that $\text{var}(\bar{X}) > 0$ for all sufficiently large n requires $\rho \geq 0$ and determine the distribution of $\bar{X}$ in the multivariate normal case.)
(b) $\bar{X}$ is consistent if $|\rho_{j-i}| \leq M\gamma^{j-i}$ with $|\gamma| < 1$.

6.68 For the situation of Example 6.84:

(a) Prove that the sequence (X_n) is stationary provided $\sigma^2 = 1/(1 - \beta^2)$.
(b) Show that $\mathbb{E}(X_k|x_0) = \beta^k x_0$. (*Hint:* Consider $\mathbb{E}[(X_k - \beta X_{k-1})|x_0]$.)
(c) Show that $\text{cov}(X_0, X_k) = \beta^k/(1 - \beta^2)$.

6.69 Under the conditions of Theorem 6.85, it follows that $\mathbb{E}[\mathbb{E}(X_k|X_0)]^2 \to 0$. There are some other interesting properties of this sequence.

(a) Show that

$$var[\mathbb{E}(X_k|X_0)] = \mathbb{E}[\mathbb{E}(X_k|X_0)]^2,$$
$$var[\mathbb{E}(X_k|X_0)] \geq var[\mathbb{E}(X_{k+1}|X_0)]] \ .$$

(*Hint*: Write $f_{k+1}(y|x) = \int f_k(y|x')f(x'|x)dx'$ and use Fubini and Jensen.)
(b) Show that

$$\mathbb{E}[var(X_k|X_0)] \leq \mathbb{E}[var(X_{k+1}|X_0)]$$

and that

$$\lim_{k \to \infty} \mathbb{E}[var(X_k|X_0)] = \sigma^2 .$$

6.9 Notes

6.9.1 Drift Conditions

Besides atoms and small sets, Meyn and Tweedie (1993) rely on another tool to check or establish various stability results, namely, *drift criteria*, which can be traced back to Lyapunov. Given a function V on $\mathcal{X}$, the *drift of V* is defined by

$$\Delta V(x) = \int V(y) \, P(x,dy) - V(x) \ .$$

(Functions V appearing in this setting are often referred to as *potentials*; see Norris 1997.) This notion is also used in the following chapters to verify the convergence properties of some MCMC algorithms (see, e.g., Theorem 7.15 or Mengersen and Tweedie 1996).

The following lemma is instrumental in deriving drift conditions for the transience or the recurrence of a chain (X_n).

Lemma 6.70. *If $C \in \mathcal{B}(\mathcal{X})$, the smallest positive function which satisfies the conditions*

(6.38) $\Delta V(x) \leq 0 \quad if \quad x \notin C, \quad V(x) \geq 1 \quad if \quad x \in C$

is given by

$$V^*(x) = P_x(\sigma_C < \infty) \ ,$$

where σ_C denotes

$$\sigma_C = \inf\{n \geq 0; x_n \in C\} \ .$$

Note that, if $x \notin C$, $\sigma_C = \tau_C$, while $\sigma_C = 0$ on C. We then have the following necessary and sufficient condition.

Theorem 6.71. *The ψ-irreducible chain (X_n) is transient if and only if there exist a bounded positive function V and a real number $r \geq 0$ such that for every x for which $V(x) > r$, we have*

(6.39) $\Delta V(x) > 0 \ .$

Proof. If $C = \{x; V(x) \leq r\}$ and M is a bound on V, the conditions (6.38) are satisfied by

$$\tilde{V}(x) = \begin{cases} (M - V(x))/(M - r) & \text{if } x \in C^c \\ 1 & \text{if } x \in C. \end{cases}$$

Since $\tilde{V}(x) < 1$ for $x \in C^c$, $V^*(x) = P_x(\tau_C < \infty) < 1$ on C^c, and this implies the transience of C, therefore the transience of (X_n). The converse can be deduced from a (partial) converse to Proposition 6.31 (see Meyn and Tweedie 1993, p. 190).   □

Condition (6.39) describes an average increase of $V(x_n)$ once a certain level has been attained, and therefore does not allow a sure return to 0 of V. The condition is thus incompatible with the stability associated with recurrence. On the other hand, if there exists a potential function V "attracted" to 0, the chain is recurrent.

Theorem 6.72. *Consider (X_n) a ψ-irreducible Markov chain. If there exist a small set C and a function V such that*

$$C_V(n) = \{x; V(x) \leq n\}$$

is a small set for every n, the chain is recurrent if

$$\Delta V(x) \leq 0 \text{ on } C^c.$$

The fact that $C_V(n)$ is small means that the function V is not bounded outside small sets. The attraction of the chain toward smaller values of V on the sets where V is large is thus a guarantee of stability for the chain. The proof of the above result is, again, quite involved, based on the fact that $P_x(\tau_C < \infty) = 1$ (see Meyn and Tweedie 1993, p. 191).

Example 6.73. (Continuation of Example 6.39) If the distribution of W_n has a finite support and zero expectation, (X_n) *is recurrent.* When considering $V(x) = |x|$ and r such that $\gamma_x = 0$ for $|x| > r$, we get

$$\Delta V(x) = \sum_{n=-r}^{r} \gamma_n (|x + n| - |x|),$$

which is equal to

$$\sum_{n=-r}^{r} \gamma_n n \text{ if } x \geq r \qquad \text{and} \qquad - \sum_{n=-r}^{r} \gamma_n n \text{ if } x \leq -r.$$

Therefore, $\Delta V(x) = 0$ for $x \notin \{-r + 1, \ldots, r - 1\}$, which is a small set. Conversely, if W_n has a nonzero mean, X_n is transient.   ‖

For Harris recurrent chains, positivity can also be related to a drift condition and to a "regularity" condition on visits to small sets.

Theorem 6.74. *If (X_n) is Harris recurrent with invariant measure π, there is equivalence between*

(a) π is finite;

(b) there exist a small set C and a positive number M_C such that

$$\sup_{x \in C} \mathbb{E}_x[\tau_C] \leq M_C \; ;$$

(c) there exist a small set C, a function V taking values in $\mathbb{R} \cup \{\infty\}$, and a positive real number b such that

$$(6.40) \qquad \Delta V(x) \leq -1 + b\mathbb{I}_C(x) \; .$$

See Meyn and Tweedie (1993, Chapter 11) for a proof and discussion of these equivalences. (If there exists V finite and bounded on C which satisfies (6.40), the chain (X_n) is necessarily Harris positive.)

The notion of a *Kendall atom* introduced in Section 6.6.2 can also be extended to non-atomic chains by defining *Kendall sets* as sets A such that

$$(6.41) \qquad \sup_{x \in A} \mathbb{E}_x\left[\sum_{k=0}^{\tau_A - 1} \kappa^k \right] < \infty \; .$$

with $\kappa > 1$. The existence of a Kendall set guarantees a *geometric drift* condition. If C is a Kendall set and if

$$V(x) = \mathbb{E}_x[\kappa^{\sigma_C}] \; ,$$

the function V satisfies

$$(6.42) \qquad \Delta V(x) \leq -\beta V(x) + b\mathbb{I}_C(x)$$

with $\beta > 0$ and $0 < b < \infty$. This condition also guarantees geometric convergence for (X_n) in the following way.

Theorem 6.75. *For a ψ-irreducible and aperiodic chain (X_n) and a small Kendall set C, there exist $R < \infty$ and $r > 1$, $\kappa > 1$ such that*

$$(6.43) \qquad \sum_{n=1}^{\infty} r^n \|K^n(x, \cdot) - \pi(\cdot)\| \leq R\, \mathbb{E}_x\left[\sum_{k=0}^{\tau_C} \kappa^k \right] < \infty$$

for almost every $x \in \mathcal{X}$.

The three conditions (6.41), (6.42) and (6.43) are, in fact, equivalent for ψ-irreducible aperiodic chains if A is a small set in (6.41) and if V is bounded from below by 1 in (6.42) (see Meyn and Tweedie 1993, pp. 354–355). The drift condition (6.42) is certainly the simplest to check in practice, even though the potential function V must be derived.

Example 6.76. (Continuation of Example 6.20) The condition $|\theta| < 1$ is necessary for the chain $X_n = \theta x_{n-1} + \varepsilon_n$ to be recurrent. Assume ε_n has a strictly positive density on $\mathbb{R}$. Define $V(x) = |x| + 1$. Then

$$\begin{aligned}
\mathbb{E}_x[V(X_1)] &= 1 + \mathbb{E}[|\theta X + \varepsilon_1|] \\
&\leq 1 + |\theta|\, |x| + \mathbb{E}[|\varepsilon_1|] \\
&= |\theta|\, V(x) + \mathbb{E}[|\varepsilon_1|] + 1 - |\theta|
\end{aligned}$$

and

$$\Delta V(x) \leq -(1 - |\theta|)\, V(x) + \mathbb{E}[|\varepsilon_1|] + 1 - |\theta|$$
$$= -(1 - |\theta|)\, \gamma V(x) + \mathbb{E}[|\varepsilon_1|] + 1 - |\theta| - (1 - \gamma)(1 - |\theta|)\, V(x)$$
$$\leq -\beta V(x) + b\, \mathbb{I}_C(x)$$

for $\beta = (1 - |\theta|)\, \gamma$, $b = \mathbb{E}[|\varepsilon_1|] + 1 - |\theta|$, and C equal to

$$C = \{x;\ V(x) < (\mathbb{E}[|\varepsilon_1|] + 1 - |\theta|)/(1 - \gamma)(1 - |\theta|)\}\ ,$$

if $|\theta| < 1$ and $\mathbb{E}[|\varepsilon_1|] < +\infty$. These conditions thus imply geometric ergodicity for AR(1) models.   ‖

Meyn and Tweedie (1994) propose, in addition, *explicit* evaluations of convergence rates r as well as explicit bounds R in connection with drift conditions (6.42), but the geometric convergence is evaluated under a norm induced by the very function V satisfying (6.42), which makes the result somewhat artificial.

There also is an equivalent form of *uniform ergodicity* involving drift, namely that (X_n) is aperiodic and there exist a small set C, a bounded potential function $V \geq 1$, and constants $0 < b < \infty$ and $\beta > 0$ such that

$$(6.44) \qquad\qquad \Delta V(x) \leq -\beta V(x) + b\, \mathbb{I}_C(x)\,, \qquad x \in \mathcal{X}\,.$$

In a practical case (see, e.g., Example 12.6), this alternative to the conditions of Theorem 6.59 is often the most natural approach.

As mentioned after Theorem 6.64, there exist alternative versions of the Central Limit Theorem based on drift conditions. Assume that there exist a function $f \geq 1$, a finite potential function V, and a small set C such that

$$(6.45) \qquad\qquad \Delta V(x) \leq -f(x) + b\, \mathbb{I}_C(x), \qquad x \in \mathcal{X},$$

and that $\mathbb{E}^\pi[V^2] < \infty$. This is exactly condition (6.44) above, with $f = V$, which implies that (6.45) holds for an uniformly ergodic chain.

Theorem 6.77. *If the ergodic chain (X_n) with invariant distribution π satisfies conditions (6.45), for every function g such that $|g| \leq f$, then*

$$\gamma_g^2 = \lim_{n \to \infty}\ n\mathbb{E}_\pi[S_n^2(\overline{g})]$$
$$= \mathbb{E}_\pi[\overline{g}^2(x_0)] + 2 \sum_{k=1}^{\infty} \mathbb{E}_\pi[\overline{g}(x_0)\overline{g}(x_k)]$$

is non-negative and finite. If $\gamma_g > 0$, the Central Limit Theorem holds for $S_n(\overline{g})$. If $\gamma_g = 0$, $\sqrt{n}S_n(\overline{g})$ almost surely goes to 0.

This theorem is definitely relevant for convergence assessment of Markov chain Monte Carlo algorithms since, when $\gamma_g^2 > 0$, it is possible to assess the convergence of the ergodic averages $S_n(g)$ to the quantity of interest $\mathbb{E}^\pi[g]$. Theorem 6.77 also suggests how to implement this monitoring through renewal theory, as discussed in detail in Chapter 12.

6.9.2 Eaton's Admissibility Condition

Eaton (1992) exhibits interesting connections, similar to Brown (1971), between the admissibility of an estimator and the recurrence of an associated Markov chain. The problem considered by Eaton (1992) is to determine whether, for a *bounded* function $g(\theta)$, a generalized Bayes estimator associated with a prior measure π is admissible under quadratic loss. Assuming that the posterior distribution $\pi(\theta|x)$ is well defined, he introduces the transition kernel

$$(6.46) \qquad K(\theta, \eta) = \int_{\mathcal{X}} \pi(\theta|x) f(x|\eta) \, dx,$$

which is associated with a Markov chain $(\theta^{(n)})$ generated as follows: The transition from $\theta^{(n)}$ to $\theta^{(n+1)}$ is done by generating first $x \sim f(x|\theta^{(n)})$ and then $\theta^{(n+1)} \sim \pi(\theta|x)$. (Most interestingly, this is also a kernel used by Markov Chain Monte Carlo methods, as shown in Chapter 9.) Note that the prior measure π is an invariant measure for the chain $(\theta^{(n)})$. For every measurable set C such that $\pi(C) < +\infty$, consider

$$V(C) = \{ h \in \mathcal{L}^2(\pi); h(\theta) \geq 0 \text{ and } h(\theta) \geq 1 \text{ when } \theta \in C \}$$

and

$$\Delta(h) = \int \int \{ h(\theta) - h(\eta) \}^2 K(\theta, \eta) \pi(\eta) \, d\theta \, d\eta.$$

The following result then characterizes admissibility for *all bounded functions* in terms of Δ and $V(C)$ (that is, independently of the estimated functions g).

Theorem 6.78. *If for every C such that $\pi(C) < +\infty$,*

$$(6.47) \qquad \inf_{h \in V(C)} \Delta(h) = 0,$$

then the Bayes estimator $\mathbb{E}^{\pi}[g(\theta)|x]$ is admissible under quadratic loss for every bounded function g.

   This result is obviously quite general but only mildly helpful in the sense that the practical verification of (6.47) for every set C can be overwhelming. Note also that (6.47) always holds when π is a proper prior distribution since $h \equiv 1$ belongs to $\mathcal{L}^2(\pi)$ and $\Delta(1) = 0$ in this case. The extension then considers approximations of 1 by functions in $V(C)$. Eaton (1992) exhibits a connection with the Markov chain $(\theta^{(n)})$, which gives a condition equivalent to Theorem 6.78. First, for a given set C, a stopping rule τ_C is defined as the first integer $n > 0$ such that $(\theta^{(n)})$ belongs to C (and $+\infty$ otherwise), as in Definition 6.10.

Theorem 6.79. *For every set C such that $\pi(C) < +\infty$,*

$$\inf_{h \in V(C)} \Delta(h) = \int_C \left\{ 1 - P(\tau_C < +\infty | \theta^{(0)} = \eta) \right\} \pi(\eta) \, d\eta.$$

Therefore, the generalized Bayes estimators of bounded functions of θ are admissible if and only if the associated Markov chain $(\theta^{(n)})$ is recurrent.

Again, we refer to Eaton (1992) for extensions, examples, and comments on this result. Note, however, that the verification of the recurrence of the Markov chain $(\theta^{(n)})$ is much easier than the determination of the lower bound of $\Delta(h)$. Hobert and Robert (1999) consider the potential of using the dual chain based on the kernel

$$(6.48) \qquad K'(x, y) = \int_{\Theta} f(y|\theta)\pi(\theta|x)d\theta$$

(see Problem 6.64) and derive admissibility results for various distributions of interest.

6.9.3 Alternative Convergence Conditions

Athreya et al. (1996) present a careful development of the basic limit theorems for Markov chains, with conditions stated that are somewhat more accessible in Markov chain Monte Carlo uses, rather than formal probabilistic properties.

Consider a time-homogeneous Markov chain (X_n) where f is the invariant density and $f_k(\cdot|\cdot)$ is the conditional density of X_k given X_0. So, in particular, $f_1(\cdot|\cdot)$ is the transition kernel. For a basic limit theorem such as Theorem 6.51, there are two conditions that are required on the transition kernel, both of which have to do with the ability of the Markov chain to visit all sets A. Assume that the transition kernel satisfies[5]: There exists a set A such that

(i) $\sum_{k=1}^{\infty} \int_A f_k(x|x_0)\,d\mu(x) > 0$ for all x_0,
(ii) $\inf_{x,y \in A} f_1(y|x) > 0$.

A set A satisfying (i) is called *accessible*, which means that from anywhere in the state space there is positive probability of eventually entering A. Condition (ii) is essentially a minorization condition. The larger the set A, the easier it is to verify (i) and the harder it is to verify (ii) . These two conditions imply that the chain is irreducible and aperiodic. It is possible to weaken (ii) to a condition that involves f_k for some $k \geq 1$; see Athreya et al. (1996).

The limit theorem of Athreya et al. (1996) can be stated as follows.

Theorem 6.80. *Suppose that the Markov chain (X_n) has invariant density $f(\cdot)$ and transition kernel $f_1(\cdot|\cdot)$ that satisfies Conditions (i) and (ii). Then*

$$(6.49) \qquad \lim_{k \to \infty} \sup_A \left| \int_A f_k(x|x_0)dx - \int_A f(x)dx \right| = 0$$

for f almost all x_0.

6.9.4 Mixing Conditions and Central Limit Theorems

In Section 6.7.2, we established a Central Limit Theorem using regeneration, which allowed us to use a typical independence argument. Other conditions, known as *mixing conditions*, can also result in a Central Limit Theorem. These mixing conditions guarantee that the dependence in the Markov chain decreases fast enough, and variables that are far enough apart are close to being independent. Unfortunately, these conditions are usually quite difficult to verify. Consider the property of α-*mixing* (Billingsley 1995, Section 27).

---

[5] The conditions stated here are weaker than those given in the first edition; we thank Hani Doss for showing us this improvement.

Definition 6.81. A sequence $X_0, X_1, X_2, \ldots$ is α-*mixing* if

$$(6.50) \qquad \alpha_n = \sup_{A,B} |P(X_n \in A, X_0 \in B) - P(X_n \in A)P(X_0 \in B)|$$

goes to 0 when n goes to infinity.

So, we see that an α-mixing sequence will tend to "look independent" if the variables are far enough apart. As a result, we would expect that Theorem 6.63 is a consequence of α-mixing. This is, in fact, the case, as every positive recurrent aperiodic Markov chain is α-mixing (Rosenblatt 1971, Section VII.3), and if the Markov chain is stationary and α-mixing, the covariances go to zero (Billingsley 1995, Section 27).

However, for a Central Limit Theorem, we need even more. Not only must the Markov chain be α-mixing, but we need the coefficient α_n to go to 0 fast enough; that is, we need the dependence to go away fast enough. One version of a Markov chain Central Limit Theorem is the following (Billingsley 1995, Section 27).

Theorem 6.82. *Suppose that the Markov chain* (X_n) *is stationary and* α-*mixing with* $\alpha_n = \mathcal{O}(n^{-5})$ *and that* $\mathbb{E}[X_n] = 0$ *and* $\mathbb{E}[X_n^{12}] < \infty$. *Then,*

$$\sigma^2 = \lim_{n \to \infty} n \operatorname{var} \bar{X}_n < \infty$$

and if $\sigma^2 > 0$, $\sqrt{n} \bar{X}_n$ *tends in law to* $\mathcal{N}(0, \sigma^2)$.

This theorem is not very useful because the condition on the mixing coefficient is very hard to verify. (Billingsley 1995 notes that the conditions are stronger than needed, but are imposed to avoid technical difficulties in the proof.) Others have worked hard to get the condition in a more accessible form and have exploited the relationship between mixing and ergodicity. Informally, if (6.50) goes to 0, dividing through by $P(X_0 \in B)$ we expect that

$$(6.51) \qquad |P(X_n \in A|X_0 \in B) - P(X_n \in A)| \to 0,$$

which looks quite similar to the assumption that the Markov chain is ergodic. (This corresponds, in fact, to a stronger type of mixing called β-*mixing*. See Bradley 1986). We actually need something stronger (see Problem 7.6), like uniform ergodicity, where there are constants M and $r < 1$ such $|P(X_n \in A|X_0 \in B) - P(X_n \in A)| \leq Mr^n$. Tierney (1994) presents the following Central Limit Theorem.

Theorem 6.83. *Let* (X_n) *be a stationary uniformly ergodic Markov chain. For any function* $h(\cdot)$ *satisfying* $\operatorname{var} h(X_i) = \sigma^2 < \infty$, *there exists a real number* τ_h *such that*

$$\sqrt{n} \frac{\sum_{i=1}^n h(X_i)/n - \mathbb{E}[h(X)]}{\tau_h} \xrightarrow{\mathcal{L}} \mathcal{N}(0, 1).$$

Other versions of the Central Limit Theorem exist. See, for example, Robert (1994), who surveys other mixing conditions and their connections with the Central Limit Theorem.

6.9.5 Covariance in Markov Chains

An application of Chebychev's inequality shows that the convergence of an average of random variables from a Markov chain can be connected to the behavior of the covariances, with a sufficient condition for convergence in probability being that the covariances go to zero.

We assume that the Markov chain is Harris positive and aperiodic, and is stationary. We also assume that the random variables of the chain have finite variance. Thus, let (X_n) be a stationary ergodic Markov chain with mean 0 and finite variance. The variance of the average of the X_i's is

$$(6.52) \quad \mathrm{var}\left(\frac{\sum_{i=0}^n X_i}{(n+1)}\right) = \frac{1}{n+1}\mathrm{var}(X_0) + \frac{2}{n+1}\sum_{k=1}^n \frac{n-k+1}{n+1}\mathrm{cov}(X_0, X_k) ,$$

so the covariance term in (6.52) will go to zero if $\sum_{k=1}^n \mathrm{cov}(X_0, X_k)/n$ goes to zero, and a sufficient condition for this is that $\mathrm{cov}(X_0, X_k)$ converges to 0 (Problem 6.38).

To see when $\mathrm{cov}(X_0, X_k)$ converges to 0, write

$$
\begin{aligned}
|\mathrm{cov}(X_0, X_k)| &= |\mathbb{E}[X_0 X_k]| \\
&= |\mathbb{E}[X_0 \mathbb{E}(X_k|X_0)]| \\
(6.53) \qquad &\leq [\mathbb{E}(X_0^2)]^{1/2}\{\mathbb{E}[\mathbb{E}(X_k|X_0)]^2\}^{1/2},
\end{aligned}
$$

where we used the Cauchy-Schwarz inequality. Since $\mathbb{E}(X_0^2) = \sigma^2$, $\mathrm{cov}(X_0, X_k)$ will go to zero if $\mathbb{E}[\mathbb{E}(X_k|X_0)]^2$ goes to 0.

Example 6.84. (Continuation of Example 6.6) Consider the AR(1) model

$$(6.54) \qquad X_k = \theta X_{k-1} + \epsilon_k, \quad k = 0, \ldots, n,$$

when the ϵ_k's are iid $\mathcal{N}(0, 1)$, θ is an unknown parameter satisfying $|\theta| < 1$, and $X_0 \sim \mathcal{N}(0, \sigma^2)$. The X_k's all have marginal normal distributions with mean zero. The variance of X_k satisfies $\mathrm{var}(X_k) = \theta^2 \mathrm{var}(X_{k-1}) + 1$ and, $\mathrm{var}(X_k) = \sigma^2$ for all k, provided $\sigma^2 = 1/(1 - \theta^2)$. This is the stationary case in which it can be shown that

$$(6.55) \qquad \mathbb{E}(X_k|X_0) = \theta^k X_0$$

and, hence, $\mathbb{E}[\mathbb{E}(X_k|X_0)]^2 = \theta^{2k}\sigma^2$, which goes to zero as long as $|\theta| < 1$. Thus, $\mathrm{var}(\bar{X})$ converges to 0. (See Problem 6.68.) ‖

Returning to (6.53), let M be a positive constant and write

$$
\begin{aligned}
\mathbb{E}[\mathbb{E}(X_k|X_0)]^2 &= \mathbb{E}[\mathbb{E}(X_k \mathbb{I}_{X_k > M}|X_0) + \mathbb{E}(X_k \mathbb{I}_{X_k \leq M}|X_0)]^2 \\
(6.56) \qquad &\leq 2\mathbb{E}[\mathbb{E}(X_k \mathbb{I}_{X_k > M}|X_0)]^2 + 2\mathbb{E}[\mathbb{E}(X_k \mathbb{I}_{X_k \leq M}|X_0)]^2.
\end{aligned}
$$

Examining the two terms on the right side of (6.56), the first term can be made arbitrarily small using the fact that X_k has finite variance, while the second term converges to zero as a consequence of Theorem 6.51. We formalize this in the following theorem.

Theorem 6.85. *If the Markov chain (X_n) is positive and aperiodic, with $\mathrm{var}(X_n) < \infty$, then $\mathrm{cov}(X_0, X_k)$ converges to 0.*

7

The Metropolis–Hastings Algorithm

This chapter is the first of a series on simulation methods based on *Markov chains*. However, it is a somewhat strange introduction because it contains a description of the most general algorithm of all. The next chapter (Chapter 8) concentrates on the more specific slice sampler, which then introduces the Gibbs sampler (Chapters 9 and 10), which, in turn, is a special case of the Metropolis–Hastings algorithm. (However, the Gibbs sampler is different in both fundamental methodology and historical motivation.)

The motivation for this reckless dive into a completely new and general simulation algorithm is that there exists no simple case of the Metropolis–Hastings algorithm that would "gently" explain the fundamental principles of the method; a global presentation does, on the other hand, expose us to the almost infinite possibilities offered by the algorithm.

Unfortunately, the drawback of this ordering is that some parts of the chapter will be completely understood only after reading later chapters. But realize that this is the pivotal chapter of the book, one that addresses the methods that radically changed our perception of simulation and opened countless new avenues of research and applications. It is thus worth reading this chapter more than once!

7.1 The MCMC Principle

It was shown in Chapter 3 that it is not necessary to directly simulate a sample from the distribution f to approximate the integral

$$\Im = \int h(x)f(x)dx \, ,$$

since other approaches like *importance sampling* can be used. While Chapter 14 will clarify the complex connections existing between importance sampling and Markov chain Monte Carlo methods, this chapter first develops a somewhat different strategy and shows that it is possible to obtain a sample $X_1, \ldots, X_n$ approximately distributed from f without directly simulating from f. The basic principle underlying the methods described in this chapter and the following ones is *to use an ergodic Markov chain with stationary distribution f.*

While we will discuss below some rather general schemes to produce valid transition kernels associated with arbitrary stationary distributions, the working principle of MCMC algorithms is thus as follows: For an arbitrary starting value $x^{(0)}$, a chain $(X^{(t)})$ is generated using a transition kernel with stationary distribution f, which ensures the convergence in distribution of $(X^{(t)})$ to a random variable from f. (Given that the chain is ergodic, the starting value $x^{(0)}$ is, in principle, unimportant.)

Definition 7.1. A *Markov chain Monte Carlo (MCMC) method* for the simulation of a distribution f is any method producing an ergodic Markov chain $(X^{(t)})$ whose stationary distribution is f.

This simple idea of using a Markov chain with limiting distribution f may sound impractical. In comparison with the techniques of Chapter 3, here we rely on more complex asymptotic convergence properties than a simple Law of Large Numbers, as we generate dependencies within the sample that slow down convergence of the approximation of $\Im$. Thus, the number of iterations required to obtain a good approximation seems a priori much more important than with a standard Monte Carlo method. The appeal to Markov chains is nonetheless justified from at least two points of view. First, in Chapter 5, we have already seen that some stochastic optimization algorithms (for example, the Robbins–Monro procedure in Note 5.5.3) naturally produce Markov chain structures. It is a general fact that the use of Markov chains allows for a greater scope than the methods presented in Chapters 2 and 3. Second, regular Monte Carlo and MCMC algorithms both satisfy the $O(1/\sqrt{n})$ convergence requirement for the approximation of $\Im$. There are thus many instances where a specific MCMC algorithm dominates, variance-wise, the corresponding Monte Carlo proposal. For instance, while importance sampling is virtually a universal method, its efficiency relies of adequate choices of the importance function and this choice gets harder and harder as the dimension increases, a practical realization of the curse of dimensionality. At a first level, some generic algorithms, like the Metropolis–Hastings algorithms, also use simulations from almost any arbitrary density g to actually generate from an equally arbitrary given density f. At a second level, however, since these algorithms allow for the dependence of g on the previous simulation, the choice of g does not require a particularly elaborate construction a priori but can take advantage of the local characteristics of the stationary distribution. Moreover, even when an Accept–Reject algorithm is available, it is sometimes more efficient to use the pair (f, g) through a Markov chain, as detailed in Section 7.4. Even if this point is not obvious at this stage, it must be stressed that the (re)discovery of Markov chain Monte Carlo methods by statisticians in the 1990s has produced considerable progress in simulation-based inference and, in particular, in Bayesian inference, since it has allowed the analysis of a multitude of models that were too complex to be satisfactorily processed by previous schemes.

7.2 Monte Carlo Methods Based on Markov Chains

Despite its formal aspect, Definition 7.1 can be turned into a working principle: the use of a chain $(X^{(t)})$ produced by a Markov chain Monte Carlo algorithm with stationary distribution f is fundamentally identical to the use of an iid sample from f in the sense that the ergodic theorem (Theorem 6.63) guarantees the (almost sure) convergence of the empirical average

$$(7.1) \qquad \mathfrak{I}_T = \frac{1}{T} \sum_{t=1}^{T} h(X^{(t)})$$

to the quantity $\mathbb{E}_f[h(X)]$. A sequence $(X^{(t)})$ produced by a Markov chain Monte Carlo algorithm can thus be employed just as an iid sample. If there is no particular requirement of independence but if, rather, the purpose of the simulation study is to examine the properties of the distribution f, there is no need for the generation of n independent chains $(X_i^{(t)})$ $(i = 1, \ldots, n)$, where only some "terminal" values $X_i^{(T_0)}$ are kept: the choice of the value T_0 may induce a bias and, besides, this approach results in the considerable waste of $n(T_0 - 1)$ simulations out of nT_0. In other words, a single realization (or *path*) of a Markov chain is enough to ensure a proper approximation of $\mathfrak{I}$ through estimates like (7.1) for the functions h of interest (and sometimes even of the density f, as detailed in Chapter 10). Obviously, handling this sequence is somewhat more arduous than in the iid case because of the dependence structure, but some approaches to the convergence assessment of (7.1) are given in Section 7.6 and in Chapter 12. Chapter 13 will also discuss strategies to efficiently produce iid samples with MCMC algorithms.

Given the principle stated in Definition 7.1, one can propose an infinite number of practical implementations as those, for instance, used in statistical physics. The Metropolis–Hastings algorithms described in this chapter have the advantage of imposing minimal requirements on the target density f and allowing for a wide choice of possible implementations. In contrast, the Gibbs sampler described in Chapters 8–10 is more restrictive, in the sense that it requires some knowledge of the target density to derive some conditional densities, but it can also be more effective than a generic Metropolis–Hastings algorithm.

7.3 The Metropolis–Hastings algorithm

Before illustrating the universality of Metropolis–Hastings algorithms and demonstrating their straightforward implementation, we first address the (important) issue of theoretical validity. Since the results presented below are valid for all types of Metropolis–Hastings algorithms, we do not include examples in this section, but rather wait for Sections 7.4 and 7.5, which present a collection of specific algorithms.

7.3.1 Definition

The Metropolis–Hastings algorithm starts with the objective *(target)* density f. A conditional density $q(y|x)$, defined with respect to the dominating measure for the model, is then chosen. The Metropolis–Hastings algorithm can be implemented in practice when $q(\cdot|x)$ is easy to simulate from and is either explicitly available (up to a multiplicative constant *independent of x*) or *symmetric*; that is, such that $q(x|y) = q(y|x)$. The target density f must be available to some extent: a general requirement is that the ratio

$$f(y)/q(y|x)$$

is known up to a constant *independent* of x.

The Metropolis–Hastings algorithm associated with the objective (target) density f and the conditional density q produces a Markov chain $(X^{(t)})$ through the following transition.

Algorithm A.24 –Metropolis–Hastings–

```
Given x⁽ᵗ⁾,
```

1. Generate $Y_t \sim q(y|x^{(t)})$.
2. Take

$$X^{(t+1)} = \begin{cases} Y_t & \text{with probability} \quad \rho(x^{(t)}, Y_t), \\ x^{(t)} & \text{with probability} \quad 1 - \rho(x^{(t)}, Y_t), \end{cases}$$

where   [A.24]

$$\rho(x, y) = \min\left\{ \frac{f(y)}{f(x)} \frac{q(x|y)}{q(y|x)}, 1 \right\}.$$

The distribution q is called the *instrumental* (or *proposal*) *distribution* and the probability $\rho(x, y)$ the *Metropolis–Hastings acceptance probability*.

This algorithm always accepts values y_t such that the ratio $f(y_t)/q(y_t|x^{(t)})$ is increased, compared with the previous value $f(x^{(t)})/(q(x^{(t)}|y_t)$. It is only in the symmetric case that the acceptance is driven by the objective ratio $f(y_t)/f(x^{(t)})$. An important feature of the algorithm [A.24] is that it may accept values y_t such that the ratio is decreased, similar to stochastic optimization methods (see Section 5.4). Like the Accept–Reject method, the Metropolis–Hastings algorithm depends only on the ratios

$$f(y_t)/f(x^{(t)}) \quad \text{and} \quad q(x^{(t)}|y_t)/q(y_t|x^{(t)})$$

and is, therefore, independent of normalizing constants, assuming, again, that $q(\cdot|x)$ is known up to a constant that is *independent* of x[1].

Obviously, the probability $\rho(x^{(t)}, y_t)$ is defined only when $f(x^{(t)}) > 0$. However, if the chain starts with a value $x^{(0)}$ such that $f(x^{(0)}) > 0$, it follows that $f(x^{(t)}) > 0$ for every $t \in \mathbb{N}$ since the values of y_t such that $f(y_t) = 0$ lead to $\rho(x^{(t)}, y_t) = 0$ and are, therefore, rejected by the algorithm. We will make the *convention* that the ratio $\rho(x, y)$ is equal to 0 when both $f(x)$ and $f(y)$ are null, in order to avoid theoretical difficulties.

There are similarities between [A.24] and the Accept–Reject methods of Section 2.3, and it is possible to use the algorithm [A.24] as an alternative to an Accept–Reject algorithm for a given pair (f, g). These approaches are compared in Section 7.4. However, a sample produced by [A.24] differs from an iid sample. For one thing, such a sample may involve repeated occurrences of the same value, since rejection of Y_t leads to repetition of $X^{(t)}$ at time $t + 1$ (an impossible occurrence in absolutely continuous iid settings). Thus, in calculating a mean such as (7.1), the Y_t's generated by the algorithm [A.24] can be associated with weights of the form m_t/T $(m_t = 0, 1, \ldots)$, where m_t counts the number of times the subsequent values have been rejected. (This makes the comparison with importance sampling somewhat more relevant, as discussed in Section 7.6 and Chapter 14.)

While [A.24] is a generic algorithm, defined for all f's and q's, it is nonetheless necessary to impose minimal regularity conditions on both f and the conditional distribution q for f to be the limiting distribution of the chain $(X^{(t)})$ produced by [A.24]. For instance, it is easier if $\mathcal{E}$, the support of f, is *connected*: an unconnected support $\mathcal{E}$ can invalidate the Metropolis–Hastings algorithm. For such supports, it is necessary to proceed on one connected component at a time and show that the different connected components of $\mathcal{E}$ are linked by the kernel of [A.24]. If the support of $\mathcal{E}$ is truncated by q, that is, if there exists $A \subset \mathcal{E}$ such that

$$\int_A f(x)dx > 0 \quad \text{and} \quad \int_A q(y|x)dy = 0, \qquad \forall x \in \mathcal{E},$$

---

[1] If we insist on this independence from x, it is because forgetting a term in $q(\cdot|x)$ that depends on x does jeopardize the validity of the whole algorithm.

the algorithm [A.24] does not have f as a limiting distribution since, for $x^{(0)} \notin A$, the chain $(X^{(t)})$ never visits A. Thus, a minimal necessary condition is that

$$\bigcup_{x \in \text{supp} f} \text{supp } q(\cdot|x) \supset \text{supp } f \ .$$

To see that f is the stationary distribution of the Metropolis chain, we first examine the Metropolis kernel more closely and find that it satisfies the detailed balance property (6.22). (See Problem 7.3 for details of the proof.)

Theorem 7.2. *Let $(X^{(t)})$ be the chain produced by [A.24]. For every conditional distribution q whose support includes $\mathcal{E}$,*

(a) the kernel of the chain satisfies the detailed balance condition with f;
(b) f is a stationary distribution of the chain.

Proof. The transition kernel associated with [A.24] is

$$(7.2) \qquad K(x,y) = \rho(x,y)q(y|x) + (1 - r(x))\delta_x(y) \ ,$$

where $r(x) = \int \rho(x,y)q(y|x)dy$ and δ_x denotes the Dirac mass in x. It is straightforward to verify that

$$\rho(x,y)q(y|x)f(x) = \rho(y,x)q(x|y)f(y)$$

(7.3)

$$(1 - r(x))\delta_x(y)f(x) = (1 - r(y))\delta_y(x)f(y) \ ,$$

which together establish detailed balance for the Metropolis–Hastings chain. Part (*b*) now follows from Theorem 6.46. □

The stationarity of f is therefore established for almost any conditional distribution q, a fact which indicates the universality of Metropolis–Hastings algorithms.

7.3.2 Convergence Properties

To show that the Markov chain of [A.24] indeed converges to the stationary distribution and that (7.1) is a convergent approximation to $\mathfrak{I}$, we need to apply further the theory developed in Chapter 6.

Since the Metropolis–Hastings Markov chain has, by construction, an invariant probability distribution f, if it is also an aperiodic Harris chain (see Definition 6.32), then the ergodic theorem (Theorem 6.63) does apply to establish a result like the convergence of (7.1) to $\mathfrak{I}$.

A sufficient condition for the Metropolis–Hastings Markov chain to be *aperiodic* is that the algorithm [A.24] allows events such as $\{X^{(t+1)} = X^{(t)}\}$; that is, that the probability of such events is not zero, and thus

(7.4) $$P\left[f(X^{(t)})\,q(Y_t|X^{(t)}) \le f(Y_t)\,q(X^{(t)}|Y_t)\right] < 1.$$

Interestingly, this condition implies that q is not the transition kernel of a reversible Markov chain with stationary distribution f.[2] (Note that q is not the transition kernel of the Metropolis–Hastings chain, given by (7.2), which *is* reversible.)

The fact that [A.24] works only when (7.4) is satisfied is not overly troublesome, since it merely states that it is useless to further perturb a Markov chain with transition kernel q if the latter already converges to the distribution f. It is then sufficient to directly study the chain associated with q.

The property of *irreducibility* of the Metropolis–Hastings chain $(X^{(t)})$ follows from sufficient conditions such as positivity of the conditional density q; that is,

(7.5) $$q(y|x) > 0 \text{ for every } (x,y) \in \mathcal{E} \times \mathcal{E},$$

since it then follows that every set of $\mathcal{E}$ with positive Lebesgue measure can be reached in a single step. As the density f is the invariant measure for the chain, the chain is *positive* (see Definition 6.35) and Proposition 6.36 implies that the chain is recurrent. We can also establish the following stronger result for the Metropolis–Hastings chain.

Lemma 7.3. *If the Metropolis–Hastings chain* $(X^{(t)})$ *is* f-*irreducible, it is Harris recurrent.*

Proof. This result can be established by using the fact that a characteristic of Harris recurrence is that the only bounded harmonic functions are constant (see Proposition 6.61).

If h is a harmonic function, it satisfies

$$h(x_0) = \mathbb{E}[h(X^{(1)})|x_0] = \mathbb{E}[h(X^{(t)})|x_0]\;.$$

Because the Metropolis–Hastings chain is positive recurrent and aperiodic, we can use Theorem 6.80, as in the discussion surrounding (6.27), and conclude that h is f-almost everywhere constant and equal to $\mathbb{E}_f[h(X)]$. To show that h is everywhere constant, write

$$\mathbb{E}[h(X^{(1)})|x_0] = \int \rho(x_0,x_1)\,q(x_1|x_0)\,h(x_1)dx_1 + (1 - r(x_0))\,h(x_0)\,,$$

and substitute $\mathbb{E}h(X)$ for $h(x_1)$ in the integral above. It follows that

$$\mathbb{E}_f[h(X)]\,r(x_0) + (1 - r(x_0))\,h(x_0) = h(x_0)\,;$$

that is, $(h(x_0) - \mathbb{E}[h(X)])\,r(x_0) = 0$ for every $x_0 \in \mathcal{E}$. Since $r(x_0) > 0$ for every $x_0 \in \mathcal{E}$, by virtue of the f-irreducibility, h is necessarily constant and the chain is Harris recurrent.   □

---

[2] For instance, (7.4) is not satisfied by the successive steps of the Gibbs sampler (see Theorem 10.13).

We therefore have the following convergence result for Metropolis–Hastings Markov chains.

Theorem 7.4. *Suppose that the Metropolis–Hastings Markov chain $(X^{(t)})$ is f-irreducible.*

(i) If $h \in L^1(f)$, then

$$\lim_{T \to \infty} \frac{1}{T} \sum_{t=1}^{T} h(X^{(t)}) = \int h(x) f(x) dx \qquad a.e.\ f.$$

(ii) If, in addition, $(X^{(t)})$ is aperiodic, then

$$\lim_{n \to \infty} \left\| \int K^n(x, \cdot) \mu(dx) - f \right\|_{TV} = 0$$

for every initial distribution μ, where $K^n(x, \cdot)$ denotes the kernel for n transitions, as in (6.5).

Proof. If $(X^{(t)})$ is f-irreducible, it is Harris recurrent by Lemma 7.3 , and part (i) then follows from Theorem 6.63 (the Ergodic Theorem). Part (ii) is an immediate consequence of Theorem 6.51. □

As the f-irreducibility of the Metropolis–Hastings chain follows from the above-mentioned positivity property of the conditional density q, we have the following immediate corollary, whose proof is left as an exercise.

Corollary 7.5. *The conclusions of Theorem 7.4 hold if the Metropolis–Hastings Markov chain $(X^{(t)})$ has conditional density $q(x|y)$ that satisfies (7.4) and (7.5).*

Although condition (7.5) may seem restrictive, it is often satisfied in practice. (Note that, typically, conditions for irreducibility involve the transition kernel of the chain, as in Theorem 6.15 or Note 6.9.3.)

We close this section with a result due to Roberts and Tweedie (1996) (see Problem 7.35) which gives a somewhat less restrictive condition for irreducibility and aperiodicity.

Lemma 7.6. *Assume f is bounded and positive on every compact set of its support $\mathcal{E}$. If there exist positive numbers ε and δ such that*

(7.6) $$q(y|x) > \varepsilon \quad \text{if} \quad |x - y| < \delta,$$

then the Metropolis–Hastings Markov chain $(X^{(t)})$ is f-irreducible and aperiodic. Moreover, every nonempty compact set is a small set.

The rationale behind this result is the following. If the conditional distribution $q(y|x)$ allows for moves in a neighborhood of $x^{(t)}$ with diameter bounded from below and if f is such that $\rho(x^{(t)}, y)$ is positive in this neighborhood, then any subset of $\mathcal{E}$ can be visited in k steps for k large enough. (This property obviously relies on the assumption that $\mathcal{E}$ is connected.)

Proof. Consider $x^{(0)}$ an arbitrary starting point and $A \subset \mathcal{E}$ an arbitrary measurable set. The connectedness of $\mathcal{E}$ implies that there exist $m \in \mathbb{N}$ and a sequence $x^{(i)} \in \mathcal{E}$ ($1 \le i \le m$) such that $x^{(m)} \in A$ and $|x^{(i+1)} - x^{(i)}| < \delta$. It is therefore possible to link $x^{(0)}$ and A through a sequence of balls with radius δ. The assumptions on f imply that the acceptance probability of a point $x^{(i)}$ of the ith ball starting from the $(i-1)$st ball is positive and, therefore, $P_{x^{(0)}}^m(A) = P(X^{(m)} \in A|X^{(0)} = x^{(0)}) > 0$. By Theorem 6.15, the f-irreducibility of $(X^{(t)})$ is established.

For an arbitrary value $x^{(0)} \in \mathcal{E}$ and for every $y \in B(x^{(0)}, \delta/2)$ (the ball with center $x^{(0)}$ and radius $\delta/2$) we have

$$
P_y(A) \ge \int_A \rho(y,z)\, q(z|y)\, dz
$$
$$
= \int_{A \cap D_y} \frac{f(z)}{f(y)}\, q(y|z)\, dz + \int_{A \cap D_y^c} q(z|y)\, dz \ ,
$$

where $D_y = \{z; f(z)q(y|z) \le f(y)q(z|y)\}$. It therefore follows that

$$
P_y(A) \ge \int_{A \cap D_y \cap B} \frac{f(z)}{f(y)}\, q(y|z)\, dz + \int_{A \cap D_y^c \cap B} q(z|y)\, dz
$$
$$
\ge \frac{\inf_B f(x)}{\sup_B f(x)} \int_{A \cap D_y \cap B} q(y|z)\, dz + \int_{A \cap D_y^c \cap B} q(z|y)\, dz
$$
$$
\ge \varepsilon\, \frac{\inf_B f(x)}{\sup_B f(x)}\, \lambda(A \cap B) \ ,
$$

where λ denotes the Lebesgue measure on $\mathcal{E}$. The balls $B(x^{(0)}, \delta/2)$ are small sets associated with uniform distributions on $B(x^{(0)}, \delta/2)$. This simultaneously implies the aperiodicity of $(X^{(t)})$ and the fact that every compact set is small. $\qquad\square$

Corollary 7.7. *The conclusions of Theorem 7.4 hold if the Metropolis–Hastings Markov chain $(X^{(t)})$ has invariant probability density f and conditional density $q(x|y)$ that satisfy the assumptions of Lemma 7.6.*

One of the most fascinating aspects of the algorithm [A.24] is its universality; that is, the fact that an arbitrary conditional distribution q with support $\mathcal{E}$ can lead to the simulation of an arbitrary distribution f on $\mathcal{E}$. On the other hand, this universality may only hold formally if the instrumental distribution q rarely simulates points in the main portion of $\mathcal{E}$; that is to say, in the region where most of the mass of the density f is located. This issue of selecting a good proposal distribution q for a given f is detailed in Section 7.6.

Since we have provided no examples so far, we now proceed to describe two particular approaches used in the literature, with some probabilistic properties and corresponding examples. Note that a complete classification of the Metropolis–Hastings algorithms is impossible, given the versatility of the method and the possibility of creating even more hybrid methods (see, for instance, Roberts and Tweedie 1995, 2004 and Stramer and Tweedie 1999b).

7.4 The Independent Metropolis–Hastings Algorithm

7.4.1 Fixed Proposals

This method appears as a straightforward generalization of the Accept–Reject method in the sense that the instrumental distribution q is independent of $X^{(t)}$ and is denoted g by analogy. The algorithm [A.24] will then produce the following transition from $x^{(t)}$ to $X^{(t+1)}$.

Algorithm A.25 –Independent Metropolis–Hastings–

Given $x^{(t)}$

  1 Generate $Y_t \sim g(y)$.
  2 Take   [A.25]

$$X^{(t+1)} = \begin{cases} Y_t & \text{with probability} \quad \min\left\{ \dfrac{f(Y_t)\, g(x^{(t)})}{f(x^{(t)})\, g(Y_t)}, 1 \right\} \\ x^{(t)} & \text{otherwise.} \end{cases}$$

Although the Y_t's are generated independently, the resulting sample is not iid: for instance, the probability of acceptance of Y_t depends on $X^{(t)}$ (except in the trivial case when $f = g$).

The convergence properties of the chain $(X^{(t)})$ follow from properties of the density g in the sense that $(X^{(t)})$ is irreducible and aperiodic (thus, ergodic according to Corollary 7.5) if and only if g is almost everywhere positive on the support of f. Stronger properties of convergence like geometric and uniform ergodicity are also clearly described by the following result of Mengersen and Tweedie (1996).

Theorem 7.8. *The algorithm [A.25] produces a uniformly ergodic chain if there exists a constant M such that*

(7.7) $$f(x) \le M g(x), \quad \forall x \in \text{supp } f.$$

In this case,

$$(7.8) \qquad \|K^n(x, \cdot) - f\|_{TV} \leq 2\left(1 - \frac{1}{M}\right)^n,$$

where $\| \cdot \|_{TV}$ denotes the total variation norm introduced in Definition 6.47. On the other hand, if for every M, there exists a set of positive measure where (7.7) does not hold, $(X^{(t)})$ is not even geometrically ergodic.

Proof. If (7.7) is satisfied, the transition kernel satisfies

$$K(x, x') \geq g(x') \min\left\{\frac{f(x')g(x)}{f(x)g(x')}, 1\right\}$$

$$= \min\left\{f(x')\frac{g(x)}{f(x)}, g(x')\right\} \geq \frac{1}{M} f(x').$$

The set $\mathcal{E}$ is therefore small and the chain is uniformly ergodic (Theorem 6.59).

To establish the bound on $\|K^n(x, \cdot) - f\|_{TV}$, first write

$$\|K(x, \cdot) - f\|_{TV} = 2 \sup_A \left|\int_A (K(x, y) - f(y))dy\right|$$

$$= 2 \int_{\{y; f(y) \geq K(x,y)\}} (f(y) - K(x, y))dy$$

$$(7.9) \qquad \leq 2\left(1 - \frac{1}{M}\right) \int_{\{y; f(y) \geq K(x,y)\}} f(y)dy$$

$$\leq 2\left(1 - \frac{1}{M}\right).$$

We now continue with a recursion argument to establish (7.8). We can write

$$\int_A (K^2(x, y) - f(y))dy = \int_{\mathcal{E}} \left[\int_A (K(u, y) - f(y))dy\right]$$

$$(7.10) \qquad \times (K(x, u) - f(u))du,$$

and an argument like that in (7.9) leads to

$$(7.11) \qquad \|K^2(x, \cdot) - f\|_{TV} \leq 2\left(1 - \frac{1}{M}\right)^2.$$

We next write a general recursion relation

$$\int_A (K^{n+1}(x, y) - f(y))dy$$

$$(7.12) \qquad = \int_{\mathcal{E}} \left[\int_A (K^n(u, y) - f(y))dy\right] (K(x, u) - f(u))du,$$

and proof of (7.8) is established by induction (Problem 7.11).

If (7.7) does not hold, then the sets

$$D_n = \left\{ x; \frac{f(x)}{g(x)} \geq n \right\}$$

satisfy $P_f(D_n) > 0$ for every n. If $x \in D_n$, then

$$P(x, \{x\}) = 1 - \mathbb{E}_g \left[\min \left\{ \frac{f(Y)g(x)}{g(Y)f(x)}, 1 \right\} \right]$$

$$= 1 - P_g \left(\frac{f(Y)}{g(Y)} \geq \frac{f(x)}{g(x)} \right) - \mathbb{E}_g \left[\frac{f(Y)g(x)}{f(x)g(Y)} \, \mathbb{I}_{\frac{f(Y)}{g(Y)} < \frac{f(x)}{g(x)}} \right]$$

$$\geq 1 - P_g \left(\frac{f(Y)}{g(Y)} \geq n \right) - \frac{g(x)}{f(x)} \geq 1 - \frac{2}{n},$$

since Markov inequality implies that

$$P_g \left(\frac{f(Y)}{g(Y)} \geq n \right) \leq \frac{1}{n} \mathbb{E}_g \left[\frac{f(Y)}{g(Y)} \right] = \frac{1}{n}.$$

Consider a small set C such that $D_n \cap C^c$ is not empty for n large enough and $x_0 \in D_n \cap C^c$. The return time to C, τ_C, satisfies

$$P_{x_0}(\tau_C > k) \geq \left(1 - \frac{2}{n} \right)^k ;$$

therefore, the radius of convergence of the series (in κ) $\mathbb{E}_{x_0}[\kappa^{\tau_C}]$ is smaller than $n/(n-2)$ for every n, and this implies that $(X^{(t)})$ cannot be geometrically ergodic, according to Theorem 6.75. □

This particular class of Metropolis–Hastings algorithms naturally suggests a comparison with Accept–Reject methods since every pair (f, g) satisfying (7.7) can also induce an Accept–Reject algorithm. Note first that the expected acceptance probability for the variable simulated according to g is larger in the case of the algorithm [A.25].

Lemma 7.9. *If (7.7) holds, the expected acceptance probability associated with the algorithm [A.25] is at least $\frac{1}{M}$ when the chain is stationary.*

Proof. If the distribution of $f(X)/g(X)$ is absolutely continuous,[3] the expected acceptance probability is

---

[3] This constraint implies that f/g is not constant over some interval.

$$\mathbb{E}\left[\min\left\{\frac{f(Y_t)g(X^{(t)})}{f(X^{(t)})g(Y_t)},1\right\}\right] = \int \mathbb{I}_{\frac{f(y)g(x)}{g(y)f(x)}>1}\, f(x)g(y)\, dxdy$$

$$+ \int \frac{f(y)g(x)}{g(y)f(x)}\, \mathbb{I}_{\frac{f(y)g(x)}{g(y)f(x)}\leq1}\, f(x)g(y)\, dxdy$$

$$= 2\int \mathbb{I}_{\frac{f(y)g(x)}{g(y)f(x)}\geq1}\, f(x)g(y)\, dxdy$$

$$\geq 2\int \mathbb{I}_{\frac{f(y)}{g(y)}\geq\frac{f(x)}{g(x)}}\, f(x)\frac{f(y)}{M}\, dxdy$$

$$= \frac{2}{M} P\left(\frac{f(X_1)}{g(X_1)} \geq \frac{f(X_2)}{g(X_2)}\right) = \frac{1}{M}\,.$$

Since X_1 and X_2 are independent and distributed according to f, this last probability is equal to $1/2$, and the result follows.   □

Thus, the independent Metropolis–Hastings algorithm [A.25] is more efficient than the Accept–Reject algorithm [A.4] in its handling of the sample produced by g, since, on the average, it accepts more proposed values. A more advanced comparison between these approaches is about as difficult as the comparison between Accept–Reject and importance sampling proposed in Section 3.3.3, namely that the size of one of the two samples is random and this complicates the computation of the variance of the resulting estimator. In addition, the correlation between the X_i's resulting from [A.25] prohibits a closed-form expression of the joint distribution. We therefore study the consequence of the correlation on the variance of both estimators through an example. (See also Liu 1996b and Problem 7.33 for a comparison based on the eigenvalues of the transition operators in the discrete case, which also shows the advantage of the Metropolis–Hastings algorithm.)

Example 7.10. Generating gamma variables.   Using the algorithm of Example 2.19 (see also Example 3.15), an Accept–Reject method can be derived to generate random variables from the $\mathcal{G}a(\alpha,\beta)$ distribution using a Gamma $\mathcal{G}a(\lfloor\alpha\rfloor,b)$ candidate (where $\lfloor a\rfloor$ denotes the integer part of a). When $\beta = 1$, the optimal choice of b is

$$b = \lfloor\alpha\rfloor/\alpha\,.$$

The algorithms to compare are then

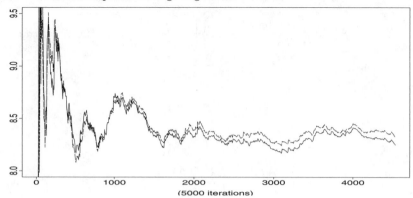

(5000 iterations)

Fig. 7.1. Convergence of Accept–Reject (solid line) and Metropolis–Hastings (dashed line) estimators to $\mathbb{E}_f[X^2] = 8.33$, for $\alpha = 2.43$ based on the same sequence $y_1, \ldots, y_{5000}$ simulated from $\mathcal{G}a(2, 2/2.43)$. The number of acceptances in [A.27] is then random. The final values of the estimators are 8.25 for [A.27] and 8.32 for [A.26].

Algorithm A.26 –Gamma Metropolis–Hastings–

1. Generate $Y_t \sim \mathcal{G}a(\lfloor \alpha \rfloor, \lfloor \alpha \rfloor / \alpha)$.
2. Take   [A.26]

$$X^{(t+1)} = \begin{cases} Y_t & \text{with probability } \varrho_t \\ x^{(t)} & \text{otherwise,} \end{cases}$$

where

$$\varrho_t = \min\left[\left(\frac{Y_t}{x^{(t)}} \exp\left\{\frac{x^{(t)} - Y_t}{\alpha}\right\}\right)^{\alpha - \lfloor \alpha \rfloor}, 1\right].$$

and

Algorithm A.27 –Gamma Accept–Reject–

1. Generate $Y \sim \mathcal{G}a(\lfloor \alpha \rfloor, \lfloor \alpha \rfloor / \alpha)$.
2. Accept $X = Y$ with probability   [A.27]

$$\left(\frac{ey \exp(-y/\alpha)}{\alpha}\right)^{\alpha - \lfloor \alpha \rfloor}$$

Note that (7.7) does apply in this particular case with $\exp(x/\alpha)/x > e/\alpha$.

A first comparison is based on a sample $(y_1, \ldots, y_n)$, of fixed size n, generated from $\mathcal{G}a(\lfloor \alpha \rfloor, \lfloor \alpha \rfloor / \alpha)$ with $x^{(0)}$ generated from $\mathcal{G}a(\alpha, 1)$. The number

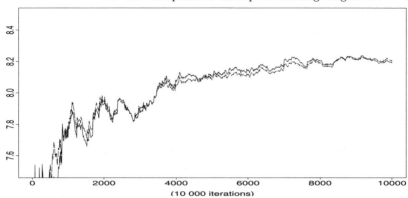

Fig. 7.2. Convergence to $\mathbb{E}_f[X^2] = 8.33$ of Accept–Reject (full line) and Metropolis–Hastings (dots) estimators for 10,000 acceptances in [A.27], the same sequence of y_i's simulated from $\mathcal{G}a(2, 2/2.43)$ being used in [A.27] and [A.26]. The final values of the estimators are 8.20 for [A.27] and 8.21 for [A.26].

t of values accepted by [A.27] is then random. Figure 7.1 describes the convergence of the estimators of $\mathbb{E}_f[X^2]$ associated with both algorithms for the same sequence of y_i's and exhibits strong agreement between the approaches, with the estimator based on [A.26] being closer to the exact value 8.33 in this case.

On the other hand, the number t of values accepted by [A.27] can be fixed and [A.26] can then use the resulting sample of random size n, $y_1, \ldots, y_n$. Figure 7.2 reproduces the comparison in this second case and exhibits a behavior rather similar to Figure 7.1, with another close agreement between estimators and, the scale being different, a smaller variance (which is due to the larger size of the effective sample).

Note, however, that both comparisons are biased. In the first case, the sample of $X^{(i)}$ produced by [A.27] does not have the distribution f and, in the second case, the sample of Y_i's in [A.26] is not iid. In both cases, this is due to the use of a stopping rule which modifies the distribution of the samples.∥

Example 7.11. Logistic Regression. We return to the data of Example 1.13, which described a logistic regression relating the failure of O-rings in shuttle flights to air temperature. We observe (x_i, y_i), $i = 1, \ldots, n$ according to the model

$$Y_i \sim \text{Bernoulli}(p(x_i)), \quad p(x) = \frac{\exp(\alpha + \beta x)}{1 + \exp(\alpha + \beta x)},$$

where $p(x)$ is the probability of an O-ring failure at temperature x. The likelihood is

$$L(\alpha, \beta | \mathbf{y}) \propto \prod_{i=1}^{n} \left(\frac{\exp(\alpha + \beta x_i)}{1 + \exp(\alpha + \beta x_i)} \right)^{y_i} \left(\frac{1}{1 + \exp(\alpha + \beta x_i)} \right)^{1 - y_i}$$

and we take the prior to be

$$\pi_\alpha(\alpha | b) \pi_\beta(\beta) = \frac{1}{b} e^\alpha e^{-e^\alpha / b} d\alpha d\beta,$$

which puts an exponential prior on $\log \alpha$ and a flat prior on β, and insures propriety of the posterior distribution (Problem 7.25). To complete the prior specification we must give a value for b, and we choose the data-dependent value that makes $\mathbb{E}\alpha = \hat{\alpha}$, where $\hat{\alpha}$ is the MLE of α. (This also insures that the prior will not have undue influence, as it is now centered near the likelihood.) It can be shown that

$$\mathbb{E}[\alpha] = \int_0^\infty \frac{1}{b} e^\alpha e^{-e^\alpha / b} d\alpha = \int_0^\infty \log(w) \frac{1}{b} e^{-w/b} dw = \log(b) - \gamma,$$

where γ is *Euler's Constant*, equal to .577216. Thus we take $\hat{b} = e^{\hat{\alpha} + \gamma}$.

The posterior distribution is proportional to $L(\alpha, \beta | \mathbf{y}) \pi(\alpha, \beta)$, and to simulate from this distribution we take an independent candidate

$$g(\alpha, \beta) = \pi_\alpha(\alpha | \hat{b}) \phi(\beta),$$

where $\phi(\beta)$ is a normal distribution with mean $\hat{\beta}$ and variance $\hat{\sigma}_\beta^2$, the MLEs. Note that although basing the prior distribution on the data is somewhat in violation of the formal Bayesian paradigm, nothing is violated if the candidate depends on the data. In fact, this will usually result in a more effective simulation, as the candidate is placed close to the target.

Generating a random variable from $g(\alpha, \beta)$ is straightforward, as it only involves the generation of a normal and an exponential random variable. If we are at the point (α_0, β_0) in the Markov chain, and we generate (α', β') from $g(\alpha, \beta)$, we accept the candidate with probability

$$\min \left\{ \frac{L(\alpha', \beta' | \mathbf{y})}{L(\alpha_0, \beta_0 | \mathbf{y})} \frac{\phi(\beta_0)}{\phi(\beta')}, 1 \right\}.$$

Figure 7.3 shows the distribution of the generated parameters and their convergence. ∥

Example 7.12. Saddlepoint tail area approximation. In Example 3.18, we saw an approximation to noncentral chi squared tail areas based on the regular and renormalized saddlepoint approximations. Such an approximation requires numerical integration, both to calculate the constant and to evaluate the tail area.

An alternative is to produce a sample $Z_1, \ldots, Z_m$, from the saddlepoint distribution, and then approximate the tail area using

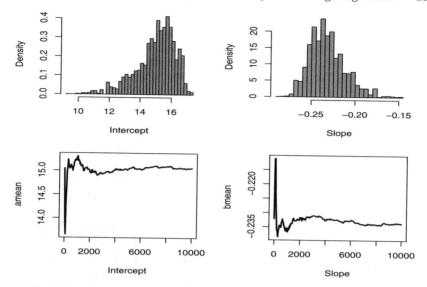

Fig. 7.3. Estimation of the slope and intercept from the Challenger logistic regression. The top panels show histograms of the distribution of the coefficients, while the bottom panels show the convergence of the means.

$$P(\bar{X} > a) = \int_{\hat{\tau}(a)}^{1/2} \left(\frac{n}{2\pi}\right)^{1/2} [K_X''(t)]^{1/2} \exp\{n\,[K_X(t) - tK_X'(t)]\}\,dt$$

$$(7.13) \qquad \approx \frac{1}{m} \sum_{i=1}^{m} \mathbb{I}[Z_i > \hat{\tau}(a)]\,,$$

where $K_X(\tau)$ is the cumulant generating function of X and $\hat{\tau}(x)$ is the solution of the saddlepoint equation $K'(\hat{\tau}(x)) = x$ (see Section 3.6.2).

Note that we are simulating from the transformed density. It is interesting (and useful) to note that we can easily derive an instrumental density to use in a Metropolis–Hastings algorithm. Using a Taylor series approximation, we find that

$$(7.14) \qquad \exp\{n\,[K_X(t) - tK_X'(t)]\} \approx \exp\left\{-nK_X''(0)\frac{t^2}{2}\right\}\,,$$

so a first choice for an instrumental density is the $\mathcal{N}(0, 1/nK_X''(0))$ distribution (see Problem 7.26 for details). Booth et al. (1999) use a Student's t approximation instead.

We can now simulate the noncentral chi squared tail areas using a normal instrumental density with $K_X''(t) = 2[p(1 - 2t) + 4\lambda]/(1 - 2t)^3$. The results are presented in Table 7.1, where we see that the approximations are quite good. Note that the same set of simulated random variables can be used for all the tail area probability calculations. Moreover, by using the Metropolis–

Hastings algorithm, we have avoided calculating the normalizing constant for the saddlepoint approximation.   ‖

Interval	Renormalized	Exact	Monte Carlo
$(36.225, \infty)$	0.0996	0.1	0.0992
$(40.542, \infty)$	0.0497	0.05	0.0497
$(49.333, \infty)$	0.0099	0.01	0.0098

Table 7.1. Monte Carlo saddlepoint approximation of a noncentral chi squared integral for $p = 6$ and $\lambda = 9$, based on $10,000$ simulated random variables.

As an aside, note that the usual classification of "Hastings" for the algorithm [A.25] is somewhat inappropriate, since Hastings (1970) considers the algorithm [A.24] in general, using random walks (Section 7.5) rather than independent distributions in his examples. It is also interesting to recall that Hastings (1970) proposes a theoretical justification of these methods for finite state-space Markov chains based on the finite representation of real numbers in a computer. However, a complete justification of this physical discretization needs to take into account the effect of the approximation in the entire analysis. In particular, it needs to be verified that the computer choice of discrete approximation to the continuous distribution has no effect on the resulting stationary distribution or irreducibility of the chain. Since Hastings (1970) does not go into such detail, but keeps to the simulation level, we prefer to study the theoretical properties of these algorithms by bypassing the finite representation of numbers in a computer and by assuming flawless pseudo-random generators, namely algorithms producing variables which are uniformly distributed on $[0, 1]$. See Roberts et al. (1995) for a theoretical study of some effects of the computer discretization.

A final note about independent Metropolis–Hastings algorithms is that they cannot be omniscient: there are settings where an independent proposal does not work well because of the complexity of the target distribution. Since the main purpose of MCMC algorithms is to provide a crude but easy simulation technique, it is difficult to imagine spending a long time on the design of the proposal distribution. This is specially pertinent in high-dimensional models where the capture of the main features of the target distribution is most often impossible. There is therefore a limitation of the independent proposal, which can be perceived as a *global* proposal, and a need to use more *local* proposals that are not so sensitive to the target distribution, as presented in Section 7.5. Another possibility, developed in Section 7.6.3, is to validate *adaptive* algorithms that learn from the ongoing performances of the current proposals to refine their construction. But this solution is delicate, both from a theoretical (*"Does ergodicity apply?"*) and an algorithmic (*"How does one tune the adaptation?"*) point of view. The following section first develops a specific kind of adaptive algorithm.

7.4.2 A Metropolis–Hastings Version of ARS

The ARS algorithm, which provides a general Accept–Reject method for log-concave densities in dimension one (see Section 2.4.2), can be generalized to the ARMS method (which stands for *Adaptive Rejection Metropolis Sampling*) following the approach developed by Gilks et al. (1995). This generalization applies to the simulation of arbitrary densities, instead of being restricted to log-concave densities as the ARS algorithm, by simply adapting the ARS algorithm for densities f that are not log-concave. The algorithm progressively fits a function g, which plays the role of a pseudo-envelope of the density f. In general, this function g does not provide an upper bound on f, but the introduction of a Metropolis–Hastings step in the algorithm justifies the procedure.

Using the notation from Section 2.4.2, take $h(x) = \log f_1(x)$ with f_1 proportional to the density f. For a sample $S_n = \{x_i, 0 \leq i \leq n+1\}$, the equations of the lines between $(x_i, h(x_i))$ and $(x_{i+1}, h(x_{i+1}))$ are denoted by $y = L_{i,i+1}(x)$. Consider

$$\tilde{h}_n(x) = \max\{L_{i,i+1}(x), \min[L_{i-1,i}(x), L_{i+1,i+2}(x)]\} ,$$

for $x_i \leq x < x_{i+1}$, with

$$
\begin{aligned}
&\tilde{h}_n(x) = L_{0,1}(x) && \text{if} \quad x < x_0, \\
&\tilde{h}_n(x) = \max[L_{0,1}(x), L_{1,2}(x)] && \text{if} \quad x_0 \leq x < x_1, \\
&\tilde{h}_n(x) = \max[L_{n,n+1}(x), L_{n-1,n}(x)] && \text{if} \quad x_n \leq x < x_{n+1}, \\
\text{and} \quad &\tilde{h}_n(x) = L_{n,n+1}(x) && \text{if} \quad x \geq x_{n+1}.
\end{aligned}
$$

The resulting proposal distribution is $g_n(x) \propto \exp\{\tilde{h}_n(x)\}$. The ARMS algorithm is based on g_n and it can be decomposed into two parts, a first step which is a standard Accept–Reject step for the simulation from the instrumental distribution

$$\psi_n(x) \propto \min\left[f_1(x), \exp\{\tilde{h}_n(x)\}\right] ,$$

based on g_n, and a second part, which is the acceptance of the simulated value by a Metropolis–Hastings procedure:

Algorithm A.28 –ARMS Metropolis–Hastings–

1. Simulate Y from $g_n(y)$ and $U \sim \mathcal{U}_{[0,1]}$
   until
   $$U \le f_1(Y)/\exp\{\tilde{h}_n(Y)\}.$$

2. Generate $V \sim \mathcal{U}_{[0,1]}$ and take   [A.28]

$$(7.15) \qquad X^{(t+1)} = \begin{cases} Y & \text{if} \quad V < \dfrac{f_1(Y)\,\psi_n(x^{(t)})}{f_1(x^{(t)})\,\psi_n(Y)} \wedge 1 \\[2mm] x^{(t)} & \text{otherwise.} \end{cases}$$

The Accept–Reject step indeed produces a variable distributed from $\psi_n(x)$ and this justifies the expression of the acceptance probability in the Metropolis–Hastings step. Note that [A.28] is a particular case of the approximate Accept–Reject algorithms considered by Tierney (1994) (see Problem 7.9). The probability (7.15) can also be written

$$\begin{cases} \min\left[1, \dfrac{f_1(Y)\exp\{\tilde{h}_n(x^{(t)})\}}{f_1(x^{(t)})\exp\{\tilde{h}_n(Y)\}}\right] & \text{if} \ \ f_1(Y) > \exp\{\tilde{h}_n(Y)\}, \\[4mm] \min\left[1, \dfrac{\exp\{\tilde{h}_n(x^{(t)})\}}{f_1(x^{(t)})}\right] & \text{otherwise,} \end{cases}$$

which implies a sure acceptance of Y when $f_1(x^{(t)}) < \exp\{\tilde{h}_n(x^{(t)})\}$; that is, when the bound is correct.

Each simulation of $Y \sim g_n$ in Step 1 of [A.28] provides, in addition, an update of S_n in $S_{n+1} = S_n \cup \{y\}$, and therefore of g_n, when Y is rejected. As in the case of the ARS algorithm, the initial S_n set must be chosen so that g_n is truly a probability density. If the support of f is not bounded from below, $L_{0,1}$ must be increasing and, similarly, if the support of f is not bounded from above, $L_{n,n+1}$ must be decreasing. Note also that the simulation of g_n detailed in Section 2.4.2 is valid in this setting.

Since the algorithm [A.28] appears to be a particular case of independent Metropolis–Hastings algorithm, the convergence and ergodicity results obtained in Section 7.4 should apply for [A.28]. This is not the case, however, because of the lack of time homogeneity of the chain (see Definition 6.4) produced by [A.28]. The transition kernel, based on g_n, can change at each step with a positive probability. Since the study of nonhomogeneous chains is quite delicate, the algorithm [A.28] can be justified only by reverting to the homogeneous case; that is, by fixing the function g_n and the set S_n after a warm-up period of length n_0. The constant n_0 need not be fixed in advance as this warm-up period can conclude when the approximation of f_1 by g_n is satisfactory, for instance when the rejection rate in Step 1 of [A.28] is sufficiently small. The algorithm [A.28] must then start with an initializing (or calibrating) step which adapts the parameters at hand (in this case, g_n) to the function f_1. This adaptive structure is generalized in Section 7.6.3.

The ARMS algorithm is useful when a precise analytical study of the density f is impossible, as, for instance, in the setup of generalized linear models. In fact, f (or f_1) needs to be computed in only a few points to initialize the algorithm, which thus does not require the search for "good" density g which approximates f. This feature should be contrasted to the cases of the independent Metropolis–Hastings algorithm and of sufficiently fast random walks as in the case of [A.29].

Example 7.13. Poisson logistic model. For the generalized linear model in Example 2.26, consider a logit dependence between explanatory and dependent (observations) variables,

$$Y_i|x_i \sim \mathcal{P}\left(\frac{\exp(bx_i)}{1 + \exp(bx_i)}\right), \qquad i = 1, \ldots, n,$$

which implies the restriction $\lambda_i < 1$ on the parameters of the Poisson distribution, $Y_i \sim \mathcal{P}(\lambda_i)$. When b has the prior distribution $\mathcal{N}(0, \tau^2)$, the posterior distribution is

$$\pi(b|\mathbf{x}, \mathbf{y}) \propto \frac{\exp\left\{\sum_i y_i(bx_i)\right\}}{\prod_i(1 + \exp(bx_i))} \exp\left\{-\sum_i \frac{e^{bx_i}}{1 + e^{bx_i}}\right\} e^{-b^2/2\tau^2}.$$

This posterior distribution $\pi(b|\mathbf{x})$ is not easy to simulate from and one can use the ARMS Metropolis–Hastings algorithm instead. ‖

7.5 Random Walks

A natural approach for the practical construction of a Metropolis–Hastings algorithm is to take into account the value previously simulated to generate the following value; that is, to consider a *local* exploration of the neighborhood of the current value of the Markov chain. This idea is already used in algorithms such as the simulated annealing algorithm [A.19] and the stochastic gradient method given in (5.4).

Since the candidate g in algorithm [A.24] is allowed to depend on the current state $X^{(t)}$, a first choice to consider is to simulate Y_t according to

$$Y_t = X^{(t)} + \varepsilon_t,$$

where ε_t is a random perturbation with distribution g, independent of $X^{(t)}$. In terms of the algorithm [A.24], $q(y|x)$ is now of the form $g(y - x)$. The Markov chain associated with q is a *random walk* (see Example 6.39) on $\mathcal{E}$.

δ	0.1	0.5	1.0
Mean	0.399	−0.111	0.10
Variance	0.698	1.11	1.06

Table 7.2. Estimators of the mean and the variance of a normal distribution $\mathcal{N}(0,1)$ based on a sample obtained by a Metropolis–Hastings algorithm using a random walk on $[-\delta, \delta]$ (15,000 simulations).

The convergence results of Section 7.3.2 naturally apply in this particular case. Following Lemma 7.6, if g is positive in a neighborhood of 0, the chain $(X^{(t)})$ is f-irreducible and aperiodic, therefore ergodic. The most common distributions g in this setup are the uniform distributions on spheres centered at the origin or standard distributions like the normal and the Student's t distributions. All these distributions usually need to be scaled; we discuss this problem in Section 7.6. At this point, we note that the choice of a *symmetric function* g (that is, such that $g(-t) = g(t)$), leads to the following original expression of [A.24], as proposed by Metropolis et al. (1953).

Algorithm A.29 –Random walk Metropolis–Hastings–

Given $x^{(t)}$,

1. Generate $Y_t \sim g(|y - x^{(t)}|)$.
2. Take   [A.29]

$$X^{(t+1)} = \begin{cases} Y_t & \text{with probability } \min\left\{1, \dfrac{f(Y_t)}{f(x^{(t)})}\right\} \\ x^{(t)} & \text{otherwise.} \end{cases}$$

Example 7.14. A random walk normal generator. Hastings (1970) considers the generation of the normal distribution $\mathcal{N}(0,1)$ based on the uniform distribution on $[-\delta, \delta]$. The probability of acceptance is then $\rho(x^{(t)}, y_t) = \exp\{(x^{(t)^2} - y_t^2)/2\} \wedge 1$. Figure 7.4 describes three samples of 15,000 points produced by this method for $\delta = 0.1, 0.5$, and 1. The corresponding estimates of the mean and variance are provided in Table 7.2. Figure 7.4 clearly shows the different speeds of convergence of the averages associated with these three values of δ, with an increasing regularity (in δ) of the corresponding histograms and a faster exploration of the support of f.   ‖

Despite its simplicity and its natural features, the random walk Metropolis–Hastings algorithm does not enjoy uniform ergodicity properties. Mengersen and Tweedie (1996) have shown that in the case where supp $f = \mathbb{R}$, this algorithm cannot produce a uniformly ergodic Markov chain on $\mathbb{R}$ (Problem 7.16). This is a rather unsurprising feature when considering the *local* character of

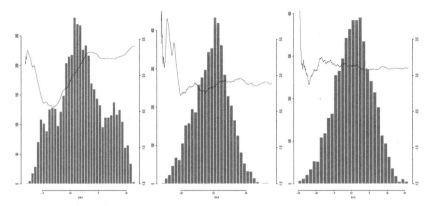

Fig. 7.4. Histograms of three samples produced by the algorithm [A.29] for a random walk on $[-\delta, \delta]$ with (a) $\delta = 0.1$, (b) $\delta = 0.5$, and (c) $\delta = 1.0$, with the convergence of the means (7.1), superimposed with scales on the right of the graphs (15,000 simulations).

the random walk proposal, centered at the current value of the Markov chain.

Although uniform ergodicity cannot be obtained with random walk Metropolis–Hastings algorithms, it is possible to derive necessary and sufficient conditions for geometric ergodicity. Mengersen and Tweedie (1996) have proposed a condition based on the *log-concavity of f* in the tails; that is, if there exist $\alpha > 0$ and x_1 such that

$$(7.16) \qquad \log f(x) - \log f(y) \geq \alpha |y - x|$$

for $y < x < -x_1$ or $x_1 < x < y$.

Theorem 7.15. *Consider a symmetric density f which is log-concave with associated constant α in (7.16) for $|x|$ large enough. If the density g is positive and symmetric, the chain $(X^{(t)})$ of [A.29] is geometrically ergodic. If f is not symmetric, a sufficient condition for geometric ergodicity is that $g(t)$ be bounded by $b \exp\{-\alpha|t|\}$ for a sufficiently large constant b.*

The proof of this result is based on the use of the drift function $V(x) = \exp\{\alpha|x|/2\}$ (see Note 6.9.1) and the verification of a geometric drift condition of the form

$$(7.17) \qquad \Delta V(x) \leq -\lambda V(x) + b\mathbb{I}_{[-x^*, x^*]}(x) \,,$$

for a suitable bound x^*. Mengersen and Tweedie (1996) have shown, in addition, that this condition on g is also necessary in the sense that if $(X^{(t)})$ is geometrically ergodic, there exists $s > 0$ such that

$$(7.18) \qquad \int e^{s|x|} f(x)dx < \infty \,.$$

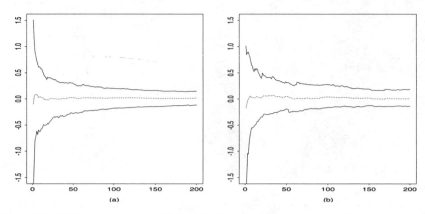

Fig. 7.5. 90% confidence envelopes of the means produced by the random walk Metropolis–Hastings algorithm [A.24] based on a instrumental distribution $\mathcal{N}(0,1)$ for the generation of (a) a normal distribution $\mathcal{N}(0,1)$ and (b) a distribution with density ψ. These envelopes are derived from 500 parallel independent chains and with identical uniform samples on both distributions.

Example 7.16. A comparison of tail effects. In order to assess the practical effect of this theorem, Mengersen and Tweedie (1996) considered two random walk Metropolis–Hastings algorithms based on a $\mathcal{N}(0,1)$ instrumental distribution for the generation of (a) a $\mathcal{N}(0,1)$ distribution and (b) a distribution with density $\psi(x) \propto (1 + |x|)^{-3}$. Applying Theorem 7.15 (see Problem 7.18), it can be shown that the first chain associated is geometrically ergodic, whereas the second chain is not. Figures 7.5(a) and 7.5(b) represent the average behavior of the sums

$$\frac{1}{T} \sum_{t=1}^{T} X^{(t)}$$

over 500 chains initialized at $x^{(0)} = 0$. The 5% and 95% quantiles of these chains show a larger variability of the chain associated with the distribution ψ, in terms of both width of the confidence region and precision of the resulting estimators.   ‖

We next look at a discrete example where Algorithm [A.29] generates a geometrically ergodic chain.

Example 7.17. Random walk geometric generation. Consider generating a geometric[4] distribution, $\mathcal{G}eo(\theta)$ using [A.29] with $(Y^{(t)})$ having transition probabilities $q(i,j) = P(Y^{(t+1)} = j | Y^{(t)} = i)$ given by

---

[4] The material used in the current example refers to the drift condition introduced in Note 6.9.1.

$$q(i,j) = \begin{cases} 1/2 & i = j-1, j+1 \text{ and } j = 1, 2, 3, \ldots \\ 1/2 & i = 0, 1 \text{ and } j = 0 \\ 0 & \text{otherwise;} \end{cases}$$

that is, q is the transition kernel of a symmetric random walk on the non-negative integers *with reflecting boundary at* 0.

Now, $X \sim \mathcal{G}eo(\theta)$ implies $P(X = x) = (1 - \theta)^x \theta$ for $x = 0, 1, 2, \ldots$. The transition matrix has a band diagonal structure and is given by

$$T = \begin{pmatrix} \frac{1+\theta}{2} & \frac{1-\theta}{2} & 0 & 0 & \cdots \\ \frac{1}{2} & \frac{\theta}{2} & \frac{1-\theta}{2} & 0 & \cdots \\ 0 & \frac{1}{2} & \frac{\theta}{2} & \frac{1-\theta}{2} & \cdots \\ & \ddots & \ddots & \ddots & \cdots \end{pmatrix}.$$

Consider the potential function $V(i) = \beta^i$ where $\beta > 1$, and recall that $\Delta V(y^{(0)}) = \mathbb{E}[V(Y^{(1)})|y^{(0)}] - V(y^{(0)})$. For $i > 0$, we have

$$\mathbb{E}[V(Y^{(1)})|Y^{(0)} = i] = \frac{1}{2}\,\beta^{i-1} + \frac{\theta}{2}\,\beta^i + \frac{1-\theta}{2}\,\beta^{i+1}$$

$$= V(i)\left(\tfrac{1}{2\beta} + \tfrac{\theta}{2} + \tfrac{1-\theta}{2}\beta\right).$$

Thus, $\Delta V(i) = V(i)(1/(2\beta) + \theta/2 - 1 + \beta(1 - \theta)/2) = V(i)g(\theta, \beta)$. For a fixed value of θ, $g(\theta, \beta)$ is minimized by $\beta = 1/\sqrt{1-\theta}$. In this case, $\Delta V(i) = (\sqrt{1-\theta} + \theta/2 - 1)V(i)$ and $\lambda = \sqrt{1-\theta} + \theta/2 - 1$ is the geometric rate of convergence. The closer θ is to 1, the faster the convergence.   ‖

Tierney (1994) proposed a modification of the previous algorithm with a proposal density of the form $g(y - a - b(x - a))$; that is,

$$y_t = a + b(x^{(t)} - a) + z_t, \qquad z_t \sim g.$$

This autoregressive representation can be seen as intermediary between the independent version ($b = 0$) and the random walk version ($b = 1$) of the Metropolis–Hastings algorithm. Moreover, when $b < 0$, $X^{(t)}$ and $X^{(t+1)}$ are negatively correlated, and this may allow for faster excursions on the surface of f if the symmetry point a is well chosen. Hastings (1970) also considers an alternative to the uniform distribution on $[x^{(t)} - \delta, x^{(t)} + \delta]$ (see Example 7.14) with the uniform distribution on $[-x^{(t)} - \delta, -x^{(t)} + \delta]$: The convergence of the empirical average to 0 is then faster in this case, but the choice of 0 as center of symmetry is obviously crucial and requires some a priori information on the distribution f. In a general setting, a and b can be calibrated during the first iterations. (See also Problem 7.23.) (See also Chen and Schmeiser 1993, 1998 for the alternative "hit-and-run" algorithm, which proceeds by generating a random direction in the space and moves the current value by a random distance along this direction.)

7.6 Optimization and Control

The previous sections have established the theoretical validity of the Metropolis–Hastings algorithms by showing that under suitable (and not very restrictive) conditions on the transition kernel, the chain produced by [A.24] is ergodic and, therefore, that the mean (7.1) converges to the expectation $\mathbb{E}_f[h(X)]$. In Sections 7.4 and 7.4, however, we showed that the most common algorithms only rarely enjoy strong ergodicity properties (geometric or uniform ergodicity). In particular, there are simple examples (see Problem 7.5) that show how slow convergence can be.

This section addresses the problem of choosing the transition kernel $q(y|x)$ and illustrates a general acceleration method for Metropolis–Hastings algorithms, which extends the conditioning techniques presented in Section 4.2.

7.6.1 Optimizing the Acceptance Rate

When considering only the classes of algorithms described in Section 7.4, the most common alternatives are to use the following:

(a) a fully automated algorithm like ARMS ([A.28]);
(b) an instrumental density g which approximates f, such that f/g is bounded for uniform ergodicity to apply to the algorithm [A.25];
(c) a random walk as in [A.29].

In case (a), the automated feature of [A.28] reduces "parameterization" to the choice of initial values, which are theoretically of limited influence on the efficiency of the algorithm. In both of the other cases, the choice of g is much more critical, as it determines the performances of the resulting Metropolis–Hastings algorithm. As we will see below, the few pieces of advice available on the choice of g are, in fact, contrary! Depending on the type of Metropolis–Hastings algorithm selected, one would want high acceptance rates in case (b) and low acceptance rates in case (c).

Consider, first, the independent Metropolis–Hastings algorithm introduced in Section 7.4. Its similarity with the Accept–Reject algorithm suggests a choice of g that maximizes the average *acceptance rate*

$$
\rho = \mathbb{E}\left[\min\left\{\frac{f(Y)\,g(X)}{f(X)\,g(Y)}, 1\right\}\right]
$$
$$
= 2P\left(\frac{f(Y)}{g(Y)} \geq \frac{f(X)}{g(X)}\right), \qquad X \sim f,\; Y \sim g,
$$

as seen[5] in Lemma 7.9. In fact, the optimization associated with the choice of g is related to the speed of convergence of $\frac{1}{T}\sum_{t=1}^{T} h(X^{(t)})$ to $\mathbb{E}_f[h(X)]$

---

[5] Under the same assumption of no point mass for the ratio $f(Y)/g(Y)$.

and, therefore, to the ability of the algorithm [A.25] to quickly explore any complexity of f (see, for example, Theorem 7.8).

If this optimization is to be generic (that is, independent of h), g should reproduce the density f as faithfully as possible, which implies the maximization of ρ. For example, a density g that is either much less or much more concentrated, compared with f, produces a ratio

$$\frac{f(y)\,g(x)}{f(x)\,g(y)} \wedge 1$$

having huge variations and, therefore, leads to a low acceptance rate.

The acceptance rate ρ is typically impossible to compute, and one solution is to use the minorization result $\rho \geq 1/M$ of Lemma 7.9 to minimize M as in the case of the Accept–Reject algorithm.

Alternatively, we can consider a more *empirical* approach that consists of choosing a parameterized instrumental distribution $g(\cdot|\theta)$ and adjusting the corresponding parameters θ based on the evaluated acceptance rate, now $\hat{\rho}(\theta)$; that is, first choose an initial value for the parameters, θ_0, and estimate the corresponding acceptance rate, $\hat{\rho}(\theta_0)$, based on m iterations of [A.25], then modify θ_0 to obtain an increase in $\hat{\rho}$.

In the simplest cases, θ_0 will reduce to a scale parameter which is increased or decreased depending on the behavior of $\hat{\rho}(\theta)$. In multidimensional settings, θ_0 can also include a position parameter or a matrix acting as a scale parameter, which makes optimizing $\rho(\theta)$ a more complex task. Note that $\hat{\rho}(\theta)$ can be obtained by simply counting acceptances or through

$$\frac{2}{m} \sum_{i=1}^{m} \mathbb{I}_{\{f(y_i)g(x_i|\theta) > f(x_i)g(y_i|\theta)\}} \,,$$

where $x_1, \ldots, x_m$ is a sample from f, obtained, for instance, from a first MCMC algorithm, and $y_1, \ldots, y_m$ is an iid sample from $g(\cdot|\theta)$. Therefore, if θ is composed of location and scale parameters, a sample $((x_1, y_1), \ldots, (x_m, y_m))$ corresponding to a value θ_0 can be used repeatedly to evaluate different values of θ by a deterministic modification of y_i, which facilitates the maximization of $\rho(\theta)$.

Example 7.18. Inverse Gaussian distribution. The *inverse Gaussian distribution* has the density

$$(7.19) \qquad f(z|\theta_1, \theta_2) \propto z^{-3/2} \exp\left\{ -\theta_1 z - \frac{\theta_2}{z} + 2\sqrt{\theta_1\theta_2} + \log \sqrt{2\theta_2} \right\}$$

on $\mathbb{R}_+$ ($\theta_1 > 0, \theta_2 > 0$). Denoting $\psi(\theta_1, \theta_2) = 2\sqrt{\theta_1\theta_2} + \log\sqrt{2\theta_2}$, it follows from a classical result on exponential families (see Brown 1986, Chapter 2, Robert 2001, Lemma 3.3.7, or Problem 1.38) that

β	0.2	0.5	0.8	0.9	1	1.1	1.2	1.5
$\hat{\rho}(\beta)$	0.22	0.41	0.54	0.56	0.60	0.63	0.64	0.71
$\mathbb{E}[Z]$	1.137	1.158	1.164	1.154	1.133	1.148	1.181	1.148
$\mathbb{E}[1/Z]$	1.116	1.108	1.116	1.115	1.120	1.126	1.095	1.115

Table 7.3. Estimation of the means of Z and of $1/Z$ for the inverse Gaussian distribution $\mathcal{IN}(\theta_1, \theta_2)$ by the Metropolis–Hastings algorithm [A.25] and evaluation of the acceptance rate for the instrumental distribution $\mathcal{G}a(\sqrt{\theta_2/\theta_1}\,\beta, \beta)$ ($\theta_1 = 1.5, \theta_2 = 2$, and $m = 5000$).

$$\mathbb{E}\left[(Z, 1/Z)\right] = \nabla\psi(\theta_1, \theta_2)$$
$$= \left(\sqrt{\frac{\theta_2}{\theta_1}}, \sqrt{\frac{\theta_1}{\theta_2}} + \frac{1}{2\theta_2}\right).$$

A possible choice for the simulation of (7.19) is the Gamma distribution $\mathcal{G}a(\alpha, \beta)$ in algorithm [A.25], taking $\alpha = \beta\sqrt{\theta_2/\theta_1}$ so that the means of both distributions coincide. Since

$$\frac{f(x)}{g(x)} \propto x^{-\alpha-1/2} \exp\left\{(\beta - \theta_1)x - \frac{\theta_2}{x}\right\},$$

the ratio f/g is bounded for $\beta < \theta_1$. The value of x which maximizes the ratio is the solution of

$$(\beta - \theta_1)x^2 - \left(\alpha + \frac{1}{2}\right)x + \theta_2 = 0\;;$$

that is,

$$x_\beta^* = \frac{(\alpha + 1/2) - \sqrt{(\alpha + 1/2)^2 + 4\theta_2(\theta_1 - \beta)}}{2(\beta - \theta_1)}.$$

The analytical optimization (in β) of

$$M(\beta) = (x_\beta^*)^{-\alpha-1/2} \exp\left\{(\beta - \theta_1)x_\beta^* - \frac{\theta_2}{x_\beta^*}\right\}$$

is not possible, although, in this specific case the curve $M(\beta)$ can be plotted for given values of θ_1 and θ_2 and the optimal value β^* can be approximated numerically. Typically, the influence of the choice of β must be assessed empirically; that is, by approximating the acceptance rate ρ via the method described above.

Note that a new sample $(y_1, \ldots, y_m)$ must be simulated for every new value of β. Whereas $y \sim \mathcal{G}a(\alpha, \beta)$ is equivalent to $\beta y \sim \mathcal{G}a(\alpha, 1)$, the factor α depends on β and it is not possible to use the same sample for several values of β. Table 7.3 provides an evaluation of the rate ρ as a function of β and gives estimates of the means of Z and $1/Z$ for $\theta_1 = 1.5$ and $\theta_2 = 2$. The constraint on the ratio f/g then imposes $\beta < 1.5$. The corresponding theoretical values are respectively 1.155 and 1.116, and the optimal value of β is $\beta^* = 1.5$.   ‖

The *random walk* version of the Metropolis–Hastings algorithm, introduced in Section 7.5, requires a different approach to acceptance rates, given the dependence of the instrumental distribution on the current state of the chain. In fact, a high acceptance rate does not necessarily indicate that the algorithm is moving correctly since it may indicate that the random walk is moving too slowly on the surface of f. If $x^{(t)}$ and y_t are close, in the sense that $f(x^{(t)})$ and $f(y_t)$ are approximately equal, the algorithm [A.29] leads to the acceptance of y with probability

$$\min\left(\frac{f(y_t)}{f(x^{(t)})}, 1\right) \simeq 1 \ .$$

A higher acceptance rate may therefore correspond to a slower convergence as the moves on the support of f are more limited. In the particular case of multimodal densities whose modes are separated by zones of extremely small probability, the negative effect of limited moves on the surface of f clearly shows. While the acceptance rate is quite high for a distribution g with small variance, the probability of jumping from one mode to another may be arbitrarily small. This phenomenon occurs, for instance, in the case of mixtures of distributions (see Section 9.7.1) and in overparameterized models (see, e.g., Tanner and Wong 1987 and Besag et al. 1995). In contrast, if the average acceptance rate is low, the successive values of $f(y_t)$ tend to be small compared with $f(x^{(t)})$, which means that the random walk moves quickly on the surface of f since it often reaches the "borders" of the support of f (or, at least, that the random walk explores regions with low probability under f).

The above analysis seems to require an advanced knowledge of the density of interest, since an instrumental distribution g with too narrow a range will slow down the convergence rate of the algorithm. On the other hand, a distribution g with a wide range results in a waste of simulations of points outside the range of f without improving the probability of visiting all of the modes of f. It is unfortunate that an automated parameterization of g cannot guarantee uniformly optimal performances for the algorithm [A.29], and that the rules for choosing the rate presented in Note 7.8.4 are only heuristic.

7.6.2 Conditioning and Accelerations

Similar[6] to the Accept–Reject method, the Metropolis–Hastings algorithm does not take advantage of the total set of random variables that are generated. Lemma 7.9 shows that the "rate of waste" of these variables y_t is lower than for the Accept–Reject method, but it still seems inefficient to ignore the

---

[6] This section presents material related to nonparametric Rao–Blackwellization, as in Section 4.2, and may be skipped on a first reading.

rejected y_t's. As the rejection mechanism relies on an independent uniform random variable, it is reasonable to expect that the rejected variables bring, although indirectly, some relevant information on the distribution f. As in the conditioning method introduced in Section 4.2, the *Rao–Blackwellization* technique applies in the case of the Metropolis–Hastings algorithm. (Other approaches to Metropolis–Hastings acceleration can be found in Green and Han 1992, Gelfand and Sahu 1994, or McKeague and Wefelmeyer 2000.)

First, note that a sample produced by the Metropolis–Hastings algorithm, $x^{(1)}, \ldots, x^{(T)}$, is based on two samples, $y_1, \ldots, y_T$ and $u_1, \ldots, u_T$, with $y_t \sim q(y|x^{(t-1)})$ and $u_t \sim \mathcal{U}_{[0,1]}$. The mean (7.1) can then be written

$$\delta^{MH} = \frac{1}{T} \sum_{t=1}^{T} h(x^{(t)}) = \frac{1}{T} \sum_{t=1}^{T} \sum_{i=1}^{t} \mathbb{I}_{x^{(t)}}$$

$$= \frac{1}{T} \sum_{t=1}^{T} h(y_t) \sum_{i=t}^{T} \mathbb{I}_{x^{(i)} = y_t}$$

and the conditional expectation

$$\delta^{RB} = \frac{1}{T} \sum_{t=1}^{T} h(y_t) \, \mathbb{E} \left[\sum_{i=t}^{T} \mathbb{I}_{X^{(i)} = y_t} \middle| y_1, \ldots, y_T \right]$$

$$= \frac{1}{T} \sum_{t=1}^{T} h(y_t) \left(\sum_{i=t}^{T} P(X^{(i)} = y_t | y_1, \ldots, y_T) \right)$$

dominates the empirical mean, δ^{MH}, under quadratic loss. This is a consequence of the Rao–Blackwell Theorem (see Lehmann and Casella 1998, Section 1.7), resulting from the fact that δ^{RB} integrates out the variation due to the uniform sample.

The practical interest of this alternative to δ^{MH} is that the probabilities $P(X^{(i)} = y_t | y_1, \ldots, y_T)$ can be explicitly computed. Casella and Robert (1996) have established the two following results, which provide the weights for $h(y_t)$ in δ^{RB} both for the independent Metropolis–Hastings algorithm and the general Metropolis–Hastings algorithm. In both cases, the computational complexity of these weights is of order $\mathcal{O}(T^2)$, which is a manageable order of magnitude.

Consider first the case of the independent Metropolis–Hastings algorithm associated with the instrumental distribution g. For simplicity's sake, assume that $X^{(0)}$ is simulated according to the distribution of interest, f, so that the chain is stationary, and the mean (7.1) can be written

$$\delta^{MH} = \frac{1}{T+1} \sum_{t=0}^{T} h(x^{(t)}) \,,$$

with $x^{(0)} = y_0$. If we denote

n	10	25	50	100
h_1	50.11	49.39	48.27	46.68
h_2	42.20	44.75	45.44	44.57

Table 7.4. Decrease (in percentage) of squared error risk associated with δ^{RB} for the evaluation of $\mathbb{E}[h_i(X)]$, evaluated over 7500 simulations for different sample sizes n. (*Source*: Casella and Robert 1996).

$$w_i = \frac{f(y_i)}{g(y_i)}, \qquad \rho_{ij} = \frac{w_j}{w_i} \wedge 1 \qquad (0 \le i < j),$$

$$\zeta_{ii} = 1, \qquad \zeta_{ij} = \prod_{t=i+1}^{j} (1 - \rho_{it}) \qquad (i < j),$$

we have the following theorem, whose proof is left to Problem 7.31.

Theorem 7.19. *The estimator δ^{RB} can be written*

$$\delta^{RB} = \frac{1}{T+1} \sum_{i=0}^{T} \varphi_i \, h(y_i),$$

where

$$\varphi_i = \tau_i \sum_{j=i}^{T} \zeta_{ij} \,,$$

and the conditional probability $\tau_i = P(X^{(i)} = y_i | y_0, y_1, \ldots, y_T)$, $i = 0, \ldots, T$, is given by $\tau_0 = 1$ and $\tau_i = \sum_{j=0}^{i-1} \tau_j \, \zeta_{j(i-1)} \rho_{ji}$ for $i > 0$.

The computation of ζ_{ij} for a fixed i requires $(T - i)$ multiplications since $\zeta_{i(j+1)} = \zeta_{ij}(1 - \rho_{i(j+1)})$; therefore, the computation of all the ζ_{ij}'s require $T(T+1)/2$ multiplications. The derivations of τ_i and φ_i are of the same order of complexity.

Example 7.20. Rao–Blackwellization improvement for a $\mathcal{T}_3$ simulation. Suppose the target distribution is $\mathcal{T}_3$ and the instrumental distribution is Cauchy, $\mathcal{C}(0,1)$. The ratio f/g is bounded, which ensures a geometric rate of convergence for the associated Metropolis–Hastings algorithm. Table 7.4 illustrates the improvement brought by δ^{RB} for some functions of interest $h_1(x) = x$ and $h_2(x) = \mathbb{I}_{(1.96,+\infty)}(x)$, whose (exact) expectations $\mathbb{E}[h_i(X)]$ ($i = 1, 2$) are 0 and 0.07, respectively. Over the different sample sizes selected for the experiment, the improvement in mean square error brought by δ^{RB} is of the order 50%.   ‖

We next consider the general case, with an arbitrary instrumental distribution $q(y|x)$. The dependence between Y_i and the set of previous variables Y_j $(j < i)$ (since $X^{(i-1)}$ can be equal to $Y_0, Y_1, \ldots,$ or Y_{i-1}) complicates the expression of the joint distribution of Y_i and U_i, which cannot be obtained in closed form for arbitrary n. In fact, although $(X^{(t)})$ is a Markov chain, (Y_t) is not.

Let us denote

$$\rho_{ij} = \frac{f(y_j)/q(y_j|y_i)}{f(y_i)/q(y_i|y_j)} \wedge 1 \qquad (j > i),$$

$$\overline{\rho}_{ij} = \rho_{ij}q(y_{j+1}|y_j), \quad \underline{\rho}_{ij} = (1 - \rho_{ij})q(y_{j+1}|y_i) \qquad (i < j < T),$$

$$\zeta_{jj} = 1, \quad \zeta_{jt} = \prod_{l=j+1}^{t} \underline{\rho}_{jl} \qquad (i < j < T),$$

$$\tau_0 = 1, \quad \tau_j = \sum_{t=0}^{j-1} \tau_t \zeta_{t(j-1)} \, \overline{\rho}_{tj}, \quad \tau_T = \sum_{t=0}^{T-1} \tau_t \zeta_{t(T-1)} \rho_{tT} \qquad (i < T),$$

$$\omega_T^i = 1, \quad \omega_i^j = \overline{\rho}_{ji}\omega_{i+1}^i + \underline{\rho}_{tj}\omega_{i+1}^j \qquad (0 \le j < i < T).$$

Casella (1996) derives the following expression for the weights of $h(y_i)$ in δ^{RB}. We again leave the proof as a problem (Problem 7.32).

Theorem 7.21. *The estimator* δ^{RB} *satisfies*

$$\delta^{RB} = \frac{\sum_{i=0}^{T} \varphi_i \, h(y_i)}{\sum_{i=0}^{T-1} \tau_i \, \zeta_{i(T-1)}},$$

with $(i < T)$

$$\varphi_i = \tau_i \left[\sum_{j=i}^{T-1} \zeta_{ij}\omega_{j+1}^i + \zeta_{i(T-1)}(1 - \rho_{iT}) \right]$$

and $\varphi_T = \tau_T$.

Although these estimators are more complex than in the independent case, the complexity of the weights is again of order $\mathcal{O}(T^2)$ since the computations of $\overline{\rho}_{ij}, \underline{\rho}_{ij}, \zeta_{ij}, \tau_i,$ and ω_j^i involve $T(T+1)/2$ multiplications. Casella and Robert (1996) give algorithmic advice toward easier and faster implementation.

Example 7.22. (Continuation of Example 7.20) Consider now the simulation of a T_3 distribution based on a random walk with perturbations distributed as $\mathcal{C}(0, \sigma^2)$. The choice of σ determines the acceptance rate for the Metropolis–Hastings algorithm: When $\sigma = 0.4$, it is about 0.33, and when $\sigma = 3.0$, it increases to 0.75.

As explained in Section 7.6.1, the choice $\sigma = 0.4$ is undoubtedly preferable in terms of efficiency of the algorithm. Table 7.5 confirms this argument,

n	10	25	50	100
h_1	10.7	8.8	7.7	7.7
	(1.52)	(0.98)	(0.63)	(0.3)
$\sigma = 0.4$ h_2	23.6	25.2	25.8	25.0
	(0.02)	(0.01)	(0.006)	(0.003)
h_1	0.18	0.15	0.11	0.07
	(2.28)	(1.77)	(1.31)	(0.87)
$\sigma = 3.0$ h_2	0.99	0.94	0.71	1.19
	(0.03)	(0.02)	(0.014)	(0.008)

Table 7.5. Improvement brought by δ^{RB} (in %) and quadratic risk of the empirical average (in parentheses) for different sample sizes and $50,000$ simulations of the random walk based on $\mathcal{C}(0, \sigma^2)$ (*Source*: Casella and Robert 1996).

since the quadratic risk of the estimators (7.1) is larger for $\sigma = 3.0$. The gains brought by δ^{RB} are smaller, compared with the independent case. They amount to approximately 8% and 25% for $\sigma = 0.4$ and 0.1% and 1% for $\sigma = 3$. Casella and Robert (1996) consider an additional comparison with an importance sampling estimator based on the same sample $y_1, \ldots, y_n$.   ‖

7.6.3 Adaptive Schemes

Given the range of situations where MCMC applies, it is unrealistic to hope for a *generic* MCMC sampler that would function in every possible setting. The more generic proposals like random walk Metropolis–Hastings algorithms are known to fail in large dimension and disconnected supports, because they take too long to explore the space of interest (Neal 2003). The reason for this impossibility theorem is that, in realistic problems, the complexity of the distribution to simulation is the very reason why MCMC is used! So it is difficult to ask for a prior opinion about this distribution, its support or the parameters of the proposal distribution used in the MCMC algorithm: intuition is close to void in most of these problems.

However, the performances of off-the-shelf algorithms like the random walk Metropolis–Hastings scheme bring information about the distribution of interest and, thus, should be incorporated in the design of better and more powerful algorithms. The problem is that we usually miss the time to train the algorithm on these previous performances and are looking for the Holy Grail of automated MCMC procedures! While it is natural to think that the information brought by the first steps of an MCMC algorithm should be used in later steps, there is a severe catch: using the whole past of the "chain" implies that this is not a Markov chain any longer. Therefore, usual convergence theorems do not apply and the validity of the corresponding algorithms is questionable. Further, it may be that, in practice, such algorithms do degenerate to point masses because of a too-rapid decrease in the variation of their proposal.

Example 7.23. t-distribution. Consider a $\mathcal{T}(\nu, \theta, 1)$ sample $(x_1, \ldots, x_n)$ with ν known. Assume in addition a flat prior $\pi(\theta) = 1$ on θ as in a noninformative environment. While the posterior distribution can be easily computed at an arbitrary value of θ, direct simulation and computation from this posterior is impossible. In a Metropolis–Hastings framework, we could fit a normal proposal from the empirical mean and variance of the previous values of the chain,

$$\mu_t = \frac{1}{t} \sum_{i=1}^{t} \theta^{(i)} \quad \text{and} \quad \sigma_t^2 = \frac{1}{t} \sum_{i=1}^{t} (\theta^{(i)} - \mu_t)^2.$$

Notwithstanding the dependence on the past, we could then use the Metropolis–Hastings acceptance probability

$$\prod_{j=2}^{n} \left[\frac{\nu + (x_j - \theta^{(t)})^2}{\nu + (x_j - \xi)^2} \right]^{-(\nu+1)/2} \frac{\exp -(\mu_t - \theta^{(t)})^2/2\sigma_t^2}{\exp -(\mu_t - \xi)^2/2\sigma_t^2},$$

where ξ is the proposed value from $\mathcal{N}(\mu_t, \sigma_t^2)$. The invalidity of this scheme (because of the dependence on the whole sequence of $\theta^{(i)}$'s till iteration t) is illustrated in Figure 7.6: when the range of the initial values is too small, the sequence of $\theta^{(i)}$'s cannot converge to the target distribution and concentrates on too small a support. But the problem is deeper, because even when the range of the simulated values is correct, the (long-term) dependence on past values modifies the distribution of the sequence. Figure 7.7 shows that, for an initial variance of 2.5, there is a bias in the histogram, even after $25,000$ iterations and stabilization of the empirical mean and variance.   ‖

Even though the Markov chain is converging *in distribution* to the target distribution (when using a proper, i.e., time-homogeneous, updating scheme), using past simulations to create a nonparametric approximation to the target distribution does not work either. For instance, Figure 7.8 shows the output of an adaptive scheme in the setting of Example 7.23 when the proposal distribution is the Gaussian kernel based on earlier simulations. A very large number of iterations is not sufficient to reach an acceptable approximation of the target distribution.

The overall message is thus that one should not *constantly* adapt the proposal distribution on the past performances of the simulated chain. Either the adaptation must cease after a period of *burn in* (not to be taken into account for the computations of expectations and quantities related to the target distribution), or the adaptive scheme must be theoretically assessed in its own right. This latter path is not easy and only a few examples can be found (so

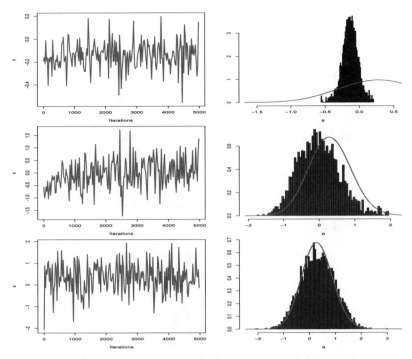

Fig. 7.6. Output of the adaptive scheme for the t-distribution posterior with a sample of 10 $x_j \sim \mathcal{T}_3$ and initial variances of *(top)* 0.1, *(middle)* 0.5, and *(bottom)* 2.5. The left column plots the sequence of $\theta^{(i)}$'s while the right column compares its histogram against the true posterior distribution *(with a different scale for the upper graph).*

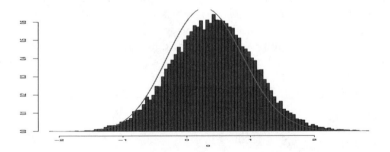

Fig. 7.7. Comparison of the distribution of an adaptive scheme sample of 25,000 points with initial variance of 2.5 and of the target distribution.

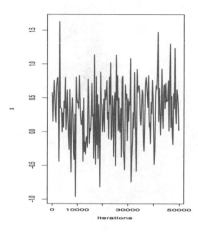
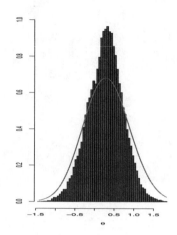

Fig. 7.8. Sample produced by $50,000$ iterations of a nonparametric adaptive MCMC scheme and comparison of its distribution with the target distribution.

far) in the literature. See, e.g., Gilks et al. (1998), who use regeneration to create block independence and preserve Markovianity on the paths rather than on the values (see also Sahu and Zhigljavsky 1998 and Holden 1998), Haario et al. (1999, 2001), who derive a proper[7] adaptation scheme in the spirit of Example 7.23 by using a Gaussian proposal and a ridge-like correction to the empirical variance,

$$\Sigma_t = s\mathrm{Cov}(\theta^{(0)}, \ldots, \theta^{(t)}) + s\epsilon I_d\,,$$

where s and ϵ are constant, and Andrieu and Robert (2001), who propose a more general framework of valid adaptivity based on stochastic optimization and the Robbin–Monro algorithm. (The latter actually embeds the chain of interest $\theta^{(t)}$ in a larger chain $(\theta^{(t)}, \xi^{(t)})$ that also includes the parameter of the proposal distribution as well as the gradient of a performance criterion.) We will again consider adaptive algorithms in Chapter 14, with more accessible theoretical justifications.

7.7 Problems

7.1 Calculate the mean of a Gamma$(4.3, 6.2)$ random variable using
   (a) Accept-Reject with a Gamma$(4, 7)$ candidate.
   (b) Metropolis–Hastings with a Gamma$(4, 7)$ candidate.
   (c) Metropolis–Hastings with a Gamma$(5, 6)$ candidate.

---

[7] Since the chain is not Markov, the authors need to derive an ergodic theorem on their own.

In each case monitor the convergence.

7.2 Student's $\mathcal{T}_\nu$ density with ν degrees of freedom is given by

$$f(x|\nu) = \frac{\Gamma\left(\frac{\nu+1}{2}\right)}{\Gamma\left(\frac{\nu}{2}\right)} \frac{1}{\sqrt{\nu\pi}} \left(1 + \frac{x^2}{\nu}\right)^{-(\nu+1)/2}.$$

Calculate the mean of a t distribution with 4 degrees of freedom using a Metropolis–Hastings algorithm with candidate density
(a) $N(0,1)$
(b) t with 2 degrees of freedom.
Monitor the convergence of each.

7.3 Complete some details of Theorem 7.2:
(a) To establish (7.2), show that

$$\begin{aligned}
K(x, A) &= P(X_{t+1} \in A|X_t = x) \\
&= P(Y \in A \text{ and } X_{t+1} = Y|X_t = x) + P(x \in A \text{ and } X_{t+1} = x|X_t = x) \\
&= \int_A q(y|x)\varrho(x,y)dy + \int_{\mathcal{Y}} \mathbb{I}(x \in A)(1 - \varrho(x,y))q(y|x)dy,
\end{aligned}$$

where $q(y|x)$ is the instrumental density and $\varrho(x,y) = P(X_{t+1} = y|X_t = x)$. Take the limiting case $A = \{y\}$ to establish (7.2).
(b) Establish (7.3). Notice that $\delta_y(x)f(y) = \delta_x(y)f(x)$.

7.4 For the transition kernel,

$$X^{(t+1)}|x^{(t)} \sim \mathcal{N}(\rho x^{(t)}, \tau^2)$$

gives sufficient conditions on ρ and τ for the stationary distribution π to exist. Show that, in this case, π is a normal distribution and that (7.4) occurs.

7.5 (Doukhan et al. 1994) The algorithm presented in this problem is used in Chapter 12 as a benchmark for slow convergence.
(a) Prove the following result:

Lemma 7.24. *Consider a probability density g on $[0,1]$ and a function $0 < \rho < 1$ such that*

$$\int_0^1 \frac{g(x)}{1 - \rho(x)}\, dx < \infty.$$

The Markov chain with transition kernel

$$K(x, x') = \rho(x)\, \delta_x(x') + (1 - \rho(x))\, g(x')\,,$$

where δ_x is the Dirac mass at x, has stationary distribution

$$f(x) \propto g(x)/(1 - \rho(x)).$$

(b) Show that an algorithm for generating the Markov chain associated with Lemma 7.24 is given by

Algorithm A.30 –Repeat or Simulate–

1. Take $X^{(t+1)} = x^{(t)}$ with probability $\rho(x^{(t)})$
2. Else, generate $X^{(t+1)} \sim g(y)$.   [A.30]

(c) Highlight the similarity with the Accept–Reject algorithm and discuss in which sense they are complementary.

7.6 (Continuation of Problem 7.5) Implement the algorithm of Problem 7.5 when g is the density of the $\mathcal{Be}(\alpha + 1, 1)$ distribution and $\rho(x) = 1 - x$. Give the expression of the stationary distribution f. Study the acceptance rate as α varies around 1. (*Note:* Doukhan et al. 1994 use this example to derive β-mixing chains which do not satisfy the Central Limit Theorem.)

7.7 (Continuation of Problem 7.5) Compare the algorithm [A.30] with the corresponding Metropolis–Hastings algorithm; that is, the algorithm [A.25] associated with the same pair (f, g). (*Hint:* Take into account the fact that [A.30] simulates only the y_t's which are not discarded and compare the computing times when a recycling version as in Section 7.6.2 is implemented.)

7.8 Determine the distribution of Y_t given $y_{t-1}, \ldots$ in [A.25].

7.9 (Tierney 1994) Consider a version of [A.25] based on a "bound" M on f/g that is not a uniform bound; that is, $f(x)/g(x) > M$ for some x.

(a) If an Accept–Reject algorithm uses the density g with acceptance probability $f(y)/Mg(y)$, show that the resulting variables are generated from

$$\tilde{f}(x) \propto \min\{f(x), Mg(x)\},$$

instead of f.

(b) Show that this error can be corrected, for instance by using the Metropolis–Hastings algorithm:

> 1. Generate $Y_t \sim \tilde{f}$.
> 2. Accept with probability
>
> $$P(X^{(t+1)} = y_t | x^{(t)}, y_t) = \begin{cases} \min\left\{1, \dfrac{f(y_t)g(x^{(t)})}{g(y_t)f(x^{(t)})}\right\} & \text{if } \dfrac{f(y_t)}{g(y_t)} > M \\ \min\left\{1, \dfrac{Mg(x^{(t)})}{f(x^{(t)})}\right\} & \text{otherwise.} \end{cases}$$

to produce a sample from f.

7.10 The inequality (7.8) can also be established using *Orey's inequality* (See Problem 6.42 for a slightly different formulation) For two transitions P and Q,

$$\|P^n - Q^n\|_{TV} \leq 2P(X_n \neq Y_n), \qquad X_n \sim P^n, \quad Y_n \sim Q^n.$$

Deduce that when P is associated with the stationary distribution f and when X_n is generated by [A.25], under the condition (7.7),

$$\|P^n - f\|_{TV} \leq \left(1 - \frac{1}{M}\right)^n.$$

Hint: Use a coupling argument based on

$$X^n = \begin{cases} Y^n & \text{with probability } 1/M \\ Z^n \sim \dfrac{g(z) - f(z)/M}{1 - 1/M} & \text{otherwise.} \end{cases}$$

7.11 Complete the proof of Theorem 7.8:

(a) Verify (7.10) and prove (7.11). (*Hint:* By (7.9), the inner integral is immediately bounded by $1 - \frac{1}{M}$. Then repeat the argument for the outer integral.)

(b) Verify (7.12) and prove (7.8).

7.12 In the setup of Hastings (1970) uniform-normal example (see Example 7.14):

(a) Study the convergence rate (represented by the 90% interquantile range) and the acceptance rate when δ increases.

(b) Determine the value of δ which minimizes the variance of the empirical average. (*Hint:* Use a simulation experiment.)

7.13 Show that, for an arbitrary Metropolis–Hastings algorithm, every compact set is a small set when f and q are positive and continuous everywhere.

7.14 (Mengersen and Tweedie 1996) With respect to Theorem 7.15, define

$$A_x = \{y;\ f(x) \le f(y)\} \quad \text{and} \quad B_x = \{y;\ f(x) \ge f(y)\}.$$

(a) If f is symmetric, show that $A_x = \{|y| < |x|\}$ for $|x|$ larger than a value x_0.

(b) Define x_1 as the value after which f is log-concave and $x^* = x_0 \vee x_1$. For $V(x) = \exp s|x|$ and $s < \alpha$, show that

$$\frac{\mathbb{E}[V(X_1)|x_0 = x]}{V(x)} \le 1 + \int_0^x \left[e^{s(y-x)} - 1 \right] g(x - y) dy$$

$$+ \int_x^{2x} e^{-\alpha(y-x)} \left[e^{s(y-x)} - 1 \right] g(x - y) dy$$

$$+ 2 \int_x^{\infty} g(y) dy\ .$$

(c) Show that

$$\int_0^x \left(e^{-sy} - 1 + e^{-(\alpha-s)y} - e^{-\alpha y} \right) g(y) dy$$

$$= - \int_0^x \left[1 - e^{-sy} \right] \left[1 - e^{-(\alpha-s)y} \right] g(y) dy$$

and deduce that (7.17) holds for $x > x^*$ and x^* large enough.

(d) For $x < x^*$, show that

$$\frac{\mathbb{E}[V(X_1)|x_0 = x]}{V(x)} \le 1 + 2 \int_{x^*}^{\infty} g(y) dy + 2e^{sx^*} \int_0^{x^*} g(z) dz$$

and thus establish the theorem.

7.15 Examine whether the following distributions are log-concave in the tails: Normal, log-normal, Gamma, Student's t, Pareto, Weibull.

7.16 The following theorem is due to Mengersen and Tweedie (1996).

Theorem 7.25. *If the support of f is not compact and if g is symmetric, the chain $(X^{(t)})$ produced by [A.29] is not uniformly ergodic.*

Assume that the chain satisfies Doeblin's condition (Theorem 6.59).

(a) Take x_0 and $A_0 =]-\infty, x_0]$ such that $\nu(A_0) > \varepsilon$ and consider the *unilateral* version of the random walk, with kernel

$$K^-(x, A) = \frac{1}{2} \mathbb{I}_A(x) + \int_{A \cap]-\infty, x]} g(x - y)\, dy\ ;$$

that is, the random walk which only goes to the left. Show that for $y > x_0$,

$$P^m(y, A_0) \leq P_y(\tau \leq m) \leq P_y(\tau^- \leq m),$$

where τ and τ^- are the return times to A_0 for the chain $(X^{(t)})$ and for K^-, respectively,

(b) For y sufficiently large to satisfy

$$(K^-)^m(y, A_0) = (K^-)^m(0,] - \infty, x_0 - y]) < \frac{\delta}{m},$$

show that

$$P_y(\tau^- \leq m) \leq \sum_{j=1}^{m} (K^-)^j(y, A_0) \leq m(K^-)^m(y, A_0),$$

contradicting Doeblin's condition and proving the theorem.

(c) Formulate a version of Theorem 7.25 for higher-dimensional chains.

7.17 Mengersen and Tweedie (1996) also establish the following theorem:

Theorem 7.26. *If g is continuous and satisfies*

$$\int |x| \, g(x) dx < \infty,$$

the chain $(X^{(t)})$ of [A.29] is geometrically ergodic if and only if

$$\bar{\varphi} = \lim_{x \to \infty} \frac{d}{dx} \log f(x) < 0.$$

(a) To establish sufficiency, show that for $x < y$ large enough, we have

$$\log f(y) - \log f(x) = \int_x^y \frac{d}{dt} \log f(t) dt \leq \frac{\bar{\varphi}}{2} (y - x).$$

Deduce that this inequality ensures the log-concavity of f and, therefore, the application of Theorem 7.15.

(b) For necessity, suppose $\bar{\varphi} = 0$. For every $\delta > 0$, show that you can choose x large enough so that $\log f(x + z) - \log f(x) \geq -\delta z$, $z > 0$ and, therefore, $f(x + z) \exp(\delta z) \geq f(x)$. By integrating out the z's, show that

$$\int_x^\infty f(y) \, e^{\delta y} \, dy = \infty,$$

contradicting condition (7.18).

7.18 For the situation of Example 7.16, show that

$$\lim_{x \to \infty} \frac{d}{dx} \log \varphi(x) = -\infty \quad \text{and} \quad \lim_{x \to \infty} \frac{d}{dx} \log \psi(x) = 0,$$

showing that the chain associated with φ is geometrically ergodic and the chain associated with ψ is not.

7.19 Verify that the transition matrix associated with the geometric random walk in Example 7.17 is correct and that $\beta = \theta^{-1/2}$ minimizes λ_β.

7.20 The Institute for Child Health Policy at the University of Florida studies the effects of health policy decisions on children's health. A small portion of one of their studies follows.

The overall health of a child (**metq**) is rated on a 1–3 scale, with 3 being the worst. Each child is in an HMO[8] (variable **np**, 1=nonprofit, −1=for profit). The dependent variable of interest (y_{ij}) is the use of an emergency room (**erodds**, 1=used emergency room, 0=did not). The question of interest is whether the status of the HMO affects the emergency room choice.

(a) An appropriate model is the logistic regression model,

$$\text{logit}(p_{ij}) = a + bx_i + cz_{ij}, \quad i = 1, \ldots, k, \quad j = 1, \ldots, n_i,$$

where x_i is the HMO type, z_{ij} is the health status of the child, and p_{ij} is the probability of using an emergency room. Verify that the likelihood function is

$$\prod_{i=1}^{k} \prod_{j=1}^{n_i} \left(\frac{\exp(a + bx_i + cz_{ij})}{1 + \exp(a + bx_i + cz_{ij})} \right)^{y_{ij}} \left(\frac{1}{1 + \exp(a + bx_i + cz_{ij})} \right)^{1 - y_{ij}}.$$

(Here we are only distinguishing between for-profit and non-profit, so $k = 2$.)

(b) Run a standard GLM on these data[9] and get the estimated mean and variance of a, b, and c.

(c) Use normal candidate densities with mean and variance at the GLM estimates in a Metropolis–Hastings algorithm that samples from the likelihood. Get histograms of the parameter values.

x	4	7	8	9	10	11	12
y	2,10	4,22	16	10	18,26	17,28	14,20
					34		24,28

x	13	14	15	16	17	18	19
y	26,34	26,36	20,26	32,40	32,40	42,56	36,46
	34,46	60,80	54		50	76,84	68

x	20	22	23	24	25
y	32,48	66	54	70,92	85
	52,56,64			93,120	

Table 7.6. Braking distances of 50 cars, $x =$ speed (mph), $y =$ distance to stop (feet).

7.21 The famous "braking data" of Tukey (1977) is given in Table 7.6. It is thought that a good model for this dataset is a quadratic model

---

[8] A person who joins an *HMO* (for Health Maintenance Organization) obtains their medical care through physicians belonging to the HMO.

[9] Available as **LogisticData.txt** on the book website.

$$y_{ij} = a + bx_i + cx_i^2 + \varepsilon_{ij}, \quad i = 1, \ldots, k, \quad j = 1, \ldots n_i.$$

If we assume that $\varepsilon_{ij} \sim N(0, \sigma^2)$, independent, then the likelihood function is

$$\left(\frac{1}{\sigma^2}\right)^{N/2} e^{\frac{1}{2\sigma^2} \sum_{ij}(y_{ij} - a - bx_i - cx_i^2)^2},$$

where $N = \sum_i n_i$. We can view this likelihood function as a posterior distribution of $a, b, c,$ and σ^2, and we can sample from it with a Metropolis–Hastings algorithm.

(a) Get estimates of $a, b, c,$ and σ^2 from a usual linear regression.

(b) Use the estimates to select a candidate distribution. Take normals for a, b, c and inverted gamma for σ^2.

(c) Make histograms of the posterior distributions of the parameters. Monitor convergence.

(d) Robustness considerations could lead to using an error distribution with heavier tails. If we assume that $\varepsilon_{ij} \sim t(0, \sigma^2)$, independent, then the likelihood function is

$$\left(\frac{1}{\sigma^2}\right)^{N/2} \prod_{ij} \left(1 + \frac{(y_{ij} - a - bx_i - cx_i^2)^2}{\nu}\right)^{-(\nu+1)/2},$$

where ν is the degrees of freedom. For $\nu = 4$, use Metropolis–Hastings to sample $a, b, c,$ and σ^2 from this posterior distribution. Use either normal or t candidates for $a, b, c,$ and either inverted gamma or half-t for σ^2.

(*Note:* See Problem 11.4 for another analysis of this dataset.)

7.22 The *traveling salesman problem* is a classic in combinatoric and operations research, where a salesman has to find the shortest route to visit each of his N customers.

(a) Show that the problem can be described by (i) a permutation σ on $\{1, \ldots, N\}$ and (ii) a distance $d(i, j)$ on $\{1, \ldots, N\}$.

(b) Deduce that the traveling salesman problem is equivalent to minimization of the function

$$H(\sigma) = \sum_i d(i, \sigma(i)).$$

(c) Propose a Metropolis–Hastings algorithm to solve the problem with a simulated annealing scheme (Section 5.2.3).

(d) Derive a simulation approach to the solution of $Ax = b$ and discuss its merits.

7.23 Check whether a negative coefficient b in the random walk $Y_t = a + b(X^{(t)} - a) + Z_t$ induces a negative correlation between the $X^{(t)}$'s. Extend to the case where the random walk has an ARCH-like structure,

$$Y_t = a + b(X^{(t)} - a) + \exp(c + d(X^{(t)} - a)^2)Z_t.$$

7.24 Implement the Metropolis–Hastings algorithm when f is the normal $\mathcal{N}(0, 1)$ density and $q(\cdot|x)$ is the uniform $\mathcal{U}[-x - \delta, -x + \delta]$ density. Check for negative correlation between the $X^{(t)}$'s when δ varies.

7.25 Referring to Example 7.11

(a) Verify that $\log \alpha$ has an exponential distribution.

(b) Show that the posterior distribution is proper, that is

$$\int L(\alpha, \beta | \mathbf{y}) \pi(\alpha, \beta) d\alpha d\beta < \infty .$$

(c) Show that $\mathbb{E}\alpha = \log b - \gamma$, where γ is Euler's constant.

7.26 Referring to the situation of Example 7.12:

(a) Use Taylor series to establish the approximations

$$K_X(t) \approx K_X(0) + K'_X(0)t + K''_X(0)t^2/2$$
$$tK'_X(t) \approx t \left[K'_X(0) + K''_X(0)t \right]$$

and hence (7.14).

(b) Write out the Metropolis–Hastings algorithm that will produce random variables from the saddlepoint distribution.

(c) Apply the Metropolis saddlepoint approximation to the noncentral chi squared distribution and reproduce the tail probabilities in Table 7.1.

7.27 Given a Cauchy $\mathcal{C}(0, \sigma)$ instrumental distribution:

(a) Experimentally select σ to maximize (i) the acceptance rate when simulating a $\mathcal{N}(0,1)$ distribution and (ii) the squared error when estimating the mean (equal to 0).

(b) Same as (a), but when the instrumental distribution is $\mathcal{C}(x^{(t)}, \sigma)$.

7.28 Show that the Rao–Blackwellized estimator δ^{RB} does not depend on the normalizing factors in f and g.

7.29 Reproduce the experiment of Example 7.20 in the case of a Student's $\mathcal{T}_7$ distribution.

7.30 In the setup of the Metropolis–Hastings algorithm [A.24], the Y_t's are generated from the distributions $q(y|x^{(t)})$. Assume that $Y_1 = X^{(1)} \sim f$.

(a) Show that the estimator

$$\delta_0 = \frac{1}{T} \sum_{t=1}^{T} \frac{f(y_t)}{q(y_t | x^{(t)})} h(y_t)$$

is an unbiased estimator of $\mathbb{E}_f[h(X)]$.

(b) Derive, from the developments of Section 7.6.2, that the Rao–Blackwellized version of δ_0 is

$$\delta_1 = \frac{1}{n} \left\{ h(x_1) + \frac{f(y_2)}{q(y_2|x_1)} h(y_2) \right.$$
$$\left. + \frac{\sum_{i=3}^{T} \sum_{j=1}^{i-1} f(y_i) \delta_j \zeta_{j(i-2)} (1 - \rho_{j(i-1)}) \omega_i^j}{\sum_{i=1}^{T-1} \delta_i \zeta_{i(T-1)}} h(y_i) \right\} .$$

(c) Compare δ_1 with the Rao–Blackwellized estimator of Theorem 7.21 in the case of a $\mathcal{T}_3$ distribution for the estimation of $h(x) = \mathbb{I}_{x>2}$.

7.31 Prove Theorem 7.19 as follows:

(a) Use the properties of the Markov chain to show that the conditional probability τ_i can be written as

$$\tau_i = \sum_{j=0}^{i-1} P(X_i = y_i | X_{i-1} = y_j) \, P(X_{i-1} = y_j) = \sum_{j=0}^{i-1} \rho_{ji} \, P(X_{i-1} = y_j).$$

(b) Show that

$$P(X_{i-1} = y_j) = P(X_j = y_j, X_{j+1} = y_j, \ldots, X_{i-1} = y_j)$$
$$= (1 - \rho_{j(j+1)}) \cdots (1 - \rho_{j(i-1)})$$

and, hence, establish the expression for the weight φ_i.

7.32 Prove Theorem 7.21 as follows:

(a) As in the independent case, the first step is to compute $P(X_j = y_i | y_0, y_1, \ldots, y_n)$. The event $\{X_j = y_i\}$ can be written as the set of all the i-tuples $(u_1, \ldots, u_i)$ leading to $\{X_i = y_i\}$, of all the $(j - i)$-tuples $(u_{i+1}, \ldots, u_j)$ corresponding to the rejection of $(y_{i+1}, \ldots, y_j)$ and of all the $(n - j)$-tuples $u_{j+1}, \ldots, u_n$ following after $X_j = y_i$.
Define $B_0^1 = \{u_1 > \rho_{01}\}$ and $B_1^1 = \{u_1 < \rho_{01}\}$, and let $B_k^t(u_1, \ldots, u_t)$ denote the event $\{X_t = y_k\}$. Establish the relation

$$B_k^t(u_1, \ldots, u_t) = \bigcup_{m=0}^{k-1} \left[B_m^{k-1}(u_1, \ldots, u_{k-1}) \right.$$
$$\left. \cap \{u_k < \rho_{mk}, u_{k+1} > \rho_{k(t+1)}, \ldots, u_t > \rho_{kt}\} \right],$$

and show that

$$\{x_j = y_i\} = \bigcup_{k=0}^{i-1} \left[B_k^{i-1}(u_1, \ldots, u_{i-1}) \right.$$
$$\left. \cap \{u_i < \rho_{ki}, u_{i+1} > \rho_{i(i+1)}, \ldots, u_j > \rho_{ij}\} \right].$$

(b) Let $p(u_1, \ldots, u_T, y_1, \ldots, y_T) = p(\mathbf{u}, \mathbf{y})$ denote the joint density of the U_i's and the $Y_i's$. Show that $\tau_i = \int_A p(\mathbf{u}, \mathbf{y}) du_1 \cdots du_i$, where $A = \bigcup_{k=0}^{i-1} B_k^{i-1}(u_1, \ldots, u_{i-1}) \cap \{u_i < \rho_{ki}\}$.

(c) Show that $\omega_{j+1}^i = \int_{\{x_j = y_i\}} p(\mathbf{u}, \mathbf{y}) du_{j+1} \cdots du_T$ and, using part (b), establish the identity

$$\tau_i \prod_{t=i+1}^{j} (1 - \rho_{it}) q(y_{t+1} | y_i) \omega_{j+1}^i = \tau_i \prod_{t=i+1}^{j} \underline{\rho}_{it} \omega_{j+1}^i = \tau_i \, \zeta_{ij} \, \omega_{j+1}^i.$$

(d) Verify the relations

$$\tau_i = \sum_{t=0}^{i-1} \tau_i \, \zeta_{t(i-1)j} \bar{\rho}_{ti} \quad \text{and} \quad \omega_{j+1}^i = \omega_{j+2}^{j+1} \underline{\rho}_{i(j+1)} + \omega_{j+2}^i \bar{\rho}_{i(j+1)},$$

which provide a recursion relation on the ω_j^i's depending on acceptance or rejection of y_{j+1}. The case $j = T$ must be dealt with separately, since there is no generation of y_{T+1} based on $q(y | x_T)$. Show that ω_T^i is equal to $\rho_{iT} + (1 - \rho_{iT}) = 1$.

(e) The probability $P(X_j = y_i)$ can be deduced from part (d) by computing the marginal distribution of $(Y_1, \ldots, Y_T)$. Show that $1 = \sum_{i=0}^{T} P(X_T = y_i) = \sum_{i=0}^{T-1} \tau_i \zeta_{i(T-1)}$, and, hence, the normalizing constant for part (d) is $\left(\sum_{i=0}^{T-1} \tau_i \zeta_{i(T-1)} \right)^{-1}$, which leads to the expression for φ.

7.33 (Liu 1996b) Consider a finite state-space $\mathcal{X} = \{1, \ldots, m\}$ and a Metropolis–Hastings algorithm on $\mathcal{X}$ associated with the stationary distribution $\pi = (\pi_1, \ldots, \pi_m)$ and the proposal distribution $p = (p_1, \ldots, p_m)$.

(a) For $\omega_i = \pi_i/p_i$ and $\lambda_k = \sum_{i=k}^{m}(p_i - \pi_i/\omega_k)$, express the transition matrix of the Metropolis–Hastings algorithm as $K = G + \mathbf{e}p^T$, where $\mathbf{e} = (1, \ldots, 1)^T$. Show that G is upper triangular with diagonal elements the λ_k's.

(b) Deduce that the eigenvalues of G and K are the λ_k's.

(c) Show that for $||p|| = \sum |p_i|$,

$$||K^n(x, \cdot) - \pi||^2 \leq \frac{\lambda_1^{2n}}{\pi(x)},$$

following a result by Diaconis and Hanlon (1992).

7.34 (Ó Ruanaidh and Fitzgerald 1996) Given the model

$$y_i = A_1 e^{\lambda_1 t_i} + A_2 e^{\lambda_2 t_i} + c + \epsilon_i,$$

where $\epsilon_i \sim \mathcal{N}(0, \sigma^2)$ and the t_i's are known observation times, study the estimation of $(A_1, A_2, \lambda_1, \lambda_2, \sigma)$ by recursive integration (see Section 5.2.4) with particular attention to the Metropolis–Hastings implementation.

7.35 (Roberts 1998) Take f to be the density of the $\mathcal{E}xp(1)$ distribution and g_1 and g_2 the densities of the $\mathcal{E}xp(0.1)$ and $\mathcal{E}xp(5)$ distributions, respectively. The aim of this problem is to compare the performances of the independent Metropolis–Hastings algorithms based on the pairs (f, g_1) and (f, g_2).

(a) Compare the convergences of the empirical averages for both pairs, based on 500 replications of the Markov chains.

(b) Show that the pair (f, g_1) leads to a geometrically ergodic Markov chain and (f, g_2) does not.

7.36 For the Markov chain of Example 7.28, define

$$\xi^{(t)} = \mathbb{I}_{[0,13]}(\theta^{(t)}) + 2\mathbb{I}_{[13,\infty)}(\theta^{(t)}).$$

(a) Show that $(\xi^{(t)})$ is not a Markov chain.

(b) Construct an estimator of the pseudo-transition matrix of $(\xi^{(t)})$.

7.37 Show that the transition associated with the acceptance probability (7.20) also leads to f as invariant distribution, for every symmetric function s. (*Hint:* Use the reversibility equation.)

7.38 Show that the Metropolis–Hastings algorithm is, indeed, a special case of the transition associated with the acceptance probability (7.20) by providing the corresponding $s(x, y)$.

7.39 (Peskun 1973) Let $\mathbb{P}_1$ and $\mathbb{P}_2$ be regular (see Problems 6.9 and 6.10), reversible stochastic matrices with the same stationary distribution π on $\{1, \ldots, m\}$. Show that if $\mathbb{P}_1 \leq \mathbb{P}_2$ (meaning that the off-diagonal elements are smaller in the first case) for every function h,

$$\lim_{N \to \infty} \text{var}\left[\sum h(X_1^{(t)})\right] / N \geq \lim_{N \to \infty} \text{var}\left[\sum h(X_2^{(t)})\right] / N,$$

where $(X_i^{(t)})$ is a Markov chain with transition matrix $\mathbb{P}_i$ $(i = 1, 2)$. (*Hint:* Use Kemeny and Snell 1960 result on the asymptotic variance in Problem 6.50.)

7.40 (Continuation of Problem 7.39) Deduce from Problem 7.39 that for a given instrumental matrix Q in a Metropolis–Hastings algorithm, the choice

$$p_{ij}^* = q_{ij}\left(\frac{\pi_j q_{ji}}{\pi_i q_{ij}} \wedge 1\right)$$

is optimal among the transitions such that

$$p_{ij} = \frac{q_{ij}s_{ij}}{1 + \dfrac{\pi_i q_{ij}}{\pi_j q_{ji}}} = q_{ij}\alpha_{ij}\,,$$

where $s_{ij} = s_{ji}$ and $0 \le \alpha_{ij} \le 1$. (*Hint:* Give the corresponding α_{ij} for the Metropolis–Hastings algorithm and show that it is maximal for $i \ne j$. Tierney 1998, Mira and Geyer 1998, and Tierney and Mira 1998 propose extensions to the continuous case.)

7.41 Show that f is the stationary density associated with the acceptance probability (7.20).

7.42 In the setting of Example 7.28, implement the simulated annealing algorithm to find the maximum of the likelihood $L(\theta|x_1, x_2, x_3)$. Compare with the performances based on $\log L(\theta|x_1, x_2, x_3)$.

7.43 (Winkler 1995) A *Potts model* is defined on a set S of "sites" and a finite set G of "colors" by its energy

$$H(x) = -\sum_{(s,t)} \alpha_{st}\mathbb{I}_{x_s = x_t}, \qquad x \in G^S,$$

where $\alpha_{st} = \alpha_{ts}$, the corresponding distribution being $\pi(x) \propto \exp(H(x))$. An additional structure is introduced as follows: "Bonds" b are associated with each pair (s,t) such that $\alpha_{st} > 0$. These bonds are either active ($b = 1$) or inactive ($b = 0$).

(a) Defining the joint distribution

$$\mu(x, b) \propto \prod_{b_{st}=0} q_{st} \prod_{b_{st}=1} (1 - q_{st})\mathbb{I}_{x_s = x_t},$$

with $q_{st} = \exp(\alpha_{st})$, show that the marginal of μ in x is π. Show that the marginal of μ in b is

$$\mu(b) \propto |G|^{c(b)} \prod_{b_{st}=0} q_{st} \prod_{b_{st}=1} (1 - q_{st}),$$

where $c(b)$ denotes the number of *clusters* (the number of sites connected by active bonds).

(b) Show that the *Swendson–Wang* (1987) algorithm

```
1. Take b_st = 0 if x_s ≠ x_t and, for x_s = x_t,

         ⎧ 1  with probability 1 − q_st
   b_st = ⎨
         ⎩ 0  otherwise.

2. For every cluster, choose a color at random on G.
```

leads to simulations from π (*Note:* This algorithm is acknowledged as accelerating convergence in image processing.)

7.44 (McCulloch 1997) In a generalized linear mixed model, assume the link function is $h(\xi_i) = x_i'\beta + z_i'\mathbf{b}$, and further assume that $\mathbf{b} = (b_1, \ldots, b_I)$ where $\mathbf{b} \sim f_\mathbf{b}(\mathbf{b}|D)$. (Here, we assume φ to be unknown.)

(a) Show that the usual (incomplete-data) likelihood is

$$L(\theta, \varphi, D|y) = \int \prod_{i=1}^{n} f(y_i|\theta_i) f_\mathbf{b}(\mathbf{b}|D) d\mathbf{b} .$$

(b) Denote the complete data by $\mathbf{w} = (\mathbf{y}, \mathbf{b})$, and show that

$$\log L_W = \sum_{i=1}^{n} \log f(y_i|\theta_i) + \log f_\mathbf{b}(b_i|D) .$$

(c) Show that the EM algorithm, given by

> 1. Choose starting values $\beta^{(0)}$, $\varphi^{(0)}$, and $D^{(0)}$.
> 2. Calculate (expectations evaluated under $\beta^{(m)}$, $\varphi^{(m)}$, and $D^{(m)}$) $\beta^{(m+1)}$ and $\varphi^{(m+1)}$, which maximize $\mathbb{E}[\log f(y_i|\theta_i, u, \beta, \varphi)|y]$.
> 3. $D^{(m+1)}$ maximizes $\mathbb{E}[f_\mathbf{b}(\mathbf{b}|D)|y]$.
> 4. Set m to $m+1$.

converges to the MLE.

The next problems (7.45–7.49) deal with Langevin diffusions, as introduced in Section 7.8.5.

7.45 Show that the naïve discretization of (7.22) as $dt = \sigma^2$, $dL_t = X^{(t+\sigma^2)} - X^{(t)}$, and $dB_t = B_{t+\sigma^2} - B_t$ does lead to the representation (7.23).

7.46 Consider f to be the density of $\mathcal{N}(0, 1)$. Show that when $\sigma = 2$ in (7.23), the limiting distribution of the chain is $\mathcal{N}(0, 2)$.

7.47 Show that (7.27) can be directly simulated as

$$\theta_2 \sim \mathcal{N}\left(\frac{y}{3}, \frac{5}{6}\right), \qquad \theta_1|\theta_2 \sim \mathcal{N}\left(\frac{2y - \theta_2}{5}, \frac{4}{5}\right) .$$

7.48 Show that when (7.24) exists and is larger (smaller) than 1 (-1) at $-\infty$ (∞), the random walk (7.23) is transient.

7.49 (Stramer and Tweedie 1999a) Show that the following stochastic differential equation still produces f as the stationary distribution of the associated process:

$$dL_t = \sigma(L_t)\nabla \log f(L_t) + b(L_t)dt,$$

when

$$b(x) = \frac{1}{2}\nabla \log f(x)\sigma^2(x) + \sigma(x)\nabla\sigma(x).$$

Give a discretized version of this differential equation to derive a Metropolis–Hastings algorithm and apply to the case $\sigma(x) = \exp(\omega|x|)$.

7.8 Notes

7.8.1 Background of the Metropolis Algorithm

The original Metropolis algorithm was introduced by Metropolis et al. (1953) in a setup of optimization on a discrete state-space, in connection with particle physics:

the paper was actually published in the *Journal of Chemical Physics*. All the authors of this seminal paper were involved at a very central level in the Los Alamos research laboratory during and after World War II. At the beginning of the war, desk calculators (operated mostly by the wives of the scientists working in Los Alamos) were used to evaluate the behavior of nuclear explosives, to be replaced by the end of the war by one of the first computers under the impulsion of von Neumann. Both a physicist and a mathematician, Metropolis, who died in Los Alamos in 1999, came to this place in April 1943 and can be considered (with Stanislaw Ulam) to be the father of Monte Carlo methods. They not only coined the name "Monte Carlo methods", but also ran the first computer Monte Carlo experiment on the MANIAC[10] computer in 1948![11] Also a physicist, Marshall Rosenbluth joined the Los Alamos Lab later, in 1950, where he worked on the development of the H bomb till 1953. Edward Teller is the most controversial character of the group: as early as 1942, he was one of the first scientists to work on the Manhattan Project that led to the production of the A bomb. Almost as early, he became obsessed with the hydrogen (H) bomb that he eventually managed to design with Stanislaw Ulam and better computer facilities in the early 1950s.[12] Teller's wife, Augusta (Mici) Teller, emigrated with him from Germany in the 1930s and got a Ph.D. from Pittsburgh University; she was also part of the team of "computers" operating the desk calculators in Los Alamos.

The Metropolis algorithm was later generalized by Hastings (1970) and Peskun (1973, 1981) to statistical simulation. Despite several other papers that highlighted its usefulness in specific settings (see, for example, Geman and Geman 1984, Tanner and Wong 1987, Besag 1989), the starting point for an intensive use of Markov chain Monte Carlo methods by the statistical community can be traced to the presentation of the *Gibbs sampler* by Gelfand and Smith (1990) and Smith, as explained in Casella and George (1992) or Chib (1995).

The gap of more than 30 years between Metropolis et al. (1953) and Gelfand and Smith (1990) can be partially attributed to the lack of appropriate computing power, as most of the examples now processed by Markov chain Monte Carlo algorithms could not have been treated previously.

As shown by Hastings (1970), Metropolis–Hastings algorithms are a special case of a more general class of algorithms whose transition is associated with the acceptance probability

$$(7.20) \qquad \varrho(x, y) = \frac{s(x, y)}{1 + \dfrac{f(x)q(y|x)}{f(y)q(x|y)}} \ ,$$

where s is an arbitrary positive symmetric function such that $\varrho(x, y) \leq 1$ (see also Winkler 1995). The particular case $s(x, y) = 1$ is also known as the *Boltzman algorithm* and is used in simulation for particle physics, although Peskun (1973) has shown that, in the discrete case, the performance of this algorithm is always suboptimal compared to the Metropolis–Hastings algorithm (see Problem 7.40).

---

[10] MANIAC stands for *Mathematical Analyzer, Numerical Integrator and Computer*.

[11] Although Metropolis attributes the original idea to Enrico Fermi, 15 years earlier!

[12] To end up on a somber note, Edward Teller later testified against Oppenheimer in the McCarthy trials and, much later, was a fervent proponent of the "Star Wars" defense system under the Reagan administration.

For more details on the historical developments, see Hitchcock (2003).

7.8.2 Geometric Convergence of Metropolis–Hastings Algorithms

The sufficient condition (7.6) of Lemma 7.6 for the irreducibility of the Metropolis–Hastings Markov chain is particularly well adapted to *random walks*, with transition densities $q(y|x) = g(y - x)$. It is, indeed, enough that g is positive in a neighborhood of 0 to ensure the ergodicity of [A.24] (see Section 7.5 for a detailed study of these methods). On the other hand, convergence results stronger than the simple ergodic convergence of (7.1) or than the (total variation) convergence of $\|P^n_{x^{(0)}} - f\|_{TV}$ are difficult to derive without introducing additional conditions on f and q. For instance, it is impossible to establish *geometric convergence* without a restriction to the discrete case or without considering particular transition densities, since Roberts and Tweedie (1996) have come up with chains which are not geometrically ergodic. Defining, for every measure ν and every ν-measurable function h, the *essential supremum*

$$\operatorname{ess}_\nu \sup\, h(x) = \inf\{w; \nu(h(x) > w) = 0\}\,,$$

they established the following result, where $\bar\rho$ stands for the average acceptance probability.

Theorem 7.27. *If the marginal probability of acceptance satisfies*

$$\operatorname{ess}_f \sup\, (1 - \bar\rho(x)) = 1,$$

the algorithm [A.24] is not geometrically ergodic.

Therefore, if $\bar\rho$ is not bounded from below on a set of measure 1, a geometric speed of convergence cannot be guaranteed for [A.24]. This result is important, as it characterizes Metropolis–Hastings algorithms that are weakly convergent (see the extreme case of Example 12.10); however, it cannot be used to establish nongeometric ergodicity, since the function $\bar\rho$ is almost always intractable.

When the state space $\mathcal{E}$ is a small set (see Chapter 6), Roberts and Polson (1994) note that the chain $(X^{(t)})$ is uniformly ergodic. However, this is rarely the case when the state-space is uncountable. It is, in fact, equivalent to *Doeblin's condition*, as stated by Theorem 6.59. Chapter 9 exhibits examples of continuous Gibbs samplers for which uniform ergodicity holds (see Examples 10.4 and 10.17), but Section 7.5 has shown that in the particular case of random walks, uniform ergodicity almost never holds, even though this type of move is a natural choice for the instrumental distribution.

7.8.3 A Reinterpretation of Simulated Annealing

Consider a function E defined on a finite set $\mathcal{E}$ with such a large cardinality that a minimization of E based on the comparison of the values of $E(\mathcal{E})$ is not feasible. The simulated annealing technique (see Section 5.2.3) is based on a conditional density q on $\mathcal{E}$ such that $q(i|j) = q(j|i)$ for every $(i, j) \in \mathcal{E}^2$. For a given value $T > 0$, it produces a Markov chain $(X^{(t)})$ on $\mathcal{E}$ by the following transition:

1. **Generate** ζ_t **according to** $q(\zeta|x^{(t)})$.

2. Take

$$X^{(t+1)} = \begin{cases} \zeta_t & \text{with probability} \quad \exp\left(\{E(X^{(t)}) - E(\zeta_t)\}/T\right) \wedge 1 \\ X^{(t)} & \text{otherwise.} \end{cases}$$

As noted in Section 5.2.3, the simulated value ζ_t is automatically accepted when $E(\zeta_t) \leq E(X^{(t)})$. The fact that the simulated annealing algorithm may accept a value ζ_t with $E(\zeta_t)$ larger than $E(X^{(t)})$ is a very positive feature of the method, since it allows for escapes from the attraction zone of local minima of E when T is large enough. The simulated annealing algorithm is actually a Metropolis–Hastings algorithm with stationary distribution $f(x) \propto \exp(-E(x)/T)$, provided that the matrix of the $q(i|j)$'s generates an irreducible chain. Note, however, that the theory of time-homogeneous Markov chains presented in Chapter 6 does not cover the extension to the case when T varies with t and converges to 0 "slowly enough" (typically in $1/\log t$).

7.8.4 Reference Acceptance Rates

Roberts et al. (1997) recommend the use of instrumental distributions with *acceptance rate close to $1/4$ for models of high dimension and equal to $1/2$ for the models of dimension 1 or 2*. This heuristic rule is based on the asymptotic behavior of an *efficiency criterion* equal to the ratio of the variance of an estimator based on an iid sample and the variance of the estimator (3.1); that is,

$$\left[1 + 2 \sum_{k>0} \text{corr}\left(X^{(t)}, X^{(t+k)}\right)\right]^{-1}$$

in the case $h(x) = x$. When f is the density of the $\mathcal{N}(0,1)$ distribution and g is the density of a Gaussian random walk with variance σ, Roberts et al. (1997) have shown that the optimal choice of σ is 2.4, with an asymmetry in the efficiency in favor of large values of σ. The corresponding acceptance rate is

$$\rho = \frac{2}{\pi} \arctan\left(\frac{2}{\sigma}\right) ,$$

equal to 0.44 for $\sigma = 2.4$. A second result by Roberts et al. (1997), based on an approximation of $X^{(t)}$ by a Langevin diffusion process (see Section 7.8.5) when the dimension of the problem goes to infinity, is that the acceptance probability converges to 0.234 (approximately $1/4$). An equivalent version of this empirical rule is to *take the scale factor in g equal to $2.38/\sqrt{d}\,\Sigma$*, where d is the dimension of the model and Σ is the asymptotic variance of $X^{(t)}$. This is obviously far from being an absolute optimality result since this choice is based on the particular case of the normal distribution, which is not representative (to say the least) of the distributions usually involved in the Markov chain Monte Carlo algorithms. In addition, Σ is never known in practice.

The implementation of this heuristic rule follows the principle of an *algorithmic calibration*, first proposed by Müller (1991); that is, of successive modifications of scale factors followed by estimations of Σ and of the acceptance rate until this rate is close to $1/4$ and the estimation of Σ remains stable. Note, again, that the

σ	0.1	0.2	0.5	1.0	5.0	8.0	10.0	12.0
ρ_σ	0.991	0.985	0.969	0.951	0.893	0.890	0.891	0.895
t_σ	106.9	67.1	32.6	20.4	9.37	9.08	9.15	9.54
h_1	41.41	44.24	44.63	43.76	42.59	42.12	42.92	42.94
h_2	0.035	0.038	0.035	0.036	0.036	0.036	0.035	0.036
h_3	0.230	0.228	0.227	0.226	0.228	0.229	0.228	0.230

Table 7.7. Performances of the algorithm [A.29] associated with (7.21) for $x_1 = -8$, $x_1 = 8$, and $x_3 = 17$ and the random walk $\mathcal{C}(0, \sigma^2)$. These performances are evaluated via the acceptance probability ρ_σ, the interjump time, t_σ, and the empirical variance associated with the approximation of the expectations $\mathbb{E}^\pi[h_i(\theta)]$ (20,000 simulations).

convergence results of Metropolis–Hastings algorithms only apply for these adaptive versions when the different hyperparameters of the instrumental distribution are fixed: As long as these parameters are modified according to the simulation results, the resulting Markov chain is heterogeneous (or is not a Markov chain anymore if the parameters depend on the whole history of the chain). This cautionary notice signifies that, in practice, the use of a (true) Metropolis–Hastings algorithm must be preceded by a calibration step which determines an acceptable range for the simulation hyperparameters.

Example 7.28. Cauchy posterior distribution. Consider X_1, X_2, and X_3 iid $\mathcal{C}(\theta, 1)$ and $\pi(\theta) = \exp(-\theta^2/100)$. The posterior distribution on θ, $\pi(\theta|x_1, x_2, x_3)$, is proportional to

$$(7.21) \qquad e^{-\theta^2/100}[(1 + (\theta - x_1)^2)(1 + (\theta - x_2)^2)(1 + (\theta - x_3)^2)]^{-1},$$

which is trimodal when x_1, x_2, and x_3 are sufficiently spaced out, as suggested by Figure 1.1 for $x_1 = 0$, $x_2 = 5$, and $x_3 = 9$. This distribution is therefore adequate to test the performances of the Metropolis–Hastings algorithm in a unidimensional setting. Given the dispersion of the distribution (7.21), we use a random walk based on a Cauchy distribution $\mathcal{C}(0, \sigma^2)$. (Chapter 9 proposes an alternative approach via the Gibbs sampler for the simulation of (7.21).)

Besides the probability of acceptance, ρ_σ, a parameter of interest for the comparison of algorithms is the *interjump* time, t_σ; that is, the average number of iterations that it takes the chain to move to a different mode. (For the above values, the three regions considered are $(-\infty, 0]$, $(0, 13]$, and $(13, +\infty)$.) Table 7.7 provides, in addition, an evaluation of [A.29], through the values of the standard deviation of the random walk σ, in terms of the variances of some estimators of $\mathbb{E}^\pi[h_i(\theta)]$ for the functions

$$h_1(\theta) = \theta, \quad h_2(\theta) = \left(\theta - \frac{17}{3}\right)^2 \quad \text{and} \quad h_3(\theta) = \mathbb{I}_{[4,8]}(\theta).$$

The means of these estimators for the different values of σ are quite similar and equal to 8.96, 0.063, and 0.35, respectively. A noteworthy feature of this example is that the probability of acceptance never goes under 0.88 and, therefore, the goal of

Roberts et al. (1997) cannot be attained with this choice of instrumental distribution, whatever σ is. This phenomenon is quite common in practice.   ‖

7.8.5 Langevin Algorithms

Alternatives to the random walk Metropolis–Hastings algorithm can be derived from *diffusion theory* as proposed in Grenander and Miller (1994) and Phillips and Smith (1996). The basic idea of this approach is to seek a *diffusion equation* (or a *stochastic differential equation*) which produces a *diffusion* (or continuous-time process) with stationary distribution f and then discretize the process to implement the method. More specifically, the *Langevin diffusion* L_t is defined by the stochastic differential equation

$$(7.22) \qquad dL_t = dB_t + \frac{1}{2}\nabla \log f(L_t)dt,$$

where B_t is the standard *Brownian motion*; that is, a random function such that $B_0 = 0$, $B_t \sim \mathcal{N}(0, \omega^2 t)$, $B_t - B_{t'} \sim \mathcal{N}(0, \omega^2|t - t'|)$ and $B_t - B_{t'}$ is independent of $B_{t'}$ ($t > t'$). (This process is the limit of a simple random walk when the magnitude of the steps, Δ, and the time between steps, τ, both approach zero in such a way that

$$\lim \Delta/\sqrt{\tau} = \omega.$$

See, for example, Ethier and Kurtz 1986 , Resnick 1994, or Norris 1997. As stressed by Roberts and Rosenthal (1998), the Langevin diffusion (7.22) is the only non-explosive diffusion which is reversible with respect to f.

The actual implementation of the diffusion algorithm involves a discretization step where (7.22) is replaced with the random walk like transition

$$(7.23) \qquad x^{(t+1)} = x^{(t)} + \frac{\sigma^2}{2}\nabla \log f(x^{(t)}) + \sigma\varepsilon_t,$$

where $\varepsilon_t \sim \mathcal{N}_p(0, I_p)$ and σ^2 corresponds to the discretization size (see Problem 7.45). Although this naïve discretization aims at reproducing the convergence of the random walk to the Brownian motion, the behavior of the Markov chain (7.23) may be very different from that of the diffusion process (7.22). As shown by Roberts and Tweedie (1995), the chain (7.23) may well be transient! Indeed, a sufficient condition for this transience is that the limits

$$(7.24) \qquad \lim_{x \to \pm\infty} \sigma^2 \nabla \log f(x)|x|^{-1}$$

exist and are larger than 1 and smaller than -1 at $-\infty$ and $+\infty$, respectively, since the moves are then necessarily one-sided for large values of $|x^{(t)}|$. Note also the strong similarity between (7.23) and the stochastic gradient equation inspired by (5.4),

$$(7.25) \qquad x^{(t+1)} = x^{(t)} + \frac{\sigma}{2}\nabla \log f(x^{(t)}) + \sigma\varepsilon_t \ .$$

As suggested by Besag (1994), a way to correct this negative behavior is to treat (7.23) as a regular Metropolis–Hastings instrumental distribution; that is, to accept the new value Y_t with probability

$$\frac{f(Y_t)}{f(x^{(t)})} \cdot \frac{\exp\left\{-\left\|Y_t - x^{(t)} - \frac{\sigma^2}{2}\nabla \log f(x^{(t)})\right\|^2 \bigg/ 2\sigma^2\right\}}{\exp\left\{-\left\|x^{(t)} - Y_t - \frac{\sigma^2}{2}\nabla \log f(Y_t)\right\|^2 \bigg/ 2\sigma^2\right\}} \wedge 1.$$

The corresponding Metropolis–Hastings algorithm will not necessarily outperform the regular random walk Metropolis–Hastings algorithm, since Roberts and Tweedie (1995) show that the resulting chain is not geometrically ergodic when $\nabla \log f(x)$ goes to 0 at infinity, similar to the random walk, but the (basic) ergodicity of this chain is ensured.

Roberts and Rosenthal (1998) give further results about the choice of the scaling factor σ, which should lead to an acceptance rate of 0.574 to achieve optimal convergence rates in the special case where the components of x are uncorrelated[13] under f. Note also that the proposal distribution (7.23) is rather natural from a Laplace approximation point of view since it corresponds to a second-order approximation of f. Indeed, by a standard Taylor expansion,

$$\log f(x^{(t+1)}) = \log f(x^{(t)}) + (x^{(t+1)} - x^{(t)})'\nabla \log f(x^{(t)})$$
$$+ \frac{1}{2}(x^{(t+1)} - x^{(t)})' \left[\nabla'\nabla \log f(x^{(t)})\right](x^{(t+1)} - x^{(t)}),$$

the random walk type approximation to $f(x^{(t+1)})$ is

$$f(x^{(t+1)}) \propto \exp\left\{(x^{(t+1)} - x^{(t)})'\nabla \log f(x^{(t)})\right.$$
$$\left. - \frac{1}{2}(x^{(t+1)} - x^{(t)})'H(x^{(t)})(x^{(t+1)} - x^{(t)})\right\}$$
$$\propto \exp\left\{-\frac{1}{2}(x^{(t+1)})'H(x^{(t)})x^{(t+1)}\right.$$
$$\left. + (x^{(t+1)})'\left[\nabla \log f(x^{(t)}) + H(x^{(t)})x^{(t)}\right]\right\}$$
$$\propto \exp\left\{-\frac{1}{2}[x^{(t+1)} - x^{(t)} - (H(x^{(t)}))^{-1}\nabla \log f(x^{(t)})]'H(x^{(t)})\right.$$
$$\left. \times [x^{(t+1)} - x^{(t)} - (H(x^{(t)}))^{-1}\nabla \log f(x^{(t)})]\right\},$$

where $H(x^{(t)}) = -\nabla'\nabla \log f(x^{(t)})$ is the Hessian matrix. If we simplify this approximation by replacing $H(x^{(t)})$ with $\sigma^{-2}I_p$, the Taylor approximation then leads to the random walk with a *drift* term

(7.26) $$x^{(t+1)} = x^{(t)} + \sigma^2\nabla \log f(x^{(t)}) + \sigma\varepsilon_t.$$

From an exploratory point of view, the addition of the gradient of $\log f$ is relevant, since it should improve the moves toward the modes of f, whereas only requiring a minimum knowledge of this density function (in particular, constants are not needed). Note also that, in difficult settings, exact gradients can be replaced by numerical derivatives.

---

[13] The authors also show that this corresponds to a variance of order $p^{-1/3}$, whereas the optimal variance for the Metropolis–Hastings algorithm is of order p^{-1} (see Roberts et al. 1997).

Stramer and Tweedie (1999b) start from the lack of uniform minimal performances of (7.23) to build up modifications (see Problem 7.49) which avoid some pathologies of the basic Langevin Metropolis–Hastings algorithm. They obtain general geometric and uniformly ergodic convergence results.

Example 7.29. Nonidentifiable normal model. To illustrate the performance of the Langevin diffusion method in a more complex setting, consider the nonidentifiable model

$$Y \sim \mathcal{N}\left(\frac{\theta_1 + \theta_2}{2}, 1\right)$$

associated with the exchangeable prior $\theta_1, \theta_2 \sim \mathcal{N}(0, 1)$. The posterior distribution

$$(7.27) \qquad \pi(\theta_1, \theta_2 | y) \propto \exp\left(-\frac{1}{2}\left\{\theta_1^2 \frac{5}{4} + \theta_2^2 \frac{5}{4} + \frac{\theta_1 \theta_2}{2} - (\theta_1 + \theta_2)y\right\}\right)$$

is then well defined (and can be simulated directly, see Problem 7.47), with a ridge structure due to the nonidentifiability. The Langevin transition is then based on the proposal

$$\mathcal{N}_2\left(\frac{\sigma^2}{2}\left(\frac{2y - 2\theta_2^{(t)} - 5\theta_1^{(t)}}{8}, \frac{4y - 2\theta_1^{(t)} - 5\theta_2^{(t)}}{8}\right) + (\theta_1^{(t)}, \theta_2^{(t)}), \sigma^2 I_2\right).$$

A preliminary calibration then leads to $\sigma = 1.46$ for an acceptance rate of $1/2$. As

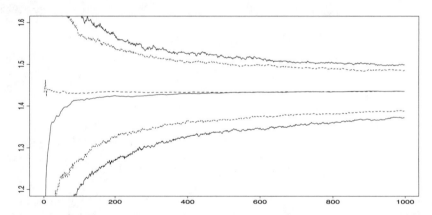

Fig. 7.9. Convergence of the empirical averages for the Langevin Metropolis–Hastings (full line) and iid simulation (dashes) of (θ_1, θ_2) for the estimation of $\mathbb{E}[\theta_1]$ and corresponding 90% equal tail confidence intervals when $y = 4.3$. The final intervals are $[1.373, 1.499]$ and $[1.388, 1.486]$ for the Langevin and exact algorithms, respectively.

seen in Figure 7.9, the 90% range is larger for the output of the Langevin algorithm than for an iid sample in the approximation of $\mathbb{E}[\theta_1 | y]$, but the ratio of these ranges is only approximately 1.3, which shows a moderate loss of efficiency in using the Langevin approximation.   ‖

8

The Slice Sampler

He'd heard stories that shredded documents could be reconstructed. All it took was patience: colossal patience.
—Ian Rankin, *Let it Bleed*

While many of the MCMC algorithms presented in the previous chapter are both generic and universal, there exists a special class of MCMC algorithms that are more model dependent in that they exploit the local conditional features of the distributions to simulate. Before starting the general description of such algorithms, gathered under the (somewhat inappropriate) name of *Gibbs sampling*, we provide in this chapter a simpler introduction to these special kind of MCMC algorithms. We reconsider the *fundamental theorem of simulation* (Theorem 2.15) in light of the possibilities opened by MCMC methodology and construct the corresponding *slice sampler*.

The previous chapter developed simulation techniques that could be called "generic," since they require only a limited amount of information about the distribution to be simulated. For example, the generic algorithm ARMS (Note 7.4.2) aims at reproducing the density f of this distribution in an automatic manner where only the numerical value of f at given points matters. However, Metropolis–Hastings algorithms can achieve higher levels of efficiency when they take into account the specifics of the target distribution f, in particular through the calibration of the acceptance rate (see Section 7.6.1). Moving even further in this direction, the properties and performance of the method presented in this chapter are closely tied to the distribution f.

8.1 Another Look at the Fundamental Theorem

Recall from Section 2.3.1 that the generation from a distribution with density $f(x)$ is equivalent to uniform generation on the subgraph of f,

$$\mathscr{S}(f) = \{(x, u); 0 \leq u \leq f(x)\},$$

whatever the dimension of x, and f need only be known up to a normalizing constant.

From the development at the beginning of Chapter 7, we can consider the possibility of using a Markov chain with stationary distribution equal to this uniform distribution on $\mathscr{S}(f)$ as an approximate way to simulate from f. A natural solution is to use a *random walk* on $\mathscr{S}(f)$, since a random walk on a set $\mathscr{A}$ usually results in a stationary distribution that is the uniform distribution on $\mathscr{A}$ (see Examples 6.39, 6.40, and 6.73).

There are many ways of implementing a random walk on this set, but a natural solution is to go one direction at a time, that is, to move iteratively along the u-axis and then along the x-axis. Furthermore, we can use uniform moves on both directions, since, as formally shown below, the associated Markov chain on $\mathscr{S}(f)$ does not require a Metropolis–Hastings correction to have the uniform distribution on $\mathscr{S}(f)$ as stationary distribution. Starting from a point (x, u) in $\{(x, u) : 0 < u < f(x)\}$, the move along the u-axis will correspond to the conditional distribution

(8.1) $$U|X = x \sim \mathcal{U}(\{u : u \leq f(x)\}),$$

resulting in a change from point (x, u) to point (x, u'), still in $\mathscr{S}(f)$, and then the move along the x-axis to the conditional distribution

(8.2) $$X|U = u' \sim \mathcal{U}(\{x : u' \leq f(x)\}),$$

resulting in a change from point (x, u') to point (x', u').

This set of proposals is the basis chosen for the original *slice sampler* of Neal (1997) (published as Neal 2003) and Damien et al. (1999), which thus uses a 2-step uniform random walk over the subgraph. We inaccurately call it the *2D slice sampler* to distinguish it from the general slice sampler defined in Section 8.2, even though the dimension of x is arbitrary.

Algorithm A.31 –2D slice sampler–

At iteration t, simulate

1. $u^{(t+1)} \sim \mathcal{U}_{[0, f(x^{(t)})]}$;   [A.31]
2. $x^{(t+1)} \sim \mathcal{U}_{A^{(t+1)}}$, with

$$A^{(t+1)} = \{x : f(x) \geq u^{(t+1)}\}.$$

From (8.1) it is also clear that $x^{(t)}$ is always part of the set $A^{(t+1)}$, which is thus nonempty. Moreover, the algorithm remains valid if $f(x) = Cf_1(x)$, and we use f_1 instead of f (see Problem 8.1). This is quite advantageous in settings where f is an unnormalized density like a posterior density.

As we will see in more generality in Chapter 10, the validity of [A.31] as an MCMC algorithm associated with f_1 stems from the fact that both steps 1. and 2. in [A.31] successively preserve the uniform distribution on the subgraph of f: first, if $x^{(t)} \sim f(x)$ and $u^{(t+1)} \sim \mathcal{U}_{[0, f_1(x^{(t)})]}$, then

$$\left(x^{(t)}, u^{(t+1)}\right) \sim f(x) \frac{\mathbb{I}_{[0, f_1(x)]}(u)}{f_1(x)} \propto \mathbb{I}_{0 \leq u \leq f_1(x)}.$$

Second, if $x^{(t+1)} \sim \mathcal{U}_{A^{(t+1)}}$, then

$$\left(x^{(t)}, u^{(t+1)}, x^{(t+1)}\right) \sim f(x^{(t)}) \frac{\mathbb{I}_{[0, f_1(x^{(t)})]}(u^{(t+1)})}{f_1(x^{(t)})} \frac{\mathbb{I}_{A^{(t+1)}}(x^{(t+1)})}{\mathrm{mes}(A^{(t+1)})},$$

where $\mathrm{mes}(A^{(t+1)})$ denotes the (generally Lebesgue) measure of the set $A^{(t+1)}$. Thus

$$\left(x^{(t+1)}, u^{(t+1)}\right) \sim C \int \mathbb{I}_{0 \leq u \leq f_1(x)} \frac{\mathbb{I}_{f_1(x^{(t+1)}) \geq u}}{\mathrm{mes}(A^{(t+1)})} \, dx$$

$$= C \mathbb{I}_{0 \leq u \leq f_1(x^{(t+1)})} \int \frac{\mathbb{I}_{u \leq f_1(x)}}{\mathrm{mes}(A^{(t+1)})} \, dx$$

$$\propto \mathbb{I}_{0 \leq u \leq f_1(x^{(t+1)})}$$

and the uniform distribution on $\mathscr{S}(f)$ is indeed stationary for both steps.

Example 8.1. Simple slice sampler. Consider the density $f(x) = \frac{1}{2} e^{-\sqrt{x}}$ for $x > 0$. While it can be directly simulated from (Problem 8.2), it also yields easily to the slice sampler. Indeed, applying (8.1) and (8.2), we have

$$U|x \sim \mathcal{U}\left(0, \frac{1}{2} e^{-\sqrt{x}}\right), \quad X|u \sim \mathcal{U}\left(0, [\log(2u)]^2\right).$$

We implement the sampler to generate $50,000$ variates, and plot them along with the density in Figure 8.1, which shows that the agreement is very good. The performances of the slice sampler may however deteriorate when $\sqrt{x}$ is replaced with $x^{1/d}$ for d large enough, as shown in Roberts and Tweedie (2004). ‖

Example 8.2. Truncated normal distribution The distribution represented in Figure 8.2 is a truncated normal distribution $\mathcal{N}_-^+(3, 1)$, restricted to the interval $[0, 1]$,

$$f(x) \propto f_1(x) = \exp\{-(x+3)^2/2\} \, \mathbb{I}_{[0,1]}(x).$$

As mentioned previously, the naïve simulation of a normal $\mathcal{N}(3, 1)$ random variable until the outcome is in $[0, 1]$ is suboptimal because there is only a 2% probability of this happening. However, devising and optimizing the

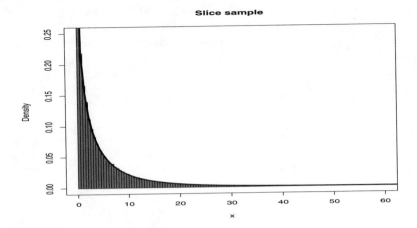

Fig. 8.1. The slice sampler histogram and density for Example 8.1.

Accept–Reject algorithm of Example 2.20 can be costly if the algorithm is to be used only a few times. The slice sampler [A.31] applied to this problem is then associated with the horizontal slice

$$A^{(t+1)} = \{y; \ \exp\{-(y+3)^2/2\} \geq u \, f_1(x^{(t)})\} \,.$$

If we denote $\omega^{(t)}$ the value $u \, f_1(x^{(t)})$, the slice is also given by

$$A^{(t+1)} = \{y \in [0,1]; \ (y+3)^2 \leq -2 \log(\omega^{(t)})\} \,,$$

which is an interval of the form $[0, \gamma^{(t)}]$. Figure 8.2 shows the first ten steps of [A.31] started from $x^{(0)} = 0.25$.   ‖

This algorithm will work well only if the exploration of the subgraph of f_1 by the corresponding random walk is fast enough; if we take the above example of the truncated normal, this is the case. Whatever the value of $x^{(t)}$, the next value $x^{(t+1)}$ can be anything in $[0,1]$. This property actually holds for all f's: given that, when $\omega^{(t+1)}$ is close to 0, the set $A^{(t+1)}$ is close to the entire support of f and, formally, every value $x^{(t+1)}$ in the support of f can be simulated in one step. (We detail below some more advanced results about convergence properties of the slice sampler.)

Figure 8.3 illustrates the very limited dependence on the starting value for the truncated normal example: the three graphs represent a sample of values of $x^{(10)}$ when $(x^{(0)}, \omega^{(1)})$ is in the upper left hand corner, the lower right hand corner and the middle of the subgraph, respectively. To study the effect of the starting value, we always used the same sequence of uniforms for the three starting points. As can be seen from the plot, the samples are uniformly spread out over the subgraph of f and, more importantly, almost identical!

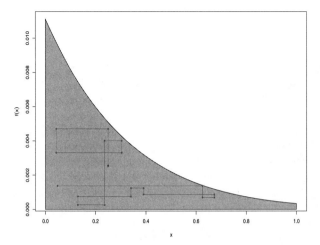

Fig. 8.2. First ten steps of the slice sampler for the truncated normal distribution $\mathcal{N}(3,1)$, restricted to the interval $[0,1]$.

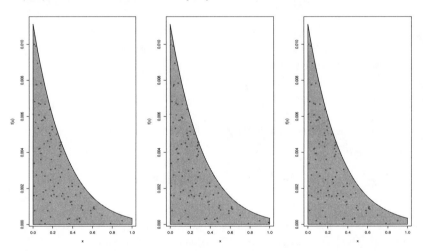

Fig. 8.3. Comparison of three samples obtained by ten iterations of the slice sampler starting from $(.01, .01)$ *(left)*, $(.99, .001)$ *(center)* and $(.25, .025)$ *(right)* and based on the same pool of uniform variates.

In this simple example, ten iterations are thus enough to cancel the effect of the starting value.

The major (practical) difficulty with the slice sampler is with the simulation of the uniform distribution $\mathcal{U}_{A^{(t+1)}}$, since the determination of the set of y's such that $f_1(y) \geq \omega$ can be intractable if f_1 is complex enough. We thus consider an extension to [A.31] in the next section, but call attention to an extension by Neal (2003) covered in Note 8.5.1.

8.2 The General Slice Sampler

As indicated above, sampling uniformly from the slice $A^{(t)} = \{x; f_1(x) \geq \omega^{(t)}\}$ may be completely intractable, even with the extension of Note 8.5.1. This difficulty persists as the dimension of x gets larger. However, there exists a generalization of the 2D slice sampler [A.31] that partially alleviates this difficulty by introducing multiple slices.

This general slice sampler can be traced back to the auxiliary variable algorithm of Edwards and Sokal (1988) applied to the Ising model of Example 5.8 (see also Swendson and Wang 1987, Wakefield et al. 1991, Besag and Green 1993, Damien and Walker 1996, Higdon 1996, Neal 1997, Damien et al. 1999, and Tierney and Mira 1999). It relies upon the decomposition of the density $f(x)$ as

$$f(x) \propto \prod_{i=1}^{k} f_i(x),$$

where the f_i's are positive functions, but *not necessarily densities*. For instance, in a Bayesian framework with a flat prior, the $f_i(x)$ may be chosen as the individual likelihoods.

This decomposition can then be associated with k auxiliary variables ω_i, rather than one as in the fundamental theorem, in the sense that each $f_i(x)$ can be written as an integral

$$f_i(x) = \int \mathbb{I}_{0 \leq \omega_i \leq f_i(x)} \, d\omega_i,$$

and that f is the marginal distribution of the joint distribution

(8.3) $$(x, \omega_1, \ldots, \omega_k) \sim p(x, \omega_1, \ldots, \omega_k) \propto \prod_{i=1}^{k} \mathbb{I}_{0 \leq \omega_i \leq f_i(x)}.$$

This particular *demarginalization* of f (Section 5.3.1) introduces a larger dimensionality to the problem and induces a generalization of the random walk of Section 8.1 which is to have uniform proposals one direction at a time. The corresponding generalization of [A.31] is thus as follows.

Algorithm A.32 –Slice Sampler–

At iteration $t+1$, simulate

1. $\omega_1^{(t+1)} \sim \mathcal{U}_{[0, f_1(x^{(t)})]}$;

   $\vdots$ [A.32]

k. $\omega_k^{(t+1)} \sim \mathcal{U}_{[0, f_k(x^{(t)})]}$;

k+1. $x^{(t+1)} \sim \mathcal{U}_{A^{(t+1)}}$, with

$$A^{(t+1)} = \{y; \, f_i(y) \geq \omega_i^{(t+1)}, \, i = 1, \ldots, k\}.$$

Example 8.3. A 3D slice sampler. Consider the density proportional to

$$(1 + \sin^2(3x))\,(1 + \cos^4(5x))\,\exp\{-x^2/2\}.$$

The corresponding functions are, for instance, $f_1(x) = (1+\sin^2(3x))$, $f_2(x) = (1 + \cos^2(5x))$, and $f_3(x) = \exp\{-x^2/2\}$. In an iteration of the slice sampler, three uniform $\mathcal{U}([0,1])$ u_1, u_2, u_3 are generated and the new value of x is uniformly distributed over the set

$$\left\{x : |x| \le \sqrt{-2\log\omega_3}\right\} \cap \left\{x : \sin^2(3x) \ge 1 - \omega_1\right\} \cap \left\{x : \cos^4(5x) \ge 1 - \omega_2\right\},$$

which is made of one or several intervals depending on whether $\omega_1 = u_1 f_1(x)$ and $\omega_2 = u_2 f_2(x)$ are larger than 1. Figure 8.4 shows how the sample produced by the algorithm fits the target density. ‖

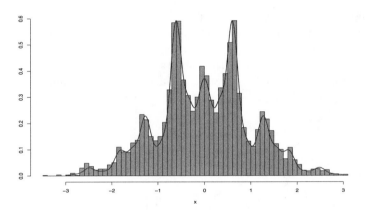

Fig. 8.4. Histogram of a sample produced by $5,000$ iterations of a 3D slice sampler and superposition of the target density proportional to $(1 + \sin^2(3x))\,(1 + \cos^4(5x))\,\exp\{-x^2/2\}$.

In Section 5.3.1, the completion was justified by a *latent variable* representation of the model. Here, the auxiliary variables are more artificial, but there often is an obvious connection in (natural) latent variable models.

Example 8.4. Censored data models. As noted in Section 5.3.1, censored data models (Example 5.14) can be associated with a missing data structure. Consider

$$y^* = (y_1^*, y_2^*) = (y \wedge r, \mathbb{I}_{y < r}), \quad \text{with } y \sim f(y|\theta) \text{ and } r \sim h(r),$$

so the observation y is censored by the random variable r. If we observe $(y_1^*, \ldots, y_n^*)$, the density of $y_i^* = (y_{1i}^*, y_{2i}^*)$ is then

(8.4) $$\int_{y_{1i}^*}^{+\infty} f(y|\theta)\, dy \, h(y_{1i}^*)\, \mathbb{I}_{y_{2i}^*=0} + \int_{y_{1i}^*}^{+\infty} h(r)\, dr \, f(y_{1i}^*|\theta)\, \mathbb{I}_{y_{2i}^*=1} \;.$$

In the cases where (8.4) cannot be explicitly integrated, the likelihood and posterior distribution associated with this model may be too complex to be used. If θ has the prior distribution π, the posterior distribution satisfies

$$\pi(\theta|y_1^*, \ldots, y_n^*) \propto \pi(\theta) \prod_{\{i:y_{2i}^*=0\}} \left\{ h(y_{1i}^*) \int_{y_{1i}^*}^{+\infty} f(y|\theta) dy \right\}$$

$$\times \prod_{\{i:y_{2i}^*=1\}} \left\{ f(y_{1i}^*|\theta) \int_{y_{1i}^*}^{+\infty} h(r) dr \right\}$$

$$\propto \pi(\theta) \prod_{\{i:y_{2i}^*=0\}} \int_{y_{1i}^*}^{+\infty} f(y|\theta) dy \prod_{\{i:y_{2i}^*=1\}} f(y_{1i}^*|\theta) \;.$$

If the *uncensored* model leads to explicit computations for the posterior distribution of θ, we will see in the next chapter that a logical completion is to reconstruct the original data, $y_1, \ldots, y_n$, conditionally on the observed y_i^*'s $(i = 1, \ldots, n)$, and then implement the Gibbs sampler on the two groups θ and the unobserved y's. At this level, we can push the demarginalization further to first represent the above posterior as the marginal of

$$\pi(\theta) \prod_{\{i:y_{2i}^*=0\}} f(y_{1i}|\theta)\, \mathbb{I}_{y_{1i}>y_{1i}^*} \prod_{\{i:y_{2i}^*=1\}} f(y_{1i}^*|\theta)$$

and write this distribution itself as the marginal of

$$\mathbb{I}_{0\le\omega_0\le\pi(\theta)} \prod_{\{i:y_{2i}^*=0\}} \left\{ \mathbb{I}_{y_{1i}>y_{1i}^*}\, \mathbb{I}_{0\le\omega_i\le f(y_{1i}|\theta)} \right\} \prod_{\{i:y_{2i}^*=1\}} \mathbb{I}_{0\le\omega_i\le f(y_{1i}^*|\theta)} \;,$$

adding to the unobserved y_{1i}'s the basic auxiliary variables ω_i $(0 \le i \le n)$. ∥

Although the representation (8.3) of f takes us farther away from the fundamental theorem of simulation, the validation of this algorithm is the same as in Section 8.1: each of the $k + 1$ steps in [A.32] preserves the distribution (8.3). This will also be the basis of the Gibbs sampler in Chapters 9 and 10.

While the basic appeal in using this generalization is that the set $A^{(t+1)}$ may be easier to compute as the intersection of the slices

$$A_i^{(t+1)} = \{y;\ f_i(y) \ge \omega_i^{(t+1)}\} \;,$$

there may still be implementation problems. As k increases, the determination of the set $A^{(t+1)}$ usually gets increasingly complex. Neal (2003) develops proposals in parallel to improve the slice sampler in this multidimensional setting, using "over-relaxed" and "reflective slice sampling" procedures. But these proposals are too specialized to be discussed here and, also, they require careful calibration and cannot be expected to automatically handle all settings. Note also that, in missing variable models, the number of auxiliary variables (that is, of slices) increases with the number of observations and may create deadlocks for large data sizes.

8.3 Convergence Properties of the Slice Sampler

Although, in the following chapters, we will discuss in detail the convergence of the Gibbs sampler, of which the slice sampler is a special case, we present in this section some preliminary results. These are mostly due to Roberts and Rosenthal (1998) and and show that the convergence rate of the slice sampler can be evaluated quantitatively.

First, note that, for the 2D slice sampler [A.31], if we denote by $\mu(\omega)$ the Lebesgue measure of the set

$$\mathscr{S}(\omega) = \{y; \, f_1(y) \geq \omega\} \,,$$

the transition kernel is such that

$$
\begin{aligned}
\Pr\left(f_1(X^{(t+1)}) \leq \eta | f_1(X^{(t)}) = v\right) &= \int\int \mathbb{I}_{w \leq f_1(x) \leq \eta} \, \mathbb{I}_{0 \leq w \leq v} \, \frac{dw \, dx}{v \, \mu(w)} \\
&= \frac{1}{v} \int_0^{\eta \wedge v} \frac{\mu(w) - \mu(\eta)}{\mu(w)} \, dw \\
&= \frac{1}{v} \int_0^v \max\left(1 - \frac{\mu(\eta)}{\mu(w)}, 0\right) dw \,.
\end{aligned}
$$

The properties of the chain are thus entirely characterized by the measure μ and, moreover, they are equivalent for $(X^{(t+1)})$ and $(f_1(X^{(t+1)}))$ which also is a Markov chain in this case (Problems 8.9 and 8.10).

Under boundedness conditions, Tierney and Mira (1999) established the following uniform ergodicity result.

Lemma 8.5. *If f_1 is bounded and* $\mathrm{supp} f_1$ *is bounded, the slice sampler [A.31] is uniformly ergodic.*

Proof. Without loss of generality, assume that f_1 is bounded by 1 and that the support $\mathrm{supp} f_1$ is equal to $(0, 1)$. To establish uniform ergodicity, it is necessary and sufficient to prove that Doeblin's condition (Theorem 6.59) holds, namely that the whole space $\mathrm{supp} f_1$ is a small set. This follows from the fact that

$$\xi(v) = \Pr\left(f_1(X^{(t+1)}) \leq \eta | f_1(X^{(t)}) = v\right)$$

is decreasing in v for all η. In fact, when $v \geq \eta$,

$$\xi(v) = \frac{1}{v} \int_0^\eta \frac{\mu(w) - \mu(\eta)}{\mu(w)} \, dw$$

is clearly decreasing in v. When $v \leq \eta$, $\xi(v)$ is the average of the function

$$\left(1 - \frac{\mu(\eta)}{\mu(W)}\right)$$

when W is uniformly distributed over $(0, v)$. Since $\mu(\omega)$ is decreasing in ω, this average is equally decreasing in v. The maximum of $\xi(v)$ is thus

$$\lim_{v \to 0} \xi(v) = \lim_{v \to 0} \left(1 - \frac{\mu(\eta)}{\mu(v)}\right) = 1 - \mu(\eta)$$

by L'Hospital's rule, while the minimum of $\xi(v)$ is

$$\lim_{v \to 1} \xi(v) = \int_0^\eta \frac{\mu(w) - \mu(\eta)}{\mu(w)} \, dw \, .$$

Since we are able to find nondegenerate upper and lower bounds on the cdf of the transition kernel, we can derive Doeblin's condition and thus uniform ergodicity.   □

In a more general setting, Roberts and Rosenthal (1998) exhibit a *small set* associated with the 2D slice sampler.

Lemma 8.6. *For any $\epsilon^* > \epsilon_\star$, $\{x; \epsilon_\star < f_1(x) < \epsilon^*\}$ is a small set, that is,*

$$P(x, \cdot) \geq \frac{\epsilon_\star}{\epsilon^*} \nu(\cdot) \text{ where } \nu(A) = \frac{1}{\epsilon_\star} \int_0^{\epsilon_\star} \frac{\lambda(A \cap \{y; f_1(y) > \epsilon\})}{\mu(\epsilon)} d\,\epsilon$$

and λ denotes Lebesgue measure.

Proof. For $\epsilon_\star < x < \epsilon^*$,

$$\Pr(x, A) = \frac{1}{x} \int_0^x \frac{\lambda(A \cap \{y; f_1(y) > \epsilon\})}{\mu(\epsilon)} d\epsilon$$

$$\geq \frac{1}{\epsilon^*} \int_0^{\epsilon_\star} \frac{\lambda(A \cap \{y; f_1(y) > \epsilon\})}{\mu(\epsilon)} d\epsilon \, .$$

We thus recover the minorizing measure given in the lemma.   □

If f_1 is bounded, ϵ^* can be chosen as the maximum of f_1 and the set $\mathscr{S}(\epsilon_\star)$ is a small set. Roberts and Rosenthal (1998) also derive a drift condition (Note 6.9.1), as follows.

Lemma 8.7. *Assume*

(i) f_1 is bounded by 1,
(ii) the function μ is differentiable,

(iii) $\mu'(y)y^{1+1/\alpha}$ *is non-increasing for $y < \epsilon^\star$ and an $\alpha > 1$.*

Then, for $0 < \beta < \min(\alpha - 1, 1)/\alpha$,

$$V(x) = f_1(x)^{-\beta}$$

is a drift function, that is, it satisfies the drift condition (6.42) on the sets $\mathscr{S}(\epsilon^\star)$.

Proof. Recall that $\mathscr{S}(\epsilon^\star) = \{y; \, f_1(y) \geq \epsilon^\star\}$. If $f_1(x) \leq \epsilon^\star$, the image of V by the slice sampler transition kernel satisfies

$$
\begin{aligned}
KV(x) &= \frac{1}{f_1(x)} \int_0^{f_1(x)} \frac{1}{\mu(\omega)} \int_{\mathscr{S}(\omega)} f_1(y)^{-\beta} \, dy \, d\omega \\
&= \frac{1}{f_1(x)} \int_0^{f_1(x)} \frac{1}{\mu(\omega)} \int_\omega^\infty z^{-\beta}(-\mu'(z)) \, dz \, d\omega \\
&\leq \frac{1}{f_1(x)} \int_0^{f_1(x)} \frac{\int_y^{\epsilon^\star} z^{-\beta}(-\mu'(z)) \, dz}{\int_y^{\epsilon^\star} (-\mu'(z)) \, dz} \, d\omega \\
&\leq \frac{1}{f_1(x)} \int_0^{f_1(x)} \frac{\int_y^{\epsilon^\star} z^{-(1+\beta+1/\alpha)} \, dz}{\int_y^{\epsilon^\star} z^{-(1+1/\alpha)} \, dz} \, d\omega \\
&= \frac{1}{1+\alpha\beta} \frac{1}{f_1(x)} \left(\int_0^{f_1(x)} \omega^{-\beta} \, d\omega + \int_0^{f_1(x)} \frac{\epsilon^{\star -1/\alpha}(\omega^{-\beta} - \epsilon^{\star -\beta})}{\omega^{-1/\alpha} - \epsilon^{\star -1/\alpha}} \, d\omega \right) \\
&\leq \frac{V(x)}{(1-\beta)(1+\alpha\beta))} + \frac{\alpha\beta\epsilon^{\star -\beta}}{1+\alpha\beta}
\end{aligned}
$$

(see Problem 8.12 for details).

When $f_1(x) \geq \epsilon^\star$, the image $KV(x)$ is non-increasing with $f_1(x)$ and, therefore,

$$KV(x) \leq KV(x_0) = -\epsilon^{\star\beta} \frac{1 + \alpha\beta(1-\beta)}{(1+\alpha\beta)(1-\beta)}$$

for x_0 such that $f_1(x_0) = \epsilon^\star$.

Therefore, outside $\mathscr{S}(\epsilon^\star)$,

$$KV(x) \leq \frac{1}{(1-\beta)(1+\alpha\beta))} + \frac{\alpha\beta(\epsilon_\star/\epsilon^\star)^\beta}{1+\alpha\beta} V(x) = \lambda V(x)$$

for any $\epsilon_\star < \epsilon^\star$ and, on $\mathscr{S}(\epsilon^\star)$,

$$KV(x) \leq b = -\epsilon^{\star\beta} \frac{1 + \alpha\beta(1-\beta)}{(1+\alpha\beta)(1-\beta)} - \lambda,$$

which establishes the result. $\square$

Therefore, under the conditions of Lemma 8.7, the Markov chain associated with the slice sampler is geometrically ergodic, as follows from Theorem 6.75. In addition, Roberts and Rosenthal (1998) derive explicit bounds on the total variation distance, as in the following example, the proof of which is beyond the scope of this book.

Example 8.8. Exponential $\mathcal{E}xp(1)$ distribution   If $f_1(x) = \exp(-x)$, Roberts and Rosenthal (1998) show that, for $n > 23$,

$$(8.5) \qquad ||K^n(x, \cdot) - f(\cdot)||_{TV} \leq .054865 \, (0.985015)^n \, (n - 15.7043) \,.$$

This implies, for instance, that, when $n = 530$, the total variation distance between K^{530} and f is less than 0.0095. While this figure is certainly over-conservative, given the smoothness of f_1, it is nonetheless a very reasonable bound on the convergence time. $\qquad\qquad ||$

Roberts and Rosenthal (1998) actually show a much more general result: *For any density such that $\epsilon\mu'(\epsilon)$ is non-increasing, (8.5) holds for all x's such that $f(x) \geq .0025 \sup f(x)$.* In the more general case of the (product) slice sampler [A.32], Roberts and Rosenthal (1998) also establish geometric ergodicity under stronger conditions on the functions f_i.

While unidimensional log-concave densities satisfy the condition on μ (Problem 8.15), Roberts and Rosenthal (2001) explain why multidimensional slice samplers may perform very poorly through the following example.

Example 8.9. A poor slice sampler. Consider the density $f(x) \propto \exp\{-||x||\}$ in $\mathbb{R}^d$. Simulating from this distribution is equivalent to simulating the radius $z = |x|$ from

$$\eta_d(z) \propto z^{d-1} e^{-z}\,, \qquad z > 0\,,$$

or, by a change of variable $u = z^d$, from

$$\pi_d(u) \propto e^{-u^{1/d}}\,, \qquad u > 0\,.$$

If we run the slice sampler associated with π_d, the performances degenerate as d increases. This sharp decrease in the performances is illustrated by Figure 8.5: for $d = 1, 5$, the chain mixes well and the autocorrelation function decreases fast enough. This is not the case for $d = 10$ and even less for $d = 50$, where mixing is slowed down so much that convergence does not seem possible! $\qquad\qquad ||$

Roberts and Rosenthal (2001) take advantage of this example to propose an alternative to the slice sampler called the *polar slice sample*, which relates more to the general Gibbs sampler of next chapter (see Problem 9.21).

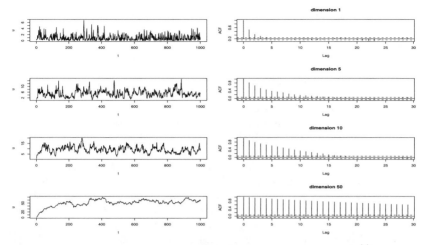

Fig. 8.5. Rawplots and autocorrelation functions for the series $(z^{(t)})$ generated by the slice samplers associated with $f_1(z) = z^{d-1}\,e^{-z}$ for $d = 1, 5, 10, 50$.

8.4 Problems

8.1 Referring to Algorithm [A.31] and equations (8.1) and (8.2):
  (a) Show that the stationary distribution of the Markov chain [A.31] is the uniform distribution on the set $\{(x, u) : 0 < u < f(x)\}$.
  (b) Show that the conclusion of part (a) remains the same if we use f_1 in [A.31], where $f(x) = C f_1(x)$.

8.2 In the setup of Example 8.1, show that the cdf associated with the density $\exp(-\sqrt{x})$ on $\mathbb{R}_+$ can be computed in closed form. (*Hint:* Make the change of variable $z = \sqrt{x}$ and do an integration by parts on $z \exp(-z)$.)

8.3 Consider two unnormalized versions of a density f, f_1 and f_2. By implementing the slice sampling algorithm [A.31] on both f_1 and f_2, show that the chains $(x^{(t)})_t$ produced by both versions are exactly the same if they both start from the same value $x^{(0)}$ and use the same uniform $u^{(t)} \sim \mathcal{U}([0, 1])$.

8.4 As a generalization of the density in Example 8.1, consider the density $f(x) \propto \exp\{-x^d\}$, for $d < 1$. Write down a slice sampler algorithm for this density, and evaluate its performance for $d = .1, .25, .4$.

8.5 Show that a possible slice sampler associated with the standard normal density, $f(x) \propto \exp(-x^2/2)$, is associated with the two conditional distributions

$$(8.6) \quad \omega|x \sim \mathcal{U}_{[0, \exp(-x^2/2)]}\,, \qquad X|\omega \sim \mathcal{U}\left([-\sqrt{-2\log(\omega)}, \sqrt{-2\log(\omega)}]\right)\,.$$

Compare the performances of this slice sampler with those of an iid sampler from $\mathcal{N}(0, 1)$ by computing the empirical cdf at 0, .67, .84, 1.28, 1.64, 1.96, 2.33, 2.58, 3.09, and 3.72 for two samples of same size produced under both approaches. (Those figures correspond to the .5, .75, .8, .9, .95, .99, .995, .999 and .9999 quantiles, respectively.)

8.6 Reproduce the comparison of Problem 8.5 in the case of (a) the gamma distribution and (b) the Poisson distribution.

8.7 Consider the mixture distribution (1.3), where the p_j's are unknown and the f_j's are known densities. Given a sample $x_1, \ldots, x_n$, examine whether or not a slice sampler can be constructed for the associated posterior distribution.

8.8 (Neal 2003) Consider the following hierarchical model ($i = 1, \ldots, 10$)

$$x_i | v \sim \mathcal{N}(0, e^v), \qquad v \sim \mathcal{N}(0, 3).$$

(a) When simulating this distribution from a independent Metropolis–Hastings algorithm with $\mathcal{N}(0, 1)$ proposals on all variables, show that the algorithm fails to recover the proper distribution for v by looking at the smaller values of v.

(b) Explain why this poor behavior occurs by considering the acceptance probability when v is smaller than -5.

(c) Evaluate the performance of the corresponding random Metropolis–Hastings algorithm which updates one component at a time with a Gaussian proposal centered at the current value and variance equal to one. Explain why the problem now occurs for the larger values of v.

(d) Compare with a single-variable slice sampling, that is, slice sampling applied to the eleven full conditional distributions of v given the x_i's and of the x_i's given the x_j's and v.

8.9 (Roberts and Rosenthal 1998) Show that, for the slice sampler [A.31], $(f_1(X^{(t)}))$ is a Markov chain with the same convergence properties as the original chain $(X^{(t)})$.

8.10 (Roberts and Rosenthal 1998) Using the notation of Section 8.3, show that, if f_1 and $\tilde{f}_1$ are two densities on two spaces $\mathcal{X}$ and $\tilde{\mathcal{X}}$ such that $\mu(w) = \tilde{\mu}(aw)$ for all w's, the kernels of the chains $(f_1(X^{(t)}))$ and $(\tilde{f}_1(\tilde{X}^{(t)}))$ are the same, even when the dimensions of $\mathcal{X}$ and $\tilde{\mathcal{X}}$ are different.

8.11 (Roberts and Rosenthal 1998) Consider the distribution on $\mathbb{R}^d$ with density proportional to

$$f_0(x) \, \mathbb{I}_{\{y \in \mathcal{X}; \, f_1(y) \geq w\}}(x).$$

Let T be a differentiable injective transformation, with Jacobian J.

(a) Show that sampling x from this density is equivalent to sampling $z = T(x)$ from

$$f_0^T(z) = f_0(T^{-1}(z))/J(T^{-1}(z)) \, \mathbb{I}_{T(\{y \in \mathcal{X}; \, f_1(y) \geq w\})}(x),$$

and deduce that $P_{f_0}(x, A) = P_{f_0^T}(T(x), T(A))$.

(b) Show that the transformation $T(x) = (T_1(x), x_2, \ldots, x_d)$, where $T_1(x) = \int_0^{x_1} f_0(t, x_2, \ldots, x_d) \, dt$, is such that $f_0(T^{-1}(z))/J(T^{-1}(z))$ is constant over the range of T.

8.12 (Roberts and Rosenthal 1998) In the proof of Lemma 8.7, when $f_1(x) \leq \epsilon^\star$:

(a) Show that the first equality follows from the definition of V.

(b) Show that the second equality follows from

$$\mu(w) = \int_w^\infty (-\mu'(z)) \, dz.$$

(c) Show that the first inequality follows from the fact that $z^{-\beta}$ is decreasing and the ratio can be expressed as an expectation.

(d) Show that, if g_1 and g_2 are two densities such that $g_1(x)/g_2(x)$ is increasing, and if h is also increasing, then $\int h(x)g_1(x)dx \geq \int h(x)g_2(x)dx$. Deduce the second inequality from this general result.

(e) Establish the third equality.

(f) Show that

$$\frac{z^{-\beta} - \epsilon^{\star-\beta}}{z^{-1/\alpha} - \epsilon^{\star-1/\alpha}}$$

is an increasing function of $z < \epsilon^{\star}$ when $\beta\alpha < 1$. Derive the last inequality.

8.13 (Roberts and Rosenthal 1998) Show that $\mu(y)'y^{1+1/\alpha}$ is non-increasing in the following two cases:

(a) $\mathcal{X} = \mathbb{R}^+$ and $f_1(x) \propto \exp{-\gamma x}$ for x large enough;

(b) $\mathcal{X} = \mathbb{R}^+$ and $f_1(x) \propto x^{-\delta}$ for x large enough.

8.14 (Roberts and Rosenthal 1998) In Lemma 8.7, show that, if $\epsilon^{\star} = 1$, then the bound b on $KV(x)$ when $x \in \mathscr{S}(\epsilon^{\star})$ simplifies into $b = \frac{\alpha\beta(1-\epsilon_{\star}^{\beta})}{(1+\alpha\beta)}$.

8.15 (Roberts and Rosenthal 1998) Show that, if the density f on $\mathbb{R}$ is log-concave, then $\omega\mu(\omega)'$ is non-increasing. (*Hint:* Show that, if $\mu^{-1}(\omega)$ is log-concave, then $\omega\mu(\omega)'$ is non-increasing.)

8.16 (Roberts and Rosenthal 1998) Show that, if the density f in $\mathbb{R}^d$ is such that

$$y\, D(y;\theta)^{d-1} \frac{\partial}{\partial y} D(y;\theta)$$

is non-increasing for all θ's and $D(y;\theta) = \sup\{t > 0; f(t\theta) \leq y\}$, then $\omega\mu(\omega)'$ is non-increasing.

8.17 Examine whether or not the density on $\mathbb{R}_+$ defined as $f(u) \propto \exp{-u^{1/d}}$ is log-concave. Show that $f(x) \propto \exp{-\|x\|}$ is log-concave in any dimension d.

8.18 In the general strategy proposed by Neal (2003) and discussed in Note 8.5.1, $\mathscr{I}_t$ can be shrunk during a rejection sampling scheme as follows: if ξ is rejected as not belonging to $\mathscr{A}_t$, change L_t to ξ if $\xi < x^{(t)}$ and R_t to ξ if $\xi > x^{(t)}$. Show that this scheme produces an acceptable interval $\mathscr{I}_t$ at the end of the rejection scheme.

8.19 In the stepping-out procedure described in Note 8.5.1, show that choosing ω too small may result in an non-irreducible Markov chain when $A^{(t)}$ is not connected. Show that the doubling procedure avoids this difficulty due to the random choice of the side of doubling.

8.20 Again referring to Note 8.5.1, when using the same scale ω, compare the expansion rate of the intervals in the stepping-out and the doubling procedures, and show that the doubling procedure is faster.

8.5 Notes

8.5.1 Dealing with Difficult Slices

After introducing the term "slice sampling" in Neal (1997), Neal (2003) proposed improvements to the standard slice sampling algorithm [A.31]. In particular, given the frequent difficulty in generating exactly a uniform $\mathcal{U}_{A^{(t+1)}}$, he suggests to implement this slice sampler in a univariate setting by replacing the "slice" $A^{(t)}$ with an interval $\mathscr{I}_t = (L_t, R_t)$ that contains most of the slice. He imposes the condition on $\mathscr{I}_t$ that the set $\mathscr{A}_t$ of x's in $A^{(t)} \cap \mathscr{I}_t$ where the probability of constructing $\mathscr{I}_t$ starting from x is the same as the probability of constructing $\mathscr{I}_t$ starting from $x^{(t-1)}$ can be constructed easily. This property is essential to ensure that the stationary distribution is the uniform distribution on the slice $A^{(t)}$. (We refer the reader to Neal

2003 for more details about the theoretical validity of his method, but warn him/her that utmost caution must be exercised in the construction of $\mathscr{I}_t$ for detailed balance to hold!) As noted in Neal (2003) and Mira and Roberts (2003), irreducibility of the procedure may be a problem (Problem 8.19).

Besides the obvious choices when $\mathscr{I}_t$ is $A^{(t)}$ and when it is the entire range of the (bounded) support of f, an original solution of Neal (2003), called the "stepping-out" procedure, is to create an interval of length ω containing $x^{(t-1)}$ and to expand this interval in steps of size ω till both ends are outside the slice $A^{(t)}$. Similarly, his "doubling procedure" consists in the same random starting interval of length ω whose length is doubled (leftwise or rightwise at random) recursively till both ends are outside the slice. While fairly general, this scheme also depends on a scale parameter ω that may be too large for some values of $x^{(t)}$ and too small for others.

Once an acceptable interval $\mathscr{I}_t$ has been produced, a point ξ is drawn at random on $\mathscr{I}_t$: if it belongs to $\mathscr{A}_t$, it is accepted. Otherwise, repeated sampling must be used, with the possible improvement on $\mathscr{I}_t$ gathered from previous failed attempts (see Problem 8.18).

The Two-Stage Gibbs Sampler

All that mattered was that slowly, by degrees, by left and right then left
and right again, he was guiding them towards the destination.
—Ian Rankin, *Hide and Seek*

The previous chapter presented the slice sampler, a special case of a Markov
chain algorithm that did not need an Accept–Reject step to be valid, seemingly
because of the uniformity of the target distribution. The reason why the slice
sampler works is, however, unrelated to this uniformity and we will see in this
chapter a much more general family of algorithms that function on the same
principle. This principle is that of using the true conditional distributions
associated with the target distribution to generate from that distribution.

In order to facilitate the link with Chapter 8, we focus in this chapter on
the two-stage Gibbs sampler before covering in Chapter 10 the more general
case of the multi-stage Gibbs sampler. There are several reasons for this dichotomy. First, the two-stage Gibbs sampler can be derived as a generalization
of the slice sampler. The two algorithms thus share superior convergence properties that do not apply to the general multistage Gibbs sampler. A second
reason is that the two-stage Gibbs sampler applies naturally in a wide range of
statistical models that do not call for the generality of Chapter 10. Lastly, the
developments surrounding the Gibbs sampler are so numerous that starting
with a self-contained algorithm like the two-stage Gibbs sampler provides a
gentler entry to the topic.

9.1 A General Class of Two-Stage Algorithms

9.1.1 From Slice Sampling to Gibbs Sampling

Instead of a density $f_X(x)$ as in Chapter 8, consider now a joint density $f(x, y)$
defined on an arbitrary product space, $\mathscr{X} \times \mathscr{Y}$. If we use the fundamental

theorem of simulation (Theorem 2.15) in this setup, we simulate a uniform distribution on the set

$$\mathcal{S}(f) = \{(x, y, u) : 0 \leq u \leq f(x, y)\} \, .$$

Since we now face a three-component setting, a natural implementation of the random walk principle is to move uniformly in one component at a time. This means that, starting at a point (x, y, u) in $\mathcal{S}(f)$, we generate

(i) X along the x-axis from the uniform distribution on $\{x : u \leq f(x, y)\}$,
(ii) Y along the y-axis from the uniform distribution on $\{y : u \leq f(x', y)\}$,
(iii) U along the u-axis from the uniform distribution on $[0, f(x', y')]$.

(Note that this is different from the 3D slice sampler of Example 8.3.)

There are two important things to note:

(1) Generating from the uniform distribution on $\{x : u \leq f(x, y)\}$ is equivalent to generating from the uniform distribution

$$\{x : f_{X|Y}(x|y) \geq u/f_Y(y)\},$$

where $f_{X|Y}$ and f_Y denote the conditional and marginal distributions of X given Y and of Y, respectively, that is,

$$(9.1) \qquad f_Y(y) = \int f(x, y) dx \quad \text{and} \quad f_{X|Y}(x|y) = \frac{f(x, y)}{f_Y(y)} \, .$$

(2) The sequence of uniform generations along the three axes does *not* need to be done in the same order x-y-u all the time for the Markov chain to remain stationary with stationary distribution the uniform on $\mathcal{S}(f)$.

Thus, for example, simulations along the x and the u axes can be repeated several times before moving to the simulation along the y axis. If we put these two remarks together and consider the limiting case where the X and the U simulations are repeated an *infinite* number of times before moving to the Y simulation, we end up with a simulation (in X) of

$$X \sim f_{X|Y}(x|y) \, ,$$

by virtue of the 2D slice sampler. Consider now the same repetition of Y and U simulations with X fixed at its latest value: in the limiting case, this produces a simulation (in Y) of

$$Y \sim f_{Y|X}(y|x) \, .$$

Assuming that both of these conditional distributions can be simulated, we can therefore implement the limiting case of the slice sampler and still maintain the stationarity of the uniform distribution. In addition, the simulation of the

U's gets somehow superfluous since we are really interested in the generation from $f(x, y)$, rather than from the uniform on $\mathscr{S}(f)$.

Although this is *not* the way it was originally derived, this introduction points out the strong link between the slice sampler and the two-stage Gibbs sampler, sometimes called *Data Augmentation* (Tanner and Wong 1987). It also stresses how the two-stage Gibbs sampler takes further advantage of the knowledge on the distribution f, compared with the slice sampler that uses only the numerical values of $f(x)$. This is reflected in the fact that each step of the two-stage Gibbs sampler amounts to an infinity of steps of a special slice sampler. (Note that this does not mean that the two-stage Gibbs sampler always does better than any slice sampler.)

9.1.2 Definition

The algorithmic implementation of the two-stage Gibbs sampler is thus straightforward. If the random variables X and Y have joint density $f(x, y)$, the two-stage Gibbs sampler generates a Markov chain (X_t, Y_t) according to the following steps:

Algorithm A.33 –Two-stage Gibbs sampler–

> Take $X_0 = x_0$
> For $t = 1, 2, \ldots$, generate
>
> 1. $Y_t \sim f_{Y|X}(\cdot|x_{t-1})$;   [A.33]
> 2. $X_t \sim f_{X|Y}(\cdot|y_t)$.

where $f_{Y|X}$ and $f_{X|Y}$ are the conditional distributions associated with f, as in (9.1).

We note here that not only is the sequence (X_t, Y_t) a Markov chain, but also each subsequence (X_t) and (Y_t) is a Markov chain. For example, the chain (X_t) has transition density

$$K(x, x^*) = \int f_{Y|X}(y|x) f_{X|Y}(x^*|y) dy,$$

which indeed depends on the past only through the last value of (X_t). (Note the similarity to Eaton's transition (6.46).) In addition, it is also easy to show that f_X is the stationary distribution associated with this (sub)chain, since

$$
\begin{aligned}
f_X(x') &= \int f_{X|Y}(x'|y) f_Y(y) dy \\
&= \int f_{X|Y}(x'|y) \int f_{Y|X}(y|x) f_X(x) dx \, dy \\
&= \int \left[\int f_{X|Y}(x'|y) f_{Y|X}(y|x) dy \right] f_X(x) dx \ . \\
&= \int K(x, x') f_X(x) dx \ .
\end{aligned}
$$

(9.2)

Example 9.1. Normal bivariate Gibbs. For the special case of the bivariate normal density,

$$(9.3) \qquad (X, Y) \sim \mathcal{N}_2\left(0, \begin{pmatrix} 1 & \rho \\ \rho & 1 \end{pmatrix}\right),$$

the Gibbs sampler is
`Given` y_t`, generate`

$$(9.4) \qquad \begin{aligned} X_{t+1} \mid y_t &\sim \mathcal{N}(\rho y_t,\ 1 - \rho^2), \\ Y_{t+1} \mid x_{t+1} &\sim \mathcal{N}(\rho x_{t+1},\ 1 - \rho^2). \end{aligned}$$

The Gibbs sampler is obviously not necessary in this particular case, as iid copies of (X, Y) can be easily generated using the Box–Muller algorithm (see Example 2.8). Note that the corresponding marginal Markov chain in X is defined by the AR(1) relation

$$X_{t+1} = \rho^2 X_t + \sigma \epsilon_t, \qquad \epsilon_t \sim \mathcal{N}(0, 1),$$

with $\sigma^2 = 1 - \rho^2 + \rho^2(1 - \rho^2) = 1 - \rho^4$. As shown in Example 6.43, the stationary distribution of this chain is indeed the normal distribution $\mathcal{N}\left(0, \frac{1-\rho^4}{1-\rho^4}\right)$.   ‖

This motivation of the two-stage Gibbs sampler started with a joint distribution $f(x, y)$. However, what we saw in Chapter 8 was exactly the opposite justification. There, we started with a marginal density $f_X(x)$ and constructed (or *completed*) a joint density to aid in simulation where the second variable Y (that is, U in slice sampling) is an *auxiliary variable* that is not directly relevant from the statistical point of view. This connection will be detailed in Section 10.1.2 for the general Gibbs sampler, but we can point out at this stage that there are many settings where a natural completion of $f_X(x)$ into $f(x, y)$ does exist. One such setting is the domain of *missing data models*, introduced in Section 5.3.1,

$$f(x|\theta) = \int_{\mathscr{Z}} g(x, z|\theta)\, dz,$$

as, for instance, the mixtures of distributions (Examples 1.2 and 5.19).

Data Augmentation was introduced independently (that is, unrelated to Gibbs sampling) by Tanner and Wong (1987), and is, perhaps, more closely related to the EM algorithm of Dempster et al. (1977) (Section 5.3.2) and methods of stochastic restoration (see Note 5.5.1). It is even more related to recent versions of EM such as ECM and MCEM (see Meng and Rubin 1991, 1992 and Rubin, Liu and Rubin 1994, and Problem 9.11).

Example 9.2. Gibbs sampling on mixture posterior. Consider a *mixture of distributions*

$$(9.5) \qquad\qquad \sum_{j=1}^{k} p_j\, f(x|\xi_j)\,,$$

where $f(\cdot|\xi)$ belongs to an exponential family

$$f(x|\xi) = h(x)\, \exp\{\xi \cdot x - \psi(\xi)\}$$

and ξ is distributed from the associated conjugate prior

$$\pi(\xi|\alpha_0, \lambda) \propto \exp\{\lambda(\xi \cdot \alpha_0 - \psi(\xi))\}\,, \qquad \lambda > 0, \quad \alpha_0 \in \mathcal{X}\,,$$

while

$$(p_1, \ldots, p_k) \sim \mathcal{D}_k(\gamma_1, \ldots, \gamma_k)\,.$$

Given a sample $(x_1, \ldots, x_n)$ from (9.5), we can associate with every observation an indicator variable $z_i \in \{1, \ldots, k\}$ that indicates which component of the mixture is associated with x_i (see Problems 5.8–5.10). The demarginalization (or *completion*) of model (9.5) is then

$$Z_i \sim \mathcal{M}_k(1; p_1, \ldots, p_k), \qquad x_i|z_i \sim f(x|\xi_{z_i})\,.$$

Thus, considering $x_i^* = (x_i, z_i)$ (instead of x_i) entirely eliminates the mixture structure since the likelihood of the completed model is

$$\ell(p, \xi|x_i^*, \ldots, x_i^*) \propto \prod_{i=1}^{n} p_{z_i}\, f(x_i|\xi_{z_i})$$

$$= \prod_{j=1}^{k} \prod_{i; z_i=j} p_j\, f(x_i|\xi_j)\,.$$

(This latent structure is also exploited in the original implementation of the EM algorithm; see Section 5.3.2.) The two steps of the Gibbs sampler are

Algorithm A.34 –Mixture Posterior Simulation–

1. Simulate $Z_i\ (i = 1, \ldots, n)$ from

$$P(Z_i = j) \propto p_j\, f(x_i|\xi_j) \qquad (j = 1, \ldots, k)$$

and compute the statistics   [A.34]

$$n_j = \sum_{i=1}^{n} \mathbb{I}_{z_i=j}\,, \qquad n_j \bar{x}_j = \sum_{i=1}^{n} \mathbb{I}_{z_i=j} x_i\,.$$

2. Generate $(j = 1, \ldots, k)$

$$\xi \sim \pi\left(\xi \left| \frac{\lambda_j \alpha_j + n_j \bar{x}_j}{\lambda_j + n_j}, \lambda_j + n_j \right.\right),$$

$$p \sim \mathcal{D}_k(\gamma_1 + n_1, \ldots, \gamma_k + n_k)\,.$$

As an illustration, consider the same setting as Example 5.19, namely a normal mixture with two components with equal known variance and fixed weights,

$$(9.6) \qquad p\mathcal{N}(\mu_1, \sigma^2) + (1 - p)\mathcal{N}(\mu_2, \sigma^2).$$

We assume in addition a normal $\mathcal{N}(0, 10\sigma^2)$ prior distribution on both means μ_1 and μ_2. Generating directly from the posterior associated with a sample $\mathbf{x} = (x_1, \ldots, x_n)$ from (9.6) quickly turns impossible, as discussed for instance in Diebolt and Robert (1994) and Celeux et al. (2000), because of a combinatoric explosion in the number of calculations, which grow as $\mathcal{O}(2^n)$.

As for the EM algorithm (Problems 5.8–5.10), a natural completion of (μ_1, μ_2) is to introduce the (unobserved) component indicators z_i of the observations x_i, namely,

$$P(Z_i = 1) = 1 - P(Z_i = 2) = p \qquad \text{and} \qquad X_i|Z_i = k \sim \mathcal{N}(\mu_k, \sigma^2).$$

The completed distribution is thus

$$\pi(\mu_1, \mu_2, \mathbf{z}|\mathbf{x}) \propto \exp\{-(\mu_1^2 + \mu_2^2)/20\sigma^2\} \prod_{z_i=1} p \exp\{-(x_i - \mu_1)^2/2\sigma^2\} \times$$

$$\prod_{z_i=2} (1 - p) \exp\{-(x_i - \mu_2)^2/2\sigma^2\}.$$

Since μ_1 and μ_2 are independent, given $(\mathbf{z}, \mathbf{x})$, with distributions $(j = 1, 2)$, the conditional distributions are

$$\mathcal{N}\left(\sum_{z_i=j} x_i/(.1 + n_j), \sigma^2/(.1 + n_j)\right),$$

where n_j denotes the number of z_i's equal to j. Similarly, the conditional distribution of $\mathbf{z}$ given (μ_1, μ_2) is a product of binomials, with

$$P(Z_i = 1|x_i, \mu_1, \mu_2)$$

$$= \frac{p \exp\{-(x_i - \mu_1)^2/2\sigma^2\}}{p \exp\{-(x_i - \mu_1)^2/2\sigma^2\} + (1 - p)\exp\{-(x_i - \mu_2)^2/2\sigma^2\}}.$$

Figure 9.1 illustrates the behavior of the Gibbs sampler in that setting, with a simulated dataset of 500 points from the $.7\mathcal{N}(0, 1) + .3\mathcal{N}(2.7, 1)$ distribution. The representation of the MCMC sample after 15,000 iterations is quite in agreement with the posterior surface, represented via a grid on the (μ_1, μ_2) space and some contours; while it may appear to be too concentrated around one mode, the second mode represented on this graph is much lower since there is a difference of at least 50 in log-posterior values. However, the Gibbs sampler may also fail to converge, as described in Diebolt and Robert (1994) and illustrated in Figure 12.16. When initialized at a local mode of the likelihood, the magnitude of the moves around this mode may be too limited to allow for exploration of further modes (in a reasonable number of iterations).   ‖

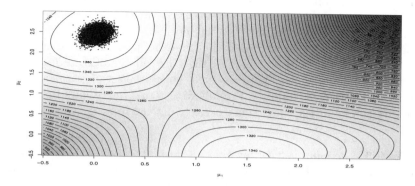

Fig. 9.1. Gibbs sample of $5,000$ points for the mixture posterior against the posterior surface.

9.1.3 Back to the Slice Sampler

Just as the two-stage Gibbs sampler can be seen as a limiting case of the slice sampler in a three-coordinate problem, the slice sampler can be interpreted as a special case of two-stage Gibbs sampler when the joint distribution is the uniform distribution on the subgraph $\mathscr{S}(f)$. From this point of view, the slice sampler starts with $f_X(x)$ and creates a joint density $f(x, u) = \mathbb{I}(0 < u < f_X(x))$. The associated conditional densities are

$$f_{X|U}(x|u) = \frac{\mathbb{I}(0 < u < f_X(x))}{\int \mathbb{I}(0 < u < f_X(x))dx} \quad \text{and} \quad f_{U|X}(u|x) = \frac{\mathbb{I}(0 < u < f_X(x))}{\int \mathbb{I}(0 < u < f_X(x))du},$$

which are exactly those used in the slice sampler. Therefore, the X sequence is also a Markov chain with transition kernel

$$K(x, x') = \int f_{X|U}(x'|u)f_{U|X}(u|x)du$$

and stationary density $f_X(x)$.

What the slice sampler tells us is that we can induce a Gibbs sampler for any marginal distribution $f_X(x)$ by creating a joint distribution that is, formally, arbitrary. Starting from $f_X(x)$, we can take any conditional density $g(y|x)$ and create a Gibbs sampler with

$$f_{X|Y}(x|y) = \frac{g(y|x)f_X(x)}{\int g(y|x)f_X(x)dx} \quad \text{and} \quad f_{Y|X}(y|x) = \frac{g(y|x)f_X(x)}{\int g(y|x)f_X(x)dy}.$$

9.1.4 The Hammersley–Clifford Theorem

A most surprising feature of the Gibbs sampler is that the conditional distributions contain sufficient information to produce a sample from the joint distribution. (This is the case for both two-stage and multi-stage Gibbs; see also

Section 10.1.3.) By comparison with maximization problems, this approach is akin to maximizing an objective function successively in every direction of a given basis. It is well known that this optimization method does not necessarily lead to the global maximum, but may end up in a saddlepoint.

It is, therefore, somewhat remarkable that the full conditional distributions perfectly summarize the joint density, although the set of marginal distributions obviously fails to do so. The following result then shows that the joint density can be directly and constructively derived from the conditional densities.

Theorem 9.3. *The joint distribution associated with the conditional densities* $f_{Y|X}(y|x)$ *and* $f_{X|Y}(x|y)$ *has the joint density*

$$f(x,y) = \frac{f_{Y|X}(y|x)}{\int \left[f_{Y|X}(y|x)/f_{X|Y}(x|y) \right] dy}.$$

Proof. Since $f(x,y) = f_{Y|X}(y|x)f_X(x) = f_{X|Y}(x|y)f_Y(y)$ we have

$$(9.7) \qquad \int \frac{f_{Y|X}(y|x)}{f_{X|Y}(x|y)} dy = \int \frac{f_Y(y)}{f_X(x)} dy = \frac{1}{f_X(x)},$$

and the result follows.   □

This derivation of $f(x,y)$ obviously requires the existence and computation of the integral (9.7). However, this result clearly demonstrates the fundamental feature that the two conditionals are sufficiently informative to recover the joint density. Note, also, that this theorem makes the implicit assumption that the joint density $f(x,y)$ exists. (See Section 10.4.3 for a discussion of what happens when this assumption is not satisfied.)

9.2 Fundamental Properties

A particularly nice feature of the two-stage Gibbs sampler is that this algorithm lends itself to componentwise study, because the associated sequences $(X^{(t)})$ and $(Y^{(t)})$ are Markov chains. This decomposition into two Markov chains enables us to more thoroughly evaluate the properties of the two-stage Gibbs sampler.

9.2.1 Probabilistic Structures

We have already seen in Section 9.1 that the individual subchains are both Markov chains. We next state their formal properties.

A sufficient condition for irreducibility of the Gibbs Markov chain is the following condition, introduced by Besag (1974) (see also Section 10.1.3). We state it in full generality as it also applies to the general case of Chapter 10.

Definition 9.4. Let $(Y_1, Y_2, \ldots, Y_p) \sim g(y_1, \ldots, y_p)$, where $g^{(i)}$ denotes the marginal distribution of Y_i. If $g^{(i)}(y_i) > 0$ for every $i = 1, \ldots, p$, implies that $g(y_1, \ldots, y_p) > 0$, then g satisfies the *positivity condition*.

Thus, the support of g is the Cartesian product of the supports of the $g^{(i)}$'s. Moreover, it follows that the conditional distributions will not reduce the range of possible values of Y_i when compared with g. In this case, two arbitrary Borel subsets of the support can be joined in a single iteration of [A.33]. (Recall that *strong irreducibility* is introduced in Definition 6.13.)

Lemma 9.5. *Each of the sequences $(X^{(t)})$ and $(Y^{(t)})$ produced by [A.33] is a Markov chain with corresponding stationary distributions*

$$f_X(x) = \int f(x, y)\, dy \quad and \quad f_Y(y) = \int f(x, y)\, dx \,.$$

If the positivity constraint on f holds, then both chains are strongly irreducible.

Proof. The development of (9.2) shows that each chain is, individually, a Markov chain. Under the positivity constraint, if f is positive, $f_{X|Y}(x|y)$ is positive on the (projected) support of f and every Borel set of $\mathcal{X}$ can be visited in a single iteration of [A.33], establishing the strong irreducibility. This development also applies to $(Y^{(t)})$. □

This elementary reasoning shows, in addition, that if only the chain $(X^{(t)})$ is of interest and if the condition $f_{X|Y}(x|y) > 0$ holds for every pair (X', Y), irreducibility is satisfied. As shown further in Section 9.2.3, the "dual" chain $(Y^{(t)})$ can be used to establish some probabilistic properties of $(X^{(t)})$.

Convergence of the two-stage Gibbs sampler will follow as a special case of the general Gibbs sampler introduced in Chapter 10. However, for completeness we state the following convergence result here, whose proof follows the same lines as Lemma 7.3 and Theorem 7.4.

Theorem 9.6. *Under the positivity condition, if the transition kernel*

$$K((x, y), (x', y')) = f_{X|Y}(x'|y) f_{Y|X}(y'|x')$$

is absolutely continuous with respect to the dominating measure, the chain $(X^{(t)}, Y^{(t)})$ is Harris recurrent and ergodic with stationary distribution f.

Theorem 9.6 implies convergence for most two-stage Gibbs samplers since the kernel $K(x, y)$ will be absolutely continuous in most setups, and there will be no point mass to worry about. In addition, the specific features of the two-stage Gibbs sampler induces special convergence properties (interleaving, duality) described in the following sections.

Typical illustrations of the two-stage Gibbs sampler are in missing variable models, where one chain usually is on a finite state space.

Example 9.7. Grouped counting data. For 360 consecutive time units, consider recording the number of passages of individuals, per unit time, past some sensor. This can be, for instance, the number of cars observed at a crossroad or the number of leucocytes in a region of a blood sample. Hypothetical results are given in Table 9.1. This table, therefore, involves a grouping of the

Number of passages	0	1	2	3	4 or more
Number of observations	139	128	55	25	13

Table 9.1. Frequencies of passage for 360 consecutive observations.

observations with four passages and more. If we assume that every observation is a Poisson $\mathcal{P}(\lambda)$ random variable, the likelihood of the model corresponding to Table 9.1 is

$$\ell(\lambda|x_1,\ldots,x_5) \propto e^{-347\lambda}\lambda^{128+55\times2+25\times3}\left(1-e^{-\lambda}\sum_{i=0}^{3}\frac{\lambda^i}{i!}\right)^{13},$$

for $x_1 = 139,\ldots,x_5 = 13$.

For $\pi(\lambda) = 1/\lambda$ and $\mathbf{y} = (y_1,\ldots,y_{13})$, vector of the 13 units larger than 4, it is possible to complete the posterior distribution $\pi(\lambda|x_1,\ldots,x_5)$ in $\pi(\lambda, y_1,\ldots,y_{13}|x_1,\ldots,x_5)$ and to propose the Gibbs sampling algorithm associated with the two components λ and $\mathbf{y}$,

Algorithm A.35 –Poisson–Gamma Gibbs Sampler–

Given $\lambda^{(t-1)}$,

1. Simulate $Y_i^{(t)} \sim \mathcal{P}(\lambda^{(t-1)})\,\mathbb{I}_{y\geq4}$ $(i = 1,\ldots,13)$
2. Simulate   [A.35]

$$\lambda^{(t)} \sim \mathcal{G}a\left(313 + \sum_{i=1}^{13} y_i^{(t)}, 360\right).$$

Figure 9.2 describes the convergence of the Rao–Blackwellized estimator (see Section 4.2 for motivation and Sections 9.3 and 10.4.2 for justification)

$$\delta_{rb} = \frac{1}{T}\sum_{t=1}^{T}\mathbb{E}\left[\lambda|x_1,\ldots,x_5,y_1^{(t)},\ldots,y_{13}^{(t)}\right]$$

$$= \frac{1}{360T}\sum_{t=1}^{T}\left(313 + \sum_{i=1}^{13} y_i^{(t)}\right),$$

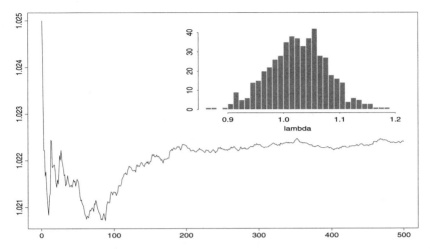

Fig. 9.2. Evolution of the estimator δ_{rb} against the number of iterations of $[A.35]$ and (insert) histogram of the sample of $\lambda^{(t)}$'s for 500 iterations.

to $\delta^{\pi} = 1.0224$, the Bayes estimator of λ, along with an histogram of the $\lambda^{(t)}$'s simulated from $[A.35]$. The convergence to the Bayes estimator is particularly fast in this case, a fact related to the feature that δ_{rb} only depends on the finite state-space chain $(y^{(t)})$,   ‖

Example 9.8. Grouped multinomial data. Tanner and Wong (1987) consider the multinomial model

$$X \sim \mathcal{M}_5\left(n; a_1\mu + b_1, a_2\mu + b_2, a_3\eta + b_3, a_4\eta + b_4, c(1 - \mu - \eta)\right),$$

with

$$0 \le a_1 + a_2 = a_3 + a_4 = 1 - \sum_{i=1}^{4} b_i = c \le 1,$$

where the $a_i, b_i \ge 0$ are known, based on genetic considerations. This model is equivalent to a sampling from

$$Y \sim \mathcal{M}_9\left(n; a_1\mu, b_1, a_2\mu, b_2, a_3\eta, b_3, a_4\eta, b_4, c(1 - \mu - \eta)\right),$$

where some observations are aggregated (and thus missing),

$$X_1 = Y_1 + Y_2, \ X_2 = Y_3 + Y_4, \ X_3 = Y_5 + Y_6, \ X_4 = Y_7 + Y_8, \ X_5 = Y_9.$$

A natural prior distribution on (μ, η) is the Dirichlet prior $\mathcal{D}(\alpha_1, \alpha_2, \alpha_3)$,

$$\pi(\mu, \eta) \propto \mu^{\alpha_1 - 1}\eta^{\alpha_2 - 1}(1 - \eta - \mu)^{\alpha_3 - 1},$$

where $\alpha_1 = \alpha_2 = \alpha_3 = 1/2$ corresponds to the noninformative case. The posterior distributions $\pi(\mu|x)$ and $\pi(\eta|x)$ are not readily available, as shown in Problem 9.10. If we define $\mathbf{Z} = (Z_1, Z_2, Z_3, Z_4) = (Y_1, Y_3, Y_5, Y_7)$, the completed posterior distribution can be defined as

$$
\pi(\eta, \mu | x, \mathbf{z}) = \pi(\eta, \mu | y)
$$
$$
\propto \mu^{\alpha_1 - 1} \eta^{\alpha_2 - 1} (1 - \eta - \mu)^{\alpha_3 - 1} \mu^{z_1} \mu^{z_2} \eta^{z_3} \eta^{z_4} (1 - \eta - \mu)^{x_5}
$$
$$
= \mu^{z_1 + z_2 + \alpha_1 - 1} \eta^{z_3 + z_4 + \alpha_2 - 1} (1 - \eta - \mu)^{x_5 + \alpha_3 - 1}.
$$

Thus,

$$
(\mu, \eta, 1 - \mu - \eta) | x, \mathbf{z} \sim \mathcal{D}(z_1 + z_2 + \alpha_1, z_3 + z_4 + \alpha_2, x_5 + \alpha_3).
$$

Moreover,

$$
Z_i | x, \mu, \eta \sim \mathcal{B}\left(x_i, \frac{a_i \mu}{a_i \mu + b_i}\right) \qquad (i = 1, 2),
$$
$$
Z_i | x, \mu, \eta \sim \mathcal{B}\left(x_i, \frac{a_i \eta}{a_i \eta + b_i}\right) \qquad (i = 3, 4).
$$

Therefore, if $\theta = (\mu, \eta)$, this completion provides manageable conditional distributions $g_1(\theta|y)$ and $g_2(\mathbf{z}|x, \theta)$. ‖

Example 9.9. Capture–recapture uniform model. In the setup of *capture–recapture models* seen in Examples 2.25 and 5.22, the simplest case corresponds to the setting where the size N of the entire population is unknown and each individual has a probability p of being captured in every capture experiment, whatever its past history. For two successive captures, a sufficient statistic is the triplet (n_{11}, n_{10}, n_{01}), where n_{11} is the number of individuals which have been captured twice, n_{10} the number of individuals captured only in the first experiment, and n_{01} the number of individuals captured only in the second experiment. Writing $n = n_{10} + n_{01} + 2n_{11}$, the total number of captures, the likelihood of this model is given by

$$
\ell(N, p | n_{11}, n_{10}, n_{01}) \propto \binom{N}{n_{11}\ n_{10}\ n_{01}} p^n (1 - p)^{N-n}.
$$

Castledine (1981) calls this a *uniform* likelihood (see also Wolter 1986 or George and Robert 1992).

It is easy to see that the likelihood function factors through n and $n' = n_{10} + n_{11} + n_{01}$, the total number of different individuals captured in the experiments. If $\pi(p, N)$ corresponds to a Poisson distribution $\mathcal{P}(\lambda)$ on N and a uniform distribution $\mathcal{U}_{[0,1]}$ on p,

$$
\pi(p, N | n, n') \propto \frac{N!}{(N - n')!} p^n (1 - p)^{N-n} \frac{e^{-\lambda} \lambda^N}{N!}.
$$

implies that

$$(N - n')|p, n, n' \sim \mathcal{P}(\lambda) \,,$$
$$p|N, n, n' \sim \mathcal{B}e(n + 1, N - n + 1) \,,$$

and therefore that the two-stage Gibbs sampling is available in this setup (even though direct computation is possible; see Problems 9.12 and 9.13). ‖

9.2.2 Reversible and Interleaving Chains

The two-stage Gibbs sampler was shown by Liu et al. (1994) to have a very strong structural property. The two (sub)chains, called, say, the *chain of interest*, $(X^{(t)})$, and the *instrumental* (or *dual*) *chain*, $(Y^{(t)})$, satisfy a duality property they call *interleaving*. This property is mostly characteristic of two-stage Gibbs samplers.

Definition 9.10. Two Markov chains $(X^{(t)})$ and $(Y^{(t)})$ are said to be *conjugate to each other with the interleaving property* (or *interleaved*) if

(i) $X^{(t)}$ and $X^{(t+1)}$ are independent conditionally on $Y^{(t)}$;
(ii) $Y^{(t-1)}$ and $Y^{(t)}$ are independent conditionally on $X^{(t)}$; and
(iii) $(X^{(t)}, Y^{(t-1)})$ and $(X^{(t)}, Y^{(t)})$ are identically distributed under stationarity.

In most cases where we have interleaved chains, there is more interest in one of the chains. That the property of interleaving is always satisfied by two-stage Gibbs samplers is immediate, as shown below. Note that the (global) chain $(X^{(t)}, Y^{(t)})$ is not necessarily (time-)reversible.

Lemma 9.11. *Each of the chains $(X^{(t)})$ and $(Y^{(t)})$ generated by a two-stage Gibbs sampling algorithm is reversible, and the chain $(X^{(t)}, Y^{(t)})$ satisfies the interleaving property.*

Proof. We first establish reversibility for each chain, a property that is independent of interleaving: it follows from the detailed balance condition (Theorem 6.46). Consider (X_0, Y_0) distributed from the stationary distribution g, with respective marginal distributions $g^{(2)}(x)$ and $g^{(1)}(y)$. Then, if

$$K_1(x_0, x_1) = \int g_2(y_0|x_0) g_1(x_1|y_0) dy_0$$

denotes the transition kernel of $(X^{(t)})$,

$$g^{(2)}(x_0) \, K_1(x_0, x_1) = g^{(2)}(x_0) \int g_2(y_0|x_0) \, g_1(x_1|y_0) \, dy_0$$

$$= \int g(x_0, y_0) \, g_1(x_1|y_0) \, dy_0$$

$$[g^{(2)}(x_0) = \int g(x_0, y_0) dy_0]$$

$$= \int g(x_0, y_0) \, \frac{g_2(y_0|x_1) \, g^{(2)}(x_1)}{g^{(1)}(y_0)} \, dy_0$$

$$\left[g_1(x_1|y_0) = \frac{g_2(y_0|x_1) g^{(2)}(x_1)}{g^{(1)}(y_0)} \right]$$

$$= g^{(2)}(x_1) \, K_1(x_1, x_0),$$

where the last equality follows by integration. Thus, the reversibility of the $(X^{(t)})$ chain is established, and a similar argument applies for the reversibility of the $(Y^{(t)})$ chain; that is, if K_2 denotes the associated kernel, the detailed balance condition

$$g^{(1)}(y_0) \, K_2(y_0, y_1) = g^{(1)}(y_1) \, K_2(y_1, y_0)$$

is satisfied.

Turning to the interleaving property of Definition 9.10, the construction of each chain establishes (i) and (ii) directly. To see that property (iii) is satisfied, we note that the joint cdf of X_0 and Y_0 is

$$P(X_0 \le x, Y_0 \le y) = \int_{-\infty}^{x} \int_{-\infty}^{y} g(v, u) du dv$$

and the joint cdf of X_1 and Y_0 is

$$P(X_1 \le x, Y_0 \le y) = \int_{-\infty}^{x} \int_{-\infty}^{y} g_1(v|u) g^{(1)}(u) du dv.$$

Since

$$g_1(v|u) g^{(1)}(u) = \int g_1(v|u) g(v', u) dv'$$

$$= \int g_1(v|u) g_1(v'|u) g^{(1)}(u) dv'$$

$$= g(v, u),$$

the result follows, and $(X^{(t)})$ and $(Y^{(t)})$ are interleaved chains. □

To ensure that the entire chain $(X^{(t)}, Y^{(t)})$ is reversible, an additional step is necessary in [A.33].

Algorithm A.36 –Reversible Two-Stage Gibbs–

Given $y_2^{(t)}$,

1. Simulate $W \sim g_1(w|y_2^{(t)})$;
2. Simulate $Y_2^{(t+1)} \sim g_2(y_2|w)$;   [A.36]
3. Simulate $Y_1^{(t+1)} \sim g_1(y_1|y_2^{(t+1)})$.

In this case,

$$
\begin{aligned}
(Y_1, Y_2, Y_1', Y_2') &\sim g(y_1, y_2) \int g_1(w|y_2)\, g_2(y_2'|w)\, dw\, g_1(y_1'|y_2') \\
&= g(y_1', y_2') \int \frac{g_2(y_2'|w)}{g^1(y_2')}\, g(w, y_2)\, dw\, g_1(y_1|y_2) \\
&= g(y_1', y_2') \int g_1(w|y_2')\, g_2(y_2|w)\, dw\, g_1(y_1|y_2)
\end{aligned}
$$

is distributed as (Y_1', Y_2', Y_1, Y_2).

9.2.3 The Duality Principle

The well-known concept of Rao–Blackwellization (Sections 4.2, 7.6.2, and to come in Section 9.3) is based on the fact that a conditioning argument, using variables other than those directly of interest, can result in an improved procedure. In the case of interleaving Markov chains, this phenomenon goes deeper. We will see in this section that this phenomenon is more fundamental than a mere improvement of the variance of some estimators, as it provides a general technique to establish convergence properties for the chain of interest $(X^{(t)})$ based on the instrumental chain $(Y^{(t)})$, even when the latter is unrelated with the inferential problem. Diebolt and Robert (1993, 1994) have called this use of the dual chain the *Duality Principle* when they used it in the setup of mixtures of distributions. While it has been introduced in a two-stage Gibbs sampling setup, this principle extends to other Gibbs sampling methods since it is sufficient that the chain of interest $(X^{(t)})$ be generated conditionally on another chain $(Y^{(t)})$, which supplies the probabilistic properties.

Theorem 9.12. *Consider a Markov chain $(X^{(t)})$ and a sequence $(Y^{(t)})$ of random variables generated from the conditional distributions*

$$
X^{(t)}|y^{(t)} \sim \pi(x|y^{(t)})\,, \qquad Y^{(t+1)}|x^{(t)}, y^{(t)} \sim f(y|x^{(t)}, y^{(t)})\,.
$$

If the chain $(Y^{(t)})$ is ergodic (geometrically or uniformly ergodic) and if $\pi_{y^{(0)}}^t$ denotes the distribution of $(X^{(t)})$ associated with the initial value $y^{(0)}$, the norm $\|\pi_{y^{(0)}}^t - \pi\|_{TV}$ goes to 0 when t goes to infinity (goes to 0 at a geometric or uniformly bounded rate).

Proof. The transition kernel is

$$K(y, y') = \int_{\mathcal{X}} \pi(x|y) \, f(y'|x, y) \, dx.$$

If $f_{y^{(0)}}^t(y)$ denotes the marginal density of $Y^{(t)}$, then the marginal of $X^{(t)}$ is

$$\pi_{y^{(0)}}^t(x) = \int_{\mathcal{Y}} \pi(x|y) \, f_{y^{(0)}}^t(y) \, dy$$

and convergence in the $(X^{(t)})$ chain can be tied to convergence in the $(Y^{(t)})$ chain by the following. The total variation in the $(X^{(t)})$ chain is

$$
\begin{aligned}
\|\pi_{y^{(0)}}^t - \pi\|_{TV} &= \frac{1}{2} \int_{\mathcal{Y}} \left| \pi_{y^{(0)}}^t(x) - \pi(x) \right| \, dx \\
&= \frac{1}{2} \int_{\mathcal{X}} \left| \int_{\mathcal{Y}} \pi(x|y) \left(f_{y^{(0)}}^t(y) - f(y) \right) \, dy \right| \, dx \\
&\leq \frac{1}{2} \int_{\mathcal{X} \times \mathcal{Y}} \left| f_{y^{(0)}}^t(y) - f(y) \right| \pi(x|y) \, dx dy \\
&= \|f_{y^{(0)}}^t - f\|_{TV},
\end{aligned}
$$

so the convergence properties of $\|f_{y^{(0)}}^t - f\|_{TV}$ can be transfered to $\|\pi_{y^{(0)}}^t - \pi\|_{TV}$. Both sequences have in the same speed of convergence since

$$\|f_{y^{(0)}}^{t+1} - f\|_{TV} \leq \|\pi_{y^{(0)}}^t - \pi\|_{TV} \leq \|f_{y^{(0)}}^t - f\|_{TV}.$$

$\square$

Note that this setting contains as a particular case *hidden Markov models*, where $Y^{(t+1)}|x^{(t)}, y^{(t)} \sim f(y|y^{(t)})$ is not observed. These models will be detailed in Section 14.3.2.

The duality principle is even easier to state, and stronger, when $(Y^{(t)})$ is a finite state-space Markov chain.

Theorem 9.13. *If $(Y^{(t)})$ is a finite state-space Markov chain, with state-space $\mathcal{Y}$, such that*

$$P(Y^{(t+1)} = k|y^{(t)}, x) > 0, \qquad \forall \, k \in \mathcal{Y}, \forall x \in \mathcal{X},$$

the sequence $(X^{(t)})$ derived from $(Y^{(t)})$ by the transition $\pi(x|y^{(t)})$ is uniformly ergodic.

Notice that this convergence result does not impose any constraint on the transition $\pi(x|y)$, which, for instance, is not necessarily everywhere positive.

Proof. The result follows from Theorem 9.12. First, write

$$p_{ij} = \int P(Y^{(t+1)} = j | x, y^{(t)} = i) \, \pi(x | y^{(t)} = i) \, dx, \qquad i, j \in \mathcal{Y}.$$

If we define the lower bound $\rho = \min_{i,j \in \mathcal{Y}} p_{ij} > 0$, then

$$P(Y^{(t)} = i | y^{(0)} = i) \geq \rho \sum_{k \in \mathcal{Y}} P(Y^{(t-1)} = k | y^{(0)} = i) = \rho \,,$$

and hence

$$\sum_{t=1}^{\infty} P(Y^{(t)} = i | y^{(0)} = i) = \infty$$

for every state i of $\mathcal{Y}$. Thus, the chain is positive recurrent with limiting distribution

$$q_i = \lim_{t \to \infty} P(Y^{(t)} = i | y^{(0)} = i).$$

Uniform ergodicity follows from the finiteness of $\mathcal{Y}$ (see, e.g., Billingsley 1968)
□

The statement of Theorems 9.12 and 9.13 is complicated by the fact that $(X^{(t)})$ is not necessarily a Markov chain and, therefore, the notions of ergodicity and of geometric ergodicity do not apply to this sequence. However, there still exists a limiting distribution π for $(X^{(t)})$, obtained by transformation of the limiting distribution of $(Y^{(t)})$ by the transition $\pi(x|y)$. The particular case of two-stage Gibbs sampling allows for a less complicated formulation.

Corollary 9.14. *For two interleaved Markov chains, $(X^{(t)})$ and $(Y^{(t)})$, if $(X^{(t)})$ is ergodic (geometrically ergodic), then $(Y^{(t)})$ is ergodic (geometrically ergodic).*

Example 9.15. (Continuation of Example 9.8) The vector $(z_1, \ldots, z_4)$ takes its values in a finite space of size $(x_1 + 1) \times (x_2 + 1) \times (x_3 + 1) \times (x_4 + 1)$ and the transition is strictly positive. Corollary 9.14 therefore implies that the chain $(\eta^{(t)}, \mu^{(t)})$ is uniformly ergodic. ‖

Finally, we turn to the question of rate of convergence, and find that the duality principle still applies, and the rates transfer between chains.

Proposition 9.16. *If $(Y^{(t)})$ is geometrically convergent with compact state-space and with convergence rate ρ, there exists C_h such that*

(9.8)
$$\left\| \mathbb{E}[h(X^{(t)}) | y^{(0)}] - \mathbb{E}^\pi[h(X)] \right\| < C_h \, \rho^t \,,$$

for every function $h \in \mathcal{L}_1(\pi(\cdot | x))$ uniformly in $y^{(0)} \in \mathcal{Y}$.

Proof. For $h(x) = (h_1(x), \ldots, h_d(x))$,

$$\left\| \mathbb{E}[h(X^{(t)})|y^{(0)}] - \mathbb{E}^\pi[h(X)] \right\|^2$$

$$= \sum_{i=1}^{d} \left(\mathbb{E}[h_i(X^{(t)})|y^{(0)}] - \mathbb{E}^\pi[h_i(X)] \right)^2$$

$$= \sum_{i=1}^{d} \left(\int h_i(x)\{\pi^t(x|y^{(0)}) - \pi(x)\}\, dx \right)^2$$

$$= \sum_{i=1}^{d} \left(\int h_i(x)\, \pi(x|y) \int \{f^t(y|y^{(0)}) - f(y)\}\, dy dx \right)^2$$

$$\leq \sum_{i=1}^{d} \sup_{y \in \mathcal{Y}} \mathbb{E}[|h_i(X)|\,|y]^2\, 4\, \|f_{y^{(0)}}^t - f\|_{TV}^2$$

$$\leq 4d \max_i \sup_{y \in \mathcal{Y}} \mathbb{E}[|h_i(X)|\,|y]^2\, \|f_{y^{(0)}}^t - f\|_{TV}^2\,.$$

$\square$

Unfortunately, this result has rather limited consequences for the study of the sequence $(X^{(t)})$ since (9.8) is a *average* property of $(X^{(t)})$, while MCMC algorithms such as [A.33] only produce *a single* realization from $\pi_{z^{(0)}}^t$. We will see in Section 12.2.3 a more practical implication of the Duality Principle since Theorem 9.12 may allow a control of convergence by renewal.

9.3 Monotone Covariance and Rao–Blackwellization

In the spirit of the Duality Principle, Rao–Blackwellization exhibits an interesting difference between statistical perspectives and simulation practice, in the sense that the approximations used in the estimator do not (directly) involve the chain of interest. As shown in Section 4.2 and Section 7.6.2, conditioning on a subset of the simulated variables may produce considerable improvement upon the standard empirical estimator in terms of variance, by a simple "recycling" of the rejected variables (see also Section 3.3.3). Two-stage Gibbs sampling and its generalization of Chapter 10 do not permit this kind of recycling since every simulated value is accepted (Theorem 10.13). Nonetheless, Gelfand and Smith (1990) propose a type of conditioning christened *Rao–Blackwellization* in connection with the Rao–Blackwell Theorem (see Lehmann and Casella 1998, Section 1.7) and defined as *parametric* Rao–Blackwellization by Casella and Robert (1996) to differentiate from the form studied in Sections 4.2 and 7.6.2.

For $\mathbf{Y} = (Y_1, Y_2) \sim g(y_1, y_2)$, *Rao–Blackwellization* is based on the marginalization identity

$$g^{(1)}(y_1) = \int g_1(y_1|y_2)\, g^{(2)}(y_2)\, dy_2\,.$$

It thus replaces

$$\delta_0 = \frac{1}{T} \sum_{t=1}^{T} h\left(y_1^{(t)}\right)$$

with

$$\delta_{rb} = \frac{1}{T} \sum_{t=1}^{T} \mathbb{E}\left[h(Y_1)|y_2^{(t)}\right].$$

Both estimators converge to $\mathbb{E}[h(Y_1)]$ and, under the stationary distribution, they are both unbiased. A simple application of the identity $\mathrm{var}(U) = \mathrm{var}(\mathbb{E}[U|V]) + \mathbb{E}[\mathrm{var}(U|V)]$ implies that

(9.9) $$\mathrm{var}\left(\mathbb{E}\left[h(Y_1)|Y_2^{(t)}\right]\right) \leq \mathrm{var}(h(Y_1)).$$

This led Gelfand and Smith (1990) to suggest the use of δ_{rb} instead of δ_0. However, inequality (9.9) is insufficient to conclude on the domination of δ_{rb} when compared with δ_0, as it fails to take into account the correlation between the $Y^{(t)}$'s. The domination of δ_0 by δ_{rb} can therefore be established in only a few cases; Liu et al. (1994) show in particular that it holds for the two-stage Gibbs sampler. (See also Geyer 1995 for necessary conditions.)

We establish the domination result in Theorem 9.19, but we first need some preliminary results, beginning with a representation lemma yielding the interesting result that covariances are positive in an interleaved chain.

Lemma 9.17. *If $h \in \mathcal{L}_2(g_2)$ and if $(X^{(t)})$ is interleaved with $(Y^{(t)})$, then*

$$\mathrm{cov}\left(h(Y^{(1)}), h(Y^{(2)})\right) = \mathrm{var}(\mathbb{E}[h(Y)|X]).$$

Proof. Assuming, without loss of generality, that $\mathbb{E}_{g_2}[h(Y)] = 0$,

$$\begin{aligned}
\mathrm{cov}\left(h(Y^{(1)}), h(Y^{(2)})\right) &= \mathbb{E}\left[h(Y^{(1)})h(Y^{(2)})\right] \\
&= \mathbb{E}\left\{\mathbb{E}\left[h(Y^{(1)})|X^{(2)}\right]\mathbb{E}\left[h(Y^{(2)})|X^{(2)}\right]\right\} \\
&= \mathbb{E}\left[\mathbb{E}\left[h(Y^{(1)})|X^{(2)}\right]^2\right] = \mathrm{var}(\mathbb{E}[h(Y)|X])\,,
\end{aligned}$$

where the second equality follows from iterating the expectation and using the conditional independence of the interleaving property. The last equality uses reversibility (that is, condition (iii)) of the interleaved chains. □

Proposition 9.18. *If $(Y^{(t)})$ is a Markov chain with the interleaving property, the covariances*

$$\mathrm{cov}(h(Y^{(1)}), h(Y^{(t)}))$$

are positive and decreasing in t for every $h \in \mathcal{L}_2(g_2)$.

Proof. Lemma 9.17 implies, by induction, that

$$\text{cov}\left(h(Y^{(1)}), h(Y^{(t)})\right) = \mathbb{E}\left[\mathbb{E}[h(Y)|X^{(2)}]\,\mathbb{E}[h(Y)|X^{(t)}]\right]$$

(9.10)
$$= \text{var}(\mathbb{E}[\cdots \mathbb{E}[\mathbb{E}[h(Y)|X]|Y]\cdots]),$$

where the last term involves $(t-1)$ conditional expectations, alternatively in Y and in X. The decrease in t directly follows from the inequality on conditional expectations, by virtue of the representation (9.10) and the inequality (9.9). □

The result on the improvement brought by Rao–Blackwellization then easily follows from Proposition 9.18.

Theorem 9.19. *If $(X^{(t)})$ and $(Y^{(t)})$ are two interleaved Markov chains, with stationary distributions f_X and f_Y, respectively, the estimator δ_{rb} dominates the estimator δ_0 for every function h with finite variance under both f_X and f_Y.*

Proof. Again assuming $\mathbb{E}[h(X)] = 0$, and introducing the estimators

$$(9.11) \qquad \delta_0 = \frac{1}{T} \sum_{t=1}^{T} h(X^{(t)}), \qquad \delta_{rb} = \frac{1}{T} \sum_{t=1}^{T} \mathbb{E}[h(X)|Y^{(t)}],$$

it follows that

$$\text{var}(\delta_0) = \frac{1}{T^2} \sum_{t,t'} \text{cov}\left(h(X^{(t)}), h(X^{(t')})\right)$$

(9.12)
$$= \frac{1}{T^2} \sum_{t,t'} \text{var}(\mathbb{E}[\cdots \mathbb{E}[h(X)|Y]\cdots])$$

and

$$\text{var}(\delta_{rb}) = \frac{1}{T^2} \sum_{t,t'} \text{cov}\left(\mathbb{E}[h(X)|Y^{(t)}], \mathbb{E}[h(X)|Y^{(t')}]\right)$$

(9.13)
$$= \frac{1}{T^2} \sum_{t,t'} \text{var}(\mathbb{E}[\cdots \mathbb{E}[\mathbb{E}[h(X)|Y]|X]\cdots]),$$

according to the proof of Proposition 9.18, with $|t-t'|$ conditional expectations in the general term of (9.12) and $|t - t'| + 1$ in the general term of (9.13). It is then sufficient to compare $\text{var}(\delta_0)$ with $\text{var}(\delta_{rb})$ term by term to conclude that $\text{var}(\delta_0) \geq \text{var}(\delta_{rb})$. □

One might question whether Rao–Blackwellization will always result in an appreciable variance reduction, even as the sample size (or the number of Monte Carlo iterations) increases. This point was addressed by Levine (1996), who formulated this problem in terms of the *asymptotic relative efficiency* (ARE) of δ_0 with respect to its Rao–Blackwellized version δ_{rb}, given in (9.11), where the pairs $(X^{(t)}, Y^{(t)})$ are generated from a bivariate Gibbs sampler. The ARE is a ratio of the variances of the limiting distributions for the two estimators, which are given by

$$(9.14) \qquad \sigma_{\delta_0}^2 = \text{var}(h(X^{(0)})) + 2\sum_{k=1}^{\infty} \text{cov}(h(X^{(0)}), h(X^{(k)}))$$

and

$$(9.15) \qquad \begin{aligned} \sigma_{\delta_{rb}}^2 &= \text{var}(\mathbb{E}[h(X)|Y]) \\ &+ 2\sum_{k=1}^{\infty} \text{cov}(\mathbb{E}[h(X^{(0)})|Y^{(0)}], \mathbb{E}[h(X^{(k)}|Y^{(k)}]). \end{aligned}$$

Levine (1996) established that the ratio $\sigma_{\delta_0}^2/\sigma_{\delta_1}^2 \geq 1$, with equality if and only if $\text{var}(h(X)) = \text{cov}(\mathbb{E}[h(X)|Y]) = 0$.

Example 9.20. (Continuation of Example 9.1) For the Gibbs sampler (9.4), it can be shown (Problem 9.5) that $\text{cov}(X^{(0)}, X^{(k)}) = \rho^{2k}$, for all k, and

$$\sigma_{\delta_0}^2/\sigma_{\delta_{rb}}^2 = \frac{1}{\rho^2} > 1.$$

So, if ρ is small, the amount of improvement, which is independent of the number of iterations, can be substantial. ‖

9.4 The EM–Gibbs Connection

As mentioned earlier, the EM algorithm can be seen as a precursor of the two-stage Gibbs sampler in missing data models (Section 5.3.1), in that it similarly exploits the conditional distribution of the missing variables. The connection goes further, as seen below.

Recall from Section 5.3.2 that, if $\mathbf{X} \sim g(x|\theta)$ is the observed data, and we augment the data with $\mathbf{z}$, where $\mathbf{Z} \sim f(\mathbf{x}, \mathbf{z}|\theta)$, then we have the complete-data and incomplete-data likelihoods

$$L^c(\theta|\mathbf{x}, \mathbf{z}) = f(\mathbf{x}, \mathbf{z}|\theta) \text{ and } L(\theta|\mathbf{x}) = g(\mathbf{x}|\theta),$$

with the missing data density

$$k(\mathbf{z}|\mathbf{x}, \theta) = \frac{L^c(\theta|\mathbf{x}, \mathbf{z})}{L(\theta|\mathbf{x})}.$$

If we can normalize the complete-data likelihood in θ (this is the only condition for the equivalence mentioned above), that is, if $\int L^c(\theta|\mathbf{x}, \mathbf{z})d\theta < \infty$, then define

$$L^*(\theta|\mathbf{x}, \mathbf{z}) = \frac{L^c(\theta|\mathbf{x}, \mathbf{z})}{\int L^c(\theta|\mathbf{x}, \mathbf{z})d\theta}$$

and create the two-stage Gibbs sampler:

(9.16)   1. $\mathbf{z}|\theta \sim k(\mathbf{z}|\mathbf{x}, \theta)$
           2. $\theta|\mathbf{z} \sim L^*(\theta|\mathbf{x}, \mathbf{z})$.

Note the direct connection to an EM algorithm based on L^c and k. The "E" step in the EM algorithm calculates the expected value of the log-likelihood over $\mathbf{z}$, often by calculating $\mathbb{E}(\mathbf{Z}|\mathbf{x}, \theta)$ and substituting in the log-likelihood. In the Gibbs sampler this step is replaced with generating a random variable from the density k. The "M" step of the EM algorithm then takes as the current value of θ the maximum of the expected complete-data log-likelihood. In the Gibbs sampler this step is replaced by generating a value of θ from L^*, the normalized complete-data likelihood.

Example 9.21. Censored data Gibbs. For the censored data example considered in Example 5.14, the distribution of the missing data is

$$Z_i \sim \frac{\phi(z - \theta)}{1 - \Phi(a - \theta)}$$

and the distribution of $\theta|x, z$ is

$$L(\theta|x, z) \propto \prod_{i=1}^{m} e^{-(x_i - \theta)^2/2} \prod_{i=m+1}^{n} e^{-(z_i - \theta)^2/2},$$

which corresponds to a

$$\mathcal{N}\left(\frac{m\bar{x} + (n - m)\bar{z}}{n}, \frac{1}{n}\right)$$

distribution, and so we immediately have that L^* exists and that we can run a Gibbs sampler (Problem 9.14).   ‖

The validity of the "EM/Gibbs" sampler follows in a straightforward manner from its construction. The transition kernel of the Markov chain is

$$\mathcal{K}(\theta, \theta'|\mathbf{x}) = \int_{\mathbf{z}} k(\mathbf{z}|\mathbf{x}, \theta)L^*(\theta'|\mathbf{x}, \mathbf{z}) \, d\mathbf{z}$$

and it can be shown (Problem 9.15) that the invariant distribution of the chain is the incomplete data likelihood, that is,

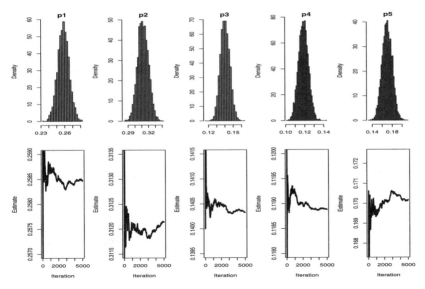

Fig. 9.3. Gibbs output for cellular phone data, 5000 iterations

$$L(\theta'|\mathbf{x}) = \int_{\Theta} \mathcal{K}(\theta, \theta'|\mathbf{x})L(\theta|\mathbf{x})d\theta.$$

Since $L(\theta'|\mathbf{x}, \mathbf{z})$ is integrable in θ, so is $L(\theta'|\mathbf{x})$, and hence the invariant distribution is a proper density. So the Markov chain is positive, and convergence follows from Theorem 9.6.

Example 9.22. Cellular phone Gibbs. As an illustration of the EM–Gibbs connection, we revisit Example 5.18, but now we use the Gibbs sampler to get our solution. From the complete data likelihood (5.18) and the missing data distribution (5.19) we have (Problem 9.16)

$$\mathbf{p}|W_1, W_2, \ldots, W_5, \sum_i X_i \sim \mathcal{D}(W_1 + 1, W_2 + 1, \ldots, W_5 + \sum_i X_i + 1)$$

$$\text{(9.17)} \qquad \sum_i X_i \sim \mathcal{N}eg\left(\sum_{i=1}^{m} n_i + m, 1 - p_5\right).$$

The results of the Gibbs iterations are shown in Figure 9.3. The point estimates agree with those of the EM algorithm (Example 5.18), $\hat{p} = (0.258, 0.313, 0.140, 0.118, 0.170)$, with the exception of $\hat{p}_5$, which is larger than the MLE. This may reflect the fact that the Gibbs estimate is a mean (and gets pulled a bit into the tail), while the MLE is a mode. Measures of error can be obtained from either the iterations or the histograms. ‖

Based on the same functions $L(\theta|\mathbf{y}, \mathbf{z})$ and $k(\mathbf{z}|\theta, \mathbf{y})$ the EM algorithm will get the ML estimator from $L(\theta|\mathbf{y})$, whereas the Gibbs sampler will get us the entire function. This likelihood implementation of the Gibbs sampler was used by Casella and Berger (1994) and is also described by Smith and Roberts (1993). A version of the EM algorithm, where the Markov chain connection is quite apparent, was given by Baum and Petrie (1966) and Baum et al. (1970).

9.5 Transition

As a natural extension of the (2D) slice sampler, the two-stage Gibbs sampler enjoys many optimality properties. In Chapter 8 we also developed the natural extension of the slice sampler to the case when the density f is easier to study when decomposed as a product of functions f_i (Section 8.2). Chapter 10 will provide the corresponding Gibbs generalization to this general slice sampler, which will cover cases when the decomposition of the simulation in two conditional simulations is not feasible any longer. This generalization obviously leads to a wider range of models, but also to fewer optimality properties.

As an introduction to the next chapter, note that if $(X, Y) = (X, (Y_1, Y_2))$, and if simulating from $f_{Y|X}$ is not directly possible, the two-stage Gibbs sampler could also be applied to this (conditional) density. This means that a sequence of successive simulations from $f(y_1|x, y_2)$ and from $f(y_2|x, y_1)$ (which is the translation of the two-stage Gibbs sampler for the conditional $f_{Y|X}$) converges to a simulation from $f_{Y|X}(y_1, y_2|x)$.

The fundamental property used in Chapter 10 is that stopping the recursion between $f(y_1|x, y_2)$ and $f(y_2|x, y_1)$ "before" convergence has no effect on the validation of the algorithm. In fact, a *single* simulation of each component is sufficient! Obviously, this feature will generalize (by a cascading argument) to an arbitrary number of components in Y.

9.6 Problems

9.1 For the Gibbs sampler [A.33]:
  (a) Show that the sequence (X_i, Y_i) is a Markov chain, as is each sequence (X_i) and (Y_i).
  (b) Show that $f_X(\cdot)$ and $f_Y(\cdot)$ are respectively the invariant densities of the X and Y sequences of [A.33].
9.2 Write a Gibbs sampler to generate standard bivariate normal random variables (with mean 0, variance 1 and correlation ρ). (Recall that if (X, Y) is standard bivariate normal, the conditional density of $X|Y = y$ is $N(\rho y, (1 - \rho^2))$. For $\rho = .3$, use the generated random variables to estimate the density of $X^2 + Y^2$ and calculate $P(X^2 + Y^2 > 2)$.
9.3 Referring to Problem 5.18, estimate p_A, p_B and p_O using a Gibbs sampler. Make a histogram of the samples.
9.4 In the case of the two-stage Gibbs sampler, the relationship between the Gibbs and Metropolis–Hastings algorithms becomes particularly clear. If we have the bivariate Gibbs sampler $X \sim f(x|y)$ and $Y \sim f(y|x)$, consider the X chain alone and show:

(a) $K(x, x') = g(x|x') = \int f(x'|y)f(y|x)dy;$

(b) $\varrho = \min\left\{\frac{f(x')/g(x'|x)}{f(x)/g(x|x')}, 1\right\}$, where $f(\cdot)$ is the marginal distribution;

(c) $f(x')/g(x'|x) = f(x)/g(x|x')$, so $\varrho = 1$ and the Metropolis–Hastings proposal is always accepted.

9.5 For the situation of Example 9.20:

(a) Verify the variance representations (9.14) and (9.15).

(b) For the bivariate normal sampler, show that $\text{cov}(X_1, X_k) = \rho^{2k}$, for all k, and $\sigma_{\delta_0}^2/\sigma_{\delta_1}^2 = 1/\rho^2$.

9.6 The monotone decrease of the correlation seen in Section 9.3 does not hold uniformly for all Gibbs samplers as shown by the example of Liu et al. (1994): For the bivariate normal Gibbs sampler of Example 9.20, and $h(x, y) = x - y$, show that

$$\text{cov}[h(X_1, Y_1), h(X_2, Y_2)] = -\rho(1 - \rho)^2 < 0.$$

9.7 Refer to the Horse Kick data of Table 2.1. Fit the loglinear model of Example 9.7 using the bivariate Gibbs sampler, along with the ARS algorithm, and estimate both $\pi(a|\mathbf{x}, \mathbf{y})$ and $\pi(b|\mathbf{x}, \mathbf{y})$. Obtain both point estimates and error bounds for a and b. Take $\sigma^2 = \tau^2 = 5$.

9.8 The data of Example 9.7 can also be analyzed as a loglinear model, where we fit $\log \lambda = a + bt$, where $t = $ number of passages.

(a) Using the techniques of Example 2.26, find the posterior distributions of a and b. Compare you answer to that given in Example 9.7. (Ignore the "4 or more", and just use the category as 4.)

(b) Find the posterior distributions of a and b, but now take the "4 or more" censoring into account, as in Example 9.7.

9.9 The situation of Example 9.7 also lends itself to the EM algorithm, similar to the Gibbs treatment (Example 9.8) and EM treatment (Example 5.21) of the grouped multinomial data problem. For the data of Table 9.1:

(a) Use the EM algorithm to calculate the MLE of λ.

(b) Compare your answer in part (a) to that from the Gibbs sampler of Algorithm [A.35].

(c) Establish that the Rao–Blackwellized estimator is correct.

9.10 In the setup of Example 9.8, the (uncompleted) posterior distribution is available as

$$\pi(\eta, \mu|x) \propto (a_1\mu + b_1)^{x_1}(a_2\mu + b_2)^{x_2}(a_3\eta + b_3)^{x_3}(a_4\eta + b_4)^{x_4}$$
$$\times (1 - \mu - \eta)^{x_5 + \alpha_3 - 1}\mu^{\alpha_1 - 1}\eta^{\alpha_2 - 1}.$$

(a) Show that the marginal distributions $\pi(\mu|x)$ and $\pi(\eta|x)$ can be explicitly computed as polynomials when the α_i's are integers.

(b) Give the marginal posterior distribution of $\xi = \mu/(1 - \eta - \mu)$. (*Note:* See Robert 1995a for a solution.)

(c) Evaluate the Gibbs sampler proposed in Example 9.8 by comparing approximate moments of μ, η, and ξ with their exact counterpart, derived from the explicit marginal.

9.11 There is a connection between the EM algorithm and Gibbs sampling, in that both have their basis in Markov chain theory. One way of seeing this is to show that the incomplete-data likelihood is a solution to the integral equation of successive substitution sampling and that Gibbs sampling can then be used to calculate the likelihood function. If $L(\theta|\mathbf{y})$ is the incomplete-data likelihood and $L(\theta|\mathbf{y},\mathbf{z})$ is the complete-data likelihood, define

$$L^*(\theta|\mathbf{y}) = \frac{L(\theta|\mathbf{y})}{\int L(\theta|\mathbf{y})d\theta}\,, \quad L^*(\theta|\mathbf{y},\mathbf{z}) = \frac{L(\theta|\mathbf{y},\mathbf{z})}{\int L(\theta|\mathbf{y},\mathbf{z})d\theta}\,,$$

assuming both integrals to be finite.

(a) Show that $L^*(\theta|\mathbf{y})$ is the solution to

$$L^*(\theta|\mathbf{y}) = \int \left[\int L^*(\theta|\mathbf{y},\mathbf{z})k(\mathbf{z}|\theta',\mathbf{y})d\mathbf{z} \right] L^*(\theta'|\mathbf{y})d\theta'\,,$$

where $k(\mathbf{z}|\theta,\mathbf{y}) = L(\theta|\mathbf{y},\mathbf{z})/L(\theta|\mathbf{y})$.

(b) Show that the sequence $\theta_{(j)}$ from the Gibbs iteration

$$\theta_{(j)} \sim L^*(\theta|\mathbf{y},\mathbf{z}_{(j-1)})\,,$$
$$\mathbf{z}_{(j)} \sim k(\mathbf{z}|\theta_{(j)},\mathbf{y})\,,$$

converges to a random variable with density $L^*(\theta|\mathbf{y})$ as j goes ∞. How can this be used to compute the likelihood function $L(\theta|\mathbf{y})$?

(*Note:* Based on the same functions $L(\theta|\mathbf{y},\mathbf{z})$ and $k(\mathbf{z}|\theta,\mathbf{y})$ the EM algorithm will get the ML estimator from $L(\theta|\mathbf{y})$, whereas the Gibbs sampler will get us the entire function. This likelihood implementation of the Gibbs sampler was used by Casella and Berger 1994 and is also described by Smith and Roberts 1993. A version of the EM algorithm, where the Markov chain connection is quite apparent, was given by Baum and Petrie 1966 and Baum et al. 1971.)

9.12 In the setup of Example 9.9, the posterior distribution of N can be evaluated by recursion.

(a) Show that

$$\pi(N) \propto \frac{(N-n_0)!\lambda^N}{N!(N-n_t)!}\,.$$

(b) Using the ratio $\pi(N)/\pi(N-1)$, derive a recursion relation to compute $\mathbb{E}^\pi[N|n_0,n_t]$.

(c) In the case $n_0 = 112$, $n_t = 79$, and $\lambda = 500$, compare the computation time of the above device with the computation time of the Gibbs sampler. (*Note:* See George and Robert 1992 for details.)

9.13 Recall that, in the setting of Example 5.22, animal i, $i = 1, 2, \ldots, n$ may be captured at time j, $j = 1, 2, \ldots, t$, in one of m locations, where the location is a multinomial random variable $H_{ij} \sim \mathcal{M}_m(\theta_1, \ldots, \theta_m)$. Given $H_{ij} = k$ ($k = 1, 2, \ldots, m$), the animal is captured with probability p_k, represented by the random variable $X \sim \mathcal{B}(p_k)$. Define $y_{ijk} = \mathbb{I}(h_{ij} = k)\mathbb{I}(x_{ijk} = 1)$.

(a) Under the conjugate priors

$$(\theta_1, \ldots, \theta_m) \sim \mathcal{D}(\lambda_1, \ldots, \lambda_m) \quad \text{and} \quad p_k \sim \mathcal{B}e(\alpha, \beta)\,,$$

show that the full conditional posterior distributions are given by

$$\{\theta_1, \ldots, \theta_m\} \sim \mathcal{D}(\lambda_1 + \Sigma_{i=1}^n \Sigma_{j=1}^t y_{ij1}, \ldots, \lambda_m + \Sigma_{i=1}^n \Sigma_{j=1}^t y_{ijm})$$

and

$$p_k \sim \mathcal{B}e(\alpha + \Sigma_{i=1}^n \Sigma_{j=1}^t x_{ijk}, \beta + n - \Sigma_{i=1}^n \Sigma_{j=1}^t x_{ijk})\,.$$

(b) Deduce that all of the full conditionals are conjugate and thus that the Gibbs sampler is straightforward to implement.

(c) For the data of Problem 5.28, estimate the θ_i's and the p_i's using the Gibbs sampler starting from the prior parameter values $\alpha = \beta = 5$ and $\lambda_i = 2$.

(*Note:* Dupuis 1995 and Scherrer 1997 discuss how to choose the prior parameter values to reflect the anticipated movement of the animals.)

9.14 Referring to Example 9.21:

(a) Show that, as a function of θ, the normalized complete-data likelihood is $\mathcal{N}((m\bar{x} + (n - m)\bar{z})/n, 1/n)$.

(b) Derive a Monte Carlo EM algorithm to estimate θ.

(c) Contrast the Gibbs sampler algorithm with the EM algorithm of Example 9.21 and the Monte Carlo EM algorithm of part (b).

9.15 Referring to Section 9.4:

(a) Show that the Gibbs sampler of (9.16) has $L(\theta|\mathbf{x})$ as stationary distribution.

(b) Show that if $L(\theta|\mathbf{x}, \mathbf{z})$ is integrable in θ, then so is $L(\theta|\mathbf{x})$, and hence the Markov chain of part(*a*) is positive.

(c) Complete the proof of ergodicity of the Markov chain. (*Hint:* In addition to Theorem 10.10, see Theorem 6.51. Theorem 9.12 may also be useful in some situations.)

9.16 Referring to Example 9.22:

(a) Verify the distributions in (9.17) for the Gibbs sampler.

(b) Compare the output of the Gibbs sampler to the EM algorithm of Example 5.18. Which algorithm do you prefer and why?

9.17 (Smith and Gelfand 1992) For $i = 1, 2, 3$, consider $Y_i = X_{1i} + X_{2i}$, with

$$X_{1i} \sim \mathcal{B}(n_{1i}, \theta_1), \qquad X_{2i} \sim \mathcal{B}(n_{2i}, \theta_2).$$

(1) Give the likelihood $L(\theta_1, \theta_2)$ for $n_{1i} = 5, 6, 4$, $n_{2i} = 5, 4, 6$, and $y_i = 7, 5, 6$.

(2) For a uniform prior on (θ_1, θ_2), derive the Gibbs sampler based on the natural parameterization.

(3) Examine whether an alternative parameterization or a Metropolis–Hastings algorithm may speed up convergence.

9.18 For the Gibbs sampler

$$X \mid y \sim \mathcal{N}(\rho y, \ 1 - \rho^2),$$
$$Y \mid x \sim \mathcal{N}(\rho x, \ 1 - \rho^2),$$

of Example 9.1:

(a) Show that for the X chain, the transition kernel is

$$K(x^*, x) = \frac{1}{2\pi(1 - \rho^2)} \int e^{-\frac{1}{2(1-\rho^2)}(x - \rho y)^2} e^{-\frac{1}{2(1-\rho^2)}(y - \rho x^*)^2} \, dy.$$

(b) Show that $X \sim \mathcal{N}(0, 1)$ is the invariant distribution of the X chain.

(c) Show that $X|x^* \sim \mathcal{N}(\rho^2 x^*, 1 - \rho^4)$. (*Hint:* Complete the square in the exponent of part (a).)

(d) Show that we can write $X_k = \rho^2 X_{k-1} + U_k$, $k = 1, 2, \ldots$, where the U_k are iid $\mathcal{N}(0, 1 - \rho^4)$ and that $\text{cov}(X_0, X_k) = \rho^{2k}$, for all k. Deduce that the covariances go to zero.

9.19 In the setup of Example 9.23, show that the likelihood function is not bounded and deduce that, formally, there is no maximum likelihood estimator. (*Hint:* Take $\mu_{j_0} = x_{i_0}$ and let τ_{j_0} go to 0.)

9.20 A model consists in the partial observation of normal vectors $Z = (X, Y) \sim \mathcal{N}_2(0, \Sigma)$ according to a mechanism of random censoring. The corresponding data are given in Table 9.2.

x	1.17	−0.98	0.18	0.57	0.21	−	−	−
y	0.34	−1.24	−0.13	−	−	−0.12	−0.83	1.64

Table 9.2. Independent observations of $Z = (X, Y) \sim \mathcal{N}_2(0, \Sigma)$ with missing data (denoted −).

(a) Show that inference can formally be based on the likelihood

$$\prod_{i=1}^{3} \left\{ |\Sigma|^{-1/2} e^{-z_i^t \Sigma^{-1} z_i/2} \right\} \sigma_1^{-2} e^{-(x_4^2 + x_5^2)/2\sigma_1^2} \sigma_2^{-3} e^{-(y_6^2 + y_7^2 + y_8^2)/2\sigma_2^2}.$$

(b) Show that the choice of the prior distribution $\pi(\Sigma) \propto |\Sigma|^{-1}$ leads to difficulties given that σ_1 and σ_2 are isolated in the likelihood.

(c) Show that the missing components can be simulated through the following algorithm.

Algorithm A.37 −Normal Completion−

1. Simulate

$$X_i^* \sim \mathcal{N}\left(\rho \frac{\sigma_1}{\sigma_2} y_i, \sigma_1^2(1 - \rho^2)\right) \qquad (i = 6, 7, 8),$$

$$Y_i^* \sim \mathcal{N}\left(\rho \frac{\sigma_2}{\sigma_1} x_i, \sigma_2^2(1 - \rho^2)\right) \qquad (I = 4, 5).$$

2. Generate

$$\Sigma^{-1} \sim \mathcal{W}_2(8, X^{-1}),$$

with $X = \sum_{i=1}^{8} z_i^* z_i^{*t}$, the dispersion matrix of the completed data.

to derive the posterior distribution of the quantity of interest, ρ.

(d) Propose a Metropolis–Hastings alternative based on a slice sampler.

9.21 Roberts and Rosenthal (2003) derive the *polar slice sampler* from the decomposition $\pi(x) \propto f_0(x) f_1(x)$ of a target distribution $\pi(x)$.

(a) Show that the Gibbs sampler whose two steps are (i) to simulate U uniformly on $(0, f_1(x))$ and (ii) to simulate X from $f_0(x)$ restricted to the set of x's such that $f_1(x) > u$ is a valid MCMC algorithm with target distribution π.

(b) Show that, if f_0 is constant, this algorithm is simply the slice sampler of Algorithm [A.31] in Chapter 8. (It is also called the *uniform slice sampler* in Roberts and Rosenthal 2003.)

(c) When $x \in \mathbb{R}^d$, using the case $f_0(x) = |x|^{-(d-1)}$ and $f_1(x) = |x|^{d-1} \pi(x)$ is called the *polar slice sampler*. Show that using the polar slice sampler on a d-dimensional log-concave target density is equivalent to a uniform slice sampler on a corresponding one-dimensional log-concave density.

(d) Illustrate this property in the case when $\pi(x) = \exp -|x|$.

9.22 Consider the case of a mixture of normal distributions,

$$\tilde{f}(x) = \sum_{j=1}^k p_j \frac{e^{-(x-\mu_j)^2/(2\tau_j^2)}}{\sqrt{2\pi}\,\tau_j}.$$

(a) Show that the conjugate distribution on (μ_j, τ_j) is

$$\mu_j | \tau_j \sim \mathcal{N}\left(\alpha_j, \tau_j^2/\lambda_j\right), \qquad \tau_j^2 \sim \mathcal{IG}\left(\frac{\lambda_j + 3}{2}, \frac{\beta_j}{2}\right).$$

(b) Show that two valid steps of the Gibbs sampler are as follows.

Algorithm A.38 –Normal Mixture Posterior Simulation–

1. Simulate $(i = 1, \ldots, n)$

$$Z_i \sim P(Z_i = j) \propto p_j \, \exp\left\{-(x_i - \mu_j)^2/(2\tau_j^2)\right\} \tau_j^{-1}$$

and compute the statistics $(j = 1, \ldots, k)$ [A.38]

$$n_j = \sum_{i=1}^n \mathbb{I}_{z_i=j}, \quad n_j \bar{x}_j = \sum_{i=1}^n \mathbb{I}_{z_i=j} x_i, \quad s_j^2 = \sum_{i=1}^n \mathbb{I}_{z_i=j}(x_i - \bar{x}_j)^2.$$

2. Generate

$$\mu_j | \tau_j \sim \mathcal{N}\left(\frac{\lambda_j \alpha_j + n_j \bar{x}_j}{\lambda_j + n_j}, \frac{\tau_j^2}{\lambda_j + n_j}\right),$$

$$\tau_j^2 \sim \mathcal{IG}\left(\frac{\lambda_j + n_j + 3}{2}, \frac{\lambda_j \alpha_j^2 + \beta_j + s_j^2 - (\lambda_j + n_j)^{-1}(\lambda_j \alpha_j + n_j \bar{x}_j)^2}{2}\right),$$

$$p \sim \mathcal{D}_k(\gamma_1 + n_1, \ldots, \gamma_k + n_k).$$

(*Note:* Robert and Soubiran 1993 use this algorithm to derive the maximum likelihood estimators by recursive integration (see Section 5.2.4), showing that the Bayes estimators converge to the local maximum, which is the closest to the initial Bayes estimator.)

9.23 Referring to Section 9.7.2, for the factor ARCH model of (9.18):

(a) Propose a noninformative prior distribution on the parameter $\theta = (\alpha, \beta, a, \Sigma)$ that leads to a proper posterior distribution.

(b) Propose a completion step for the latent variables based on $f(y_t^* | y_t, y_{t-1}^*, \theta)$. (*Note:* See Diebold and Nerlove 1989, Gouriéroux et al. 1993, Kim et al. 1998, and Billio et al. 1998 for different estimation approaches to this model.)

9.24 Check whether a negative coefficient b in the random walk $Y_t = a + b(X^{(t)} - a) + Z_t$ induces a negative correlation between the $X^{(t)}$'s. Extend to the case where the random walk has an ARCH structure,

$$Y_t = a + b(X^{(t)} - a) + \exp(c + d(X^{(t)} - a)^2)Z_t.$$

9.25 (Diebolt and Robert 1990a,b)

 (a) Show that, for a mixture of distributions from exponential families, there exist conjugate priors. (*Hint:* See Example 9.2.)

 (b) For the conjugate priors, show that the posterior expectation of the mean parameters of the components can be written in a closed form.

 (c) Show that the convergence to stationarity is geometric for all the chains involved in the Gibbs sampler for the mixture model.

 (d) Show that Rao–Blackwellization applies in the setup of normal mixture models and that it theoretically improves upon the naïve average. (*Hint:* Use the Duality Principle.)

9.26 (Roeder and Wasserman 1997) In the setup of normal mixtures of Example 9.2:

 (a) Derive the posterior distribution associated with the prior $\pi(\mu, \tau)$, where the τ_j^2's are inverted Gamma $\mathcal{IG}(\nu, A)$ and $\pi(\mu_j | \mu_{j-1}, \tau)$ is a left-truncated normal distribution

$$\mathcal{N}\left(\mu_{j-1}, B(\tau_j^{-2} + \tau_{j-1}^{-2}))^{-1}\right),$$

 except for $\pi(\mu_1) = 1/\mu_1$. Assume that the constant B is known and $\pi(A) = 1/A$.

 (b) Show that the posterior is always proper.

 (c) Derive the posterior distribution using the noninformative prior

$$\pi(\mu, \tau) = \tau^{-1}, \quad p, q_j \sim \mathcal{U}_{[0,1]}, \quad \sigma_j \sim \mathcal{U}_{[0,1]}, \quad \theta_j \sim \mathcal{N}(0, \zeta^2).$$

 and compare.

9.27 Consider the following mixture of uniforms,[1]

$$p\mathcal{U}_{[\lambda, \lambda+1]} + (1-p)\mathcal{U}_{[\mu, \mu+1]},$$

 and an ordered sample $x_1 \leq \cdots \leq x_{n_1} \leq \cdots \leq x_n$ such that $x_{n_1} + 1 < x_{n_1+1}$.

 (a) Show that the chain $(\lambda^{(t)}, \mu^{(t)})$ associated with the Gibbs sampler corresponding to [A.34] is not irreducible.

 (b) Show that the above problem disappears if $\mathcal{U}_{[\mu, \mu+1]}$ is replaced with $\mathcal{U}_{[\mu, \mu+0.5]}$ (for n large enough).

9.28 (Billio et al. 1999) A *dynamic desequilibrium* model is defined as the observation of

$$Y_t = \min(Y_{1t}^*, Y_{2t}^*),$$

 where the Y_{it}^* are distributed from a parametric joint model, $f(y_{1t}^*, y_{2t}^*)$.

 (a) Give the distribution of (Y_{1t}^*, Y_{2t}^*) conditional on Y_t.

 (b) Show that a possible completion of the model is to first draw the regime (1 versus 2) and then draw the missing component.

 (c) Show that when $f(y_{1t}^*, y_{2t}^*)$ is Gaussian, the above steps can be implemented without approximation.

9.7 Notes

9.7.1 Inference for Mixtures

Although they may seem to apply only for some very particular sets of random phenomena, *mixtures of distributions* (9.5) are of wide use in practical modeling.

---

[1] This problem was suggested by Eric Moulines.

However, as already noticed in Examples 1.2 and 3.7, they can be challenging from an inferential point of view (that is, when estimating the parameters p_j and ξ_j). Everitt (1984), Titterington et al. (1985), MacLachlan and Basford (1988), West (1992), Titterington (1996), Robert (1996a), and Marin et al. (2004) all provide different perspectives on mixtures of distributions, discuss their relevance for modeling purposes, and give illustrations of their use in various setups of (9.5).

We assume, without a considerable loss of generality, that $f(\cdot|\xi)$ belongs to an exponential family

$$f(x|\xi) = h(x) \, \exp\{\xi \cdot x - \psi(\xi)\} \, ,$$

and we consider the associated conjugate prior on ξ (see Robert 2001, Section 3.3)

$$\pi(\xi|\alpha_0, \lambda) \propto \exp\{\lambda(\xi \cdot \alpha_0 - \psi(\xi))\} \, , \qquad \lambda > 0, \quad \alpha_0 \in \mathcal{X}.$$

For the mixture (9.5), it is therefore possible to associate with each component $f(\cdot|\xi_j)$ $(j = 1, \ldots, k)$ a conjugate prior $\pi(\xi_j|\alpha_j, \lambda_j)$. We also select for $(p_1, \ldots, p_k)$ the standard Dirichlet conjugate prior; that is,

$$(p_1, \ldots, p_k) \sim \mathcal{D}_k(\gamma_1, \ldots, \gamma_k) \, .$$

Given a sample $(x_1, \ldots, x_n)$ from (9.5), and conjugate priors on ξ and $(p_1, \ldots, p_k)$ (see Robert 2001, Section 3.3), the posterior distribution associated with this model is formally explicit (see Problem 9.25). However, it is virtually useless for large, or even moderate, values of n. In fact, the posterior distribution,

$$\pi(p, \xi|x_1, \ldots, x_n) \propto \prod_{i=1}^{n} \left\{ \sum_{j=1}^{k} p_j \, f(x_i|\xi_j) \right\} \, \pi(p, \xi),$$

is better expressed as a sum of k^n terms which correspond to the different allocations of the observations x_i to the components of (9.5). Although each term is conjugate, the number of terms involved in the posterior distribution makes the computation of the normalizing constant and of posterior expectations totally infeasible for large sample sizes (see Diebolt and Robert 1990a). (In a simulation experiment, Casella et al. 1999 actually noticed that very few of these terms carry a significant posterior weight, but there is no manageable approach to determine which terms are relevant and which are not.) The complexity of this model is such that there are virtually no other solutions than using the Gibbs sampler (see, for instance, Smith and Makov 1978 or Bernardo and Giròn 1986, 1988, for pre-Gibbs approximations).

The solution proposed by Diebolt and Robert (1990c,b, 1994), Lavine and West (1992), Verdinelli and Wasserman (1992), and Escobar and West (1995) is to take advantage of the missing data structure inherent to (9.5), as in Example 9.2.

Good performance of the Gibbs sampler is guaranteed by the above setup since the *Duality Principle* of Section 9.2.3 applies. One can also deduce geometric convergence and a Central Limit Theorem. Moreover, Rao–Blackwellization is justified (see Problem 9.25).

The practical implementation of [A.34] might, however, face serious convergence difficulties, in particular because of the phenomenon of the "absorbing component" (Diebolt and Robert 1990b, Mengersen and Robert 1996, Robert 1996a). When only a small number of observations are allocated to a given component j_0, the following probabilities are quite small:

(1) The probability of allocating new observations to the component j_0.
(2) The probability of reallocating, to another component, observations already allocated to j_0.

Even though the chain $(z^{(t)}, \xi^{(t)})$ corresponding to [A.34] is irreducible, the practical setting is one of an almost-absorbing state, which is called a *trapping state* as it requires an enormous number of iterations in [A.34] to escape from this state. In the extreme case, the probability of escape is below the minimal precision of the computer and the trapping state is truly absorbing, due to computer "rounding errors."

This problem can be linked with a potential difficulty of this model, namely that it does not allow a noninformative (or improper) Bayesian approach, and therefore necessitates the elicitation of the hyperparameters γ_j, α_j and λ_j. Moreover, a vague choice of these parameters (taking for instance, $\gamma_j = 1/2$, $\alpha_j = \alpha_0$, and small λ_j's) often has the effect of increasing the occurrence of trapping states (Chib 1995).

Example 9.23. (Continuation of Example 9.2) Consider the case where $\lambda_j \ll 1$, $\alpha_j = 0$ $(j = 1, \ldots, k)$, and where a *single observation* x_{i_0} is allocated to j_0. Using the algorithm [A.38], we get the approximation

$$\mu_{j_0} | \tau_{j_0} \sim \mathcal{N}\left(x_{i_0}, \tau_{j_0}^2\right) , \qquad \tau_{j_0}^{-2} \sim \mathcal{IG}(2, \beta_j/2) ,$$

so $\tau_{j_0}^2 \approx \beta_j/4 \ll 1$ and $\mu_{j_0} \approx x_{i_0}$. Therefore,

$$P\left(Z_{i_0} = j_0\right) \propto p_{j_0}\tau_{j_0}^{-1} \gg p_{j_1} \, e^{-\left(x_{i_0} - \mu_{j_1}\right)^2/2\tau_{j_1}^2} \, \tau_{j_1}^{-1} \propto P\left(Z_{i_0} = j_1\right)$$

for $j_1 \neq j_0$. On the other hand, if $i_1 \neq i_0$, it follows that

$$P\left(Z_{i_1} = j_0\right) \propto p_{j_0} \, e^{-\left(x_{i_1} - \mu_{j_0}\right)^2/2\tau_{j_0}^2} \, \tau_{j_0}^{-1}$$
$$\ll p_{j_1} \, e^{-\left(x_{i_1} - \mu_{j_1}\right)^2/2\tau_{j_1}^2} \, \tau_{j_1}^{-1} \propto P\left(Z_{i_1} = j_1\right) ,$$

given the very rapid decrease of $\exp\{-t^2/2\tau_{j_0}^2\}$.   ‖

An attempt at resolution of the paradox of trapping states may be to blame the Gibbs sampler, which moves too slowly on the likelihood surface, and to replace it by an Metropolis–Hastings algorithm with wider moves, as developed by Celeux et al. (2000). The trapping phenomenon is also related to the lack of a maximum likelihood estimator in this setup, since the likelihood is not bounded. (See Problem 9.19 or Lehmann and Casella 1998.)

9.7.2 ARCH Models

Together with the techniques of Section 5.5.4 and Chapter 7, the Gibbs sampler can be implemented for estimation in an interesting class of missing data models that are used in Econometrics and Finance.

A Gaussian ARCH (<u>A</u>uto <u>R</u>egressive <u>C</u>onditionally <u>H</u>eteroscedastic) model is defined, for $t = 2, \ldots, T$, by

(9.18)
$$\begin{cases} Z_t = (\alpha + \beta Z_{t-1}^2)^{1/2}\varepsilon_t^*, \\ X_t = aZ_t + \varepsilon_t, \end{cases}$$

where $a \in \mathbb{R}^p$, $\varepsilon_t^* \overset{iid}{\sim} \mathcal{N}(0,1)$, and $\varepsilon_t \overset{iid}{\sim} \mathcal{N}_p(0, \Sigma)$ independently. Let $\theta = (a, \beta, \sigma)$ denote the parameter of the model and assume, in addition, that $Z_1 \sim \mathcal{N}(0,1)$. The sample is denoted by $x^T = (x_1, \ldots, x_T)$, meaning that the z_t's are not observed. (For details on the theory and use of ARCH models, see the review papers by Bollerslev et al. 1992 and Kim et al. 1998, or the books by Enders 1994 and Gouriéroux 1996.)

For simplicity and identifiability reasons, we consider the special case

$$(9.19) \qquad \begin{cases} Z_t = (1 + \beta Z_{t-1}^2)^{1/2} \varepsilon_t^*, & \varepsilon_t^* \sim \mathcal{N}(0,1), \\ X_t = a Z_t + \varepsilon_t, & \varepsilon_t \sim \mathcal{N}_p(0, \sigma^2 I_p). \end{cases}$$

Of course, the difficulty in estimation under this model comes from the fact that only the x_t's are observed.

However, in missing data models such as this one (see also Billio et al. 1998), the likelihood function $L(\theta|x)$ can be written as the marginal of $f(x, z|\theta)$ and the likelihood ratio is thus available in the form

$$\frac{L(\theta|x)}{L(\eta|x)} = \mathbb{E}_\eta \left[\frac{f(x, Z|\theta)}{f(x, Z|\eta)} \middle| x \right], \quad Z \sim f(x, z|\eta),$$

and can be approximated by

$$(9.20) \qquad \hat{h}_\eta(\theta) = \frac{1}{m} \sum_{i=1}^{m} \frac{f(x, z_i|\theta)}{f(x, z_i|\eta)}, \quad Z_i \sim f(z|x, \eta).$$

(See Section 5.5.4 and Problem 5.11.)

	True θ	Starting value η	Approximate maximum likelihood estimate $\hat{\theta}$
a_1	-0.2	-0.153	-0.14
a_2	0.6	0.43	0.42
β	0.8	0.86	0.99
σ^2	0.2	0.19	0.2

Table 9.3. Estimation result for the factor ARCH model (9.19) with a simulated sample of size $T = 100$ and a Bayes estimate as starting value. (*Source:* Billio et al. 1998.)

The approximation of the likelihood ratio (9.20) is then based on the simulation of the missing data $Z^T = (Z_1, \ldots, Z_T)$ from

$$\tilde{f}(z^T | x^T, \theta) \propto f(z^T, x^T | \theta)$$

$$(9.21) \qquad \propto \sigma^{-2T} \exp\left\{ -\sum_{t=1}^{T} ||x_t - a z_t||^2 / 2\sigma^2 \right\} e^{-z_1^2/2} \prod_{t=2}^{T} \frac{e^{-z_t^2/2(1+\beta z_{t-1}^2)}}{(1 + \beta z_{t-1}^2)^{1/2}},$$

whose implementation using a Metropolis–Hastings algorithm (see Chapter 7) is detailed in Problem 9.24. Given a sample $(z_1^T, \ldots, z_m^T)$ simulated from (9.21), with $\theta = \eta$, and an observed sample x^T, the approximation (9.20) is given by

$$(9.22) \qquad \frac{1}{m} \sum_{i=1}^{m} \frac{f(z_i^T, x^T | \theta)}{f(z_i^T, x^T | \eta)},$$

where $f(z_i^T, x^T | \theta)$ is defined in (9.21).

Billio et al. (1998) consider a simulated sample of size $T = 100$ with $p = 2$, $a = (-0.2, 0.6)$, $\beta = 0.8$, and $\sigma^2 = 0.2$. The above approximation method is quite sensitive to the value of η and a good choice is the noninformative Bayes estimate associated with the prior $\pi(a, \beta, \sigma) = 1/\sigma$, which can be obtained by a Metropolis–Hastings algorithm (see Problem 9.24). Table 9.3 gives the result of the maximization of (9.22) for $m = 50,000$. The maximum likelihood estimator $\hat{\beta}$ is far from the true value β, but the estimated log-likelihood ratio at $(\hat{a}, \hat{\beta}, \hat{\sigma})$ is 0.348, which indicates that the likelihood is rather flat in this region.

The Multi-Stage Gibbs Sampler

In this place he was found by Gibbs, who had been sent for him in some haste. He got to his feet with promptitude, for he knew no small matter would have brought Gibbs in such a place at all.
—G.K. Chesterton, *The Innocence of Father Brown*

After two chapters of preparation on the slice and two-stage Gibbs samplers, respectively, we are now ready to envision the entire picture for the Gibbs sampler. We describe the general method in Section 10.1, whose theoretical properties are less complete than for the two-stage special case (see Section 10.2): The defining difference between that sampler and the multi-stage version considered here is that the interleaving structure of the two-stage chain does not carry over. Some of the consequences of interleaving are the fact that the individual subchains are also Markov chains, and the Duality Principle and Rao-Blackwellization hold in some generality. None of that is true here, in the multi-stage case. Nevertheless, the multi-stage Gibbs sampler enjoys many optimality properties, and still might be considered the workhorse of the MCMC world. The remainder of this chapter deals with implementation considerations, many in connection with the important role of the Gibbs sampler in Bayesian Statistics.

10.1 Basic Derivations

10.1.1 Definition

The following definition is a rather natural extension of what we saw in Section 8.2 for the general slice sampler. Suppose that for some $p > 1$, the random variable $\mathbf{X} \in \mathcal{X}$ can be written as $\mathbf{X} = (X_1, \ldots, X_p)$, where the X_i's are either uni- or multidimensional. Moreover, suppose that we can simulate from the corresponding univariate conditional densities $f_1, \ldots, f_p$, that is, we can simulate

$$X_i | x_1, x_2, \ldots, x_{i-1}, x_{i+1}, \ldots, x_p \sim f_i(x_i | x_1, x_2, \ldots, x_{i-1}, x_{i+1}, \ldots, x_p)$$

for $i = 1, 2, \ldots, p$. The associated *Gibbs sampling* algorithm (or *Gibbs sampler*) is given by the following transition from $X^{(t)}$ to $X^{(t+1)}$:

Algorithm A.39 –The Gibbs Sampler–

Given $\mathbf{x}^{(t)} = (x_1^{(t)}, \ldots, x_p^{(t)})$, generate

1. $X_1^{(t+1)} \sim f_1(x_1 | x_2^{(t)}, \ldots, x_p^{(t)})$;

2. $X_2^{(t+1)} \sim f_2(x_2 | x_1^{(t+1)}, x_3^{(t)}, \ldots, x_p^{(t)})$,   [A.39]

   $\vdots$

p. $X_p^{(t+1)} \sim f_p(x_p | x_1^{(t+1)}, \ldots, x_{p-1}^{(t+1)})$.

The densities $f_1, \ldots, f_p$ are called the *full conditionals*, and a particular feature of the Gibbs sampler is that these are the only densities used for simulation. Thus, even in a high-dimensional problem, *all of the simulations may be univariate*, which is usually an advantage.

Example 10.1. Autoexponential model. The *autoexponential model* of Besag (1974) has been found useful in some aspects of spatial modeling. When $y \in \mathbb{R}_+^3$, the corresponding density is

$$f(y_1, y_2, y_3) \propto \exp\{-(y_1 + y_2 + y_3 + \theta_{12} y_1 y_2 + \theta_{23} y_2 y_3 + \theta_{31} y_3 y_1)\},$$

with known $\theta_{ij} > 0$. The full conditional densities are exponential. For example,

$$Y_3 | y_1, y_2 \sim \mathcal{E}xp\left(1 + \theta_{23} y_2 + \theta_{31} y_1\right).$$

They are thus very easy to simulate from. In contrast, the other conditionals and the marginal distributions have forms such as

$$f(y_2 | y_1) \propto \frac{\exp\{-(y_1 + y_2 + \theta_{12} y_1 y_2)\}}{1 + \theta_{23} y_2 + \theta_{31} y_1},$$

$$f(y_1) \propto e^{-y_1} \int_0^{+\infty} \frac{\exp\{-y_2 - \theta_{12} y_1 y_2\}}{1 + \theta_{23} y_2 + \theta_{31} y_1} \, dy_2,$$

which cannot be simulated easily.   ‖

Example 10.2. Ising model. For the *Ising model* of Example 5.8, where

$$f(s) \propto \exp\left\{-H\sum_i s_i - J\sum_{(i,j)\in\mathcal{N}} s_i s_j\right\}, \qquad s_i \in \{-1,1\},$$

and where $\mathcal{N}$ denotes the neighborhood relation for the network, the full conditionals are given by

$$f(s_i|s_{j\neq i}) = \frac{\exp\{-Hs_i - Js_i\sum_{j:(i,j)\in\mathcal{N}} s_j\}}{\exp\{-H - J\sum_j s_{j:(i,j)\in\mathcal{N}}\} + \exp\{H + J\sum_j s_{j:(i,j)\in\mathcal{N}}\}}$$

$$(10.1) \qquad = \frac{\exp\{-(H + J\sum_{j:(i,j)\in\mathcal{N}} s_j)(s_i + 1)\}}{1 + \exp\{-2(H + J\sum_{j:(i,j)\in\mathcal{N}} s_j)\}}.$$

It is therefore particularly easy to implement the Gibbs sampler [A.39] for these conditional distributions by successively updating each node i of the network. (See Swendson and Wang 1987 and Swendson et al. 1992 for improved algorithms in this case, as introduced in Problem 7.43.)   ‖

Although the Gibbs sampler is, formally, a special case of the Metropolis–Hastings algorithm (or rather a combination of Metropolis–Hastings algorithms applied to different components; see Theorem 10.13), the Gibbs sampling algorithm has a number of distinct features:

(i) The acceptance rate of the Gibbs sampler is uniformly equal to 1. Therefore, every simulated value is accepted and the suggestions of Section 7.6.1 on the optimal acceptance rates do not apply in this setting. This also means that convergence assessment for this algorithm should be treated differently than for Metropolis–Hastings techniques.

(ii) The use of the Gibbs sampler implies limitations on the choice of instrumental distributions and requires a prior knowledge of some analytical or probabilistic properties of f.

(iii)The Gibbs sampler is, by construction, multidimensional. Even though some components of the simulated vector may be artificial for the problem of interest, or unnecessary for the required inference, the construction is still at least two-dimensional.

(iv)The Gibbs sampler does not apply to problems where the number of parameters varies, as in Chapter 11, because of the obvious lack of irreducibility of the resulting chain.

10.1.2 Completion

Following the mixture method discussed in Sections 5.3.1 and 8.2, the Gibbs sampling algorithm can be generalized by a "demarginalization" or completion construction.

Definition 10.3. Given a probability density f, a density g that satisfies

$$\int_Z g(x, z)\, dz = f(x)$$

is called a *completion* of f.

The density g is chosen so that the full conditionals of g are easy to simulate from and the Gibbs algorithm [A.39] is implemented on g instead of f. For $p > 1$, write $y = (x, z)$ and denote the conditional densities of $g(y) = g(y_1, \ldots, y_p)$ by

$$Y_1 | y_2, \ldots, y_p \sim g_1(y_1 | y_2, \ldots, y_p),$$
$$Y_2 | y_1, y_3, \ldots, y_p \sim g_2(y_2 | y_1, y_3, \ldots, y_p),$$
$$\vdots$$
$$Y_p | y_1, \ldots, y_{p-1} \sim g_p(y_p | y_1, \ldots, y_{p-1}).$$

The move from $Y^{(t)}$ to $Y^{(t+1)}$ is then defined as follows.

Algorithm A.40 –Completion Gibbs Sampler–

Given $(y_1^{(t)}, \ldots, y_p^{(t)})$, simulate

1. $Y_1^{(t+1)} \sim g_1(y_1 | y_2^{(t)}, \ldots, y_p^{(t)})$,
2. $Y_2^{(t+1)} \sim g_2(y_2 | y_1^{(t+1)}, y_3^{(t)}, \ldots, y_p^{(t)})$,   [A.40]
   $\vdots$
p. $Y_p^{(t+1)} \sim g_p(y_p | y_1^{(t+1)}, \ldots, y_{p-1}^{(t+1)})$.

The two-stage Gibbs sampler of Chapter 9 obviously corresponds to the particular case of [A.40], where f is completed in g and x in $y = (y_1, y_2)$ such that both conditional distributions $g_1(y_1 | y_2)$ and $g_2(y_2 | y_1)$ are available. (Again, note that both y_1 and y_2 can be either scalars or vectors.)

In cases when such completions seem necessary (for instance, when every conditional distribution associated with f is not explicit, or when f is unidimensional), there exist an infinite number of densities for which f is a marginal density. We will not discuss this choice in terms of optimality, because, first, there are few practical results on this topic (as it is similar to finding an optimal density g in the Metropolis–Hastings algorithm) and, second, because there exists, in general, a *natural* completion of f in g. As already pointed out in Chapter 9, *missing data models* (Section 5.3.1) provide a series of examples of natural completion.

In principle, the Gibbs sampler does not require that the completion of f into g and of x in $y = (x, z)$ should be related to the problem of interest. Indeed, there are settings where the vector z has no meaning from a statistical point of view and is only a useful device, that is, an *auxiliary variable* as in (2.8).

Example 10.4. Cauchy–Normal posterior distribution. As shown in Chapter 2 (Section 2.2), Student's t distribution can be generated as a mixture of a normal distribution by a chi squared distribution. This decomposition is useful when the expression $[1 + (\theta - \theta_0)^2]^{-\nu}$ appears in a more complex distribution. Consider, for instance, the density

$$f(\theta|\theta_0) \propto \frac{e^{-\theta^2/2}}{[1 + (\theta - \theta_0)^2]^\nu}.$$

This is the posterior distribution resulting from the model

$$X|\theta \sim \mathcal{N}(\theta, 1) \quad \text{and} \quad \theta \sim \mathcal{C}(\theta_0, 1)$$

(see also Example 12.12). A similar function arises in the estimation of the parameter of interest in a linear calibration model (see Example 1.16).

The density $f(\theta|\theta_0)$ can be written as the marginal density

$$f(\theta|\theta_0) \propto \int_0^\infty e^{-\theta^2/2} \, e^{-[1+(\theta-\theta_0)^2]\,\eta/2} \, \eta^{\nu-1} \, d\eta$$

and can, therefore, be completed as

$$g(\theta, \eta) \propto e^{-\theta^2/2} \, e^{-[1+(\theta-\theta_0)^2]\,\eta/2} \, \eta^{\nu-1},$$

which leads to the conditional densities

$$g_1(\eta|\theta) = \left(\frac{1 + (\theta - \theta_0)^2}{2}\right)^\nu \frac{\eta^{\nu-1}}{\Gamma(\nu)} \, \exp\left\{-[1 + (\theta - \theta_0)^2]\,\eta/2\right\},$$

$$g_2(\theta|\eta) = \sqrt{\frac{1+\eta}{2\pi}} \, \exp\left\{-\left(\theta - \frac{\eta\theta_0}{1+\eta}\right)^2 \frac{1+\eta}{2}\right\},$$

that is, to a gamma and a normal distribution on η and θ, respectively:

$$\eta|\theta \sim \mathcal{G}a\left(\nu, \frac{1 + (\theta - \theta_0)^2}{2}\right), \qquad \theta|\eta \sim \mathcal{N}\left(\frac{\theta_0\eta}{1+\eta}, \frac{1}{1+\eta}\right).$$

Note that the parameter η is completely meaningless for the problem at hand but serves to facilitate computations. (See Problem 10.7 for further properties.) ‖

The Gibbs sampler of [A.40] has also been called the Gibbs sampler with *systematic scan* (or systematic sweep), as the path of iteration is to proceed systematically in one direction. Such a sampler results in a non-reversible Markov chain, but we can construct a reversible Gibbs sampler with *symmetric scan*. The following algorithm guarantees the reversibility of the chain $(Y^{(t)})$ (see Problem 10.12).

Algorithm A.41 –Reversible Gibbs Sampler–

Given $(y_2^{(t)}, \ldots, y_p^{(t)})$, generate

1. $Y_1^* \sim g_1(y_1 | y_2^{(t)}, \ldots, y_p^{(t)})$
2. $Y_2^* \sim g_2(y_2 | y_1^*, y_3^{(t)}, \ldots, y_p^{(t)})$

$\vdots$

p-1. $Y_{p-1}^* \sim g_{p-1}(y_{p-1} | y_1^*, \ldots, y_{p-2}^*, y_p^{(t)})$

p. $Y_p^{(t+1)} \sim g_p(y_p | y_1^*, \ldots, y_{p-1}^*)$ [A.41]

p+1. $Y_{p-1}^{(t+1)} \sim g_{p-1}(y_{p-1} | y_1^*, \ldots, y_{p-2}^*, y_p^{(t+1)})$

$\vdots$

2p-1. $Y_1^{(t+1)} \sim g_1(y_1 | y_2^{(t+1)}, \ldots, y_p^{(t+1)})$

An[1] alternative to [A.41] has been proposed by Liu et al. (1995) and is called Gibbs sampling with *random scan*, as the simulation of the components of y is done in a random order following each transition. For the setup of [A.40], the modification [A.42] produces a reversible chain with stationary distribution g, and every simulated value is used.

Algorithm A.42 –Random Sweep Gibbs Sampler–

1. Generate a permutation $\sigma \in \mathcal{G}_p$;
2. Simulate $Y_{\sigma_1}^{(t+1)} \sim g_{\sigma_1}(y_{\sigma_1} | y_j^{(t)}, j \neq \sigma_1)$;   [A.42]

$\vdots$

p+1. Simulate $Y_{\sigma_p}^{(t+1)} \sim g_{\sigma_p}(y_{\sigma_p} | y_j^{(t+1)}, j \neq \sigma_p)$.

This algorithm improves upon [A.41], which only uses one simulation out of two.[2] (Recall that, following Theorem 6.65, the reversibility of $(Y^{(t)})$ allows application of the Central Limit Theorem.)

10.1.3 The General Hammersley–Clifford Theorem

We have already seen the Hammersley–Clifford Theorem in the special case of the two-stage Gibbs sampler (Section 9.1.4). The theorem also holds in the general case and follows from similar manipulations of the conditional distri-

---

[1] Step p is only called once, since repeating it would result in a complete waste of time!

[2] Obviously, nothing prevents the use of all the simulations in integral approximations.

butions.[3] (Hammersley and Clifford 1970; see also Besag 1974 and Gelman and Speed 1993).

To extend Theorem 9.3 to an arbitrary p, we need the condition of *positivity* given in Definition 9.4). We then have the following result, which specifies the joint distribution up to a proportionality constant, as in the two-stage Theorem 9.3.

Theorem 10.5. Hammersley–Clifford. *Under the positivity condition, the joint distribution g satisfies*

$$g(y_1, \ldots, y_p) \propto \prod_{j=1}^{p} \frac{g_{\ell_j}(y_{\ell_j} | y_{\ell_1}, \ldots, y_{\ell_{j-1}}, y'_{\ell_{j+1}}, \ldots, y'_{\ell_p})}{g_{\ell_j}(y'_{\ell_j} | y_{\ell_1}, \ldots, y_{\ell_{j-1}}, y'_{\ell_{j+1}}, \ldots, y'_{\ell_p})}$$

for every permutation ℓ on $\{1, 2, \ldots, p\}$ and every $y' \in \mathcal{Y}$.

Proof. For a given $y' \in \mathcal{Y}$,

$$\begin{aligned}
g(y_1, \ldots, y_p) &= g_p(y_p | y_1, \ldots, y_{p-1}) g^p(y_1, \ldots, y_{p-1}) \\
&= \frac{g_p(y_p | y_1, \ldots, y_{p-1})}{g_p(y'_p | y_1 \ldots, y_{p-1})} g(y_1, \ldots, y_{p-1}, y'_p) \\
&= \frac{g_p(y_p | y_1, \ldots, y_{p-1})}{g_p(y'_p | y_1, \ldots, y_{p-1})} \frac{g_{p-1}(y_{p-1} | y_1, \ldots, y_{p-2}, y'_p)}{g_{p-1}(y'_{p-1} | y_1, \ldots, y_{p-2}, y'_p)} \\
&\quad \times g(y_1, \ldots, y'_{p-1}, y'_p) \,.
\end{aligned}$$

A recursion argument then shows that

$$g(y_1, \ldots, y_p) = \prod_{j=1}^{p} \frac{g_j(y_j | y_1, \ldots, y_{j-1}, y'_{j+1}, \ldots, y'_p)}{g_j(y'_j | y_1, \ldots, y_{j-1}, y'_{j+1}, \ldots, y'_p)} g(y'_1, \ldots, y'_p).$$

The proof is identical for an arbitrary permutation ℓ. □

The extension of Theorem 10.5 to the non-positive case is more delicate and requires additional assumptions, as shown by Example 10.7. Besag (1994) proposes a formal generalization which is not always relevant in the setup of Gibbs sampling algorithms. Hobert et al. (1997) modify Besag's (1994) condition to preserve the convergence properties of these algorithms and show, moreover, that the connectedness of the support of g is essential for [A.40] to converge under every regular parameterization of the model. (See Problems 10.5 and 10.8.)

This result is also interesting from the general point of view of MCMC algorithms, since it shows that the density g is known up to a multiplicative

---

[3] Clifford and Hammersley never published their result. Hammersley (1974) justifies their decision by citing the impossibility of extending this result to the non-positive case, as shown by Moussouris (1974) through a counterexample.

constant when the conditional densities $g_i(y_i|y_{j\neq i})$ are available. It is therefore possible to compare the Gibbs sampler with alternatives like Accept–Reject or Metropolis–Hastings algorithms. This also implies that the Gibbs sampler is never *the single available method* to simulate from g and that it is always possible to include Metropolis–Hastings steps in a Gibbs sampling algorithm, following an hybrid strategy developed in Section 10.3.

10.2 Theoretical Justifications

10.2.1 Markov Properties of the Gibbs Sampler

In this section we investigate the convergence properties of the Gibbs sampler in the multi-stage case. The first thing we show is that most Gibbs samplers satisfy the minimal conditions necessary for ergodicity. We have mentioned that Gibbs sampling is a special case of the Metropolis–Hastings algorithm (this notion is formalized in Theorem 10.13). Unfortunately, however, this fact is not much help to us in the current endeavor.

We will work with the general Gibbs sampler [A.40] and show that the Markov chain $(Y^{(t)})$ converges to the distribution g and the subchain $(X^{(t)})$ converges to the distribution f. It is important to note that although $(Y^{(t)})$ is, by construction, a Markov chain, the subchain $(X^{(t)})$ is, typically, not a Markov chain, except in the particular case of two-stage Gibbs (see Section 9.2).

Theorem 10.6. *For the Gibbs sampler of [A.40], if $(Y^{(t)})$ is ergodic, then the distribution g is a stationary distribution for the chain $(Y^{(t)})$ and f is the limiting distribution of the subchain $(X^{(t)})$.*

Proof. The kernel of the chain $(Y^{(t)})$ is the product

$$(10.2)\qquad K(\mathbf{y},\mathbf{y}') = g_1(y_1'|y_2,\ldots,y_p)$$
$$\times\, g_2(y_2'|y_1',y_3,\ldots,y_p)\cdots g_p(y_p'|y_1',\ldots,y_{p-1}').$$

For the vector $\mathbf{y} = (y_1,y_2,\ldots,y_p)$, let $g^i(y_1,\ldots,y_{i-1},y_{i+1},\ldots,y_p)$ denote the marginal density of the vector $\mathbf{y}$ with y_i integrated out. If $\mathbf{Y}\sim g$ and A is measurable under the dominating measure, then

$$P(\mathbf{Y}'\in A) = \int \mathbb{I}_A(\mathbf{y}')\,K(\mathbf{y},\mathbf{y}')\,g(\mathbf{y})\,d\mathbf{y}'d\mathbf{y}$$
$$= \int \mathbb{I}_A(\mathbf{y}')\,[g_1(y_1'|y_2,\ldots,y_p)\cdots g_p(y_p'|y_1',\ldots,y_{p-1}')]$$
$$\times\,[g_1(y_1|y_2,\ldots,y_p)g^1(y_2,\ldots,y_p)]\,dy_1\cdots dy_pdy_1'\ldots dy_p'$$
$$= \int \mathbb{I}_A(\mathbf{y}')\,g_2(y_2'|y_1',\ldots,y_p)\cdots g_p(y_p'|y_1',\ldots,y_{p-1}')$$
$$\times\, g(y_1',y_2,\ldots,y_p)dy_2\cdots dy_pdy_1'\cdots dy_p'\,,$$

where we have integrated out y_1 and combined $g_1(y_1'|y_2,\ldots,y_p)g^1(y_2,\ldots,y_p)$
$= g(y_1',y_2,\ldots,y_p)$. Next, write $g(y_1',y_2,\ldots,y_p) = g(y_2|y_1',y_3,\ldots,y_p)g^2(y_1',y_3,$
$\ldots,y_p)$, and integrate out y_2 to obtain

$$P(\mathbf{Y}' \in A) = \int \mathbb{I}_A(\mathbf{y}')\, g_3(y_3'|y_1',y_2',\ldots,y_p)\cdots g_p(y_p'|y_1',\ldots,y_{p-1}')$$
$$\times\, g(y_1',y_2',y_3,\ldots,y_p)\, dy_3 \cdots dy_p dy_1' \cdots dy_p' \,.$$

By continuing in this fashion and successively integrating out the y_i's, the
above probability is

$$P(\mathbf{Y}' \in A) = \int_A g(y_1',\ldots,y_p')\, d\mathbf{y}',$$

showing that g is the stationary distribution. Therefore, Theorem 6.50 implies
that $Y^{(t)}$ is asymptotically distributed according to g and $X^{(t)}$ is asymptotically distributed according to the marginal f, by integration.   □

A shorter proof can be based on the stationarity of g after *each* step in the
p steps of $[A.40]$ (see Theorem 10.13).

We now consider the irreducibility of the Markov chain produced by the
Gibbs sampler. The following example shows that a Gibbs sampler need not
be irreducible:

Example 10.7. Nonconnected support. Let $\mathcal{E}$ and $\mathcal{E}'$ denote the disks
of $\mathbb{R}^2$ with radius 1 and respective centers $(1,1)$ and $(-1,-1)$. Consider the
distribution with density

$$(10.3) \qquad f(x_1,x_2) = \frac{1}{2\pi}\left\{\mathbb{I}_{\mathcal{E}}(x_1,x_2) + \mathbb{I}_{\mathcal{E}'}(x_1,x_2)\right\},$$

which is shown in Figure 10.1. From this density we cannot produce an irreducible chain through $[A.40]$, since the resulting chain remains concentrated
on the (positive or negative) quadrant on which it is initialized. (Note that a
change of coordinates such as $z_1 = x_1 + x_2$ and $z_2 = x_1 - x_2$ is sufficient to
remove this difficulty.)   ‖

Recall the *positivity condition*, given in Definition 9.4, and the consequence
that two arbitrary Borel subsets of the support can be joined in a single
iteration of $[A.40]$. We therefore have the following theorem, whose proof is
left to Problem 10.11.

Theorem 10.8. *For the Gibbs sampler of $[A.40]$, if the density g satisfies the
positivity condition, it is irreducible.*

Unfortunately, the condition of Theorem 10.8 is often difficult to verify,
and Tierney (1994) gives a more manageable condition which we will use
instead.

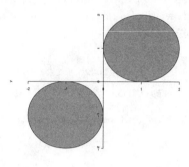

Fig. 10.1. Support of the function $f(x_1, x_2)$ of (10.3)

Lemma 10.9. *If the transition kernel associated with [A.40] is absolutely continuous with respect to the dominating measure, the resulting chain is Harris recurrent.*

The proof of Lemma 10.9 is similar to that of Lemma 7.3, where Harris recurrence was shown to follow from irreducibility. Here, the condition on the Gibbs transition kernel yields an irreducible chain, and Harris recurrence follows. Once the irreducibility of the chain associated with [A.40] is established, more advanced convergence properties can be established.

This condition of absolute continuity on the kernel (10.2) is satisfied by most decompositions. However, if one of the i steps ($1 \le i \le p$) is replaced by a simulation from an Metropolis–Hastings algorithm, as in the hybrid algorithms described in Section 10.3, absolute continuity is lost and it will be necessary to either study the recursion properties of the chain or to introduce an additional simulation step to guarantee Harris recurrence.

From the property of Harris recurrence, we can now establish a result similar to Theorem 7.4.

Theorem 10.10. *If the transition kernel of the Gibbs chain $(Y^{(t)})$ is absolutely continuous with respect to the measure μ,*

(i) If $h_1, h_2 \in L^1(g)$ with $\int h_2(y)dg(y) \ne 0$, then

$$\lim_{n \to \infty} \frac{\sum_{t=1}^{T} h_1(Y^{(t)})}{\sum_{t=1}^{T} h_2(Y^{(t)})} = \frac{\int h_1(y)dg(y)}{\int h_2(y)dg(y)} \quad a.e. \ g.$$

(ii) If, in addition, $(Y^{(t)})$ is aperiodic, then, for every initial distribution μ,

$$\lim_{n \to \infty} \left\| \int K^n(y, \cdot)\mu(dx) - g \right\|_{TV} = 0.$$

We also state a more general condition, which follows from Lemma 7.6 by allowing for moves in a minimal neighborhood.

Lemma 10.11. *Let* $\mathbf{y} = (y_1, \ldots, y_p)$ *and* $\mathbf{y}' = (y_1', \ldots, y_p')$ *and suppose there exists* $\delta > 0$ *for which* $\mathbf{y}, \mathbf{y}' \in supp(g)$, $|\mathbf{y} - \mathbf{y}'| < \delta$ *and*

$$g_i(y_i | y_1, \ldots, y_{i-1}, y_{i+1}', \ldots, y_p') > 0, \quad i = 1, \ldots, p.$$

If there exists $\delta' < \delta$ *for which almost every pair* $(\mathbf{y}, \mathbf{y}') \in supp(g)$ *can be connected by a finite sequence of balls with radius* δ' *having the (stationary) measure of the intersection of two consecutive balls positive, then the chain produced by* [A.40] *is irreducible and aperiodic.*

The laborious formulation of this condition is needed to accommodate settings where the support of g is not connected, as in Example 10.7, since a necessary irreducibility condition is that two connected components of $supp(g)$ can be linked by the kernel of [A.40]. The proof of Lemma 10.11 follows from arguments similar to those in the proof of Lemma 7.6.

Corollary 10.12. *The conclusions of Theorem 10.10 hold if the Gibbs sampling Markov chain* $(Y^{(t)})$ *has conditional densities* $g_i(y_i | y_1, \ldots, y_{i-1}, y_{i+1}', \ldots, y_p')$ *that satisfy the assumptions of Lemma 10.11.*

10.2.2 Gibbs Sampling as Metropolis–Hastings

We now examine the exact relationship between the Gibbs sampling and Metropolis–Hastings algorithms by considering [A.40] as the composition of p Markovian kernels. (As mentioned earlier, this representation is not sufficient to establish irreducibility since each of these separate Metropolis–Hastings algorithms does not produce an irreducible chain, given that it only modifies one component of y. In fact, these kernels are *never irreducible*, since they are constrained to subspaces of lower dimensions.)

Theorem 10.13. *The Gibbs sampling method of* [A.40] *is equivalent to the composition of* p *Metropolis–Hastings algorithms, with acceptance probabilities uniformly equal to* 1.

Proof. If we write [A.40] as the composition of p "elementary" algorithms which correspond to the p simulation steps from the conditional distribution, it is sufficient to show that each of these algorithms has an acceptance probability equal to 1. For $1 \leq i \leq p$, the instrumental distribution in step i of [A.40] is given by

$$\begin{aligned} q_i(y'|y) = &\, \delta_{(y_1, \ldots, y_{i-1}, y_{i+1}, y_p)}(y_1', \ldots, y_{i-1}', y_{i+1}', y_p') \\ &\times g_i(y_i' | y_1, \ldots, y_{i-1}, y_{i+1}, y_p) \end{aligned}$$

and the ratio defining the probability $\rho(y, y')$ is therefore

$$\frac{g(y') \, q_i(y|y')}{g(y) \, q_i(y'|y)} = \frac{g(y') \, g_i(y_i|y_1,\ldots,y_{i-1},y_{i+1},y_p)}{g(y) \, g_i(y_i'|y_1,\ldots,y_{i-1},y_{i+1},y_p)}$$

$$= \frac{g_i(y_i'|y_1,\ldots,y_{i-1},y_{i+1},y_p) \, g_i(y_i|y_1,\ldots,y_{i-1},y_{i+1},y_p)}{g_i(y_i|y_1,\ldots,y_{i-1},y_{i+1},y_p) \, g_i(y_i'|y_1,\ldots,y_{i-1},y_{i+1},y_p)}$$

$$= 1.$$

□

Note that Gibbs sampling is not the only MCMC algorithm to enjoy this property. As noted in Section 7.2, every kernel q associated with a reversible Markov chain with invariant distribution g has an acceptance probability uniformly equal to 1 (see also Barone and Frigessi 1989 and Liu and Sabbati 1999). However, the (global) acceptance probability for the vector $(y_1,\ldots,y_p)$ is usually different from 1 and a direct processing of [A.40] as a particular Metropolis–Hastings algorithm leads to a positive probability of rejection.

Example 10.14. (Continuation of Example 10.1) Consider the two-dimensional autoexponential model

$$g(y_1,y_2) \propto \exp\{-y_1 - y_2 - \theta_{12}y_1y_2\} \, .$$

The kernel associated with [A.40] is then composed of the conditional densities

$$K(y,y') = g_1(y_1'|y_2) \, g_2(y_2'|y_1'),$$

with

$$g_1(y_1|y_2) = (1 + \theta_{12}y_2) \, \exp\left\{-(1 + \theta_{12}y_2) \, y_1\right\},$$
$$g_2(y_2|y_1) = (1 + \theta_{12}y_1) \, \exp\left\{-(1 + \theta_{12}y_1) \, y_2\right\}.$$

The ratio

$$\frac{g(y_1',y_2')}{g(y_1,y_2)} \frac{K((y_1',y_2'),(y_1,y_2))}{K((y_1,y_2),(y_1',y_2'))} = \frac{(1 + \theta_{12}y_2')(1 + \theta_{12}y_1)}{(1 + \theta_{12}y_2)(1 + \theta_{12}y_1')} \, e^{\theta_{12}(y_2'y_1 - y_1'y_2)}$$

is thus different from 1 for almost every vector (y_1,y_2,y_1',y_2').   ‖

In this global analysis of [A.40], it is possible to reject some of the values produced by the sequence of steps 1,..., p of this algorithm. This version of the Metropolis–Hastings algorithm could then be compared with the original Gibbs sampler. No full-scale study has been yet undertaken in this direction, except for the modification introduced by Liu (1995, 1996a) and presented in Section 10.3. The fact that the first approach allows for rejections seems beneficial when considering the results of Gelman et al. (1996), but the Gibbs sampling algorithm cannot be evaluated in this way.

10.2.3 Hierarchical Structures

To conclude this section, we investigate a structure for which Gibbs sampling is particularly well adapted, that of *hierarchical models*. These are structures in which the distribution f can be decomposed as $(I \geq 1)$

$$f(x) = \int f_1(x|z_1)f_2(z_1|z_2)\cdots f_I(z_I|z_{I+1})f_{I+1}(z_{I+1})\,dz_1\cdots dz_{I+1},$$

for either structural or computational reasons. Such models naturally appear in the Bayesian analysis of complex models, where the diversity of the prior information or the variability of the observations may require the introduction of several levels of prior distributions (see Wakefield et al. 1994, Robert 2001, Chapter 10, Bennett et al. 1996, Spiegelhalter et al. 1996, and Draper 1998). This is, for instance, the case of Example 10.17, where the first level represents the exchangeability of the parameters λ_i, and the second level represents the prior information. In the particular case where the prior information is sparse, hierarchical modeling is also useful since diffuse (or noninformative) distributions can be introduced at various levels of the hierarchy.

The following two examples illustrate situations where hierarchical models are particularly useful (see also Problems 10.29–10.36).

Example 10.15. Animal epidemiology. Research in animal epidemiology sometimes uses data from groups of animals, such as litters or herds. Such data may not follow some of the usual assumptions of independence, etc., and, as a result, variances of parameter estimates tend to be larger (this phenomenon is often referred to as "overdispersion"). Schukken et al. (1991) obtained counts of the number of cases of clinical mastitis[4] in dairy cattle herds over a one year period.

If we assume that, in each herd, the occurrence of mastitis is a Bernoulli random variable, and if we let X_i, $i = 1,\ldots,m$, denote the number of cases in herd i, it is then reasonable to model $X_i \sim \mathcal{P}(\lambda_i)$, where λ_i is the underlying rate of infection in herd i. However, there is lack of independence here (mastitis is infectious), which might manifest itself as overdispersion. To account for this, Schukken et al. (1991) put a gamma prior distribution on the Poisson parameter. A complete hierarchical specification is

$$X_i \sim \mathcal{P}(\lambda_i),$$
$$\lambda_i \sim \mathcal{G}a(\alpha, \beta_i),$$
$$\beta_i \sim \mathcal{G}a(a, b),$$

where α, a, and b are specified. The posterior density of λ_i, $\pi(\lambda_i|\mathbf{x}, \alpha)$, can now be simulated via the Gibbs sampler

$$\lambda_i \sim \pi(\lambda_i|\mathbf{x}, \alpha, \beta_i) = \mathcal{G}a(x_i + \alpha, 1 + \beta_i),$$
$$\beta_i \sim \pi(\beta_i|\mathbf{x}, \alpha, a, b, \lambda_i) = \mathcal{G}a(\alpha + a, \lambda_i + b).$$

---

[4] Mastitis is an inflammation usually caused by infection.

Figure 10.2 shows selected estimates of λ_i and β_i. For more details see Problem 10.10 or Eberly and Casella (2003).   ‖

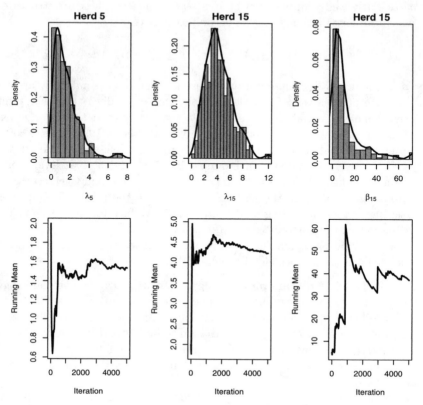

Fig. 10.2. Histograms, density estimates and running mean estimates for three selected parameters of the mastitis data (Example 10.15).

Example 10.16. Medical models. Part of the concern of the study of *pharmacokinetics* is the modeling of the relationship between the dosage of a drug and the resulting concentration in the blood. (More generally, pharmacokinetics studies the different interactions of a drug and the body.) Gilks et al. (1993) introduce an approach for estimating pharmacokinetic parameters that uses the traditional mixed-effects model and nonlinear structure, but which is also robust to the outliers common to clinical trials. For a given dose d_i administered at time 0 to patient i, the measured log concentration in the blood at time t_{ij}, X_{ij}, is assumed to follow a Student's t distribution

$$\frac{X_{ij} - \log g_{ij}(\lambda_i)}{\sigma\sqrt{n/(n-2)}} \sim \mathcal{T}_n,$$

where $\lambda_i = (\log C_i, \log V_i)'$ is a vector of parameters for the ith individual, σ^2 is the measurement error variance, and g_{ij} is given by

$$g_{ij}(\lambda_i) = \frac{d_i}{V_i} \exp\left(-\frac{C_i \, t_{ij}}{V_i}\right).$$

(The parameter C_i represents *clearance* and V_i represents *volume* for patient i.) Gilks et al. (1993) then complete the hierarchy by specifying an noninformative prior on σ, $\pi(\sigma) = 1/\sigma$ and

$$\lambda_i \sim \mathcal{N}(\theta, \Sigma),$$

$$\theta \sim \mathcal{N}(\tau_1, T_1), \quad \text{and} \quad \Sigma^{-1} \sim \mathcal{W}_2(\tau_2, T_2),$$

where the values of τ_1, T_1, τ_2, and T_2 are specified. Conjugate structures can then be exhibited for most parameters by using the Dickey's decomposition of the Student's t distribution (see Example 4.5); that is, by associating to each x_{ij} an (artificial) variable ω_{ij} such that

$$X_{ij}|\omega_{ij} \sim \mathcal{N}\left(\log g_{ij}(\lambda_i), \omega_{ij}\sigma^2 \frac{n}{n-2}\right).$$

Using this completion, the full conditional distributions on the C_i's and θ are normal, while the full conditional distributions on σ^2 and Σ are inverse gamma and inverse Wishart, respectively. The case of the V_i's is more difficult to handle since the full conditional density is proportional to
(10.4)

$$\exp\left(\frac{-1}{2}\left\{\sum_j \left(\log V_i + \frac{C_i t_{ij}}{V_i} - \mu_i\right)^2 \bigg/ \xi + (\log V_i - \gamma_i)^2 / \zeta\right\}\right),$$

where the hyperparameters μ_i, ξ, γ_i, and ζ depend on the other parameters and x_{ij}. Gilks et al. (1993) suggest using an Accept–Reject algorithm to simulate from (10.4), by removing the $C_i t_{ij}/V_i$ terms. Another possibility is to use a Metropolis–Hastings step, as described in Section 10.3.3.   ‖

For some cases of hierarchical models, it is possible to show that the associated Gibbs chains are *uniformly ergodic*. Typically, this can only be accomplished on a case-by-case study, as in the following example. Schervish and Carlin (1992) studies the weaker property of *geometric convergence* of [A.40] in some detail, which requires conditions on the kernel that can be difficult to assess in practice.

Example 10.17. Nuclear pump failures. Gaver and O'Muircheartaigh (1987) introduced a model that is frequently used (or even overused) in the Gibbs sampling literature to illustrate various properties (see, for instance, Gelfand and Smith 1990, Tanner 1996, or Guihenneuc-Jouyaux and Robert

Pump	1	2	3	4	5	6	7	8	9	10
Failures	5	1	5	14	3	19	1	1	4	22
Time	94.32	15.72	62.88	125.76	5.24	31.44	1.05	1.05	2.10	10.48

Table 10.1. Numbers of failures and times of observation of 10 pumps in a nuclear plant (*Source*: Gaver and O'Muircheartaigh 1987).

1998). This model describes multiple failures of pumps in a nuclear plant, with the data given in Table 10.1. The modeling is based on the assumption that the failures of the ith pump follow a Poisson process with parameter λ_i ($1 \leq i \leq 10$). For an observed time t_i, the number of failures p_i is thus a Poisson $\mathcal{P}(\lambda_i t_i)$ random variable. The associated prior distributions are ($1 \leq i \leq 10$)

$$\lambda_i \overset{iid}{\sim} \mathcal{G}a(\alpha, \beta), \qquad \beta \sim \mathcal{G}a(\gamma, \delta),$$

with $\alpha = 1.8$, $\gamma = 0.01$, and $\delta = 1$ (see Gaver and O'Muircheartaigh 1987 for a motivation of these numerical values). The joint distribution is thus

$$\pi(\lambda_1, \ldots, \lambda_{10}, \beta | t_1, \ldots, t_{10}, p_1, \ldots, p_{10})$$

$$\propto \prod_{i=1}^{10} \left\{ (\lambda_i t_i)^{p_i} e^{-\lambda_i t_i} \lambda_i^{\alpha-1} e^{-\beta \lambda_i} \right\} \beta^{10\alpha} \beta^{\gamma-1} e^{-\delta\beta}$$

$$\propto \prod_{i=1}^{10} \left\{ \lambda_i^{p_i+\alpha-1} e^{-(t_i+\beta)\lambda_i} \right\} \beta^{10\alpha+\gamma-1} e^{-\delta\beta}$$

and a natural decomposition[5] of π in conditional distributions is

$$\lambda_i | \beta, t_i, p_i \sim \mathcal{G}a(p_i + \alpha, t_i + \beta) \qquad (1 \leq i \leq 10),$$

$$\beta | \lambda_1, \ldots, \lambda_{10} \sim \mathcal{G}a\left(\gamma + 10\alpha, \delta + \sum_{i=1}^{10} \lambda_i\right).$$

The transition kernel on β associated with [A.40] and this decomposition satisfies (see Problem 10.3)

$$K(\beta, \beta') = \int \frac{(\beta')^{\gamma+10\alpha-1}}{\Gamma(10\alpha+\gamma)} \left(\delta + \sum_{i=1}^{10} \lambda_i\right)^{\gamma+10\alpha} \exp\left\{-\beta'\left(\delta + \sum_{i=1}^{10} \lambda_i\right)\right\}$$

$$\times \prod_{i=1}^{10} \frac{(t_i + \beta)^{p_i+\alpha}}{\Gamma(p_i + \alpha)} \lambda_i^{p_i+\alpha-1} \exp\{-(t_i + \beta)\lambda_i\} \, d\lambda_1 \ldots d\lambda_{10}$$

$$(10.5) \qquad \geq \frac{\delta^{\gamma+10\alpha}(\beta')^{\gamma+10\alpha-1}}{\Gamma(10\alpha+\gamma)} e^{-\delta\beta'} \prod_{i=1}^{10} \left(\frac{t_i}{t_i + \beta'}\right)^{p_i+\alpha}.$$

---

[5] This decomposition reflects the hierarchical structure of the model.

This minorization by a positive quantity which does not depend on β implies that the entire space $(\mathbb{R}_+)$ is a small set for the transition kernel; thus, by Theorem 6.59, the chain (β^t) is uniformly ergodic (see Definition 6.58).

As shown in Section 9.2.3, uniform ergodicity directly extends to the dual chain $\lambda^t = (\lambda_1^t, \ldots, \lambda_{10}^t)$.   ‖

Note that Examples 10.15 and 10.17 are cases of two-stage Gibbs samplers, which means that $(X^{(t)})$ and $(Z^{(t)})$ are interleaved Markov chains (Definition 9.10). As noted earlier, the interesting features of interleaving disappear for $p \geq 3$, because the subchains $(Y_1^{(t)}), \ldots, (Y_p^{(t)})$ are not Markov chains, although the vector $(\mathbf{Y}^{(t)})$ is. Therefore, there is no transition kernel associated with $(Y_i^{(t)})$, and the study of uniform ergodicity can only cover a grouped vector of $(p-1)$ of the p components of $(\mathbf{Y}^{(t)})$ since the original kernel cannot, in general, be bounded uniformly from below. If we denote $\mathbf{z}_1 = (y_2, \ldots, y_p)$, the transition from $\mathbf{z}_1$ to $\mathbf{z}_1'$ has the following kernel:

$$K_1(\mathbf{z}_1, \mathbf{z}_1') = \int_{\mathcal{Y}_1} g_1(y_1'|\mathbf{z}_1) g_2(y_2'|y_1', y_3, \ldots, y_p) \cdots$$

(10.6)
$$\times g_p(y_p'|y_1', y_2', \ldots, y_{p-1}') \, dy_1' \, .$$

While some setups result in a uniform bound on $K_1(\mathbf{z}_1, \mathbf{z}_1')$, it is often impossible to achieve a uniform minorization of K. For example, it is impossible to bound the transition kernel of the autoexponential model (Example 10.1) from below.

10.3 Hybrid Gibbs Samplers

10.3.1 Comparison with Metropolis–Hastings Algorithms

A comparison between a Gibbs sampling method and an arbitrary Metropolis–Hastings algorithm would seem a priori to favor Gibbs sampling since it derives its conditional distributions from the true distribution f, whereas a Metropolis–Hastings kernel is, at best, based on a approximation of this distribution f. In particular, we have noted several times that Gibbs sampling methods are, by construction, more straightforward than Metropolis–Hastings methods, since they cannot have a "bad" choice of the instrumental distribution and, hence, they avoid useless simulations (*rejections*). Although these algorithms can be formally compared, seeking a ranking of these two main types of MCMC algorithms is not only illusory but also somewhat pointless.

However, we do stress in this section that the availability and apparent objectivity of Gibbs sampling methods are not necessarily compelling arguments. If we consider the Gibbs sampler of Theorem 10.13, the Metropolis–Hastings algorithms which underly this method are not valid on an individual basis since they do not produce irreducible Markov chains $(Y^{(t)})$. Therefore, only a combination of a sufficient number of Metropolis–Hastings algorithms can ensure the validity of the Gibbs sampler. This composite structure is also a weakness of the method, since a decomposition of the joint distribution f given a particular system of coordinates does not necessarily agree with the form of f. Example 10.7 illustrates this incompatibility in a pathological case: A wrong choice of the coordinates traps the corresponding Gibbs sampling in one of the two connected components of the support of f. Hills and Smith (1992, 1993) also propose examples where an incorrect parameterization of the model significantly increases the convergence time for the Gibbs sampler. As seen in Note 9.7.1 in the particular case of mixtures of distributions, parameterization influences the performances of the Gibbs sampler to the extent of getting into a trapping state.

To draw an analogy, let us recall that when a function $v(y_1, \ldots, y_p)$ is maximized one component at a time, the resulting solution is not always satisfactory since it may correspond to a saddlepoint or to a local maximum of v. Similarly, the simulation of a single component at each iteration of $[A.40]$ restricts the possible excursions of the chain $(Y^{(t)})$ and this implies that Gibbs sampling methods are generally *slow* to converge, since they are slow to explore the surface of f.

This intrinsic defect of the Gibbs sampler leads to phenomena akin to convergence to local maxima in optimization algorithms, which are expressed by strong attractions to the closest local modes and, in consequence, to difficulties in exploring the entire range of the support of f.

Example 10.18. Two-dimensional mixture. Consider a two-dimensional mixture of normal distributions,

$$(10.7) \qquad p_1 \mathcal{N}_2(\mu_1, \Sigma_1) + p_2 \mathcal{N}_2(\mu_2, \Sigma_2) + p_3 \mathcal{N}_2(\mu_3, \Sigma_3),$$

given in Figure 10.3 as a gray-level image.[6] Both unidimensional conditionals are also mixtures of normal distributions and lead to a straightforward Gibbs sampler. The first 100 steps of the associated Gibbs sampler are represented on Figure 10.3; they show mostly slow moves along the two first components of the mixture and a single attempt to reach the third component, which is too far in the tail of the conditional. Note that the numerical values chosen for this illustration are such that the third component has a 31% probability mass in (10.7). (Each step of the Gibbs sampler is given in the graph, which explains for the succession of horizontal and vertical moves.)   ‖

10.3.2 Mixtures and Cycles

The drawbacks of Metropolis–Hastings algorithms are different from those of the Gibbs sampler, as they are more often related to a bad agreement between

---

[6] As opposed to the other mixture examples, this example considers simulation from (10.7), which is a more artificial setting, given that direct simulation is straightforward.

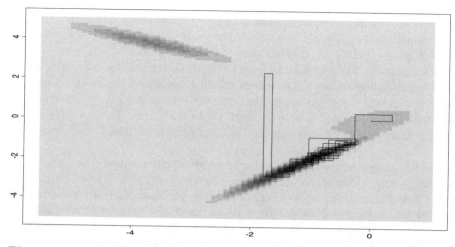

Fig. 10.3. Successive (full conditional) moves of the Gibbs chain on the surface of the stationary distribution (10.7), represented by gray levels (darker shades mean higher elevations).

f and the instrumental distribution. Moreover, the freedom brought by Metropolis–Hastings methods sometimes allows for remedies to these drawbacks through the modification of some scale (parameters or hyperparameters are particularly useful).

Compared to the Gibbs sampler, a failing of Metropolis–Hastings algorithms is to miss the finer details of the distribution f, if the simulation is "too coarse." However, following Tierney (1994), a way to take advantage of both algorithms is to implement a *hybrid* approach which uses both Gibbs sampling and Metropolis–Hastings algorithms.

Definition 10.19. An *hybrid MCMC algorithm* is a Markov chain Monte Carlo method which simultaneously utilizes both Gibbs sampling steps and Metropolis–Hastings steps. If $K_1, \ldots, K_n$ are kernels which correspond to these different steps and if $(\alpha_1, \ldots, \alpha_n)$ is a probability distribution, a *mixture* of $K_1, K_2, \ldots, K_n$ is an algorithm associated with the kernel

$$\tilde{K} = \alpha_1 K_1 + \cdots + \alpha_n K_n$$

and a *cycle* of $K_1, K_2, \ldots, K_n$ is the algorithm with kernel

$$K^* = K_1 \circ \cdots \circ K_n,$$

where "$\circ$" denotes the composition of functions.

Of course, this definition is somewhat ambiguous since Theorem 10.13 states that the Gibbs sampling is already a composition of Metropolis–Hastings kernels; that is, a cycle according to the above definition. Definition 10.19

must therefore be understood as processing heterogeneous MCMC algorithms, where the chain under study is a subchain $(Y^{(kt)})_t$ of the chain produced by the algorithm.

From our initial perspective concerning the speed of convergence of the Gibbs sampler, Definition 10.19 leads us to consider modifications of the initial algorithm where, every m iterations, [A.40] is replaced by a Metropolis–Hastings step with larger dispersion or, alternatively, at each iteration, this Metropolis–Hastings step is selected with probability $1/m$. These modifications are particularly helpful to escape "trapping effects" related to local modes of f.

Hybrid procedures are valid from a theoretical point of view, when the *heterogeneity* of the chains generated by cycles is removed by considering only the subchains $(Y^{(mt)})_t$ (although the entire chain should be exploited). A composition of kernels associated with an identical stationary distribution f leads to a kernel with stationary distribution f. The irreducibility and aperiodicity of $\tilde{K}$ directly follows from the irreducibility and the aperiodicity of one of the i kernels K_i. In the case of a *cycle*, we already saw that the irreducibility of K^* for the Gibbs sampler does not require the irreducibility of its component kernels and a specific study of the algorithm at hand may sometimes be necessary. Tierney (1994) also proposes sufficient conditions for uniform ergodicity.

Proposition 10.20. *If K_1 and K_2 are two kernels with the same stationary distribution f and if K_1 produces a uniformly ergodic Markov chain, the mixture kernel*

$$\tilde{K} = \alpha K_1 + (1 - \alpha)K_2 \qquad (0 < \alpha < 1)$$

is also uniformly ergodic. Moreover, if $\mathcal{X}$ is a small set for K_1 with $m = 1$, the kernel cycles $K_1 \circ K_2$ and $K_2 \circ K_1$ are uniformly ergodic.

Proof. If K_1 produces a uniformly ergodic Markov chain, there exists $m \in \mathbb{N}$, $\varepsilon_m > 0$, and a probability measure ν_m such that K_1 satisfies

$$K_1^m(x, A) \geq \varepsilon_m \nu_m(A), \qquad \forall x \in \mathcal{X}, \quad \forall A \in \mathcal{B}(\mathcal{X}).$$

Therefore, we have the minorization condition

$$(\alpha K_1 + (1 - \alpha)K_2)^m(x, A) \geq \alpha^m K_1^m(x, A) \geq \alpha^m \varepsilon_m \nu_m(A),$$

which, from Theorem 6.59, establishes the uniform ergodicity of the mixture kernel.

If $\mathcal{X}$ is a small set for K_1 with $m = 1$, we have the minorizations

$$(K_1 \circ K_2)(x, A) = \int_A \int_{\mathcal{X}} K_2(x, dy)\, K_1(y, dz)$$

$$\geq \varepsilon_1 \nu_1(A) \int_{\mathcal{X}} K_2(x, dy) = \varepsilon_1 \nu_1(A)$$

and

$$(K_2 \circ K_1)(x, A) = \int_A \int_{\mathcal{X}} K_1(x, dy)\, K_2(y, dz)$$
$$\geq \varepsilon_1 \int_A \int_{\mathcal{X}} \nu_1(dy) K_2(x, dy) = \varepsilon_1 (K_2 \circ \nu_1)(A).$$

From Theorem 6.59, both cycles are therefore uniformly ergodic. $\qquad \square$

These results are not only formal since it is possible (see Theorem 7.8) to produce a uniformly ergodic kernel from an independent Metropolis–Hastings algorithm with instrumental distribution g such that f/g is bounded. Hybrid MCMC algorithms can therefore be used to impose uniform ergodicity in an almost automatic way.

The following example, due to Nobile (1998), shows rather clearly how the introduction of a Metropolis–Hastings step in the algorithm speeds up the exploration of the support of the stationary distribution.

Example 10.21. Probit model. A (dichotomous) probit model is defined by random variables D_i $(1 \leq i \leq n)$ such that $(1 \leq j \leq 2)$

(10.8) $$P(D_i = 1) = 1 - P(D_i = 0) = P(Z_i \geq 0)$$

with $Z_i \sim \mathcal{N}(R_i\beta, \sigma^2)$, $\beta \in \mathbb{R}$, R_i being a covariate. (Note that the z_i's are *not* observed. This is a special case of a *latent variable model*, as mentioned in Sections 1.1 and 5.3.1. See also Problem 10.14.) For the prior distribution

$$\sigma^{-2} \sim \mathcal{G}a(1.5, 1.5)\,, \qquad \beta|\sigma \sim \mathcal{N}(0, 10^2)\,,$$

Figure 10.4 plots the $20,000$ first iterations of the Gibbs chain $(\beta^{(t)}, \sigma^{(t)})$ against some contours of the true posterior distribution. (See Example 12.1 for further details.) The exploration is thus very poor, since the chain does not even reach the region of highest posterior density. (A reason for this behavior is that the likelihood is quite uninformative about (β, σ), providing only a lower bound on β/σ, as explained in Nobile 1998. See also Problem 10.14.)

A hybrid alternative, proposed by Nobile (1998), is to insert a Metropolis–Hastings step after each Gibbs cycle. The proposal distribution merely rescales the current value of the Markov chain $y^{(t)}$ by a random scale factor c, drawn from an exponential $\mathcal{E}xp(1)$ distribution (which is similar to the "hit-and-run" method of Chen and Schmeiser (1993). The rescaled value $cy^{(t)}$ is then accepted or rejected by a regular Metropolis–Hastings scheme to become the current value of the chain. Figure 10.5 shows the improvement brought by this hybrid scheme, since the MCMC sample now covers most of the support of the posterior distribution for the same number of iterations as in Figure 10.4. $\qquad \|$

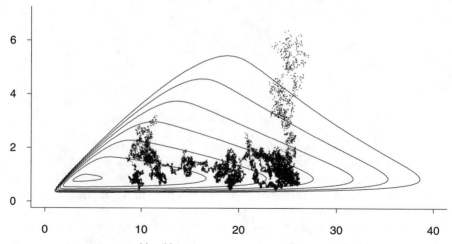

Fig. 10.4. Sample of $(\beta^{(t)}, \sigma^{(t)})$ obtained by the Gibbs sampler plotted with some contours of the posterior distribution of (β, σ) for $20,000$ observations from (10.8), when the chain is started at $(\beta, \sigma) = (25, 5)$.

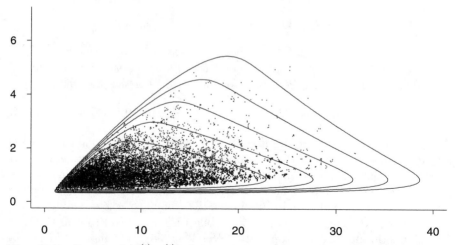

Fig. 10.5. Sample of $(\beta^{(t)}, \sigma^{(t)})$ obtained by an hybrid MCMC algorithm plotted with the posterior distribution for the same $20,000$ observations as in Figure 10.4 and the same starting point.

10.3.3 Metropolizing the Gibbs Sampler

Hybrid MCMC algorithms are often useful at an elementary level of the simulation process; that is, when some components of [A.40] cannot be easily simulated. Rather than looking for a customized algorithm such as Accept–Reject in each of these cases or for alternatives to Gibbs sampling, there is a compromise suggested by Müller (1991, 1993) (sometimes called "Metropolis-within-Gibbs"). In any step i of the algorithm [A.40] with a difficult simulation from $g_i(y_i|y_j, j \neq i)$, substitute a simulation from an instrumental distribution q_i. In the setup of [A.40], Müller's (1991) modification is

Algorithm A.43 –Hybrid MCMC–

For $i = 1, \ldots, p$, given $(y_1^{(t+1)}, \ldots, y_{i-1}^{(t+1)}, y_i^{(t)}, \ldots, y_p^{(t)})$:

1. Simulate

$$\tilde{y}_i \sim q_i(y_i|y_1^{(t+1)}, \ldots, y_i^{(t)}, y_{i+1}^{(t)}, \ldots, y_p^{(t)})$$

2. Take

[A.43]

$$y_i^{(t+1)} = \begin{cases} y_i^{(t)} & \text{with probability } 1 - \rho, \\ \tilde{y}_i & \text{with probability } \rho, \end{cases}$$

where

$$\rho = 1 \wedge \left\{ \frac{\left(\dfrac{g_i(\tilde{y}_i|y_1^{(t+1)}, \ldots, y_{i-1}^{(t+1)}, y_{i+1}^{(t)}, \ldots, y_p^{(t)})}{q_i(\tilde{y}_i|y_1^{(t+1)}, \ldots, y_{i-1}^{(t+1)}, y_i^{(t)}, y_{i+1}^{(t)}, \ldots, y_p^{(t)})} \right)}{\left(\dfrac{g_i(y_i^{(t)}|y_1^{(t+1)}, \ldots, y_{i-1}^{(t+1)}, y_{i+1}^{(t)}, \ldots, y_p^{(t)})}{q_i(y_i^{(t)}|y_1^{(t+1)}, \ldots, y_{i-1}^{(t+1)}, \tilde{y}_i, y_{i+1}^{(t)}, \ldots, y_p^{(t)})} \right)} \right\} .$$

An important point about this substitution is that the above Metropolis–Hastings step is only used *once* in an iteration from [A.40]. The modified step thus produces a *single* simulation $\tilde{y}_i$ instead of trying to approximate $g_i(y_i|y_j, j \neq i)$ more accurately by producing T simulations from q_i. The reasons for this choice are twofold: First, the resulting hybrid algorithm is valid since g is its stationary distribution (see Section 10.3.2 or Problems 10.13 and 10.15.). Second, Gibbs sampling also leads to an approximation of g. To provide a more "precise" approximation of $g_i(y_i|y_j, j \neq i)$ in [A.43] does not necessarily lead to a better approximation of g and the replacement of g_i by q_i may even be beneficial for the speed of excursion of the chain on the surface of g.[7] (See also Chen and Schmeiser 1998.)

When several Metropolis–Hastings steps appear in a Gibbs algorithm, either because of some complex conditional distributions (see Besag et al. 1995) or because of convergence contingencies, a method proposed by Müller (1993) is to run a single acceptance step after the p conditional simulations. This approach is more time-consuming in terms of simulation (since it may result in the rejection of the p simulated components), but the resulting algorithm

---

[7] The multi-stage Gibbs sampler is itself an illustration of this, as announced in Section 9.5.

can be written as a simple Metropolis–Hastings method instead of a (hybrid) combination of Metropolis–Hastings algorithms. Moreover, it produces an approximation $q(y)$ of the distribution $g(y)$ rather than *local* approximations of the conditional distributions of the subvectors y_i.

We conclude this section with a surprising result of Liu (1995), who shows that, in a discrete state setting, Gibbs sampling can be improved by Metropolis–Hastings steps. The improvement is expressed in terms of a reduction of the variance of the empirical mean of the $h(y^{(t)})$'s for the Metropolis–Hastings approach. This modification, called *Metropolization* by Liu (1995), is based on the following result, established by Peskun (1973), where $T_1 \ll T_2$ means that the non-diagonal elements of T_2 are larger that those of T_1 (see Problem 7.39 for a proof).

Lemma 10.22. *Consider two reversible Markov chains on a countable state-space, with transition matrices T_1 and T_2 such that $T_1 \ll T_2$. The chain associated with T_2 dominates the chain associated with T_1 in terms of variances.*

Given a conditional distribution $g_i(y_i|y_j, j \neq i)$ on a discrete space, the modification proposed by Liu (1995) is to use an additional Metropolis–Hastings step.

Algorithm A.44 –Metropolization of the Gibbs Sampler–

Given $y^{(t)}$,

1. Simulate $z_i \neq y_i^{(t)}$ with probability

$$\frac{g_i(z_i|y_j^{(t)}, j \neq i)}{1 - g_i(y_i^{(t)}|y_j^{(t)}, j \neq i)} \; ;$$

2. Accept $y_i^{(t+1)} = z_i$ with probability   [A.44]

$$\frac{1 - g_i(y_i^{(t)}|y_j^{(t)}, j \neq i)}{1 - g_i(z_i|y_j^{(t)}, j \neq i)} \wedge 1.$$

The probability of moving from $y_i^{(t)}$ to a different value is then necessarily higher in [A.44] than in the original Gibbs sampling algorithm and Lemma 10.22 implies the following domination result.

Theorem 10.23. *The modification [A.44] of the Gibbs sampler is more efficient in terms of variances.*

Example 10.24. (Continuation of Example 9.8) For the aggregated multinomial model of Tanner and Wong (1987), the completed variables z_i

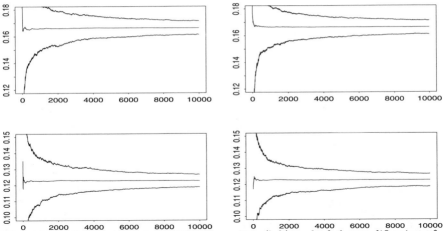

Fig. 10.6. Comparison of the Gibbs sampling *(left)* and of the modification of Liu (1995) *(right)*, for the estimation of $\mathbb{E}[\mu|x]$ *(top)* and of $\mathbb{E}[\eta|x]$ *(bottom)*. The 90% confidence regions on both methods have been obtained on 500 parallel chains.

$(i = 1, \ldots, 4)$ take values in $\{0, 1, \ldots, x_i\}$ and can, therefore, be simulated from [A.44], that is, from a binomial distribution

$$\mathcal{B}\left(x_i, \frac{a_i\mu}{a_i\mu + b_i}\right) \quad \text{for } i = 1, 2 \quad \text{and} \quad \mathcal{B}\left(x_i, \frac{a_i\eta}{a_i\eta + b_i}\right) \quad \text{for } i = 3, 4,$$

until the simulated value $\tilde{z}_i$ is different from $z_i^{(t)}$. This new value $\tilde{z}_i$ is accepted with probability $(i = 1, 2)$

$$\frac{1 - \dbinom{z_i^{(t)}}{x_i}\left(\dfrac{a_i\mu}{a_i\mu + b_i}\right)^{z_i^{(t)}}\left(\dfrac{b_i}{a_i\mu + b_i}\right)^{x_i - z_i^{(t)}}}{1 - \dbinom{\tilde{z}_i}{x_i}\left(\dfrac{a_i\mu}{a_i\mu + b_i}\right)^{\tilde{z}_i}\left(\dfrac{b_i}{a_i\mu + b_i}\right)^{x_i - \tilde{z}_i}}$$

and $(i = 3, 4)$

$$\frac{1 - \dbinom{z_i^{(t)}}{x_i}\dfrac{(a_i\eta)^{z_i^{(t)}}\, b_i^{x_i - z_i^{(t)}}}{(a_i\eta + b_i)^{x_i}}}{1 - \dbinom{\tilde{z}_i}{x_i}\dfrac{(a_i\mu)^{\tilde{z}_i}\, b_i^{x_i - \tilde{z}_i}}{(a_i\eta + b_i)^{x_i}}}.$$

Figure 10.6 describes the convergence of estimations of μ and η under the two simulation schemes for the following data:

$$(a_1, a_2, a_3, a_4) = (0.06, 0.14, 0.11, 0.09),$$
$$(b_1, b_2, b_3, b_4) = (0.17, 0.24, 0.19, 0.20),$$
$$(x_1, x_2, x_3, x_4, x_5) = (9, 15, 12, 7, 8).$$

n	10	100	1000	10,000
μ	0.302	0.0984	0.0341	0.0102
	0.288	0.0998	0.0312	0.0104
η	0.234	0.0803	0.0274	0.00787
	0.234	0.0803	0.0274	0.00778

Table 10.2. 90% interquantile ranges for the original Gibbs sampling *(top)* and the modification of Liu (1995) *(bottom)*, for the estimation of $\mathbb{E}[\mu|x]$ and $\mathbb{E}[\eta|x]$.

As shown by Table 10.2, the difference on the variation range of the approximations of both $\mathbb{E}[\mu|x]$ and $\mathbb{E}[\eta|x]$ is quite minor, in particular for the estimation of η, and are not always larger for the original Gibbs sampler.   ‖

10.4 Statistical Considerations

10.4.1 Reparameterization

Convergence of both Gibbs sampling and Metropolis–Hastings algorithms may suffer from a poor choice of parameterization (see the extreme case of Example 10.7). As a result of this, following Hills and Smith (1992), the MCMC literature has considered changes in the parameterization of a model as a way to speed up convergence in a Gibbs sampler. (See Gilks and Roberts 1996, for a more detailed review.) It seems, however, that most efforts have concentrated on the improvement of specific models, resulting in a lack of a general methodology for the choice of a "proper" parameterization. Nevertheless, the overall advice is to try to make the components "as independent as possible."

As noted in Example 10.18, convergence performances of the Gibbs sampler may be greatly affected by the choice of the coordinates. For instance, if the distribution g is a $\mathcal{N}_2(0, \Sigma)$ distribution with a covariance matrix Σ such that its eigenvalues satisfy $\lambda_{\min}(\Sigma) \ll \lambda_{\max}(\Sigma)$ and its eigenvectors correspond to the first and second diagonals of $\mathbb{R}^2$, the Gibbs sampler based on the conditionals $g(x_1|x_2)$ and $g(x_2|x_1)$ is very slow to explore the entire range of the support of g. However, if $y_1 = x_1 + x_2$ and $y_2 = x_1 - x_2$ is the selected parameterization (which corresponds to the coordinates in the eigenvector basis) the Gibbs sampler will move quite rapidly over the support of g. (Hobert et al. 1997 have shown that the influence of the parameterization on convergence performances may be so drastic that the chain can be irreducible for some parameterizations and not irreducible for others.)

Similarly, the geometry of the selected instrumental distribution in a Metropolis–Hastings algorithm must somehow correspond to the geometry of the support of g for good acceptance rates to be obtained. In particular, as pointed out by Hills and Smith (1992), if a normal second-order approximation to the density g is used, the structure of the Hessian matrix will matter in a Gibbs implementation, whereas the simplification brought by a Laplace approximation must be weighted against the influence of the parameterization. For instance, the value of the approximation based on the Taylor expansion

$$h(\theta) = h(\hat{\theta}) + \frac{1}{2}(\theta - \hat{\theta})^t (\nabla \nabla^t h)|_{\theta = \hat{\theta}}(\theta - \hat{\theta}) ,$$

with $h(\theta) = \log g(\theta)$, depends on the choice of the parameterization.

Gelfand and Sahu (1994), Gelfand et al. (1995, 1996) have studied the effects of parameterization on different linear models.

Example 10.25. Random effects model.   Consider the simple random effects model

$$Y_{ij} = \mu + \alpha_i + \varepsilon_{ij} , \qquad i = 1, \ldots, I, \quad j = 1, \ldots, J,$$

where $\alpha_i \sim \mathcal{N}(0, \sigma_\alpha^2)$ and $\varepsilon_{ij} \sim \mathcal{N}(0, \sigma_y^2)$. For a flat prior on μ, the Gibbs sampler implemented for the $(\mu, \alpha_1, \ldots, \alpha_I)$ parameterization exhibits high correlation and consequent slow convergence if $\sigma_y^2/(IJ\sigma_\alpha^2)$ is large (see Problem 10.22). On the other hand, if the model is rewritten as the hierarchy

$$Y_{ij} \sim \mathcal{N}(\eta_i, \sigma_y^2), \qquad \eta_i \sim \mathcal{N}(\mu, \sigma_\alpha^2),$$

the correlations between the η_i's and between μ and the η_i's are lower (see Problem 10.23).

Another approach is suggested by Vines and Gilks (1994) and Vines et al. (1995), who eliminate the unidentifiability feature by so-called *sweeping*; that is, by writing the model as

$$Y_{ij} = \nu + \varphi_i + \varepsilon_{ij},$$

with $(1 \leq i \leq I, 1 \leq j \leq J)$

$$\sum_i \varphi_i = 0, \qquad \varphi_i \sim \mathcal{N}\left(0, \sigma_\alpha^2 (1 - (1/I))\right) ,$$

and $\text{cov}(\varphi_i, \varphi_j) = -\sigma_\alpha^2/I$. This choice leads to even better correlation structures since the parameter ν is independent of the φ_i's while $\text{corr}(\varphi_i, \varphi_j) = -1/I$, a posteriori.   ‖

The end of this section deals with a special feature of mixture estimation (see Example 9.2), whose relevance to MCMC algorithms is to show that a change in the parameterization of a model can accelerate convergence.

As a starting point, consider the paradox of trapping states, discussed in Section 9.7.1. We can reassess the relevance of the selected prior distribution, which may not endow the posterior distribution with sufficient stability. The

conjugate distribution used in Example 9.2 ignores the specific nature of the
mixture (9.5) since it only describes the completed model, and creates an independence between the components of the mixture. An alternative proposed
in Mengersen and Robert (1996) is to relate the various components through
a common reference, namely a scale, location, or scale *and* location parameter.

Example 10.26. (Continuation of Example 9.2) For instance, normal
distributions are location–scale distributions. Each component can thus be
expressed in terms of divergence from a global location, μ, and a global scale,
τ. Using somewhat abusive (but convenient) notation, we write

$$(10.9) \qquad \sum_{j=1}^{k} p_j \, \mathcal{N}(\mu + \tau\theta_j, \tau^2\sigma_j^2),$$

also requiring that $\theta_1 = 0$ and $\sigma_1 = 1$, which avoids overparameterization.
Although the model (10.9) is an improvement over (9.5), it is still unable to
ensure proper stability of the algorithm.[8] The solution proposed in Robert
and Mengersen (1999) is, instead, to express each component as a perturba-
tion of the previous component, which means using the location and scale
parameters of the previous component as a *local reference*. Starting from the
normal distribution $\mathcal{N}(\mu, \tau^2)$, the two-component mixture

$$p\mathcal{N}(\mu, \tau^2) + (1 - p)\mathcal{N}(\mu + \tau\theta, \tau^2\sigma^2)$$

is thus modified into

$$p\mathcal{N}(\mu, \tau^2) + (1-p)q\mathcal{N}(\mu + \tau\theta, \tau^2\sigma^2) + (1-p)(1-q)\mathcal{N}(\mu + \tau\theta + \tau\sigma\varepsilon, \tau^2\sigma^2\omega^2) \,,$$

a three-component mixture.

The k component version of the reparameterization of (9.5) is

$$p\mathcal{N}(\mu, \tau^2) + (1 - p) \cdots (1 - q_{k-2}) \, \mathcal{N}(\mu + \ldots + \tau \cdots \sigma_{k-2}\theta_{k-1}, \tau^2 \ldots \sigma_{k-1}^2)$$

$$(10.10) + \sum_{j=1}^{k-2} (1 - p) \cdots (1 - q_{j-1})q_j \, \mathcal{N}(\mu + \cdots + \tau \cdots \sigma_{j-1}\theta_j, \tau^2 \cdots \sigma_j^2) \,.$$

The mixture (10.10) being invariant under permutations of indices, we can
impose an *identifiability constraint*, for instance,

$$(10.11) \qquad \sigma_1 \leq 1, \ldots, \sigma_{k-1} \leq 1 \,.$$

The prior distribution can then be modified to

$$(10.12) \qquad \pi(\mu, \tau) = \tau^{-1}, \ p, q_j \sim \mathcal{U}_{[0,1]}, \ \sigma_j \sim \mathcal{U}_{[0,1]}, \ \theta_j \sim \mathcal{N}(0, \zeta^2) \,.$$

---

[8] It reproduces the feature of the original parameterization, namely of independence
between the parameters conditional on the global location–scale parameter.

The representation (10.10) allows the use of an improper distribution on (μ, τ), and (10.11) justifies the use of a uniform distribution on the σ_i's (Robert and Titterington 1998). The influence of the only hyperparameter, ζ, on the resulting estimations of the parameters is, moreover, moderate, if present. (See Problems 9.26 and 10.39.)   ‖

Example 10.27. Acidity level in lakes. In a study of acid neutralization capacity in lakes, Crawford et al. (1992) used a Laplace approximation to the normal mixture model

$$(10.13) \qquad \tilde{f}(x) = \sum_{j=1}^{3} p_j \, \frac{e^{-(x-\mu_j)^2/2w_j^2}}{\sqrt{2\pi} \, w_j}.$$

This model can also be fit with a Gibbs sampling algorithm based on the reparameterization of Example 10.26 (see Problem 10.39), and Figures 10.7 and 10.8 illustrate the performances of this algorithm on a benchmark example introduced in Crawford et al. (1992). Figure 10.7 shows the (lack of) evolution of the estimated density based on the averaged parameters from the Gibbs sampler when the number of iterations T increases. Figure 10.8 gives the corresponding convergence of these averaged parameters and shows that the stability is less obvious at this level. This phenomenon occurs quite often in mixture settings and is to be blamed on the weak identifiability of these models, for which quite distinct sets of parameters lead to very similar densities, as shown by Figure 10.7.   ‖

In the examples above, the reparameterization of mixtures and the corresponding correction of prior distributions result in a higher stability of Gibbs sampler algorithms and, in particular, with the disappearance of the phenomenon of trapping states

The identifiability constraint used in (10.11) has an advantage over equivalent constraints, such as $p_1 \geq p_2 \geq \cdots \geq p_k$ or $\mu_1 \leq \mu_2 \leq \cdots \leq \mu_k$. It not only automatically provides a compact support for the parameters but also allows the use of a uniform prior distribution. However, it sometimes slows down convergence of the Gibbs sampler.

Although the new parameterization helps the Gibbs sampler to distinguish between the homogeneous components of the sample (that is, to identify the observations with the same indicator variable), this reparameterization can encounter difficulties in moving from homogeneous subsamples. When the order of the component indices does not correspond to the constraint (10.11), the probability of simultaneously permuting all the observations of two ill-ordered but homogeneous subsamples is very low. One solution in the normal case is to identify the homogeneous components without imposing the constraint (10.11). The uniform prior distribution $\mathcal{U}_{[0,1]}$ on σ_j must be replaced by $\sigma_j \sim \frac{1}{2}\mathcal{U}_{[0,1]} + \frac{1}{2}\mathcal{P}a(2,1)$, which is equivalent to assuming that either σ_j or

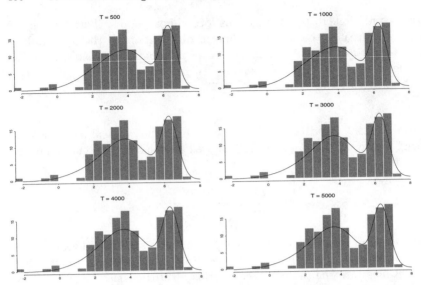

Fig. 10.7. Evolution of the estimation of the density (10.13) for three components and T iterations of a Gibbs sampling algorithm. The estimations are based on 149 observations of acidity levels for lakes in the American Northeast, used in Crawford et al. (1992) and represented by the histograms.

σ_j^{-1} is distributed from a $\mathcal{U}_{[0,1]}$ distribution. The simulation steps for σ_j are modified, but there exists a direct Accept–Reject algorithm to simulate from the conditional posterior distribution of the σ_j's. An alternative solution is to keep the identifiability constraints and to use a hybrid Markov chain Monte Carlo algorithm, where, in every U iterations, a random permutation of the values of z_i is generated via a Metropolis–Hastings scheme:

Algorithm A.45 –Hybrid Allocation Mixture Estimation–

> 0. Simulate Z_i $(i = 1, \ldots, n)$.
> 1. Generate a random permutation ψ on $\{1, 2, \ldots, k\}$ and derive
>
> $$\tilde{z} = (\psi(z_i))_i \quad \text{and} \quad \tilde{\xi} = (\tilde{p}, \tilde{\mu}, \tilde{w}) = (p_{\psi(j)}, \mu_{\psi(j)}, w_{\psi(j)})_j.$$
> $$[A.45]$$
>
> 2. Generate ξ' conditionally on $(\tilde{\xi}, \tilde{z})$ by a standard Gibbs sampler iteration.
> 3. Accept $(\xi', \tilde{z})$ with probability
>
> $$(10.14) \qquad \frac{\pi(\xi', \tilde{z})\pi_1(\psi^{-1}(z^-)|\xi') \, \pi_2(\xi|z^-, \psi(\xi'))}{\pi(\xi, z^-)\pi_1(\psi^{-1}(\tilde{z})|\xi) \, \pi_2(\xi'|\tilde{z}, \tilde{\xi})} \wedge 1 \,.$$

where $\pi(\xi, z)$ denotes the distribution of (ξ, z),

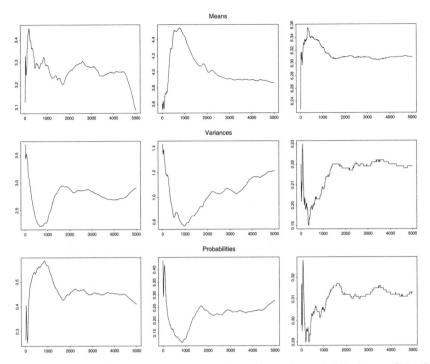

Fig. 10.8. Convergence of estimators of the parameters of the mixture (10.13) for the same iterations as in Figure 10.7.

$$\pi(\xi, z) \propto \prod_{i=1}^{n} \left(\sum_{j=1}^{k} p_j \frac{e^{-(\mu_j - x_i)^2/(2w_j^2)}}{w_j} \, \mathbb{I}_{z_i=j} \right) \pi(\xi) \,,$$

where $\pi_1(z|\xi)$ and $\pi_2(\xi'|z, \xi)$ are the conditional distributions used in the Gibbs sampler and z^- is the previous value of the allocation vector. Note that, in (10.14), $\pi_1(\psi^{-1}(\tilde z)|\xi) = \pi_1(\tilde z|\tilde\xi)$.

This additional Metropolis–Hastings step may result in delicate computations in terms of manipulation of indices but it is of the same order of complexity as an iteration of the Gibbs sampler since the latter also requires the computation of the sums $\sum_{j=1}^{k} p_j f(x_i|\xi_j)$, $i = 1, \dots, n$. The value of U in [A.45] can be arbitrarily fixed (for instance, at 50) and later modified depending on the average acceptance rate corresponding to (10.14). A high acceptance rate means that the Gibbs sampler lacks a sufficient number of iterations to produce a stable allocation to homogeneous classes; a low acceptance rate suggests reduction of U so as to more quickly explore the possible permutations.

10.4.2 Rao-Blackwellization

Some of the Rao-Blackwellization results of Section 9.3 carry over to the multi-stage Gibbs sampler. For example, Liu et al. (1995) are able to extend Proposition 9.18 to the multi-stage Gibbs sampler in the case of the *random Gibbs sampler*, where every step only updates a single component of y. In the setup of [A.40], define a multinomial distribution $\sigma = (\sigma_1, \ldots, \sigma_p)$.

Algorithm A.46 –Random Gibbs Sampler–

Given $y^{(t)}$,

1. Select a component $\nu \sim \sigma$.   [A.46]
2. Generate $Y_\nu^{(t+1)} \sim g_\nu(y_\nu|y_j^{(t)}, j \neq \nu)$ and take

$$y_j^{(t+1)} = y_j^{(t)} \qquad \text{for} \quad j \neq \nu.$$

Note that, although [A.46] only generates one component of y at each iteration, the resulting chain is strongly irreducible (Definition 6.13) because of the random choice of ν. It also satisfies the following property.

Proposition 10.28. *The chain $(Y^{(t)})$ generated by [A.46] has the property that for every function $h \in \mathcal{L}_2(g)$, the covariance $\mathrm{cov}(h(Y^{(0)}), h(Y^{(t)}))$ is positive and decreasing in t.*

Proof. Assume again $\mathbb{E}_g[h(Y)] = 0$; then

$$\mathbb{E}[h(Y^{(0)})h(Y^{(t)})] = \sum_{v=1}^{p} \sigma_v \, \mathbb{E}[\mathbb{E}[h(Y^{(0)})h(Y^{(1)})|\nu = v, (Y_j^{(0)}, j \neq v)]]$$

$$= \sum_{v=1}^{p} \sigma_v \, \mathbb{E}[\mathbb{E}[h(Y^{(0)})|(Y_j^{(0)}, j \neq v)]^2]$$

$$= \mathbb{E}[\mathbb{E}[h(Y)|\nu, (Y_j, j \neq \nu)]^2],$$

due to the reversibility of the chain and the independence between $Y^{(0)}$ and $Y^{(1)}$, conditionally on ν and $(y_j^{(0)}, j \neq v)$. A simple recursion implies

$$\mathbb{E}[h(Y^{(0)})h(Y^{(t)})] = \mathrm{var}(\mathbb{E}[\cdots \mathbb{E}[\mathbb{E}[h(Y)|\nu, (Y_j, j \neq \nu)]|Y]\cdots]),$$

where the second term involves t conditional expectations, successively in $(\nu, (y_j, j \neq \nu))$ and in Y.   □

This proof suggests choosing a distribution σ that more heavily weights components with small $\mathbb{E}[h(Y)|(y_j, j \neq v)]^2$, so the chain will typically visit components where $g_v(y_v|y_j, j \neq v)$ is not too variable. However, the opposite choice seems more logical when considering the *speed* of convergence to the stationary distribution, since components with high variability should be visited more often in order to accelerate the exploration of the support of g. This dichotomy between acceleration of the convergence to the stationary distribution and reduction of the empirical variance is typical of MCMC methods.

Another substantial benefit of Rao–Blackwellization is an elegant method for the approximation of densities of different components of y. Since

$$\frac{1}{T} \sum_{t=1}^{T} g_i(y_i | y_j^{(t)}, j \neq i)$$

is unbiased and converges to the marginal density $g_i(y_i)$ if these conditional densities are available in closed form, it is unnecessary (and inefficient) to use nonparametric density estimation methods such as *kernel methods* (see Fan and Gijbels 1996 or Wand and Jones 1995). This property can also be used in extensions of the Riemann sum method (see Section 4.3) to setups where the density f needs to be approximated (see Problem 10.17).

Another consequence of Proposition 9.18 is a justification of the technique of *batch sampling* proposed in some MCMC algorithms (Geyer 1992, Raftery and Lewis 1992, Diebolt and Robert 1994). Batch sampling involves subsampling the sequence $(Y^{(t)})_t$ produced by a Gibbs sampling method into $(Y^{(ks)})_s$ ($k > 1$), in order to decrease the dependence between the points of the sample. However, Lemma 12.2 describes a negative impact of subsampling on the variance of the corresponding estimators.

10.4.3 Improper Priors

This section discusses a particular danger resulting from careless use of Metropolis–Hastings algorithms, in particular the Gibbs sampler. We know that the Gibbs sampler is based on conditional distributions derived from $f(x_1, \ldots, x_q)$ or $g(y_1, \ldots, y_p)$. What is particularly insidious is that these conditional distributions may be well defined and may be simulated from, but may not correspond to any joint distribution g; that is, the function g given by Lemma 10.5 is not integrable. The same problem may occur when using a proportionality relation as $\pi(\theta|x) \propto \pi(\theta)f(x|\theta)$ to derive a Metropolis–Hastings algorithm for $\pi(\theta)f(x|\theta)$.

This problem is not a *defect* of the Gibbs sampler, nor even a simulation[9] problem, but rather a problem of carelessly using the Gibbs sampler in a situation for which the underlying assumptions are violated. It is nonetheless important to warn the user of MCMC algorithms against this danger, because it corresponds to a situation often encountered in Bayesian noninformative (or *"default"*) modelings.

The construction of the Gibbs sampler directly from the conditional distributions is a strong incentive to bypass checking for the propriety of g,

---

[9] The "distribution" g does not exist in this case.

especially in complex setups. In particular, the function g may well be un-defined in a "generalized" Bayesian approach where the prior distribution is "improper," in the sense that it is a σ-finite measure instead of being a regu-lar probability measure (see Berger 1985, Section 3.3, or Robert 2001, Section 1.5).

Example 10.29. Conditional exponential distributions. The following model was used by Casella and George (1992) to point out the difficulty of assessing the impropriety of a posterior distribution through the conditional distributions. The pair of conditional densities

$$X_1|x_2 \sim \mathcal{E}xp(x_2) \,, \qquad X_2|x_1 \sim \mathcal{E}xp(x_1)$$

are well defined and are functionally compatible in the sense that the ratio

$$\frac{f(x_1|x_2)}{f(x_2|x_1)} = \frac{x_1 \exp(-x_1\,x_2)}{x_2 \exp(-x_2\,x_1)} = \frac{x_1}{x_2}$$

separates into a function of x_1 and a function of x_2. By virtue of Bayes theo-rem, this is a necessary condition for the existence of a joint density $f(x_1, x_2)$ corresponding to these conditional densities. However, recall that Theorem 9.3 requires that

$$\int \frac{f(x_1|x_2)}{f(x_2|x_1)} dx_1 < \infty,$$

but the function x^{-1} cannot be integrated on $\mathbb{R}_+$. Moreover, if the joint density, $f(x_1, x_2)$, existed, it would have to satisfy $f(x_1|x_2) = f(x_1, x_2)/\int f(x_1, x_2)dx_1$. The only function $f(x_1, x_2)$ satisfying this equation is

$$f(x_1, x_2) \propto \exp(-x_1 x_2).$$

Thus, these conditional distributions do not correspond to a joint probability distribution (see Problem 10.24).   ‖

Example 10.30. Normal improper posterior. Consider $X \sim \mathcal{N}(\theta, \sigma^2)$. A Bayesian approach when there is little prior information on the parameters θ and σ is Jeffreys (1961). This approach is based on the derivation of the prior distribution of (θ, σ) from the Fisher information of the model as $I(\theta, \sigma)^{1/2}$, the square root being justified by invariance reasons (as well as information theory considerations; see Lehmann and Casella 1998 or Robert 2001, Section 3.5.3). Therefore, in this particular case, $\pi(\theta, \sigma) = \sigma^{-2}$ and the posterior distribution follows by a (formal) application of Bayes' Theorem,

$$\pi(\theta, \sigma|x) = \frac{f(x|\theta, \sigma)\, \pi(\theta, \sigma)}{\int f(x|\zeta)\, \pi(\zeta)\, d\zeta} \,.$$

It is straightforward to check that

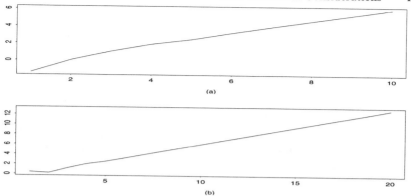

Fig. 10.9. First iterations of the chains (a) $\log(|\theta^{(t)}|)$ and (b) $\log(\sigma^{(t)})$ for a divergent Gibbs sampling algorithm.

$$\int f(x|\zeta)\,\pi(\zeta)\,d\zeta = \int \frac{1}{\sqrt{2\pi}\sigma}\,e^{-(x-\theta)^2/(2\sigma^2)}\,\frac{d\theta\,d\sigma}{\sigma^2} = \infty,$$

and, hence, $\pi(\theta,\sigma|x)$ is not defined.

However, this impropriety may never be detected as it is common Bayesian practice to consider only the proportionality relation between $\pi(\theta,\sigma|x)$ and $f(x|\theta,\sigma)\,\pi(\theta,\sigma)$. Similarly, a derivation of a Gibbs sampler for this model is also based on the proportionality relation since the conditional distributions $\pi(\theta|\sigma,x)$ and $\pi(\sigma|\theta,x)$ can be obtained directly as

$$\pi(\theta|\sigma,x) \propto e^{-(\theta-x)^2/(2\sigma^2)}, \qquad \pi(\sigma|\theta,x) \propto \sigma^{-3}\,e^{-(\theta-x)^2/(2\sigma^2)},$$

which lead to the full conditional distributions $\mathcal{N}(x,\sigma^2)$ and $\mathcal{E}xp((\theta-x)^2/2)$ for θ and σ^{-2}, respectively.

This practice of omitting the computation of normalizing constants is justified as long as the joint distribution does exist and is particularly useful because these constants often cannot be obtained in closed form. However, the use of improper prior distributions implies that the derivation of conditional distributions from the proportionality relation is not always justified. In the present case, the Gibbs sampler associated with these two conditional distributions is

1. Simulate $\theta^{(t+1)} \sim \mathcal{N}(x,\sigma^{(t)2})$;
2. Simulate $\varepsilon \sim \mathcal{E}xp\left((\theta^{(t+1)} - x)^2/2\right)$ and take
   $\sigma^{(t+1)} = 1/\sqrt{\varepsilon}.$

For an observation $x = 0$, the behavior of the chains $(\theta^{(t)})$ and $(\sigma^{(t)})$ produced by this algorithm is shown in Figure 10.9, which exhibits an extremely fast divergence[10] to infinity (both graphs are plotted in log scales).   ‖

The setup of improper posterior distributions and of the resulting behavior of the Gibbs sampler can be evaluated from a theoretical point of view, as in Hobert and Casella (1996, 1998). For example, the associated Markov chain cannot be positive recurrent since the σ-finite measure associated with the conditional distributions given by Lemma 10.5 is invariant. However, the major task in such settings is to come up with indicators to flag that something is wrong with the stationary measure, so that the experimenter can go back to his/her prior distribution and check for propriety.

Given the results of Example 10.30, it may appear that a simple graphical monitoring is enough to exhibit deviant behavior of the Gibbs sampler. However, this is not the case in general and there are many examples, some of which are published (see Casella 1996), where the output of the Gibbs sampler seemingly does not differ from a positive recurrent Markov chain. Often, this takes place when the divergence of the posterior density occurs "at 0," that is, at a specific point whose immediate neighborhood is rarely visited by the chain. Hobert and Casella (1996) illustrate this seemingly acceptable behavior on an example initially treated in Gelfand et al. (1990).

Example 10.31. Improper random effects posterior. Consider a random effects model,

$$Y_{ij} = \beta + U_i + \varepsilon_{ij}, \qquad i = 1, \ldots, I, \; j = 1, \ldots, J,$$

where $U_i \sim \mathcal{N}(0, \sigma^2)$ and $\varepsilon_{ij} \sim \mathcal{N}(0, \tau^2)$. The Jeffreys (improper) prior for the parameters β, σ, and τ is

$$\pi(\beta, \sigma^2, \tau^2) = \frac{1}{\sigma^2 \tau^2} .$$

The conditional distributions

$$U_i | y, \beta, \sigma^2, \tau^2 \sim \mathcal{N}\left(\frac{J(\bar{y}_i - \beta)}{J + \tau^2 \sigma^{-2}}, (J\tau^{-2} + \sigma^{-2})^{-1} \right) ,$$

$$\beta | u, y, \sigma^2, \tau^2 \sim \mathcal{N}(\bar{y} - \bar{u}, \tau^2 / JI) ,$$

$$\sigma^2 | u, \beta, y, \tau^2 \sim \mathcal{IG}\left(I/2, (1/2) \sum_i u_i^2 \right) ,$$

$$\tau^2 | u, \beta, y, \sigma^2 \sim \mathcal{IG}\left(IJ/2, (1/2) \sum_{i,j} (y_{ij} - u_i - \beta)^2 \right) ,$$

---

[10] The impropriety of this posterior distribution is hardly surprising from a statistical point of view since this modeling means trying to estimate both parameters of the distribution $\mathcal{N}(\theta, \sigma^2)$ from a single observation without prior information.

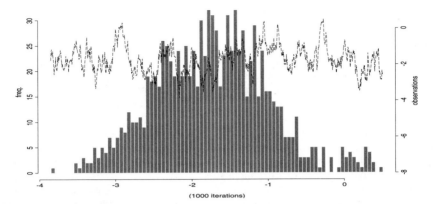

Fig. 10.10. Sequence $(\beta^{(t)})$ and corresponding histogram for the random effects model. The measurement scale for the $\beta^{(t)}$'s is on the right and the scale of the histogram is on the left.

are well defined and a Gibbs sampling can be easily implemented in this setting. Figure 10.10 provides the sequence of the $\beta^{(t)}$ produced by $[A.40]$ and the corresponding histogram for 1000 iterations. The trend of the sequence and the histogram do not indicate that the corresponding "joint distribution" does not exist (Problem 10.25).   ‖

Under some regularity conditions on the transition kernel, Hobert and Casella (1996) have shown that if there exist a positive function b, $\varepsilon > 0$ and a compact set C such that $b(x) < \varepsilon$ for $x \in C^c$, the chain $(y^{(t)})$ satisfies

$$(10.15) \qquad \liminf_{t \to +\infty} \frac{1}{t} \sum_{s=1}^{t} b(y^{(s)}) = 0.$$

A drawback of the condition (10.15) is the derivation of the function b, whereas the monitoring of a lim inf is delicate in a simulation experiment. Other references on the analysis of improper Gibbs samplers can be found in Besag et al. (1995) and in the subsequent discussion (see, in particular, Roberts and Sahu 1997), as well as in Natarajan and McCulloch (1998).

10.5 Problems

10.1 Propose a direct Accept–Reject algorithm for the conditional distribution of Y_2 given y_1 in Example 10.1. Compare with the Gibbs sampling implementation in a Monte Carlo experiment. (*Hint:* Run parallel chains and compute 90% confidence intervals for both methods.)

10.2 Devise and implement a simulation algorithm for the Ising model of Example 10.2.

10.3 In the setup of Example 10.17:

(a) Evaluate the numerical value of the boundary constant ϵ derived from (10.5), given the data in Table 10.1.

(b) Establish the lower bound in (10.5). (*Hint*: Replace $\sum \lambda_i$ by 0 in the integrand [but not in the exponent] and do the integration.)

10.4 For the data of Table 10.1

(a) Estimate $\lambda_1, \ldots, \lambda_{10}$ using the model of Example 10.17.

(b) Estimate a and b in the loglinear model $\log \lambda = a + bt$, using a Gibbs sampler and the techniques of Example 2.26.

(c) Compare the results of the analyses.

10.5 (Besag 1994) Consider a distribution f on a finite state-space $\mathcal{X}$ of dimension k and conditional distributions $f_1, \ldots, f_k$ such that for every $(x, y) \in \mathcal{X}^2$, there exist an integer m and a sequence $x_0 = x, \ldots, x_m = y$ where x_i and x_{i+1} only differ in a single component and $f(x_i) > 0$.

(a) Show that this condition extends Hammersley–Clifford Theorem (Theorem 10.5) by deriving $f(y)$ as

$$f(y) = f(x) \prod_{i=0}^{m-1} \frac{f(x_{i+1})}{f(x_i)}.$$

(b) Deduce that irreducibility and ergodicity holds for the associated Gibbs sampler.

(c) Show that the same condition on a continuous state-space is not sufficient to ensure ergodicity of the Gibbs sampler. (*Hint:* See Example 10.7.)

10.6 Compare the usual demarginalization of the Student's t distribution discussed in Example 10.4 with an alternative using a slice sampler, by computing the empirical cdf at several points of interest for both approaches and the same number of simulations. (*Hint:* Use two uniform dummy variables.)

10.7 A bound similar to the one in Example 10.17, established in Problem 10.3, can be obtained for the kernel of Example 10.4; that is, show that

$$K(\theta, \theta') = \int_0^\infty \sqrt{\frac{1+\eta}{2\pi}} \exp\left\{ -\left(\theta' - \frac{\theta_0 \eta}{1+\eta} \right)^2 \frac{1+\eta}{2} \right\} \left(\frac{1 + (\theta - \theta_0)^2}{2} \right)^\nu$$

$$\times \frac{\eta^{\nu-1}}{\Gamma(\nu)} \exp\left\{ \frac{-\eta}{2} (1 + (\theta - \theta_0)^2) \right\} d\eta$$

$$\geq \left[1 + (\theta' - \theta_0)^2 \right]^{-\nu} \frac{e^{-(\theta')^2/2}}{\sqrt{2\pi}}.$$

(*Hint*: Establish that

$$\exp\left\{ -\left(\theta' - \frac{\theta_0 \eta}{1+\eta} \right)^2 \frac{1+\eta}{2} \right\} \geq \exp\left\{ -\frac{1}{2}(\theta')^2 - \frac{\eta}{2}(\theta'^2 - 2\theta'\theta_0 + \theta_0^2) \right\}$$

and integrate.)

10.8 (Hobert et al. 1997) Consider a distribution f on $\mathbb{R}^k$ to be simulated by Gibbs sampling. A one-to-one transform Ψ on $\mathbb{R}^k$ is called a parameterization. The convergence of the Gibbs sampler can be jeopardized by a bad choice of parameterization.

(a) Considering that a Gibbs sampler can be formally associated with the full conditional distributions for every choice of a parameterization, show that there always exist parameterizations such that the Gibbs sampler fails to converge.

(b) The fact that the convergence depends on the parameterization vanishes if (i) the support of f is arcwise connected, that for every $(x, y) \in (\text{supp}(f))^2$, there exists a continuous function φ on $[0, 1]$ with $\varphi(0) = x$, $\varphi(1) = y$, and $\varphi([0, 1]) \subset \text{supp}(f)$ and (ii) the parameterizations are restricted to be continuous functions.

(c) Show that condition (i) in (b) is necessary for the above property to hold. (*Hint:* See Example 10.7.)

(d) Show that a rotation of $\pi/4$ of the coordinate axes eliminates the irreducibility problem in Example 10.7.

10.9 In the setup of Example 5.13, show that the Gibbs sampling simulation of the ordered normal means θ_{ij} can either be done in $I \times J$ conditional steps or in only two conditional steps. Conduct an experiment to compare both approaches.

10.10 The clinical mastitis data described in Example 10.15 is the number of cases observed in 127 herds of dairy cattle (the herds are adjusted for size). The data are given in Table 10.3.

0	0	0	0	0	0	0	1	1	1	1	1	1	1	1
1	1	1	1	2	2	2	2	2	2	2	2	3	3	3
3	3	3	3	3	3	4	4	4	4	4	4	4	4	5
5	5	5	5	5	5	5	6	6	6	6	6	6	6	6
6	7	7	7	7	7	7	8	8	8	8	8	9	9	9
10	10	10	10	11	11	11	11	11	11	11	12	12	12	12
13	13	13	13	13	14	14	15	16	16	16	16	17	17	17
18	18	18	19	19	19	19	20	20	21	21	22	22	22	22
23	25	25	25	25	25	25								

Table 10.3. Occurrences of clinical mastitis in 127 herds of dairy cattle. (*Source:* Schukken et al. 1991.)

(a) Verify the conditional distributions for λ_i and β_i.

(b) For these data, implement the Gibbs sampler and plot the posterior density $\pi(\lambda_1|\mathbf{x}, \alpha)$. (Use the values $\alpha = .1, a = b = 1$.)

(c) Make histograms and monitor the convergence of λ_5, λ_{15}, and β_{15}.

(d) Investigate the sensitivity of your answer to the specification of α, a and b.

10.11 (a) For the Gibbs sampler of [A.40], if the Markov chain $(Y^{(t)})$ satisfies the positivity condition (see Definition 9.4), show that the conditional distributions $g(y_i|y_1, y_2, \ldots, y_{i-1}, y_{i+1}, \ldots, y_p)$ will not reduce the range of possible values of Y_i when compared with g.

(b) Prove Theorem 10.8.

10.12 Show that the Gibbs sampler of Algorithm [A.41] is reversible. (Consider a generalization of the proof of Algorithm [A.36].)

10.13 For the hybrid algorithm of [A.43], show that the resulting chain is a Markov chain and verify its stationary distribution. Is the chain reversible?

10.14 Show that the model (10.8) is not identifiable in (β, σ). (*Hint:* Show that it only depends on the ratio β/σ.)

10.15 If $K(x, x')$ is a Markov transition kernel with stationary distribution g, show that the Metropolis–Hastings algorithm where, at iteration t, the proposed value $y_t \sim K(x^{(t)}, y)$ is accepted with probability

$$\frac{f(y_t)}{f(x^{(t)})} \frac{g(x^{(t)})}{g(y_t)} \wedge 1,$$

provides a valid MCMC algorithm for the stationary distribution f. (*Hint:* Use detailed balance.)

10.16 For the algorithm of Example 7.13, an obvious candidate to simulate $\pi(a|\mathbf{x}, \mathbf{y}, b)$ and $\pi(b|\mathbf{x}, \mathbf{y}, a)$ is the ARS algorithm of Section 2.4.2.

(a) Show that

$$\log \pi(a|\mathbf{x}, \mathbf{y}, b) = \sum_i y_i a - \sum_i y_i \log\left(1 + e^{a+bx_i}\right) - \sum_i \frac{e^{a+bx_i}}{1 + e^{a+bx_i}} - \frac{a^2}{2\sigma^2},$$

and, thus,

$$\frac{\partial}{\partial a} \log \pi(a|\mathbf{x}, \mathbf{y}, b) = \sum_i y_i - \sum_i y_i \frac{e^{a+bx_i}}{1 + e^{a+bx_i}} - \sum_i \frac{e^{a+bx_i}}{(1 + e^{a+bx_i})^2} - \frac{a}{\sigma^2}$$

and

$$\frac{\partial^2}{\partial a^2} \log \pi(a|\mathbf{x}, \mathbf{y}, b) = -\sum_i \frac{y_i e^{a+bx_i}}{(1 + e^{a+bx_i})} - \sum_i \frac{e^{a+bx_i} - e^{2(a+bx_i)}}{(1 + e^{a+bx_i})^3} - \sigma^{-2}$$

$$= -\sum_i \frac{e^{a+bx_i}}{(1 + e^{a+bx_i})^3} \left\{(1 + y_i) - (1 - y_i)e^{a+bx_i}\right\} - \sigma^{-2}.$$

Argue that this last expression is not always negative, so the ARS algorithm cannot be applied.

(b) Show that

$$\frac{\partial^2}{\partial b^2} \log \pi(b|\mathbf{x}, \mathbf{y}, a)$$

$$= -\sum_i \frac{e^{a+bx_i}}{(1 + e^{a+bx_i})^3} \left\{(1 + y_i) - (1 - y_i)e^{a+bx_i}\right\} x_i^2 - \tau^{-2}$$

and deduce there is also no log-concavity in the b direction.

(c) Even though distributions are not log-concave, they can be simulated with the ARMS algorithm [A.28]. Give details on how to do this.

10.17 Show that the Riemann sum method of Section 4.3 can be used in conjunction with Rao–Blackwellization to cover multidimensional settings. (*Note:* See Philippe and Robert 1998a for details.)

10.18 The *tobit* model is used in econometrics (see Tobin 1958). It is based on a transform of a normal variable $y_i^* \sim \mathcal{N}(x_i^t \beta, \sigma^2)$ $(i = 1, \dots, n)$ by truncation

$$y_i = \max(y_i^*, 0).$$

Show that the following algorithm provides a valid approximation of the posterior distribution of (β, σ^2).

1. Simulate $y_i^* \sim \mathcal{N}_-(x_i^t\beta, \sigma^2, 0)$ if $y_i = 0$.
2. Simulate $(\beta, \sigma) \sim \pi(\beta, \sigma | y^*, x)$ with   [$A31$]

$$\pi(\beta, \sigma | y^*, x) \propto \sigma^{-n} \exp\left\{ -\sum_i (y_i^* - x_i^t\beta)^2 / 2\sigma^2 \right\} \pi(\beta, \sigma).$$

10.19 Let (X_n) be a reversible stationary Markov chain (see Definition 6.44) with $\mathbb{E}[X_i] = 0$.

(a) Show that the distribution of $X_k | X_1$ is the same as the distribution of $X_1 | X_k$.

(b) Show that the covariance between alternate random variables is positive. More precisely, show that $\text{cov}(X_0, X_{2\nu}) = \mathbb{E}[\mathbb{E}(X_0 | X_\nu)^2] > 0$.

(c) Show that the covariance of alternate random variables is decreasing; that is, show that $\text{cov}(X_0, X_{2\nu}) \geq \text{cov}(X_0, X_{2(\nu+1)})$. (*Hint*: Use the fact that

$$\mathbb{E}[\mathbb{E}(X_0 | X_\nu)^2] = \text{var}[\mathbb{E}(X_0 | X_\nu)] \geq \text{var}[\mathbb{E}\{\mathbb{E}(X_0 | X_\nu) | X_{\nu+1}\}],$$

and show that this latter quantity is $\text{cov}(X_0, X_{2(\nu+1)})$.)

10.20 Show that a Gibbs sampling kernel with more than two full conditional steps cannot produce interleaving chains.

10.21 In the setup of Example 10.2:

(a) Consider a grid in $\mathbb{R}^2$ with the simple nearest-neighbor relation $\mathcal{N}$ (that is, $(i, j) \in \mathcal{N}$ if and only if $\min(|i_1 - j_1|, |i_2 - j_2|) = 1$). Show that a Gibbs sampling algorithm with only two conditional steps can be implemented in this case.

(b) Implement the Metropolization scheme of Liu (1995) discussed in Section 10.3.3 and compare the results with those of part (a).

10.22 (Gelfand et al. 1995) In the setup of Example 10.25, show that the original parameterization leads to the following correlations:

$$\rho_{\mu, \alpha_i} = \left(1 + \frac{I\sigma_y^2}{J\sigma_\alpha^2} \right)^{-1/2}, \qquad \rho_{\alpha_i, \alpha_j} = \left(1 + \frac{I\sigma_y^2}{J\sigma_\alpha^2} \right)^{-1},$$

for $i \neq j$.

10.23 (Continuation of Problem 10.22) For the hierarchical parameterization, show that the correlations are

$$\rho_{\mu, \eta_i} = \left(1 + \frac{IJ\sigma_\alpha^2}{\sigma_y^2} \right)^{-1/2}, \qquad \rho_{\eta_i, \eta_j} = \left(1 + \frac{IJ\sigma_\alpha^2}{\sigma_y^2} \right)^{-1},$$

for $i \neq j$.

10.24 In Example 10.29, we noted that a pair of conditional densities is *functionally compatible* if the ratio $f_1(x|y)/f_2(y|x) = h_1(x)/h_2(y)$, for some functions h_1 and h_2. This is a necessary condition for a joint density to exist, but not a sufficient condition. If such a joint density does exist, the pair f_1 and f_2 would be *compatible*.

(a) Formulate the definitions of compatible and functionally compatible for a set of densities $f_1, \ldots, f_m$.

(b) Show that if $f_1, \ldots, f_m$ are the full conditionals from a hierarchical model, they are functionally compatible.
(c) Prove the following theorem, due to Hobert and Casella (1998), which shows there cannot be any stationary probability distribution for the chain to converge to unless the densities are compatible.

Theorem 10.32. *Let $f_1, \ldots, f_m$ be a set of functionally compatible conditional densities on which a Gibbs sampler is based. The resulting Markov chain is positive recurrent if and only if $f_1, \ldots, f_m$ are compatible.*

(*Note:* Compatibility of a set of densities was investigated by Besag 1974, Arnold and Press 1989, and Gelman and Speed 1993.)

10.25 For the situation of Example 10.31,
(a) Show that the full "posterior distribution" of $(\beta, \sigma^2, \tau^2)$ is

$$\pi(\beta, \sigma^2, \tau^2|y) \propto \sigma^{-2-I} \, \tau^{-2-IJ} \, \exp\left\{-\frac{1}{2\tau^2} \sum_{i,j} (y_{ij} - \bar{y}_i)^2\right\}$$

$$\times \exp\left\{-\frac{J \sum_i (\bar{y}_i - \beta)^2}{2(\tau^2 + J\sigma^2)}\right\} (J\tau^{-2} + \sigma^{-2})^{-I/2}.$$

(*Hint:* Integrate the (unobservable) random effects u_i.)
(b) Integrate out β and show that the marginal posterior density of $(\sigma^2, \tau^2|)$ is

$$\pi(\sigma^2, \tau^2|y) \propto \frac{\sigma^{-2-I} \, \tau^{-2-IJ}}{(J\tau^{-2} + \sigma^{-2})^{I/2}} (\tau^2 + J\sigma^2)^{1/2}$$

$$\times \exp\left\{-\frac{1}{2\tau^2} \sum_{i,j} (y_{ij} - \bar{y}_i)^2 - \frac{J}{2(\tau^2 + J\sigma^2)} \sum_i (\bar{y}_i - \bar{y})^2\right\}.$$

(c) Show that the full posterior is not integrable since, for $\tau \neq 0$, $\pi(\sigma^2, \tau^2|y)$ behaves like σ^{-2} in a neighborhood of 0.

10.26 (Chib 1995) Consider the approximation of the marginal density $m(x) = f(x|\theta)\pi(\theta)/\pi(\theta|x)$, where $\theta = (\theta_1, \ldots, \theta_B)$ and the full conditionals $\pi(\theta_r|x, \theta_s, s \neq r)$ are available.
(a) In the case $B = 2$, show that an appropriate estimate of the marginal loglikelihood is

$$\hat{\ell}(x) = \log \, f(x|\theta_1^*) + \log \, \pi_1(\theta_1^*) - \log \left\{\frac{1}{T} \sum_{t=1}^T \pi(\theta_1^*|x, \theta_2^{(t)})\right\},$$

where θ_1^* is an arbitrary point, assuming that $f(x|\theta) = f(x|\theta_1)$ and that $\pi_1(\theta_1)$ is available.
(b) In the general case, rewrite the posterior density as

$$\pi(\theta|x) = \pi_1(\theta_1|x)\pi_2(\theta_2|x, \theta_1) \cdots \pi_B(\theta_B|x, \theta_1, \ldots, \theta_{B-1}).$$

Show that an estimate of $\pi_r(\theta_r^*|x, \theta_1^*, \ldots, \theta_{r-1}^*)$ is

$$\hat{\pi}_r(\theta_r^*|x, \theta_s^*, s < r) = \frac{1}{T} \sum_{t=1}^T \pi(\theta_r^*|x, \theta_1^*, \ldots, \theta_{r-1}^*, \theta_{r+1}^{(t)}, \ldots, \theta_B^{(t)}),$$

where the $\theta_\ell^{(t)}$ ($\ell > r$) are simulated from the full conditionals $\pi_\ell(\theta_\ell|\theta_1^*, \ldots, \theta_r^*, \theta_{r+1}, \ldots, \theta_B)$.

(c) Deduce that an estimate of the joint posterior density is $\prod_{r=1}^{B} \hat{\pi}_r(\theta_r^*|x, \theta_s^*,$ $s < r)$ and that an estimate of the marginal loglikelihood is

$$\hat{\ell}(x) = \log \ f(x|\theta^*) + \log \ \pi(\theta^*) - \sum_{r=1}^{B} \log \ \hat{\pi}_r(\theta_r^*|x, \theta_s^*, s < r)$$

for an arbitrary value θ^*.

(d) Discuss the computational cost of this method as a function of B.

(e) Extend the method to the approximation of the predictive density $f(y|x) = \int f(y|\theta)\pi(\theta|x)d\theta$.

10.27 (Fishman 1996) Given a contingency table with cell sizes N_{ij}, row sums $N_{i\cdot}$, and column sums $N_{\cdot j}$, the chi squared statistics for independence is

$$\chi^2 = \sum_{(i,j)} \frac{(N_{ij} - N_{i\cdot}N_{\cdot j}/N)^2}{N_{i\cdot}N_{\cdot j}/N}.$$

Assuming fixed margins $N_{i\cdot}$ and $N_{\cdot j}$ ($i = 1, \ldots, I, j = 1, \ldots, J$), the goal is to simulate the distribution of χ^2. Design a Gibbs sampling experiment to simulate a contingency table under fixed margins. (*Hint:* Show that the vector to be simulated is of dimension $(I-1)(J-1)$.) (*Note:* Alternative methods are described by Aldous 1987 and Diaconis and Sturmfels 1998.)

10.28 The one-way random effects model is usually written

$$Y_{i,j} = \mu + \alpha_i + \varepsilon_{ij}, \quad \alpha_i \sim \mathcal{N}(0, \sigma_\alpha^2), \quad \varepsilon_{ij} \sim \mathcal{N}(0, \sigma^2), \quad j = 1, \ldots, n_i, \quad i = 1, \ldots, k.$$

(a) Show it can also be written as the hierarchical model

$$Y_{i,j}|\alpha_i \sim \mathcal{N}(\mu + \alpha_i, \sigma^2)$$
$$\alpha_i \sim \mathcal{N}(0, \sigma_\alpha^2).$$

(b) Show the joint density is

$$\left(\frac{1}{\sigma_\alpha^2}\right)^{k/2} \exp \frac{1}{2\sigma_\alpha^2} \sum_i \alpha_i^2 \left(\frac{1}{\sigma^2}\right)^{Nk/2} \exp \frac{1}{2\sigma_\alpha^2} \sum_{ij} (y_{ij} - \mu - \alpha_i)^2.$$

(*Note:* This is a complete data likelihood if we consider the α_i's to be missing data.)

(c) Show that the full conditionals are given by

$$\alpha_i|\mathbf{y}, \mu, \sigma_\alpha^2, \sigma^2 \sim N\left(\frac{n_i\sigma_\alpha^2}{n_i\sigma_\alpha^2 + \sigma^2}(\bar{y}_i - \mu), \frac{\sigma_\alpha^2\sigma^2}{n_i\sigma_\alpha^2 + \sigma^2}\right),$$

$$\mu|\mathbf{y}, \alpha, \sigma_\alpha^2, \sigma^2 \sim N\left(\bar{y} - \bar{\alpha}, \frac{\sigma^2}{Nk}\right),$$

$$\sigma_\alpha^2|\mathbf{y}, \alpha, \mu, \sigma^2 \sim \mathcal{IG}\left(\frac{k}{2} - 1, \frac{1}{2}\sum_i \alpha_i^2\right),$$

$$\sigma^2|\mathbf{y}, \alpha, \mu, \sigma_\alpha^2 \sim \mathcal{IG}\left(\frac{Nk}{2} - 1, \frac{1}{2}\sum_{ij}(y_{ij} - \mu - \alpha_i)^2\right),$$

where $N = \sum_i n_i$ and $\mathcal{IG}(a, b)$ is the inverted gamma distribution (see Appendix A).

(d) The following data come from a large experiment to assess the precision of estimation of chemical content of turnip greens, where the leaves represent a random effect (Snedecor and Cochran 1971). Run a Gibbs sampler to estimate the variance components, plot the histograms and monitor the convergence of the chain.

Leaf	% Ca
1	3.28 3.09 3.03 3.03
2	3.52 3.48 3.38 3.38
3	2.88 2.80 2.81 2.76
4	3.34 3.38 3.24 3.26

Table 10.4. Calcium concentration (%) in turnip greens (*Source:* Snedecor and Cochran 1971).

10.29 $\dagger^{11}$(Gelfand et al. 1990) For a population of 30 rats, the weight y_{ij} of rat i is observed at age x_j and is associated with the model

$$Y_{ij} \sim \mathcal{N}(\alpha_i + \beta_i(x_j - \bar{x}), \sigma^2).$$

(a) Give the Gibbs sampler associated with this model and the prior

$$\alpha_i \sim \mathcal{N}(\alpha_c, \sigma_\alpha^2), \qquad \beta_i \sim \mathcal{N}(\beta_c, \sigma_\beta^2),$$

with almost flat hyperpriors

$$\alpha_c, \beta_c \sim \mathcal{N}(0, 10^4), \quad \sigma_c^{-2}, \sigma_\alpha^{-2}, \sigma_\beta^{-2} \sim \mathcal{G}a(10^{-3}, 10^{-3}).$$

(b) Assume now that

$$Y_{ij} \sim \mathcal{N}(\beta_{1i} + \beta_{2i}x_j, \sigma_c^2),$$

with $\beta_i = (\beta_{1i}, \beta_{2i}) \sim \mathcal{N}_2(\mu_\beta, \Omega_\beta)$. Using a Wishart hyperprior $\mathcal{W}(2, R)$ on Ω_β, with

$$R = \begin{pmatrix} 200 & 0 \\ 0 & 0.2 \end{pmatrix},$$

give the corresponding Gibbs sampler.

(c) Study whether the original assumption of independence between β_{1i} and β_{2i} holds.

10.30 $\dagger$(Spiegelhalter et al. 1995a,b) Binomial observations

$$R_i \sim \mathcal{B}(n_i, p_i), \qquad i = 1, \ldots, 12,$$

correspond to mortality rates for cardiac surgery on babies in hospitals. When the failure probability p_i for hospital i is modeled by a random effect structure

$$\text{logit}(p_i) \sim \mathcal{N}(\mu, \tau^2),$$

---

[11] Problems with this dagger symbol are studied in detail in the BUGS manual of Spiegelhalter et al. (1995a,b) . Corresponding datasets can also be obtained from this software (see Note 10.6.2).

with almost flat hyperpriors

$$\mu \sim \mathcal{N}(0, 10^4), \quad \tau^{-2} \sim \mathcal{G}a(10^{-3}, 10^{-3}),$$

examine whether a Gibbs sampler can be implemented. (*Hint:* Consider the possible use of ARS as in Section 2.4.2.)

10.31 †(Boch and Aitkin 1981) Data from the Law School Aptitude Test (LSAT) corresponds to multiple-choice test answers $(y_{j1}, \ldots, y_{j5})$ in $\{0, 1\}^5$. The y_{jk}'s are modeled as $\mathcal{B}(p_{jk})$, with $(j = 1, \ldots, 1000, k = 1, \ldots, 5)$

$$\text{logit}(p_{jk}) = \theta_j - \alpha_k, \qquad \theta_j \sim \mathcal{N}(0, \sigma^2).$$

(This is the *Rasch model*.) Using vague priors on the α_k's and σ^2, give the marginal distribution of the probability P_i to answer $i \in \{0, 1\}^5$. (*Hint:* Show that P_i is the posterior expectation of the probability $P_{i|\theta}$ conditional on ability level.)

10.32 †(Dellaportas and Smith 1993) Observations t_i on survival time are related to covariates z_i by a Weibull model

$$T_i | z_i \sim \mathcal{W}e(r, \mu_i), \qquad \mu_i = \exp(\beta^t z_i),$$

with possible censoring. The prior distributions are

$$\beta_j \sim \mathcal{N}(0, 10^4), \qquad r \sim \mathcal{G}a(1, 10^{-4}).$$

(a) Construct the associated Gibbs sampler and derive the posterior expectation of the median, $\log(2 \exp(-\beta^t z_i))^{1/r}$.
(b) Compare with an alternative implementation using a slice sampler [A.32]

10.33 †(Spiegelhalter et al. 1995a,b) In the study of the effect of a drug on a heart disease, the number of contractions per minute for patient i is recorded before treatment (x_i) and after treatment (y_i). The full model is $X_i \sim \mathcal{P}(\lambda_i)$ and for uncured patients, $Y_i \sim \mathcal{P}(\beta \lambda_i)$, whereas for cured patients, $y_i = 0$.
(a) Show that the conditional distribution of Y_i given $t_i = x_i + y_i$ is $\mathcal{B}(t_i, \beta/(1 + \beta))$ for the uncured patients.
(b) Express the distribution of the Y_i's as a mixture model and derive the Gibbs sampler.

10.34 †(Breslow and Clayton 1993) In the modeling of breast cancer cases y_i according to age x_i and year of birth d_i, an exchangeable solution is

$$\begin{aligned} Y_i &\sim \mathcal{P}(\mu_i), \\ \log(\mu_i) &= \log(d_i) + \alpha_{x_i} + \beta_{d_i}, \\ \beta_k &\sim \mathcal{N}(0, \sigma^2). \end{aligned}$$

(a) Derive the Gibbs sampler associated with almost flat hyperpriors on the parameters α_j, β_k, and σ.
(b) Breslow and Clayton (1993) consider a dependent alternative where for $k = 3, \ldots, 11$ we have

$$(10.16) \qquad \beta_k | \beta_1, \ldots, \beta_{k-1} \sim \mathcal{N}(2\beta_{k-1} - \beta_{k-2}, \sigma^2),$$

while $\beta_1, \beta_2 \sim \mathcal{N}(0, 10^5 \sigma^2)$. Construct the associated Gibbs sampler and compare with the previous results.

(c) An alternative representation of (10.16) is

$$\beta_k | \beta_j, j \neq k \sim \mathcal{N}(\bar{\beta}_k, n_k \sigma^2).$$

Determine the value of $\bar{\beta}_k$ and compare the associated Gibbs sampler with the previous implementation.

10.35 †(Dobson 1983) The effect of a pesticide is tested against its concentration x_i on n_i beetles, $R_i \sim \mathcal{B}(n_i, p_i)$ of which are killed. Three generalized linear models are in competition:

$$p_i = \frac{\exp(\alpha + \beta x_i)}{1 + \exp(\alpha + \beta x_i)},$$
$$p_i = \Phi(\exp(\alpha + \beta x_i)),$$
$$p_i = 1 - \exp(-\exp(\alpha + \beta x_i));$$

that is, the logit, probit, and log-log models, respectively. For each of these models, construct a Gibbs sampler and compute the expected posterior *deviance*; that is, the posterior expectation of

$$D = 2 \left(\sum_{i=1}^n \hat{\ell}_i - \sum_{i=1}^n \ell_i \right),$$

where

$$\ell_i = r_i \log(p_i) + (n_i - p_i) \log(1 - p_i), \qquad \hat{\ell}_i = \max_{p_i} \ell_i.$$

10.36 †(Spiegelhalter et al. 1995a,b) Consider a standard Bayesian ANOVA model $(i = 1, \ldots, 4, j = 1, \ldots, 5)$

$$Y_{ij} \sim \mathcal{N}(\mu_{ij}, \sigma^2),$$
$$\mu_{ij} = \alpha_i + \beta_j,$$
$$\alpha_i \sim \mathcal{N}(0, \sigma_\alpha^2),$$
$$\beta_j \sim \mathcal{N}(0, \sigma_\beta^2),$$
$$\sigma^{-2} \sim \mathcal{G}a(a, b),$$

with $\sigma_\alpha^2 = \sigma_\beta^2 = 5$ and $a = 0$, $b = 1$. Gelfand et al. (1992) impose the constraints $\alpha_1 > \ldots > \alpha_4$ and $\beta_1 < \cdots < \beta_3 > \cdots > \beta_5$.

(a) Give the Gibbs sampler for this model. (*Hint:* Use the optimal truncated normal Accept–Reject algorithm of Example 2.20.)

(b) Change the parameterization of the model as $(i = 2, \ldots, 4, j = 1, 2, 4, 5)$

$$\alpha_i = \alpha_{i-1} + \epsilon_i, \qquad \epsilon_i > 0,$$
$$\beta_j = \beta_3 - \eta_j, \qquad \eta_j > 0,$$

and modify the prior distribution in

$$\alpha_1 \sim \mathcal{N}(0, \sigma_\alpha^2), \quad \beta_3 \sim \mathcal{N}(0, \sigma_\alpha^2), \quad \epsilon_i, \eta_j \sim \mathcal{G}a(0, 1).$$

Check whether the posterior distribution is well defined and compare the performances of the corresponding Gibbs sampler with the previous implementation.

10.37 In the setup of Note 10.6.3:

(a) Evaluate the time requirements for the computation of the exact formulas for the $\mu_i - \mu_{i+1}$.

(b) Devise an experiment to test the maximal value of $x_{(n)}$ which can be processed on your computer.

(c) In cases when the exact value can be computed, study the convergence properties of the corresponding Gibbs sampler.

10.38 In the setup of Note 10.6.3, we define the *canonical moments* of a distribution and show that they can be used as a representation of this distribution.

(a) Show that the two first moments μ_1 and μ_2 are related by the following two inequalities:

$$\mu_1^2 \leq \mu_2 \leq \mu_1 \,,$$

and that the sequence (μ_k) is monotonically decreasing to 0.

(b) Consider a kth-degree polynomial

$$P_k(x) = \sum_{i=0}^{k} a_i x^i.$$

Deduce from

$$\int_0^1 P_k^2(x)\, dG(x)\, dx \geq 0$$

that

(10.17)
$$a^t C_k a \geq 0, \qquad \forall a \in \mathbb{R}^{k+1},$$

where

$$C_k = \begin{pmatrix} 1 & \mu_1 & \mu_2 & \cdots & \mu_k \\ \mu_1 & \mu_2 & \mu_3 & \cdots & \mu_{k+1} \\ \cdots & \cdots & \cdots\cdots & & \\ \mu_k & \mu_{k+1} & & \cdots & \mu_{2k} \end{pmatrix}$$

and $a^t = (a_0, a_1, \ldots, a_k)$.

(c) Show that for every distribution g, the moments μ_k satisfy

$$\begin{vmatrix} 1 & \mu_1 & \mu_2 & \cdots & \mu_k \\ \mu_1 & \mu_2 & \mu_3 & \cdots & \mu_{k+1} \\ \cdots & \cdots & \cdots\cdots & & \cdots \\ \mu_k & \mu_{k+1} & & \cdots & \mu_{2k} \end{vmatrix} \geq 0.$$

(*Hint:* Interpret this as a property of C_k.)

(d) Using inequalities similar to (10.17) for the polynomials $t(1-t)P_k^2(t)$, $tP_k^2(t)$, and $(1-t)P_k^2(t)$, derive the following inequalities on the moments of G:

$$\begin{vmatrix} \mu_1 - \mu_2 & \mu_2 - \mu_3 & \cdots & \mu_{k-1} - \mu_k \\ \mu_2 - \mu_3 & \mu_3 - \mu_4 & \cdots & \mu_k - \mu_{k+1} \\ \cdots & \cdots & \cdots & \cdots \\ \mu_{k-1} - \mu_k & & \cdots & \mu_{2k-1} - \mu_{2k} \end{vmatrix} \geq 0 \,,$$

$$\begin{vmatrix} \mu_1 & \mu_2 & \cdots & \mu_k \\ \mu_2 & \mu_3 & \cdots & \mu_{k+1} \\ \cdots & \cdots & \cdots & \cdots \\ \mu_k & \mu_{k+1} & \cdots & \mu_{2k-1} \end{vmatrix} \geq 0 \,,$$

$$\begin{vmatrix} 1 - \mu_1 & \mu_1 - \mu_2 & \cdots & \mu_{k-1} - \mu_k \\ \mu_1 - \mu_2 & \mu_2 - \mu_3 & \cdots & \mu_k - \mu_{k+1} \\ \cdots & \cdots & \cdots & \cdots \\ \mu_{k-1} - \mu_k & & \cdots & \mu_{2k-2} - \mu_{2k-1} \end{vmatrix} \geq 0 \,.$$

(e) Show that the bounds in parts (c) and (d) induce a lower (upper) bound $\underline{c}_{2k}$ ($\bar{c}_{2k}$) on μ_{2k} and that part (c) [(d)] induces a lower (upper) bound $\underline{c}_{2k-1}$ ($\bar{c}_{2k-1}$) on μ_{2k-1}.

(f) Defining p_k as

$$p_k = \frac{c_k - \underline{c}_k}{\bar{c}_k - \underline{c}_k},$$

show that the relation between $(p_1, ..., p_n)$ and $(\mu_1, ..., \mu_n)$ is one-to-one for every n and that the p_i are independent.

(g) Show that the inverse transform is given by the following recursive formulas. Define

$$q_i = 1 - p_i, \quad \zeta_1 = p_1, \quad \zeta_i = p_i q_{i-1} \quad (i \geq 2).$$

then

$$\begin{cases} S_{1,k} = \zeta_1 + \cdots + \zeta_k & (k \geq 1), \\ S_{j,k} = \sum_{i=1}^{k-j+1} \zeta_i S_{j-1,i+j-1} & (j \geq 2), \\ c_n = S_{n,n}. \end{cases}$$

(*Note:* See Dette and Studden 1997 for details and useful complements on canonical moments.)

10.39 The modification of [A.38] corresponding to the reparameterization discussed in Example 10.26 only involves a change in the generation of the parameters. In the case $k = 2$, show that it is given by

Simulate

$$p|\mu, \theta, \sigma, \tau \sim \mathcal{B}e(n_1 + 1, n_2 + 1);$$

$$\theta|\mu, \sigma, \tau, p \sim \mathcal{N} \left(\frac{(\bar{x}_2 - \mu)}{\tau} \left[1 + \frac{\sigma^2 \zeta^{-2}}{n_2} \right]^{-1}, \frac{\sigma^2}{n_2 + \sigma^2 \xi^{-2}} \right);$$

$$\mu|\theta, \sigma, \tau, , p \sim \mathcal{N} \left(\frac{n_1 \bar{x}_1 + \sigma^{-2} n_2 (\bar{x}_2 - \tau\theta)}{n_1 + n_2 \sigma^{-2}}, \frac{\tau^2 \sigma^2}{n_2 \sigma^2 + n_2} \right);$$

(10.18) $\quad \sigma^{-2}|\mu, \theta, \tau, p \sim \mathcal{G}a \left(\frac{n_2 - 1}{2}, \frac{s_2^2 + n_2 (\bar{x}_2 - \mu - \tau\theta)^2}{2\tau^2} \right) \mathbb{I}_{\sigma^2 < 1};$

$$\tau^{-2}|\mu, \theta, \sigma, p \sim \mathcal{G}a \left(\frac{n}{2}, \frac{1}{2} \left(s_1^2 + n_1 (\bar{x}_1 - \mu)^2 + \frac{s_2^2}{\sigma^2} + \frac{n_2 (\bar{x}_2 - \mu)^2}{n_2 \zeta^2 + \sigma^2} \right) \right).$$

10.40 (Gruet et al. 1999) As in Section 10.4.1, consider a reparameterization of a mixture of exponential distributions, $\sum_{j=1}^{k} p_j \, \mathcal{E}xp(\lambda_j)$.

(a) Use the identifiability constraint, $\lambda_1 \geq \lambda_2 \geq \cdots \geq \lambda_k$, to write the mixture with the "cascade" parameterization

$$p \, \mathcal{E}xp(\lambda) + \sum_{j=1}^{k-1} (1-p) \cdots q_j \, \mathcal{E}xp(\lambda\sigma_1 \cdots \sigma_j),$$

with $q_{k-1} = 1$, $\sigma_1 \leq 1, \ldots, \sigma_{k-1} \leq 1$.

(b) For the prior distribution $\pi(\lambda) = \frac{1}{\lambda}$, p, $q_j \sim \mathcal{U}_{[0,1]}$, $\sigma_j \sim \mathcal{U}_{[0,1]}$, show that the corresponding posterior distribution is always proper.

(c) For the prior of part (b), show that [A.34] leads to the following algorithm:

Algorithm A.47 –Exponential Mixtures–

> **2. Generate**
>
> $$\lambda \sim \mathcal{G}a\left(n, n_0\overline{x}_0 + \sigma_1 n_1\overline{x}_1 + \cdots + \sigma_1 \cdots \sigma_{k-1} n_{k-1}\overline{x}_{k-1}\right),$$
>
> $$\sigma_1 \sim \mathcal{G}a\left(n_1 + \cdots + n_{k-1}, \lambda\left\{n_1\overline{x}_1 + \sigma_2 n_2\overline{x}_2 + \cdots + \right.\right.$$
> $$\left.\left. \sigma_2 \cdots \sigma_{k-1} n_{k-1}\overline{x}_{k-1}\right\}\right) \mathbb{I}_{\sigma_1 \leq 1},$$
>
> $$\vdots$$
>
> $$\sigma_{k-1} \sim \mathcal{G}a\left(n_{k-1}, \lambda n_{k-1}\overline{x}_{k-1}\right) \mathbb{I}_{\sigma_{k-1} \leq 1},$$
> $$p \sim \mathcal{B}e(n_0 + 1, n - n_0 + 1),$$
>
> $$\vdots$$
>
> $$q_{k-2} \sim \mathcal{B}e(n_{k-2} + 1, n_{k-1} + 1),$$

where $n_0, n_1, \ldots, n_{k-1}$ denote the size of subsamples allocated to the components $\mathcal{E}xp(\lambda)$, $\mathcal{E}xp(\lambda\sigma_1)$, ..., $\mathcal{E}xp(\lambda\sigma_1 \ldots \sigma_{k-1})$ and $n_0\overline{x}_0, n_1\overline{x}_1, \ldots, n_{k-1}\overline{x}_{k-1}$ are the sums of the observations allocated to these components.

10.6 Notes

10.6.1 A Bit of Background

Although somewhat removed from statistical inference in the classical sense and based on earlier techniques used in statistical Physics, the landmark paper by Geman and Geman (1984) brought Gibbs sampling into the arena of statistical application. This paper is also responsible for the name *Gibbs sampling*, because it implemented this method for the Bayesian study of *Gibbs random fields* which, in turn, derive their name from the physicist Josiah Willard Gibbs (1839–1903). This original implementation of the Gibbs sampler was applied to a discrete image processing problem and did not involve completion.

The work of Geman and Geman (1984) built on that of Metropolis et al. (1953) and Hastings (1970) and his student Peskun (1973), influenced Gelfand and Smith (1990) to write a paper that sparked new interest in Bayesian methods, statistical computing, algorithms, and stochastic processes through the use of computing algorithms such as the Gibbs sampler and the Metropolis–Hastings algorithm. It is interesting to see, in retrospect, that earlier papers had proposed similar solutions but did not find the same response from the statistical community. Among these, one may quote Besag (1974, 1986), Besag and Clifford (1989), Broniatowski et al. (1984), Qian and Titterington (1990), and Tanner and Wong (1987).

10.6.2 The BUGS Software

The acronym BUGS stands for *Bayesian inference using Gibbs sampling*. This software was developed by Spiegelhalter et al. (1995a,b,c) at the MRC Biostatistics Unit in Cambridge, England. As shown by its name, it has been designed to take advantage of the possibilities of the Gibbs sampler in Bayesian analysis. BUGS includes a language which is C or R like and involves declarations about the model, the data, and the prior specifications, for single or multiple levels in the prior modeling. For instance, for the benchmark nuclear pump failures dataset of Example 10.17, the model and priors are defined by

```
for (i in 1:N) {
    theta[i] ~ dgamma(alpha,beta)
    lambda[i]~ theta[i] * t[i]
    x[i]     ~ dpois(lambda[i])
}
alpha ~ dexp(1.0)
beta  ~ dgamma(0.1,1.0)
```

(see Spiegelhalter et al. 1995b, p.9). Most standard distributions are recognized by BUGS (21 are listed in Spiegelhalter et al. 1995a), which also allows for a large range of transforms. BUGS also recognizes a series of commands like compile, data, out, and stat. The output of BUGS is a table of the simulated values of the parameters after an open number of warmup iterations, the batch size being also open.

A major restriction of this software is the use of the conjugate priors or, at least, log-concave distributions for the Gibbs sampler to apply. However, more complex distributions can be handled by discretization of their support and assessment of the sensitivity to the discretization step. In addition, improper priors are not accepted and must be replaced by proper priors with small precision, like dnorm(0,0.0001), which represents a normal modeling with mean 0 and *precision* (inverse variance) 0.0001.

The BUGS manual (Spiegelhalter et al. 1995a) is quite informative and well written.[12] In addition, the authors have compiled a most helpful example manual (Spiegelhalter et al. 1995b,1996), which exhibits the ability of BUGS to deal with an amazing number of models, including meta-analysis, latent variable, survival analysis, nonparametric smoothing, model selection and geometric modeling, to name a few. (Some of these models are presented in Problems 10.29–10.36.) The BUGS software is also compatible with the convergence diagnosis software CODA presented in Note 12.6.2.

10.6.3 Nonparametric Mixtures

Consider $X_1, \ldots, X_n$ distributed from a mixture of geometric distributions,

---

[12] At the present time, that is, Spring 2004, the BUGS software is available as freeware on the Web site http://www.mrc-bsu.cam.ac.uk/bugs for a wide variety of platforms.

$$X_1, \ldots, X_n \sim \int_0^1 \theta^x (1-\theta) \, dG(\theta), \qquad X_i \in \mathbb{N},$$

where G is an arbitrary distribution on $[0,1]$. In this nonparametric setup, the likelihood can be expressed in terms of the moments

$$\mu_i = \int_0^1 \theta^i dG(\theta), \qquad i = 1, \ldots$$

since G is then identified by the μ_i's. The likelihood can be written

$$L(\mu_1, \mu_2, \ldots | x_1, \ldots, x_n) = \prod_{i=1}^n (\mu_{x_i} - \mu_{1+x_i}).$$

A direct Bayesian modeling of the μ_i's is impossible because of the constraint between the moments, such as $\mu_i \geq \mu_1^i$ ($i \geq 2$), which create dependencies between the different moments (Problem 10.38). The *canonical moments* technique (see Olver 1974 and Dette and Studden 1997, can overcome this difficulty by expressing the μ_i's as transforms of a sequence (p_j) on $[0,1]$ (see Problem 10.38). Since the p_j's are not constrained, they can be modeled as uniform on $[0,1]$. The connection between the μ_i's and the p_j's is given by recursion equations. Let $q_i = 1 - p_i$, $\zeta_1 = p_1$, and $\zeta_i = p_i q_{i+1}$, ($i \geq 2$), and define

$$\begin{aligned} S_{1,k} &= \zeta_1 + \cdots + \zeta_k, & (k \geq 1) \\ S_{j,k} &= \textstyle\sum_{u=1}^{k-j+1} \zeta_u S_{j-1,u+j-1}, & (j \geq 2). \end{aligned}$$

It is then possible to show that $\mu_j - \mu_{j+1} = q_1 S_{j,j}$ (see Problem 10.38).

From a computational point of view, the definition of the μ_i's via recursion equations complicates the exact derivation of Bayes estimates, and they become too costly when $\max x_i > 5$. This setting where numerical complexity prevents the analytical derivation of Bayes estimators can be solved via Gibbs sampling.

The complexity of the relations $S_{j,k}$ is due to the action of the sums in the recursion equations; for instance,

$$\mu_3 - \mu_4 = q_1 p_2 q_2 \{ p_1 q_2 (p_1 q_2 + p_2 q_3) + p_2 q_3 (p_1 q_2 + p_2 q_3 + p_3 q_4) \}.$$

This complexity can be drastically reduced if, through a demarginalization (or completion) device, every sum $S_{j,k}$ in the recursion equation is replaced by one of its terms $\zeta_u S_{j-1,u+j-1}$ ($1 \leq u \leq k - j + 1$). In fact, $\mu_k - \mu_{k+1}$ is then a product of p_i's and q_j's, which leads to a beta distribution on the parameters p_i. To achieve such simplification, the expression

$$\begin{aligned} P(X_i = k) = \mu_k - \mu_{k+1} &= \zeta_1 S_{k-1,k} \\ &= \zeta_1 \{ \zeta_1 S_{k-2,k-1} + \zeta_2 S_{k-2,k} \} \end{aligned}$$

can be interpreted as a marginal distribution of the X_i by introducing $Z_1^i \in \{0,1\}$ such that

$$\begin{aligned} P(X_i = k, Z_1^i = 0) &= \zeta_1^2 S_{k-2,k-1}, \\ P(X_i = k, Z_1^i = 1) &= \zeta_1 \zeta_2 S_{k-2,k}. \end{aligned}$$

Then, in a similar manner, introduce $Z_2^i \in \{0,1,2\}$ such that the density of (X_i, Z_1^i, Z_2^i) (with respect to counting measure) is

$$f(x_i, z_1^i, z_2^i) = \zeta_1 \zeta_{z_1^i + 1} \zeta_{z_2^i + 1} S_{x_i - 3, x_i - 2 + z_2^i}.$$

The replacement of the $S_{k,j}$'s thus requires the introduction of $(k-1)$ variables Z_s^i for each observation $x_i = k$. Once the model is completed by the Z_s^i's, the posterior distribution of p_j,

$$\prod_{i=1}^{n} (q_1 p_1 q_2 p_{z_1^i + 1} q_{z_1^i + 2} \cdots q_{z_{x_i - 1}^i + 2}),$$

is a product of beta distributions on the p_j's, which are easily simulated.

Similarly, the distribution of Z_s^i conditionally on p_j and the other dummy variables Z_r^i $(r \neq s)$ is given by

$$\pi(z_s^i | p, z_{s-1}^i = v, z_{s+1}^i = w) \propto p_w q_{w+1} \mathbb{I}_{z_s^i = w-1} + \cdots + p_{v+2} q_{v+3} \mathbb{I}_{z_s^i = v+1}.$$

The Gibbs sampler thus involves a large number of (additional) steps in this case, namely $1 + \sum_i (x_i - 1)$ simulations, since it imposes the "local" generation of the z_s^i's. In fact, an arbitrary grouping of z_s^i would make the simulation much more difficult, except for the case of a division of $z^i = (z_1^i, \ldots, z_{x_i-1}^i)$ into two subvectors corresponding to the odd and even indices, respectively.

Suppose that the parameter of interest is $\mathbf{p} = (p_1, \ldots, p_{K+1})$, where K is the largest observation. (The distribution of p_i for indices larger than $K+1$ is unchanged by the observations. See Robert 2001.) Although $\mathbf{p}$ is generated conditionally on the complete data (x_i, z^i) $(i = 1, \ldots, n)$, this form of Gibbs sampling is not a Data Augmentation scheme since the z^i's are not simulated conditionally on θ, but rather one component at a time, from the distribution

$$f(z_s^i | \mathbf{p}, z_{s-1}^i = v, z_{s+1}^i = w) \propto p_w q_{w+1} \mathbb{I}_{z_s^i = w-1} + \cdots + p_{v+2} q_{v+3} \mathbb{I}_{z_s^i = v+1}.$$

However, this complexity does not prevent the application of Theorem 9.13 since the sequence of interest is generated conditionally on the z_s^i's. Geometric convergence thus applies.

10.6.4 Graphical Models

Graphical models use graphs to analyze statistical models. They have been developed mainly to represent conditional independence relations, primarily in the field of *expert systems* (Whittaker 1990, Spiegelhalter et al. 1993). The Bayesian approach to these models, as a way to incorporate model uncertainty, has been aided by the advent of MCMC techniques, as stressed by Madigan and York (1995) an expository paper on which this note is based.

The construction of a graphical model is based on a collection of independence assumptions represented by a graph. We briefly recall here the essentials of graph theory and refer to Lauritzen (1996) for details. A *graph* is defined by a set of *vertices* or *nodes*, $\alpha \in \mathcal{V}$, which represents the random variables or factors under study, and by a set of *edges*, $(\alpha, \beta) \in \mathcal{V}^2$, which can be ordered (the graph is then said to be *directed*) or not (the graph is *undirected*). For a directed graph, α is a *parent* of β if (α, β) is an edge (and β is then a *child* of α).[13] Graphs are also often assumed to

---

[13] Directed graphs can be turned into undirected graphs by adding edges between nodes which share a child and dropping the directions.

be *acyclic*; that is, without directed paths linking a node α with itself. This leads to the notion of *directed acyclic graphs*, introduced by Kiiveri and Speed (1982), often represented by the acronym DAG.

For the construction of probabilistic models on graphs, an important concept is that of a *clique*. A *clique* C is a maximal subset of nodes which are all joined by an edge (in the sense that there is no subset containing C and satisfying this condition). An ordering of the cliques of an undirected graph $(C_1, \ldots, C_n)$ is *perfect* if the nodes of each clique C_i contained in a previous clique are all members of one previous clique (these nodes are called the *separators*, $\alpha \in S_i$). In this case, the joint distribution of the random variable V taking values in $\mathcal{V}$ is

$$p(V) = \prod_{v \in V} p(v | \mathcal{P}(v)) ,$$

where $\mathcal{P}(v)$ denotes the parents of v. This can also be written as

(10.19)
$$p(V) = \frac{\prod_{i=1}^{n} p(C_i)}{\prod_{i=1}^{n} p(S_i)} ,$$

and the model is then called *decomposable*; see Spiegelhalter and Lauritzen (1990), Dawid and Lauritzen (1993) or Lauritzen (1996). As stressed by Spiegelhalter et al. (1993), the representation (10.19) leads to a *principle of local computation*, which enables the building of a prior distribution, or the simulation from a conditional distribution on a single clique. (In other words, the distribution is *Markov with respect to the undirected graph*, as shown by Dawid and Lauritzen 1993.) The appeal of this property for a Gibbs implementation is then obvious.

When the densities or probabilities are parameterized, the parameters are denoted by θ_A for the marginal distribution of $V \in A$, $A \subset \mathcal{V}$. (In the case of discrete models, $\theta = \theta_V$ may coincide with p itself; see Example 10.33.) The prior distribution $\pi(\theta)$ must then be compatible with the graph structure: Dawid and Lauritzen (1993) show that a solution is of the form

(10.20)
$$\pi(\theta) = \frac{\prod_{i=1}^{n} \pi_i(\theta_{C_i})}{\prod_{i=1}^{n} \tilde{\pi}_i(\theta_{S_i})} ,$$

thus reproducing the clique decomposition (10.19).

Example 10.33. Discrete event graph. Consider a decomposable graph such that the random variables corresponding to all the nodes of $\mathcal{V}$ are discrete. Let $w \in W$ be a possible value for the vector of these random variables and $\theta(w)$ be the associated probability. For the perfect clique decomposition $(C_1, \ldots, C_n)$, $\theta(w_i)$ denotes the marginal probability that the subvector $(v, v \in C_i)$ takes the value w_i ($\in W_i$) and, similarly, $\theta(w_i^s)$ is the probability that the subvector $(v, v \in S_i)$ takes the value w_i^s when $(S_1, \ldots, S_n)$ is the associated sequence of separators. In this case,

$$\theta(w) = \frac{\prod_{i=1}^{n} \theta(w_i)}{\prod_{i=1}^{n} \theta(w_i^s)}.$$

As illustrated by Madigan and York (1995), a Dirichlet prior can be constructed on $\theta_W = (\theta(w), w \in W)$, which leads to genuine Dirichlet priors on the $\theta_{W_i} = (\theta(w_i), w_i \in W_i)$, under the constraint that the Dirichlet weights are identical over the intersection of two cliques. Dawid and Lauritzen (1993) demonstrate that this prior is unique, given the marginal priors on the cliques. $\parallel$

Example 10.34. Graphical Gaussian model. Giudici and Green (1999) provide another illustration of prior specification in the case of a *graphical Gaussian model*, $\mathbf{X} \sim \mathcal{N}_p(0, \Sigma)$, where the precision matrix $K = \{k_{ij}\} = \Sigma^{-1}$ must comply with the conditional independence relations on the graph. For instance, if $\mathbf{X}_v$ and $\mathbf{X}_w$ are independent given the rest of the graph, then $k_{vw} = 0$. The likelihood can then be factored as

$$f(\mathbf{x}|\Sigma) = \frac{\prod_{i=1}^{n} f(\mathbf{x}_{C_i}|\Sigma^{C_i})}{\prod_{i=1}^{n} f(\mathbf{x}_{S_i}|\Sigma^{S_i})},$$

with the same clique and separator notations as above, where $f(\mathbf{x}_C|\Sigma^C)$ is the normal $\mathcal{N}_{p_C}(0, \Sigma^C)$ density, following the decomposition (10.19). The prior on Σ can be chosen as the conjugate inverse Wishart priors on the Σ^{C_i}'s, under some compatibility conditions. $\parallel$

Madigan and York (1995) discuss an MCMC approach to model choice and model averaging in this setup, whereas Dellaportas and Forster (1996) and Giudici and Green (1999) implement reversible jump algorithms for determining the probable graph structures associated with a given dataset, the latter under a Gaussian assumption.

11

Variable Dimension Models and Reversible Jump Algorithms

"We're wasting our time," he said. "The one thing we need is the one thing we'll never get."
—Ian Rankin, *Resurrection Men*

While the previous chapters have presented a general class of MCMC algorithms, there exist settings where they are not general enough. A particular case of such settings is that of variable dimension models. There, the parameter (and simulation) space is not well defined, being a finite or denumerable collection of unrelated subspaces. To have an MCMC algorithm moving within this collection of spaces requires more advanced tools, if only because of the associated measure theoretic subtleties. Section 11.1 motivates the use of variable dimension models in the setup of Bayesian model choice and model comparison, while Section 11.2 presents the general theory of reversible jump algorithms, which were tailored for these models. Section 11.3 examines further algorithms and methods related to this issue.

11.1 Variable Dimension Models

In general, a variable dimension model is, to quote Peter Green, a *"model where one of the things you do not know is the number of things you do not know"*. This means that the statistical model under consideration is not defined precisely enough for the dimension of the parameter space to be fixed. As detailed below in Section 11.1.1, this setting is closely associated with *model selection*, a collection of statistical procedures that are used at the early state of a statistical analysis, namely, when the model to be used is not yet fully determined.

In addition to model construction, there are other situations where several models are simultaneously considered. For instance, this occurs in model

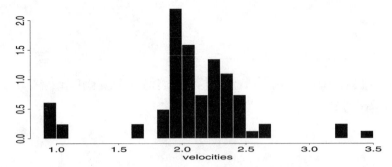

Fig. 11.1. Velocity *(km/second)* of galaxies in the Corona Borealis Region.

checking, model comparison, model improvement, and model pruning, with many areas of application: variable selection in generalized linear models, change point determination in signal processing, object recognition in image analysis, coding identification in DNA analysis, dependence reconstruction in expert systems, and so on.

Example 11.1. Mixture modeling. Consider the dataset represented by the histogram of Figure 11.1, and also provided in Table 11.1 (see Problem 11.7). It consists of the velocities of 82 galaxies previously analyzed by Roeder (1992) and is often used as a benchmark example in mixture analysis (see, e.g., Chib 1995, Phillips and Smith 1996, Raftery 1996, Richardson and Green 1997, or Robert and Mengersen 1999).

A probabilistic model considered for the representation of this dataset is a Gaussian mixture model,

$$(11.1) \qquad \mathcal{M}_k : x_i \sim \sum_{j=1}^{k} p_{jk} \mathcal{N}(\mu_{jk}, \sigma_{jk}^2) \qquad (i = 1, \ldots, 82),$$

but the index k, that is, the number of components in the mixture (or of clusters of galaxies in the sample), is under debate. It cannot therefore be arbitrarily fixed to, say, $k = 7$ for the statistical analysis of this dataset.   ‖

11.1.1 Bayesian Model Choice

While the concept of a variable dimension model is loosely defined, we can give a more formal definition which mostly pertains to the important case of model selection.

Definition 11.2. A Bayesian *variable dimension model* is defined as a collection of models $(k = 1, \ldots, K)$,

$$\mathcal{M}_k = \{ f(\cdot|\theta_k); \ \theta_k \in \Theta_k \} ,$$

associated with a collection of priors on the parameters of these models,

$$\pi_k(\theta_k) \, ,$$

and a prior distribution on the indices of these models,

$$\{\varrho(k) \, , k = 1, \dots, K\} \, .$$

Note that we will also use the more concise notation

(11.2) $$\pi(k, \theta_k) = \varrho(k)\pi_k(\theta_k) \, .$$

This is a density, in the sense that $\varrho(k)$ is a density with respect to the counting measure on $\mathbb{N}$, while $\pi_k(\theta_k)$ is typically a density with respect to Lebesgue measure on Θ_k. The function (11.2) is then a density with respect to Lebesgue measure on the union of spaces,

$$\Theta = \bigcup_k \{k\} \times \Theta_k \, .$$

From a Bayesian perspective, this representation of the problem implies that inference is formally complete! Indeed, once the prior and the model are defined, the model selected from the dataset $\mathbf{x}$ is determined from the posterior probabilities

$$p(\mathcal{M}_i|\mathbf{x}) = \frac{\varrho(i) \displaystyle\int_{\Theta_i} f_i(x|\theta_i)\pi_i(\theta_i)d\theta_i}{\displaystyle\sum_j \varrho(j) \int_{\Theta_j} f_j(x|\theta_j)\pi_j(\theta_j)d\theta_j}$$

by either taking the model with largest $p(\mathcal{M}_i|x)$ or using *model averaging* through

$$\sum_j p(\mathcal{M}_i|\mathbf{x}) \int_{\Theta_j} f_j(x|\theta_j)\pi_j(\theta_j|\mathbf{x}) \, d\theta_j \, ,$$

where $\mathbf{x}$ denotes the observed dataset, as a *predictive* distribution, even though more sophisticated decision theoretic perspectives could be adopted (see Robert, 2001, Chapter 7).

11.1.2 Difficulties in Model Choice

There are several kinds of difficulties with this formal resolution of the model choice issue. While the definition of a prior distribution on the parameter space

$$\Theta = \bigcup_k \{k\} \times \Theta_k$$

does not create any problem, thanks to the decomposition of $\pi(\theta)$ into $\varrho(k)\pi_k(\theta_k)$, a first difficulty is that at the inferential level model choice is a complex notion. The setting does not clearly belong to either the estimation or the testing domain.

As an example, when considering model choice as an estimation problem, there is a tendency to overfit the model by selecting a model $\mathcal{M}_i$ with a large number of parameters, and overfitting can only be countered with priors distributions $\varrho(j)$ that depend on the sample size (see Robert, 2001, Chapter 7). Similarly, when adopting the perspective that model choice is a special case of testing between models, the subsequent inference on the most "likely" model fails to account for the selection process and its inherent error. This vagueness, central to the model choice formulation, will translate into a requirement for a many faceted prior distribution.

We must stress here that our understanding of the Bayesian model choice issue is that we must chose a completely new set of parameters for each model Θ_k and set the parameter space as the union of the model parameter spaces Θ_k, even though some parameters may have a similar meaning in two different models.

Example 11.3. Order of an AR(p) model. Recall that an AR(p) model (Example 6.6) is given as the *autoregressive* representation of a time series,

$$\mathcal{M}_p \; : \; X_t = \sum_{i=1}^{p} \theta_{ip} X_{t-i} + \sigma_p \varepsilon_t \; .$$

When comparing an AR(p) and an AR$(p+1)$ model, it could be assumed that the first p autoregressive coefficients of $\theta_{i(p+1)}$ would be the same as the θ_{ip}'s, that is, that an AR(p) model is simply an AR$(p+1)$ model with an extra zero coefficient.

We note that it is important to consider the coefficients for each of the models as an entire set of coefficients, and not individually. This is not only because the models are different but, more importantly, because the best fitting AR$(p+1)$ model is not necessarily a modification of the best fitting AR(p) model (obtained by adding an extra term, that is, a non-zero $\theta_{(p+1)(p+1)}$). Moreover, from a Bayesian point of view, the parameters θ_{ip}'s are not independent a posteriori. Similarly, even though the variance σ_p^2 has the same formal meaning for all values of p, we insist on using a different variance parameter for each value of p, hence the notation σ_p^2.   ‖

However, many statisticians prefer to use some parameters that are common to all models, in order to reduce model and computational complexity (Problems 4.1 and 4.2), and also to enforce Occam's parsimony requirement (see Note 11.5.1).[1] As we will see below, the reversible jump technique of Sec-

---

[1] At another level, this alternative often allows for a resolution of some testing difficulties associated with improper priors; see Berger and Pericchi (1998) and Robert (2001, Chapter 6) for details.

tion 11.2 is based upon this assumption of partially exchangeable parameters between models, since it uses proposal distributions that modify only a part of the parameter vector to move between models. (The centering technique of Brooks et al. 2003b relies on the same assumption.)

More central to the theme of this book, there also are *computational* difficulties related to variable dimension models. For instance, the number of models in competition may well be infinite. Even when the number of models is finite, there is additional complexity in representing, or simulating from, the posterior distribution (11.2) in that a sampler must move both *within* and *between* models Θ_k. While the former (move) pertains to previous developments in Chapters 7–10, the latter (move) requires a deeper measure-theoretic basis to ensure the overall validity of the correct MCMC moves, that is, to preserve $\pi(\theta|x)$ as the stationary distribution of the simulated Markov chain on Θ.

Lastly, we mention that there is an enormous, and growing, literature on the topic of model selection, from both a frequentist and a Bayesian point of view. Starting with the seminal paper of Akaike (1974), which defined one of the first model selection criterion (now known as AIC), a major focus of that methodology is to examine the properties of the model selection procedure, and to insure, in some sense, that the correct model will be selected if the number of observations is infinite. (This is known as *consistency* in model selection.) However, many of our concerns are different from those in the model selection literature, and we will not go in details of the many model selection criteria that are available. A good introduction to the many facets of model selection is the collection edited by Lahiri (2001).

11.2 Reversible Jump Algorithms

There have been several earlier approaches in the literature to deal with variable dimension models using, for instance, birth-and-death processes (Ripley 1977, Geyer and Møller 1994) or pseudo-priors (Carlin and Chib 1995, see Problem 11.9), but the general formalization of this problem has been presented by Green (1995). Note that, at this stage, regular Gibbs sampling is impossible when considering distributions of the form (11.2): if one conditions on k, then $\theta_k \in \Theta_k$, and if one conditions on θ_k, then k cannot move. Therefore a standard Gibbs sampler cannot provide moves *between* models Θ_k without further modification of the setting.

11.2.1 Green's Algorithm

If we let $x = (k, \theta_k)$, the solution proposed by Green (1995) is based on a *reversible* transition kernel K, that is, a kernel satisfying

$$\int_A \int_B K(x, dy)\pi(x)dx = \int_B \int_A K(y, dx)\pi(y)dy$$

for all $A, B \subset \Theta$ and for some invariant density π (see Section 6.5.3). To see more clearly[2] how this condition can be satisfied and how a proper reversible kernel K can be constructed, we decompose K according to the model in which it proposes a move: for the model $\mathcal{M}_m$, q_m denotes a transition measure on $\mathcal{M}_m$, and ρ_m is the corresponding probability of accepting this move (or *jump*). The decomposition of the kernel is thus

$$K(x, B) = \sum_m \int_B \rho_m(x, y') q_m(x, dy') + \omega(x) \mathbb{I}_B(x),$$

where

$$\omega(x) = 1 - \sum_m (\rho_m q_m)(x, \Theta_m)$$

represents the probability of no move.

Typically, and mostly for practicality's sake, the jumps are limited to moves from $\mathcal{M}_m$ to models with dimensions close to the dimension of Θ_k, possibly including $\mathcal{M}_m$: constructing a sensible proposal q_m for a move from $x = (m, \theta_m)$ to $y = (m', \theta_{m'})$ is generally too difficult when $\mathcal{M}_m$ and $\mathcal{M}_{m'}$ differ by many dimensions. The definition of ρ_m (and the verification of the reversibility assumption) relies on the following assumption: *The joint measure $\pi(dx) q_m(x, dy)$ must be absolutely continuous with respect to a symmetric measure $\xi_m(dx, dy)$ on $\Theta \times \Theta$.* If $g_m(x, y)$ denotes the density of $q_m(x, dy) \pi(dx)$ against this dominating measure $\xi_m(dx, dy)$ and if ρ_m is written in the usual Metropolis–Hastings form

$$\rho_m(x, y) = \min\left\{ 1, \frac{g_m(y, x)}{g_m(x, y)} \right\},$$

then reversibility is ensured by the symmetry of the measure ξ_m:

$$\int_A \int_B \rho_m(x, y) q_m(x, dy) \pi(dx) = \int_A \int_B \rho_m(x, y) g_m(x, y) \xi_m(dx, dy)$$
$$= \int_A \int_B \rho_m(y, x) g_m(y, x) \xi_m(dy, dx)$$
$$= \int_A \int_B \rho_m(y, x) q_m(y, dx) \pi(dy),$$

as $\rho_m(x, y) g_m(x, y) = \rho_m(y, x) g_m(y, x)$ by construction.

The main difficulty of this approach lies in the determination of the measure ξ_m, given the symmetry constraint. If the jumps are decomposed into moves between pairs of models, $\mathcal{M}_{k_1}$ and $\mathcal{M}_{k_2}$, the (clever!) idea of Green

---

[2] The construction of the reversible jump kernel is slightly involved. For an alternate description of this technique, which might be easier to understand, we refer the reader to Section 11.2.2. There the reversible jump algorithm is justified using ordinary Metropolis–Hastings arguments.

(1995) is to supplement each of the spaces Θ_{k_1} and Θ_{k_2} with adequate artificial spaces in order to create a *bijection* between them.

For instance, if $\dim(\Theta_{k_1}) > \dim(\Theta_{k_2})$ and if the move from Θ_{k_1} to Θ_{k_2} can be represented by a deterministic transformation of $\theta^{(k_1)}$

$$\theta^{(k_2)} = T(\theta^{(k_1)}),$$

Green (1995) imposes a *dimension matching* condition which is that the opposite move from Θ_{k_2} to Θ_{k_1} is concentrated on the curve

$$\{\theta^{(k_1)} \,:\, \theta^{(k_2)} = T(\theta^{(k_1)})\}\,.$$

In the general case, if $\theta^{(k_1)}$ is completed by a simulation $u_1 \sim g_1(u_1)$ into $(\theta^{(k_1)}, u_1)$ and $\theta^{(k_2)}$ by $u_2 \sim g_2(u_2)$ into $(\theta^{(k_2)}, u_2)$ so that the mapping between $(\theta^{(k_1)}, u_1)$ and $(\theta^{(k_2)}, u_2)$ is a bijection,

$$(11.3) \qquad\qquad (\theta^{(k_2)}, u_2) = T(\theta^{(k_1)}, u_1),$$

the probability of acceptance for the move from $\mathcal{M}_{k_1}$ to $\mathcal{M}_{k_2}$ is then

$$(11.4) \qquad \min\left(\frac{\pi(k_2, \theta^{(k_2)})}{\pi(k_1, \theta^{(k_1)})}\, \frac{\pi_{21} g_2(u_2)}{\pi_{12} g_1(u_1)}\, \left|\frac{\partial T(\theta^{(k_1)}, u_1)}{\partial(\theta^{(k_1)}, u_1)}\right|, 1\right),$$

involving the Jacobian of the transform (11.3), the probability π_{ij} of choosing a jump to $\mathcal{M}_{k_j}$ while in $\mathcal{M}_{k_i}$, and g_i, the density of u_i. This proposal satisfies the detailed balance condition and the symmetry assumption of Green (1995) if the move from $\mathcal{M}_{k_2}$ to $\mathcal{M}_{k_1}$ also satisfies (11.3) with $u_2 \sim g_2(u_2)$.

The pseudo-code representation of Green's (1995) algorithm is thus as follows.

Algorithm A.48 —Green's Algorithm—

At iteration t, if $x^{(t)} = (m, \theta_m^{(t)})$,

1 Select model $\mathcal{M}_n$ with probability π_{mn}
2 Generate $u_{mn} \sim \varphi_{mn}(u)$
3 Set $(\theta_n, v_{nm}) = T_{mn}(\theta_m^{(t)}, u_{mn})$
4 Take $\theta_n^{(t)} = \theta_n$ with probability   [A.48]

$$\min\left(\frac{\pi(n, \theta_n)}{\pi(m, \theta_m^{(t)})}\, \frac{\pi_{nm}\varphi_{nm}(v_{nm})}{\pi_{mn}\varphi_{mn}(u_{mn})}\, \left|\frac{\partial T_{mn}(\theta_m^{(t)}, u_{mn})}{\partial(\theta_n^{(t)}, u_{mn})}\right|, 1\right),$$

and take $\theta_m^{(t+1)} = \theta_m^{(t)}$ otherwise.

As pointed out by Green (1995), the density π does not need to be normalized, but the different component densities $\pi_k(\theta_k)$ must be known up to the same constant.

Example 11.4. A linear Jacobian. To illustrate the procedure, Green (1995) considers the toy example of switching between the parameters $(1, \theta)$ and $(2, \theta_1, \theta_2)$ using the following moves:

(i) To go from $(2, \theta_1, \theta_2)$ to $(1, \theta)$, set $\theta = (\theta_1 + \theta_2)/2$.
(ii) To go from $(1, \theta)$ to $(2, \theta_1, \theta_2)$, generate a random variable $u \sim g(u)$ and set $\theta_1 = \theta - u$ and $\theta_2 = \theta + u$.

These moves represent one-to-one transformations of variables in $\mathbb{R}^2$; that is,

$$\left(\frac{\theta_1 + \theta_2}{2}, \frac{\theta_2 - \theta_1}{2}\right) = T_{21}(\theta_1, \theta_2), \qquad (\theta - u, \theta + u) = T_{12}(\theta, u),$$

with corresponding Jacobians

$$\frac{\partial T_{21}(\theta_1, \theta_2)}{\partial(\theta_1, \theta_2)} = \frac{1}{2}, \qquad \frac{\partial T_{12}(\theta, u)}{\partial(\theta, u)} = 2.$$

The acceptance probability for a move from $(1, \theta)$ to $(2, \theta_1, \theta_2)$ is thus

$$\min\left(2\,\frac{\pi(2, \theta_1, \theta_2)}{\pi(1, \theta)}\,\frac{\pi_{21}}{\pi_{12}g(u)}, 1\right),$$

where $u = (\theta_2 - \theta_1)/2$. $\|$

11.2.2 A Fixed Dimension Reassessment

While the above development is completely valid from a mathematical point of view, we now redefine Green's algorithm via a *saturation* scheme that provides better intuition for the determination of the acceptance probability. When considering a specific move from $\mathcal{M}_m$ to $\mathcal{M}_n$, that is, from $\theta_m \in \Theta_m$ to $\theta_n \in \Theta_n$, where $d_m = \dim \Theta_m < \dim \Theta_n = d_n$, we can indeed describe an equivalent move in *a fixed dimension setting*.

As described above, the central feature of Green's algorithm is to add an auxiliary variable $u_{mn} \in \mathcal{U}_{mn}$ to θ_m so that $\Theta_m \times \mathcal{U}_{mn}$ and Θ_n are in bijection (one-to-one) relation. (We consider the special case where θ_n needs not be completed.) Using a *regular* Metropolis–Hastings scheme, to propose a move from the pair (θ_m, u_{mn}) to θ_n is like proposing to do so when the corresponding stationary distributions are $\pi(m, \theta_m)\varphi_{mn}(u_{mn})$ and $\pi(n, \theta_n)$, respectively, and when the proposal distribution is *deterministic*, since

$$\theta_n = T_{mn}(\theta_m, u_{mn}).$$

This is an unusual setting for Metropolis–Hastings moves, because of its deterministic feature, but it can solved by the following approximation: Consider that the move from (θ_m, u_{mn}) to θ_n proceeds by generating

$$\theta_n \sim \mathcal{N}_{d_n}\left(T_{mn}(\theta_m, u_{mn}), \varepsilon I\right), \qquad \varepsilon > 0,$$

and that the reciprocal proposal is to take (θ_m, u_{mn}) as the T_{mn}-inverse transform of a normal $\mathcal{N}_{d_n}(\theta_n, \varepsilon I)$. (This is feasible since T_{mn} is a bijection.) This reciprocal proposal then has the density

$$\frac{\exp\left\{-\left[\theta_n - T_{mn}(\theta_m, u_{mn})\right]^2 / 2\varepsilon\right\}}{(2\pi\varepsilon)^{d_n/2}} \times \left|\frac{\partial T_{mn}(\theta_m, u_{mn})}{\partial(\theta_m, u_{mn})}\right|$$

by the Jacobian rule. Therefore, the Metropolis–Hastings acceptance ratio for this regular move is

$$1 \wedge \left(\frac{\pi(n, \theta_n)}{\pi(m, \theta_m)\varphi_{mn}(u_{mn})} \left|\frac{\partial T_{mn}(\theta_m, u_{mn})}{\partial(\theta_m, u_{mn})}\right| \right.$$
$$\left. \times \frac{\exp\left\{-\left[\theta_n - T_{mn}(\theta_m, u_{mn})\right]^2 / 2\varepsilon\right\} / (2\pi\varepsilon)^{d_n/2}}{\exp\left\{-\left[\theta_n - T_{mn}(\theta_m, u_{mn})\right]^2 / 2\varepsilon\right\} / (2\pi\varepsilon)^{d_n/2}}\right)$$

and the normal densities cancel as in a regular random walk proposal. If we take into account the probabilities of the moves between $\mathcal{M}_m$ and $\mathcal{M}_n$, we thus end up with

$$1 \wedge \left(\frac{\pi(n, \theta_n)\pi_{nm}}{\pi(m, \theta_m)\varphi_{mn}(u_{mn})\pi_{mn}} \left|\frac{\partial T_{mn}(\theta_m, u_{mn})}{\partial(\theta_m, u_{mn})}\right|\right).$$

Since this probability does not depend on ε, we can let ε go to zero and obtain the equivalent of the ratio (11.4). The reversible jump algorithm can thus be reinterpreted as a sequence of local fixed dimensional moves between the models $\mathcal{M}_k$ (Problem 11.5).

11.2.3 The Practice of Reversible Jump MCMC

The dimension matching transform, T_{mn}, while incredibly flexible, can be quite difficult to create and much more difficult to optimize; one could almost say this universality in the choice of T_{mn} is a drawback with the method. In fact, the total freedom left by the reversible jump principle about the choice of the jumps, which are often referred to as *split* and *merge* moves in embedded models, creates a potential opening for inefficiency and requires tuning steps which may be quite demanding. As also mentioned in a later chapter, this is a setting where *wealth is a mixed blessing*, if only because the total lack of direction in the choice of the jumps may result in a lengthy or even impossible calibration of the algorithm.

Example 11.5. Linear versus quadratic regression. Instead of choosing a particular regression model, we can use the reversible jump algorithm to do *model averaging*.

Suppose that the two candidate models are the linear and quadratic regression models; that is,

$$y_i = \beta_0 + \beta_1 x_i + \varepsilon_i \quad \text{and} \quad y_i = \beta_0 + \beta_1 x_i + \beta_2 x_i^2 + \varepsilon_i .$$

If we represent either regression by $\mathbf{y} = \mathbf{X}\beta + \varepsilon$, where $\varepsilon \sim \mathcal{N}(0, \sigma^2 I)$, the least squares estimate $\hat{\beta} = (\mathbf{X}'\mathbf{X})^{-1}\mathbf{X}'\mathbf{y}$ has distribution

$$\hat{\beta} \sim \mathcal{N}(\beta, \sigma^2 (\mathbf{X}'\mathbf{X})^{-1}) .$$

Using normal prior distributions will result in normal posterior distributions, and the reversible jump algorithm will then be jumping between a two-dimensional and three-dimensional normal distribution.

To jump between these models, it seems sensible to first transform to *orthogonal coordinates,* as a jump that is made by simply adding or deleting a coefficient will not affect the fit of the other coefficients. We thus find an orthogonal matrix $\mathbf{P}$ and diagonal matrix D_λ satisfying

$$\mathbf{P}' \left(\mathbf{X}'\mathbf{X} \right) \mathbf{P} = D_\lambda .$$

The elements of D_λ, λ_i, are the eigenvalues of $\mathbf{X}'\mathbf{X}$ and the columns of $\mathbf{P}$ are its eigenvectors. We then write $\mathbf{X}^* = \mathbf{X}\mathbf{P}$ and $\alpha = \mathbf{P}'\beta$, and we work with the model $\mathbf{y} = \mathbf{X}^*\alpha + \varepsilon$.

If each α_i has a normal prior distribution, $\alpha_i \sim \mathcal{N}(0, \tau^2)$, its posterior density, denoted by f_i, is $\mathcal{N}(b_i, b_i \sigma^2)$, where $b_i = \frac{\lambda_i \tau^2}{\lambda_i \tau^2 + \sigma^2}$. The possible moves are as follows:

(i) linear $\to$ linear: $(\alpha_0, \alpha_1) \to (\alpha_0', \alpha_1')$, where $(\alpha_0', \alpha_1') \sim f_0 f_1$,
(ii) linear $\to$ quadratic: $(\alpha_0, \alpha_1) \to (\alpha_0, \alpha_1, \alpha_2')$, where $\alpha_2' \sim f_2$,
(iii) quadratic $\to$ quadratic: $(\alpha_0, \alpha_1, \alpha_2) \to (\alpha_0', \alpha_1', \alpha_2')$, where $(\alpha_0', \alpha_1', \alpha_2') \sim f_0 f_1 f_2$,
(iv) quadratic $\to$ linear: $(\alpha_0, \alpha_1, \alpha_2) \to (\alpha_0', \alpha_1')$, where $(\alpha_0', \alpha_1') = (\alpha_0, \alpha_1)$.

The algorithm was implemented on simulated data with move probabilities π_{ij} all taken to be $1/4$ and a prior probability of $1/2$ on each regression model. The resulting fits are given in Figure 11.2. It is interesting to note that when the model is quadratic, the reversible jump fit is close to that of quadratic least squares, but it deviates from quadratic least squares when the underlying model is linear. (See Problem 11.3 for more details.) $\parallel$

Example 11.6. Piecewise constant densities. Consider a density f on $[0, 1]$ of the form

$$f(x) = \sum_{i=1}^{k} \omega_i \mathbb{I}_{[a_i, a_{i+1})}(x) ,$$

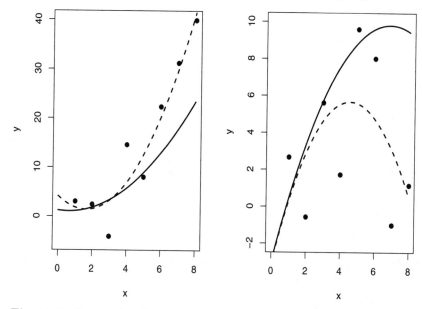

Fig. 11.2. Comparison of quadratic least squares (dashed line) and reversible jump (solid line) fits. In the left panel the underlying model is quadratic, and in the right panel the underlying model is linear.

with

$$a_1 = 0, \quad a_{k+1} = 1, \quad \text{and} \quad \sum_{i=1}^{k} \omega_i(a_{i+1} - a_i) = 1.$$

This model corresponds, for instance, a special kind of nonparametric density estimation, where the estimator is a step function.

Assuming all parameters unknown (including k), define $p_i = \omega_i(a_{i+1} - a_i)$, so these are probabilities. Let the prior distribution be

$$\pi(k, p^{(k)}, a^{(k)}) = \frac{\lambda^k e^{-\lambda}}{k!} \frac{\Gamma(k/2) p_1^{-1/2} \cdots p_k^{-1/2}}{\Gamma(1/2)^k} (k-1)! \mathbb{I}_{a_2 \leq \ldots \leq a_k},$$

where $p^{(k)} = (p_1, \ldots, p_k)$ and $a^{(k)} = (a_2, \ldots, a_k)$, which implies a Poisson distribution on k, $\mathcal{P}(\lambda)$, a uniform distribution on $\{a_2, \ldots, a_k\}$, and a Dirichlet distribution $D_k(1/2, \ldots, 1/2)$ on the weights p_i of the components $\mathcal{U}_{[a_i, a_{i+1}]}$ of f. (Note that the density integrates to 1 over Θ.)

For a sample $x_1, \ldots, x_n$, the posterior distribution is

$$\pi(k, p^{(k)}, a^{(k)} | x_1, \ldots, x_n) \propto \frac{\lambda^k}{k} \frac{\Gamma(k/2)}{\Gamma(1/2)^k} p_1^{n_1 - 1/2} \cdots p_k^{n_k - 1/2}$$
$$\times \prod_{i=1}^{k} (a_{i+1} - a_i)^{-n_i} \mathbb{I}_{a_2 \leq \ldots \leq a_k},$$

where n_j is the number of observations between a_j and a_{j+1}. We can, for instance, restrict the moves to jumps only between neighboring models; that is, models with one more or one less component in the partition of $[0, 1]$. We then represent the jump from model Θ_k to model Θ_{k-1} as a random choice of i $(1 \leq i \leq k - 1)$, and the aggregation of the ith and $(i + 1)$st components as

$$a_2^{(k-1)} = a_2^{(k)}, \ldots, a_i^{(k-1)} = a_i^{(k)}, \ a_{i+1}^{(k-1)} = a_{i+2}^{(k)}, \ldots, a_{k-1}^{(k-1)} = a_k^{(k)}$$

and

$$p_1^{(k-1)} = p_1^{(k)}, \ldots, p_i^{(k-1)} = (p_i^{(k)} + p_{i+1}^{(k)}), \ldots, p_{k-1}^{(k-1)} = p_k^{(k)}.$$

For reasons of symmetry, the opposite (upward) jump implies choosing a component i at random and breaking it by the procedure

1. Generate $u_1, u_2 \sim \mathcal{U}_{[0,1]}$;
2. Take $a_i^{(k)} = a_i^{(k-1)}$, $a_{i+1}^{(k)} = u_1 a_i^{(k-1)} + (1 - u_1)a_{i+1}^{(k-1)}$ and $a_{i+2}^{(k)} = a_{i+1}^{(k-1)}$;
3. Take $p_i^{(k)} = u_2 p_i^{(k-1)}$, and $p_{i+1}^{(k)} = (1 - u_2)p_i^{(k-1)}$.

The other quantities remain identical up to a possible index shift. The weight corresponding to the jump from model Θ_k to model Θ_{k+1} is then

$$\rho_m = \min \left\{ 1, \frac{\pi(k + 1, p^{(k+1)}, a^{(k+1)})}{\pi(k, p^{(k)}, a^{(k)})} \left| \frac{\partial(p^{(k+1)}, a^{(k+1)})}{\partial(p^{(k)}, a^{(k)}, u_1, u_2)} \right| \right\}.$$

As an illustration, take $k = 3$ and consider the jump from model Θ_3 to model Θ_4. The transformation is given by

$$\begin{pmatrix} p_1^{(3)} \\ p_2^{(3)} \\ p_3^{(3)} \\ a_2^{(3)} \\ a_3^{(3)} \\ u_1 \\ u_2 \end{pmatrix} \longrightarrow \begin{pmatrix} p_1 \\ u_2 p_2 \\ (1 - u_2)p_2 \\ p_3 \\ a_2 \\ u_1 a_2 + (1 - u_1)a_3 \\ a_3 \end{pmatrix} = \begin{pmatrix} p_1^{(4)} \\ p_2^{(4)} \\ p_3^{(4)} \\ p_4^{(4)} \\ a_2^{(4)} \\ a_3^{(4)} \\ a_4^{(4)} \end{pmatrix}$$

with Jacobian

$$\left| \frac{\partial(p^{(4)}, a^{(4)})}{\partial(p^{(3)}, a^{(3)}, u_1, u_2)} \right| = \begin{vmatrix} 1 & 0 & 0 & 0 & 0 & 0 & 0 \\ 0 & u_2 & 1 - u_2 & 0 & 0 & 0 & 0 \\ 0 & 0 & 0 & 1 & 0 & 0 & 0 \\ 0 & 0 & 0 & 0 & 1 & u_1 & 0 \\ 0 & 0 & 0 & 0 & 0 & 1 - u_1 & 1 \\ 0 & 0 & 0 & 0 & 0 & a_2^{(3)} - a_3^{(3)} & 0 \\ 0 & p_2^{(3)} & p_2^{(3)} & 0 & 0 & 0 & 0 \end{vmatrix}$$

$$= \left| p_2^{(3)}(a_2^{(3)} - a_3^{(3)}) \right|.$$

(See Problem 11.6 for extensions.)   ‖

These two examples are fairly straightforward, and there are much more realistic illustrations in the literature. We will see two of these below for mixtures of distributions and AR(p) models, respectively. See Green (1995), Denison et al. (1998), and Holmes et al. (2002) for other illustrations. In particular, Cappé et al. (2003, 2004), provide in great detail the implementation of the reversible jump algorithm for a *hidden Markov model* (see Section 14.3.2).

Example 11.7 is quite popular in the reversible jump literature, in that Richardson and Green (1997) dealt with a realistic mixture model, providing sensible answers with seemingly straightforward proposals.

Example 11.7. (Continuation of Example 11.1) If we consider a model $\mathcal{M}_k$ to be the k-component normal mixture distribution $\mathcal{M}_k$ in (11.1), moves between models involve changing the number of components in the mixture and thus adding new components or removing older components.

A first possibility considered in Richardson and Green (1997) is to restrict the moves to models with one *more* or one *less* component, that is, from $\mathcal{M}_k$ to $\mathcal{M}_{k+1}$ or $\mathcal{M}_{k-1}$, and to create a reversible birth-and-death process using the prior on the components (assumed to be independent) as a proposal for the birth step.

The *birth step* associated with the move from $\mathcal{M}_k$ to $\mathcal{M}_{k+1}$ consists in adding a new normal component in the mixture. The new $(k+1)$-component normal mixture distribution is then made of the previous k-component normal mixture distribution and a new component, with weight $p_{(k+1)(k+1)}$, and mean and variance, $\mu_{(k+1)(k+1)}$ and $\sigma^2_{(k+1)(k+1)}$. These parameters can be generated in many ways, but a natural solution is to simulate $p_{(k+1)(k+1)}$ from the marginal distribution of $(p_{1(k+1)}, \ldots, p_{(k+1)(k+1)})$ (which typically is a Dirichlet $\mathcal{D}(\alpha_1, \ldots, \alpha_{k+1})$) and then simulate $\mu_{(k+1)(k+1)}$ and $\sigma^2_{(k+1)(k+1)}$ from the corresponding prior distributions. Obviously, constraining the weights to sum to one implies that the weights of the k component mixture, $p_{1(k)}, \ldots, p_{k(k)}$, have to be multiplied by $(1 - p_{(k+1)(k+1)})$ to obtain the weights of the $(k+1)$ mixture, $p_{1(k+1)}, \ldots, p_{k(k+1)}$. The parameters of the additional component are denoted $u_{k(k+1)}$ with corresponding proposal distribution $\varphi_{k(k+1)}(u_{k(k+1)})$.

It follows from the reversibility constraint that the *death step* is necessarily the (deterministic) opposite. We remove one of the k components and renormalize the other weights. In the case of the Dirichlet distribution, the corresponding acceptance probability for a birth step is then, following (11.4),

$$\min\left(\frac{\pi_{(k+1)k}}{\pi_{k(k+1)}} \frac{(k+1)!}{k!} \frac{\pi_{k+1}(\theta_{k+1})}{\pi_k(\theta_k)(k+1)\varphi_{k(k+1)}(u_{k(k+1)})}, 1\right)$$

$$(11.5) \qquad = \min\left(\frac{\pi_{(k+1)k}}{\pi_{k(k+1)}} \frac{\varrho(k+1)}{\varrho(k)} \frac{\ell_{k+1}(\theta_{k+1})(1 - p_{k+1})^{k-1}}{\ell_k(\theta_k)}, 1\right),$$

where ℓ_k denotes the likelihood of the k component mixture model $\mathcal{M}_k$, if we use the exact prior distributions (Problem 11.11). Note that the factorials $k!$ and $(k+1)!$ in the above probability appear as the numbers of ways of ordering the k and $k + 1$ components of the mixtures (Problem 11.12): the ratio cancels with $1/(k + 1)$, which is the probability of selecting a particular component for the death step. If $\pi_{(k+1)k} = \pi_{k(k+1)}$ and $\varrho(k) = \lambda^k/k!$, the birth acceptance ratio thus simplifies into

$$\min\left(\frac{\lambda(1 - p_{k+1})^{k-1}}{k + 1} \frac{\ell_{k+1}(\theta_{k+1})}{\ell_k(\theta_k)}, 1\right).$$

While this proposal can work well in some setting, as in Richardson and Green (1997) when the prior is calibrated against the data, it can also be inefficient, that is, leading to a high rejection rate, if the prior is vague, since the birth proposals are not tuned properly. A second proposal, central to the solution of Richardson and Green (1997), is to devise more local jumps between models.

In this proposal, the upward move from $\mathcal{M}_k$ to $\mathcal{M}_{k+1}$ is called a *split* move. In essence, it replaces a component, say the jth, with two components centered at this earlier component. The split parameters are thus created under a moment condition

(11.6)
$$\begin{aligned}
p_{jk} &= p_{j(k+1)} + p_{(j+1)(k+1)}, \\
p_{jk}\mu_{jk} &= p_{j(k+1)}\mu_{j(k+1)} + p_{(j+1)(k+1)}\mu_{(j+1)(k+1)}, \\
p_{jk}\sigma_{jk}^2 &= p_{j(k+1)}\sigma_{j(k+1)}^2 + p_{(j+1)(k+1)}\sigma_{(j+1)(k+1)}^2.
\end{aligned}$$

The downward move, called a *merge*, is obtained directly as (11.6) by the reversibility constraint.

The split move satisfying (11.6) can, for instance, be obtained by generating the auxiliary variable $u_{k(k+1)}$ as $u_1, u_2, u_3 \sim \mathcal{U}(0, 1)$, and then taking

$$\begin{aligned}
p_{j(k+1)} &= u_1 p_{jk}, & p_{(j+1)(k+1)} &= (1 - u_1)p_{jk}, \\
\mu_{j(k+1)} &= u_2 \mu_{jk}, & \mu_{(j+1)(k+1)} &= \frac{p_{jk} - p_{j(k+1)} u_2}{p_{jk} - p_{j(k+1)}} \mu_{jk}, \\
\sigma_{j(k+1)}^2 &= u_3 \sigma_{jk}^2, & \sigma_{(j+1)(k+1)} &= \frac{p_{jk} - p_{j(k+1)} u_3}{p_{jk} - p_{j(k+1)}} \sigma_{jk}^2.
\end{aligned}$$

The Jacobian of the split transform is thus

$$\begin{pmatrix}
u_1 & 1 - u_1 & \cdots & & \cdots & & \cdots & & \cdots \\
p_{jk} & -p_{jk} & \cdots & & \cdots & & \cdots & & \cdots \\
0 & 0 & u_2 & \frac{p_{jk} - p_{j(k+1)} u_2}{p_{jk} - p_{j(k+1)}} & \cdots & & \cdots & & \\
0 & 0 & \mu_{jk} & \frac{-p_{j(k+1)}}{p_{jk} - p_{j(k+1)}} \mu_{jk} & \cdots & & \cdots & & \\
0 & 0 & 0 & 0 & u_3 & \frac{p_{jk} - p_{j(k+1)} u_3}{p_{jk} - p_{j(k+1)}} \\
0 & 0 & 0 & 0 & \sigma_{jk}^2 & \frac{-p_{j(k+1)}}{p_{jk} - p_{j(k+1)}} \sigma_{jk}^2
\end{pmatrix},$$

with a block diagonal structure that does not require the upper part of the derivatives to be computed. The absolute value of the determinant of this matrix is then

$$p_{jk} \times |\mu_{jk}| \frac{p_{jk}}{p_{jk} - p_{j(k+1)}} \times \sigma_{jk}^2 \frac{p_{jk}}{p_{jk} - p_{j(k+1)}} = \frac{p_{jk}^3}{(1-u_1)^2} |\mu_{jk}|\sigma_{jk}^2,$$

and the corresponding acceptance probability is

$$\min\left(\frac{\pi_{(k+1)k}}{\pi_{k(k+1)}} \frac{\varrho(k+1)}{\varrho(k)} \frac{\pi_{k+1}(\theta_{k+1})\ell_{k+1}(\theta_{k+1}}{\pi_k(\theta_k)\ell_k(\theta_k)} \frac{p_{jk}^3}{(1-u_1)^2} |\mu_{jk}|\sigma_{jk}^2, 1 \right).$$

This is one possible jump proposal following Richardson and Green (1997), and others could be implemented as well with a similar or higher efficiency. Note also that the implementation of Richardson and Green (1997) is slightly different from the one presented above in that, for them, only *adjacent* components can be merged together. (Adjacency is defined in terms of the order on the means of the components.) There is no theoretical reason for doing so, but the authors consider that merging only adjacent components may result in an higher efficiency (always in terms of acceptance probability): merging components that are far apart is less likely to be accepted, even though this may create a component with a larger variance that could better model the tails of the sample. We stress that, when this constraint is adopted, reversibility implies that the split move must be restricted accordingly: only adjacent components can be created at the split stage.

The implementation of Richardson and Green (1997) also illustrates another possibility with the reversible jump algorithm, namely, the possibility of including fixed dimensional moves in addition to the variable dimensional moves. This hybrid structure is completely valid, with a justification akin to the Gibbs sampler, that is, as a composition of several MCMC steps. Richardson and Green (1997) use fixed dimension moves to update the hyperparameters of their model, as well as the missing variables associated with the mixture. (As shown, for instance, in Cappé et al., 2003, this completion step is not necessary for the algorithm to work.)

Figures 11.3–11.5 illustrate the implementation of this algorithm for the Galaxy dataset presented in Figure 11.1, also used by Richardson and Green (1997). In Figure 11.3, the MCMC output on the number of components k is represented as a histogram on k, and the corresponding sequence of k's. The prior used on k is a uniform distribution on $\{1, \ldots, 20\}$: as shown by the lower plot, most values of k are explored by the reversible jump algorithm, but the upper bound does not appear to be unduly restrictive since the $k^{(t)}$'s hardly ever reach this upper limit.

Figure 11.4 illustrates one appealing feature of the MCMC experiment, namely that conditioning the output on the most likely value of k (the posterior mode equal to 3 here) is possible. The nine graphs in this figure show the joint variation of the three types of parameters, as well as the stability of the Markov chain over the $1,000,000$ iterations; the cumulative averages are quite stable, almost from the start.

The density plotted on top of the histogram in Figure 11.5 is another good illustration of the inferential possibilities offered by reversible jump al-

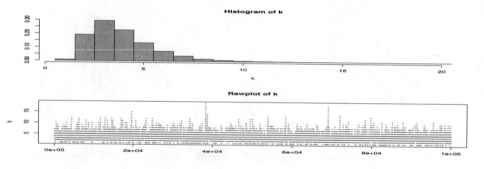

Fig. 11.3. Histogram and raw plot of $100,000$ k's produced by a reversible jump MCMC algorithm for the Galaxy dataset of Figure 11.1.

gorithms, as a case of *model averaging*: this density is obtained as the average over iterations t of

$$\sum_{j=1}^{k^{(t)}} p_{jk}^{(t)} \mathcal{N}(\mu_{jk}^{(t)}, (\sigma_{jk}^{(t)})^2),$$

which approximates the posterior expectation $\mathbb{E}[f(y|\theta)|\mathbf{x}]$, where $\mathbf{x}$ denotes the data $x_1, \ldots, x_{82}$. (This can also be seen as a Bayesian equivalent of a kernel estimator.)   ‖

Example 11.8. (Continuation of Example 11.3) An AR(p) model

$$(11.7) \qquad\qquad X_t = \sum_{i=1}^{p} \theta_{ip} X_{t-i} + \sigma_p \varepsilon_t$$

is often restricted by a stationarity condition on the process (X_t) (see, e.g., Robert, 2001, Section 4.5). While this constraint can be expressed either on the coefficients θ_{ip} of the model or on the partial auto-correlations, a convenient (from a prior point of view) representation is to use the lag-polynomial associated with the model $\mathcal{M}_p$,

$$\prod_{i=1}^{p} (1 - \lambda_i B)\, X_t = \epsilon_t, \quad \epsilon_t \sim \mathcal{N}(0, \sigma^2),$$

and to constrain the inverse roots,[3] λ_i, to stay within the unit circle if complex and within $[-1, 1]$ if real (see Barnett et al. 1996, Huerta and West 1999, and Robert, 2001, Section 4.5.2). A natural prior in this setting is then to use uniform priors for the real and complex roots λ_j, that is,

---

[3] In order to simplify notation, we have refrained from using a double index on the λ_i's, which should be, strictly speaking, the λ_{ip}'s.

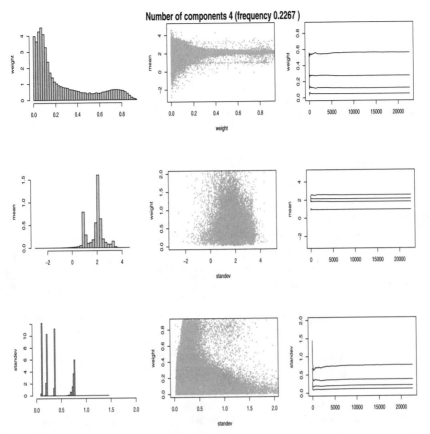

Fig. 11.4. Reversible jump MCMC output on the parameters of the model $\mathcal{M}_3$ for the Galaxy dataset, obtained by conditioning on $k = 3$. *The left column gives the histogram of the weights, means, and variances; the middle column the scatterplot of the pairs weights–means, means–variances, and variances–weights; the right column plots the cumulated averages (over iterations) for the weights, means, and variances.*

$$\pi_p(\boldsymbol{\lambda}) = \frac{1}{\lfloor p/2 \rfloor + 1} \prod_{\lambda_i \in \mathbb{R}} \frac{1}{2} \mathbb{I}_{|\lambda_i| < 1} \prod_{\lambda_i \notin \mathbb{R}} \frac{1}{\pi} \mathbb{I}_{|\lambda_i| < 1} \,,$$

where $\lfloor p/2 \rfloor + 1$ is the number of different values of r_p.

Note that this factor, while unimportant for a fixed p setting, must be included within the posterior distribution when using reversible jump since it does not vanish in the acceptance probability of a move between models $\mathcal{M}_p$ and $\mathcal{M}_q$. If it is omitted in the acceptance probability of the reversible jump move, this results in a modification of the prior probability of each model $\mathcal{M}_p$, from $\varrho(p)$ to $\varrho(p)(\lfloor p/2 \rfloor + 1)$. See also Vermaak et al. (2003) for a similar phenomenon.

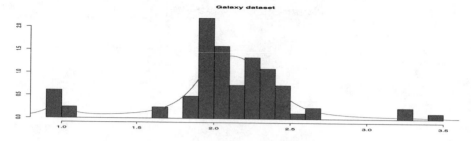

Fig. 11.5. Fit of the dataset by the averaged density, $\mathbb{E}[f(y|\theta)|\mathbf{x}]$

The most basic choice for a reversible jump algorithm in this framework is to use the uniform priors as proposals for a birth-and-death scheme where the birth moves are either from $\mathcal{M}_p$ to $\mathcal{M}_{p+1}$, for the creation of a real root λ_{p+1}, or from $\mathcal{M}_p$ to $\mathcal{M}_{p+2}$, for the creation of two conjugate complex roots λ_{p+1} and $\bar{\lambda}_{p+1}$. The corresponding death moves are then completely determined: from $\mathcal{M}_{p+1}$ to $\mathcal{M}_p$ with the deletion of a real root (if any) and from $\mathcal{M}_{p+2}$ to $\mathcal{M}_p$ with the deletion of two conjugate complex roots (if any).

As in the birth-and-death proposal for Example 11.7, the acceptance probability simplifies quite dramatically since it is, for example,

$$\min\left(\frac{\pi_{(p+1)p}}{\pi_{p(p+1)}}\frac{(r_p+1)!}{r_p!}\frac{\lfloor p/2\rfloor+1}{\lfloor (p+1)/2\rfloor+1}\frac{\ell_{p+1}(\theta_{p+1})}{\ell_p(\theta_p)},1\right)$$

in the case of a move from $\mathcal{M}_p$ to $\mathcal{M}_{p+1}$. (As for the above mixture example, the factorials are related to the possible choices of the created and the deleted roots.)

Figure 11.6 presents some views of the corresponding reversible jump MCMC algorithm. Besides the ability of the algorithm to explore a range of values of k, it also shows that Bayesian inference using these tools is much richer, since it can, for instance, condition on or average over the order k, mix the parameters of different models and run various tests of these parameters. A last remark on this graph is that both the order and the value of the parameters are well estimated, with a characteristic trimodality on the histograms of the θ_i's, even when conditioning on k different from 3, the value used for the simulation. We refer the reader to Vermaak et al. (2003) and Ehlers and Brooks (2003) for different analyses of this problem using either the partial autocorrelation representation (and the Durbin-Levinson recursive algorithm) or the usual AR(p) parameters without stationarity constraint.   ‖

For any given variable dimension model, there exist an infinity of possible reversible jump algorithms. Compared with the fixed dimension case of, for instance, Chapter 7, there is more freedom and less structure, because the between model moves cannot rely on an Euclidean structure common to both models (unless they are embedded). Brooks et al. (2003b) tried to provide

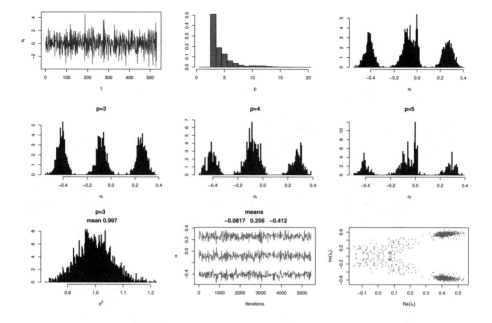

Fig. 11.6. Output of a reversible jump algorithm based on an AR(3) simulated dataset of 530 points *(upper left)* with true parameters θ_i $(-0.1, 0.3, -0.4)$ and $\sigma = 1$. The first histogram is associated with k, the following histograms are associated with the θ_i's, for different values of k, and of σ^2. The final graph is a scatterplot of the complex roots (for iterations where there were complex root(s)). The penultimate graph plots the evolution over the iterations of $\theta_1, \theta_2, \theta_3$. *(Source: Robert 2004.)*

general strategies for the construction of efficient reversible jump algorithms by setting up rules to calibrate these jumps. While their paper opens too many new perspectives to be entirely discussed here, let us mention a *scaling* proposal that relates to the calibrating issue of Section 7.6.

Brooks et al. (2003b) assume that a transform T_{mn} that moves the completion of $\mathcal{M}_m$ into the completion of $\mathcal{M}_n$ has been chosen, as well as a *centering function*, $c_{mn}(\theta_m)$, which is a special value of θ_n corresponding to θ_m. For instance, in the setup of Example 11.1, if $\theta_p = (\theta_1, \ldots, \theta_p, \sigma)$, a possible centering function is

$$c_{p(p+1)}(\theta_p) = (\theta_1, \ldots, \theta_p, 0, \sigma).$$

Once both T_{mn} and c_{mn} have been chosen, the calibration of the remaining parameters of the proposal transformation, that is, of the parameters of the distribution φ_{mn} of u_{mn} such that $(\theta_m, u_{nm}) = T_{mn}(\theta_m, u_{mn})$, is based on the constraint that the probability (11.4) of moving from θ_m to $c_{mn}(\theta_m)$ is *equal to one*. Again, in the setup of Example 11.1, if θ_p is completed by

$u_{p(p+1)} = \theta_{p+1} \sim \mathcal{N}(0, \tau^2)$, the scale τ of this completion can be determined by this constraint which, in Brooks et al.'s (2003b) case becomes

$$\tau^2 = \sigma_a^2 \left(\frac{\pi_{p(p+1)}}{\pi_{(p+1)p}} \right)^2,$$

where σ_a^2 is the variance of the prior distribution on the θ_i's.

This calibration scheme can be generalized to other settings by imposing further conditions on the acceptance probability for moving from θ_m to $c_{mn}(\theta_m)$, as detailed in Brooks et al. (2003b). Obviously, it cannot work in all settings, because this acceptance probability often depends on the specific value of θ_m, and it does not necessarily provide sensible results (as shown for instance in Robert 2004). The second difficulty with this approach is that it is completely dependent on the choice of the transform and the centering function. Brooks et al. (2003b) suggest choosing c_{mn} so that

$$\ell_n(c_{mn}(\theta_m)) = \ell_m(\theta_m),$$

but this is not always possible nor solvable.

11.3 Alternatives to Reversible Jump MCMC

Since it first appeared, reversible jump MCMC has had a vast impact on variable dimension Bayesian inference, especially model choice, but, as mentioned earlier, there also exist other approaches to the problem of variable dimension models.

11.3.1 Saturation

First, Brooks et al. (2003b) reassess the reversible jump methodology through a global *saturation scheme* already used in Section 11.2.2. More precisely, they consider a series of models $\mathcal{M}_k$ with corresponding parameter spaces Θ_k $(k = 1, \ldots, R)$ such that $\max_k \dim(\Theta_k) = n_{\max} < \infty$. For each model $\mathcal{M}_k$, the parameter θ_k is then completed with an auxiliary variable $u_k \sim \varphi_k(u_k)$ such that all (θ_k, u_k)'s are one-to-one transforms of one another. The authors define, in addition, rv's ω_k independently distributed from ψ, that are used to move between models. For instance, in the setting of Example 11.8, the saturated proposal on $\mathcal{M}_k$ may be

(11.8) $(\sigma_k + \omega_0, \theta_1 + \omega_1, \ldots, \theta_k + \omega_k, u_k), \quad u_k \in \mathbb{R}^{n_{\max} - k},$

if there is no stationarity constraint imposed on the model.

Brooks et al. (2003b) then assign the following joint auxiliary prior distribution to a parameter in Θ_k,

$$\pi(k,\theta_k)\, q_k(u_k) \prod_{i=1}^{n_{\max}} \psi(\omega_i).$$

Within this augmented (or *saturated*) framework, there is no varying dimension anymore since, for all models, the entire vector (θ_k, u_k, ω) is of fixed dimension. Therefore, moves between models can be defined just as freely as moves between points of each model. See also Besag (2000) and Godsill (2003) for a similar development, which differs from the full saturation scheme of Carlin and Chib (1995) (Problem 11.9).

Brooks et al. (2003b) propose a three-stage MCMC update as follows.

Algorithm A.49 —Saturation Algorithm—

At iteration t

a. Update the current value of the parameter, θ_k, within model $\mathcal{M}_k$
b. Update u_k and ω conditional on θ_k
c. Update the model from $\mathcal{M}_k$ into $\mathcal{M}_{k'}$ using the bijection

$$(\theta_{k'}, u_{k'}) = T_{kk'}(\theta_k, u_k | \omega).$$

Note that, for specific models, saturation schemes appear rather naturally. For instance, Green (1995) considered a time series model with change-points, $Y_t \sim p(y_k - \theta_k)$, where θ_t changes values according to a geometric jump scheme (see Problems 11.15, 11.17 and 11.18). Whatever the number of change-points in the series, a reparameterization of the model by the missing data, composed of the indicators X_t of whether or not a change occurs at time t, creates a fixed dimension model. (See Cappé et al., 2003, for a similar representation of the semi-Markov jump process introduced in Note 14.6.3.)

Example 11.9. (Continuation of Example 11.8)   Within model $\mathcal{M}_k$, for the AR(k) parameters $\theta_1, \ldots, \theta_k$ and σ_k, the move in step a can be any Metropolis–Hastings step that preserves the posterior distribution (conditional on k), like a random walk proposal on the real and complex roots of the lag-polynomial. The remaining u_k can then be updated via an AR(1) move $u'_k = \lambda u_k + \sqrt{1 - \lambda^2}\,\epsilon$ if the proper stationary distribution $q_k(u_k)$ is a $\mathcal{N}(0,1)$ distribution. As noted in Brooks et al. (2003b), the ω_k's are not really necessary when the u_k's are independent. In the move from $\mathcal{M}_k$ to $\mathcal{M}_{k+1}$, Brooks et al. (2003b) suggest using (11.8), or a combination of the non-saturated approach with an auxiliary variable where $\theta_{k+1} = \sigma u_{k,k+1}$. If u_k is updated symmetrically, that is, if $u_{k,k+1} = \theta_{k+1}/\sigma$ when moving from $\mathcal{M}_{k+1}$ to $\mathcal{M}_k$, the algorithm provides the chain with some memory of previous values *within* each model and should thus facilitate moves *between* models.   ‖

The motivation of the saturation scheme by Brooks et al. (2003b) is not a simplification of the reversible jump algorithm, since the acceptance probabilities remain fundamentally the same in step c of Algorithm [A.49], but rather a "memory" over the moves that improve between-model moves. (For instance, the components of the u_k's may be correlated, as shown in Brooks et al. 2003b.) The possibilities open by this range of possible extensions is quite exciting but wealth is a mixed blessing in that the choice of such moves requires a level of expertise that escapes the layman.

11.3.2 Continuous-Time Jump Processes

Reversible jump algorithms operate in discrete time, but similar algorithms may be formulated in *continuous time*. This has a long history (for MCMC), dating back to Preston (1976), Ripley (1987), Geyer and Møller (1994). Here we focus on Stephens's (2000) methodology, called *birth-and-death MCMC* which he developed for mixture models.

The idea at the core of this continuous time approach is to build a *Markov jump process*, to move between models $\mathcal{M}_k$, using births (to increase the dimension) and deaths (to decrease the dimension), as well as other dimension-changing moves like the split/combine moves used in Example 11.7 (see Cappé et al. 2003). The Markov jump process is such that whenever it reaches some state θ, it stays there for a exponential $\mathcal{E}xp(\lambda(\theta))$ time with intensity $\lambda(\theta)$ depending on θ, and, after expiration of this holding time, jumps to a new state according to a transition kernel. More precisely, the various moves to other models are all associated with exponential waiting times depending on the current state of the Markov chain and the actual move corresponds to the smallest waiting time (which is still exponentially distributed; see Problem 11.20).

An important feature that distinguishes this algorithm from the previous MCMC algorithms in discrete time is the *jump structure:* whenever a jump occurs, the corresponding move *is always accepted*. Therefore, what replaces the acceptance probability of reversible jump methods are the differential holding times in each state. In particular, implausible configurations, that is, configurations with small $L_k(\theta_k)\pi(k, \theta_k)$, die quickly.

To ensure that a Markov jump process has an invariant density proportional to $L_k(\theta_k)\pi(k, \theta_k)$ on $\mathcal{M}_k$, that is, the posterior distribution based on the prior $\pi(k, \theta_k)$ and the likelihood function $L_k(\theta_k)$, it is sufficient (although not necessary) that the local detailed balance equations

$$(11.9) \qquad L_k(\theta_k)\pi(k, \theta_k)\lambda_{k\ell}(\theta_k, \theta_\ell) = L_\ell(\theta_\ell)\pi(\ell, \theta_\ell)\lambda_{\ell k}(\theta_\ell, \theta_k)$$

hold. Here $\lambda_{k\ell}(\theta_k, \theta_\ell)$ denotes the intensity of moving from state $\theta_k \in \Theta_k$ to $\theta_\ell \in \Theta_\ell$. (The proof is beyond the scope of this book, but Cappé et al. 2003 provide an intuitive justification based on the approximation of the continuous time process by a sequence of reversible jump algorithms.)

Algorithm A.50 —Continuous-Time MCMC—

At time t, for $\theta^{(t)} = (k, \theta_k)$

1. For $\ell \in \{k_1, \ldots, k_m\}$, generate $u_{k\ell} \sim \varphi_{k\ell}(u_{k\ell})$ so that

$$(\theta_\ell, u_{\ell k}) = T_{k\ell}(\theta_k, u_{k\ell})$$

and compute the intensity

(11.10) $\displaystyle \lambda_{k\ell} = \frac{L_\ell(\theta_\ell)\pi(\ell, \theta_\ell)}{L_k(\theta_k)\pi(k, \theta_k)} \frac{\varphi_{\ell k}(u_{\ell k})}{\varphi_{k\ell}(u_{k\ell})} \left| \frac{\partial T_{k\ell}(\theta_k, u_{k\ell})}{\partial(\theta_k, u_{k\ell})} \right|.$

2. Compute the full intensity

$$\lambda = \lambda_{kk_1} + \cdots + \lambda_{kk_m} \qquad [A.50].$$

3. Generate the jumping time as $t + V$, with $V \sim \mathcal{E}xp(\lambda)$.
4. Select the jump move as k_i with probability λ_{kk_i}/λ.
5. Set time to $t = t + V$ and take $\theta^{(t)} = (k_i, \theta_{k_i})$.

Moves that update model parameters (or hyperparameters) without changing its dimension may also be incorporated as additional jump processes, that is, by introducing corresponding (fixed) intensities for these moves in the sum of Step 2. above. Note the strong similarity of (11.10) with the reversible jump acceptance probability (11.4). Recall that the intensity of an exponential distribution is the inverse expectation: this means that (11.10) is inversely proportional to the waiting time until the jump to (ℓ, θ_ℓ) and that this waiting time will be small if either $L_\ell(\theta_\ell)$ is large or if $L_k(\theta_k)$ is small. The latter case corresponds to a previous move to an unlikely value of θ.

The continuous time structure of [A.50] is rather artificial. Rather than being simulated as in Step 3., using a regular Rao–Blackwellization argument, it can just as well be implemented as a sequence of iterations, where each value (k, θ_k) is weighted by the average waiting time λ^{-1}.

The following example illustrates this generalization of Stephens (2000) in the setting of mixtures.

Example 11.10. (Continuation of Example 11.1.) For the following representation of the normal mixture model,

$$Y_t \sim \sum_{i=1}^{k} \omega_i \mathcal{N}(\mu_i, \sigma_i^2) \Big/ \sum_{i=1}^{k} \omega_i \,,$$

Cappé et al. (2003) have developed split-and-combine moves for continuous time schemes, reproducing Richardson and Green (1997) within this context. The *split* move for a given component θ_ℓ of the k component vector θ_k is to split this component as to give rise to a new parameter vector with $k + 1$ components, defined as

$$\left(\theta_1, \ldots, \theta_{\ell-1}, \theta_{\ell+1}, \ldots, \theta_k, T(\theta_\ell, u_{k(k+1)})\right),$$

where T is a differentiable one-to-one mapping that produces two new components $(\theta'_\ell, \theta'_{k+!})$ and $u_{k(k+1)} \sim \varphi(u)$. Note that the representation of the weights via the ω_i's is not identifiable but avoids the renormalizing step in the jump from $\mathcal{M}_k$ to $\mathcal{M}_{k+1}$ or $\mathcal{M}_{k-1}$: the θ_ℓ's are then a priori independent.

We assume that the mapping T is symmetric in the sense that

(11.11) $$P(T(\theta, u) \in B' \times B'') = P(T(\theta, u) \in B'' \times B')$$

for all $B', B'' \subset \mathbb{R} \times \mathbb{R}_+^2$.

We denote the *total* splitting intensity by $\eta(\theta_k)$, which is the sum over $\ell = 1, \ldots, k$ of the intensities associated with a split of θ_ℓ in $(\theta'_\ell, \theta'_{k+1})$, assuming that, in a split move, each component ℓ is chosen with equal probability $1/k$. In the basic scheme, all the split intensities can be chosen as constants, that is, η/k.

Conversely, the *combine* intensity, that is, the intensity of the waiting time until the combination of a pair of components of θ_k (there are $k(k-1)/2$ such pairs), can be derived from the local balance equation (11.9). For the mixture model and a move from θ_{k+1} to θ_k, with a combination of two components of θ_{k+1} into θ_ℓ and an auxiliary variable $u_{k(k+1)}$, the combine intensity $\lambda_{(k+1)k}$ is thus given by

$$\mathrm{L}_k(\theta_k)\pi(k, \theta_k)\, k!\, \frac{\eta(\theta_k)}{k}\, 2\varphi(u_{k(k+1)}) \left| \frac{\partial T(\theta_\ell, u_{k(k+1)})}{\partial(\theta_\ell, u_{k(k+1)})} \right|$$
$$= \mathrm{L}_{k+1}(\theta_{k+1})\pi(k+1, \theta_{k+1})\,(k+1)!\, \lambda_{(k+1)k}.$$

As previously seen in Example 11.7, the factorials arise because of the ordering of components and $\eta(\theta)/k$ is the rate of splitting a *particular* component as $\eta(\theta)$ is the *overall* split rate. The factor 2 is a result of the symmetry assumption (11.11): in a split move, a component θ can be split into the pair (θ', θ'') as well as into the pair (θ'', θ') in reverse order, but these configurations are equivalent. However, the two ways of getting there are typically associated with different values of u and possibly also with different densities $\varphi(u)$; the symmetry assumption is precisely what assures that the densities at these two values of u coincide and hence we may replace the sum of two densities, that we would otherwise be required to compute, by the factor 2. We could proceed without such symmetry but would then need to consider the densities of u when combining the pairs (θ', θ'') and (θ'', θ'), respectively, separately.

Thus, the intensity corresponding to the combination of two components of θ_{k+1}, (θ', θ''), is

(11.12) $$\frac{\mathrm{L}_k(\theta_k)\pi(k, \theta_k)}{\mathrm{L}_{k+1}(\theta_{k+1})\pi(k+1, \theta_{k+1})}\, \frac{\eta(\theta_k)}{(k+1)k}\, 2\varphi(u_{k(k+1)}) \left| \frac{\partial T(\theta', \theta'')}{\partial(\theta, u)} \right|.$$

Cappé et al. (2003) also compared the discrete and continuous time approaches to the mixture model of Example 11.10 and they concluded that the differences between both algorithms are very minor, with the continuous time approach generally requiring more computing time. $\|$

11.4 Problems

11.1 In the situation of Example 11.3, we stress that the best fitting $AR(p+1)$ model is not necessarily a modification of the best fitting $AR(p)$ model. Illustrate this fact by implementing the following experiment.

(a) Set $p = 3$ and parameter values $\theta_i = i$, $i = 1, \ldots, 3$.

(b) Simulate two data sets, one from an $AR(3)$ model and one from and $AR(1)$ model.

For each data set:

(c) Estimate the $AR(3)$ model parameters θ_i and σ_i for $i = 1, \ldots, 3$, using independent priors $\theta_i \sim \mathcal{N}(0, \tau^2)$ and $\sigma_i^{-2} \sim \mathcal{G}a(1, 1)$

(d) Estimate the $AR(1)$ model parameters θ_1 and σ_1. Then for $i = 2, \ldots, p$, estimate θ_i conditional on the $\theta_{i-1}, \theta_{i-2}, \ldots$ under independent priors $\theta_{ip} \sim \mathcal{N}(0, \tau^2)$, $\sigma_1^{-2} \sim \mathcal{G}a(1, 1)$ for both simulated datasets.

(e) Compare the results of (b) and (c).

11.2 In the situation of Example 7.13, if a second explanatory variable, z, is potentially influential in

$$\mathbb{E}[Y | x, z] = \frac{\exp(a + bx + cz)}{1 + \exp(a + bx + cz)},$$

the two competing models are $\Theta_1 = \{1\} \times \mathbb{R}^2$ and $\Theta_2 = \{2\} \times \mathbb{R}^3$, with $\theta_1 = (a, b)$ and $\theta_2 = (a, b, c)$. Assume the jump from Θ_2 to Θ_1 is a (deterministic) transform of $(1, a, b, c)$ to $(2, a, b)$. Show that the jump from Θ_1 to Θ_2 in a reversible jump algorithm must necessarily preserve the values of a and b and thus only modify c. Write out the details of the reversible jump algorithm.

11.3 For the situation of Example 11.5, show that the four possible moves are made with the following probabilities:

(i) $(\alpha_0, \alpha_1) \rightarrow (\alpha_0', \alpha_1')$, with probability $\pi_{11} \min\left(1, \frac{f_0(\alpha_0') f_1(\alpha_1')}{f_0(\alpha_0) f_1(\alpha_1)}\right)$,

(ii) $(\alpha_0, \alpha_1) \rightarrow (\alpha_0, \alpha_1, \alpha_2')$, with probability $\pi_{12} \min\left(1, \frac{f_2(\alpha_2')}{f_2(\alpha_2)}\right)$,

(iii) $(\alpha_0, \alpha_1, \alpha_2) \rightarrow (\alpha_0', \alpha_1', \alpha_2')$ with probability $\pi_{21} \min\left(1, \frac{f_0(\alpha_0') f_1(\alpha_1') f_2(\alpha_2')}{f_0(\alpha_0) f_1(\alpha_1) f_2(\alpha_2)}\right)$,

(iv) $(\alpha_0, \alpha_1, \alpha_2) \rightarrow (\alpha_0', \alpha_1')$ with probability $\pi_{22} \min\left(1, \frac{f_0(\alpha_0) f_1(\alpha_1)}{f_0(\alpha_0) f_1(\alpha_1) f_2(\alpha_2)}\right)$.

11.4 Similar to the situation of Example 11.5, the data in Table 7.6, given in Problem 7.21, are a candidate for model averaging between a linear and quadratic regression. This "braking data" (Tukey 1977) is the distance needed to stop (y) when an automobile travels at a particular speed (x) and brakes.

(a) Fit the data using a reversible jump algorithm that jumps between a linear and quadratic model. Use normal priors with relatively large variances.

(b) Assess the robustness of the fit to specifications of the π_{ij}. More precisely, choose values of the π_{ij} so that the algorithm spends 25%, 50%, and 75% of the time in the quadratic space. How much does the overall model average change?

11.5 Consider a sequence of models $\mathcal{M}_k$, $k = 1, \ldots, K$, with corresponding parameters $\theta_k \in \Theta_k$. For a given reversible jump algorithm, the auxiliary variables

$$u_{mn} \sim \varphi_{mn}(u)$$

are associated with completion transforms T_{mn} such that

$$(\theta_m, u_{mn}) = T_{mn}(\theta_n, u_{nm}).$$

(a) Show that the *saturated* model made of the distribution of the θ_k's completed with all the u_{mn}'s,

$$\pi(k, \theta_k) \prod_{mn} \varphi_{mn}(u_{mn}),$$

is a fixed dimension model with the correct marginal on (k, θ_k).

(b) Show that the reversible jump move is not a Metropolis–Hastings proposal compatible with this joint distribution.

(c) Explain why the u_{mn}'s do not need to be taken into account in moves that do not involve both models $\mathcal{M}_n$ and $\mathcal{M}_m$. (*Hint:* Include an additional pair (u_{mn}, u_{nm}) and check that they vanish from the probability ratio.)

11.6 Consider the model introduced in Example 11.6.

(a) Show that the weight of the jump from Θ_{k+1} to Θ_k involves the Jacobian

$$\left| \frac{\partial(p^{(k)}, a^{(k)}, u_1, u_2)}{\partial(p^{(k+1)}, a^{(k+1)})} \right|.$$

(b) Compute the Jacobian for the jump from Θ_4 to Θ_3. (The determinant can be calculated by row-reducing the matrix to triangular form, or using the fact that $\begin{vmatrix} A & B \\ C & D \end{vmatrix} = |A| \times |D - CA^{-1}B|$.)

11.7 For the model of Example 11.6:

(a) Derive general expressions for the weight of the jump from Θ_k to Θ_{k+1} and from Θ_{k+1} to Θ_k.

(b) Select a value of the Poisson parameter λ and implement a reversible jump algorithm to fit a density estimate to the data of Table 11.1. (*Note:* Previous investigation and subject matter considerations indicate that there may be between three and seven modes. Use this information to choose an appropriate value of λ.)

(c) Investigate the robustness of your density estimate to variations in λ.

11.8 (Chib 1995) Consider a posterior distribution $\pi(\theta_1, \theta_2, \theta_3 | x)$ such that the three full conditional distributions $\pi(\theta_1 | \theta_2, \theta_3, x)$, $\pi(\theta_2 | \theta_1, \theta_3, x)$, and $\pi(\theta_3 | \theta_1, \theta_2, x)$ are available.

(a) Show that

$$\log m(x) = \log f(x|\hat{\theta}) + \log \pi(\hat{\theta}) - \log \pi(\hat{\theta}_3 | \hat{\theta}_1, \hat{\theta}_2, x)$$
$$- \log \pi(\hat{\theta}_2 | \hat{\theta}_1, x) - \log \pi(\hat{\theta}_1 | x),$$

where $(\hat{\theta}_1, \hat{\theta}_2, \hat{\theta}_3)$ are arbitrary values of the parameters that may depend on the data.

9172	9350	9483	9558	9775	10227
10406	16084	16170	18419	18552	18600
18927	19052	19070	19330	19343	19349
19440	19473	19529	19541	19547	19663
19846	19856	19863	19914	19918	19973
19989	20166	20175	20179	20196	20215
20221	20415	20629	20795	20821	20846
20875	20986	21137	21492	21701	21814
21921	21960	22185	22209	22242	22249
22314	22374	22495	22746	22747	22888
22914	23206	23241	23263	23484	23538
23542	23666	23706	23711	24129	24285
24289	24366	24717	24990	25633	26960
26995	32065	32789	34279		

Table 11.1. Velocity *(km/second)* of galaxies in the Corona Borealis Region. (*Source:* Roeder 1992.)

(b) Show that $\pi(\theta_1|x)$ can be approximated by

$$\hat{\pi}(\theta_1|x) = \frac{1}{T} \sum_{t=1}^{T} \pi(\theta_1, \theta_2^{(t)}, \theta_3^{(t)}|x),$$

where the $(\theta_1^{(t)}, \theta_2^{(t)}, \theta_3^{(t)})$'s are generated by Gibbs sampling.

(c) Show that $\pi(\theta_2|\hat{\theta}_1, x)$ can be approximated by

$$\hat{\pi}(\theta_2|\hat{\theta}_1, x) = \frac{1}{T} \sum_{t=1}^{T} \pi(\theta_2|\hat{\theta}_1, \theta_3^{(t)}, x),$$

where the $(\theta_2^{(t)}, \theta_3^{(t)})$'s are generated by Gibbs sampling from the conditional distributions $\pi(\theta_2|\hat{\theta}_1, \theta_3^{(t-1)}, x)$ and $\pi(\theta_3|\hat{\theta}_1, \theta_2^{(t)}, x)$, that is, with θ_1 being kept equal to $\hat{\theta}_1$.

(d) Apply this marginalization method to the setup of Example 11.1 and the data in Table 11.1.

(e) Evaluate the computing cost of this method.

(*Note:* See Chib and Jeliazkov 2001 for an extension to larger numbers of components.)

11.9 A saturation scheme due to Carlin and Chib (1995) is to consider all models at once. Given $\mathcal{M}_k$ $(k = 1, \cdots, K)$ with corresponding priors $\pi_k(\theta_k)$, and prior weights ρ_k, the parameter space is

$$\Theta = \{1, \ldots, K\} \times \prod_{k=1}^{K} \Theta_k.$$

(a) If μ denotes the model indicator, show that the posterior distribution is

$$\pi(\mu, \theta_1, \ldots, \theta_K|x) \propto \rho_\mu f_\mu(x|\theta_\mu) \prod_{k=1}^{K} \pi_k(\theta_k).$$

(b) Show that

$$m(x|\mu = j) = \int f_j(x|\theta_j)\pi(\theta_1,\ldots,\theta_K|\mu = j)\, d\theta = \int f_j(x|\theta_j)\pi_j(\theta_j)\, d\theta_j$$

does not depend on the $\pi_k(\theta_k)$'s for $k \neq j$.

(c) Deduce that, when in model $\mathcal{M}_j$, that is, when $\mu = j$, the parameters θ_k for $k \neq j$ can be simulated using arbitrary distributions $\tilde{\pi}_k(\theta_k|\mu = j)$. (*Note:* Those are called *pseudo-priors* by Carlin and Chib 1995.)

(d) Construct a corresponding Gibbs sampler on $(\mu, (\theta_1,\ldots,\theta_K))$, where μ is generated from

$$P(\mu = j|x, \theta_1,\ldots,\theta_K) \propto \rho_j f_j(x|\theta_j)\pi_j(\theta) \prod_{k\neq j} \tilde{\pi}_k(\theta_k|\mu = j)\,.$$

(e) Comment on the costliness of the method when K is large, namely, on the drawback that all models $\mathcal{M}_k$ must be considered at every stage.

11.10 (Continuation of Problem 11.9) On a dataset of n pine trees, the grain (strength) y_i is regressed against either the wood density x_i,

$$\mathcal{M}_1: \quad y_i = \alpha + \beta x_i + \sigma\varepsilon_i\,,$$

or a modified (resin adapted) density, z_i,

$$\mathcal{M}_2: \quad y_i = \gamma + \delta z_i + \tau\varepsilon_i\,.$$

Both model parameters are associated with conjugate priors,

$$\begin{pmatrix}\alpha\\\beta\end{pmatrix},\begin{pmatrix}\gamma\\\delta\end{pmatrix} \sim \mathcal{N}\left(\begin{pmatrix}3000\\185\end{pmatrix},\begin{bmatrix}10^6 & 0\\0 & 10^4\end{bmatrix}\right)\,, \quad \sigma^2, \tau^2 \sim \mathcal{IG}(a,b)\,,$$

with $(a, b) = (1, 1)$. For the pseudo-priors

$$\alpha|\mu = 2 \sim \mathcal{N}(3000, 52^2)\,, \quad \beta|\mu = 2 \sim \mathcal{N}(185, 12^2)\,,$$
$$\gamma|\mu = 1 \sim \mathcal{N}(3000, 43^2)\,, \quad \delta|\mu = 1 \sim \mathcal{N}(185, 9^2)\,,$$

and $\sigma^2, \tau^2 \sim \mathcal{IG}(a, b)$, describe the saturated algorithm of Carlin and Chib (1995). (*Note:* For their dataset, the authors were forced to use disproportionate weights, $\rho_1 = .9995$ and $\rho_2 = .0005$ in order to force visits to the model $\mathcal{M}_1$.)

11.11 In the setting of Example 11.7, for the birth and death moves,

(a) If $(p_1,\ldots,p_k) \sim \mathcal{D}(\alpha_1,\ldots,\alpha_k)$, show that $p_k \sim \mathcal{B}e(\alpha_k, \alpha_1 + \ldots + \alpha_{k-1})$.

(b) Deduce that, if

$$\pi_k(\theta_k) = \prod_{j=1}^k \pi^c(\mu_{jk}, \sigma_{jk}) \times \frac{\Gamma(\alpha_1 + \ldots + \alpha_k)}{\Gamma(\alpha_1)\cdots\Gamma(\alpha_k)}\, p_{1k}^{\alpha_1 - 1}\cdots p_{kk}^{\alpha_k - 1}\,,$$

the reversible jump acceptance ratio is indeed (11.5) when $p_{(k+1)(k+1)}$ is simulated from $\mathcal{B}e(\alpha_k, \alpha_1 + \ldots + \alpha_{k-1})$ and $(\mu_{(k+1)(k+1)}, \sigma_{(k+1)(k+1)})$ from π^c.

11.12 Show that a mixture model $\mathcal{M}_k$ as in Example 11.7 is invariant under permutation of its components and deduce that the parameter θ_k is identifiable up to a permutation of its components.

11.13 In the setup of the AR(p) model of Example 11.3, it is preferable to impose stationarity (or causality) on the model (11.7) through the constraint $|\lambda_i| < 1$ on the roots λ_i, as described in Example 11.8. The case when some roots are at the boundary of this constrained space $|\lambda_i| = 1$ is sometimes singled out in economics and finance, as an indicator of a special behavior of the time series. Such roots are called *unit roots*.

(a) By considering the particular case $p = 1$ and $\lambda_1 = 1$, show that the model (11.7) is not stationary at the boundary.

(b) Show that the prior distribution defined in Example 11.8 does not put weight on the boundary.

(c) If we denote by u_p^r and u_p^c the number of real and complex unit roots in $\mathcal{M}_p$, respectively, write the uniform prior on the λ_i's when there are u_p^r real and u_p^c complex unit roots.

(d) Derive a reversible jump algorithm in the spirit of Example 11.8, by introducing moves within $\mathcal{M}_p$ that change either u_p^r or u_p^c.

11.14 A switching AR model is an autoregressive model where the mean can switch between two values as follows.

$$X_t|x_{t-1}, z_t, z_{t-1} \sim \mathcal{N}(\mu_{z_t} + \varphi(x_{t-1} - \mu_{z_{t-1}}), \sigma^2)$$
$$Z_t|z_{t-1} \sim \rho_{z_{t-1}}\mathbb{I}_{z_{t-1}}(z_t) + (1 - \rho_{z_{t-1}})\mathbb{I}_{1-z_{t-1}}(z_t),$$

where $Z_t \in \{0, 1\}$ is not observed, with $z_0 = 0$ and $x_0 = \mu_0$.

(a) Construct a data augmentation algorithm by showing that the conditional distributions of the z_t's ($1 < t < T$) are

$$P(Z_t|z_{t-1}, z_{t+1}, x_t, x_{t-1}, x_{t+1}) \propto$$

$$\exp\left\{-\frac{1}{2\sigma^2}\left[(x_t - \mu_{z_t} - \varphi(x_{t-1} - \mu_{z_{t-1}}))^2\right.\right.$$

$$\left.\left.+(x_{t+1} - \mu_{z_{t+1}} - \varphi(x_t - \mu_{z_t}))^2\right]\right\}$$

$$\times \left(\rho_{z_{t-1}}\mathbb{I}_{z_{t-1}}(z_t) + (1 - \rho_{z_{t-1}})\mathbb{I}_{1-z_{t-1}}(z_t)\right)$$

$$\times \left(\rho_{z_t}\mathbb{I}_{z_t}(z_{t+1}) + (1 - \rho_{z_t})\mathbb{I}_{1-z_t}(z_{t+1})\right),$$

with appropriate modifications for the limiting cases Z_1 and Z_T.

(b) Under the prior

$$(\mu_1 - \mu_0) \sim \mathcal{N}(0, \varsigma^2), \quad \pi(\varphi, \sigma^2) = 1/\sigma, \quad \rho_0, \rho_1 \sim \mathcal{U}_{[0,1]},$$

show that the parameters can be generated as

$$\mu_0 \sim \mathcal{N}\left([n_{00}(1-\varphi)(\overline{y}_{00} - \varphi\tilde{y}_{00}) + n_{01}\varphi(\mu_1 - \overline{y}_{01} + \varphi\tilde{y}_{01})\right.$$
$$+n_{10}(\varphi\mu_1 + \overline{y}_{10} - \varphi\tilde{y}_{10}) + \varsigma^{-2}\sigma^2\mu_1]\left[n_{00}(1-\varphi)^2 + n_{01}\varphi^2\right.$$
$$\left.+n_{10} + \varsigma^2\sigma^2\right]^{-1}, \sigma^2[n_{00}(1-\varphi)^2 + n_{01}\varphi^2 + n_{10} + \varsigma^2\sigma^2]^{-1}\right),$$

$$\mu_1 \sim \mathcal{N}\left([n_{11}(1-\varphi)(\overline{y}_{11} - \varphi\tilde{y}_{11}) + n_{10}\varphi(\mu_0 - \overline{y}_{10} + \varphi\tilde{y}_{10})\right.$$
$$+n_{01}(\varphi\mu_0 + \overline{y}_{01} - \varphi\tilde{y}_{01}) + \varsigma^{-2}\sigma^2\mu_0]\left[n_{11}(1-\varphi)^2 + n_{10}\varphi^2\right.$$
$$\left.+n_{01} + \varsigma^2\sigma^2\right]^{-1}, \sigma^2[n_{11}(1-\varphi)^2 + n_{10}\varphi^2 + n_{01} + \varsigma^2\sigma^2]^{-1}\right),$$

$$\varphi \sim \mathcal{N}\left(\frac{\sum_{t=1}^{T}(y_t - \mu_{z_t})(y_{t-1} - \mu_{z_{t-1}})}{\sum_{t=1}^{T}(y_{t-1} - \mu_{z_{t-1}})^2}, \frac{\sigma^2}{\sum_{t=1}^{T}(y_{t-1} - \mu_{z_{t-1}})^2}\right),$$

$$\sigma^2 \sim \mathcal{IG}\left(\frac{n+1}{2}, \frac{1}{2}\sum_{t=1}^{T}(y_t - \mu_{z_t} - \varphi(y_{t-1} - \mu_{z_{t-1}}))^2\right),$$

$$\rho_0 \sim \mathcal{B}e(n_{00} + 1, n_{01} + 1),$$

$$\rho_1 \sim \mathcal{B}e(n_{11} + 1, n_{10} + 1),$$

where n_{ij} denotes the number of jumps from i to j and

$$n_{ij}\bar{y}_{ij} = \sum_{\substack{z_{t-1}=i \\ z_t=j}} y_t, \qquad n_{ij}\tilde{y}_{ij} = \sum_{\substack{z_{t-1}=i \\ z_t=j}} y_{t-1}.$$

(c) Generalize this algorithm to a birth-and-death reversible jump algorithm when the number of means μ_i is unknown.

11.15 Consider a sequence $X_1, \ldots, X_n$ such that

$$X_i \sim \begin{cases} \mathcal{N}(\mu_1, \tau^2) & \text{if } i \le i_0 \\ \mathcal{N}(\mu_2, \tau^2) & \text{if } i > i_0. \end{cases}$$

(a) When i_0 is uniformly distributed on $\{1, \ldots, n-1\}$, determine whether the improper prior $\pi(\mu_1, \mu_2, \tau) = 1/\tau$ leads to a proper posterior.

(b) Let $\mathcal{M}_k$ denote the model with change-point at $i = k$. Derive a Monte Carlo algorithm for this setup.

(c) Use an importance sampling argument to recycle the sample produced by the algorithm in part (b) when the prior is

$$\pi(\mu_1, \mu_2, \tau) = \exp\left(-(\mu_1 - \mu_2)^2/\tau^2\right)/\tau^2.$$

11.16 Here we look at an alternative formulation for the situation of Problem 11.15.

(a) Show that an alternative implementation is to use $(n-1)$ latent variables z_i which are the indicators of change-points.

(b) Extend the analysis to the case where the number of change-points is unknown.

11.17 The analysis of counting data is sometimes associated with a change-point representation. Consider a Poisson process

$$Y_i \sim \mathcal{P}(\theta t_i) \ (i \le \tau), \qquad Y_i \sim \mathcal{P}(\lambda t_i) \qquad (i > \tau),$$

with

$$\theta \sim \mathcal{G}a(\alpha_1, \beta_1), \quad \lambda \sim \mathcal{G}a(\alpha_2, \beta_2),$$

and

$$\beta_1 \sim \mathcal{G}a(\delta_1, \varepsilon_1), \quad \beta_2 \sim \mathcal{G}a(\delta_2, \varepsilon_2),$$

where α_i, ϵ_i, and δ_i are assumed to be known $(i = 1, 2)$.

(a) Check that the posterior distribution is well defined.

(b) Show that the following algorithm

Decade/Year	0 1 2 3 4 5 6 7 8 9
1850	− 4 5 4 1 0 4 3 4 0
1860	6 3 3 4 0 2 6 3 3 5
1870	4 5 3 1 4 4 1 5 5 3
1880	4 2 5 2 2 3 4 2 1 3
1890	2 2 1 1 1 1 3 0 0 1
1900	0 1 1 0 0 3 1 0 3 2
1910	2 0 1 1 1 0 1 0 1 0
1920	0 0 2 1 0 0 0 1 1 0
1930	2 3 3 1 1 2 1 1 1 1
1940	2 4 2 0 0 0 1 4 0 0
1950	0 1 0 0 0 0 0 1 0 0
1960	1 0 1 − − − − − − −

Table 11.2. Yearly number of mining accidents in England for the years 1851 to 1962, from Maguire et al. (1952) and Jarrett (1979) datasets. (*Source*: Carlin et al. 1992.)

Algorithm A.51 –Changepoint Poisson Model–

1. Generate $(1 \leq k \leq n)$

$$\tau \sim P(\tau = k) \propto \exp\left\{(\lambda - \theta)\sum_{i=1}^{k} t_i\right\} \exp\left\{\log(\theta/\lambda)\sum_{i=1}^{k} y_i\right\}.$$

2. Generate   [A.51]

$$\theta \sim \mathcal{G}a\left(\alpha_1 + \sum_{i=1}^{\tau} y_i, \beta_1 + \sum_{i=1}^{\tau} t_i\right),$$

$$\lambda \sim \mathcal{G}a\left(\alpha_2 + \sum_{i=\tau+1}^{n} y_i, \beta_2 + \sum_{i=\tau+1}^{n} t_i\right).$$

3. Generate

$$\beta_1 \sim \mathcal{G}a(\delta_1 + \alpha_1, \theta + \varepsilon_1), \quad \beta_2 \sim \mathcal{G}a(\delta_2 + \alpha_2, \lambda + \varepsilon_2).$$

is a valid Gibbs sampler.

(c) Show that the model can also be represented by associating with each observation y_i a latent variable $Z_i \in \{0, 1\}$ which is the indicator of the changepoint. Deduce that this representation does not modify the above algorithm.

(d) For the data in Table 11.2, which summarizes the number of mining accidents in England from 1851 to 1962, from 1851 to 1962, and for the hyperparameters $\alpha_1 = \alpha_2 = 0.5$, $\delta_1 = \delta_2 = 0$, and $\varepsilon_1 = \varepsilon_2 = 1$, apply the above algorithm to estimate the change-point. (*Note:* Raftery and Akman 1986 obtain estimates of k located between 1889 and 1892.)

11.18 Barry and Hartigan (1992, 1993) study another change-point model where the number ν of change-points is random. They introduce the notion of a *partition product*; assuming that the probability that ν is equal to k and that the change-points are in $1 < i_1 < \ldots < i_k < n$, this is given by $\rho(k, i_1, \ldots, i_k) \propto c_{1 i_1} c_{i_1 i_2} \ldots c_{i_k n}$, where

$$c_{ij} = \begin{cases} (j-i)^{-3} & \text{for } 1 < i < j < n \\ (j-i)^{-2} & \text{for } i = 1 \text{ or } j = n \\ n^{-1} & \text{for } i = 1 \text{ and } j = n. \end{cases}$$

The distributions on the observables X_i and their mean ξ_{i_j} are normal ($i_{j-1} \le i < i_j$)

$$X_i \sim \mathcal{N}(\xi_{i_j}, \sigma^2), \qquad \xi_{i_j} \sim \mathcal{N}\left(\mu_0, \frac{\sigma_0^2}{i_j - i_{j-1}}\right),$$

while the prior distributions on the parameters σ, σ_0, μ_0, and p are

$$\pi(\sigma^2) = 1/\sigma^2, \qquad \frac{\sigma_0^2}{\sigma_0^2 + \sigma^2}\Big|\sigma \sim \mathcal{U}_{[0, w_0]}, \qquad \pi(\mu_0) = 1 \text{ and } p \sim \mathcal{U}_{[0, p_0]},$$

with $p_0 \le 1$ and $w_0 \le 1$.

(a) Show that the posterior distribution of $\nu, i_1, \ldots, i_\nu, \xi_{i_1}, \ldots, \xi_{i_\nu}, \xi_n$ is

$$\pi(\nu, i_1, \ldots, \xi_n | \sigma, \sigma_0, p, \mu_0, x_1, \ldots, x_n) \propto c_{1 i_1} \ldots c_{i_\nu n}$$
$$\times \exp\left\{-\frac{(i_1 - 1)(\xi_{i_1} - \mu_0)^2}{2\sigma_0^2}\right\} \cdots \exp\left\{-\frac{(n - i_\nu)(\xi_n - \mu_0)^2}{2\sigma_0^2}\right\}$$
$$\times \prod_{j=1}^{\nu+1} \prod_{i_{j-1} \le i < i_j} \exp\left\{-\frac{(x_i - \xi_{i_j})^2}{2\sigma^2}\right\}$$
$$\times (i_1 - 1)^{1/2} \ldots (i_j - i_{j-1})^{1/2} \ldots (n - i_\nu)^{1/2}.$$

(b) Show that integrating ξ_{i_j} out leads to

$$\pi(\nu, i_1, \ldots, i_\nu | \sigma, \sigma_0, p, \mu_0, x_1, \ldots, x_n) \propto c_{1 i_1} \ldots c_{i_\nu n}$$
$$\times \prod_{j=1}^{\nu+1} \exp\left\{-\sum_{i=i_{j-1}}^{i_j - 1} \frac{(x_i - \bar{x}_{i_{j-1}})^2}{2\sigma^2} - \frac{(i_j - i_{j-1})(\bar{x}_{i_{j-1}} - \mu_0)^2}{2(\sigma^2 + \sigma_0^2)}\right\}.$$

(c) Deduce that the posterior distribution is well-defined.

(d) Construct a reversible jump algorithm for the simulation of $(\nu, i_1, \ldots, i_\nu)$.

11.19 (a) We now extend the model of Problem 11.17 to the case when the number of change-points, k, is unknown. We thus have change-points $1 < \tau_1 < \cdots < \tau_k \le n$ such that, for $\tau_i \le j \le \tau_i - 1$, $Y_j \sim \mathcal{P}(\theta_{i+1} t_j)$. Using the priors

$$\theta_i \sim \mathcal{G}a(\alpha, \beta_i), \qquad \beta_i \sim \mathcal{G}a(\delta, \epsilon),$$

a uniform prior on $(\tau_1, \ldots, \tau_k)$, and $k \sim \mathcal{P}(\lambda)$, derive a reversible jump algorithm for the estimation of k.

(b) Show that the latent variable structure described in (c) can be extended to this setup and that it avoids the variable dimension representation. Derive a Gibbs sampler corresponding to this latent variable structure.

11.20 Show that, if $X_1, \ldots, X_n$ are iid $\mathcal{E}xp(\lambda)$, the random variable $\min(X_1, \ldots, X_n)$ is distributed from an $\mathcal{E}xp(n\lambda)$ distribution. Generalize to the case when the X_i's are $\mathcal{E}xp(\lambda_i)$.

11.21 A typical setup of *convolution* consists in the observation of

$$Y_t = \sum_{j=0}^{k} h_j X_{t-j} + \varepsilon_t , \qquad t = 1, \ldots, n,$$

where $\varepsilon_t \sim \mathcal{N}(0, \sigma^2)$, the X_i's are not observed but have finite support, $\mathcal{X} = \{s_1, \ldots, s_m\}$, and where the parameters of interest h_j are unknown. The discrete variables X_t are iid with distribution $\mathcal{M}_m(\theta_1, \ldots, \theta_m)$ on $\mathcal{X}$, the probabilities θ_i are distributed from a Dirichlet prior distribution, $\mathcal{D}_m(\gamma_1, \ldots, \gamma_m)$, and the coefficients h_j are from a conjugate distribution $h = (h_0, \ldots, h_k) \sim \mathcal{N}_{k+1}(\mu, \Sigma)$.
(a) Show that the completed likelihood is

$$L(x, h, \sigma) \propto \sigma^{-n} \exp\left\{ -\frac{1}{2\sigma^2} \sum_{t=1}^{n} \left(y_t - \sum_{j=0}^{k} h_j x_{t-j} \right)^2 \right\}$$

and examine whether or not the EM algorithm can be implemented in this setting.
(b) Verify that the completion of Y_t in (Y_t, X_t) can be done as follows:
Algorithm A.52 –Deconvolution Completion–

    1. Generate $(t = 1-k, \ldots, n)$

$$X_t \sim P(x_t = s_\ell) \propto \frac{\theta_\ell s_\ell}{\sigma^2} \exp\left\{ -s_\ell^2 \sum_{j=0}^{k} h_j^2 / 2\sigma^2 \right.$$

$$\left. + \sum_{j=0}^{k} h_j \left(y_{t+j} - \sum_{\substack{i=0 \\ i \neq j}}^{k} h_i x_{t+j-i} \right) \right\}$$

    and derive $\hspace{6cm}$ [A.52]

$$X = \begin{pmatrix} x_1 & x_0 & \ldots & x_{1-k} \\ & & \ldots & \\ x_n & x_{n-1} & \ldots & x_{n-k} \end{pmatrix}, \qquad \hat{h} = (X^t X)^{-1} X^t y ,$$

where $y = (y_1, \ldots, y_n)$.
(c) Verify that the update of the parameters (h, σ^2) is given by
    2. Generate

$$h \sim \mathcal{N}_{k+1} \left((\Sigma^{-1} + \sigma^{-2} X^t X)^{-1} (\Sigma^{-1}\mu + \sigma^{-2} X^t X \hat{h}), (\Sigma^{-1} + \sigma^{-2} X^t X)^{-1} \right),$$

$$\sigma^2 \sim \mathcal{IG}\left(\frac{n}{2}, \frac{1}{2} \sum_{t=1}^{n} \left(y_t - \sum_{j=0}^{k} h_j x_{t-j} \right)^2 \right),$$

while the step for hyperparameters θ_ℓ $(\ell = 1, \ldots, m)$ corresponds to
    3. Generate

$$(\theta_1, \ldots, \theta_m) \sim \mathcal{D}\left(\gamma_1 + \sum_{t=1-k}^{n} \mathbb{I}_{x_t = s_1}, \ldots, \gamma_m + \sum_{t=1-k}^{n} \mathbb{I}_{x_t = s_m} \right)$$

(d) When k is unknown, with a prior $k \sim \mathcal{P}(\lambda)$, develop a reversible jump algorithm for estimating k, using the above completion and a simple birth-death proposal based on moves from $\mathcal{M}_k$ to $\mathcal{M}_{k-1}$ and to $\mathcal{M}_{k+1}$.

(*Note:* See Kong et al. 1994, Gassiat 1995, Liu and Chen 1995, and Gamboa and Gassiat 1997 for more detailed introductions to deconvolution models.)

11.22 (Neal 1999) Consider a two-layer neural network (or *perceptron*), introduced in Note 5.5.2, where

$$\begin{cases} h_k = \tanh\left(\alpha_{k0} + \sum_j \alpha_{kj} x_j\right), & k = 1, \ldots, p, \\ \mathbb{E}[Y_\ell | h] = \beta_{\ell 0} + \sum_k v_{\ell k} h_k, & \ell = 1, \ldots, n. \end{cases}$$

Propose a reversible jump Monte Carlo method when the number of hidden units p is unknown, based on observations (x_j, y_j) $(j = 1, \ldots, n)$.

11.5 Notes

11.5.1 Occam's Razor

Associated with the issues of model choice and variable selection, the so-called *Occam's Razor rule* is often used as a philosophical argument in favor of parsimony, that is, simpler models. William d'Occam or d'Ockham (ca. 1290– ca. 1349) was a English theologian from Oxford who worked on the bases of empirical induction and, in particular, posed the principle later called *Occam's Razor*, which excludes a plurality of reasons for a phenomenon if they are not supported by some experimentation (see Adams 1987). This principle,

Pluralitas non est ponenda sine neccesitate,

(meaning *Entities are not to be multiplied without necessity*), is sometimes invoked as a *parsimony principle* to choose the simplest between two equally possible explanations, and its use is recurrent in the (recent) Bayesian literature. However, it does not provide a working principle per se. (At a more anecdotal level, Umberto Eco's *The Name of the Rose* borrows from Occam to create the character William of Baskerville.) See also Wolpert (1992, 1993) for an opposite perspective on Occam's Razor and Bayes factors.

12

Diagnosing Convergence

"Why does he insist that we must have a diagnosis? Some things are not meant to be known by man."
—Susanna Gregory, *An Unholy Alliance*

12.1 Stopping the Chain

In previous chapters, we have presented the theoretical foundations of MCMC algorithms and showed that, under fairly general conditions, the chains produced by these algorithms are ergodic, or even geometrically ergodic. While such developments are obviously necessary, they are nonetheless insufficient from the point of view of the implementation of MCMC methods. They do not directly result in methods of controlling *the* chain produced by an algorithm (in the sense of a *stopping rule* to guarantee that the number of iterations is sufficient). In other words, while necessary as mathematical proofs of the validity of the MCMC algorithms, general convergence results do not tell us when to stop these algorithms and produce our estimates. For instance, the mixture model of Example 10.18 is fairly well behaved from a theoretical point of view, but Figure 10.3 indicates that the number of iterations used is definitely insufficient.

Perhaps the only sure way of guaranteeing convergence is through the types of calculations discussed in Example 8.8 and Section 12.2.1, where a bound on the total variation distance between the nth-order transition kernel $K^n(x, \cdot)$ and the stationary distribution is given. Then, for a specified total variation distance, the needed value of n can be solved for. Unfortunately, such calculations are usually quite difficult, and are only feasible in relatively simple settings. Thus, for generally applicable convergence assessment strategies, we are left with empirical methods.

Example 12.1. Probit model revisited. The probit model defined in Example 10.21 is associated with the posterior distribution

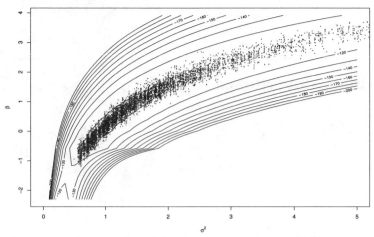

Fig. 12.1. Contour plot of the log-posterior distribution for a probit sample of $1,000$ observations, along with $1,000$ points of an MCMC sample.

$$\prod_{i=1}^{n} \Phi(r_i\beta/\sigma)^{d_i} \Phi(-r_i\beta/\sigma)^{1-d_i} \times \pi(\beta, \sigma^2),$$

where $\pi(\beta, \sigma^2)$ is the prior distribution and the pairs (r_i, d_i) the observations. In the special case where

$$\pi(\beta, \sigma^2) = \sigma^{-4} \exp\{-1/\sigma^2\} \exp\{-\beta^2/50\},$$

a contour plot of the log-posterior distribution is given in Figure 12.1, along with the last $1,000$ points of an MCMC sample after $100,000$ iterations. This MCMC sample is produced via a simple Gibbs sampler on the posterior distribution where β and σ^2 are alternatively simulated by normal and log-normal random walk proposals and accepted by a one-dimensional Metropolis–Hastings step.

While Example 10.21 was concerned with convergence difficulties with the Gibbs sampler, this (different) implementation does not seem to face the same problem, at least judging from a simple examination of Figure 12.1, since the simulated values coincide with the highest region of the log-posterior. Obviously, this is a very crude evaluation and a more refined assessment is necessary before deciding whether using the MCMC sample represented in Figure 12.1 in ergodic averages, or increasing the number of simulations to achieve a more reliable approximation.   ‖

The goal of this chapter is to present, in varying amounts of detail, a catalog of the numerous monitoring methods (or *diagnostics*) proposed in the literature, in connection with the review papers of Cowles and Carlin (1996), Robert (1995a), Brooks (1998b), Brooks and Roberts (1998), and Mengersen et al. (1999). Some of the techniques presented in this chapter have withstood the test of time, and others are somewhat exploratory in nature. We are, however, in the situation of describing a sequence of noncomparable techniques with widely varying degrees of theoretical justification and usefulness.[1]

12.1.1 Convergence Criteria

From a general point of view, there are three (increasingly stringent) types of convergence for which assessment is necessary:

(i) Convergence to the Stationary Distribution

This criterion considers convergence of the chain $\theta^{(t)}$ to the stationary distribution f (or *stationarization*), which seems to be a minimal requirement for an algorithm that is supposed to approximate simulation from f! Unfortunately, it seems that this approach to convergence issues is not particularly fruitful. In fact, from a theoretical point of view, f is only the *limiting* distribution of $\theta^{(t)}$. This means that stationarity is only achieved asymptotically.[2]

However, the original implementation of the Gibbs sampler was based on the generation of n independent initial values $\theta_i^{(0)}$ ($i = 1, \ldots, n$), and the storage of only the last simulation $\theta_i^{(T)}$ in each chain. While intuitive to some extent—the larger T is, the closer $\theta_i^{(T)}$ is to the stationary distribution—this criterion is missing the point. If μ_0 is the (initial) distribution of $\theta^{(0)}$, then the $\theta_i^{(T)}$'s are all distributed from $f_{\mu_0}^T$. In addition, this also results in a waste of resources, as most of the generated variables are discarded.

If we, instead, consider only a single realization (or *path*) of the chain $(\theta^{(t)})$, the question of *convergence* to the limiting distribution is not really relevant! Indeed, it is possible[3] to obtain the initial value $\theta^{(0)}$ from the distribution f, and therefore to act as if the chain is already in its stationary regime from the start, meaning that $\theta^{(0)}$ belongs to an area of likely (enough) values for f.

This seeming dismissal of the first type of control may appear rather cavalier, but we do think that convergence to f per se is not the major issue for most MCMC algorithms in the sense that the chain truly produced by the

---

[1] Historically, there was a flurry of papers at the end of the 90s concerned with the development of convergence diagnoses. This flurry has now quieted down, the main reason being that no criterion is absolutely foolproof, as we will see later in this chapter. As we will see again in the introduction of Chapter 13, the only way of being certain the algorithm has converged is to use iid sampling!

[2] This perspective obviously oversimplifies the issue. As already seen in the case of renewal and coupling, there exist finite instances where the chain is known to be in the stationary distribution (see Section 12.2.3).

[3] We consider a standard statistical setup where the support of f is approximately known. This may not be the case for high-dimensional setups or complex structures where the algorithm is initialized at random.

algorithm often behaves like a chain initialized from f. The issues at stake are rather the speed of exploration of the support of f and the degree of correlation between the $\theta^{(t)}$'s. This is not to say that stationarity should not be tested at all. As we will see in Section 12.2.2, regardless of the starting distribution, the chain may be slow to explore the different regions of the support of f, with lengthy stays in some regions (for example, the modes of the distribution f). A stationarity test may be useful in detecting such difficulties.

(ii) Convergence of Averages

Here, as in regular Monte Carlo settings, we are concerned with convergence of the empirical average

(12.1)
$$\frac{1}{T} \sum_{t=1}^{T} h(\theta^{(t)})$$

to $\mathbb{E}_f[h(\theta)]$ for an arbitrary function h. This type of convergence is most relevant in the implementation of MCMC algorithms. Indeed, even when $\theta^{(0)} \sim f$, the exploration of the complexity of f by the chain $(\theta^{(t)})$ can be more or less lengthy, depending on the transition chosen for the algorithm. The purpose of the convergence assessment is, therefore, to determine whether the chain has exhibited all the features of f (for instance, all the modes). Brooks and Roberts (1998) relate this convergence to the *mixing* speed of the chain, in the informal sense of a strong (or weak) dependence on initial conditions and of a slow (or fast) exploration of the support of f (see also Asmussen et al. 1992). A formal version of convergence monitoring in this setup is the convergence assessment of Section 12.1. While the ergodic theorem guarantees the convergence of this average from a theoretical point of view, the relevant issue at this stage is to determine a minimal value for T which justifies the approximation of $\mathbb{E}_f[h(\theta)]$ by (12.1) for a given level of accuracy.

(iii) Convergence to iid Sampling

This convergence criterion measures how close a sample $(\theta_1^{(t)}, \ldots, \theta_n^{(t)})$ is to being iid. Rather than approximating integrals such as $\mathbb{E}_f[h(\theta)]$, the goal is to produce variables θ_i which are (quasi-)independent. While the solution based on parallel chains mentioned above is not satisfactory, an alternative is to use *subsampling* (or *batch sampling*) to reduce correlation between the successive points of the Markov chain. This technique, which is customarily used in numerical simulation (see, for instance, Schmeiser 1989) subsamples the chain $(\theta^{(t)})$ with a batch size k, considering only the values $\eta^{(t)} = \theta^{(kt)}$. If the covariance $\text{cov}_f(\theta^{(0)}, \theta^{(t)})$ decreases monotonically with t (see Section 9.3), the motivation for subsampling is obvious. In particular, if the chain $(\theta^{(t)})$ satisfies an interleaving property (see Section 9.2.2), subsampling is justified.

However, checking for the monotone decrease of $\text{cov}_f(\theta^{(0)}, \theta^{(t)})$—which also justifies Rao–Blackwellization (see Section 9.3)—is not always possible and, in some settings, the covariance oscillates with t, which complicates the choice of k.

We note that subsampling necessarily leads to losses in efficiency with regard to the second convergence goal. In fact, as shown by MacEachern and Berliner (1994), it is always preferable to use the entire sample for the approximation of $\mathbb{E}_f[h(\theta)]$. Nonetheless, for convergence assessment, subsampling may be beneficial (see, e.g., Robert et al. 1999).

Lemma 12.2. *Suppose $h \in \mathcal{L}^2(f)$ and $(\theta^{(t)})$ is a Markov chain with stationary distribution f. If*

$$\delta_1 = \frac{1}{Tk} \sum_{t=1}^{Tk} h(\theta^{(t)}) \quad and \quad \delta_k = \frac{1}{T} \sum_{\ell=1}^{T} h(\theta^{(k\ell)}),$$

the variance of δ_1 satisfies

$$\text{var}(\delta_1) \leq \text{var}(\delta_k),$$

for every $k > 1$.

Proof. Define $\delta_k^1, \ldots, \delta_k^{k-1}$ as the shifted versions of $\delta_k = \delta_k^0$; that is,

$$\delta_k^i = \frac{1}{T} \sum_{t=1}^{T} h(\theta^{(tk-i)}), \qquad i = 0, 1, \ldots, k-1 .$$

The estimator δ_1 can then be written as $\delta_1 = \frac{1}{k} \sum_{i=0}^{k-1} \delta_k^i$, and hence

$$
\begin{aligned}
\text{var}(\delta_1) &= \text{var}\left(\frac{1}{k} \sum_{i=0}^{k-1} \delta_k^i \right) \\
&= \text{var}(\delta_k^0)/k + \sum_{i \neq j} \text{cov}(\delta_k^i, \delta_k^j)/k^2 \\
&\leq \text{var}(\delta_k^0)/k + \sum_{i \neq j} \text{var}(\delta_k^0)/k^2 \\
&= \text{var}(\delta_k) ,
\end{aligned}
$$

where the inequality follows from the Cauchy–Schwarz inequality

$$|\text{cov}(\delta_k^i, \delta_k^j)| \leq \text{var}(\delta_k^0).$$

$\square$

In the remainder of the chapter, we consider only independence issues in cases where they have bearing on the control of the chain, as in renewal theory (see Section 12.2.3). Indeed, for an overwhelming majority of cases where MCMC algorithms are used, independence is not a necessary feature.

12.1.2 Multiple Chains

Aside from distinguishing between convergence to stationarity (Section 12.2) and convergence of the average (Section 12.3), we also distinguish between the methods involving the parallel simulation of M independent chains $(\theta_m^{(t)})$ $(1 \leq m \leq M)$ and those based on a single "on-line" chain. The motivation of the former is intuitively sound. By simulating several chains, variability and dependence on the initial values are reduced and it should be easier to control convergence to the stationary distribution by comparing the estimation, using different chains, of quantities of interest. The dangers of a naïve implementation of this principle should be obvious, namely that the slower chain governs convergence and that the choice of the initial distribution is extremely important in guaranteeing that the different chains are well dispersed.

Many multiple-chain convergence diagnostics are quite elaborate (Gelman and Rubin 1992, Liu et al. 1992) and seem to propose convergence evaluations that are more robust than single-chain methods. Geyer (1992) points out that this robustness may be illusory from several points of view. In fact, good performances of these parallel methods require a degree of a priori knowledge on the distribution f in order to construct an initial distribution on $\theta_m^{(0)}$ which takes into account the features of f (modes, shape of high density regions, etc.). For example, an initial distribution which is too concentrated around a local mode of f does not contribute significantly more than a single chain to the exploration of f. Moreover, slow algorithms, like Gibbs sampling used in highly nonlinear setups, usually favor single chains, in the sense that a unique chain with MT observations and a slow rate of mixing is more likely to get closer to the stationary distribution than M chains of size T, which will presumably stay in the neighborhood of the starting point with higher probability.

An additional practical drawback of parallel methods is that they require a modification of the original MCMC algorithm to deal with the processing of parallel outputs. (See Tierney 1994 and Raftery and Lewis 1996 for other criticisms.) On the other hand, single-chain methods suffer more severely from the defect that *"you've only seen where you've been,"* in the sense that the part of the support of f which has not been visited by the chain at time T is almost impossible to detect. Moreover, a single chain may present probabilistic pathologies which are possibly avoided by parallel chains. (See the example of Figure 12.16, as opposed to the sequential importance sampling resolution of Section 14.4.4.)

As discussed in Chapter 14, and in particular in the population Monte Carlo section, Section 14.4, there exist alternative ways of taking advantage of parallel chains to improve and study convergence, in particular for a better assessment of the entire support of f. These algorithms outside the MCMC area can also be used as benchmarks to test the convergence of MCMC algorithms.

12.1.3 Monitoring Reconsidered

We agree with many authors[4] that it is somewhat of an illusion to think we can control the flow of a Markov chain and assess its convergence behavior from a few realizations of this chain. There always are settings (transition kernels) which, for most realizations, will invalidate an arbitrary indicator (whatever its theoretical justification) and the randomness inherent to the nature of the problem prevents any categorical guarantee of performance. The heart of the difficulty is the key problem of statistics, where the uncertainty due to the observations prohibits categorical conclusions and final statements. Far from being a failure acknowledgment, these remarks only aim at warning the reader about the *relative* value of the indicators developed below. As noted by Cowles and Carlin (1996), it is simply inconceivable, in the light of recent results, to envision *automated stopping rules*. Brooks and Roberts (1998) also stress that the prevalence of a given control method strongly depends on the model and on the inferential problem under study. It is, therefore, even more crucial to develop robust and general evaluation methods which extend and complement the present battery of stopping criteria. One goal is the development of "convergence diagnostic spreadsheets," in the sense of computer graphical outputs, which would graph several different features of the convergence properties of the chain under study (see Cowles and Carlin 1996, Best et al. 1995, Robert 1997, 1998, or Robert et al. 1999).

The criticisms presented in the wake of the techniques proposed below serve to highlight the incomplete aspect of each method. They do not aim at preventing their use but rather to warn against a selective interpretation of their results.

12.2 Monitoring Convergence to the Stationary Distribution

12.2.1 A First Illustration

As a first approach to assessing how close the Markov chain is to stationarity, one might try to obtain a bound on $\|K^n(x, \cdot) - f\|_{TV}$, the total variation difference between the nth step transition kernel and the stationary distribution (see Section 6.6). For example, in the case of the independent Metropolis–Hastings algorithm, we know that $\|K^n(x, \cdot) - f\|_{TV} \leq 2(1 - M^{-1})^n$ (Theorem 7.8). Using an argument based on drift conditions (Note 6.9.1), Jones and Hobert (2001) were able to obtain bounds on a number of geometrically ergodic chains. However, although their results apply to a wide class of

---

[4] To borrow from the injunction of Hastings (1970), *"even the simplest of numerical methods may yield spurious results if insufficient care is taken in their use... The setting is certainly no better for the Markov chain methods and they should be used with appropriate caution."*

Markov chains, the calculations needed to obtain the analytical bounds can be prohibitively difficult.

A natural empirical approach to convergence control is to draw pictures of the output of simulated chains, in order to detect deviant or non-stationary behaviors. For instance, as in Gelfand and Smith 1990, a first plot is to draw the sequence of the $\theta^{(t)}$'s against t. However, this plot is only useful for strong non-stationarities of the chain.

Example 12.3. (Continuation of Example 12.1) For the MCMC sample produced in Example 12.1, simple graphs may be enough to detect a lack of convergence. For instance, Figures 12.2 and 12.3 represent the evolution of β along 1000 and 10,000 iterations of the Gibbs sampler. They clearly show that the time scale required for the convergence of the MCMC algorithm is much bigger than these quantities. It is only at the scale of 100,000 iterations (Figure 12.4) that the chain seems to take a more likely turn towards stationarity. But things are not so clear-cut when looking at the last simulated values: Figure 12.5 presents the same heterogeneities as Figure 12.3. This indicates that the Markov chain, rather than being slow to converge to the stationary distribution, is slow to explore the support of the posterior distribution; that is, it is mixing very slowly. This feature can also be checked on Figure 12.6, which indicates the values of the log-posterior at consecutive points of the Markov chain.

Another interesting feature, related to this convergence issue, is that some aspects of the posterior may be captured more quickly than others, reinforcing our point that the fundamental problem is mixing rather than convergence *per se*. When looking at the marginal posterior distribution of β/σ, the stabilization of the empirical distribution is much faster, as shown by the comparison of the first and of the last 5,000 iterations (out of 100,000) in Figure 12.7. This is due to the fact that β/σ is an identifiable parameter for the probit model $P(D = 1|R) = \Phi(R\beta/\sigma)$. (For the simulated dataset, the true value of β/σ is 0.89, which agrees quite well with the MCMC sample being distributed around 0.9.) ‖

12.2.2 Nonparametric Tests of Stationarity

Standard nonparametric tests, such as Kolmogorov–Smirnov or Kuiper tests (Lehmann 1975), can be applied in the stationarity assessment of a single output of the chain $\theta^{(t)}$. In fact, when the chain is stationary, $\theta^{(t_1)}$ and $\theta^{(t_2)}$ have the same marginal distribution for arbitrary times t_1 and t_2. Given an MCMC sample $\theta^{(1)}, \ldots, \theta^{(T)}$, it thus makes sense to compare the distributions of the two halves of this sample, $(\theta^{(1)}, \ldots, \theta^{(T/2)})$ and $(\theta^{(T/2+1)}, \ldots, \theta^{(T)})$. Since usual nonparametric tests are devised and calibrated in terms of iid samples, there needs to be a correction for the correlation between the $\theta^{(t)}$'s.

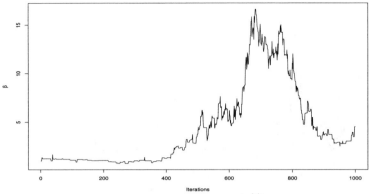

Fig. 12.2. Evolution of the Markov chain $(\beta^{(t)})$ along the first $1,000$ iterations when $(\theta^{(0)}, \sigma^{(0)}) = (1, 2)$.

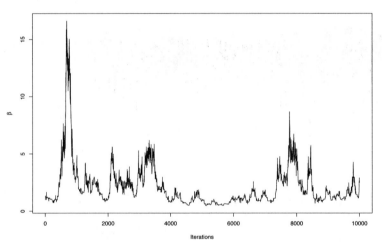

Fig. 12.3. Evolution of the Markov chain $(\beta^{(t)})$ along the first $10,000$ iterations.

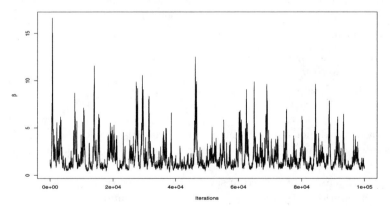

Fig. 12.4. Evolution of the Markov chain $(\beta^{(t)})$ along $100,000$ iterations.

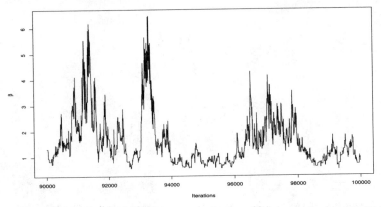

Fig. 12.5. Evolution of the Markov chain $(\beta^{(t)})$ along the last $10,000$ iterations, out of $100,000$.

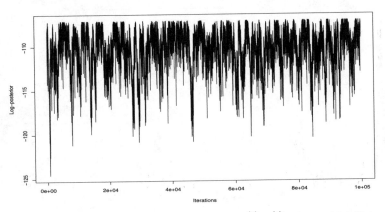

Fig. 12.6. Evolution of the value of $\pi(\beta^{(t)}, \sigma^{(t)})$ along $100,000$ iterations.

This correction can be achieved by the introduction of a batch size G leading to the construction of two (quasi-) independent samples. For each of the two halves above, select subsamples $(\theta_1^{(G)}, \theta_1^{(2G)}, \ldots)$ and $(\theta_2^{(G)}, \theta_2^{(2G)}, \ldots)$. Then, for example, the *Kolmogorov–Smirnov statistic* is

$$(12.2) \qquad K = \frac{1}{M} \sup_\eta \left| \sum_{g=1}^{M} \mathbb{I}_{(0,\eta)}(\theta_1^{(gG)}) - \sum_{g=1}^{M} \mathbb{I}_{(0,\eta)}(\theta_2^{(gG)}) \right|$$

in the case of a one-dimensional chain. For multidimensional chains, (12.2) can be computed on either a function of interest or on each component of the vector $\theta^{(t)}$.

The statistic K can be processed in several ways to derive a stopping rule. First, under the stationarity assumption as M goes to infinity, the limiting distribution of $\sqrt{M}\,K$ has the cdf

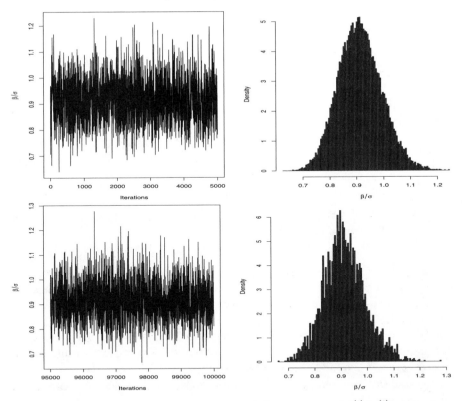

Fig. 12.7. Comparison of the distribution of the sequence of $\beta^{(t)}/\sigma^{(t)}$'s over the first $5,000$ iterations (top) and the last $5,000$ iterations (bottom), along $100,000$ iterations.

$$(12.3) \qquad R(x) = 1 - \sum_{k=1}^{\infty} (-1)^{k-1} e^{-2k^2 x^2} \, ,$$

which can be easily approximated by a finite sum (see Problem 12.2). The corresponding p-value can therefore be computed for each T until it gets above a given level. (An approximation of the 95% quantile, $q_\alpha = 1.36$ (for $M \geq 100$), simplifies this stage.) Of course, to produce a valid inference, we must take into account the sequential nature of the test and the fact that K is computed as the infimum over all components of $\theta^{(t)}$ of the corresponding values of (12.2).

An exact derivation of the level of the derived test is quite difficult given the correlation between the $\theta^{(t)}$'s and the influence of the subsampling mechanism. Another use of (12.2), which is more graphic, is to represent the sample of $\sqrt{M}\ K_T$'s against T and to check visually for a stable distribution around small values. (See also Brooks et al. 2003a, for another approach using the Kolmogorov–Smirnov test)

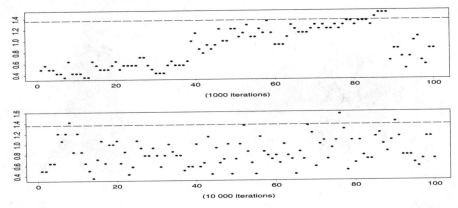

Fig. 12.8. Plot of 100 Kolmogorov–Smirnov statistics for $T = 1000$ and $T = 10,000$ iterations. The dotted line corresponds to the 95% level.

Obviously, an assessment of stationarity based on a single chain is open to criticism: In cases of strong attraction from a local mode, the chain will most likely behave as if it was simulated from the restriction of f to the neighborhood of this mode and thus lead to a convergence diagnosis (this is the *"you've only seen where you've been"* defect mentioned in Section 12.1.2). However, in more intermediate cases, where the chain $(\theta^{(t)})$ stays for a while in the neighborhood of a mode before visiting another modal region, the subsamples $(\theta_1^{(t)})$ and $(\theta_2^{(t)})$ should exhibit different features until the chain explores every modal region.[5]

Example 12.4. Nuclear pump failures. In the model of Gaver and O'Muircheartaigh (1987), described in Example 10.17, we consider the subchain $(\beta^{(t)})$ produced by the algorithm. Figure 12.8 gives the values of the Kolmogorov–Smirnov statistics K for $T = 1000$ and $T = 10,000$ iterations, with $M = 100$ and 100 values of $10K_T$. Although both cases lead to similar proportions of about 95% values under the level 1.36, the first case clearly lacks the required homogeneity, since the statistic is almost monotone in t. This behavior may correspond to the local exploration of a modal region for $t \leq 400$ and to the move to another region of importance for $400 \leq t \leq 800$. ‖

12.2.3 Renewal Methods

Chapter 6 has presented the basis of *renewal theory* in Section 6.3.2 through the notion of *small set* (Definition 6.19). This theory is then used in Section 6.7 for a direct derivation of many limit theorems (Theorems 6.63 and 6.64). It is also possible to take advantage of this theory for the purpose(s) of convergence control, either through small sets as in Robert (1995a) or through the

---

[5] We may point out that, once more, it is a matter of proper scaling of the algorithm.

alternative of Mykland et al. (1995), which is based on a less restrictive representation of renewal called *regeneration*. (See also Gilks et al. 1998, Sahu and Zhigljavsky 1998, 2003 for a related updating technique.) The use of small sets is obviously relevant in the control of convergence of averages, as presented in Section 12.3.3, and as a correct discretization technique; see Section 12.4.2, but it is above all a central notion for the convergence to stationarity.

As we will see in Chapter 13, renewal is a key to the derivation of perfect sampling schemes (see Section 13.2.4), in which we can obtain exact samples from the stationary distribution. This is done using *Kac's representation of stationary distributions* (Section 6.5.2). The difficulties with minorization conditions, namely the derivation of the minorizing measure and the very small probabilities of renewal, remain, however, and restrict the practicality of this approach for stationarity assessment.

Mykland et al. (1995) replaced small sets and the corresponding minorization condition with a generalization through functions s such that

$$(12.4) \qquad P(\theta^{(t+1)} \in B | \theta^{(t)}) \geq s(\theta^{(t)})\, \nu(B), \qquad \theta^{(t)} \in \Theta,\ B \in \mathcal{B}(\Theta)\,.$$

Small sets thus correspond to the particular case $s(x) = \varepsilon \mathbb{I}_C(x)$. If we define

$$r(\theta^{(t)}, \xi) = \frac{s(\theta^{(t)})\, \nu(\xi)}{K(\theta^{(t)}, \xi)}\,,$$

then each time t is a renewal time with probability $r(\theta^{(t-1)}, \xi)$.

Based on the regeneration rate, Mykland et al. (1995) propose a graphical convergence assessment. When an MCMC algorithm is in its stationary regime and has correctly explored the support of f, the regeneration rate must remain approximately constant (because of stationarity). Thus, they plot an approximation of the regeneration rate,

$$\hat{r}_T = \frac{1}{T} \sum_{t=1}^{T} r(\theta^{(t)}, \theta^{(t+1)})\,,$$

against T. When the normalizing constant of ν is available they instead plot

$$\tilde{r}_T = \frac{1}{T} \sum_{t=1}^{T} s(\theta^{(t)})\,.$$

An additional recommendation of the authors is to smooth the graph of $\hat{r}_T$ by nonparametric techniques, but the influence of the smoothing parameter must be assessed to avoid overoptimistic conclusions (see Section 12.6.1 for a global criticism of functional nonparametric techniques).

Metropolis–Hastings algorithms provide a natural setting for the implementation of this regeneration method, since they allow for complete freedom in the choice of the transition kernel. Problems related to the Dirac mass in the kernel can be eliminated by considering only its absolutely continuous part,

$$\rho(\theta^{(t)}, \xi) \, q(\xi|\theta^{(t)}) = \min\left\{\frac{f(\xi)}{f(\theta^{(t)})} \, q(\theta^{(t)}|\xi), q(\xi|\theta^{(t)})\right\},$$

using the notation of [A.24]. The determination of s and ν is facilitated in the case of a so-called *pseudo-reversible* transition; that is, when there exists a positive function $\tilde{f}$ with

$$(12.5) \qquad\qquad \tilde{f}(\theta) \, q(\xi|\theta) = \tilde{f}(\xi) \, q(\theta|\xi) \, .$$

Equation (12.5) thus looks like a detailed balance condition for a reversible Markov chain, but $\tilde{f}$ is not necessarily a probability density.

Lemma 12.5. *Suppose that q satisfies (12.5) and for the transition induced by q, the functions s_q and ν_q satisfy (12.4). Let $w(\theta) = f(\theta)/\tilde{f}(\theta)$ and for every $c > 0$, define the function*

$$s(\theta) = s_q(\theta) \, \min\left\{\frac{c}{w(\theta)}, 1\right\}$$

and the density

$$\nu(\theta) = \nu_q(\theta) \, \min\left\{\frac{w(\theta)}{c}, 1\right\}.$$

then, for the Metropolis–Hastings algorithm associated with q, $s(\theta)$ and $\nu(\theta)$ satisfy (12.4).

Proof. Since

$$P(\theta^{(t+1)} \in B|\theta^{(t)}) \geq \int_B \rho(\theta^{(t)}, \xi) \, q(\xi|\theta^{(t)}) \, d\xi,$$

we have

$$P(\theta^{(t+1)} \in B|\theta^{(t)}) \geq \int_B \min\left\{\frac{w(\theta^{(t)})}{w(\xi)}, 1\right\} q(\xi|\theta^{(t)}) \, d\xi$$

$$\geq \min\left\{\frac{w(\theta^{(t)})}{c}, 1\right\} \int_B \min\left\{\frac{c}{w(\xi)}, 1\right\} s_q(\theta^{(t)}) \, \nu_q(\xi) \, d\xi$$

$$= s(\theta^{(t)}) \int_B \nu(\xi) \, d\xi.$$

$\square$

In the particular case of an *independent* Metropolis–Hastings algorithm, $q(\xi|\theta) = g(\xi)$ and (12.5) applies for $\tilde{f} = g$. Therefore, $s_q \equiv 1$, $\nu_q = g$, and

$$\nu(\xi) \propto g(\xi) \, \min\left\{\frac{f(\xi)}{cg(\xi)}, 1\right\} \propto \min\left\{f(\xi), cg(\xi)\right\},$$

which behaves as a truncation of the instrumental distribution g depending on the true density f. If $\xi \sim g$ is accepted, the probability of regeneration is then

$$(12.6) \qquad r(\theta^{(t)}, \xi) = \begin{cases} \dfrac{c}{w(\theta^{(t)}) \wedge w(\xi)} & \text{if } w(\xi) \wedge w(\theta^{(t)}) > c \\[2mm] \dfrac{w(\theta^{(t)}) \vee w(\xi)}{c} & \text{if } w(\xi) \vee w(\theta^{(t)}) < c \\[2mm] 1 & \text{otherwise.} \end{cases}$$

Note that c is a free parameter. Therefore, it can be selected (calibrated) to optimize the renewal probability (or the mixing rate).

For a *symmetric* Metropolis–Hastings algorithm, $q(\xi|\theta) = q(\theta|\xi)$ implies that $\tilde{f} \equiv 1$ satisfies (12.5). The parameters s_q and ν_q are then determined by the choice of a set D and of a value $\tilde{\theta}$ in the following way:

$$\nu_q(\xi) \propto q(\theta|\tilde{\theta})\, \mathbb{I}_D(\xi)\,, \qquad s_q(\theta) = \inf_{\xi \in D} \frac{q(\xi|\theta)}{q(\theta|\xi)}\,.$$

The setting is therefore less interesting than in the independent case since D and $\tilde{\theta}$ have first to be found, as in a regular minorization condition. Note that the choice of $\tilde{\theta}$ can be based on an initial simulation of the algorithm, using the mode or the median of $w(\theta^{(t)})$.

Example 12.6. (Continuation of Example 12.4) In the model of Gaver and O'Muircheartaigh (1987), $D = \mathbb{R}_+$ is a small set for the chain $(\beta^{(t)})$, which is thus uniformly ergodic. Indeed,

$$K(\beta, \beta') \geq \frac{\delta^{\gamma+10\alpha}(\beta')^{\gamma+10\alpha-1}}{\Gamma(10\alpha + \gamma)} e^{-\delta\beta'} \prod_{i=1}^{10} \left(\frac{t_i}{t_i + \beta'} \right)^{p_i + \alpha}$$

(see Example 10.17). The regeneration (or renewal) probability is then

$$(12.7) \qquad r(\beta, \beta') = \frac{\delta^{\gamma+10\alpha}(\beta')^{\gamma+10\alpha-1}}{K(\beta, \beta')\, \Gamma(\gamma + 10\alpha)} e^{-\delta\beta'} \prod_{i=1}^{10} \left(\frac{t_i}{t_i + \beta'} \right)^{p_i + \alpha},$$

where $K(\beta, \beta')$ can be approximated by

$$\frac{1}{M} \sum_{m=1}^{M} \frac{\left(\delta + \sum_{i=1}^{10} \lambda_{im} \right)^{\gamma+10\alpha} (\beta')^{\gamma+10\alpha-1} e^{-\beta'\left(\delta + \sum_{i=1}^{10} \lambda_{im}\right)}}{\Gamma(\gamma + 10\alpha)},$$

with $\lambda_{im} \sim \mathcal{G}a(p_i + \alpha, t_i + \beta)$. Figure 12.9 provides the plot of $r(\beta^{(t)}, \beta^{(t+1)})$ against t, and the graph of the averages $\hat{r}_T$, based on only the first four observations (pumps) in the dataset. (This dataset illustrates the problem with perfect sampling mentioned at the beginning of this section. The renewal

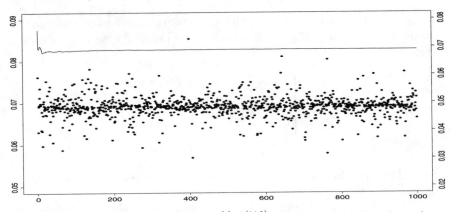

Fig. 12.9. Plot of the probabilities $r(\beta^{(t)}, \beta^{(t+1)})$ for the pump failure data when $\mathbb{R}_+$ is chosen as the small set. Superimposed (and scaled to the right), is the average $\hat{r}_T$.

probabilities associated with 10 observations are quite small, resulting in the product in (12.7) being very small, and limiting the practical use.) Mykland et al. (1995) thus replace $\mathbb{R}_+$ with the interval $D = [2.3, 3.1]$ to achieve renewal probabilities of reasonable magnitude.   ‖

Although renewal methods involve a detailed study of the chain $(\theta^{(t)})$ and may require modifications of the algorithms, as in the hybrid algorithms of Mykland et al. (1995) (see also Section 12.3.3), their basic independence structure brings a certain amount of robustness in the monitoring of MCMC algorithms, more than the other approaches presented in this chapter. The specificity of the monitoring obviously prevents a completely automated implementation, but this drawback cannot detract from its attractive theoretical properties, since it is one of the very few completely justified monitoring methods.

12.2.4 Missing Mass

Another approach to convergence monitoring is to assess how much of the support of the target distribution has been explored by the chain via an evaluation of

$$(12.8) \qquad \int_{\mathcal{A}} f(x)\, dx \, ,$$

if $\mathcal{A}$ denotes the support of the distribution of the chain (after a given number of iterations). This is not necessarily easy, especially in large dimensions, but Philippe and Robert (2001) propose a solution, based on Riemann sums, that operates in low dimensions.

Indeed, the Riemann approximation method of Section 4.3 provides a simple convergence diagnosis since, when f is a one-dimensional density, the quantity

(12.9)
$$\sum_{t=1}^{T-1} [\theta_{(t+1)} - \theta_{(t)}]\, f(\theta_{(t)})$$

converges to 1, even when the $\theta_{(t)}$'s are not generated from the density f. (See Definition 4.8 and Proposition 4.9.) Therefore, if the chain $(\theta_{(t)})$ has failed to explore some (significant) part of the support of f, the approximation (12.9) gives a signal of non-convergence by providing an evaluation of the mass of the region explored by the chain thus far.

Example 12.7. Bimodal target. For illustration purposes, consider the density
$$f(x) = \frac{\exp{-x^2/2}}{\sqrt{2\pi}}\, \frac{4(x - .3)^2 + .01}{4(1 + (.3)^2) + .01}.$$
As shown by Figure 12.10, this distribution is bimodal and a normal random walk Metropolis–Hastings algorithm with a small variance like .04 may face difficulties moving between the two modes. As shown in Figure 12.11, the criterion (12.9) identifies the missing mass corresponding to the second mode when the chain has not yet visited that mode. Note that once the chain enters the second modal region, that is, after the 800th iteration, the criterion (12.9) converges very quickly to 1.   ‖

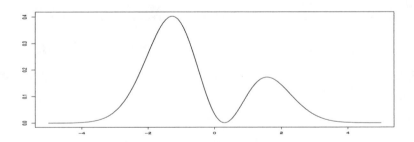

Fig. 12.10. Density $f(x) \propto \exp{-x^2/2}\,(4(x - .3)^2 + .01)$.

In multidimensional settings, it was mentioned in Section 4.3 that Riemann sums are no longer efficient estimators. However, (12.9) can still be used as a convergence assessment when Rao–Blackwellization estimates (Section 10.4.2) can give a convergent approximation to the marginal distributions. In that case, for each marginal f of interest, the Rao–Blackwellized estimate is used instead in (12.9).

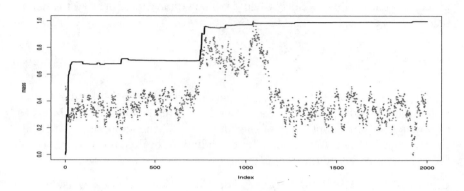

Fig. 12.11. Evolution of the missing mass criterion (12.9) and superposition of the normal random walk Metropolis–Hastings sequence, for the target density of Figure 12.10.

Example 12.8. Bivariate mixture revisited. Recall the case of the bivariate normal mixture of Example 10.18,

$$(12.10) \qquad (X,Y) \sim p\mathcal{N}_2(\mu, \Sigma) + (1-p)\mathcal{N}_2(\nu, \Sigma'),$$

where $\mu = (\mu_1, \mu_2), \nu = (\nu_1, \nu_2) \in \mathbb{R}^2$ and the covariance matrices are

$$\Sigma = \begin{pmatrix} a & c \\ c & b \end{pmatrix}, \qquad \Sigma' = \begin{pmatrix} a' & c' \\ c' & b' \end{pmatrix}.$$

In this case, the conditional distributions are also normal mixtures,

$$X|y \sim \omega_y \mathcal{N}\left(\mu_1 + \frac{(y-\mu_2)c}{b}, \frac{\det\Sigma}{b}\right) + (1-\omega_y)\mathcal{N}\left(\nu_1 + \frac{(y-\nu_2)c'}{b'}, \frac{\det\Sigma'}{b'}\right)$$

$$Y|x \sim \omega_x \mathcal{N}\left(\mu_2 + \frac{(x-\mu_1)c}{a}, \frac{\det\Sigma}{a}\right) + (1-\omega_x)\mathcal{N}\left(\nu_2 + \frac{(y-\nu_1)c'}{a'}, \frac{\det\Sigma'}{a'}\right),$$

where

$$\omega_x = \frac{p^{-1/2}\exp(-(x-\mu_1)^2/(2a))}{pa^{-1/2}\exp(-(x-\mu_1)^2/(2a)) + pa'^{-1/2}\exp(-(y-\nu_1)^2/(2a'))}$$

$$\omega_y = \frac{pb^{-1/2}\exp(-(y-\mu_2)^2/(2b))}{pb^{-1/2}\exp(-(y-\mu_2)^2/(2b)) + pb'^{-1/2}\exp(-(y-\nu_2)^2/(2b'))}.$$

They thus provide a straightforward Gibbs sampler, while the marginal distributions of X and Y are again normal mixtures,

$$X \sim p\mathcal{N}(\mu_1, a) + (1-p)\mathcal{N}(\nu_1, a')$$
$$Y \sim p\mathcal{N}(\mu_2, b) + (1-p)\mathcal{N}(\nu_2, b'),$$

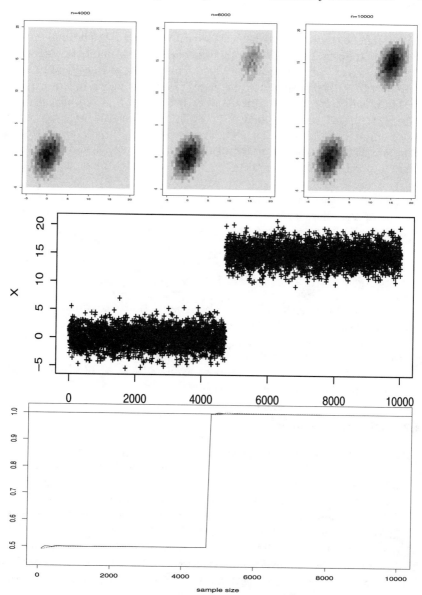

Fig. 12.12. *(top)* 2D histogram of the Markov chain of Example 12.8 after 4000, 6000 and 10,000 iterations; *(middle)* Path of the Markov chain for the first coordinate x; *(bottom)* Control curves for the bivariate mixture model, for the parameters $\mu = (0,0)$, $\nu = (15,15)$, $p = 0.5$, $\Sigma = \Sigma' = \left(\begin{smallmatrix} 3 & 1 \\ 1 & 3 \end{smallmatrix}\right)$. *(Source:* Philippe and Robert 2001.)

which can be used in the mass evaluation (12.9).

As seen in Example 10.18, when the two components of the normal mixture (12.10) are far apart, the Gibbs sampler may take a large number of iterations to jump from one component to the other. This feature is thus ideal to study the properties of the convergence diagnostic (12.9). As shown by Figure 12.12 *(top and middle)*, for the numerical values $\mu = (0,0)$, $\nu = (15,15)$, $p = 0.5$, $\Sigma = \Sigma' = \left(\begin{smallmatrix} 3 & 1 \\ 1 & 3 \end{smallmatrix}\right)$, the chain takes almost $5,000$ iterations to jump to the second component and this is exactly diagnosed by (12.9) *(bottom)*, where the evaluations (12.9) for the marginal distributions of both X and Y converge quickly to $p = 0.5$ at first and then to 1 when the second mode is visited.   ‖

A related approach is to build a nonparametric approximation of the distribution of the chain $(\theta^{(t)})$, $\hat{f}$, and to evaluate the integral (12.8) by an importance sampling argument, that is, to compute

$$\Im_T = \frac{1}{T} \sum_{t=1}^{T} \frac{f(\theta^{(t)})}{\hat{f}(\theta^{(t)})} .$$

Obviously, the convergence of $\Im_T$ to 1 also requires that f is available in closed form, inclusive of the normalizing constant. If not, multiple estimates as in Section 12.3.2 can be used. Again, using nonparametric estimates requires that the dimension of the chain be small enough.

Brooks and Gelman (1998b) propose a related assessment based on the score function, whose expectation is zero.

12.2.5 Distance Evaluations

Roberts (1992) considers convergence from a functional point of view, as in Schervish and Carlin (1992) (and only for Gibbs sampling). Using the norm induced by f defined in Section 6.6.1, he proposes an unbiased estimator of the distance $\|f_t - f\|$, where f_t denotes the marginal density of the *symmetrized* chain $\theta^{(t)}$: The symmetrized chain is obtained by adding to the steps **1.**, **2.**,...,**k.** of the Gibbs sampler the additional steps **k.**,**k-1.**,...,**1.**, as in [A.41]. This device leads to a reversible chain, creating in addition a *dual* chain $(\tilde{\theta}^{(t)})$, which is obtained by the inversion of the steps of the Gibbs sampler: Starting with $\theta^{(t)}$, $\tilde{\theta}^{(t)}$ is generated conditionally on $\theta^{(t)}$ by steps **1.**,**2.**,...,**k.**, then $\theta^{(t+1)}$ is generated conditionally on $\tilde{\theta}^{(t)}$ by steps **k.**,**k-1.**,...,**1.**.

Using m parallel chains $(\theta_\ell^{(t)})$ $(\ell = 1, \ldots, m)$ started with the same initial value $\theta^{(0)}$, Roberts (1992) shows that an unbiased estimator of $\|f_t - f\| + 1$ is

$$J_t = \frac{1}{m(m-1)} \sum_{1 \leq \ell \neq s \leq m} \frac{K_-(\tilde{\theta}_\ell^{(0)}, \theta_s^{(t)})}{f(\theta_s^{(t)})} ,$$

where K_- denotes the transition kernel for the steps **k.**,**k-1.**,...,**1.** of [A.39] (see Problem 12.6). Since the distribution f is typically only known up to a multiplicative constant, the limiting value of J_t is unknown. Thus, the diagnostic based on this unbiased estimation of $\|f_t - f\| + 1$ is to evaluate the stabilization of J_t graphically. This method requires both $K_-(\theta, \theta')$ and the normalizing constant of $K_-(\theta, \theta')$ to be known, as in the method of Ritter and Tanner (1992) .

An additional diagnostic can be derived from the convergence of

$$I_t = \frac{1}{m} \sum_{i=1}^{m} \frac{K_-\left(\tilde{\theta}_\ell^{(0)}, \theta_i^{(t)}\right)}{f\left(\theta_i^{(t)}\right)}$$

to the same limit as J_t, which is equal to 1 when the normalizing constant is available. Variations and generalizations of this method are considered by Roberts (1994), Brooks et al. (1997), and Brooks and Roberts (1998).

From a theoretical point of view, this method of evaluation of the distance to the stationary distribution is quite satisfactory, but it does not exactly meet the convergence requirements of point (ii) of Section 12.1 for several reasons:

(a) It requires parallel runs of several Gibbs sampler algorithms and thus results in a loss of efficiency in the execution (see Section 12.4).
(b) The convergence control is based on f_t, the marginal distribution of $(\theta^{(t)})$, which is typically of minor interest for Markov chain Monte Carlo purposes, when a single realization of f_t is observed, and which does not directly relate to the speed of convergence of empirical means.
(c) The computation (or the approximation) of the normalizing constant of K_- can be time-consuming and stabilizing around a mean value does not imply that the chain has explored all the modes of f (see also Section 12.4).

Liu et al. (1992) also evaluate convergence to the stationary distribution through the difference between f, the limiting distribution, and f_t, the distribution of $\theta^{(t)}$. Their method is based on an expression of the variance of $f_t(\theta)/f(\theta)$ which is close to the J_t's of Roberts (1992).

Lemma 12.9. Let $\theta_1^-, \theta_2^- \sim f_{(t-1)}$ be independent, and generate $\theta_1 \sim K(\theta_1^-, \theta_1)$, $\theta_2 \sim K(\theta_2^-, \theta_2)$. Then the quantity

$$U = \frac{f(\theta_1)}{f(\theta_2)} \frac{K(\theta_1^-, \theta_2)}{K(\theta_1^-, \theta_1)}$$

satisfies

$$\mathbb{E}_f[U] = \text{var}_f\left(\frac{f_t(\theta)}{f(\theta)}\right) + 1.$$

Proof. The independence of the θ_i^-'s implies that

$$\mathbb{E}_f\left[\frac{f(\theta_1)}{f(\theta_2)}\frac{K(\theta_1^-,\theta_2)}{K(\theta_1^-,\theta_1)}\right] = \int \frac{f(\theta_1)}{f(\theta_2)}\,K(\theta_1^-,\theta_2)\,f_{t-1}(\theta_1^-)\,d\theta_1^-\,d\theta_1 d\theta_2$$

$$= \int \frac{f_t(\theta_2)^2}{f(\theta_2)}\,d\theta_2$$

$$= 1 + \int \left(\frac{f_t(\theta_2)}{f(\theta_2)}-1\right)^2 f(\theta_2)\,d\theta_2\ ,$$

with

$$\mathbb{E}_f\left[\frac{f_t(\theta_2)}{f(\theta_2)}\right] = 1.$$

□

Given M parallel chains, each iteration t provides $M(M-1)$ values $U^{(i,j,t)}$ ($i \neq j$). These can be processed either graphically or using the method of Gelman and Rubin (1992) (see Section 12.3.4) for $M/2$ independent values $U^{(i,j,t)}$ (that is, using disjoint pairs (i,j)). Note that the computation of U does not require the computation of the normalizing constant for the kernel K.

12.3 Monitoring Convergence of Averages

12.3.1 A First Illustration

We introduce this criterion with an artificial setting where the rate of convergence can be controlled to be as slow as we wish.

Example 12.10. Pathological beta generator. As demonstrated in Problem 7.5 (Algorithm [A.30]), a Markov chain $(X^{(t)})$ such that

(12.11) $$X^{(t+1)} = \begin{cases} Y \sim \mathcal{B}e(\alpha+1,1) & \text{with probability } x^{(t)} \\ x^{(t)} & \text{otherwise} \end{cases}$$

is associated with the stationary distribution

$$f(x) \propto x^{\alpha+1-1}/\{1-(1-x)\} = x^{\alpha-1}\ ,$$

that is, the beta distribution $\mathcal{B}e(\alpha,1)$. (Note that a Metropolis–Hastings algorithm based on the same pair (f,g) would lead to accept y with probability $x^{(t)}/y$, which is larger than $x^{(t)}$ but requires a simulation of y before deciding on its acceptance.)

This scheme has interesting theoretical properties in that it produces a Markov chain that does not satisfy the assumptions of the Central Limit Theorem. (Establishing this is beyond our scope. See, for example, Doukhan et al. 1994.) A consequence of this is the "slow" convergence of empirical

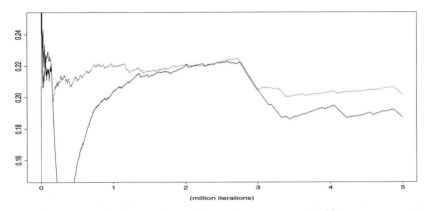

Fig. 12.13. Convergence of the empirical average of $h(x^{(t)})$ to $\alpha = 0.2$ for the algorithm (12.11) (solid lines) and the Metropolis–Hastings algorithm (dotted lines).

averages, since they are not controlled by the CLT. For instance, for $h(x) = x^{1-\alpha}$, Figure 12.13 describes the evolution of the empirical average of the $h(x^{(t)})$'s to α (see Problem 12.4), a convergence that is guaranteed by the ergodic theorem (Proposition 6.63). As it obvious from the curve, convergence is not achieved after 5×10^6 iterations. Figure 12.13 also provides a comparison with a realization of the Metropolis–Hastings algorithm based on the same pair. Both graphs correspond to identical beta generations of y_t in (12.11), with a higher probability of acceptance for the Metropolis–Hastings algorithm. The graph associated with the Metropolis–Hastings algorithm, while more regular that the graph for the transition (12.11), is also very slow to converge. The final value is still closer to the exact expectation $\alpha = 0.20$.

A simple explanation (due to Jim Berger) of the bad behavior of both algorithms is the considerable difference between the distributions $\mathcal{Be}(0.2, 1)$ and $\mathcal{Be}(1.2, 1)$. In fact, the ratio of the corresponding cdf's is $x^{0.2}/x^{1.2} = 1/x$ and, although the quantile at level 10% is 10^{-5} for $\mathcal{Be}(0.2, 1)$, the probability of reaching the interval $[0, 10^{-5}]$ under the $\mathcal{Be}(1.2, 1)$ distribution is 10^{-6}. Therefore, the number of simulations from $\mathcal{Be}(1.2, 1)$ necessary to obtain an adequate coverage of this part of the support of $\mathcal{Be}(0.2, 1)$ is enormous. Moreover, note that the probability of leaving the interval $[0, 10^{-5}]$ using [A.30] is less than 10^{-6}, and it is of the same order of magnitude for the corresponding Metropolis–Hastings algorithm. ‖

Thus, as in the first case, graphical outputs can detect obvious problems of convergence of the empirical average. To proceed further in this evaluation, Yu (1994) and Yu and Mykland (1998) propose to use cumulative sums (CUSUM), graphing the partial differences

$$(12.12) \qquad D_T^i = \sum_{t=1}^{i} [h(\theta^{(t)}) - S_T], \qquad i = 1, \dots, T,$$

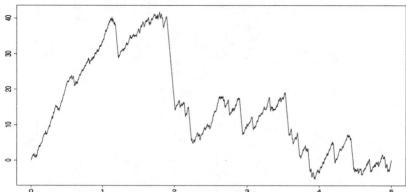

Fig. 12.14. Evolution of the D_t^i criterion for the chain produced in Example 12.10 and for $\alpha = 0.2$ and $h(x) = x^{0.8}$ (first axis millions of iterations).

where

$$(12.13) \qquad\qquad S_T = \frac{1}{T} \sum_{t=1}^{T} h(\theta^{(t)})$$

is the final average. These authors derive a qualitative evaluation of the mixing speed of the chain and the correlation between the $\theta^{(t)}$'s: When the mixing of the chain is high, the graph of D_T^i is highly irregular and concentrated around 0. Slowly mixing chains (that is, chains with a slow pace of exploration of the stationary distribution) produce regular graphs with long excursions away from 0. Figure 12.14 contains the graph which corresponds to the dataset of Example 12.10, exhibiting a slow convergence behavior already indicated in Figure 12.13. (See Brooks 1998c for a more quantitative approach to CUSUM M.)

Figure 12.15 provides the CUSUM analysis of the MCMC sample of Example 12.1 for the parameters $\beta^{(t)}$, $\sigma^{(t)}$ and $\beta^{(t)}/\sigma^{(t)}$. Even more than in the raw-plots, the two first parameters do not exhibit a strong stationarity, since the CUSUM plots are very regular and spend the entire sequence above 0. We can also note the strong similarity between the CUSUM plots of $\beta^{(t)}$ and $\sigma^{(t)}$, due to the fact that only $\beta^{(t)}/\sigma^{(t)}$ is identifiable from the model. On the other hand, and as expected, the CUSUM plot for $\beta^{(t)}/\sigma^{(t)}$ is more in agreement with a Brownian motion sequence.

Similarly, for the case of the mixture over the means of Example 9.2, Figure 12.16 exhibits a situation when the Gibbs sampler cannot escape the attraction of the local (and much lower) mode. The corresponding CUSUMs for both means μ_1 and μ_2 in Figure 12.17 are nonetheless close enough to the golden standard of the Brownian motion. This difficulty is common to most "on-line" methods; that is, to diagnoses based on a *single* chain. It is almost impossible to detect the existence of other modes of f (or of other unexplored regions of the space with positive probability). (Exceptions are the Riemann

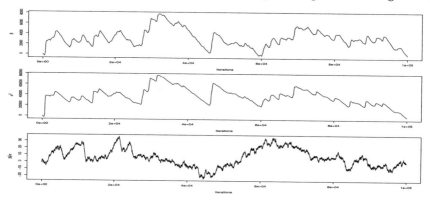

Fig. 12.15. Evolution of the CUSUM criterion D_t^i for the chains $\beta^{(t)}$ (top), $\sigma^{(t)}$ (middle) and $\beta^{(t)}/\sigma^{(t)}$ (bottom) of Example 12.1.

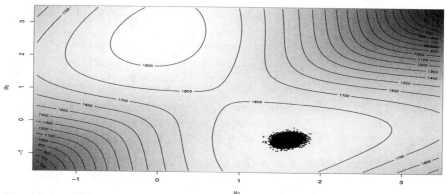

Fig. 12.16. Gibbs sample for the two mean mixture model, when initialized close to the second and lower mode, for true values $\mu_1 = 0$ and $\mu_2 = 2.5$, over the log-likelihood surface.

sum indicator (12.9) which "estimates" 1 and can detect probability losses in the region covered by the sample, and the related method developed in Brooks 1998a.)

12.3.2 Multiple Estimates

In most cases, the graph of the raw sequence $(\theta^{(t)})$ is unhelpful in the detection of stationarity or convergence. Indeed, it is only when the chain has explored different regions of the state-space during the observation time that lack of stationarity can be detected. (Gelman and Rubin 1992 also illustrate the use of the raw sequence graphs when using chains in parallel with quite distinct supports.)

Given some quantities of interest $\mathbb{E}_f[h(\theta)]$, a more helpful indicator is the behavior of the averages (12.13) in terms of T. A necessary condition

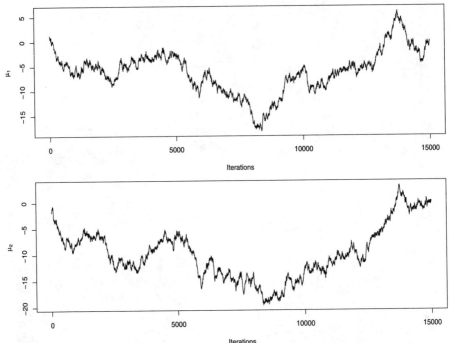

Fig. 12.17. CUSUMS for both means μ_1 and μ_2 corresponding to the sample of Figure 12.16.

for convergence is the stationarity of the sequence (S_T), even though the stabilizing of this sequence may correspond to the influence of a single mode of the density f, as shown by Gelman and Rubin (1992).

Robert (1995a) proposes a more robust approach to graphical convergence monitoring, related to control variates (Section 4.4.2). The idea is to simultaneously use several convergent estimators of $\mathbb{E}_f[h(\theta)]$ based on the same chain $(\theta^{(t)})$, until all estimators coincide (up to a given precision). The most common estimation techniques in this setup are, besides the empirical average S_T, the *conditional* (or *Rao–Blackwellized*) version of this average, either in its nonparametric version (Section 7.6.2) for Metropolis–Hastings algorithms, or in its parametric version (Section 9.3) for Gibbs sampling,

$$S_T^C = \frac{1}{T} \sum_{t=1}^{T} \mathbb{E}[h(\theta)|\eta^{(t)}] \,,$$

where $(\eta^{(t)}, \theta^{(t)})$ is the Markov chain produced by the algorithm.

A second technique providing convergent estimators is *importance sampling* (Section 3.3). If the density f is available up to a constant,[6] the impor-

---

[6] As it is, for instance, in all cases where the Gibbs sampler applies, as shown by the Hammersley–Clifford theorem (Section 10.1.3).

tance sampling alternative is

$$S_T^P = \sum_{t=1}^{T} w_t \, h(\theta^{(t)}) \, ,$$

where $w_t \propto f(\theta^{(t)})/g_t(\theta^{(t)})$ and g_t is the true density used for the simulation of $\theta^{(t)}$. In particular, in the case of Gibbs sampling,

$$g_t(\theta^{(t)}) \propto f_1(\theta_1^{(t)}|\theta_2^{(t-1)}, \ldots, \theta_k^{(t-1)}) \cdots f_k(\theta_k^{(t)}|\theta_1^{(t)}, \ldots, \theta_{k-1}^{(t)}) \, .$$

If, on the other hand, the chain $(\theta^{(t)})$ is produced by a Metropolis–Hastings algorithm, the variables actually simulated, $\eta^{(t)} \sim q(\eta|\theta^{(t-1)})$, can be recycled through the estimator

$$S_T^{MP} = \sum_{t=1}^{T} w_t \, h(\eta^{(t)})$$

with $w_t \propto f(\eta^{(t)})/q(\eta^{(t)}|\theta^{(t-1)})$.

The variance decomposition

$$(12.14) \qquad \operatorname{var}\left(\sum_t w_t X_t\right) = \sum_t \operatorname{var}(w_t X_t)$$

does not directly apply to the estimators S_T^P and S_T^{MP} since the weights w_t are known only up to a multiplicative constant and are, therefore, normalized by the inverse of their sum. However, it can be assumed that the effect of this normalization on the correlation vanishes when T is large (Problem 12.14).

More importantly, importance sampling removes the correlation between the terms in the sum. In fact, whether or not the $\theta^{(t)}$s are correlated, the importance-weighted terms will always be uncorrelated. So, for importance sampling estimators, the variance of the sum will equal the sum of the variances of the individual terms.

Lemma 12.11. *Let $(Z^{(t)})$ a Markov chain with transition kernel q. Then*

$$\operatorname{var}\left(\sum_{t=1}^{T} h(Z^{(t)}) \frac{f(Z^{(t)})}{q(Z^{(t)}|Z^{(t-1)})}\right) = \sum_{t=1}^{T} \operatorname{var}\left(h(Z^{(t)}) \frac{f(Z^{(t)})}{q(Z^{(t)}|Z^{(t-1)})}\right) ,$$

provided these quantities are well defined.

Proof. Assume, without loss of generality, that $\mathbb{E}_f[h(Z)] = 0$. If we define $w_t = f(Z^{(t)})/q(Z^{(t)}|Z^{(t-1)})$, the covariance between $w_t h(Z^{(t)})$ and $w_{t+r} h(Z^{(t+r)})$ is

$$\mathbb{E}\left[w_t h(Z^{(t)}) w_{t+r} h(Z^{(t+r)})\right]$$
$$= \mathbb{E}\left[\mathbb{E}\left(w_t h(Z^{(t)})|Z^{(t+r-1)}\right) \mathbb{E}\left(w_{t+r} h(Z^{(t+r)})|Z^{(t+r-1)}\right)\right],$$

where we have iterated the expectation and used the Markov property of conditional independence. The second conditional expectation is

$$\mathbb{E}\left(w_{t+r}h(Z^{(t+r)})|Z^{(t+r-1)}\right) = \int h(x)\frac{f(x)}{q(x|Z^{(t+r-1)})}\,q(x|Z^{(t+r-1)})\,dx$$
$$= \mathbb{E}_f h(X) = 0,$$

showing that the covariance is zero. $\square$

The consequences of (12.14) on convergence assessment are twofold. First, they indicate that, up to second order, S_T^P and S_T^{MP} behave as in the independent case, and thus allow for a more traditional convergence control on these quantities. Second, (12.14) implies that the variance of S_T^P (or of S_T^{MP}), when it exists, decreases at speed $1/T$ in *stationarity settings*. Thus, non-stationarity can be detected if the decrease of the variations of S_T^P does not fit in a confidence parabola of order $1/\sqrt{T}$. Note also that the density f can sometimes be replaced (in w_t) by an approximation. In particular, in settings where Rao–Blackwellization applies,

$$(12.15) \qquad\qquad \hat{f}_T(\theta) = \frac{1}{T}\sum_{t=1}^{T} f_1(\theta|\eta^{(t)})$$

provides an *unbiased estimator* of f (see Wei and Tanner 1990a and Tanner 1996). A parallel chain $(\eta^{(t)})$ should then be used to ensure the independence of $\hat{f}_T$ and of $h(\theta^{(t)})$.

The fourth estimator based on $(\theta^{(t)})$ is the *Riemann approximation* (Section 4.3); that is,

$$(12.16) \qquad\qquad S_T^R = \sum_{t=1}^{T-1} [\theta_{(t+1)} - \theta_{(t)}]\, h(\theta_{(t)})\, f(\theta_{(t)}) ,$$

which estimates $\mathbb{E}_f[h(\theta)]$, where $\theta_{(1)} \leq \cdots \leq \theta_{(T)}$ denotes the *ordered chain* $(\theta^{(t)})_{1\leq t\leq T}$. This estimator is mainly studied in the iid case (see Proposition 4.9), but it can, nonetheless, be included as an alternative estimator in the present setup, since its performances tends to be superior to those of the previous estimators.

The main drawback of S_T^R lies in the unidimensionality requirement, the quality of multidimensional extensions of the Riemann approximation decreasing quickly with the dimension (see Yakowitz et al. 1978). When h involves only one component of θ, the marginal distribution of this component should therefore be used or replaced with an approximation like (12.15), which increases the computation burden, and is not always available. For extensions and alternatives, see Robert (1995a) and Philippe and Robert (2001).

Example 12.12. Cauchy posterior distribution. For the posterior distribution of Example 7.18,

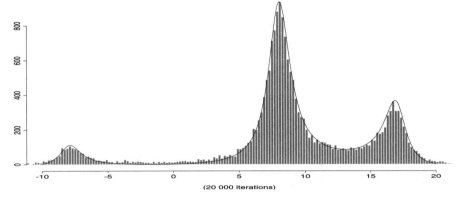

Fig. 12.18. Comparison of the density (12.17) and the histogram from a sample of 20,000 points simulated by Gibbs sampling, for $x_1 = -8$, $x_2 = 8$, $x_3 = 17$, and $\sigma = 50$.

(12.17)
$$\pi(\theta|x_1, x_2, x_3) \propto e^{-\theta^2/2\sigma^2} \big[(1 + (\theta - x_1)^2)$$
$$\times\ (1 + (\theta - x_2)^2)(1 + (\theta - x_3)^2)\big]^{-1},$$

a Gibbs sampling algorithm which approximates this distribution can be derived by demarginalization, namely by introducing three artificial variables, $\eta_1, \eta_2,$ and η_3, such that

$$\pi(\theta, \eta_1, \eta_2, \eta_3|x_1, x_2, x_3) \propto e^{-\theta^2/2\sigma^2}\, e^{-(1+(\theta-x_1)^2)\eta_1/2}$$
$$\times\, e^{-(1+(\theta-x_2)^2)\eta_2/2}\, e^{-(1+(\theta-x_3)^2)\eta_3/2}.$$

In fact, similar to the t distribution (see Example 10.1), expression (12.17) appears as the marginal distribution of $\pi(\theta, \eta_1, \eta_2, \eta_3|x_1, x_2, x_3)$ and the conditional distributions of the Gibbs sampler $(i = 1, 2, 3)$

$$\eta_i|\theta, x_i \sim \mathcal{E}xp\left(\frac{1 + (\theta - x_i)^2}{2}\right),$$
$$\theta|x_1, x_2, x_3, \eta_1, \eta_2, \eta_3 \sim \mathcal{N}\left(\frac{\eta_1 x_1 + \eta_2 x_2 + \eta_3 x_3}{\eta_1 + \eta_2 + \eta_3 + \sigma^{-2}}, \frac{1}{\eta_1 + \eta_2 + \eta_3 + \sigma^{-2}}\right),$$

are easy enough to simulate. Figure 12.18 illustrates the efficiency of this algorithm by exhibiting the agreement between the histogram of the simulated $\theta^{(t)}$'s and the true posterior distribution (12.17).

For simplicity's sake, we introduce

$$\mu(\eta_1, \eta_2, \eta_3) = \frac{\eta_1 x_1 + \eta_2 x_2 + \eta_3 x_3}{\eta_1 + \eta_2 + \eta_3 + \sigma^{-2}}$$

and

$$\tau^{-2}(\eta_1, \eta_2, \eta_3) = \eta_1 + \eta_2 + \eta_3 + \sigma^{-2}.$$

If the function of interest is $h(\theta) = \exp(-\theta/\sigma)$ (with known σ), the different approximations of $\mathbb{E}_\pi[h(\theta)]$ proposed above can be derived. The conditional version of the empirical mean is

$$S_T^C = \frac{1}{T} \sum_{t=1}^{T} \exp\left\{-\mu\left(\eta_1^{(t)}, \eta_2^{(t)}, \eta_3^{(t)}\right) + \tau^2\left(\eta_1^{(t)}, \eta_2^{(t)}, \eta_3^{(t)}\right)/2\right\},$$

importance sampling is associated with the weights

$$w_t \propto \frac{\exp\left\{-\dfrac{(\theta^{(t)})^2}{2\sigma^2} + \dfrac{\left(\theta^{(t)} - \mu\left(\eta_1^{(t)}, \eta_2^{(t)}, \eta_3^{(t)}\right)\right)^2}{2\tau^2\left(\eta_1^{(t)}, \eta_2^{(t)}, \eta_3^{(t)}\right)}\right\}}{\tau\left(\eta_1^{(t)}, \eta_2^{(t)}, \eta_3^{(t)}\right) \prod_{i=1}^{3}[1 + (x_i - \theta^{(t)})^2]},$$

and the Riemann approximation is

$$S_T^R = \frac{\displaystyle\sum_{t=1}^{T-1}(\theta_{(t+1)} - \theta_{(t)})\, e^{-\theta_{(t)}/\sigma - \theta_{(t)}^2/(2\sigma^2)} \prod_{i=1}^{3}[1 + (x_i - \theta_{(t)})^2]^{-1}}{\displaystyle\sum_{t=1}^{T-1}(\theta_{(t+1)} - \theta_{(t)})\, e^{-\theta_{(t)}^2/(2\sigma^2)} \prod_{i=1}^{3}[1 + (x_i - \theta_{(t)})^2]^{-1}},$$

where $\theta_{(1)} \leq \cdots \leq \theta_{(T)}$ denotes the ordered sample of the $\theta^{(t)}$'s.

Figure 12.19 graphs the convergence of the four estimators versus T. As is typical, S_T and S_T^C are similar almost from the start (and, therefore, their difference cannot be used to diagnose [lack of] convergence). The strong stability of S_T^R, which is the central value of S_T and S_T^C, indicates that convergence is achieved after a small number of iterations. On the other hand, the behavior of S_T^P indicates that importance sampling does not perform satisfactorily in this case and is most likely associated with an infinite variance. (The ratio $f(\theta)/f_1(\theta|\eta)$ corresponding to the Gibbs sampling usually leads to an infinite variance (see Section 3.3.2), since f is the marginal of $f_1(\theta|\eta)$ and has generally fatter tails than f_1. For example, this is the case when f is a t density and $f_1(\cdot|\eta)$ is a normal density with variance η.)   ‖

Example 12.13. (Continuation of Example 12.10) The chain $(X^{(t)})$ produced by (12.11) allows for a conditional version when $h(x) = x^{1-\alpha}$ since

$$\mathbb{E}\left[\left(X^{(t+1)}\right)^{1-\alpha}\Big|x^{(t)}\right] = \left(1 - x^{(t)}\right)\left(x^{(t)}\right)^{1-\alpha} + x^{(t)}\,\mathbb{E}[Y^{1-\alpha}]$$

$$= \left(1 - x^{(t)}\right)\left(x^{(t)}\right)^{1-\alpha} + x^{(t)}\,\frac{1+\alpha}{2}.$$

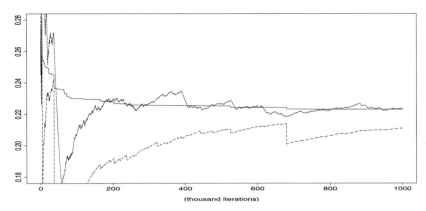

(thousand iterations)

Fig. 12.19. Convergence of four estimators S_T (solid line), S_T^C (dotted line), S_T^R (mixed) and S_T^P (long dashes, below) of the expectation under (12.17) of $h(\theta) = \exp(-\theta/\sigma)$, for $\sigma^2 = 50$ and $(x_1, x_2, x_3) = (-8, 8, 17)$. The graphs of S_T and S_T^C are identical. The final values are 0.845, 0.844, 0.828, and 0.845 for S_T, S_T^C, S_T^P, and S_T^R respectively.

The importance sampling estimator based on the simulated $y^{(t)}$'s is

$$S_T^P = \sum_{t=1}^{T} \frac{(y^{(t)})^{\alpha-1}}{(y^{(t)})^\alpha}(y^{(t)})^{1-\alpha} \bigg/ \sum_{t=1}^{T} \frac{(y^{(t)})^{\alpha-1}}{(y^{(t)})^\alpha}$$

$$= \sum_{t=1}^{T} (y^{(t)})^{-\alpha} \bigg/ \sum_{t=1}^{T} (y^{(t)})^{-1}$$

and the Riemann approximation based on the same $y^{(t)}$'s is

$$S_T^R = \sum_{t=1}^{T-1} \left(y_{(t+1)} - y_{(t)}\right) \bigg/ \sum_{t=1}^{T-1} \left(y_{(t+1)} - y_{(t)}\right) y_{(t)}^{\alpha-1}$$

$$= \frac{y_{(T)} - y_{(1)}}{\sum_{t=1}^{T} \left(y_{(t+1)} - y_{(t)}\right) y_{(t)}^{\alpha-1}}.$$

Figure 12.20 gives an evaluation of these different estimators, showing the incredibly slow convergence of (12.1), since this experiment uses 1 million iterations. Note that the importance sampling estimator S_T^P approaches the true value $\alpha = 0.2$ faster than S_T and S_T^R (S_T^C being again indistinguishable from S_T), although the repeated jumps in the graph of S_T^P do not indicate convergence. Besides, eliminating the first 200, 000 iterations of the chain does not stabilize convergence, which shows that the very slow mixing of the algorithm is responsible for this phenomenon rather than a lack of convergence to stationarity. ‖

Both examples highlight the limitations of the method of multiple estimates:

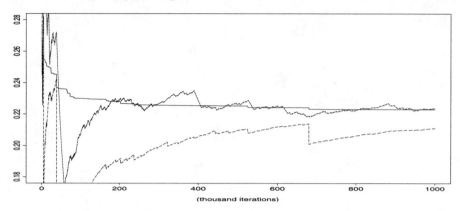

Fig. 12.20. Convergence of four estimators S_T (solid line), S_T^C (dotted line), S_T^R (mixed dashes) and S_T^P (long dashes, below) of $\mathbb{E}[(X^{(t)})^{0.8}]$ for the $\mathcal{Be}(0.2, 1)$ distribution, after elimination of the first $200,000$ iterations. The graphs of S_T and S_T^C are identical. The final values are 0.224, 0.224, 0.211, and 0.223 for S_T, S_T^C, S_T^R, and S_T^P respectively, to compare with a theoretical value of 0.2.

(1) The method does not always apply (parametric Rao–Blackwellization and Riemann approximation are not always available).
(2) It is intrinsically conservative (since the speed of convergence is determined by the slower estimate).
(3) It cannot detect missing modes, except when the Riemann approximation can be implemented.
(4) It is often the case that Rao–Blackwellization cannot be distinguished from the standard average and that importance sampling leads to an infinite variance estimator.

12.3.3 Renewal Theory

As mentioned in Section 12.2.3, it is possible to implement small sets as a theoretical control, although their applicability is more restricted than for other methods. Recall that Theorem 6.21, which was established by Asmussen (1979), guarantees the existence of small sets for ergodic Markov chains. Assuming $(\theta^{(t)})$ is strongly aperiodic, there exist a set C, a positive number $\varepsilon < 1$, and a probability measure ν such that

$$(12.18) \qquad P(\theta^{(t+1)} \in B|\theta^{(t)}) \geq \varepsilon\nu(B)\,\mathbb{I}_C(\theta^{(t)})\,.$$

Recall also that the *augmentation technique* of Athreya and Ney (1978) consists in a modification of the transition kernel K into a kernel associated with the transition

$$\theta^{(t+1)} = \begin{cases} \xi \sim \nu & \text{with probability } \varepsilon \\ \xi \sim \dfrac{K(\theta^{(t)}, \xi) - \varepsilon\nu(\xi)}{1 - \varepsilon} & \text{with probability } (1 - \varepsilon), \end{cases}$$

when $\theta^{(t)} \in C$. Thus, every passage in C results in an independent generation from ν with probability ε. Denote by $\tau_C(k)$ the time of the $(k+1)$st visit to C associated with an independent generation from ν; $(\tau_C(k))$ is then a sequence of renewal times. Therefore, the partial sums

$$S_k = \sum_{t=\tau_C(k)+1}^{\tau_C(k+1)} h(\theta^{(t)})$$

are iid in the stationary regime. Under conditions (6.31) in Proposition 6.64, it follows that the S_k's satisfy a version of the Central Limit Theorem,

$$\frac{1}{\sqrt{K}} \sum_{k=1}^{K} (S_k - \mu_C \, \mathbb{E}_f[h(\theta)]) \xrightarrow{\mathcal{L}} \mathcal{N}(0, \sigma_C^2) \,,$$

with K the number of renewals out of T iterations and $\mu_C = \mathbb{E}_f[\tau_C(2) - \tau_C(1)]$. Since these variables are iid, it is even possible to estimate σ_C^2 by the usual estimator

$$(12.19) \qquad \tilde{\sigma}_C^2 = \frac{1}{K} \sum_{k=1}^{K} \left(S_k - \frac{\lambda_k}{\tau_C(K)} \sum_{\ell=1}^{K} S_\ell \right)^2 \,,$$

where $\lambda_k = \tau_C(k+1) - \tau_C(k)$.

These probabilistic results, although elementary, have nonetheless a major consequence on the monitoring of MCMC algorithms since they lead to an estimate of the limit variance γ_h^2 of the Central Limit Theorem (Proposition 6.64), as well as a stopping rule for these algorithms. This result thus provides an additional criterion for convergence assessment, since, *for different small sets C, the ratios $K_C \tilde{\sigma}_C^2 / T$ must converge to the same value*. Again, this criterion is not foolproof since it requires sufficiently dispersed small sets.

Proposition 12.14. *When the Central Limit Theorem applies to the Markov chain under study, the variance γ_h^2 of the limiting distribution associated with the sum (12.1) can be estimated by*

$$K_C \tilde{\sigma}_C^2 / T \,,$$

where K_C is the number of renewal events before time T and $\tilde{\sigma}_C^2$ is given by (12.19).

Proof. The Central Limit Theorem result

$$\frac{1}{\sqrt{T}} \sum_{t=1}^{T} (h(\theta^{(t)}) - \mathbb{E}_f[h(\theta)]) \xrightarrow{\mathcal{L}} \mathcal{N}(0, \gamma_h^2)$$

can be written

$$\frac{1}{\sqrt{T}} \sum_{t=1}^{\tau_C(1)} (h(\theta^{(t)}) - \mathbb{E}_f[h(\theta)]) + \frac{1}{\sqrt{T}} \sum_{k=1}^{K_C} (S_k - \lambda_k \, \mathbb{E}_f[h(\theta)])$$

$$(12.20) \qquad + \frac{1}{\sqrt{T}} \sum_{t=\tau_C(K_C)+1}^{K} (h(\theta^{(t)}) - \mathbb{E}_f[h(\theta)]) \xrightarrow{\mathcal{L}} \mathcal{N}(0, \gamma_h^2).$$

Under the conditions

$$\mathbb{E}_f[\lambda_1^2] < \infty \qquad \text{and} \qquad \mathbb{E}_f\left[\left(\sum_{t=1}^{\tau_C(1)} h(\theta^{(t)})\right)^2\right] < \infty \, ,$$

which are, in fact, the sufficient conditions (6.31) for the Central Limit Theorem (see Proposition 6.64), the first and the third terms of (12.20) almost surely converge to 0. Since

$$\frac{1}{\sqrt{K_C}} \sum_{k=1}^{K_C} (S_k - \lambda_k \, \mathbb{E}_f[h(\theta)]) \xrightarrow{\mathcal{L}} \mathcal{N}(0, \sigma_C^2)$$

and T/K_C almost surely converges to μ_C, the average number of excursions between two passages in C, it follows that

$$\frac{1}{\sqrt{T}} \sum_{k=1}^{K_C} (h(\theta^{(t)}) - \lambda_k \, \mathbb{E}_f[h(\theta)]) \xrightarrow{\mathcal{L}} \mathcal{N}\left(0, \frac{\sigma_C^2}{\mu_C}\right) ,$$

which concludes the proof.   □

The fundamental factor in this monitoring method is the triplet (C, ε, ν) which must be determined for every application and which (strongly) depends on the problem at hand. Moreover, if the renewal probability $\varepsilon\pi(C)$ is too small, this approach is useless as a practical control device. (See Section 6.7.2.3 and Hobert et al. 2002 for further details.)

Latent variable and missing data setups often lead to this kind of difficulty as ε decreases as a power of the number of observations. The same problem often occurs in high-dimensional problems (see Gilks et al. 1998). Mykland et al. (1995) eliminate this difficulty by modifying the algorithm, as detailed in Section 12.3.3. If we stick to the original version of the algorithm, the practical construction of (C, ε, ν) seems to imply a detailed study of the chain $(\theta^{(t)})$ and of its transition kernel. Examples 10.1 and 10.17 in Section 10.2.1 have, however, shown that this study can be conducted for some Gibbs samplers and we see below how a more generic version can be constructed for an important class of problems.

Example 12.15. (Continuation of Example 12.12) The form of the posterior distribution (12.17) suggests using a small set C equal to $[r_1, r_2]$ with

$x_2 \in [r_1, r_2]$, $x_1 < r_1$, and $x_3 > r_2$ (if $x_1 < x_2 < x_3$). This choice leads to the following bounds:

$$\rho_{11} = r_1 - x_1 < |\theta - x_1| < \rho_{12} = r_1 - x_1,$$

$$0 < |\theta - x_2| < \rho_{22} = \max(r_2 - x_2, x_2 - r_1),$$

$$\rho_{31} = x_3 - r_2 < |\theta - x_3| < \rho_{32} = x_3 - r_1,$$

for $\theta \in C$, which give ε and ν. In fact,

$$
\begin{aligned}
K(\theta, \theta') &\geq \int_{\mathbb{R}^3_+} \frac{1}{\sqrt{2\pi}} \, \tau(\eta_1, \eta_2, \eta_3)^{-1} \\
&\quad \times \exp\left\{-(\theta' - \mu(\eta_1, \eta_2, \eta_3))^2 / (2\tau^2(\eta_1, \eta_2, \eta_3))\right\} \\
&\quad \times \frac{1 + \rho_{11}^2}{2} \exp\{-(1 + \rho_{12}^2)\eta_1/2\} \frac{1}{2} \exp\{-(1 + \rho_{22}^2)\eta_2/2\} \\
&\quad \times \frac{1 + \rho_{31}^2}{2} \exp\{-(1 + \rho_{32}^2)\eta_3/2\} \, d\eta_1 \, d\eta_2 \, d\eta_3 \\
&= \frac{1 + \rho_{11}^2}{1 + \rho_{12}^2} \frac{1}{1 + \rho_{22}^2} \frac{1 + \rho_{31}^2}{1 + \rho_{32}^2} \, \nu(\theta') = \varepsilon\nu(\theta') \,,
\end{aligned}
$$

where ν is the marginal density of θ for

$$
\begin{aligned}
(\theta, \eta_1, \eta_2, \eta_3) &\sim \mathcal{N}\left(\mu(\eta_1, \eta_2, \eta_3), \tau^2(\eta_1, \eta_2, \eta_3)\right) \\
&\quad \times \mathcal{E}xp\left(\frac{1 + \rho_{12}^2}{2}\right) \mathcal{E}xp\left(\frac{1 + \rho_{22}^2}{2}\right) \mathcal{E}xp\left(\frac{1 + \rho_{32}^2}{2}\right).
\end{aligned}
$$

One can thus evaluate the frequency $1/\mu_C$ of visits to C and calibrate r_1 and r_2 to obtain optimal renewal probabilities.   ‖

Another difficulty with this method is that the splitting technique requires the generation of variables distributed as

(12.21)
$$R(\theta^{(t)}, \xi) = \frac{K(\theta^{(t)}, \xi) - \varepsilon\nu(\xi)}{1 - \varepsilon},$$

called the *residual kernel*. It is actually enough to know the ratio $K(\theta^{(t)}, \xi)/\varepsilon\nu(\xi)$ since, following from Lemma 2.27, the algorithm

Algorithm A.53 –Negative Weight Mixture Simulation–

1. Simulate $\xi \sim K(\theta^{(t)}, \xi)$.

2. Accept ξ with probability $1 - \dfrac{\varepsilon\nu(\xi)}{K(\theta^{(t)}, \xi)}$,   [A.53]

   otherwise go to 1.

provides simulations from the distribution (12.21). Since the densities $K(\theta, \xi)$ and ν are in general unknown, it is necessary to use approximations, as in Section 12.2.5.

Example 12.16. (Continuation of Example 12.15) The integral forms of $K(\theta, \xi)$ and of $\nu(\theta)$ (that is, their representation as marginals of other distributions) allow for a Rao–Blackwellized approximation. Indeed,

$$\hat{K}(\xi) = \frac{1}{M} \sum_{i=1}^{M} \varphi(\xi | \eta_1^i, \eta_2^i, \eta_3^i)$$

and

$$\hat{\nu}(\xi) = \frac{1}{M} \sum_{i=1}^{M} \varphi(\xi | \tilde{\eta}_1^i, \tilde{\eta}_2^i, \tilde{\eta}_3^i) \;,$$

with

$$\varphi(\xi | \eta_1, \eta_2, \eta_3)$$
$$= \exp\left\{ -\left(\xi - \mu(\eta_1, \eta_2, \eta_3)\right)^2 / (2\tau^2(\eta_1, \eta_2, \eta_3)) \right\} \tau^{-1}(\eta_1, \eta_2, \eta_3)$$

and

$$\eta_j^i \sim \mathcal{E}xp\left(\frac{1 + (\xi - x_j)^2}{2} \right), \quad \tilde{\eta}_j^i \sim \mathcal{E}xp\left(\frac{1 + \rho_{j2}^2}{2} \right) \quad (j = 1, 2, 3)$$

provide convergent (in M) approximations of $K(\theta, \xi)$ and $\nu(\xi)$, respectively, by the usual Law of Large Numbers and, therefore, suggest using the approximation

$$\frac{K(\theta, \xi)}{\nu(\xi)} \sim \frac{\sum_{i=1}^{M} \varphi(\xi | \eta_1^i, \eta_2^i, \eta_3^i)}{\sum_{i=1}^{M} \varphi(\xi | \tilde{\eta}_1^i, \tilde{\eta}_2^i, \tilde{\eta}_3^i)} \;.$$

The variables η_j^i must be simulated at every iteration of the Gibbs sampler, whereas the variables $\tilde{\eta}_j^i$ can be simulated only once when starting the algorithm.

Table 12.1 gives the performances of the diagnostic for different values of r, when $C = [x_2 - r, x_2 + r]$. The decrease of the bound ε as a function of r is quite slow and indicates a high renewal rate. It is thus equal to 11% for $r = 0.54$. The stabilizing of the variance γ_h^2 occurs around of 1160 for $h(x) = x$, but the number of simulations before stabilizing is huge, since one million iterations do not provide convergence to the same quantity, except for values of r between 0.32 and 0.54.   ‖

Mykland et al. (1995) provide a simple and clever way of avoiding the residual kernel altogether. Suppose that $\theta^{(t)} \in C$. Instead of drawing $\delta_t \sim \mathcal{B}(\varepsilon)$ and then simulating $\theta^{(t+1)}$ conditional on $(\delta_t, \theta^{(t)})$ (which entails simulation

r	0.1	0.21	0.32	0.43	0.54	0.65	076	0.87	0.98	1.09
ε	0.92	0.83	0.73	0.63	0.53	0.45	0.38	0.31	0.26	0.22
μ_C	25.3	13.9	10.5	9.6	8.8	9.6	9.8	10.4	11.4	12.7
γ_h^2	1135	1138	1162	1159	1162	1195	1199	1149	1109	1109

Table 12.1. Estimation of the asymptotic variance γ_h^2 for $h(\theta) = \theta$ and renewal control for $C = [x_2 - r, x_2 + r]$ with $x_1 = -8$, $x_2 = 8$, $x_3 = 17$, and $\sigma^2 = 50$ (1 million simulations).

from $R(\theta^{(t)}, \cdot)$ if $\delta_t = 0$), one first simulates $\theta^{(t+1)}$ conditional on $\theta^{(t)}$ (in the usual way) and then simulates δ_t conditional on $(\theta^{(t)}, \theta^{(t+1)})$. Calculation of $\Pr(\delta_t = 1 | \theta^{(t)}, \theta^{(t+1)})$ is straightforward if the densities K and ν are known (Problem 12.11).

The previous example shows how *conservative* this method can be and, thus, how many iterations it requires. In addition, the method does not guarantee convergence since the small sets can always omit important parts of the support of f.

A particular setup where the renewal method behaves quite satisfactorily is the case of Markov chains with *finite state-space*. In fact, the choice of a small set is then immediate: If Θ can be written as $\{i : i \in I\}$, $C = \{i_0\}$ is a small set, with i_0 the modal state under the stationary distribution $\tilde{\pi} = (\pi_i)_{i \in I}$. If $\mathbb{P} = (p_{ij})$ denotes the transition matrix of the chain,

$$C = \{i_0\}, \qquad \varepsilon = 1, \qquad \nu = (p_{i_0 i})_{i \in I},$$

and renewal occurs *at each visit* in i_0. Considering in parallel other likely states under $\tilde{\pi}$, we then immediately derive a convergence indicator.

Example 12.17. A finite Markov chain. Consider $\Theta = \{0, 1, 2, 3\}$ and

$$\mathbb{P} = \begin{pmatrix} 0.26 & 0.04 & 0.08 & 0.62 \\ 0.05 & 0.24 & 0.03 & 0.68 \\ 0.11 & 0.10 & 0.08 & 0.71 \\ 0.08 & 0.04 & 0.09 & 0.79 \end{pmatrix}.$$

The stationary distribution is then $\tilde{\pi} = (0.097, 0.056, 0.085, 0.762)$, with mean 2.51. Table 12.2 compares estimators of γ_h^2, with $h(\theta) = \theta$, for the four small sets $C = \{i_0\}$ and shows that stabilizing is achieved for $T = 500,000$. ||

This academic example illustrates how easily the criterion can be used in finite environments. Finite state-spaces are, nonetheless, far from trivial since, as shown by the *Duality Principle* (Section 9.2.3), one can use the simplicity of some subchains to verify convergence for the Gibbs sampler. Therefore, renewal can be created for a (possibly artificial) subchain $(z^{(t)})$ and later

T/i_0	0	1	2	3
5000	1.19	1.29	1.26	1.21
500,000	1.344	1.335	1.340	1.343

Table 12.2. Estimators of γ_h^2 for $h(x) = x$, obtained by renewal in i_0.

applied to the parameter of interest if the Duality Principle applies. (See also Chauveau et al. 1998 for a related use of the Central Limit Theorem for convergence assessment.)

Example 12.18. Grouped multinomial model. The Gibbs sampler corresponding to the grouped multinomial model of Example 9.8,

$$X \sim \mathcal{M}_5(n; a_1\mu + b_1, a_2\mu + b_2, a_3\eta + b_3, a_4\eta + b_4, c(1 - \mu - \eta)) ,$$

actually enjoys a data augmentation structure and satisfies the conditions leading to the Duality Principle. Moreover, the missing data $(z_1, \ldots, z_4)$ has finite support since

$$Z_i \sim \mathcal{B}\left(x_i, \frac{a_i\mu}{a_i\mu + b_i}\right) \quad (i = 1, 2), \qquad Z_i \sim \mathcal{B}\left(x_i, \frac{a_i\eta}{a_i\eta + b_i}\right) \quad (i = 3, 4).$$

Therefore, the support of $(Z^{(t)})$ is of size $(x_1+1) \times \cdots \times (x_4+1)$. A preliminary simulation indicates that $(0, 1, 0, 0)$ is the modal state for the chain $(Z^{(t)})$, with an average excursion time of 27.1. The second most frequent state is $(0, 2, 0, 0)$, with a corresponding average excursion time of 28.1. In order to evaluate the performances of renewal control, we also used the state $(1, 1, 0, 0)$, which appears, on average, every 49.1 iterations. Table 12.3 describes the asymptotic variances obtained for the functions

$$h_1(\mu, \eta) = \mu - \eta, \quad h_2(\mu, \eta) = \mathbb{I}_{\mu > \eta}, \quad h_3(\mu, \eta) = \frac{\mu}{1 - \mu - \eta} .$$

The estimator of h_3 is obtained by Rao–Blackwellization,

$$\frac{1}{T} \sum_{t=1}^{T} \frac{0.5 + z_1^{(t)} + z_2^{(t)}}{x_5 - 0.5} ,$$

which increases the stability of the estimator and reduces its variance.

The results thus obtained are in good agreement for the three states under study, even though they cannot rigorously establish convergence.   ‖

i	$\mathbb{E}^{\pi}[h_i(\mu,\eta)|x]$	$\hat{\gamma}_i^2(1)$	$\hat{\gamma}_i^2(2)$	$\hat{\gamma}_i^2(3)$
h_1	5.10^{-4}	0.758	0.720	0.789
h_2	0.496	1.24	1.21	1.25
h_3	0.739	1.45	1.41	1.67

Table 12.3. Approximations by Gibbs sampling of posterior expectations and evaluation of variances $\hat{\gamma}_i^2$ by the renewal method for three different states (1 for $(0,1,0,0)$, 2 for $(0,2,0,0)$, and 3 for $(1,1,0,0)$) in the grouped multinomial model $(500,000$ iterations).

12.3.4 Within and Between Variances

The control strategy devised by Gelman and Rubin (1992) starts with the derivation of a distribution μ related with the modes of f, which are supposedly obtained by numerical methods. They suggest using a mixture of Student's t distributions centered around the identified modes of f, the scale being derived from the second derivatives of f at these modes. With the possible addition of an importance sampling step (see Problem 3.18), they then generate M chains $(\theta_m^{(t)})$ $(1 \le m \le M)$. For *every quantity of interest* $\xi = h(\theta)$, the stopping rule is based on the difference between a weighted estimator of the variance and the variance of estimators from the different chains.

Define

$$B_T = \frac{1}{M} \sum_{m=1}^{M} (\bar{\xi}_m - \bar{\xi})^2 \, ,$$

$$W_T = \frac{1}{M} \sum_{m=1}^{M} s_m^2 = \frac{1}{M} \sum_{m=1}^{M} \frac{1}{T} \sum_{t=1}^{T} (\xi_m^{(t)} - \bar{\xi}_m)^2 \, ,$$

with

$$\bar{\xi}_m = \frac{1}{T} \sum_{t=1}^{T} \xi_m^{(t)}, \qquad \bar{\xi} = \frac{1}{M} \sum_{m=1}^{M} \bar{\xi}_m \, ,$$

and $\xi_m^{(t)} = h(\theta_m^{(t)})$. The quantities B_T and W_T represent the *between-* and *within-chain* variances. A first estimator of the posterior variance of ξ is

$$\hat{\sigma}_T^2 = \frac{T-1}{T} W_T + B_T \, .$$

Gelman and Rubin (1992) compare $\hat{\sigma}_T^2$ and W_T, which are asymptotically equivalent (Problem 12.15). Gelman (1996) notes that $\hat{\sigma}_T^2$ overestimates the variance of $\xi_m^{(t)}$ because of the large dispersion of the initial distribution, whereas W_T underestimates this variance, as long as the different sequences $(\xi_m^{(t)})$ remain concentrated around their starting values.

The recommended criterion of Gelman and Rubin (1992) is to monitor

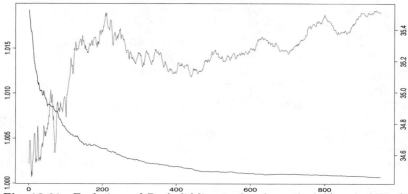

Fig. 12.21. Evolutions of R_T (solid lines and scale on the left) and of W_T (dotted lines and scale on the right) for the posterior distribution (12.17) and $h(\theta) = \theta$ ($M = 100$).

$$R_T = \frac{\hat{\sigma}_T^2 + \frac{B_T}{M}}{W_T} \frac{\nu_T}{\nu_T - 2}$$

$$= \left(\frac{T-1}{T} + \frac{M+1}{M} \frac{B_T}{W_T} \right) \frac{\nu_T}{\nu_T - 2} \, ,$$

where $\nu_T = 2(\hat{\sigma}_T^2 + \frac{B_T}{M})^2 / W_T$ and the approximate distribution of R_T is derived from the approximation $TB_T/W_T \sim \mathcal{F}(M-1, 2W_T^2/\varpi_T)$ with

$$\varpi_T = \frac{1}{M^2} \left[\sum_{m=1}^{M} s_m^4 - \frac{1}{M} \left(\sum_{m=1}^{M} s_m^2 \right)^2 \right].$$

(The approximation ignores the variability due to $\nu_T/(\nu_T - 2)$.) The stopping rule is based either on testing that the mean of R_T is equal to 1 or on confidence intervals on R_T.

Example 12.19. (Continuation of Example 12.12) Figure 12.21 describes the evolution of R_T for $h(\theta) = \theta$ and $M = 100$ and 1000 iterations. As the scale of the graph of R_T is quite compressed, one can conclude there is convergence after about 600 iterations. (The 50 first iterations have been eliminated.) On the other hand, the graph of W_T superimposed in this figure does not exhibit the same stationarity after 1000 iterations. However, the study of the associated histogram (see Figure 12.18) shows that the distribution of $\theta^{(t)}$ is stationary after a few hundred iterations. In this particular case, the criterion is therefore conservative, showing that the method can be difficult to calibrate. ‖

This method has enjoyed wide usage, in particular because of its simplicity and its connections with the standard tools of linear regression. However, we point out the following:

(a) Gelman and Rubin (1992) also suggest removing the first half of the simulated sample to reduce the dependence on the initial distribution μ. By comparison with a single-chain method, the number of wasted simulations is thus (formally) multiplied by M.

(b) The accurate construction of the initial distribution μ can be quite delicate and time-consuming. Also, in some models, the number of modes is too great to allow for a complete identification and important modes may be missed.

(c) The method relies on normal approximations, whereas the MCMC algorithms are used in settings where these approximations are, at best, difficult to satisfy and, at worst, not valid. The use of Student's t distributions by Gelman and Rubin (1992) does not remedy this. More importantly, there is no embedded test for the validity of this approximation.

(d) The criterion is *unidimensional*; therefore, it gives a poor evaluation of the correlation between variables and the necessary slower convergence of the joint distribution. Moreover, the stopping rule must be modified for every function of interest, with very limited recycling of results obtained for other functions. (Brooks and Gelman 1998a studied multidimensional extensions of this criterion based on the same approximations and Brooks and Giudici 2000 introduced a similar method in the special case of reversible jump algorithms.)

12.3.5 Effective Sample Size

Even in a stationary setting, there is an obvious difference in the use of the empirical average when compared with the standard Monte Carlo estimate of Chapter 3. Indeed, using the empirical average (12.13) as the estimator of

$$\int h(\theta) \, f(\theta) \, d\theta \,,$$

we cannot associate the standard variance estimator

$$(12.22) \qquad \hat{\nu}_T = \frac{1}{T^2} \sum_{t=1}^{T} \left(h(\theta^{(t)}) - S_T \right)^2$$

to this estimator, due to the correlations amongst the $\theta^{(t)}$'s. As mentioned above for Gelman and Rubin (1992) criterion, (12.22) corresponds to the "within" variance in Section 12.3.4, and underestimates the true variance of the estimator S_T. A first and obvious solution is to use batch sampling as in Section 12.2.2, but this is costly and the batch size still has to be determined. A more standard, if still approximate, approach is to resort to the *effective sample size*, $\hat{T}^S$, which gives the size of an iid sample with the same variance as the current sample and thus indicates the loss in efficiency due to the use of a Markov chain. This value is computed as

$$\hat{T}^S = T/\kappa(h),$$

where $\kappa(h)$ is the autocorrelation time associated with the sequence $h(\theta^{(t)})$,

$$\kappa(h) = 1 + 2 \sum_{t=1}^{\infty} \text{corr}\left(h(\theta^{(0)}), h(\theta^{(t)})\right).$$

Replacing T with $\hat{T}^S$ then leads to a more reliable estimate of the variance,

$$\tilde{\nu}_T = \frac{1}{T \times \hat{T}^S} \sum_{t=1}^{T} \left(h(\theta^{(t)}) - S_T\right)^2.$$

Estimating $\kappa(h)$ is also a delicate issue, but there exist some approximations in the literature, as in discussed in Section 12.6.1. Also, the software CODA (Section 12.6.2) contains a procedure that computes the autocorrelation. This notion of effective sample size can be found in Liu and Chen (1995) in the special case of importance resampling (Section 14.3.1).

12.4 Simultaneous Monitoring

12.4.1 Binary Control

Raftery and Lewis (1992) (see also Raftery and Banfield 1991) attempt to reduce the study of the convergence of the chain $(\theta^{(t)})$ to the study of the convergence of a *two-state* Markov chain, where an explicit analysis is possible. They then evaluate three parameters for the control of convergence, namely, k, the minimum "batch" (subsampling) step, t_0, the number of "warm-up" iterations necessary to achieve stationarity (to eliminate the effect of the starting value), and T, total number of iterations "ensuring" convergence (giving a chosen precision on the empirical average). The control is understood to be at the level of the derived two-state Markov chain (rather than for the original chain of interest).

The binary structure at the basis of the diagnosis is derived from the chain $\theta^{(t)}$ by

$$Z^{(t)} = \mathbb{I}_{\theta^{(t)} \le \underline{\theta}},$$

where $\underline{\theta}$ is an arbitrary value in the support of f. Unfortunately, the sequence $(Z^{(t)})$ *does not form a Markov chain*, even in the case where $\theta^{(t)}$ has a finite support (see Problem 6.56). Raftery and Lewis (1992) determined the batch size k by testing if $(Z^{(kt)})$ is a Markov chain against the alternative that $(Z^{(kt)})$ is a *second order* Markov chain (that is, that the vector $(Z^{(kt)}, Z^{(k(t+1))})$ is a Markov chain). This determination of k therefore has limited theoretical foundations and it makes sense, when accounting for the efficiency loss detailed in Lemma 12.2, to suggest working with the complete sequence $Z^{(t)}$ and not trying to justify the Markov approximation.

If $(Z^{(t)})$ is treated as a homogeneous Markov chain, with (pseudo-) transition matrix

$$\mathbb{P} = \begin{pmatrix} 1 - \alpha & \alpha \\ \beta & 1 - \beta \end{pmatrix},$$

it converges to the stationary distribution (Problem 6.51)

$$P(Z^{(\infty)} = 0) = \frac{\beta}{\alpha + \beta}, \qquad P(Z^{(\infty)} = 1) = \frac{\alpha}{\alpha + \beta}.$$

It is therefore possible to determine the *warm-up* size by requiring that

(12.23) $$\left| P(Z^{(t_0)} = i | z^{(0)} = j) - P(Z^{(\infty)} = i) \right| < \varepsilon$$

for $i, j = 0, 1$. Raftery and Lewis (1992) show that this is equivalent to (Problem 12.16)

(12.24) $$t_0 \geq \log \left(\frac{(\alpha + \beta)\varepsilon}{\alpha \vee \beta} \right) \Big/ \log |1 - \alpha - \beta|.$$

The sample size related with the (acceptable) convergence of

$$\delta_T = \frac{1}{T} \sum_{t=t_0}^{t_0+T} z^{(t)}$$

to $\dfrac{\alpha}{\alpha + \beta}$ can be determined by a normal approximation of δ_T, with variance

$$\frac{1}{T} \frac{(2 - \alpha - \beta) \, \alpha\beta}{(\alpha + \beta)^3} .$$

If, for instance, we require

$$P\left(\left| \delta_T - \frac{\alpha}{\alpha + \beta} \right| < q \right) \geq \varepsilon',$$

the value of T is (Problem 12.18)

(12.25) $$T \geq \frac{\alpha\beta(2 - \alpha - \beta)}{q^2(\alpha + \beta)^3} \, \Phi^{-1}\left(\frac{\varepsilon' + 1}{2} \right) .$$

This analysis relies on knowledge of the parameters (α, β). These are unknown in most settings of interest and must be estimated from the simulation of a test sample. Based on the independent case, Raftery and Lewis (1992) suggest using a sample size which is at least

$$T_{\min} \geq \Phi^{-1}\left(\frac{\varepsilon' + 1}{2} \right)^2 \frac{\alpha\beta}{(\alpha + \beta)^2} \, q^{-1} .$$

Table 12.4. Evolution of initializing and convergence times, and parameters (α, β) estimated after $t_0 + T$ iterations obtained from the previous round, in the $\mathcal{Be}(\alpha, 1)/\mathcal{Be}(\alpha + 1, 1)$ example.

Round	t_0	$T (\times 10^3)$	$\hat{\alpha}$	$\hat{\beta}$
1	55	431	0.041	0.027
2	85	452	0.040	0.001
3	111	470	0.041	0.029
4	56	442	0.041	0.029
5	56	448	0.040	0.027
6	58	458	0.041	0.025
7	50	455	0.042	0.025
8	50	452	0.041	0.028

This method, called *binary control*, is quite popular, in particular because programs are available (in Statlib) and also because its implementation is quite easy.[7] However, there are drawbacks to using it as a convergence indicator:

(a) The preliminary estimation of the coefficients α and β requires a chain $(\theta^{(t)})$ which is already (almost) stationary and which has, we hope, sufficiently explored the characteristics of the distribution f. If α and β are not correctly estimated, the validity of the method vanishes.

(b) The approach of Raftery and Lewis (1992) is intrinsically *unidimensional* and, hence, does not assess the correlations between components. It can thus conclude there is convergence, based on the marginal distributions, whereas the joint distribution is not correctly estimated.

(c) Once α and β are estimated, the stopping rules are independent of the model under study and of the selected MCMC algorithm, as shown by formulas (12.24) and (12.25). This generic feature is appealing for an automated implementation, but it cannot guarantee global efficiency.

Example 12.20. (Continuation of Example 12.13) For the chain with stationary distribution $\mathcal{Be}(0.2, 1)$, the probability that $X^{0.8}$ is less than 0.2, with $\mathbb{E}[X^{0.8}] = 0.2$, can be approximated from the Markov chain and the two-state (pseudo-)chain is directly derived from

$$Z^{(t)} = \mathbb{I}_{(X^{(t)})^{0.8} < 0.2} \cdot$$

Based on a preliminary sample of size $50,000$, the initial values of α and β are $\alpha_0 = 0.0425$ and $\beta_0 = 0.0294$. These approximations lead to $t_0 = 55$ and $T = 430,594$ for $\varepsilon = q = 0.01$ and $\varepsilon' = 0.99$. If we run the algorithm for $t_0 + T$ iterations, the estimates of α and β are then $\alpha^* = 0.0414$ and $\beta^* = 0.027$.

---

[7] In the example below, the batch size k is fixed at 1 and the method has been directly implemented in C, instead of calling the Statlib program.

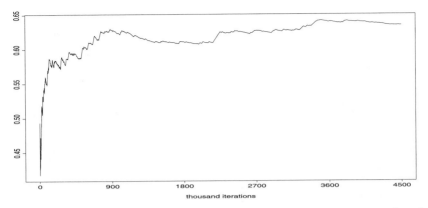

Fig. 12.22. Convergence of the mean $\hat{\varrho}_t$ for a chain generated from [$A.30$].

Repeating this recursive evaluation of (t_0, T) as a function of $(\hat{\alpha}, \hat{\beta})$ and the update of $(\hat{\alpha}, \hat{\beta})$ after $(t_0 + T)$ additional iterations, we obtain the results in Table 12.4. These exhibit a relative stability in the evaluation of (α, β) (except for the second iteration) and, thus, of the corresponding (t_0, T)'s. However, as shown by Figure 12.22, after $4,500,000$ iterations, the approximation of $P(x^{0.8} \leq 0.2)$ is equal to 0.64, while the true value is $0.2^{0.2/0.8}$ (that is, 0.669). This erroneous convergence is not indicated by the binary control method in this pathological setup.   ‖

12.4.2 Valid Discretization

A fundamental problem with the binary control technique of Raftery and Lewis (1992) in Section 12.4.1 is that it relies on an approximation, namely that the discretized sequence $(z^{(t)})$ is a Markov chain. Guihenneuc-Jouyaux and Robert (1998) have shown that there exists a rigorous discretization of Markov chains which produces Markov chains. The idea at the basis of this discretization is to use several disjoint *small sets* A_i $(i = 1, \ldots, k)$ for the chain $(\theta^{(t)})$, with corresponding parameters (ϵ_i, ν_i), and to subsample only at renewal times τ_n $(n > 1)$. The τ_n's are defined as the successive instants when the Markov chain enters one of these small sets with splitting, that is, by

$$\tau_n = \inf\{\, t > \tau_{n-1};\; \exists\, 1 \leq i \leq k,\; \theta^{(t-1)} \in A_i \text{ and } \theta^{(t)} \sim \nu_i \,\}.$$

(Note that the A_i's $(i = 1, \ldots, k)$ need not be a partition of the space.)

The discretized Markov chain is then derived from the finite-valued sequence

$$\eta^{(t)} = \sum_{i=1}^{k} i\mathbb{I}_{A_i}(\theta^{(t)})$$

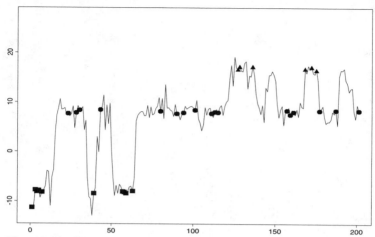

Fig. 12.23. Discretization of a continuous Markov chain, based on three small sets. The renewal events are represented by triangles for B, circles for C, and squares for D, respectively. (*Source:* Guihenneuc-Jouyaux and Robert 1998.)

as $(\xi^{(n)}) = (\eta^{(\tau_n)})$. The resulting chain is then described by the sequence of small sets encountered by the original chain $(\theta^{(t)})$ at renewal times. It can be shown that the sequence $(\xi^{(n)})$ is a homogeneous Markov chain on the finite state-space $\{1, \dots, k\}$ (Problem 12.21).

Example 12.21. (Continuation of Example 12.12) Figure 12.23 illustrates discretization for the subchain $(\theta^{(t)})$, with three small sets $C = [7.5, 8.5]$, derived in Section 12.15, and $B = [-8.5, -7.5]$ and $D = [17.5, 18.5]$, which can be constructed the same way. Although the chain visits the three sets quite often, renewal occurs with a much smaller frequency, as shown by the symbols on the sequence.   ‖

This result justifies control of Markov chains through their discretized counterparts. Guihenneuc-Jouyaux and Robert (1998) propose further uses of the discretized chain, including the evaluation of mixing rates. (See Cowles and Rosenthal 1998 for a different approach to this evaluation, based on the drift conditions of Note 6.9.1, following the theoretical developments of Rosenthal 1995.)

12.5 Problems

12.1 The witch's hat distribution

$$\pi(\theta|y) \propto \left\{ (1 - \delta)\, \sigma^{-d} e^{-\|y - \theta\|^2/(2\sigma^2)} + \delta \right\} \mathbb{I}_C(\theta), \qquad y \in \mathbb{R}^d,$$

when θ is restricted to the unit cube $C = [0, 1]^d$, has been proposed by Matthews (1993) as a calibration *benchmark* for MCMC algorithms.

(a) Construct an algorithm which correctly simulates the witch's hat distribution. (*Hint:* Show that direct simulation is possible.)

(b) The choice of δ, σ, and d can lead to arbitrarily small probabilities of either escaping the attraction of the mode or reaching the mode. Find sets of parameters (δ, σ, y) for which these two phenomena occur.

12.2 Establish that the cdf of the statistic (12.2) is (12.3).

12.3 Reproduce the experiment of Example 12.10 in the cases $\alpha = 0.9$ and $\alpha = 0.99$.

12.4 In the setup of Example 12.10:

(a) Show that $\mathbb{E}[X^{1-\alpha}] = \alpha$ when $X \sim \mathcal{B}e(\alpha, 1)$.

(b) Show that $P(X^{1-\alpha} < \varepsilon) = \varepsilon^{\alpha/1-\alpha}$.

(c) Show that the Riemann approximation S_T^R of (12.16) has a very specific shape in this case, namely that it is equal to the normalizing constant of f.

12.5 (Liu et al. 1992) Show that if $(\theta_1^{(t)})$ and $(\theta_2^{(t)})$ are independent Markov chains with transition kernel K and stationary distribution π,

$$\omega_t = \frac{\pi(\theta_1^{(t)})K(\theta_1^{(t-1)}, \theta_2^{(t)})}{\pi(\theta_2^{(t)})K(\theta_2^{(t-1)}, \theta_1^{(t)})}$$

is an unbiased estimator of the L_2 distance between π and

$$\pi^t(\theta) = \int K(\eta, \theta)\pi^{t-1}(\eta)d\eta ,$$

in the sense that $\mathbb{E}[\omega_t] = 1 + \mathrm{var}\dfrac{\pi^t(\theta)}{\pi(\theta)}$.

12.6 (Roberts 1994) Show that if $(\theta_1^{(t)})$ and $(\theta_2^{(t)})$ are independent Markov chains with transition kernel K and stationary distribution π,

$$\chi^t = \frac{K(\theta_1^{(0)}, \theta_2^{(2t-1)})}{\pi(\theta_2^{(2t-1)})}$$

satisfies

(a) $\mathbb{E}[\chi^t] = \int \frac{\pi_1^t(\theta)\pi_2^t(\theta)}{\pi(\theta)}d\theta \xrightarrow{t\to\infty} 1$,

(b) $\lim_{t\to\infty} \mathrm{var}(\chi_t) = \int (K(\theta_1^{(0)}, \theta) - \pi(\theta))^2 \frac{1}{\pi(\theta)}d\theta$,

where π_i^t denotes the density of $\theta_i^{(t)}$. (*Hint:* Use the equilibrium relation

$$K(\theta_2^{(2t-2)}, \theta_2^{(2t-1)})\pi(\theta_2^{(2t-2)}) = K(\theta_2^{(2t-1)}, \theta_2^{(2t-2)})\pi(\theta_2^{(2t-1)})$$

to prove part (a).)

12.7 (Continuation of Problem 12.6) Define

$$I_t = \frac{1}{m(m-1)}\sum_{i\neq j}\chi_{ij}^t, \quad J_t = \frac{1}{m}\sum_{i=1}^{m}\chi_{ii}^t,$$

with

$$\chi_{ij}^t = \frac{K(\theta_i^{(0)}, \theta_j^{(2t-1)})}{\pi(\theta_j^{(2t-1)})} ,$$

based on m parallel chains $(\theta_j^{(0)})$ $(j = 1, \ldots, m)$.

(a) Show that if $\theta_j^{(0)} \sim \pi$, $\mathbb{E}[I_t] = \mathbb{E}[J_t] = 1$ and that for every initial distribution on the $\theta_j^{(0)}$'s, $\mathbb{E}[I_t] \leq \mathbb{E}[J_t]$.

(b) Show that

$$\mathrm{var}(I_t) \leq \mathrm{var}(J_t)/(m-1) .$$

12.8 (Dellaportas 1995) Show that

$$\mathbb{E}_g \left[\min \left(1, \frac{f(x)}{g(x)} \right) \right] = \int |f(x) - g(x)| dx .$$

Derive an estimator of the L_1 distance between the stationary distribution π and the distribution at time t, π^t.

12.9 (Robert 1995a) Give the marginal distribution of $\xi = h_3(\mu, \nu)$ in Example 12.18 by doing the following:

(i) Show that the Jacobian of the transform of (μ, ν) in (ξ, ν) is $\frac{1-\eta}{(1+\xi)^2}$.

(ii) Express the marginal density of ξ as a $(x_1 + x_2)$ degree polynomial in $\xi/(1+\xi)$.

(iii) Show that the weights ϖ_j of this polynomial can be obtained, up to a multiplicative factor, in closed form.

Deduce that a Riemann sum estimate of $\mathbb{E}[h_3(\mu, \nu)]$ is available in this case.

12.10 For the model of Example 12.18, show that a small set is available in the (μ, η) space and derive the corresponding renewal probability.

12.11 In the setup of Section 12.3.3, if δ_t is the $\mathcal{B}(\epsilon)$ random variable associated with renewal for the minorization condition (12.18), show that the distribution of δ_t conditional on $(\theta^{(t)}, \theta^{(t+1)})$ is

$$\mathrm{Pr}\left(\delta_t = 1 | \theta^{(t)}, \theta^{(t+1)} \right) \propto \epsilon \nu(\theta^{(t+1)}),$$

$$\mathrm{Pr}\left(\delta_t = 0 | \theta^{(t)}, \theta^{(t+1)} \right) \propto K(\theta^{(t)}, \theta^{(t+1)}) - \epsilon \nu(\theta^{(t+1)}).$$

12.12 Given a mixture distribution $p\mathcal{N}(\mu, 1) + (1-p)\mathcal{N}(\theta, 1)$, with conjugate priors

$$p \sim \mathcal{U}_{[0,1]} \ , \ \mu \sim \mathcal{N}(0, \tau^2) \ , \ \theta \sim \mathcal{N}(0, \tau^2),$$

show that $[\underline{p}, \bar{p}] \times [\underline{\mu}, \bar{\mu}] \times [\underline{\theta}, \bar{\theta}]$ is a small set and deduce that ε in (12.18) decreases as a power of the sample size.

12.13 For the estimator S_T^P given in Example 12.13:

(a) Show that the variance of S_T^P in Example 12.13 is infinite.

(b) Propose an estimator of the variance of S_T^P (when it exists) and derive a convergence diagnostic.

(c) Check whether the importance sampling S_T^P is (i) available and (ii) with finite variance for Examples 10.17 and 9.8.

12.14 (a) Verify the decomposition

$$\mathrm{var}(X) = \mathrm{var}(\mathbb{E}[X|Y]) + \mathbb{E}[\mathrm{var}(X|Y)].$$

(b) Show that, up to second order, S_T^P and S_T^{MP} behave as in the independent case. That is, $\mathrm{var}S_T^P \approx \sum_t w_t \mathrm{var}h(\theta^{(t)}) + \mathrm{O}(1/t)$.

12.15 Referring to Section 12.3.4:

(a) Show that $\hat{\sigma}_T^2$ and W_T have the same expectation.

(b) Show that $\hat{\sigma}_T^2$ and W_T are both asymptotically normal. Find the limiting variances.

12.16 (Raftery and Lewis 1992) Establish the equivalence between (12.23) and (12.24) by showing first that it is equivalent to

$$|1 - \alpha - \beta|^{t_0} \leq \frac{(\alpha + \beta)\varepsilon}{\alpha \vee \beta} .$$

12.17 For the transition matrix

$$\mathbb{P} = \begin{pmatrix} 1 - \alpha & \alpha \\ \beta & 1 - \beta \end{pmatrix} ,$$

show that the second eigenvalue is $\lambda = 1 - \alpha - \beta$. Deduce that

$$|P(x^{(t)} = i|x^{(0)} = j) - P(x^{(\infty)} = i)| < \varepsilon$$

for every (i, j) if $|1 - \alpha - \beta|^t \leq \varepsilon(\alpha + \beta)/\max(\alpha, \beta)$. (*Hint:* Use the representation $P(x^{(t)} = i|x^{(0)} = j) = e_j P^t e_i'$, with $e_0 = (1, 0)$ and $e_1 = (0, 1)$.)

12.18 (Raftery and Lewis 1992) Deduce (12.25) from the normal approximation of δ_T. (*Hint:* Show that

$$\Phi\left(\sqrt{T} \frac{(\alpha + \beta)^{3/2} q}{\sqrt{\alpha\beta(2 - \alpha - \beta)}}\right) \geq \frac{\varepsilon' + 1}{2} .)$$

12.19 (Raftery and Lewis 1996) Consider the logit model

$$\log \frac{\pi_i}{1 - \pi_i} = \eta + \delta_i , \quad \delta_i \sim \mathcal{N}(0, \sigma^2), \quad \sigma^{-2} \sim \mathcal{G}a(0.5, 0.2) .$$

Study the convergence of the associated Gibbs sampler and the dependence on the starting value.

12.20 Given a three-state Markov chain with transition matrix $\mathbb{P}$, determine sufficient conditions on $\mathbb{P}$ for a derived two-state chain to be a Markov chain.

12.21 Show that the sequence $(\xi^{(n)})$ defined in Section 12.4.2 is a Markov chain. (*Hint:* Show that

$$P(\xi^{(n)} = i|\xi^{(n-1)} = j, \xi^{(n-2)} = \ell, \ldots)$$
$$= \mathbb{E}_{\theta^{(0)}}\left[\mathbb{I}_{A_i}\left(\theta^{(\tau_{n-1}-1+\Delta_n)}\right)\Big|\theta^{(\tau_{n-1}-1)} \in A_j, \theta^{(\tau_{n-2}-1)} \in A_\ell, \ldots\right] ,$$

and apply the strong Markov property.)

12.22 (Tanner 1996) Show that, if $\theta^{(t)} \sim \pi^t$ and if the stationary distribution is the posterior density associated with $f(x|\theta)$ and $\pi(\theta)$, the weight

$$\omega_t = \frac{f(x|\theta^{(t)})\pi(\theta^{(t)})}{\pi^t(\theta^{(t)})}$$

converges to the marginal $m(x)$.

12.23 Apply the convergence diagnostics of CODA to the models of Problems 10.34 and 10.36.

12.24 In the setup of Example 10.17, recall that the chain $(\beta^{(t)})$ is uniformly ergodic, with lower bound

$$r(\beta') = \frac{\delta^{\gamma+10\alpha}(\beta')^{\gamma+10\alpha-1}}{\Gamma(10\alpha+\gamma)} e^{-\delta\beta'} \prod_{i=1}^{10} \left(\frac{t_i}{t_i+\beta'}\right)^{p_i+\alpha}.$$

Implement a simulation method to determine the value of ρ and to propose an algorithm to simulate from this distribution. (*Hint:* Use the Riemann approximation method of Section 4.3 or one of the normalizing constant estimation techniques of Chen and Shao (1997); see Problem 4.1.)

12.6 Notes

12.6.1 Spectral Analysis

As already mentioned in Hastings (1970), the chain $(\theta^{(t)})$ or a transformed chain $(h(\theta^{(t)}))$ can be considered from a *time series* point of view (see Brockwell and Davis 1996, for an introduction). For instance, under an adequate parameterization, we can model $(\theta^{(t)})$ as an ARMA(p,q) process, estimating the parameters p and q, and then use partially empirical convergence control methods. Geweke (1992) proposed using the *spectral density* of $h(\theta^{(t)})$,

$$S_h(w) = \frac{1}{2\pi} \sum_{t=-\infty}^{t=\infty} \text{cov}\left(h(\theta^{(0)}), h(\theta^{(t)})\right) e^{itw},$$

where i denotes the complex square root of 1 (that is,

$$e^{itw} = \cos(tw) + i\sin(tw)\Big).$$

The spectral density is related to the asymptotic variance of (12.1) since the limiting variance γ_h of Proposition 6.77 is given by

$$\gamma_h^2 = S_h^2(0).$$

Estimating S_h by nonparametric methods like *the kernel method* (see Silverman 1986). Geweke (1992) takes the first T_A observations and the last T_B observations from a sequence of length T to derive

$$\delta_A = \frac{1}{T_A} \sum_{t=1}^{T_A} h(\theta^{(t)}), \qquad \delta_B = \frac{1}{T_B} \sum_{t=T-T_B+1}^{T} h(\theta^{(t)}),$$

and the estimates σ_A^2 and σ_B^2 of $S_h(0)$ based on both subsamples, respectively. Asymptotically (in T), the difference

(12.26)
$$\frac{\sqrt{T}(\delta_A - \delta_B)}{\sqrt{\dfrac{\sigma_A^2}{\tau_A} + \dfrac{\sigma_B^2}{\tau_B}}}$$

is a standard normal variable (with $T_A = \tau_A T$, $T_B = \tau_B T$, and $\tau_A + \tau_B < 1$). We can therefore derive a convergence diagnostic from (12.26) and a determination of the size t_0 of the training sample. The values suggested by Geweke (1992) are $\tau_A = 0.1$ and $\tau_B = 0.5$.

A global criticism of this spectral approach also applies to all the methods using a nonparametric intermediary step to estimate a parameter of the model, namely that they necessarily induce losses in efficiency in the processing of the problem (since they are based on a less constrained representation of the model). Moreover, the calibration of nonparametric estimation methods (as the choice of the *window* in the kernel method) is always delicate since it is not standardized. We therefore refer to Geweke (1992) for a more detailed study of this method, which is used in some software (see Best et al. 1995, for instance). Other approaches based on spectral analysis are given in Heidelberger and Welch (1983) and Schruben et al. (1983) to test the stationarity of the sequence by Kolmogorov–Smirnov tests (see Cowles and Carlin 1996 and Brooks and Roberts 1999 for a discussion). Note that Heidelberger and Welch (1983) test stationarity via a Kolmogorov–Smirnov test, based on

$$B_T(S) = \frac{S_{[Ts]} - [Ts]\bar{\theta}}{(T\hat{\psi}(0))^{1/2}}, \qquad 0 \le s \le 1,$$

where

$$S_t = \sum_{t=1}^{t} \theta^{(t)}, \qquad \bar{\theta} = \frac{1}{T} \sum_{t=1}^{T} \theta^{(t)},$$

and $\hat{\psi}(0)$ is an estimate of the spectral density. For large T, B_T is approximately a Brownian bridge and can be tested as such. Their method thus provide the theoretical background to that of Yu and Mykland (1998) CUSUM criterion (see Section 12.3.1).

12.6.2 The CODA Software

While the methods presented in this chapter are at various stages of their development, some of the most common techniques have been aggregated in an S-Plus or R software package called CODA, developed by Best et al. (1995). While originally intended as an output processor for the BUGS software (see Section 10.6.2), this software can also be used to analyze the output of Gibbs sampling and Metropolis–Hastings algorithms. The techniques selected by Best et al. (1996) are mainly those described in Cowles and Carlin (1995) (that is, the convergence diagnostics of Gelman and Rubin (1992) (Section 12.3.4), (Section 12.6.1) Geweke (1992) (Section 12.6.1), Heidelberger and Welch (1983), Raftery and Lewis (1992) (Section 12.4.1), plus plots of autocorrelation for each variable and of cross-correlations between variables). The MCMC output must, however, be presented in a very specific format to be processed by CODA, unless it is produced by BUGS.

13

Perfect Sampling

> There were enough broken slates underfoot to show that the process was unperfect.
> —Ian Rankin, *Set in Darkness*

The previous chapters have dealt with methods that are quickly becoming "mainstream". That is, analyses using Monte Carlo methods in general, and MCMC specifically, are now part of the applied statistician's tool kit. However, these methods keep evolving, and new algorithms are constantly being developed, with some of these algorithms resulting in procedures that seem radically different from the current standards

The final two chapters of this book cover methods that are, compared with the material in the first twelve chapters, still in their beginning stages of development. We feel, however, that these methods hold promise of evolving into mainstream methods, and that they will also become standard tools for the practicing statistician.

13.1 Introduction

MCMC methods have been presented and motivated in the previous chapters as an alternative to direct simulation methods, like Accept–Reject algorithms, because those were not adequate to process complex distributions. The ultimate step in this evolution of simulation methods is to devise exact (or "perfect") sampling techniques based on these MCMC algorithms. While we seem to be closing the loop on simulation methods by recovering exact simulation techniques as in Chapter 2, we have nonetheless gained much from the previous chapters in that, as we will see, we are now able to design much more advanced Accept–Reject algorithms, that is, algorithms that could not have been devised from scratch. This comes at a cost, though, in that these novel algorithms are often quite greedy in both computing time and storage space, but this is the price to pay for using a more generic construction in the Accept–Reject algorithms. Before starting this chapter, we must stress that the following methods, while generic in their principle, are still fairly restricted in their statistical applications, with the main bulk of perfect sampling implementations being found in spatial process problems (Note 13.6.1).

Chapter 12 details the difficult task of assessing the convergence of an MCMC sampler, that is, the validity of the approximation $\theta^{(T)} \sim \pi(x)$, and, correspondingly, the determination of a "long enough" simulation time. Nonetheless, the tools presented in that chapter are mostly hints of (or lack of) convergence, and they very rarely give a crystal-clear signal that $\theta^{(T)} \sim \pi(x)$ is valid for all simulation purposes. In fact, when this happens, the corresponding MCMC algorithm ends up providing an exact simulation from the (stationary) distribution $\pi(x)$ and thus fulfills the purpose of the present chapter.

There is another connection between Chapters 12 and 13, in that the methods developed below, even when they are too costly to be implemented to deliver iid samples from π, can be used to evaluate the necessary computing time or the mixing properties of the chain, that is, as put by Fill (1998a,b), to determine *"how long is long enough?"*

Yet another connection between perfect sampling and convergence assessments techniques is that, to reduce dependence on the starting value, some authors (e.g. Gelman and Rubin 1992) have advocated running MCMC algorithms in parallel. Perfect sampling is the extreme implementation of this principle in that it considers all possible starting values simultaneously, and runs the chains until the dependence on the starting value(s) has vanished.

Following Propp and Wilson (1996), several authors have proposed devices to sample directly from the stationary distribution π (that is, algorithms such that $\theta^{(0)} \sim \pi$), at varying computational costs[1] and for specific distributions and/or transitions. The name *perfect simulation* for such techniques was coined by Kendall (1998), replacing the *exact sampling* terminology of Propp and Wilson (1996) with a more triumphant description!

The main bulk of the work on perfect simulation deals, so far, with finite state spaces; this is due, for one thing, to the greater simplicity of these spaces and, for another, to statistical physics motivations related to the Ising model (see Example 10.2). The extension to mainstream statistical problems is much more delicate, but Murdoch and Green (1998) have shown that some standard examples in continuous settings, like the nuclear pump failure model of Example 10.17 (see Problem 12.24), do allow for *perfect simulation*. Note also that in settings when the Duality Principle of Section 9.2 applies, the stationarity of the finite chain obviously transfers to the dual chain, even if the latter is continuous.

---

[1] In most cases, the computation time required to produce $\theta^{(0)}$ exceeds by orders of magnitude the computation time of a $\theta^{(t)}$ from the transition kernel.

13.2 Coupling from the Past

As discussed earlier, a major drawback of using MCMC methods for simulating a distribution π is that their validation is only asymptotic. Formally, we would need to run the transition kernel an infinite number of iterations to achieve simulation precisely from the stationary distribution π. This is obviously impossible in practice, except if we can construct a stopping time τ on the Markov chain such that the chain at that time has the same distribution as one running for an infinite number of steps,

$$\theta^{(\tau)} \sim \pi(\theta) \,.$$

Propp and Wilson (1996) were the first to come up with such a stopping time.

13.2.1 Random Mappings and Coupling

In a finite state-space $\mathcal{X}$ of size k, the method proposed by Propp and Wilson (1996) is called *coupling from the past* (CFTP). The principle is to start k Markov chains, one in each of the k possible states, far in the past at time $-T_0$ and if all these parallel chains take the same value (or *coalesce*) by a fixed time 0, say, then the starting value is obviously irrelevant. Moreover, the distribution of the chain at time 0 is the same as if we had started the iterations at time $-\infty$, given that a chain starting at time $-\infty$ would necessarily[2] take one of the k values at time $-T_0$.

   To implement this principle requires a valid technique that ensures that coalescence at time 0 can occur with a large enough probability. For instance, if we run the above k chains *independently,* the probability that they all take the same value by time 0, will decrease as a power of k, whatever T_0. We thus need to remove the (between) *independence* assumption and use a *coupling* scheme (see Section 6.6.1), which both guarantees that all chains are marginally generated from the original transition kernel *and* that two chains that *once* take the same value remain identical *forever after*. A convenient representation for coupling proceeds through the notion of *random mapping*: given that a Markov chain transition can generically represented as

$$\theta^{(t)} = \Psi(\theta^{(t)}, u_t) \,,$$

where the u_t's are iid from a (fixed) distribution (Borovkov and Foss 1992, see Problem 13.3), we can define the successive iterations of an MCMC algorithm as a sequence of random mappings on $\mathcal{X}$,

$$\Psi_1, \ldots, \Psi_t, \ldots \,,$$

---

[2] This is not an obvious result since the reverse *forward scheme,* where all chains started at time 0 from all possible values in $\mathcal{X}$ are processed till they all coincide, does not produce a simulation from π (Problems 13.1 and 13.2).

where $\Psi_t(\theta) = \Psi(\theta, u_t)$. In particular, we can write

(13.1)
$$\theta^{(t)} = \Psi_t \circ \ldots \circ \Psi_1(\theta^{(0)}),$$

which is also called a *stochastic recursive sequence*. An important feature of this sequence of random mappings is that, for a given $\theta^{(0)}$, the successive images of $\theta^{(0)}$ by the compositions $\Psi_t \circ \ldots \circ \Psi_1$ are correctly distributed from the tth transitions of the original Markov chain. Therefore, for two different starting values, $\theta_1^{(0)}$ and $\theta_2^{(0)}$, the sequences

$$\theta_1^{(t)} = \Psi_t \circ \ldots \circ \Psi_1(\theta_1^{(0)}) \quad \text{and} \quad \theta_2^{(t)} = \Psi_t \circ \ldots \circ \Psi_1(\theta_2^{(0)})$$

are both marginally distributed from the correct Markov transitions and, furthermore, since they are based on the *same* random mappings, the two sequences will be identical from the time $t^\star$ they take the same value, which is the feature we were looking for.

Example 13.1. Coalescence. For $n \in \mathbb{N}$, $\alpha > 0$ and $\beta > 0$, consider the joint distribution

$$\theta \sim \mathcal{B}e(\alpha, \beta), \quad X|\theta \sim \mathcal{B}(n, \theta),$$

with joint density

$$\pi(x, \theta) \propto \binom{n}{x} \theta^{x+\alpha-1}(1 - \theta)^{n-x+\beta-1}.$$

Although we can simulate directly from this distribution, we can nonetheless consider a Gibbs sampler adapted to this joint density with the following transition rule at time t:

1. $\theta_{t+1} \sim \mathcal{B}e(\alpha + x_t, \beta + n - x_t)$,
2. $X_{t+1} \sim \mathcal{B}(n, \theta_{t+1})$,

with corresponding transition kernel

$$K((x_t, \theta_t), (x_{t+1}, \theta_{t+1})) \propto \binom{n}{x_{t+1}} \theta^{x_{t+1}+\alpha+x_t-1}(1 - \theta)^{\beta+2n-x_t-x_{t+1}-1}.$$

This is a special case of Data Augmentation (Section 9.2): the subchain (X_t) is a Markov chain with $X_{t+1}|x_t \sim \text{BetaBin}(n, \alpha + x_t, \beta + n - x_t)$, the beta-binomial distribution.

Consider the following numerical illustration: $n = 2$, $\alpha = 2$ and $\beta = 4$. The state space is thus $\mathcal{X} = \{0, 1, 2\}$ and the corresponding transition probabilities for the subchain (X_t) are

$$
\begin{aligned}
&\Pr(0 \mapsto 0) = .583, &&\Pr(0 \mapsto 1) = .333, &&\Pr(0 \mapsto 2) = .083, \\
&\Pr(1 \mapsto 0) = .417, &&\Pr(1 \mapsto 1) = .417, &&\Pr(1 \mapsto 2) = .167, \\
&\Pr(2 \mapsto 0) = .278, &&\Pr(2 \mapsto 1) = .444, &&\Pr(2 \mapsto 2) = .278.
\end{aligned}
$$

Given a random draw $U_{t+1} \sim \mathcal{U}(0,1)$, the paths from all possible starting points are described by Figure 13.1. This graph shows, in particular, that when $u_{t+1} > .917$, $u_{t+1} < .278$, or $u_{t+1} \in (.583,.722)$, the chain X_t moves to the same value, whatever its previous value. These ranges of u_{t+1} thus ensure coalescence in one step. $\parallel$

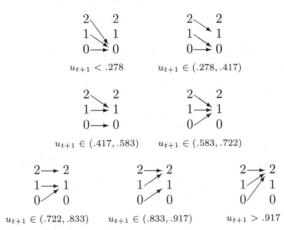

Fig. 13.1. All possible transitions for the Beta-Binomial(2,2,4) example, depending on the range of the coupling uniform variate.

As in Accept–Reject algorithms, there is a seemingly difficult point with the random mapping representation

$$\theta_t = \Psi_t(\theta_{t-1}) = \Psi(\theta_{t-1}, u_t)$$

of an MCMC transition when θ_t is generated using a random number of uniforms, because the vector u_t has no predefined length. This difficulty is easily overcome, though, by modifying Ψ into a function such that u_t is the *seed* used in the pseudo-random generator at time t (see Chapter 2). This eliminates the pretense of simulating, for each t, an infinite sequence of uniforms.

Random mappings are central to the *coupling from the past* scheme of Propp and Wilson (1996): if we find a time $T < 0$ such that the composition $\Psi_0 \circ \ldots \circ \Psi_T$ is *constant*, the value of the chain at time 0 is independent of the value of the chain at time T and, by induction, of the value of the chain at time $-\infty$. In other words, if we formally start a chain at time $-\infty$, it will end up at time 0 with the same value not matter the starting value. Before establishing more rigorously the validity of Propp and Wilson's (1996) scheme, we first state a fundamental theorem of coupling.

Theorem 13.2. *For a finite state space $\mathcal{X} = \{1,\ldots,k\}$, consider k coupled Markov chains, $(\theta_t^{(1)}),(\theta_t^{(2)}),\ldots,(\theta_t^{(k)})$, where*

(i) $\theta_0^{(j)} = j$;

(ii) $\theta_{t+1}^{(j)} = \Psi(x_t^{(j)}, u_{t+1})$, where the U_t's are mutually independent.

Then the time τ to coalescence is a random variable that depends only on $U_1, U_2, \ldots$.

An interpretation of the coalescence time τ is that it is the time at which the initial state of the chain has "worn off": the distribution of $\theta_\tau^{(j)}$ is obviously the same for all j's. A natural subsequent question is whether or not $\theta_\tau^{(j)}$ is a draw from the stationary distribution π. While it is the case that, for a *fixed* time $T^* > \tau$, $\theta_{T^*}^{(j)}$ is distributed from the stationary distribution, the fact that τ is a stopping time implies that $\theta_\tau^{(j)}$ is usually *not* distributed from π, as shown in Example 13.4 (see also Problems 13.1 and 13.2).

13.2.2 Propp and Wilson's Algorithm

The implementation of the above principle thus reduces to a search for a constant composition $\Psi_0 \circ \ldots \circ \Psi_T$: once one such $T < 0$ is found, coalescence occurs between time T and time 0, and the common value of the chains at time 0 is distributed from π. The algorithm, called *coupling from the past*, thus reads as follows:

Algorithm A.54 –Coupling from the Past–

> 1. Generate random mappings $\Psi_{-1}, \Psi_{-2}, \ldots$.
> 2. Define compositions $(t = -1, -2, \ldots)$
>
> $$\Phi_t(\theta) = \Psi_0 \circ \Psi_{-1} \circ \ldots \circ \Psi_{-t}(\theta).$$
>
> 3. Determine T such that Φ_T is constant by looking successively at $\Phi_{-1}, \Phi_{-2}, \Phi_{-4}, \ldots$
> 4. For an arbitrary value θ_0, take $\Phi_T(\theta_0)$ as a realization from π.

In Step 3 of this algorithm, the backward excursion stops[3] at one time T when Φ_T is a constant function; we are thus spared the task of running the algorithm until $-\infty$! The same applies to the simulation of the random mappings Ψ_t, which only need to be simulated till time T.

Note that, by virtue of the separate definitions of the maps Ψ_t, for a given θ_T, the value of $\theta^{(0)}$ is also given: not only does this save simulation time but, more importantly, it ensures the validity of the method.

Theorem 13.3. *The algorithm [A.54] produces a random variable distributed exactly according to the stationary distribution of the Markov chain.*

---

[3] Note that to establish that Φ_T is constant is formally equivalent to consider all possible starting values at time T and to run the corresponding coupled chains till time 0.

Proof. Following Casella et al. (2001), the proof is based on the fact that the k Markov chains do coalesce at some *finite* time into one chain, call it θ_t^*.

First, as each chain $(\theta_t^{(j)})$ starting from state $j \in \mathcal{X}$ is irreducible, and $\mathcal{X}$ is finite, there exists $N_j < \infty$ such that, for $N \geq N_j$,

$$P(\theta_N^{(j)} = \theta | \theta_0^{(j)} = j) > 0, \quad \text{for all } \theta \in \mathcal{X}.$$

Then, each chain has a positive probability of being in any state at time $N \geq \max\{N_1, N_2, \ldots, N_k\}$ and for some $\varepsilon > 0$

$$P(\theta_N^{(1)} = \theta_N^{(2)} = \cdots = \theta_N^{(k)}) > \varepsilon.$$

If we now consider the Nth iterate kernel used backward, and set

$$C_i = \{ \text{ The } k \text{ chains coalesce in } (-iN, -(i-1)N) \},$$

under the assumption that all chains are started at time $-iN$ for the associated CFTP algorithm, we have that $P(C_i) > \varepsilon$. Moreover, the C_i are independent because coalescence in $(-iN, -(i-1)N)$ depends only on $\Psi_{-iN}, \Psi_{-iN-1}, \ldots, \Psi_{-(i-1)N}$.

Finally, note that the probability that there is no coalescence after I iterations can be bounded using Bonferroni's inequality:

$$1 - \prod_{i=1}^{I}[1 - P(C_i)] < (1 - \varepsilon)^I.$$

Thus, this probability goes to 0 as I goes to ∞, showing that the probability of coalescence is 1 (Problem 13.3).

The result then follows from the fact that the CFTP algorithm starts from all possible states. This implies that the realization of a Markov chain θ_t starting at $-\infty$ will, at some time $-t$, couple with one of the CFTP chains and thereafter be equal to θ_t^*. Therefore, θ_0^* and θ_0 have the same distribution and, in particular, $\theta_0^* \sim \pi$. □

Example 13.4. (Continuation of Example 13.1). In the setting of the beta-binomial model, using the transitions summarized in Figure 13.1, we draw U_0. Suppose $U_0 \in (.833, .917)$. Then the random mapping Ψ_0 is given as follows.

$$t = -1 \quad t = 0$$

The chains have not coalesced, so we go to time $t = -2$ and draw U_{-1}, with, for instance, $U_{-1} \in (.278, 417)$. The composition $\Psi_0 \circ \Psi_{-1}$ is then as follows.

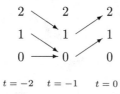

$$t = -2 \qquad t = -1 \qquad t = 0$$

The function $\Psi_0 \circ \Psi_{-1}$ is still not constant, so we go to time $t = -3$. Suppose now $U_{-2} \in (.278, .417)$. The composition $\Psi_0 \circ \Psi_{-1} \circ \Psi_{-2}$ is given by

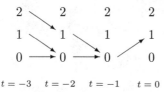

$$t = -3 \qquad t = -2 \qquad t = -1 \qquad t = 0$$

All chains have thus coalesced into $\theta_0 = 1$. We accept θ_0 as a draw from π. Note that, even though the chains have coalesced at $t = -1$, we do *not* accept $\theta_{-1} = 0$ as a draw from π.   ‖

Propp and Wilson (1996) suggested monitoring the maps only at times $-1, -2, -4, \ldots$, until coalescence occurs (at time 0), as this updating scheme has nearly optimal properties. Note that omitting some times when checking for a constant Φ_T does not invalidate the method. It simply means that we may go further backward than necessary.

13.2.3 Monotonicity and Envelopes

So far, the applicability of perfect sampling in realistic statistical problems is not obvious, given that the principle applies only to a finite state space with a manageable number of points. However, an important remark of Propp and Wilson (1996) (which has been exploited in many subsequent papers) is that the spectrum of perfect sampling can be extended to settings where a *monotonicity* structure is available, namely, when there exists a (stochastic) *ordering* $\preceq$ on the state space and when the extremal states for this ordering are known. Here, stochastic ordering is to be understood in terms of the random mappings Ψ_t, namely, that, if $\theta_1 \preceq \theta_2$, $\Psi_t(\theta_1) \leq \Psi_t(\theta_2)$ for all t's.

The existence of monotonicity structures immensely simplifies the implementation of the CFTP algorithm and correspondingly allows for extension to much more complex problems. For instance, if there exists a (stochastically) *larger* state denoted $\tilde{\mathbf{1}}$ and a (stochastically) *smaller* state $\tilde{\mathbf{0}}$, the CFTP can be run with only two chains starting from $\tilde{\mathbf{1}}$ and $\tilde{\mathbf{0}}$: when those two chains coincide at time 0, $\Phi_{-N}(\tilde{\mathbf{0}}) = \Phi_{-N}(\tilde{\mathbf{1}})$ say, the (virtual) chains starting from all possible states in $\mathcal{X}$ have also coalesced. Therefore, the intermediary chains do not require monitoring and this turns out to be a big gain if $\mathcal{X}$ is large.

Example 13.5. (Continuation of Example 13.1). Recall that the transition kernel on the (sub-)chain X_t is

$$K(x_t, x_{t+1}) \propto \binom{n}{x_{t+1}} \frac{\Gamma(\alpha + x_t + x_{t+1})\Gamma(\beta + n - x_t - x_{t+1})}{\Gamma(\alpha + x_t)\Gamma(\beta + n - x_t)}.$$

For $x_1 > x_2$,

$$\begin{aligned}
\frac{K(x_1, y)}{K(x_2, y)} &= \frac{\Gamma(\alpha + x_1 + y)\Gamma(\beta + n - x_1 - y)\Gamma(\alpha + x_2)\Gamma(\beta + n - x_2)}{\Gamma(\alpha + x_2 + y)\Gamma(\beta + n - x_2 - y)\Gamma(\alpha + x_1)\Gamma(\beta + n - x_1)} \\
&= \frac{\Gamma(\alpha + x_2)\Gamma(\beta + n - x_2)}{\Gamma(\alpha + x_1)\Gamma(\beta + n - x_1)} \frac{(\alpha + x_1 + y - 1)\cdots(\alpha + x_2 + y)}{(\beta + n - x_2 - y - 1)\cdots(\beta + n - x_1 - y)}
\end{aligned}$$

is increasing in y. This monotone ratio property implies that the transition is monotone for the usual ordering on $\mathbb{N}$ and therefore that $\tilde{\mathbf{0}} = 0$ and that $\tilde{\mathbf{1}} = n$. It is thus sufficient to monitor the chains started from 0 and from n to check for coalescence, for any value of n. This property can be observed in Figure 13.1. ‖

Example 13.6. Mixtures of distributions. Consider the simplest possible mixture structure

$$(13.2) \qquad\qquad p f_0(x) + (1 - p) f_1(x),$$

where both f_0 and f_1 are known, with a uniform prior on $0 < p < 1$. For a given sample $x_1, \ldots, x_n$, the corresponding Gibbs sampler is based on the following steps at iteration t.

1. Generate n iid $\mathcal{U}(0, 1)$ random variables $u_1^{(t)}, \ldots, u_n^{(t)}$.
2. Derive the indicator variables $z_i^{(t)}$ as $z_i^{(t)} = 0$ iff

$$u_i^{(t)} \leq \frac{p^{(t-1)} f_0(x_i)}{p^{(t-1)} f_0(x_i) + (1 - p^{(t-1)}) f_1(x_i)}$$

and compute

$$m^{(t)} = \sum_{i=1}^{n} z_i^{(t)}.$$

3. Simulate $p^{(t)} \sim \mathcal{B}e(n + 1 - m^{(t)}, 1 + m^{(t)})$.

To design a coupling strategy and a corresponding CFTP algorithm, we first notice that $m^{(t)}$ can only take $n + 1$ possible values, namely, $0, 1, \ldots, n$. We are thus de facto in a *finite* setting.

If we recall that a $\mathcal{B}e(m+1, n-m+1)$ random variable can be represented as the ratio

$$p = \sum_{i=1}^{m+1} \omega_i \Big/ \sum_{i=1}^{n+2} \omega_i$$

for $\omega_1, \ldots, \omega_{n+2}$ iid $\mathcal{E}xp(1)$, we notice in addition that, if we use for all parallel chains a common vector $(\omega_1^{(t)}, \ldots, \omega_{n+2}^{(t)})$ of iid $\mathcal{E}xp(1)$ random variables, $p^{(t)}$ is *increasing* with $m^{(t)}$. We have therefore exhibited a monotone structure associated with the Gibbs transition kernel and we thus only need to monitor two chains: one started with $m^{(-T)} = 0$ and the other with $m^{(-T)} = n$, and check whether or not they coalesce by time $t = 0$. Figure 13.2 *(left)* illustrates the sandwiching argument sketched above: if we start from all possible chains, that is, from all possible values of $m^{(t)}$, they have all coalesced by the time the two "extreme" chains have coalesced. Figure 13.2 *(right)* provides an additional (if unnecessary) checking that the stationary distribution is indeed the distribution of the chain at time $t = 0$.   ‖

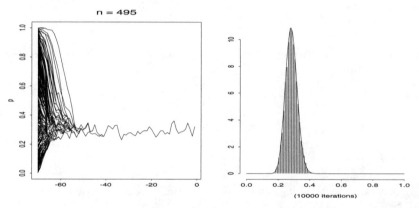

Fig. 13.2. *(left)* Simulation of $n = 495$ iid random variables from $.33\,\mathcal{N}(3.2, 3.2) + .67\,\mathcal{N}(1.4, 1.4)$ and coalescence at $t = -73$. *(right)* Iid sample from the posterior distribution produced by the CFTP algorithm and fit to the corresponding posterior density. (*Source:* Hobert et al. 1999.)

As stressed by Kendall (1998) and Kendall and Møller (1999, 2000), this monotonicity can actually be weakened into an *envelope* or *sandwiching* argument: if one can find two sequences $(\underline{\theta}_t)$ and $(\overline{\theta}_t)$, generated from a transition not necessarily associated with the target distribution π, such that, for any starting value θ_0,

(13.3) $$\underline{\theta}_t \preceq \Phi_t(\theta_0) \preceq \overline{\theta}_t,$$

the coalescence of the two bounding sequences $\underline{\theta}_t$ and $\overline{\theta}_t$ obviously implies coalescence of the (virtual) chains starting from all possible states in $\mathcal{X}$. While this strategy is generally more costly in the number of backward iterations, it nonetheless remains valid: the value produced at time 0 is distributed from π. Besides, as we will see below in Example 13.7, it may also save computing time. Note also that the upper and lower bounding sequences $\underline{\theta}_t$ and $\overline{\theta}_t$ need not necessarily evolve within $\mathcal{X}$ at every step. This is quite a useful feature when working on complex constrained spaces.

Example 13.7. (Continuation of Example 13.6.) Consider the extension of (13.2) to the 3-component case

$$(13.4) \qquad p_1 f_1(x) + p_2 f_2(x) + p_3 f_3(x), \qquad p_1 + p_2 + p_3 = 1,$$

where f_1, f_2, f_3 are known densities. If we use a (flat) Dirichlet $\mathcal{D}(1,1,1)$ prior on the parameter (p_1, p_2, p_3), an associated Gibbs sampler for a sample $x_1, \ldots, x_n$ from (13.4) is as follows.

1. Generate $u_1, \ldots, u_n \sim \mathcal{U}(0,1)$.
2. Take

$$n_1 = \sum_{i=1}^{n} \mathbb{I}\left(u_i \leq \frac{p_1 f_1(x_i)}{p_1 f_1(x_i) + p_2 f_2(x_i) + p_3 f_3(x_i)}\right),$$

$$n_2 = \sum_{i=1}^{n} \left\{ \mathbb{I}\left(u_i > \frac{p_1 f_1(x_i)}{p_1 f_1(x_i) + p_2 f_2(x_i) + p_3 f_3(x_i)}\right) \right.$$

$$\left. \times \mathbb{I}\left(u_i \leq \frac{p_1 f_1(x_i) + p_2 f_2(x_i)}{p_1 f_1(x_i) + p_2 f_2(x_i) + p_3 f_3(x_i)}\right)\right\},$$

and $n_3 = n - n_1 - n_2$.
3. Generate $(p_1, p_2, p_3) \sim \mathcal{D}(n_1 + 1, n_2 + 1, n_3 + 1)$.

Once again, since (n_1, n_2) can take only a finite number of values, the setting is *finite* and we can monitor all possible starting values of (n_1, n_2) to run our CFTP algorithm. However, there are $(n+2)(n+1)/2$ such values. The computational cost is thus in $O(n^2)$, rather than $O(n)$ for the two component case. And there is no obvious monotonicity structure, even when considering the exponential representation of the Dirichlet distribution:

$$\left(\frac{\sum_{i=1}^{n_1+1} \omega_{1i}}{\sum_{j=1}^{3} \sum_{i=1}^{n_j+1} \omega_{ji}}, \frac{\sum_{i=1}^{n_2+1} \omega_{2i}}{\sum_{j=1}^{3} \sum_{i=1}^{n_j+1} \omega_{ji}}, \frac{\sum_{i=1}^{n_3+1} \omega_{3i}}{\sum_{j=1}^{3} \sum_{i=1}^{n_j+1} \omega_{ji}}\right) \sim \mathcal{D}(n_1+1, n_2+1, n_3+1)$$

with

$$\omega_{11}, \ldots, \omega_{1(n+1)}, \omega_{21}, \ldots, \omega_{2(n+1)}, \omega_{31}, \ldots, \omega_{3(n+1)} \sim \mathcal{E}xp(1).$$

While looking only at the boundary cases $n_i = 0$ ($i = 1, 2, 3$) seemed to be the natural generalization of the case with 2 components, Hobert et al. (1999) exhibited occurrences where coalescence of the entire range of chains had not occurred at coalescence of the boundary chains.

In this setting, Hobert et al. (1999) have, however, obtained an enveloping result, namely that the image of the triangle of all possible starting values of (n_1, n_2),

$$\mathcal{T} = \{(n_1, n_2); \; n_1 + n_2 \leq n\}$$

by the corresponding Gibbs transition is contained in the *lozenge*[4]

$$\mathcal{L} = \{(n_1, n_2); \underline{n}_1 \leq n_1 \leq \overline{n}_1, \; n_2 \geq 0, \; \underline{n}_3 \leq n - n_1 - n_2 \leq \overline{n}_3\},$$

where

- $\underline{n}_1$ is the minimum of the n_1's over the images of the left border of $\mathcal{T}$;
- $\overline{n}_3$ is the n_3 coordinate of the image of $(0, 0)$;
- $\overline{n}_1$ is the n_1 coordinate of the image of $(n, 0)$;
- $\underline{n}_3$ is the minimum of the n_3's over the images of the diagonal of $\mathcal{T}$.

The reason for this property is that, for a fixed n_2,

$$\frac{p_2}{p_1} = \sum_{i=1}^{n_2+1} w_{2i} \bigg/ \sum_{i=1}^{n_1+1} w_{1i} \quad \text{and} \quad \frac{p_3}{p_1} = \sum_{i=1}^{n-n_1-n_2+1} w_{3i} \bigg/ \sum_{i=1}^{n_1+1} w_{1i}$$

are both decreasing in n_1. And this is also true for

$$m_1 = \sum_{i=1}^{n} \mathbb{I}\left(u_i \leq \left[1 + \frac{p_2 f_2(x_i) + p_3 f_3(x_i)}{p_1 f_1(x_i)}\right]^{-1}\right).$$

Besides, this boxing property is preserved over iterations, in that the image of $\mathcal{L}$ is contained in

$$\mathcal{L}' = \{(m_1, m_2); \underline{m}_1 \leq m_1 \leq \overline{m}_1, \; m_2 \geq 0, \; \underline{m}_3 \leq m_3 \leq \overline{m}_3\},$$

where

- $\underline{m}_1$ is the minimum n_1 over the images of the left border $\{n_1 = \underline{n}_1\}$;
- $\overline{m}_1$ is the maximum n_1 over the images of the right border $\{n_1 = \overline{n}_1\}$;
- $\underline{m}_3$ is the minimum n_3 over the images of the upper border $\{n_3 = \underline{n}_3\}$;
- $\overline{m}_3$ is the maximum n_3 over the images of the lower border $\{n_3 = \overline{n}_3\}$.

Therefore, checking for coalescence simplifies into monitoring the successive lozenges $\mathcal{L}$ until they reduce to a single point. Figure 13.3 represents a few successive steps of this monitoring, with the original chains provided for comparison. It illustrates the obvious point that coalescence will consume more iterations than direct monitoring of the chain, while requiring only a $O(n)$ computational effort at each iteration, compared with the brute force $O(n^2)$ effort of the direct monitoring. $\parallel$

---

[4] A *lozenge* is a figure with four equal sides and two obtuse angles and, more commonly, a diamond-shaped candy used to suppress a cough, a "cough-drop."

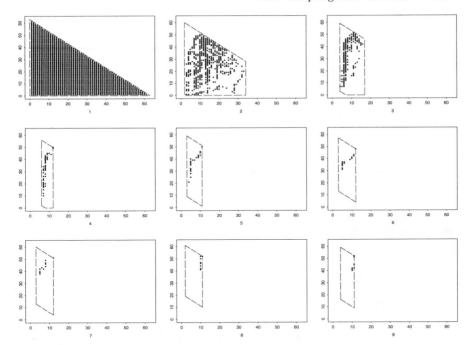

Fig. 13.3. Nine successive iterations of the coupled Gibbs sampler for the chains associated with all possible starting values (n_1, n_2), along with the evolution of the envelope $\mathcal{L}$, for $n = 63$ observations simulated from a $.12\,\mathcal{N}(1.1, 0.49) + .76\,\mathcal{N}(3.2, 0.25) + .12\,\mathcal{N}(2.5, 0.09)$ distribution. (*Source:* Hobert et al., 1999.)

Kendall (1998) also noticed that the advantages of monotonicity can be extended to *anti-monotone* settings, that is, when

$$\Psi(\theta_1, u) \preceq \Psi(\theta_2, u) \qquad \text{when} \qquad \theta_2 \preceq \theta_1 .$$

Indeed, it is sufficient to define the lower and upper bounding sequences, $(\underline{\theta}_t)$ and $(\overline{\theta}_t)$, as

$$\underline{\theta}_{t+1} = \Psi_{t+1}(\overline{\theta}_t) \qquad \text{and} \qquad \overline{\theta}_{t+1} = \Psi_{t+1}(\underline{\theta}_t),$$

since the lower bound $\underline{\theta}_t$ is transformed into an upper bound by the transform Ψ_t, and vice versa.

13.2.4 Continuous States Spaces

The above extension, while interesting, does not seem general enough to cover the important case of continuous states spaces. While Section 13.2.5 will show that there exists a (formally) generic approach to the problem, the current section illustrates some of the earlier and partial attempts to overcome the difficulty of dealing with continuous states spaces.

First, as seen in Examples 13.1 and 13.6, it is possible to devise perfect sampling algorithms in data augmentation settings (see Section 9.2.1), where one of the two chains has a finite support. In Example 13.1, once X is produced by the perfect sampling algorithm, the continuous variable θ can be generated from the conditional $\mathcal{B}e(\alpha + x, \beta + n - x)$. (This is another application of the duality principle given in Section 9.2.3.)

A more general approach to perfect sampling in the continuous case relies on the *Kac's representation of stationary distributions* (Section 6.5.2)

$$\pi(A) = \sum_{t=1}^{\infty} \Pr(N_t \in A) \Pr(T^* = t),$$

where the kernel satisfies a minorizing condition $K(x,y) \geq \varepsilon\nu(y)\mathbb{I}_C(x)$, τ_α is the associated renewal stopping rule, T^* is the *tail renewal time*

$$\Pr(T^* = t) = \frac{\Pr_\alpha(\tau_\alpha \geq t)}{\mathbb{E}_\alpha(\tau_\alpha)},$$

and N_t has the same distribution as X_t, given $X_1 \sim \nu(\cdot)$ and *no regeneration before time t.*

This was first applied by Murdoch and Green (1998) to uniformly ergodic chains (for example, Metropolis–Hastings algorithms with atoms, including the independent case (see Section 12.2.3)). When the kernel is uniformly ergodic, *Doeblin's condition* holds (Section 6.59) and the kernel satisfies $K(x,y) \geq \varepsilon\nu(y)$ for all $x \in \mathcal{X}$. Kac's representation then simplifies to

$$\pi(x) = \sum_{i=0}^{+\infty} \varepsilon(1 - \varepsilon)^i (\tilde{K}^i \nu)(x),$$

where $\tilde{K}$ denotes the residual kernel

$$\tilde{K}(x,y) = \frac{K(x,y) - \varepsilon\nu(x)}{1 - \varepsilon}.$$

The application of this representation to perfect sampling is then obvious: the backward excursion time T is distributed from a geometric distribution $\mathcal{G}eo(\varepsilon)$ and a *single* chain needs to be simulated forward from the residual kernel T times. The algorithm thus looks as follows.

Algorithm A.55 –Kac's Perfect Sampling–

1. Simulate $x_0 \sim \nu$, $\omega \sim \mathcal{G}eo(\varepsilon)$.
2. Run the transition $x_{t+1} \sim \tilde{K}(x_t, y)$ $t = 0, \cdots, \omega - 1$, and take x_ω as a realization from π.

See Green and Murdoch (1999) for other extensions, including the *multi-gamma coupling*. Hobert and Robert (2004) explore the extension of the uniformly ergodic case to the general case via Kac's representation of stationary distributions.

There is, however, a fundamental difficulty in extending CFTP from finite state spaces to general (countable or uncountable) spaces. If

$$T^* = \inf \{t \in \mathbb{N} : \Phi_t(\omega) = \Phi_t(\omega') \text{ for all } \omega, \omega'\}$$

denotes the stopping time associated with [A.54], Foss and Tweedie (1998) established the following result.

Lemma 13.8. *The stopping time T^* is almost surely finite,*

$$\Pr(T^* < \infty) = 1,$$

if and only if the Markov kernel is uniformly ergodic.

Proof. The sufficient part of the lemma is obvious: as shown above, the stopping time then follows a geometric distribution $\mathcal{G}eo(\varepsilon)$.

The necessary part is also straightforward: if C_t denotes the event that coalescence has occurred by time t, $\varepsilon_t = \Pr(C_t)$, and ν_t is the probability distribution of Φ_{t+1} conditional on C_t, then Doeblin's condition is satisfied:

$$K^{t+1}(\omega, \omega') \geq \varepsilon_t \nu_t(\omega').$$

And $\Pr(T^* < \infty) = 1$ implies there exists a $t < \infty$ such that $\varepsilon_t > 0$. □

The message contained in this result is rather unsettling: *CFTP is not guaranteed to work for Markov kernels that are not uniformly ergodic*. Thus, for instance, it is useless to try to implement [A.54] in the case of a random walk proposal on an unbounded state space. As noted in Foss and Tweedie (1998), things are even worse: *outside uniformly ergodic kernels, CFTP does not work, because the backward coupling time is either a.s. finite or a.s. infinite.*

Lemma 13.9. *If T^* is the stopping time defined in [A.54] as*

$$T^* = \inf_T \{T; \Phi_T \text{ constant}\},$$

then

$$\Pr(T^* = \infty) \in \{0, 1\}.$$

The proof of Foss and Tweedie (1998) is based on the fact that the event $\{T^* = \infty\}$ is invariant under the transition kernel and a "0–1" law applies. That is, that $\Pr(T^* = \infty)$ is either 0 or 1.

Corcoran and Tweedie (2002) point out that this does not imply that backward coupling never works for chains that are not uniformly ergodic: if bounding processes as in (13.3) are used, then coupling may occur in an almost surely finite time outside uniform ergodic setups, as exemplified in Kendall (1998). Nonetheless, situations where uniform ergodicity does not hold are more perilous in that an unmonitored CFTP may run forever.

13.2.5 Perfect Slice Sampling

The above attempts, and others in the literature, remain fragmentary in that they either are based on specialized results or suppose some advanced knowledge of the transition kernel (for example, a renewal set and its residual kernel).[5] In this section, we present a much more general approach which, even if it is not necessarily easily implementable, offers enough potential to be considered as a generic perfect sampling technique.

Recall that we introduced the slice sampling method in Chapter 8 as a general Gibbs sampling algorithm based on the use of auxiliary variables with uniform distributions of the type

$$\mathcal{U}\left([0, \pi_i(\theta)]\right).$$

In the special case of the *univariate slice sampler,* also recall that there is a single auxiliary variable, $u \sim \mathcal{U}([0, 1])$, and that the simulation of the parameter of interest is from the distribution

$$\mathcal{U}\left(\{\omega \,;\, \pi(\omega) \geq u\pi(\omega_0)\}\right).$$

In this case the slice sampler is the random walk version of the fundamental theorem of simulation, Theorem 2.15.

Mira et al. (2001) realized that the fundamental properties of the slice sampler made it an ideal tool for perfect sampling. Indeed, besides being independent of normalizing constants, the slice sampler induces a *natural order* on the space of interest, defined by π, the target distribution:

$$(13.5) \qquad \omega_1 \preceq \omega_2 \quad \text{if and only if} \quad \pi(\omega_1) \leq \pi(\omega_2).$$

This is due to the fact that if $\pi(\omega_1) \leq \pi(\omega_2)$, and if $u \sim \mathcal{U}([0, 1])$, the associated sets satisfy:

$$(13.6) \qquad \mathcal{A}_2 = \{\omega \,;\, \pi(\omega) \geq u\pi(\omega_2)\} \subset \mathcal{A}_1 = \{\omega \,;\, \pi(\omega) \geq u\pi(\omega_1)\}.$$

Therefore, if one simulates *first* the image of ω_1 as $\omega_1' \sim \mathcal{U}(\mathcal{A}_1)$, this value can be proposed as image of ω_2 and accepted if $\omega_1' \in \mathcal{A}_2$, rejected otherwise. In both cases, $\pi(\omega_1') \leq \pi(\omega_2')$, and this establishes the following result:

Lemma 13.10. *The slice sampling transition can be coupled in order to respect the natural ordering (13.5).*

There is even more to the slice sampler. In fact, as noticed by Mira et al. (2001), slice samplers induce a natural discretization of continuous state spaces. Using the inclusion (13.6), it is indeed possible to reduce the number

---

[5] Once again, we stress that we exclude from this study the fields of stochastic geometry and point processes, where the picture is quite different. See Møller (2001, 2003) and Møller and Waagepetersen (2003).

of chains considered in parallel from a continuum to a discrete and, we hope, finite set. To do this, we start from the minimal state $\tilde{\mathbf{0}} = \arg\min \pi(\omega)$ and a uniform variate u, and generate the image of $\tilde{\mathbf{0}}$ by the slice sampler transition,

$$\omega_0 \sim \mathcal{U}\left(\{\omega \, ; \, \pi(\omega) \geq u\pi(\tilde{\mathbf{0}})\}\right) .$$

This is straightforward since $\{\omega \, ; \, \pi(\omega) \geq u\pi(\tilde{\mathbf{0}})\}$ is the entire parameter set. (Obviously, this requires the parameter set to be bounded but, as noticed earlier, a parameter set can always be compactified by reparameterization, as shown in Example 13.11 below.) Once ω_0 is generated, it is acceptable not only as the image of $\tilde{\mathbf{0}}$ but also of all the starting values θ such that

$$\pi(\omega_0) \geq u\pi(\theta) .$$

The set of starting values $\{\theta; \pi(\theta) \leq \pi(\omega_0)/u\}$ is thus reduced to a single image by this coupling argument. Now, consider a value θ_1 (which can be arbitrary, see Problem 13.10) such that $\pi(\theta_1) = \pi(\omega_0)/u$ and generate its image as

$$\omega_1 \sim \mathcal{U}\left(\{\omega \, ; \, \pi(\omega) \geq u\pi(\theta_1)\}\right) .$$

Then $\pi(\omega_1) \geq \pi(\omega_0)$ and ω_1 is a valid image for all θ's such that $\pi(\omega_0)/u \leq \pi(\theta) \leq \pi(\omega_1)/u$. We have thus reduced a second continuum of starting values to a single chain. This discretization of the starting values continues until the maximum of π, if it exists, is reached, that is, when $\pi(\omega_J)/u \geq \max \pi(\omega)$. If π is unbounded, the sequence of starting values ω_j ($j \geq 0$) is infinite (but countable).

Example 13.11. Truncated normal distributions While we saw, in Example 2.20, an Accept–Reject algorithm for the simulation of univariate normal distributions, the multivariate case is more difficult and both Geweke (1991) and Robert (1993) suggested using Gibbs sampling in higher dimensions. Consider, as in Philippe and Robert (2003), a multivariate $\mathcal{N}_p(\mu, \Sigma)$ distribution restricted to the *positive quadrant*

$$\mathcal{Q}_+ = \{\mathbf{x} \in \mathbb{R}^p \, ; x_i > 0, \ i = 1, \ldots, p\} = (\mathbb{R}_+)^p ,$$

which is bounded by $(2\pi|\Sigma|)^{-p/2}$ at $\tilde{\mathbf{1}} = \mu$. While μ may well be outside the domain $\mathcal{Q}_+$, this is not relevant here (by the sandwiching argument).

The problem, in this case, is with $\tilde{\mathbf{0}}$: the minimum of the truncated normal density is achieved at ∞ and it is obviously impossible to start simulation from infinity, since this would amount to simulating from a "uniform distribution" on $\mathcal{Q}_+$. A solution is to use reparameterization: the normal distribution truncated to $\mathcal{Q}_+$ can be transformed into a distribution on $[0, 1]^d$ by the function

$$h(\mathbf{x}) = \left(\frac{x_1}{1 + x_1}, \ldots, \frac{x_p}{1 + x_p}\right) = \mathbf{z} ,$$

with the corresponding density

$$\pi(\mathbf{z}|\mu, \Sigma) \propto \exp\left\{-(h^{-1}(\mathbf{z}) - \mu)^{\mathrm{T}}\Sigma^{-1}(h^{-1}(\mathbf{z}) - \mu)/2\right\}\prod_{i=1}^{p}\frac{1}{(1 - z_i)^2}.$$

With this new parameterization, $\tilde{\mathbf{1}}$ has to be found, analytically or numerically (because of the additional Jacobian term in the density), while $\tilde{\mathbf{0}} = (1, \ldots, 1)$. In the special case when $\mu = (-2.4, 1.8)$ and

$$\Sigma = \begin{pmatrix} 1.0 & -1.2 \\ -1.2 & 4.4 \end{pmatrix}.$$

$\tilde{\mathbf{1}} = (0, 1.8/2.8)$. See Philippe and Robert (2003) for the implementation details of perfect sampling in this setting.   ‖

When π is bounded and the maximum value $\tilde{\mathbf{1}} = \arg\max \pi(\omega)$ is known (see Problem 13.14 for an extension), an associated perfect sampling algorithm thus reads as follows.

Algorithm A.56 –Perfect Monotone Slice Sampler–

At time $-T$

1. Start from $\tilde{\mathbf{0}}$.
2. Generate $u_{-T} \sim \mathcal{U}([0, 1])$.
3. Generate $\omega_0^{(-T)}$ uniformly over the parameter set.
4. Generate the images $\omega_0^{(-t)}$ as $t = -T + 1, \ldots, 0$

$$\mathcal{U}\left(\{\omega; \pi(\omega) \geq u_{-t}\pi(\omega_0^{(-t-1)})\}\right).$$

5. Set $\omega_1^{(-T-1)} = \tilde{\mathbf{1}}$.
6. For $t = -T, \ldots, 0$,
   if $\omega_0^{(-t)}$ satisfies

$$\pi(\omega_0^{(-t)}) \geq u_{-t}\pi(\omega_1^{(-t-1)}),$$

   there is coalescence at time 0;
   if not, generate $\omega_1^{(-t)}$ as

$$\mathcal{U}\left(\{\omega; \pi(\omega) \geq u_{-t}\pi(\omega_1^{(-t-1)})\}\right).$$

7. If there is no coalescence at time 0, increment T.

Example 13.12. Exponential mixtures For an exponential mixture

$$p\,\mathcal{E}xp(\lambda_0) + (1 - p)\,\mathcal{E}xp(\lambda_1),$$

where $0 < p < 1$ is known, a direct implementation of slice sampling is quite difficult, since simulation from

$$\mathcal{U}\left(\left\{(\lambda_0, \lambda_1); \prod_{i=1}^{n} [p\lambda_0 \exp(-x_i\lambda_0) + (1-p)\lambda_1 \exp(-x_i\lambda_1)] \geq \epsilon\right\}\right)$$

is impossible. However, using the missing data structure of the mixture model (see Example 9.2)

$$\mathbf{z}, \lambda_0, \lambda_1 \mid \mathbf{x} \sim \pi(\lambda_0, \lambda_1) \prod_{i=1}^{n} p^{z_i}(1-p)^{1-z_i} \lambda_{z_i} \exp(-x_i\lambda_{z_i}),$$

where $z_i \in \{0, 1\}$, it is feasible to come up with a much simpler version. The idea developed in Casella et al. (2002) is to integrate out the parameters, i.e., (λ_0, λ_1), and use the slice sampler on the marginal posterior distribution of the missing data vector, $\mathbf{z}$. This has the obvious advantage of re-expressing the problem within a finite state space.

If the prior on λ_i is a $\mathcal{G}a(\alpha_i, \beta_i)$ distribution, the marginal posterior is available in closed form,

$$(13.7) \qquad \mathbf{z} \mid \mathbf{x} \sim p^{n_0}(1-p)^{n_1} \frac{\Gamma(\alpha_0 + n_0 - 1)\Gamma(\alpha_1 + n_1 - 1)}{(\beta_0 + s_0)^{\alpha_0 + n_0}(\beta_1 + s_1)^{\alpha_1 + n_1}},$$

where $(i = 0, 1)$

$$n_i = \sum_{j=1}^{n} \mathbb{I}_{z_j=i} \quad \text{and} \quad s_i = \sum_{j=1}^{n} \mathbb{I}_{z_j=i} \, x_j.$$

It thus factors through the sufficient statistic (n_0, s_0).

In addition, the extreme values $\tilde{\mathbf{0}}$ and $\tilde{\mathbf{1}}$ are available. For every n_0, s_0 belongs to the interval $[\underline{s}_0(n_0), \overline{s}_0(n_0)]$, where

$$\underline{s}_0(n_0) = \sum_{j=1}^{n_0} x_{(j)} \quad \text{and} \quad \overline{s}_0(n_0) = \sum_{j=1}^{n_0} x_{(n+1-j)},$$

in which the $x_{(i)}$'s denote the order statistics of the sample $x_1, \ldots, x_n$, with $x_{(1)} \leq \cdots \leq x_{(n)}$. The minimum of the marginal posterior is achieved at

$$s_0^*(n_0) = \frac{(n_0 + \alpha_0)(\beta_1 + S) - (n - n_0 + \alpha_1)\beta_0}{n + \alpha_0 + \alpha_1},$$

where S is the sum of all observations, provided this value is in $[\underline{s}_0(n_0), \overline{s}_0(n_0)]$. The maximum is obviously attained at one of the two extremes. The points $\tilde{\mathbf{1}}$ and $\tilde{\mathbf{0}}$ are then obtained by $(n+1)$ comparisons over all n_0's. Note that, even though $s_0^*(n_0)$ is not necessarily a *possible* value of s_0, the sandwiching argument introduced in Section 13.2.3 validates its use for studying coalescence.

This scheme is only practical for small values of n (say $n \leq 40$), because of the large number of values of s_0, namely, $\binom{n}{n_0}$. Figure 13.4 gives an illustration of its behavior for $n = 40$.   ‖

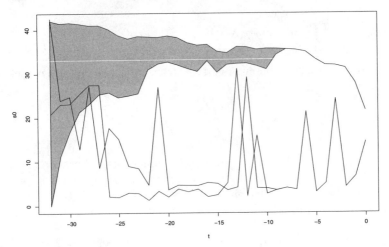

Fig. 13.4. Coupling from the past on the distribution (13.7) for a fixed $n_0 = 35$, for $n = 40$ simulated observations, with the sequence of $s_0^{(t)}$'s for both the upper and the lower chains, and the range of the marginal posterior in the background. (*Source:* Casella et al. 2002.)

13.2.6 Perfect Sampling via Automatic Coupling

In addition to slice sampling, there is another approach to perfect sampling, which is interesting from a theoretical point of view, as it highlights the connection with Accept–Reject algorithms. It is based on Breyer and Roberts (2000a) automatic coupling.

We have two chains $(\omega_0^{(t)})$ and $(\omega_1^{(t)})$, where the first chain, $(\omega_0^{(t)})$, is run following a standard Metropolis–Hastings scheme with proposal density $q(\omega'|\omega)$ and corresponding acceptance probability

$$\frac{\pi(\omega')\,q(\omega_0^{(t)}|\omega')}{\pi(\omega_0^{(t)})\,q(\omega'|\omega_0^{(t)})} \wedge 1 \,.$$

The transition for the second chain, $(\omega_1^{(t)})$, is based on the same proposed value ω', with the difference being that the proposal distribution is *independent* of the second chain and is associated with the acceptance probability

$$\frac{\pi(\omega')\,q(\omega_1^{(t)}|\omega_0^{(t)})}{\pi(\omega_1^{(t)})\,q(\omega'|\omega_0^{(t)})} \wedge 1 \,.$$

If ω' is *not accepted* by the second chain, it then stays the same, $\omega_1^{(t+1)} = \omega_1^{(t)}$. Note that, keeping in mind the perfect sampling perspective, we can use the same uniform variable u_t for both transitions, thus ensuring that the chains coalesce once they both accept a proposed value ω'.

The implications of this rudimentary coupling strategy on perfect sampling are interesting. Consider the special case of an independent proposal with density h for the first chain $(\omega_0^{(t)})$. Starting from $x_0^{(t)}$ with candidate y_t, the above acceptance probabilities simplify to

$$\varrho_0^{(t)} = \frac{\pi(y_t)\,h(x_0^{(t)})}{\pi(x_0^{(t)})\,h(y_t)} \wedge 1 \quad \text{and} \quad \varrho_1^{(t)} = \frac{\pi(y_t)\,h(x_1^{(t)})}{\pi(x_1^{(t)})\,h(y_t)} \wedge 1 .$$

We can then introduce the (total) ordering associated with π/h, that is,

$$\omega_0 \preceq \omega_1 \quad \text{if} \quad \frac{\pi(\omega_0)}{h(\omega_0)} \leq \frac{\pi(\omega_1)}{h(\omega_1)} ,$$

and observe that the above coupling scheme preserves the ordering (Problem 13.9). This property, namely that the coupling proposal preserves the ordering, was noted by Corcoran and Tweedie (2002).

If we are in a setting where there exist a minimal and a maximal element, $\tilde{0}$ and $\tilde{1}$, for the order $\preceq$, we can then run a perfect sampler based on the two chains starting from $\tilde{0}$ and $\tilde{1}$. Note that, since the choice of h is arbitrary, it can be oriented towards an easy derivation of $\tilde{0}$ and $\tilde{1}$. Nonetheless, this cannot be achieved in every setting, since the existence of $\tilde{1}$ implies that the ratio π/h is bounded, therefore that the corresponding MCMC chain is uniformly ergodic, as shown in Theorem 7.8. In the special case where the state space $\mathcal{X}$ is compact, h can be chosen as the uniform distribution on $\mathcal{X}$ and the extremal elements $\tilde{0}$ and $\tilde{1}$ are then those induced by π.

Example 13.13. (Continuation of Example 13.12) This strategy applies in the case of the mixture model, since the pair (n_0, s_0) evolves in a compact (finite) state. As an illustration, we fix the value of n_0 and simulate (exactly) s_0 from the density proportional to

$$(13.8) \qquad \frac{\Gamma(\alpha_0 + n_0)}{(\beta_0 + s_0)^{\alpha_0 + n_0}} \frac{\Gamma(\alpha_1 + n - n_0)}{(\beta_1 + S - s_0)^{\alpha_1 + n - n_0}}$$

on the finite set of the possible values of

$$s_0 = \sum_{j=1}^{n_0} x_{i_j} .$$

This means that, starting at time $-T$ from $\tilde{0}$, we generate the chain $(\omega_0^{(t)})$ forward in time according to a regular Metropolis–Hastings scheme with uniform proposal on the set of possible values of s_0. The chain $(\omega_1^{(t)})$, starting from $\tilde{1}$, is then coupled to the first chain by the above coupling scheme. If coalescence had not occurred by time zero, we increase T.

Figure 13.5 illustrates the rudimentary nature of the coupling: for almost $1,000$ iterations, coupling is rejected by $(\omega_1^{(t)})$, which remains equal to the original value $\tilde{1}$, while $(\omega_0^{(t)})$ moves with a good mixing rate. Once coupling has occurred, both chains coalesce.   ‖

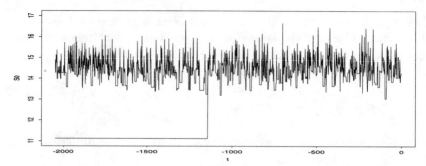

Fig. 13.5. Sample path of both chains $(\omega_0^{(t)})$ and $(\omega_1^{(t)})$ for a successful CFTP in a two-component exponential mixture simulated sample ($n = 81$ and $n_0 = 59$). (*Source:* Casella et al. 2002.)

The above example shows the fundamental structure of this automatic CFPT algorithm: it relies on a *single chain*, since the second chain is only used through its value at $\bar{1}$, that is, $\max \pi(\theta)/h(\theta)$. At the time the two chains merge, the simulated value from h satisfies

$$\pi(\omega_0^{(t)})/h(\omega_0^{(t)}) \le u_t \max\{\pi(\theta)/h(\theta)\}.$$

It is therefore *exactly* distributed from π, since this is the usual Accept–Reject step.[6] Therefore, in the case of independent proposals, the automatic CFPT is an Accept–Reject algorithm! See Corcoran and Tweedie (2002) for other and more interesting extensions. We now consider a second and much less obvious case of an Accept–Reject algorithm corresponding to a perfect sampling algorithm.

13.3 Forward Coupling

Despite the general convergence result that coalescence takes place over an almost surely finite number of backward iterations when the chain is uniformly ergodic (see Section 13.2.3), there are still problems with the CFTP algorithm, in that the running time is nonetheless potentially unbounded. Consider the practical case of a user who cannot wait for a simulation to come out and decides to abort runs longer than T_0, restarting a new CFTP each time. This abortion of long runs modifies the output of the algorithm and does not produce exact simulations from π (Problem 13.16). To alleviate this difficulty, Fill (1998a) proposed an alternative algorithm which can be interrupted while preserving the central feature of CFTP.

Fill's algorithm is a forward–backward coupling algorithm in that it considers a fixed time horizon in the future, i.e., *forward*, and then runs the

---

[6] This is a very special case when the distribution at coalescence is the stationary distribution.

Markov chain *backward* to time 0 to see if this value is acceptable. It relies on the availability of the *reverse kernel* $\tilde{K}$, namely,

$$(13.9) \qquad \tilde{K}(\omega,\omega') = \frac{\pi(\omega')K(\omega',\omega)}{\pi(\omega)},$$

which corresponds to running the associated Markov chain backward in time.[7] It also requires simulation of the variable U in the random mapping Ψ conditional on its image, that is, conditional on $\omega' = \Psi(\omega,u)$ for fixed ω and ω'.

Let $g_\Psi(\cdot|\omega,\omega')$ denote the conditional distribution of U. A first version of this algorithm for a finite state space is the following.

Algorithm A.57 –Fill's Perfect Sampling Algorithm–

1. Choose a time T and a state $x_T = z$.
2. Generate $X_{T-1}|x_T$, $X_{T-2}|x_{T-1}$, ..., $X_0|x_1$ from the reverse kernel $\tilde{K}$.
3. Generate $U_1 \sim g_\Psi(u|x_0,x_1)$, $U_2 \sim g_\Psi(u|x_1,x_2)$, ..., $U_T \sim g_\Psi(u|x_{T-1},x_T)$ and derive the random mappings $\Psi_1 = \Psi(u_1,\cdot),\ldots,\Psi_T = \Psi(u_T,\cdot)$.
4. Check whether the composition $\Psi_1 \circ \cdots \circ \Psi_T$ is constant
5. If so, then accept x_0 as a draw from π.
6. Otherwise begin again, possibly with new values of T and z.

Lemma 13.14. *The algorithm [A.57] returns an exact simulation from the stationary distribution π.*

There are several ways to prove that [A.57] is correct. Problem 13.17 gives one proof, but perhaps the most intuitive approach is to establish that [A.57] is, in essence, a rejection algorithm.

Proof. Let $C_T(z)$ denote the event that all chains have coalesced and are in state z at time T. The algorithm [A.57] is a rejection algorithm in that it generates $X_0 = x$, then accepts x as a draw from π if $C_T(z)$ has occurred. The associated proposal distribution is the T-step transition density $\tilde{K}^T(z,\cdot)$. The validity of the algorithm is established if x is accepted with probability

$$\frac{1}{M}\frac{\pi(x)}{\tilde{K}^T(z,x)}, \quad \text{where} \quad M \geq \sup_x \frac{\pi(x)}{\tilde{K}^T(z,x)}.$$

From (13.9), $\pi(x)/\tilde{K}^T(z,x) = \pi(z)/K^T(x,z)$. In addition, $P[C_T(z)] \leq K^T(x',z)$ for any x', since the probability of coalescence is smaller than the probability that the chain reaches z starting in x'. Hence $P[C_T(z)] \leq \min_{x'} K^T(x',z)$ and we have the bound

---

[7] If detailed balance (Definition 6.45) holds, $\tilde{K} = K$.

$$\max_x \frac{\pi(x)}{\tilde{K}^T(z,x)} = \max_x \frac{\pi(z)}{K^T(x,z)} = \frac{\pi(z)}{\min_{x'} K^T(x',z)} \le \frac{\pi(z)}{P[C_T(z)]}.$$

So set $M = \frac{\pi(z)}{P[C_T(z)]}$ and, to run a correct Accept–Reject algorithm, we should accept $X_0 = x$ with probability

$$\frac{1}{M}\frac{\pi(x)}{\tilde{K}^T(z,x)} = \frac{P[C_T(z)]}{\pi(z)}\frac{\pi(x)}{\tilde{K}^T(z,x)} = \frac{P[C_T(z)]}{\pi(z)}\frac{\pi(z)}{K^T(x,z)} = \frac{P[C_T(z)]}{K^T(x,z)},$$

where we have again used detailed balance. And, as detailed in Problem 13.17, we have that

$$\frac{P[C_T(z)]}{K^T(x,z)} = P[C_T(z)|x \to z],$$

where $\{x \to z\}$ denotes the event that $X_0 = x$ given that $X_T = z$. This is exactly the probability that [A.57] accepts the simulated $X_0 = x$.   □

Finally, note that the Accept–Reject algorithm is more efficient if M is smaller, so choosing z to be the state that minimizes $\pi(z)/P[C_T(z)]$ is a good choice. This will be a difficult calculation but, in running the algorithm, these probabilities can be estimated.

Example 13.15. (Continuation of Example 13.1) To illustrate the behavior of [A.57] for the Beta-binomial example, we choose $T = 3$ and $X_T = 2$. Since the chain is reversible, $\tilde{K} = K$, with probabilities given by Figure 13.1. Suppose that $(X_2, X_1, X_0) = (1, 0, 1)$. The following picture shows the corresponding transitions.

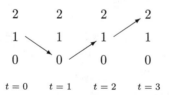

The corresponding conditional distributions on the uniforms are thus

$$U_1 \sim \mathcal{U}(0, .417), \qquad U_2 \sim \mathcal{U}(.583, .917), \quad \text{and} \quad U_3 \sim \mathcal{U}(.833, 1),$$

according to Figure 13.1. Suppose, for instance, that $U_1 \in (.278, .417)$, $U_2 \in (.833, .917)$ and $U_3 > .917$, in connection with Figure 13.1.

If we begin chains at all states, 0, 1 and 2, the trails of the different chains through time $t = 3$ are then as follows.

```
2  ＼   2   ／  2  ——▸ 2

1  ＼   1  ／   1  ／   1

0  ——▸ 0  ／    0      0

t = 0   t = 1   t = 2   t = 3
```

The three chains coalesce at time 3; so we accept $X_0 = 1$ as a draw from π. $\parallel$

The algorithm [A.57] depends on an arbitrary choice of T and X_T. This is appealing for the reasons mentioned above, namely, that the very long runs can be avoided, but one must realize that the coalescence times for CFTP and [A.57] are necessarily of the same magnitude. With the same transition kernel, we cannot expect time gains using [A.57]. If anything, this algorithm should be slower, because of the higher potential for rejection of the entire sequence if T is small and of the unnecessary runs if T is too large.

An alternative to [A.57] is to generate the U_i's unconditionally, rather than conditional on $x_i = \Psi(x_{i-1}, U)$. (Typically, $U_i \sim U(0,1)$.) This requires a preliminary rejection step, namely that only sequences of U_i's that are compatible with the path from $X_0 = x_0$ to $X_T = x$ are accepted. This does not change the distribution of the accepted X_0 and it avoids the derivation of the conditional distributions g_Ψ, but it is only practical as long as the conditional distributions are not severely constrained. This is less likely to happen if T is large.

The extensions developed for the CFTP scheme, namely monotonicity, sandwiching (Section 13.2.3), and the use of slice sampling (Section 13.2.5), obviously apply to [A.57] as well. For instance, if there exist minimum and maximum starting values, $\underline{\tilde{0}}$ and $\overline{\tilde{1}}$, the algorithm only needs to restart from $\tilde{0}$ and $\tilde{1}$, once X_0 has been generated. Note also that, in continuous settings, the conditional distributions g_Ψ will often reduce to a Dirac mass. In conclusion, once the basic forward–backward (or rather backward–forward) structure of the algorithm is understood, and its Accept–Reject nature uncovered, there is little difference in implementation and performance between [A.54] and [A.57].

13.4 Perfect Sampling in Practice

With one notable exception (see Note 13.6.1), perfect sampling has not yet become a standard tool in simulation experiments. It still excites the interest of theoreticians, maybe because of the paradoxical nature of the object, but the implementation of a perfect sampler in a given realistic problem is something far from easy. To find an adequate discretization of the starting values often requires an advanced knowledge of the transition mechanism, and also a proper choice of transition kernel (because of the uniform ergodicity requirement). For potentially universal schemes like the slice sampler (Section 13.2.5), the difficulty in implementing the univariate slice sampler reduces its appeal, and it is much more difficult to come up with monotone schemes when using the multivariate slice sampler, as illustrated by the mixture examples.

13.5 Problems

13.1 For the random walk on $\mathcal{X} = \{1, 2, .., n\}$, show that forward coalescence necessarily occurs in one of the two points 1 and n, and deduce that forward coupling does not produce a simulation from the uniform distribution on $\mathcal{X}$.

13.2 Consider a finite Markov chain with transition matrix

$$\mathbb{P} = \begin{pmatrix} .3 & .3 & .3 \\ 1 & 0 & 0 \\ 1 & 0 & 0 \end{pmatrix}.$$

(a) Show that the Markov chain is ergodic with invariant measure $(.6, .2, .2)$.

(b) Show that if three chains $x_1^{(t)}$, $x_2^{(t)}$, and $x_3^{(t)}$ are *coupled*, that is, if the three chains start from the three different states and

$$\tau = \inf\{t \geq 1; x_1^{(t)} = x_2^{(t)} = x_3^{(t)}\},$$

then the distribution of $x_i^{(\tau)}$ is not the invariant measure but is rather the Dirac mass in the first state.

13.3 (Foss and Tweedie, 1998)

(a) Establish that the stochastic recursive sequence (13.1) is a valid representation of any homogeneous Markov chain.

(b) Prove Theorem 13.2.

(c) In the proof of Theorem 13.3, we showed that coalescence occurs at some finite time. Use the Borel-Cantelli Lemma to obtain the stronger conclusion that the coalescence time is almost surely finite.

13.4 (Møller 2001) For the random mapping representation (13.1), assume that

(i) the U_t's are iid;

(ii) there exists a state ϖ such that

$$\lim_{t \to \infty} P(\Psi_t \circ \cdots \circ \Psi_0(\varpi) \in F) \to \pi(F) \qquad \forall F \subset \mathcal{X};$$

(iii) there exists an almost finite stopping time T such that

$$\Psi_0 \circ \cdots \circ \Psi_{-t}(\varpi) = \Psi_0 \circ \cdots \circ \Psi_{-T}(\varpi) \qquad \forall t \geq T.$$

Show that

$$\Psi_0 \circ \cdots \circ \Psi_{-T}(\varpi) \sim \pi.$$

(*Hint:* Use the monotone convergence theorem.)

13.5 We assume there exists some partial stochastic ordering of the state space $\mathcal{X}$ of a chain $(x^{(t)})$, denoted $\preceq$.

(a) Show that if $\tilde{\mathcal{X}}$ denotes the set of *extremal points* of $\mathcal{X}$ (that is, for every $y \in \mathcal{X}$, there exist x_1, $x_2 \in \tilde{\mathcal{X}}$ such that $x_1 \preceq y \preceq x_2$), the backward coupling algorithm needs to consider only the points of $\tilde{\mathcal{X}}$ at each step.

(b) Study the improvement brought by the above modification in the case of the transition matrix

$$\mathbb{P} = \begin{pmatrix} 0.34 & 0.22 & 0.44 \\ 0.28 & 0.26 & 0.46 \\ 0.17 & 0.31 & 0.52 \end{pmatrix}.$$

13.6 Given a sandwiching pair $\underline{\theta}_t$ and $\overline{\theta}_t$, such that $\underline{\theta}_t \preceq \Phi_t(\omega) \preceq \overline{\theta}_t$ for all ω's and all t's, show the *funneling property*

$$\Psi_{-t}\circ\cdots\Psi_{-n}(\underline{\theta}_{-n}) \preceq \Psi_{-t}\circ\cdots\Psi_{-m}(\underline{\theta}_{-m}) \preceq \Psi_{-t}\circ\cdots\Psi_{-m}(\overline{\theta}_{-m}) \preceq \Psi_{-t}\circ\cdots\Psi_{-n}(\overline{\theta}_{-n})$$

for all $t < m < n$.

13.7 Consider a transition kernel K which satisfies *Doeblin's condition*,

$$K(x,y) \geq r(x), \quad x,y \in \mathcal{X},$$

which is equivalent to uniform ergodicity (see, e.g., Theorem 6.59). As a perfect simulation technique, Murdoch and Green (1998) propose that the continuum of Markov chains coalesces into a single chain at each time with probability

$$\rho = \int r(x)dx.$$

(a) Using a coupling argument, show that if for all chains the *same uniform variable* is used to decide whether the generation from $r(x)/\rho$ occurs, then at time 0 the resulting random variable is distributed from the stationary distribution π.

(b) Deduce that the practical implementation of the method implies that a single chain is started at a random moment $-N$ in the past from the bounding probability $\rho^{-1}r(x)$ with $N \sim \mathcal{G}eo(\rho)$. Discuss why simulating from the residual kernel is not necessary.

13.8 (Hobert et al. 1999) Establish the validity of the boxing argument in Example 13.7.

13.9 (Casella et al. 2002) For the automatic coupling of Section 13.2.6, and an independent proposal with density h, show that the associated monotonicity under $\preceq$ is preserved by the associated transition when the same uniform variable u_t is used to update both chains.

(a) Show that $\varrho_0^{(t)} \geq \varrho_1^{(t)}$ when the chains are ordered.

(b) Consider the three possible cases:

    a) ω' is rejected for both chains;

    b) ω' is accepted for the first chain, but rejected by the second chain;

    c) ω' is accepted for both chains.

and deduce that the order $\preceq$ is preserved. (*Hint:* For cases (i) and (iii), this is trivial. For case (ii), the fact that ω' is rejected for the second chain implies that $\varrho_1^{(t)} < 1$, and thus that $\pi(\omega')/h(\omega') < \pi(\omega_1^{(t)})/h(\omega_1^{(t)})$.)

13.10 Show that the choice of the point ω_1 such that $\pi(\theta_1) = \pi(\omega_0)/u\}$ is immaterial for the validation of the monotone slice sampler.

13.11 In the setting of Example 13.11, find the maximum $\tilde{1}$ for the numerical values $\mu = (-2.4, 1.8)$ and $\Sigma = \begin{pmatrix} 1.0 & -1.2 \\ -1.2 & 4.4 \end{pmatrix}$.

13.12 (Mira et al. 2001) Assume that $\pi(\theta)$ can be decomposed as $\pi(\theta) \propto \pi_1(\theta)\pi_2(\theta)$. Define the slice sampler associated with the subgraph of π and determine the order under which this slice sampler is monotone.

13.13 (Mira et al. 2001) For a probability density π, define

$$Q_\pi(u) = \mu\left(\{\theta \; ; \; \pi(\theta) \geq u\}\right).$$

(a) Show that Q_π is a decreasing function.

(b) Show that Q_π characterizes the slice sampler.

(c) Show that, if π_1 and π_2 are such that $Q_{\pi_1}(u)/Q_{\pi_1}(u)$ is decreasing in u, the chains associated with the corresponding slice samplers, $(\theta_1^{(t)})$ and $(\theta_2^{(t)})$, are stochastically ordered, in that

$$\Pr\left(\pi_1(\theta_1^{(t)}) \le \omega|\theta_1^{(t-1)} = \xi\right) \ge \Pr\left(\pi_2(\theta_2^{(t)}) \le \omega|\theta_2^{(t-1)} = \xi\right).$$

(d) Examine if $Q_{\pi_1}(u)/Q_{\pi_1}(u)$ is decreasing in u when π_1/π_2 is bounded from above.

13.14 In the setting of the perfect slice sampler (Section 13.2.5), show that the maximum $\tilde{1}$ does not need to be known to start the upper chain from $\tilde{1}$, but only an upper bound on $\max \pi$.

13.15 Show that, if the slice sampler is applied to a bounded density π, it is uniformly ergodic. (*Hint:* Show that the transition kernel is

$$K(\omega, \omega') = \int_0^1 \mathbb{I}_{\pi(\omega') \ge u\pi(\omega)} \, du = \frac{\pi(\omega')}{\pi(\omega))} \wedge 1,$$

and deduce a minorizing measure if $\pi(\omega) \le M$.)
What about the converse?

13.16 A motivation for Fill's algorithm [A.57] is that interrupting [A.54] if it runs backward further than a fixed T_0 is creating a bias in the resulting sample. Show that this is the case by considering a Markov chain with transition matrix

$$\mathbb{P} = \begin{pmatrix} .1 & .4 & .3 & .2 \\ .0 & .5 & .5 & .0 \\ .2 & .0 & .8 & .0 \\ .1 & .1 & .0 & .8 \end{pmatrix}$$

and a fixed backward horizon $T_0 = 10$.

13.17 (Casella et al. 2001) Recall that $C_T(z)$ denotes the event that all chains have coalesced to state z at time T. Since [A.57] delivers a value only if $C_T(z)$ occurs, we want to establish that $P[X_0 = x|C_T(z)] = \pi(x)$.

(a) Show that

$$P[X_0 = x|C_T(z)] = \frac{P[z \to x]P[C_T(z)|x \to z]}{\sum_{x'} P[z \to x']P[C_T(z)|x' \to z]}.$$

(b) Show that, for every x',

$$P[C_T(z)|x' \to z] = \frac{P[C_T(z) \text{ and } x' \to z]}{P[x' \to z]} = \frac{P[C_T(z)]}{P[x' \to z]}.$$

(*Hint:* Use the fact that coalescence entails that, for each starting point x', the chain is in z at time T.)

(c) Note that $P[x' \to z] = K^T(x', z)$ and deduce that

$$P[X_0 = x|C_T(z)] = \frac{K^T(z, x)P[C_T(z)]/K^T(x, z)}{\sum_{x'} K^T(z, x')P[C_T(z)]/K^T(x', z)}$$
$$= \frac{\tilde{K}^T(z, x)/K^T(x, z)}{\sum_{x'} \tilde{K}^T(z, x')/K^T(x', z)}.$$

(d) Use the detailed balance condition $\pi(z)\tilde{K}^T(z,x) = \pi(x)K^T(x,z)$ to deduce

$$P[X_0 = x | C_T(z)] = \frac{\pi(x)/\pi(z)}{\sum_{x'} \pi(x')/\pi(z)} = \pi(x).$$

13.18 Wilson (2000) developed a *read-once* algorithm that goes forward only and does not require storing earlier values. For $m \in \mathbb{N}$, we define

$$\mathfrak{M}_i(\omega) = \Psi_{im} \circ \cdots \Psi_{(i-1)m+1}(\omega),$$

$\kappa_1, \kappa_2, \ldots$ as the successive indices i for which $\mathfrak{M}_i$ is a constant function,

$$\mathfrak{G}_1 = \mathfrak{M}_{\kappa_1},$$
$$\mathfrak{G}_i = \mathfrak{M}_{\kappa_i - 1} \circ \cdots \circ \mathfrak{M}_{\kappa_{i-1}} \qquad (i > 1),$$

and $\tau_i = \kappa_i - 1 - \kappa_{i-1}$.

(a) Show that, if the κ_i's are almost surely finite, and if p is the probability that $\mathfrak{M}_i$ is constant, then $p > 0$, $\mathfrak{G}_i(\omega) \sim \pi$ for all ω's and $i > 1$, and $\tau_i \sim \mathcal{G}eo(p)$.

(b) Deduce validity of the read-once algorithm of (Wilson 2000): for an arbitrary ω, take $(\mathfrak{G}_2, \ldots, \mathfrak{G}_{n+1})$ as an iid sample from π.

13.6 Notes

13.6.1 History

Maybe surprisingly for such a young topic (Propp and Wilson published their seminal paper in 1996!), perfect sampling already enjoys a wealth of reviews and introductory papers. Besides Casella et al. (2001) on which this chapter is partly based (in particular for the Beta-Binomial example), other surveys are given in Fismen (1998), Dimakos (2001), and in the large body of work of Møller and coauthors, including the book by Møller and Waagepetersen (2003). David Wilson also set up a website which catalogs papers, preprints, theses, and even the talks given about perfect sampling![8]

While our examples are statistical, there are other developments in the area of point processes and stochastic geometry, much from the work of Møller and Kendall. In particular, Kendall and Møller (2000) developed an alternative to Propp and Wilson (1996) CFTP algorithm, called *horizontal CFTP*, which mainly applies to point processes and is based on continuous time birth-and-death processes, but whose description goes rather beyond the level of this book. See also Fernández et al. (1999) for another horizontal CFTP algorithm for point processes. Berthelsen and Møller (2003) use these algorithms for nonparametric Bayesian inference on pairwise interaction point processes like the Strauss process. While the complexity of these models goes, again, beyond the level of this book, it is interesting to note that the authors exhibit an exponential increase of the mean coalescence time with the parameters of the Strauss process.

---

[8] Its current address is http://dimacs.rutgers.edu/~dbwilson/exact.html/.

13.6.2 Perfect Sampling and Tempering

Tempering, as introduced in Marinari and Parisi (1992), Geyer and Thompson (1995), and Neal (1999), appears as an *anti-simulated annealing*, in that the distribution of interest π appears as the "colder" or most concentrated distribution in a sequence (ϖ_n) of target distributions, $\varpi_N = \pi$, with the "hotter" distributions $\varpi_1, \ldots, \varpi_{N-1}$ being less and less concentrated. The purpose behind this sequence is to facilitate the moves of a Metropolis–Hastings sampler by somehow flattening the surface of the target distribution. This device is, for instance, used in Celeux et al. (2000) to impose a proper level of label switching in the simulation of the posterior distribution associated with a mixture of distributions.

Following the description of Møller and Nicholls (2004), a sequence of target distributions $(\varpi_n)_{1 \leq n \leq N^\star}$ is thus chosen such that $\varpi_{N^\star} = \pi$, ϖ_1 is a well-known distribution for which simulation is possible, and ϖ_0 is a distribution on a single atom, denoted $\mathbf{0}$. Each of the ϖ_n's is defined on a possibly different space, Ω_n. Using the auxiliary variable N, which takes values in $\{0, 1, \ldots, N^\star\}$, and defining the joint distribution

$$\mu(\theta, n) = \varrho_n \varpi_n(\theta), \qquad \sum_{n=0}^{N^\star} \varrho_n = 1,$$

a reversible jump (Section 11.2) algorithm can be constructed for this joint distribution, for instance, with moves "up," from n to $n+1$, "down," from n to $n-1$, and "fixed-level," with only θ changing. If p_U and p_D are the probabilities of moving up and down, respectively, with $1 - p_U - p_D$ the probability of choosing a fixed-level move, and if $q_{n \to n+1}(\theta'|\theta)$ and $q_{n+1 \to n}(\theta|\theta')$ denote the proposal distributions for upward and downward moves, we thus assume that the usual symmetry requirement holds for $q_{n \to n+1}$ and $q_{n+1 \to n}$, getting

$$\varrho_{n \to n+1}(\theta, \theta') = \frac{\pi_{n+1}\, \varpi_{n+1}(\theta')\, p_D\, q_{n+1 \to n}(\theta|\theta')}{\pi_n\, \varpi_n(\theta)\, p_U\, q_{n \to n+1}(\theta'|\theta)} \wedge 1$$

as the acceptance probability of an upward move. The symmetry assumption translates into

$$\varrho_{n+1 \to n}(\theta', \theta) = 1/\varrho_{n \to n+1}(\theta, \theta')$$

as the acceptance probability of a downward move.

Example 13.16. Power tempering Assuming a compact space like $[0, 1]^d$, a possible implementation of the above scheme is to define a sequence of powers $\nu_1 = 0 < \ldots < \nu_{N^\star} = 1$, with

$$\varpi_n \propto \pi^{\nu_n}.$$

These distributions are literally flattened versions of the target distribution π. The corresponding acceptance probability for a move up from n to $n+1$ is then proportional to

$$\frac{\pi_{n+1}\, p_D}{\pi_n\, p_U} \frac{\pi^{\nu_{n+1}}(\theta')}{\pi^{\nu_n}(\theta)} \frac{q_{n+1 \to n}(\theta|\theta')}{q_{n \to n+1}(\theta'|\theta)},$$

where the proportionality coefficient is the ratio of the normalizing constants of $\pi^{\nu_{n+1}}$ and π^{ν_n}. Neal (1999) bypasses this difficulty by designing an equal number of moves from n to $n+1$ and from $n+1$ to n, and by accepting the entire sequence as a single proposal, thus canceling the normalizing constants in the acceptance probability. ‖

Since the augmented chain (θ_t, N_t) is associated with the joint distribution $\mu(\theta, n)$, the subchain θ_t, conditional on $N_t = N^*$, is distributed from the target π. Therefore, monitoring the output of the MCMC sampler for only the times when $N_t = N^*$ produces a subchain converging to the target distribution π. This sounds like a costly device, since the simulations corresponding to the other values of N_t are lost, but the potential gain is in an enhanced exploration of the state space Ω_{N^*}. Note also that the other simulations can sometimes be recycled in importance sampling estimators.

There are many applications of tempering in the MCMC literature (Geyer and Thompson 1992, Müller 1999, Celeux et al. 2000). What is of most interest in connection with this chapter is that it is fairly easy to define a dominating process on $\{0, 1, \ldots, N^*\}$, as shown by Møller and Nicholls (2004). To achieve this domination, they assume in addition that

$$\varrho_{n \to n+1}(\theta, \theta') \leq \kappa_n \frac{\pi_{n+1}}{\pi_n},$$

that is, that the upward and downward transitions are bounded away from 0. The dominating process D_t is then associated with the same proposal as N_t and the acceptance probabilities

$$\alpha_{n \to n+1} = \min\left(1, \kappa_n \frac{\pi_{n+1}}{\pi_n}\right),$$

$$\alpha_{n \to n-1} = \min\left(1, \frac{\pi_{n-1}}{\kappa_{n-1} \pi_{n+1}}\right),$$

$$\alpha_{n \to n} = 1.$$

We couple N_t and D_t so that (i) the upward, downward, and fixed-level moves are chosen with the same probabilities, and (ii) the acceptance is based on the same uniform. Then,

Lemma 13.17. *The chain* (D_t) *dominates the chain* (N_t) *in that*

$$\Pr(N_{t+1} \leq D_{t+1} | N_t = n, D_t = m) = 1 \qquad \text{when} \qquad n \leq m.$$

This result obviously holds by construction of the acceptance probabilities $\alpha_{n \to n+1}$ and $\alpha_{n \to n-1}$, since those favor upward moves against downward moves. Note that (D_t) is necessarily ergodic, as a Markov chain on a finite state space, and reversible, with invariant distribution

$$\mathfrak{q}_n \propto \kappa_n \frac{\pi_{n+1}}{\pi_n}.$$

Due to this domination result, a perfect sampler is then

Algorithm A.58 –Tempered Perfect Sampling–

1. Go backward far enough to find $\tau^* < 0$ such that $D_{\tau^*} = N^*$ and $D_t = 0$ for at least one $\tau^* < t \le 0$.
2. Simulate the coupled chain (θ_t, N_t) τ^* steps, with

$$(\theta_{\tau^*}, N_{\tau^*}) = (\mathbf{0}, 0).$$

3. Take

$$(\theta_0, N_0) \sim \mu.$$

The validity of this algorithm follows from the monotonicity lemma above, in that the (virtual) chains starting from all possible values in $\{0, 1, \dots, N^*\}$ have coalesced by the time the chain starting from the largest value N^* takes the lowest value 0. Note also that $(\mathbf{0}, 0)$ is a particular renewal event for the chain (θ_t, N_t).

Example 13.18. Flour beetle mortality. As described in Carlin and Louis (2001), a logit model is constructed for a flour beetle mortality dataset, corresponding to 8 levels of dosage ω_i of an insecticide, with a_i exposed beetles and $y_i a_i$ killed beetles. The target density is

$$\pi(\mu, \sigma, \nu) \propto \pi(\mu, \sigma, \nu) \prod_{i=1}^{8} \left\{ \frac{\exp y_i \frac{\omega_i - \mu}{\sigma}}{1 + \exp \frac{\omega_i - \mu}{\sigma}} \right\}^{a_i \nu},$$

with the prior density $\pi(\mu, \sigma, \nu)$ corresponding to a normal distribution on μ, an inverted gamma distribution on σ^2 and a gamma distribution on ν.

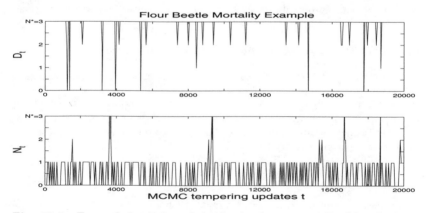

Fig. 13.6. Forward simulation of the dominating process D_t (above), along with the tempering process N_t (below). (*Source:* Møller and Nicholls 2004.)

The tempering sequence chosen in Møller and Nicholls (2004) is

$$\varpi_n(\mu, \sigma, \nu) \propto \pi(\mu, \sigma, \nu) \prod_{i=1}^{8} \left\{ \frac{\exp y_i \frac{\omega_i - \mu}{\sigma}}{1 + \exp \frac{\omega_i - \mu}{\sigma}} \middle/ \ell_i^* \right\}^{\beta_n a_i \nu},$$

with $\beta_{N^\star} = 1$, $\beta_n < \beta_{n+1}$, and

$$\ell_i^* = y_i^{y_i}(1 - y_i)^{1-y_i} ,$$

the maximum of the logit term. The move transitions $q_{n \to n+1}$ and $q_{n+1 \to n}$ are also chosen to be Dirac masses, that is, $q_{n \to n+1}(\theta'|\theta) = \mathbb{I}_\theta(\theta')$. Therefore, as $p_U = p_D$,

$$\varrho_{n \to n+1} = \frac{\pi_{n+1}}{\pi_n} \left[\prod_{i=1}^{8} \left\{ \frac{\exp y_i \frac{\omega_i - \mu}{\sigma}}{1 + \exp \frac{\omega_i - \mu}{\sigma}} \right\}^{a_i \nu} \Big/ (\ell_i^*)^{a_i \nu} \right]^{\beta_{n+1} - \beta_n} \leq \frac{\pi_{n+1}}{\pi_n}$$

and $\kappa_n = 1$. Note that, since all that matters are the probability ratios π_{n+1}/π_n, the values of the normalizing constants of the ϖ_n's do not need to be known. (The terms ℓ_i^* are also introduced to scale the various terms in the likelihood.)

In their numerical implementation, Møller and Nicholls (2004) chose $N^\star = 3$, $\beta_1 = 0$, $\beta_2 = .06$, and $\beta_3 = 0$, as well as $p_U = p_D = .333$. Figure 13.6 shows the path of the dominating process D_t, along with the tempering process N_t. Note the small number of times when $N_t = 3$, which are the times where $\pi(\mu, \sigma, \nu)$ is perfectly sampled from, and also the fact that, by construction, $N_t \leq D_t$. $\parallel$

Extensions and developments around these ideas can also be found in Brooks et al. (2002).

Iterated and Sequential Importance Sampling

"The past is important, sir," said Rebus, taking his leave.
—Ian Rankin, *The Black Book*

This chapter gives an introduction to sequential simulation methods, a collection of algorithms that build both on MCMC methods and importance sampling, with importance sampling playing a key role. We will see the relevance of importance sampling and the limitations of standard MCMC methods in many settings, as we try to make the reader aware of important and ongoing developments in this area. In particular, we present an introduction to *Population Monte Carlo* (Section 14.4), which extends these notions to a more general case, and subsumes MCMC methods

14.1 Introduction

At this stage of the book, importance sampling, as presented in Chapters 3 and 4, may appear as a precursor of MCMC methods. This is because both approaches use the ratio of *target/importance* distributions, f/g, to build approximations to the target distribution f and related integrals, and because MCMC methods appear to provide a broader framework, in particular because they can derive proposals from earlier generations. This chapter will show that this vision is not quite true. Importance sampling can also be implemented with dependent proposal[1] distributions and adaptive algorithms

---

[1] Here, we use a vocabulary slightly different from other chapters, often found in the *particle filter* literature: generations from importance distributions are often called *particles*; importance distributions will be called *proposal* distributions. Also, we restrict the use of the adjective *sequential* to settings where observations (or targets) appear according to a certain time process, rather than all at once. For other common uses of the word, we will prefer the denominations *iterated* or *repeated*.

that are much easier to build than in MCMC settings (Session 7.6.3), while providing unbiased estimators of integrals of interest (if all normalizing constants are known).

Importance sampling naturally enhances parallel sampling and there are settings, as in Example 9.2, where MCMC has a very hard time converging to the distribution of interest while importance sampling, based on identical proposals, manages to reach regions of interest for the target distribution (Section 14.4).

Importance sampling is also paramount in sequential settings, where a constant modification of the target distribution occurs with high frequency and precludes the use of standard MCMC algorithms because of computational constraints. The books by Doucet et al. (2001) and Liu (2001) are dedicated to this topic of sequential sampling methods and we thus refer the reader there for a more complete treatment of this topic.

14.2 Generalized Importance Sampling

As already stated in the above paragraph, an incorrect impression that might be drawn from the previous chapters is that importance sampling is solely a forerunner of MCMC methods, and that the latter overshadows this technique. As shown below in Section 14.4, a deeper understanding of the structures of both importance sampling and MCMC methods can lead to superior hybrid techniques that are importance sampling at the core, that is, they rely on unbiasedness at any given iteration, but borrow strength from iterated methods and MCMC proposals. We first see how iterated adaptive importance sampling is possible and why dependent steps can be embedded within importance sampling with no negative consequence on the importance sampling fundamental identity.

A first basic remark about importance sampling is that its fundamental unbiasedness property (14.1) is not necessarily jeopardized by dependences in the sample. For instance, the following result was established by MacEachern et al. (1999).

Lemma 14.1. *If f and g are two densities, with $supp(f) \subset supp(g)$, and if $\omega(x) = f(x)/g(x)$ is the associated importance weight, then for any kernel $K(x, x')$ with stationary distribution f,*

$$\int \omega(x) K(x, x') g(x) dx = f(x').$$

This invariance result is an obvious corollary of the importance sampling fundamental identity (3.9), which we repeat here:

$$(14.1) \qquad \mathbb{E}_f[h(X)] = \int_{\mathcal{X}} h(x) \frac{f(x)}{g(x)} g(x) \, dx .$$

However, its consequences are quite interesting. Lemma 14.1 implies, in particular, that MCMC transitions can be forced upon points of an importance sample with no effect on the weights. Therefore, the modification of an importance sample by MCMC transitions does not affect the unbiasedness associated with the importance estimator (3.8) (even though it may modify its variance structure), since

$$\mathbb{E}\left[\omega(X)h(X')\right] = \int \omega(x)\, h(x')\, K(x,x')\, g(x)\, dx\, dx' = \mathbb{E}_f\left[h(X)\right]$$

for any f-integrable function h. One may wonder why we need to apply an MCMC step in addition to importance sampling. The reason is that, since the MCMC kernel is tuned to the target distribution, it may correct to some extent a poor choice of importance function. This effect is not to be exaggerated, though, given that the weights do not change. For example, a point x with a large (small) weight $\omega(x) = f(x)/g(x)$, when moved to x' by the kernel $K(x,x')$, does keep its large (small) weight, even if x' is less (more) likely than x in terms of f. This drawback is linked to the fact that the kernel, rather than the proposal within the kernel, is used in Lemma 14.1. Section 14.4 will discuss more adaptive schemes that bypass this difficulty.

Although the relevance of using MCMC steps will become clearer in Section 14.3 for sequential settings, we point out at this stage that other generalizations can be found in the literature as the *dynamic weighting* of Wong and Liang (1997), where a sample of (x_t, ω_t)'s is produced, with joint distribution $g(x, \omega)$ such that

$$\int_0^\infty \omega\, g(x, \omega)\, d\omega \propto f(x)\,.$$

See Note 14.6.2 for more details on the possible implementations of this approach.

14.3 Particle Systems

14.3.1 Sequential Monte Carlo

While this issue is rather peripheral to the purpose of the book, which focuses on "mainstream" statistical models, there exist practical settings where a sequence of target distributions, $(\pi_t)_t$, is available (to some extent like normalizing or marginalizing) and needs to be approximated under severe time/storage constraints. Such settings emerged in military and safety applications. For instance, π_t may be a posterior distribution on the position and the speed of a plane at time t given some noisy measures on these quantities. The difficulty of the problem is the *time constraint:* we assume that this constraint is too tight to hope to produce a reasonable Monte Carlo approximation using standard tools, either independent (like regular importance sampling) or dependent (like MCMC algorithms) sampling.

Example 14.2. Target tracking. The original example of Gordon et al. (1993) is a tracking problem where an object (e.g., a particle, a pedestrian, or a ship) is observed through some noisy measurement of its angular position Z_t at time t. Of interest are the position (X_t, Y_t) of the object in the plane and its speed $(\dot{X}_t, \dot{Y}_t)$ (using standard physics notation). The model is then discretized as $\dot{X}_t = X_{t+1} - X_t$, $\dot{Y}_t = Y_{t+1} - y_t$, and

(14.2)
$$\dot{X}_t = \dot{X}_{t-1} + \tau \epsilon_t^x$$
$$\dot{Y}_t = \dot{Y}_{t-1} + \tau \epsilon_t^y$$
$$Z_t = \arctan(Y_t/X_t) + \eta \epsilon_t^z,$$

where ϵ_t^x, ϵ_t^y and ϵ_t^z are iid $\mathcal{N}(0,1)$ random variables. Whatever the reason for tracking this object, the distribution of interest is $\pi_t(\theta_t) = \pi(\theta_t|z_{1:t})$, where $\theta_t = (\tau, \eta, \dot{X}_t, \dot{Y}_t)$ and $z_{1:t}$ denotes (in the signal processing literature) the vector $\mathbf{z}_t = (z_1, \ldots, z_t)$. The prior distribution on this model includes the speed propagation equation (14.2), as well as priors on τ, η and the initial values $(x_0, y_0, \dot{x}_0, \dot{y}_0)$. Figure 14.1 shows a simulated sequence (x_t, y_t) as well as the corresponding z_t, in a representation inspired from the graph in Gilks and Berzuini (2001). ‖

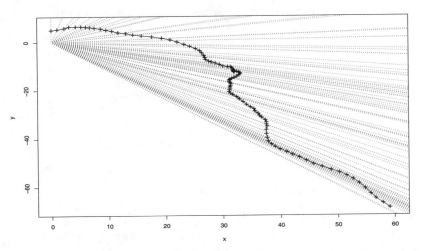

Fig. 14.1. Simulated sequence of target moves and observed angles, for $\tau = 0.2$ and $\eta = .05$.

Importance sampling seems ideally suited for this problem in that, if the distributions π_t and π_{t+1} are defined in the same space, a sample distributed (approximately) according to π_t can be recycled by importance sampling to produce a sample distributed (approximately) according to π_{t+1}. In the event that the state spaces (i.e., the supports) of π_t and π_{t+1} are not of the same dimension, it is often the case that the state space of π_{t+1} is an augmentation of the state space of π_t and thus that only a subvector needs be simulated to produce an approximate simulation from π_{t+1}. This is, in essence, the idea of *sequential importance sampling* as in, e.g., Hendry and Richard (1991) or Liu and Chen (1995).

14.3.2 Hidden Markov Models

A family of models where sequential importance sampling can be used is the family of *hidden Markov models* [abbreviated to *HMM*]. They consist of a bivariate process $(X_t, Z_t)_t$, where the subprocess (Z_t) is a homogeneous Markov chain on a state space $\mathscr{Z}$ and, conditional on (Z_t), (X_t) is a series of random variables on $\mathscr{X}$ such that the conditional distribution of X_t only depends on Z_t, as represented in Figure 14.2. When $\mathscr{Z}$ is discrete, we have, in particular,

$$(14.3) \qquad Z_t | z_{t-1} \sim P(Z_t = i | z_{t-1} = j) = p_{ji} , \qquad X_t | z_t \sim f(x | \xi_{z_t}) ,$$

where $\xi_1, \ldots, \xi_k$ denote the different values of the parameter. The process (Z_t) which is usually referred to as the *regime* or the *state* of the model, is not observable (hence, *hidden*) and inference has to be carried out only in terms of the observable process (X_t). Numerous phenomena can be modeled this way; see Note 14.6.3. For instance, Example 14.2 is a special case of an HMM where, by a switch in notation, the observable is Z_t and the hidden chain is $(\dot{X}_t, \dot{Y}_t)$.

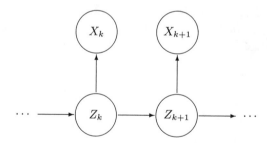

Fig. 14.2. Directed acyclic graph (DAG) representation of the dependence structure of a hidden Markov model, where (X_t) is the observable process and (Z_t) the hidden process.

When (Z_t) is a Markov chain on a discrete space, the hidden Markov model is often called, somehow illogically, a *hidden Markov chain*. In the following example, as in Example 14.2, the support of Z_t is continuous.

Example 14.3. Stochastic volatility. Stochastic volatility models are quite popular in financial applications, especially in describing series with

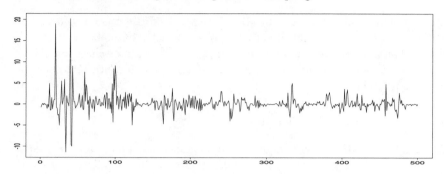

Fig. 14.3. Simulated sample (y_t) of a stochastic volatility process (14.4) with $\beta = 0$, $\sigma = 1$ and $\varphi = 0.9$. (*Source:* Mengersen et al. 1999.)

sudden changes in the magnitude of variation of the observed values (see, e.g. Jacquier et al. 1994, Kim et al. 1998). They use a *latent* linear process (Z_t), called the *volatility*, to model the variance of the observables X_t in the following way: Let $Z_0 \sim \mathcal{N}(0, \sigma^2)$ and, for $t = 1, \ldots, T$, define

$$(14.4) \qquad \begin{cases} Z_t = \varphi Z_{t-1} + \sigma \epsilon_{t-1}^* , \\ X_t = \beta e^{Z_t/2} \epsilon_t , \end{cases}$$

where ϵ_t and ϵ_t^* are iid $\mathcal{N}(0, 1)$ random variables. Figure 14.3 shows a typical stochastic volatility behavior for $\beta = 0$, $\sigma = 1$ and $\varphi = 0.9$. (For a similar model see Note 9.7.2.)   ‖

This setting is also quite representative of many applications of particle filters in state-space models,[2] in that the data (x_t) is available sequentially, and inference about z_t or $\mathbf{z}_t = (z_1, \ldots, z_t)$ (and possibly fixed parameters as those of (14.3)) is conducted with a different distribution $\pi_t(\cdot|x_1, \ldots, x_t) = \pi_t(\cdot|\mathbf{x}_t)$ at each time t, like

$$(14.5) \qquad \pi_t(\mathbf{z}_t|\mathbf{x}_t) \propto \prod_{\ell=2}^{t} p_{z_{\ell-1} z_\ell} f(x_\ell | \xi_{z_\ell})$$

and

$$(14.6) \qquad \pi_t(z_t|\mathbf{x}_t) = \sum_{\mathbf{z}_{t-1}} \pi_t((\mathbf{z}_{t-1}, z_t)|\mathbf{x}_t)$$

---

[2] The term *filtering* is usually reserved for the conditional distribution of Z_t given $(X_1, \ldots, X_t)$, while *smoothing* is used for the conditional distributions of earlier hidden states, like Z_k, $k < t$, given $(X_1, \ldots, X_t)$, and *prediction* is used for the conditional distribution of future hidden states. In most applications, like signal processing, the parameters of the model are of no interest to the analyst, whose focus is on the reconstruction of the hidden Markov process. See, e.g., Cappé et al. (2004).

in the discrete case. Note that the marginalization in (14.6) involves summing up over all possible values of $\mathbf{z}_{t-1}$, that is, p^{t-1} values when $\mathscr{Z}$ is of size p. A direct computation of this sum thus seems intractable, but there exist devices like the *forward-backward* or Baum–Welch formulas which compute this sum with complexity $O(tp^2)$, as described in Problems 14.5–14.7.

In the case of $\pi_t(\mathbf{z}_t|\mathbf{x}_t)$, when a sample of vectors $\mathbf{z}_t$ must be simulated at time t, a natural proposal (i.e., importance) distribution is, as in Gordon et al. (1993), to preserve the $(t-1)$ first components of the previous iteration, that is, to keep $\mathbf{z}_{t-1}$ unchanged, to use the predictive distribution

$$\pi_t(z_t|\mathbf{z}_{t-1},\mathbf{x}_{t-1}) = \pi(z_t|z_{t-1})$$

as an importance function for the last component z_t of $\mathbf{z}_t$, and to produce as the corresponding importance weight

$$\omega_t \propto \omega_{t-1} \frac{\pi_t(\mathbf{z}_t|\mathbf{x}_t)}{\pi_{t-1}(\mathbf{z}_{t-1}|\mathbf{x}_{t-1})\pi(z_t|z_{t-1})}$$

(14.7)
$$\propto \omega_{t-1} f(x_t|\xi_{z_t}).$$

Note that $\pi(z_t|z_{t-1})$ *is* the conditional distribution of z_t given the past of the Markov chain, which collapses to being conditional only on z_{t-1} because of the Markov property. (We have omitted the replication index in the above formulas for clarity's sake: the reader will come to realize in the next section that n particles are simulated in parallel at each time t with corresponding weights $\omega_t^{(i)}$, $i = 1, \ldots, n$.)

Although we will use these models below for illustration purposes only, we refer the reader to Cappé et al. (2004) for a detailed presentation of asymptotics, inference and computation in hidden Markov models, and Doucet et al. (2001) for additional perspectives on the computational difficulties and solutions.

14.3.3 Weight Degeneracy

There is a fundamental difficulty with a sequential application of the importance sampling technique. The importance weight at iteration t gets updated by a recursive formula of the type $\omega_t^{(i)} \propto \omega_{t-1}^{(i)} \varrho_t^{(i)}$ where $\varrho_t^{(i)}$ is, for example, an importance ratio for the additional component of the state space of π_t, like $f_t(z_t^{(i)})/g_t(z_t^{(i)})$. (See also the example of (14.7).) Therefore, in this case,

$$\omega_t^{(i)} \propto \exp\left\{\sum_{\ell=1}^{t} \log\left(f_\ell(z_\ell^{(i)})/g_\ell(z_\ell^{(i)})\right)\right\}$$

and if we take the special occurrence where g_t and f_t are both independent of t, using the Law of Large Numbers for the sum within the exponential, we see that the right hand term above is approximately

$$\exp\left\{-t\,\mathbb{E}_g\left[\log\{g(Z)/f(Z)\}\right]\right\}.$$

Since the Kullback–Leibler divergence $\mathbb{E}_g\left[\log g(Z)/f(Z)\right]$ is positive (see Note 1.6.1), the weights $\omega_t^{(i)}$ thus have a tendency to degenerate to 0. This translates into a degradation (or *attrition*) of the importance sample along iterations. When t increases the weights $\omega_t^{(i)}$ are all close to 0 except for one which is close to 1, thus only one point of the sample contributes to the evaluation of integrals, which implies that the method is useless in this extreme case.[3]

Example 14.4. (Continuation of Example 14.2) Figure 14.4 illustrates this progressive degradation of the importance sample for the model (14.2). Assume that the importance function is, in that case, the predictive distribution $\pi(\dot{x}_t, \dot{y}_t | \dot{x}_{t-1}, \dot{y}_{t-1})$. Then, if the distribution of interest is $\pi_t(\dot{\mathbf{x}}_t, \dot{\mathbf{y}}_t | z_1, \ldots, z_t)$, the corresponding importance weight is then updated as

$$\omega_t \propto \omega_{t-1}\, f(z_t | \dot{\mathbf{x}}_t, \dot{\mathbf{y}}_t) = \omega_{t-1}\, f(z_t | x_t, y_t),$$

a scheme open to the degeneracy phenomenon mentioned above. Two estimators can be derived from these weights: the *stepwise importance sampling* estimate that plots, for each time t, the value of (x_t, y_t) that corresponds to the highest importance weight ω_t, and the *final importance sampling* estimate that corresponds to the path $(\mathbf{x}_t, \mathbf{y}_t)_{1 \leq t \leq T}$ which is associated with the highest weight ω_T at the end of the observation period $(0, T)$. (As shown in Figure 14.4, the latter usually is smoother as it is more robust against diverging paths.) As time increases, the importance sample deteriorates, compared with the true path, as shown by both estimates, and it does not produce a good approximation of the last part of the true target path, for the obvious reason that, once a *single* beginning $(\mathbf{x}_t, \mathbf{y}_t)_{1 \leq t \leq T_0}$ has been chosen, there is no possible correction in future simulations. $\parallel$

14.3.4 Particle Filters

The solution to this difficulty brought up by Gordon et al. (1993) is called a *particle filter* (or *bootstrap filter*). It avoids the degeneracy problem by re-sampling at each iteration the points of the sample (or *particles*) according to their weight (and then replacing the weights by one). This algorithm thus appears as a sequential version of the SIR algorithm mentioned in Problem 3.18, where resampling is repeated at each iteration to produce a sample that is an empirical version of π_t.

---

[3] Things are even bleaker than they first appear. A weight close to 1 does not necessarily correspond to a simulated value $z_t^{(i)}$ that is important for the distribution of interest, but is more likely to correspond to a value with a relatively larger weight, albeit very small in absolute value. This is a logical consequence of the fact that the weights must sum to one.

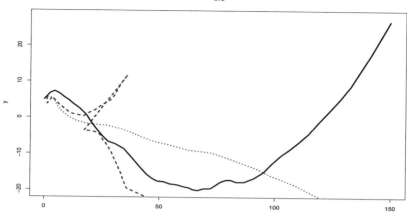

Fig. 14.4. Simulated sequence of 50 target moves (full line), for $\tau = 0.05$ and $\eta = .01$, along with the stepwise importance sampling (dashes) and the final importance sampling estimates (dots) for importance samples of size 1000.

In the particular case $\mathbf{Z}_t \sim \pi_t(\mathbf{z}_t)$, where $\mathbf{Z}_t = (\mathbf{Z}_{t-1}, Z_t)$ (as in Section 14.3.2), with importance/proposal distribution $q_t(z|\mathbf{z}_{t-1})$ that simulates only the last component of $\mathbf{Z}_t$ at time t, the algorithm of Gordon et al. (1993) reads as follows.[4]

Algorithm A.59 –Bootstrap Filter–

At time t,

1 Generate
$$\tilde{Z}_t^{(i)} \sim q_t(z|\mathbf{z}_{t-1}^{(i)}) \qquad i = 1, \ldots, n$$
and set
$$\tilde{\mathbf{z}}_t^{(i)} = (\mathbf{z}_{t-1}^{(i)}, \tilde{Z}_t^{(i)}).$$

2 Compute the importance weight
$$\omega_t^{(i)} \propto \pi_t(\tilde{\mathbf{z}}_t^{(i)})/q_t(\tilde{z}_t^{(i)}|\mathbf{z}_{t-1}^{(i)})\pi_{t-1}(\mathbf{z}_{t-1}^{(i)}).$$

3 Resample, with replacement, n particles $\mathbf{Z}_t^{(i)}$ from the $\tilde{\mathbf{z}}_t^{(i)}$'s according to the importance weights $\omega_t^{(i)}$ and set all weights to $1/n$.

Therefore, there is no direct degeneracy phenomenon, since the weights are set back to $1/n$ after each resampling step in [A.59]. The most relevant

---

[4] In the original version of the algorithm, the sampling distribution of $\tilde{Z}_t^{(i)}$ in Step 1. was restricted to the prior.

particles $\tilde{\mathbf{z}}_t^{(i)}$ are simply duplicated to the extent of their relevance and the less relevant particles disappear from the sample. This algorithm thus provides at each step an approximation to the distribution π_t by the empirical distribution (Problem 3.22)

$$(14.8) \qquad \frac{1}{n} \sum_{i=1}^{n} \mathbb{I}_{\mathbf{z}_t^{(i)}}(\mathbf{z}).$$

Obviously, degeneracy may still occur if all the particles are simulated according to a scheme that has little or no connection with the distribution π_t. In that case, the "most relevant" particle may well be completely irrelevant after a few steps.

Note that the original algorithm of Gordon et al. (1993) was designed for the simulation of the current $z_t \sim \pi_t(z)$ rather than the simulation of the *entire path* $\mathbf{z}_t$. This will make a difference in the degeneracy phenomenon studied in Section 14.3.6 and in possible improvements (as the *auxiliary particle filter* of Pitt and Shephard 1999; see Problem 14.14). This is because the marginal distribution $\pi_t(z_t)$ is usually much more complex than the joint distribution $\pi_t(\tilde{\mathbf{z}}_t)$ (see Section 14.3.2), or even sometimes unavailable, as in semi-Markov models (see Note 14.6.3).

Example 14.5. (Continuation of Example 14.2) If we use exactly the same proposal as in Example 14.4, when the target distribution is the posterior distribution of the entire path $(\mathbf{z}_t, \mathbf{y}_t)$, with the same simulated path as in Figure 14.4, the reconstituted target path in Figure 14.5 is much closer to the true path, both for the stepwise and final particle filter estimates, both of which are based on averages (rather than maximal weights).[5]   ‖

14.3.5 Sampling Strategies

The *resampling step* of Algorithm [A.59], while useful in fighting degeneracy, has a drawback. Picking the $\mathbf{z}_t^{(i)}$'s from a multinomial scheme introduces unnecessary noise to the sampling algorithm, and this noise is far from negligible since the variance of the multinomial distribution is of order $O(n)$. Several proposals can be found in the literature for the reduction of this extra Monte Carlo variation. One is the *residual sampling* of Liu and Chen (1995). Instead of resampling n points with replacement from the $\tilde{\mathbf{z}}_t^{(i)}$'s, this strategy reduces the variance by first taking $\lfloor n\omega_t^{(i)} \rfloor$ copies of $\tilde{\mathbf{z}}_t^{(i)}$, where $\lfloor x \rfloor$ denotes the integer part of x. We then sample the remaining $n - \sum_{i=1}^{n} \lfloor n\omega_t^{(i)} \rfloor$ particles from the $\tilde{\mathbf{z}}_t^{(i)}$'s, with respective probabilities proportional to $n\omega_t^{(i)} - \lfloor n\omega_t^{(i)} \rfloor$. The

---

[5] The visual fit may seem very poor judging from Figure 14.5, but one must realize that the true path and the reconstituted path are quite close from the angle point of view, since the observer is at $(0,0)$.

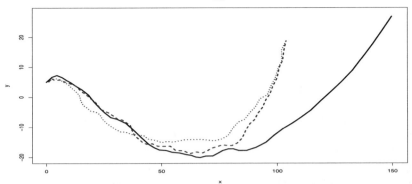

Fig. 14.5. Simulated sequence of 50 target moves, for $\tau = 0.05$ and $\eta = .01$, along with the stepwise particle filter (dashes) and the final particle filter estimates (dots), based on 1000 particles.

expected number of replications of $\tilde{\mathbf{z}}_t^{(i)}$ is still $n w_t^{(i)}$ and the sampling scheme is unbiased (Problems 14.9–14.11).

Crisan et al. (1999) propose a similar solution where they define probabilities $\varpi_i \propto N_{t-1} w_t^{(i)}$, with N_{t-1} being the previous number of particles, and they then take $N_t = \sum_{i=1}^{N_{t-1}} m_i$ with

$$m_i = \begin{cases} \lfloor N_{t-1}\, w_t^{(i)} \rfloor + 1 & \text{with probability } N_{t-1} w_t^{(i)} - \lfloor N_{t-1}\, w_t^{(i)} \rfloor, \\ \lfloor N_{t-1}\, w_t^{(i)} \rfloor & \text{otherwise.} \end{cases}$$

The drawback of this approach is that, although it is unbiased, it produces a *random number* of particles at each step and N_t thus may end up at either infinity or zero. Following Whitley (1994), Kitagawa (1996) and Carpenter et al. (1999) go even further in reducing the extra Monte Carlo variation through *systematic resampling*. They produce a vector $(m_1, \dots, m_n)$ of numbers of replications of the particles $\tilde{\mathbf{z}}_t^{(i)}$ such that $\mathbb{E}[m_i] = n\, w_t^{(i)}$ by computing the vector $(\xi_1, \dots, \xi_n)$ of the cumulative sums of $(n w_t^{(1)}, \dots, n w_t^{(n)})$, generating a *single* $u \sim \mathcal{U}([0,1])$ and then allocating the m_i's as

$$m_i = \lfloor \xi_i + u \rfloor - \lfloor \xi_{i-1} + u \rfloor, \quad i = 2, \dots, n-1$$
$$m_1 = \lfloor \xi_1 + u \rfloor, \quad m_n = n - \lfloor \xi_{n-1} + u \rfloor.$$

The amount of randomness is thus reduced to a single uniform variate. More specific results about the reduction of the variability of the estimates can be found in Chopin (2002) and Künsch (2004)

Many topical modifications of the Gordon et al. (1993) scheme can be found in the literature, and we refer the reader to Doucet et al. (2001) for a detailed coverage of the differences. Let us point out at this stage, however, the *smooth bootstrap* technique of Hürzeler and Küensch (1998) and Stavropoulos and Titterington (2001), where the empirical distribution (14.8) is replaced with a (nonparametric) kernel approximation which smoothes the proposal into a continuous distribution, as we will encounter a similar proposal in Section 14.4.

14.3.6 Fighting the Degeneracy

When considering an entire path $\tilde{\mathbf{z}}_t$ simulated by the sole addition of only the current state z_t at time t, as in Algorithm [A.59], it is not difficult to realize that as t becomes large, the algorithm must degenerate. As in repeated applications of the bootstrap (Section 1.6.2), iterated multinomial sampling can deplete the sample all the way down to a single value. That is, since the first values of the path are *never* re-simulated, the multinomial selection step at each iteration reduces the number of different values for $t < t_0$, at a rate that is faster than exponential (Chopin 2002). Therefore, if we are interested in a long path, $\tilde{\mathbf{z}}_t$, the basic particle filter algorithm simply does not work well. Note, however, that this is rarely the focus. Instead, the main interest in sequential Monte Carlo is to produce good approximations to the distribution of the process z_t at time t.

Example 14.6. (Continuation of Example 14.2) Figure 14.6 represents the range of 100 paths simulated over $T = 50$ observations of the tracking model (14.2), that is, at the end of the observation sequence, after reweighting has taken place at each observational instant. While the range of these paths is never very wide, it is much thinner at the beginning, a result of the repeated resampling at each stage of the algorithm.   ‖

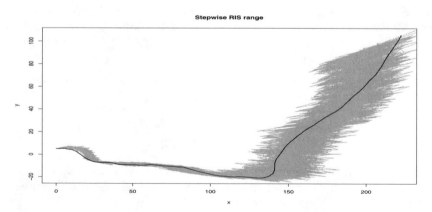

Fig. 14.6. Range of the particle filter path at the end of the observation sequence, along with the true (simulated) sequence of 50 target moves, for $\tau = 0.05$, $\eta = .01$ and 1000 particles.

If one wants to preserve the entire sample path with proper accuracy, the price to pay is, at each step, to increase the number of particles in compensation for the thinning (or *depletion*) of the sample. In general, this is not a small price, unfortunately, as the size must increase exponentially, as detailed in Section 14.3.7! In a specific setting, namely *state space models*, Doucet et al. (2000) establish an optimality result on the best proposal distribution in terms of the variance of the importance weights, conditional on the past iterations. But this optimum is not necessarily achievable, as with the regular importance optimal solution of Theorem 3.12, and can at best slow down the depletion rate.

Note that degeneracy can occur in a case that is a priori more favorable, that is, when the π_t's all have the same support and where particles associated with the previous target π_{t-1} are resampled with weights π_t/π_{t-1}. For instance, if the targets move too quickly in terms of shape or support, very few particles will survive from one iteration to the next, leading to a fast depletion of the sample.

Gilks and Berzuini (2001) have devised another solution to this problem, based on the introduction of additional MCMC moves. Their method is made of two parts, *augmentation* if the parameter space is different for π_t and π_{t-1} and *evolution*, which is the MCMC step. The corresponding algorithm reads as follows.

Algorithm A.60 –Resample–Move Particle Filter–

At iteration t,

1. **Augmentation:** For $i = 1, \ldots, n$, augment $\mathbf{z}_i^{(t-1)}$ with its missing part, as $\tilde{Z}_i^{(t)} \sim \alpha_t(\tilde{z}|\mathbf{z}_i^{(t-1)})$ and $\tilde{\mathbf{z}}_i^{(t)} = (\mathbf{z}_i^{(t-1)}, \tilde{z}_i^{(t)})$.

2. **Weighting:** Compute the corresponding importance weights for the target distribution π_t, $\omega_1^{(t)}, \ldots, \omega_n^{(t)}$, as

$$\omega_i^{(t)} = \pi_t(\tilde{\mathbf{z}}_i^{(t)}) \Big/ \alpha_t(\tilde{z}_i^{(t)}|z_i^{(t-1)}) \pi_{t-1}(\mathbf{z}_i^{(t-1)}).$$

3. **Evolution:** Generate a new iid sample $\mathbf{z}_i^{(t)}$, $i = 1, \ldots, n$, using a proposal proportional to

$$\sum_{i=1}^{n} \omega_i^{(t-1)} q_t(\mathbf{z}|\tilde{\mathbf{z}}_i^{(t-1)}),$$

where q_t is a Markov transition kernel with stationary distribution π_t.

The basis of Algorithm [A.60] is thus to pick up a point $\tilde{\mathbf{z}}_i^{(t-1)}$ according to its weight $\omega_i^{(t-1)}$, that is, via multinomial sampling, and then to implement one iteration of a Markov transition that is stable with respect to π_t. Since the regular multinomial importance sampling produces a sample that is (empirically) distributed from the target π_t, the distribution of the final sample is indeed in (empirical) agreement with π_t and is thus a valid importance sampling methodology. In other words, the importance sampling weight does not need to be adapted after the MCMC step, as already seen in Lemma 14.1.

As above, reduced variance resampling strategies can also be implemented at the resampling stage. Note that the proposal kernel q_t needs not to be known numerically, as long as simulations from q_t are possible. Gilks and Berzuini (2001) also extend this algorithm to a model choice setting with sequentially observed data, as in Section 11.2, and they use a reversible jump algorithm as part of the evolution step 3. Obviously, the success of this additional step in fighting degeneracy depends on the setup and several iterations of the MCMC step could be necessary to reach regions of acceptable values of π_t.

As noted in Clapp and Godsill (2001), the **Augmentation** step 1. in Algorithm [A.60] can itself be supplemented by an importance sampling move where the entire vector $\tilde{\mathbf{z}}_i^{(t-1)}$ may be modified through an importance distribution g (and a corresponding modification of the weight). These authors also suggested *tempering* schemes to smooth the bridge between g and the target distribution π_t. One example is to use a sequence of geometric averages

$$\pi^{[m]}(\mathbf{z}) \propto g^{\alpha_m}(\mathbf{z}) \, \pi_t^{1-\alpha_m}(\mathbf{z}) \, ,$$

where $0 < \alpha_m < 1$ increases from 0 to 1 (see Note 13.6.2 for an introduction to tempering).

14.3.7 Convergence of Particle Systems

So far, the sequential Monte Carlo methods have been discussed from a rather practical point of view, in that the convergence properties of the various estimators or approximations have been derived from those of regular importance sampling techniques, that is, mostly from the Law of Large Numbers. We must, however, take into account an additional dimension, when compared with importance sampling (as in Chapter 3), due to the iterative nature of the method. Earlier results, as those of Gilks and Berzuini (2001), assume that the sample size goes to infinity at each iteration, but this is unrealistic and contradicts the speed requirement which is fundamental to these methods. More recent results, as those of Künsch (2004), give a better understanding of how these algorithms converge and make explicit the intuition that *"there ain't no such thing as a free lunch!,"* namely, that it does not seem possible to consider a sequence of target distributions without incurring an increase in the computational expense.

Although the proofs are too advanced for the level of this book, let us point out here that Künsch (2004) shows, for state-space models, that the number of particles in the sample, N_t, has to increase exponentially fast with

t, to achieve convergence in total variation for the sequential importance sampling/particle filter method. (This result is not surprising in light of the intuitive development of Section 14.3.1.) Chopin (2004) also establishes Central Limit Theorems for both the multinomial and the residual sampling schemes, with the side result that the asymptotic variance is smaller for the residual sampling approach, again a rather comforting result. Note that for the basic sequential sampling in a fixed dimension setting where

$$\omega_t^{(i)} \propto \omega_{t-1}^{(i)} \frac{\pi_t(z_i)}{\pi_{t-1}(z_i)} \,,$$

a Central Limit Theorem also holds:

$$\sqrt{N} \left\{ \frac{\sum_{i=1}^N \omega_t^{(i)} h(Z_i)}{\sum_{i=1}^N \omega_t^{(i)}} - \mathbb{E}_{\pi_t}[h(Z)] \right\} \overset{\mathscr{L}}{\leadsto} \mathcal{N}(0, V_t^0(h)) \,,$$

where

$$V_t^0(h) = \mathbb{E}_{\pi_t} \left[\frac{\pi_t(Z)}{\pi_0(Z)} \{h(Z) - \mathbb{E}_{\pi_t}[h(Z)]\}^2 \right] \,,$$

by an application of the usual Central Limit Theorem, since the Z_i's are independent. (Establishing the Central Limit Theorems for the two other sampling schemes of Section 14.3.5 is beyond our scope.)

In comparing the above three sampling schemes, Chopin (2004) obtained the important comparison that, for a fixed state space of dimension p, and under some regularity conditions for the Central Limit Theorem to hold, the three corresponding asymptotic variances satisfy

$$V_t^0(h) = \mathrm{O}(t^{p/2-1}), \quad V_t^m(h) = \mathrm{O}(t^{p/2}), \quad \text{and} \quad V_t^r(h) = \mathrm{O}(t^{p/2}) \,,$$

where $V_t^m(h)$ and $V_t^r(h)$ denote the asymptotic variances for the estimators of $\mathbb{E}_{\pi_t}[h(X)]$ based on the multinomial and the residual resampling schemes, respectively. Therefore, resampling is more explosive in t than the mere reweighting scheme. This is due to the fact that resampling creates less depletion (thus more variation), while always sampling from the same proposal. In all cases, to fight degeneracy, that is, explosive variance, the number of particles must increase as a power of t. Crisan et al. (1999), Del Moral and Miclo (2000) and Del Moral (2001) provide alternative (if mathematically more sophisticated) entries to the study of the convergence of particle filters.

14.4 Population Monte Carlo

The population Monte Carlo (PMC) algorithm, introduced in this section, is simultaneously an iterated importance sampling scheme that produces, at each iteration, a sample approximately simulated from a target distribution *and* an *adaptive* algorithm that calibrates the proposal distribution to the target distribution along iterations. Its theoretical roots are thus within importance sampling and not within MCMC, despite its iterated features, in that the approximation to the target is valid (that is, unbiased at least to the order $O(1/n)$) at *each iteration* and does not require convergence times nor stopping rules.

14.4.1 Sample Simulation

In the previous chapters about MCMC, the stationary distribution has always been considered to be the limiting distribution of a Markov sequence (Z_t), with the practical consequence that Z_t is approximately distributed from π for t large enough. A rather straightforward extension of this perspective is to go from simulating a point distributed from π to simulating a sample of size n distributed from π or, rather, from

$$\pi^{\otimes n}(z_1, \ldots, z_n) = \prod_{i=1}^{n} \pi(z_i).$$

Implementations of this possible extension are found in Warnes (2001) and Mengersen and Robert (2003), with improvements over a naïve programming of n parallel MCMC runs. Indeed, the entire sample at iteration t can be used to design a proposal at iteration $t+1$. In Warnes (2001), for instance, a kernel estimation of the target distribution based on the sample $(z_1^{(t-1)}, \ldots, z_n^{(t-1)})$ is the proposal distribution. The difficulty with such a proposal is that multi-dimensional kernel estimators are notoriously poor. In Mengersen and Robert (2003), each point of the sample is moved using a random walk proposal that tries to avoid the other points of the sample by delayed rejection (Tierney and Mira 1998). Note that, for simulation purposes, a kernel estimator is not different from a random walk proposal. In both cases, it is more efficient to move each point of the sample separately, as the average acceptance probability of the entire sample decreases with the sample size, no matter what the proposal distribution is, using the same Kullback–Leibler argument as in Section 14.3.3. However, as we will see next, the recourse to the theory of Markov chains to justify the convergence to $\pi^{\otimes n}$ is not necessary to obtain a valid approximation of an iid sample from π.

14.4.2 General Iterative Importance Sampling

The PMC algorithm can be described in a very general framework: it is indeed possible to consider different proposal distributions *at each iteration and for each particle* with this algorithm. That is, the $Z_i^{(t)}$'s can be simulated from distributions q_{it} that may depend on past samples,

$$Z_i^{(t)} \sim q_{it}(z),$$

independently of one another (conditional on the past samples). Thus, each simulated point $z_i^{(t)}$ is allocated an importance weight

$$\varrho_i^{(t)} = \frac{\pi(z_i^{(t)})}{q_{it}(z_i^{(t)})}, \qquad i = 1, \ldots, n,$$

and approximations of the form

$$\Im_t = \frac{1}{n} \sum_{i=1}^{n} \varrho_i^{(t)} h(z_i^{(t)})$$

are then unbiased estimators of $\mathbb{E}^\pi[h(Z)]$, even when the importance distribution q_{it} depends on the entire past of the experiment. Indeed, we have

$$\begin{aligned} \mathbb{E}\left[\varrho_i^{(t)} h(z_i^{(t)})\right] &= \int \int \frac{\pi(x)}{q_{it}(x|\zeta)} h(x) q_{it}(x) \, dx \, g(\zeta) \, d\zeta \\ (14.9) \qquad &= \int \int h(x) \pi(x) \, dx \, g(\zeta) \, d\zeta = \mathbb{E}^\pi[h(X)], \end{aligned}$$

where ζ denotes the vector of past random variates that contribute to q_{it}, and $g(\zeta)$ its *arbitrary* distribution. Furthermore, assuming that the variances

$$\text{var}\left(\varrho_i^{(t)} h(Z_i^{(t)})\right)$$

exist for every $1 \leq i \leq n$, which means that the proposals q_{it} should have heavier tails than π, we have

$$(14.10) \qquad \text{var}\left(\Im_t\right) = \frac{1}{n^2} \sum_{i=1}^{n} \text{var}\left(\varrho_i^{(t)} h(Z_i^{(t)})\right),$$

due to the canceling effect of the weights $\varrho_i^{(t)}$ (Problem 14.16). In fact, even if the $Z^{(t)}$' are correlated, the importance-weighted terms will always be uncorrelated (Lemma 12.11). So, for importance sampling estimators, the variance of the sum will equal the sum of the variances of the individual terms.

Note that resampling may take place at some or even all iterations of the algorithm but that, contrary to the particle systems, there is no propagation of the weights across iterations.

As in most settings the distribution of interest π is unscaled, we instead use

$$\varrho_i^{(t)} \propto \frac{\pi(z_i^{(t)})}{q_{it}(z_i^{(t)})}, \qquad i = 1, \ldots, n,$$

scaled so that the weights $\varrho_i^{(t)}$ sum up to 1. In this case, the above unbiasedness property and the variance decomposition are both lost, although they still approximately hold. In fact, the estimation of the normalizing constant of π improves with each iteration t, since the overall average

$$(14.11) \qquad \varpi_t = \frac{1}{tn} \sum_{\tau=1}^{t} \sum_{i=1}^{n} \frac{\pi(z_i^{(\tau)})}{q_{i\tau}(z_i^{(\tau)})}$$

is a convergent estimator of the inverse of the normalizing constant. Therefore, as t increases, ϖ_t contributes less and less to the bias and variability of $\mathfrak{I}_t$ and the above properties can be considered as holding for t large enough. In addition, if ϖ_{t-1} in (14.11) is used instead of ϖ_t, that is, if

$$(14.12) \qquad \varrho_i^{(t)} = \frac{\pi(z_i^{(t)})}{\varpi_{t-1} q_{it}(z_i^{(t)})},$$

the variance decomposition (14.10) can be approximately recovered, via the same conditioning argument (Problem 14.16).

14.4.3 Population Monte Carlo

Following Iba (2000), Cappé et al. (2004) called their iterative approach *population Monte Carlo*, to stress the idea of *repeatedly* simulating an entire sample rather than *iteratively* simulating the points of an approximate sample.

Since the above section establishes that an iterated importance sampling scheme based on sample dependent proposals is fundamentally a specific kind of importance sampling, we can propose the following algorithm, which is validated by the same principles as regular importance sampling.

Algorithm A.61 –Population Monte Carlo–

> For $t = 1, \ldots, T$
>
> 1. For $i = 1, \ldots, n$,
>   - (i) Select the generating distribution $q_{it}(\cdot)$.
>   - (ii) Generate $\tilde{Z}_i^{(t)} \sim q_{it}(z)$.
>   - (iii) Compute $\varrho_i^{(t)} = \pi(z_i^{(t)})/q_{it}(\tilde{z}_i^{(t)})$.
> 2. Normalize the $\varrho_i^{(t)}$'s to sum to 1.
> 3. Resample n values from the $\tilde{z}_i^{(t)}$'s with replacement, using the weights $\varrho_i^{(t)}$, to create the sample $(Z_1^{(t)}, \ldots, Z_n^{(t)})$.

Step 1.(i) is singled out because it is an essential feature of the PMC algorithm: as demonstrated in the previous Section, the proposal distributions can be individually tailored at each step of the algorithm without jeopardizing the validity of the method. The proposals q_{it} can therefore be picked according to the performances of the previous $q_{i(t-1)}$'s (possibly in terms of survival of the values generated with a particular proposal, or a variance criterion, or even on all the previously simulated samples, if storage allows). For instance, the q_{it}'s can include, with low probability, large tails proposals as in the *defensive sampling* strategy of Hesterberg (1998) (Section 3.3.2), to ensure finite variance.

Note the formal similarly with Warnes's (2001) use of the previous sample to build a nonparametric kernel approximation to π. The main difference is that the proposal does not aim at a good approximation of π using standard nonparametric results like bandwidth selection, but may remain multiscaled over the iterations, as illustrated in Section 14.4.4 The main feature of the PMC algorithm is indeed that several scenarios can be tested in parallel and tuned along iterations, a feature that can hardly be achieved within the domain of MCMC algorithms (Section 7.6.3).

There also are similarities between the PMC algorithm and earlier proposals in the particle system literature, in particular with Algorithm [A.60], since the latter also considers iterated samples with (SIR) resampling steps based on importance weights. A major difference, though (besides the dynamic setting of moving target distributions), is that [A.60] remains an MCMC algorithm and thus needs to use Markov transition kernels with a given stationary distribution. There is also a connection with Chopin (2002), who considers iterated importance sampling with changing proposals. His setting is a special case of the PMC algorithm in a Bayesian framework, where the proposals q_{it} are the posterior distributions associated with a portion k_t of the observed dataset (and are thus independent of i and of the previous samples). As detailed in the following sections, the range of possible choices for the q_{it}'s is actually much wider.

14.4.4 An Illustration for the Mixture Model

Consider the normal mixture model of Example 5.19, that is, $p\mathcal{N}(\mu_1, 1) + (1 - p)\mathcal{N}(\mu_2, 1)$, where $p \neq 1/2$ is known, and the corresponding simulation from $\pi(\mu_1, \mu_2 | \mathbf{x})$, the posterior distribution for an iid sample $\mathbf{x} = (x_1, \ldots, x_n)$ and an arbitrary proper prior on (μ_1, μ_2). While we presented in Chapter 9 a Gibbs sampler based on a data augmentation step via the indicator variables, Celeux et al. (2003) show that a PMC sampler can be efficiently implemented *without* this augmentation step.

Given the posterior distribution

$$\pi(\mu_1, \mu_2 | \mathbf{x}) \propto \exp(-\lambda(\theta - \mu_1)^2/2\sigma^2) \exp(-\lambda(\theta - \mu_2)^2/2\sigma^2)$$
$$\prod_{i=1}^{n} \left\{ p \exp(-(x_i - \mu_1)^2/2\sigma^2) + (1 - p) \exp(-(x_i - \mu_2)^2/2\sigma^2) \right\},$$

a natural possibility is to choose a random walk for the proposal distribution (see Section 7.5). That is, starting from a sample of values of $\boldsymbol{\mu} = (\mu_1, \mu_2)$, generate random isotropic perturbations of the points of this sample.

The difficult issue of selecting the scale of the random walk (see Section 7.6), found in MCMC settings, can be bypassed by virtue of the adaptivity of the PMC algorithm. Indeed, if we take as proposals q_{it} normal distributions centered at the points of the current sample, $\mathcal{N}_2(\boldsymbol{\mu}_i^{(t)}, \sigma_{it} I_2)$, the variance factors σ_{it} can be chosen at random from a set of K scales v_k $(1 \leq k \leq K)$ ranging from, e.g., 10^3 down to 10^{-3} if this range is compatible with the range of the observations. At each iteration t of the PMC algorithm, the probability of choosing a particular scale v_k can be calibrated according to the performance of the different scales over the previous iterations. For instance, possible criterion is to select a scale proportional to its *non-degeneracy rate* on the previous iterations, that is, the percentage of points associated with v_k that survived past the resampling step 3. The reasoning behind this scheme is that, if most μ_j's associated with a given scale v_k are not resampled, the scale is not appropriate and thus should not be much used in the next iterations. However, when the survival rate is null, in order to avoid a definitive removal of the corresponding scale, the next probability ζ_k is set to a positive value ε.

In order to smooth the selection of the scales, Rao–Blackwellization should also be used in the computation of the importance weights, using as the denominator

$$\sum_k \zeta_k \varphi \left(\boldsymbol{\mu}_i^{(t)}; \boldsymbol{\mu}_i^{(t-1)}, v_k \right) ,$$

where $\varphi(\boldsymbol{\mu}; \xi, v)$ here denotes the density of the two-dimensional normal distribution with mean ξ and variance vI_2 at the vector $\boldsymbol{\mu}$.

The corresponding PMC algorithm thus looks as follows.

Algorithm A.62 –Mixture PMC algorithm–

Step 0: Initialization

For $i = 1, \ldots, n$, generate $\boldsymbol{\mu}_i^{(0)}$ from an arbitrary distribution
For $k = 1, \ldots, K$, set v_k and $\zeta_k = 1/K$.

Step t: Update

For $i = 1, \ldots, n$,
a. with probability ζ_k, take $\sigma_{it} = v_k$
b. generate

$$\boldsymbol{\mu}_i^{(t)} \sim \mathcal{N}_2 \left(\boldsymbol{\mu}_i^{(t-1)}, \sigma_{it} I_2 \right)$$

c. compute the weights

$$\varrho_i \propto \frac{\pi \left(\boldsymbol{\mu}_i^{(t)} | \mathbf{x} \right)}{\sum_k \varrho_k \varphi \left(\boldsymbol{\mu}_i^{(t)}; \boldsymbol{\mu}_i^{(t-1)}, v_k \right)} .$$

Resample the $\boldsymbol{\mu}_i^{(t)}$'s using the weights ϱ_i.
Update the ζ_k's as $\zeta_k \propto \varepsilon + r_k$ where r_k is the number of μ_i's generated with variance v_k that have been resampled in the previous step.

The performance of this algorithm is illustrated on a simulated dataset of 500 observations from $0.3\mathcal{N}(0,1) + 0.7\mathcal{N}(2,1)$. As described by Figure 14.7 through the sequence of simulated samples, the result of the experiment is that, after 8 iterations of the PMC algorithm, the simulated μ's are concentrated around the mode of interest and the scales v_k (equal to .01, .05, .1 and .5) have been selected according to their relevance, that is, with large weights for the smallest values (as also described in Figure 14.8). While the second spurious mode is visited during the first iteration of the algorithm, the relatively small value of the posterior at this mode implies that the corresponding points are not resampled at the next iteration.

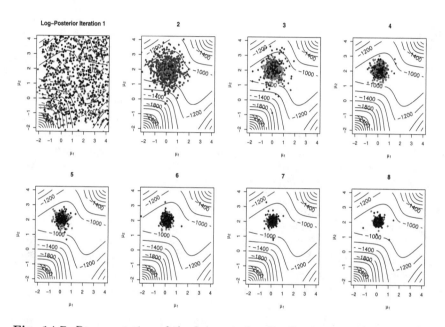

Fig. 14.7. Representation of the log-posterior distribution via contours and of a sequence of PMC samples over the first 8 iterations. The sample of 500 observations was generated from $0.3\mathcal{N}(0,1) + 0.7\mathcal{N}(2,1)$ and the prior on μ was a $\mathcal{N}(1,10)$ distribution on both means.

14.4.5 Adaptativity in Sequential Algorithms

A central feature of the PMC method is that the generality in the choice of the proposal distributions q_{it} is due to the abandonment of the MCMC framework. Indeed, were it not for the importance resampling correction, a pointwise Metropolis–Hastings algorithm would produce a parallel MCMC sampler which simply converges to the target $\pi^{\otimes n}$ in distribution. Similarly, a samplewise Metropolis–Hastings algorithm, that is, a Metropolis–Hastings

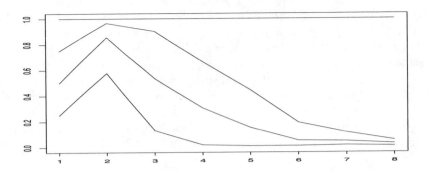

Fig. 14.8. Evolution of the cumulative weights of the four scales $v_k = .5, .1, .05, .01$ for the mixture PMC algorithm over the first 8 iterations corresponding to Figure 14.7.

algorithm aimed at the target distribution $\pi^{\otimes n}$, also produces an asymptotic approximation to this distribution, but its acceptance probability approximately decreases as a power of n. This difference is not simply a theoretical advantage since, in one example of Cappé et al. (2004), it actually occurs that a Metropolis–Hastings scheme based on the same proposal does not work well while a PMC algorithm produces correct answers.

Example 14.7. (Continuation of Example 14.3) For the stochastic volatility model (14.4), Celeux et al. (2003) consider a noninformative prior $\pi(\beta^2, \varphi, \sigma^2) = 1/(\sigma\beta)$ under the stationarity constraint $|\varphi| < 1$. Posteriors on β^2 and σ^2 are both conjugate, conditional on the z_t's, while the posterior distribution of φ is less conventional, but a standard proposal (Chib et al. 2002) is a truncated normal distribution on $]-1, 1[$ with mean and variance

$$\sum_{t=2}^{n} z_t z_{t-1} \bigg/ \sum_{t=2}^{n-1} z_t^2 \quad \text{and} \quad \sigma^2 \bigg/ \sum_{t=2}^{n-1} z_t^2 \,.$$

There have been many proposals in the literature for simulating the z_t's (see Celeux et al. 2003). For instance, one based on a Taylor expansion of the exponential is a normal distribution with mean

$$\frac{\left(1+\varphi^2\right)\mu_t/\sigma^2 + 0.5\exp\left(-\mu_t\right)y_t^2\left(1+\mu_t\right)/\beta^2 - 0.5}{\left(1+\varphi^2\right)/\sigma^2 + 0.5\exp\left(-\mu_t\right)y_t^2/\beta^2}$$

and variance

$$1\big/\big\{\left(1+\varphi^2\right)/\sigma^2 + 0.5\exp\left(-\mu_t\right)y_t^2/\beta^2\big\}\,,$$

where $\mu_t = \varphi\left(z_{t-1} + z_{t+1}\right)/(1+\varphi^2)$ is the conditional expectation of z_t given z_{t-1}, z_{t+1}.

Celeux et al. (2003) use a simulated dataset of size $n = 1000$ with $\beta^2 = 1$, $\varphi = 0.99$ and $\sigma^2 = 0.01$, in order to compare the performance of the PMC algorithm with an MCMC algorithm based on exactly the same proposals.

First, the results of the MCMC algorithm, based on $10,000$ iterations, are presented in Figures 14.9 and 14.10. The estimate of β^2 (over the last 5000 simulated values) is 0.98, while the estimate of φ is equal to 0.89 and the estimate of σ^2 is equal to 0.099. While the reconstituted volatilities are on average close to the true values, the parameter estimates are rather poor, even though the cumulative averages of Figure 14.9. do not exhibit any difficulty with convergence. Note, however, the slow mixing on β in Figure 14.9 (upper left) and, to a lesser degree, on σ^2 (middle left).

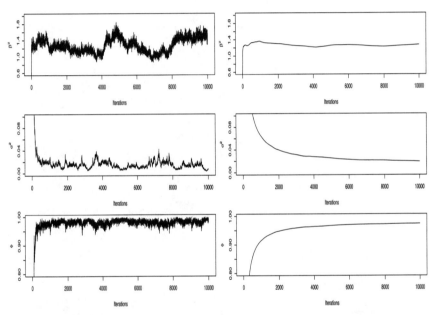

Fig. 14.9. Evolution of the MCMC samples for the three parameters *(left)* and convergence of the MCMC estimators *(right)*. (*Source:* Celeux et al. 2003.)

Then, with the same proposal distributions, Celeux et al. (2003) have iterated a PMC algorithm ten times with $M = 1000$. The results are presented in Figures 14.11 and 14.12. The estimate of φ (over the 10 iterations) is equal to 0.87, while the estimate of σ^2 is equal to 0.012 and the estimate of β^2 is equal to 1.04. These estimations are clearly closer to the true values than the ones obtained with the MCMC algorithm. (Note that the scales on Figure 14.11 (left) are much smaller than those of Figure 14.9 (right).) Moreover, Figure 14.12 provides an excellent reconstitution of the volatilities.   ‖

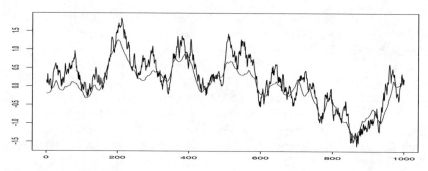

Fig. 14.10. Comparison of the true volatility *(black)* with the MCMC estimation based on the last 5000 iterations *(grey)*. (*Source:* Celeux et al. 2003.)

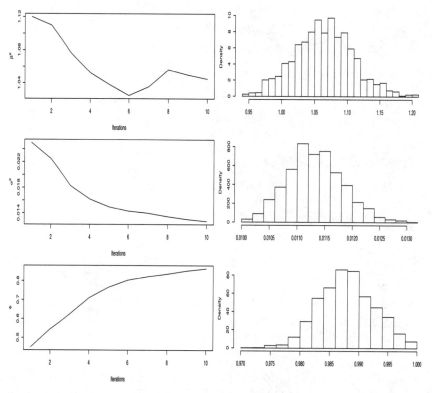

Fig. 14.11. Evolution over iterations of the PMC approximation *(left)* and 10th iteration weighted PMC sample *(right)*. (*Source:* Celeux et al. 2003.)

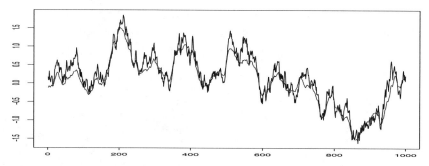

Fig. 14.12. Comparison of the true volatility *(black)* with the PMC estimation based on the 10th iteration weighted PMC sample *(grey)*. (*Source:* Celeux et al. 2003.)

As shown in Cappé et al. (2004) (see also West 1992 and Guillin et al. 2004), the PMC framework allows, in addition, for a construction of *adaptive* schemes, i.e., of proposals that correct themselves against past performances, that is much easier than in MCMC setups, as described in Section 7.6.3. Indeed, from a theoretical point of view, ergodicity is not an issue for PMC methods since the validity is obtained via importance sampling justifications and, from a practical point of view, the total freedom allowed by unrestricted parallel simulations is a major asset.

An extension of the PMC method can be found in Del Moral and Doucet (2003): their generalization is to consider two Markov kernels, K_+ and K_-, such that, given a particle at time $t-1$, $x_i^{(t-1)}$, a new particle $\tilde{x}_i^{(t)}$ is generated from $K_+(x_i^{(t-1)}, \tilde{x})$ and associated with a weight

$$\omega_i(t) \propto \frac{\pi(\tilde{x}_i^{(t)}) \, K_-(\tilde{x}_i^{(t)}, x_i^{(t-1)})}{\pi(x_i^{(t-1)}) \, K_+(x_i^{(t-1)}, \tilde{x}_i^{(t-1)})}.$$

As in Algorithm [A.61], the sample $(x_1^{(t)}, \ldots, x_n^{(t)})$ is then obtained by multinomial sampling from the $\tilde{x}_i^{(t)}$'s, using the weights $\omega_i^{(t)}$. The most intriguing feature of this extension is that the kernel K_- is irrelevant for the unbiasedness of the new sample. Indeed,

$$\mathbb{E}\left[h(\tilde{X}_i^{(t)}) \, \frac{\pi(\tilde{X}_i^{(t)}) \, K_-(\tilde{X}_i^{(t)}, X_i^{(t-1)})}{\pi(X_i^{(t-1)}) \, K_-(X_i^{(t-1)}, \tilde{X}_i^{(t-1)})}\right] = \int h(\tilde{x}) \pi(\tilde{x}) K_-(\tilde{x}, x) \, dx \, d\tilde{x}$$

$$= \mathbb{E}^\pi\left[h(X)\right],$$

whatever K_- is chosen; the only requirement is that $K_-(\tilde{x}, x)$ integrates to 1 as a function of x (which is reminiscent of Monte Carlo marginalization; see Problem 3.21).

While this scheme does provide a valid PMC algorithm, what remains to be assessed is whether *wealth is a mixed blessing*, that is, if the added value brought by the generality of the choice of the kernels K_+ and K_- is substantial. For example, Doucet et al. (2000) show that the optimal choice of K_+, in terms of conditional variance of the weights, is

$$K_+^*(x, \tilde{x}) \propto \pi(\tilde{x})K_-(\tilde{x}, x),$$

in which case

$$w_t^{(i)} \propto \frac{\int \pi(u)K_-(u, x_i^{(t-1)})du}{\pi(x_i^{(t-1)})}$$

depends only on the previous value $x_i^{(t-1)}$. This is both a positive feature, given that only resampled $\tilde{x}_i^{(t)}$ need to be simulated, and a negative feature, given that the current sample is chosen in terms of the previous sample, even though K_- is arbitrary. Note also that K_- must be chosen so that

$$\int \pi(u)K_-(u, x_i^{(t-1)})du$$

is computable. Conversely, if K_- is chosen in a symmetric way,

$$K_-^*(\tilde{x}, x) \propto \pi(x)K_+(x, \tilde{x}),$$

then

$$w_t^{(i)} \propto \frac{\pi(x_i^{(t)})}{\int \pi(u)K_+(u, x_i^{(t)})du}$$

depends only on the current value $x_i^{(t)}$. In the special case where K_+ is a Markov kernel with stationary distribution π, the integral can be computed and the importance weight is one, which is natural, but this also shows that the method is in this case nothing but parallel MCMC sampling. (See also Del Moral and Doucet 2003 for extensions.)

14.5 Problems

14.1 (MacEachern et al. 1999) Consider densities f and g such that

$$f_1(z_1) = \int f(z_1, z_2) \, dz_2 \quad \text{and} \quad g_1(z_1) = \int g(z_1, z_2) \, dz_2.$$

(a) Show that

$$\text{var}_g\left(\frac{f(Z_1, Z_2)}{g(Z_1, Z_2)}\right) \leq \text{var}_{g_1}\left(\frac{f_1(Z_1)}{g_1(Z_1)}\right),$$

where var_g denotes the variance under the joint density $g(z_1, z_2)$.

(b) Justify a Rao–Blackwellization argument that favors the integration of auxiliary variables in importance samplers.

14.2 Consider Algorithm [A.63] with $\theta = 1$ and $\delta = 0$.

(a) Show that the new weight is

$$\omega' = \begin{cases} \varrho + 1 & \text{if } u < \varrho/\varrho + 1, \\ \omega(\varrho + 1) & \text{otherwise.} \end{cases}$$

(b) Show that the expectation of ω' when integrating in $U \sim \mathcal{U}([0, 1])$ is $\omega + \varrho$.

(c) Deduce that the sequence of weights is explosive.

14.3 For the Q-move proposal of Liu et al. (2001) (see the discussion after Algorithm [A.63]) show that the average value of ω' is $a\omega(1 - \varrho) + \varrho$ if $\theta = 1$ and deduce a condition on a for the expectation of this average to increase.

14.4 Consider a hidden Markov model with hidden Markov chain (X_t) on $\{1, \ldots, \kappa\}$, associated with a transition matrix $\mathbf{P}$ and observable Y_t such that

$$Y_t | X_t = x_t \sim f(y_t | x_t) \,.$$

(a) Show the *actualization* equations are given by

$$p(x_{1:t} | y_{1:t}) = \frac{f(y_t | x_t)\, p(x_{1:t} | y_{1:(t-1)})}{p(y_t | y_{1:(t-1)})}$$

and

$$p(x_{1:t} | y_{1:(t-1)}) = \mathbb{P}_{x_{t-1} x_t} p(x_{1:(t-1)} | y_{1:(t-1)}) \,,$$

where $f(y_t | x_t)$ is the density $f_{\theta_{x_t}}(y_t)$ and $\mathbb{P}_{nm}$ denotes the (n, m)-th element of $\mathbb{P}$.

(b) Deduce from

$$p(x_{1:t} | y_{1:t}) = \frac{f(y_t | x_t)\, \mathbb{P}_{x_{t-1} x_t}}{p(y_t | y_{1:(t-1)})} p(x_{1:(t-1)} | y_{1:(t-1)})$$

that computation of the *filtering* density $p(x_{1:t} | y_{1:t})$ has the same complexity as the computation of the density $p(y_t | y_{1:(t-1)})$.

(c) Verify the *propagation* equation

$$p(x_t | y_{1:(t-1)}) = \sum_{x_{1:(t-1)}} p(x_{1:(t-1)} | y_{1:(t-1)}) \mathbb{P}_{x_{t-1} x_t} \,.$$

14.5 (Continuation of Problem 14.4) In this problem, we establish the forward–backward equations for hidden Markov chains, also called Baum–Welch formulas (Baum and Petrie 1966).

(a) Show that, if we denote $(i = 1, \ldots, \kappa)$

$$\gamma_t(i) = P(x_t = i | y_{1:T}) \qquad t \le T$$

and we set

$$\alpha_t(i) = p(y_{1:t}, x_t = i), \quad \beta_t(i) = p(y_{t+1:T}, x_t = i) \,,$$

then we have the following recursive equations

$$\begin{cases} \alpha_1(i) &= f(y_1 | x_t = i)\pi_i \\ \alpha_{t+1}(j) &= f(y_{t+1} | x_{t+1} = j) \sum_{i=1}^{\kappa} \alpha_t(i) \mathbb{P}_{ij} \end{cases}$$

$$\begin{cases} \beta_T(i) = 1 \\ \beta_t(i) = \displaystyle\sum_{j=1}^{\kappa} \mathbb{P}_{ij} f(y_{t+1}|x_{t+1} = j)\beta_{t+1}(j) \end{cases}$$

and

$$\gamma_t(i) = \alpha_t(i)\beta_t(i) \Big/ \sum_{j=1}^{\kappa} \alpha_t(j)\beta_t(j).$$

(b) Deduce that the computation of γ_t is achieved in $O(T\kappa^2)$ operations.
(c) If we denote

$$\xi_t(i,j) = P(x_t = i, x_{t+1} = j|y_{1:T}) \qquad i,j = 1,\ldots,\kappa,$$

show that

(14.13) $$\xi_t(i,j) = \frac{\alpha_t(i)a_{ij} f(y_{t+1}|x_t = j)\beta_{t+1}(j)}{\displaystyle\sum_{i=1}^{\kappa}\sum_{j=1}^{\kappa} \alpha_t(i)a_{ij} f(y_{t+1}|x_{t+1} = j)\beta_{t+1}(j)},$$

which can still be obtained in $O(T\kappa^2)$ operations.

(d) When computing the *forward* and *backward* variables, α_t and β_t, there often occur numerical problems of *overflow* or *underflow*. Show that, if the $\alpha_t(i)$'s are renormalized by $c_t = \sum_{i=1}^{\kappa} \alpha_t(i)$ at each stage t of the recursions in part (a), this renormalization, while modifying the backward variables, does not change the validity of the γ_t relations.

14.6 (Continuation of Problem 14.5)   In this problem, we introduce *backward smoothing* and the *Viterbi algorithm* for the prediction of the hidden Markov chain.

(a) Show that $p(x_s|x_{s-1}, y_{1:t}) = p(x_s|x_{s-1}, y_{s:t})$.
(b) Establish the global *backward* equations $(s = t, t-1, \ldots, 1)$

$$p(x_s|x_{s-1}, y_{1:t}) \propto \mathbb{P}_{x_{s-1}x_s} f(y_s|x_s) \sum_{x_{s+1}} p(x_{s+1}|x_s, y_{1:t}),$$

with $p(x_t|x_{t-1}, y_{1:t}) \propto \mathbb{P}_{x_{t-1}x_t} f(y_t|x_t)$, and deduce that

$$p(x_1|y_{1:t}) \propto \pi(x_1) f(y_1|x_1) \sum_{x_2} p(x_2|x_1, y_{1:t}),$$

where π is the stationary distribution of $\mathbb{P}$, can be computed in $O(t\kappa^2)$ operations.

(c) Show that the joint distribution $p(x_{1:t}|y_{1:t})$ can be computed and simulated in $O(t\kappa^2)$ operations.
(d) Show that, if the model parameters are known, the backward equations in part (a) allow for a sequential maximization of $p(x_{1:t}|y_{1:t})$. (*Note:* This algorithm comes from signal processing and is called the *Viterbi algorithm*, even though it is simply a special case of dynamic programming.)
(e) Deduce from Bayes formula that

$$p(y_{1:t}) = \frac{p(y_{1:t}|x_{1:t})p(x_{1:t})}{p(x_{1:t}|y_{1:t})}.$$

and derive from the backward equations of part (a) a representation of the observed likelihood. (*Hint:* Since the left hand side of this equation does not depend on $x_{1:t}$, one can use an arbitrary (but fixed) sequence $x^o_{1:t}$ on the right hand side.)

(f) Conclude that the likelihood of a hidden Markov model with κ states and T observations can be computed in $O(T\kappa^2)$ operations.

14.7 (Continuation of Problem 14.6) (Cappé 2001) In this problem, we show that both the likelihood and its gradient can be computed for a hidden Markov chain. We denote by $\boldsymbol{\theta}$ the parameters of the distributions of the y_t's conditional on the x_t's, and $\boldsymbol{\eta}$, the parameters of $\mathbb{P}$. We also suppose that the (initial) distribution of x_1, denoted $\varphi_1(\cdot)$, is fixed and known.

(a) We call

$$\varphi_t(i) = p(x_t = i|y_{1:t-1})$$

the *prediction filter*. Verify the *forward* equations

$$\varphi_1(j) = p(x_1 = j)$$

$$\varphi_{t+1}(j) = \frac{1}{c_t}\sum_{i=1}^{\kappa} f(y_t|x_t = i)\varphi_t(i)\mathbb{P}_{ij} \quad (t \geq 1),$$

where

$$c_t = \sum_{k=1}^{\kappa} f(y_t|x_t = k)\varphi_t(k),$$

which are based on the same principle as the backward equations.

(b) Deduce that the (log-)likelihood can be written as

$$\log p(y_{1:t}) = \sum_{r=1}^{t} \log\left[\sum_{i=1}^{\kappa} p(y_t, x_t = i|y_{1:(r-1)})\right]$$

$$= \sum_{r=1}^{t} \log\left[\sum_{i=1}^{\kappa} f(y_t|x_t = i)\varphi_t(i)\right].$$

(c) Compare this derivation with the one of question (e) in Problem 14.6, in terms of computational complexity and storage requirements.

(d) Show that the gradient against $\boldsymbol{\theta}$ of the log-likelihood is

$$\nabla_{\boldsymbol{\theta}} \log p(y_{1:t}) = \sum_{r=1}^{t} \frac{1}{c_r}$$

$$\times \sum_{i=1}^{\kappa} [\varphi_r(i)\nabla_{\boldsymbol{\theta}} f(y_r|x_r = i) + f(y_r|x_r = i)\nabla_{\boldsymbol{\theta}}\varphi_r(i)],$$

where $\nabla_{\boldsymbol{\theta}}\varphi_t(i)$ can be computed recursively as

$$\nabla_{\boldsymbol{\theta}}\varphi_{t+1}(j) = \frac{1}{c_t}\sum_{i=1}^{\kappa} (\mathbb{P}_{ij} - \varphi_{t+1}(j))$$

$$\times [\varphi_t(i)\nabla_{\boldsymbol{\theta}} f(y_t|x_t = i) + f(y_t|x_t = i)\nabla_{\boldsymbol{\theta}}\varphi_t(i)].$$

(e) Show that the gradient against η of the log-likelihood is

$$\nabla_\eta \log p(y_{1:t}) = \sum_{r=1}^{t} \frac{1}{c_r} \sum_{i=1}^{\kappa} f(y_t|x_t = i)\nabla_\eta \varphi_r(i),$$

where $\nabla_\eta \varphi_t(i)$ can be computed recursively as

$$\nabla_\eta \varphi_{t+1}(j) = \frac{1}{c_r} \sum_{i=1}^{\kappa} f(y_t|x_t = i)$$
$$\times \left[(\mathbb{P}_{ij} - \varphi_{t+1}(j))\nabla_\eta \varphi_t(i) + \varphi_t(i)\nabla_\eta \mathbb{P}_{ij} \right].$$

(f) Show that the complexity of these computations is in $O(t\kappa^2)$.

(g) Show that the storage requirements for the gradient computation are constant in t and of order $O(p\kappa)$, if p is the dimension of (θ, η).

14.8 (Continuation of Problem 14.7) In the case where $\mathbf{P}$ is known and $f(y|x)$ is the density of the normal distribution $\mathcal{N}(\mu_x, \sigma_x^2)$, explicitly write the gradient equations of part (d).

14.9 Show that residual sampling, defined in Section 14.3.5, is unbiased, that is, if $N_t^{(i)}$ denotes the number of replications of $\mathbf{x}_t^{(i)}$, then $\mathbb{E}[N_t^{(i)}] = n\omega_t^{(i)}$.

14.10 (Continuation of Problem 14.9) Show that the sampling method of Crisan et al. (1999), defined in Section 14.3.5, is unbiased. Write an explicit formula for the probability that N_t goes to 0 or to ∞ as t increases.

14.11 (Continuation of Problem 14.10) For the sampling method of Carpenter et al. (1999) defined in Section 14.3.5,

(a) Show that it is unbiased.

(b) Examine whether the variance of the m_i's is smaller than for the two above methods.

(c) Show that the simulated points are dependent.

14.12 (Continuation of Problem 14.11) Douc (2004) proposes an alternative to residual sampling called *comb sampling*. It starts as residual sampling by sampling $\lfloor n\omega_t^{(i)} \rfloor$ copies of $\tilde{\mathbf{z}}_t^{(i)}$ $(i = 1, \ldots, n)$. Denoting $N_r = n - \sum_i \lfloor n\omega_t^{(i)} \rfloor$, and $\hat{F}_n$ the empirical cdf associated with the $(\tilde{\mathbf{z}}_t^{(i)}, \omega_t^{(i)})$, the remaining N_r points are sampled by taking one point at random in each interval $(k/N_r, (k+1)/N_r)$ $(k = 0, \ldots, N_r - 1)$ and inverting this sample by $\hat{F}_n$.

(a) Show that this scheme is unbiased.

(b) Show that the simulated points are independent.

(c) Show that the corresponding variance is smaller than for the residual sampling.

14.13 For the stochastic volatility model of Example 14.3, when $\beta = 0$:

(a) Examine the identifiability of the parameters.

(b) Show that the observed likelihood satisfies

$$L(\varphi, \sigma|y_0, \ldots, y_T) = \mathbb{E}[L^c(\varphi, \sigma|y_0, \ldots, y_T, Z_0, \ldots, Z_T)|y_0, \ldots, y_T],$$

where

$$L^c(\varphi, \sigma|x_0, \ldots, x_T, z_0, \ldots, z_T) \propto \exp - \sum_{t=0}^{T} \left\{ x_t^2 e^{-z_t} + z_t \right\}/2$$

$$(\sigma)^{-T+1} \exp - \left\{ (z_0)^2 + \sum_{t=1}^{T} (z_t - \varphi z_{t-1})^2 \right\}/2(\sigma)^2 .$$

(c) Examine whether or not an EM algorithm can be implemented in this case.

(d) For a noninformative prior, $\pi(\varphi, \sigma) = 1/\sigma$, determine whether or not the posterior distribution is well-defined.

(e) Examine the data augmentation algorithm for drawing Bayesian inference on this model, and show that the simulation of the volatilities Z_t cannot be done using a standard distribution.

(f) Derive an Metropolis–Hastings modification based on an approximation of the full conditional

$$f(z_t|z_{t-1}, z_{t+1}, y^t, \varphi, \sigma) \propto \exp\left\{-(z_t - \varphi z_{t-1})^2/2\sigma^2 \right.$$
$$\left. -(z_{t+1} - \varphi z_t)^2/2\sigma^2 - z_t/2 - x_t^2 e^{-z_t}/2\right\}.$$

(*Hint:* For instance, the expression $z_t/2 + x_t^2 e^{-z_t}/2$ in the exponential can be approximated by $(z_t - \log(x_t^2))^2/2$, leading to the Metropolis–Hastings proposal

$$\mathcal{N}\left(\frac{[\varphi z_{t-1} + \varphi z_{t+1}]c^{-2} + \log(x_t^2)/2}{(1+\varphi^2)c^{-2} + 1/2}, \frac{1}{(1+\varphi^2)c^{-2} + 1/2}\right).$$

See Shephard and Pitt (1997) for another proposal.)

14.14 In the case of an hidden Markov model defined in Section 14.3.2, Pitt and Shephard (1999) propose a modification of the bootstrap filter [A.59] called the *auxiliary particle filter*.

(a) Show that, if $z_{t-1}^{(1)}, \ldots, z_{t-1}^{(n)}$ is the sample of simulated missing variables at time $t - 1$, with weights $w_{t-1}^{(1)}, \ldots, w_{t-1}^{(n)}$, the distribution

$$\hat{\pi}(z_t|x_1, \ldots, x_t) \propto f(x_t|z_t) \sum_i w_{t-1}^{(i)} \pi(z_t|z_{t-1}^{(i)})$$

provides an unbiased approximation to the predictive distribution $\pi(z_t|x_1, \ldots, x_t)$.

(b) Show that sampling from the joint distribution

$$\hat{\pi}(z_t, \kappa|x_1, \ldots, x_t) \propto f(x_t|z_t) w_{t-1}^{(\kappa)} \pi(z_t|z_{t-1}^{(\kappa)}),$$

with $\kappa = 1, \ldots, n$, is equivalent to a simulation from $\hat{\pi}$ of part (a).

(c) Deduce that this representation has the appealing feature of reweighting the particles $z_t|z_{t-1}^{(i)}$ by the density of the current observable, $f(x_t|z_t)$, and that it is thus more efficient than the standard bootstrap filter.

14.15 Using the same decomposition as in (14.9), and conditioning on a random variable ζ that determines both proposal distributions q_{it} and q_{jt}, show that $\varrho_i^{(t)} h(x_i^{(t)})$ and $\varrho_j^{(t)} h(x_j^{(t)})$ are uncorrelated and deduce (14.10). (The argument is similar that of Lemma 12.11.)

14.16 (a) For the PMC algorithm [A.61] show that when the normalizing constant of π is unknown and replaced with the estimator $1/\varpi_{t-1}$, as in (14.12), the variance decomposition (14.10) approximately holds for large t's.

(b) From the decomposition (Problem 12.14)

$$\text{var}(\mathfrak{I}_t) = \text{var}_\zeta(\mathbb{E}[\mathfrak{I}_t|\zeta])\mathbb{E}_\zeta[\text{var}(\mathfrak{I}_t|\zeta)],$$

show that the term $\text{var}_\zeta(\mathbb{E}[\mathfrak{I}_t|\zeta])$ is of order $O(1/t)$ and is thus negligible.

(c) Show that the conditional variance $\text{var}(\mathfrak{I}_t|\zeta)$ satisfies

$$\text{var}(\mathfrak{I}_t|\zeta) = \sum_i \text{var}(\varrho_i^{(t)} h(x_i^{(t)})|\zeta) \,.$$

14.17 Consider a hidden Markov Poisson model: the observed x_t's depend on an unobserved Markov chain (Z_t) such that $(i, j = 1, 2)$

$$X_t | z_t \sim \mathcal{P}(\lambda_{z_t}), \qquad P(Z_t = i | z_{t_1} = j) = p_{ji} \,.$$

(a) When $k = 2$ and a noninformative prior,

$$\pi(\lambda_1, \lambda_2, p_{11}, p_{22}) = \frac{1}{\lambda_1} \mathbb{I}_{\lambda_2 < \lambda_1} \,,$$

is used, construct the Gibbs sampler for this model. (*Hint:* Use the hidden Markov chain (Z_t).)

(b) For a prior distribution $(j = 1, \ldots, k)$

$$\lambda_j \sim \mathcal{G}a(\alpha_j, \beta_j) \,,$$

show that the parameter simulation step in the Gibbs sampler can be chosen as the generation of

$$\lambda_j \sim \mathcal{G}a(\alpha_j + n_j \bar{x}_j, \beta_j + n_j)$$

and

$$p_j \sim \mathcal{D}(\gamma_1 + n_{j1}, \ldots, \gamma_k + n_{jk}) \,,$$

where

$$n_{ji} = \sum_{t=2}^n \mathbb{I}_{z_t = i} \mathbb{I}_{z_{t-1} = j}, \quad n_j = \sum_{t=1}^n \mathbb{I}_{z_t = j} \quad \text{and} \quad n_j \bar{x}_j = \sum_{t=1}^n \mathbb{I}_{z_t = j} x_t.$$

(c) Design a reversible jump algorithm (Chapter 11) for the extension of the above model to the case when $k \sim \mathcal{P}(\xi)$.

(*Note:* Robert and Titterington (1998) detail the Bayesian estimation of such a hidden Markov Poisson model via Gibbs sampling.)

14.18 Show that a semi-Markov chain, as introduced in Note 14.6.3, is a Markov chain in the special case where the duration in each state is a geometric random variable.

14.19 Evaluate the degeneracy of a regular random walk on a $n \times n$ grid by the following simulation experiment: for an increasing sequence of values of (n, p), give the empirical frequency of random walks of length p that are *not* self-avoiding.

14.20 Establish Lemma 14.8 by showing that, if $\theta > 0$, and if g_+ denote the distribution of the augmented variable (X, W) after the step, then

$$\int_0^\infty w' \, g_+(x', w') \, dw' = 2(1 + \delta) c_0 f(x') \,,$$

where c_0 denotes the proportionality constant in (14.14). If $\theta = 0$, show that

$$\int_0^\infty w' \, g_+(x', w') \, dw' = (1 + \delta) \, c_0 f(x') \,,$$

and conclude that (14.14) holds.

14.21 If the initial distribution of (X, ω) satisfies (14.14), show that the R-moves preserve the equilibrium equation (14.14). (*Hint:* If g_+ denotes the distribution of the augmented variable (X, W) after the step, show that

$$\int \omega' \, g_+(x', \omega') \, d\omega' = 2cf(x') \, .\Big)$$

14.6 Notes

14.6.1 A Brief History of Particle Systems

The realization of the possibilities of iterating importance sampling is not new: in fact, it is about as old as Monte Carlo methods themselves! It can be found in the molecular simulation literature of the 50's, as in Hammersley and Morton (1954), Rosenbluth and Rosenbluth (1955) and Marshall (1965).[6] Hammersley and colleagues proposed such a method to simulate a self-avoiding random walk (Madras and Slade 1993) on a grid, due to huge inefficiency in regular importance sampling and rejection techniques (Problem 14.19). Although this early implementation occurred in particle physics, the use of the term "particle" only dates back to Kitagawa (1996), while Carpenter et al. (1999) coined the term "particle filter". In signal processing, early occurrences of a "particle filter" can be traced back to Handschin and Mayne (1969).

14.6.2 Dynamic Importance Sampling

This generalization of regular importance sampling, which incorporates the use of an (importance) weight within an MCMC framework, is due to Wong and Liang (1997), followed by Liu et al. (2001) and Liang (2002). At each iteration t, or for each index t of a sample, the current state X_t is associated with a weight ω_t, in such a way that the joint distribution of (X_t, ω_t), say $g(x, \omega)$, satisfies

$$(14.14) \qquad \int_0^\infty \omega \, g(x, \omega) \, d\omega \propto f(x) \, ,$$

where f is the target distribution.

Then, for a function h such that $\mathbb{E}_f[|h(X)|] < \infty$, the importance sampling identity (14.1) generalizes into

$$\mathbb{E}_g[h(X) \, W]/\mathbb{E}_g[W] = \mathbb{E}_f[h(X)]$$

and the weighted average

$$\tilde{\mathfrak{J}}_T = \sum_{t=1}^T \omega_t \, h(X_t) \Big/ \sum_{t=1}^T \omega_t$$

is a convergent estimator of the integral

---

[6] So we find, once more, Hammersley and some of the authors of the Metropolis et al. (1953) paper at the forefront of a simulation advance!

$$\mathfrak{I} = \int h(x)f(x)dx.$$

Obviously, regular importance sampling is a special case where X_t is marginally distributed from g and, conditional on x_t, ω is deterministic, with $\omega \propto f(x)/g(x)$. A more general scheme provided by Liang (2002) is based on a transition kernel K as follows.

Algorithm A.63 –Liang's Dynamic Importance Sampling–

At iteration t, given (x_t, ω_t),

1 Generate $y \sim K(x_t, y)$ and compute

$$\varrho = \omega \, \frac{f(y)K(y, x_t)}{f(x_t)K(x_t, y)}.$$

2 Generate $u \sim \mathcal{U}(0, 1)$ and take

$$(x_{t+1}, \omega_{t+1}) = \begin{cases} (y, (1+\delta)\varrho/a) & \text{if } u < a, \\ (x_t, (1+\delta)\omega_t/(1-a)) & \text{otherwise,} \end{cases}$$

where $a = \varrho/(\varrho + \theta(x_t, \omega_t))$, and θ and $\delta > 0$ are both either constant or independent random variables.

The intuition behind this construction is that the increases in the weights allow for moves that would not be allowed by regular Metropolis–Hastings algorithms. But the drawback of this construction is that the importance weights diverge (in t), as shown in Problem 14.2. (Note that the ratio $\varrho/(\varrho+\theta)$ is Boltzman's ratio (7.20).)

In comparison, Liu et al. (2001) use a more Metropolis-like ratio, in the so-called *Q-move* alternative, where

$$(x_{t+1}, \omega_{t+1}) = \begin{cases} (y, \theta \vee \varrho) & \text{if } u < 1 \wedge \varrho/\theta, \\ (x_t, a\omega_t) & \text{otherwise,} \end{cases}$$

with $a \geq 1$ either a constant or an independent random variable. In the spirit of Peskun (1973) (Lemma 10.22) Liang (2002) also establishes an inequality. That is, the acceptance probability is lower for his algorithm compared with a Metropolis–Hastings move.

Liang (2002) distinguishes between the *R-move*, where $\delta = 0$ and $\theta \equiv 1$, yielding

$$(x_{t+1}, \omega_{t+1}) = \begin{cases} (y, \varrho + 1) & \text{if } u < \varrho/(\varrho + 1) \\ (x_t, \omega_t(\varrho + 1)) & \text{otherwise,} \end{cases}$$

and the *W-move*, where $\theta \equiv 0$, $a = 1$ and

$$(x_{t+1}, \omega_{t+1}) = (y, \varrho).$$

Wong and Liang (1997) also mention the *M-move*, which is essentially the result of MacEachern et al. (1999): in the above algorithm, when f is the stationary distribution of the kernel K, $\varrho = \omega_t$ and $(x_{t+1}, \omega_{t+1}) = (y, \omega_t)$. Note that all moves are identical when $\theta \equiv 0$.

As one can see in the simple case of the W-move, these schemes are not valid MCMC transitions in that f is *not* the marginal distribution of the stationary distribution associated with the transition.

The general result for Liang's Algorithm [A.63] is as follows.

Proposition 14.8. *If the initial distribution of (X, ω) satisfies (14.14), one iteration of [A.63] preserves the equilibrium equation (14.14).*

Liu et al. (2001) show that the Q-move is only approximately correctly weighted, the approximation being related to the condition that the average acceptance probability

$$\int 1 \wedge \omega \frac{f(z)K(z,x)}{f(x)K(x,z)} \Big/ \theta K(x,z) dz$$

be almost independent from ω. See Liu et al. (2001) and Liang (2002) for a detailed study of the instability of the weights ω along iterations.

14.6.3 Hidden Markov Models

Hidden Markov models and their generalizations have enjoyed an extremely widespread use in applied problems. For example, in Econometrics, regression models and switching time series are particular cases of (14.3) (see Goldfeld and Quandt 1973, Albert and Chib 1993, McCulloch and Tsay 1994, Shephard 1994, Chib 1996, Billio and Monfort 1998). Hidden Markov chains also appear in character and speech processing (Juang and Rabiner 1991), in medicine (Celeux and Clairambault 1992, Guihenneuc-Jouyaux et al. 1998) in genetics (Churchill 1989, 1995), in engineering (Cocozza-Thivent and Guédon 1990) and in neural networks (Juang 1984). We refer the reader to MacDonald and Zucchini (1997) and Cappé et al. (2004) for additional references on the applications of Hidden Markov modeling.

Decoux (1997) proposes one of the first extensions to *hidden semi-Markov chains*, where the observations remain in a given state during a random number of epochs following a Poisson distribution and then move to another state (see Problem 14.18). Guihenneuc-Jouyaux and Richardson (1996) also propose a Markov chain Monte Carlo algorithm for the processing of a Markov process on a finite state space which corresponds to successive degrees of seropositivity.

There are many potential applications of the hidden semi-Markov modeling to settings where the sojourns in each state are too variable or too long to be modeled as a Markov process. For instance, in the modeling of DNA as a sequence of Markov chains on $\{A, C, G, T\}$, it is possible to have thousands of consecutive bases in the same (hidden) state; to model this with a hidden Markov chain is unrealistic as it leads to very small transitions with unlikely magnitudes like 10^{-89}.

One particular application of this modeling can be found for a specific neurobiological model called the *ion channel*, which is a formalized representation of ion exchanges between neurons as neurotransmission regulators in neurobiology. Ion channels can be in one of several states, each state corresponding to a given electric intensity. These intensities are only indirectly observed, via so-called *patch clamp recordings*, which are intensity variations. The observables $(y_t)_{1 \le t \le T}$ are thus directed by a hidden Gamma (indicator) process $(x_t)_{1 \le t \le T}$,

$$y_t | x_t \sim \mathcal{N}(\mu_{x_t}, \sigma^2),$$

with the hidden process such that

$$d_{j+1} = t_{j+1} - t_j \sim \mathcal{G}a(s_i, \lambda_i)$$

if $x_t = i$ for $t_j \leq t < t_{j+1}$. References to the *ion channel* model are Ball et al. (1999), Hodgson (1999), and Hodgson and Green (1999). Cappé et al. (2004) re-examine this model via the population Monte Carlo technique.

A

Probability Distributions

We recall here the density and the two first moments of most of the distributions used in this book. An exhaustive review of probability distributions is provided by Johnson and Kotz (1972), or the more recent Johnson and Hoeting (2003), Johnson et al. (1994, 1995). The densities are given with respect to Lebesgue or counting measure depending on the context.

A.1. **Normal Distribution, $\mathcal{N}_p(\theta, \Sigma)$**

($\theta \in \mathbb{R}^p$ and Σ is a $(p \times p)$ symmetric positive definite matrix.)

$$f(\mathbf{x}|\theta, \Sigma) = (\det \Sigma)^{-1/2}(2\pi)^{-p/2}e^{-(\mathbf{x}-\theta)^t \Sigma^{-1}(\mathbf{x}-\theta)/2}.$$

$\mathbb{E}_{\theta,\Sigma}[\mathbf{X}] = \theta$ and $\mathbb{E}_{\theta,\Sigma}[(\mathbf{X} - \theta)(\mathbf{X} - \theta)^t] = \Sigma$.
When Σ is not positive definite, the $\mathcal{N}_p(\theta, \Sigma)$ distribution has no density with respect to Lebesgue measure on $\mathbb{R}^p$. For $p = 1$, the *log-normal* distribution is defined as the distribution of e^X when $X \sim \mathcal{N}(\theta, \sigma^2)$.

A.2. **Gamma Distribution, $\mathcal{G}a(\alpha, \beta)$**

($\alpha, \beta > 0$.)

$$f(x|\alpha, \beta) = \frac{\beta^\alpha}{\Gamma(\alpha)}x^{\alpha-1}e^{-\beta x}\mathbb{I}_{[0,+\infty)}(x).$$

$\mathbb{E}_{\alpha,\beta}[X] = \alpha/\beta$ and $\mathrm{var}_{\alpha,\beta}(X) = \alpha/\beta^2$.
Particular cases of the Gamma distribution are the *Erlang distribution*, $\mathcal{G}a(\alpha, 1)$, the *exponential distribution* $\mathcal{G}a(1, \beta)$ (denoted by $\mathcal{E}xp(\beta)$), and the *chi squared distribution*, $\mathcal{G}a(\nu/2, 1/2)$ (denoted by χ^2_ν). (Note also that the opposite convention is sometimes adopted for the parameter, namely that $\mathcal{G}a(\alpha, \beta)$ may also be noted as $\mathcal{G}a(\alpha, 1/\beta)$. See, e.g., Berger 1985.)

A.3. **Beta Distribution, $\mathcal{B}e(\alpha, \beta)$**

($\alpha, \beta > 0$.)

$$f(x|\alpha, \beta) = \frac{x^{\alpha-1}(1 - x)^{\beta-1}}{B(\alpha, \beta)}\mathbb{I}_{[0,1]}(x),$$

where

$$B(\alpha, \beta) = \frac{\Gamma(\alpha)\Gamma(\beta)}{\Gamma(\alpha + \beta)}.$$

$\mathbb{E}_{\alpha,\beta}[X] = \alpha/(\alpha + \beta)$ and $\text{var}_{\alpha,\beta}(X) = \alpha\beta/[(\alpha + \beta)^2(\alpha + \beta + 1)]$.
The beta distribution can be obtained as the distribution of $Y_1/(Y_1 + Y_2)$
when $Y_1 \sim \mathcal{G}a(\alpha, 1)$ and $Y_2 \sim \mathcal{G}a(\beta, 1)$.

A.4. **Student's t Distribution, $\mathcal{T}_p(\nu, \theta, \Sigma)$**

($\nu > 0$, $\theta \in \mathbb{R}^p$, and Σ is a $(p \times p)$ symmetric positive-definite matrix.)

$$f(\mathbf{x}|\nu, \theta, \Sigma) = \frac{\Gamma((\nu + p)/2)/\Gamma(\nu/2)}{(\det \Sigma)^{1/2}(\nu\pi)^{p/2}}\left[1 + \frac{(\mathbf{x} - \theta)^t\Sigma^{-1}(\mathbf{x} - \theta)}{\nu}\right]^{-(\nu+p)/2}.$$

$\mathbb{E}_{\nu,\theta,\Sigma}[\mathbf{X}] = \theta$ ($\nu > 1$) and $\mathbb{E}_{\theta,\Sigma}[(\mathbf{X} - \theta)(\mathbf{X} - \theta)^t] = \nu\Sigma/(\nu - 2)$ ($\nu > 2$).
When $p = 1$, a particular case of Student's t distribution is the *Cauchy*
distribution, $\mathcal{C}(\theta, \sigma^2)$, which corresponds to $\nu = 1$. Student's t distribution
$\mathcal{T}_p(\nu, 0, I)$ can be derived as the distribution of $\mathbf{X}/Z$ when $\mathbf{X} \sim \mathcal{N}_p(0, I)$
and $\nu Z^2 \sim \chi^2_\nu$.

A.5. **Fisher's F Distribution, $\mathcal{F}(\nu, \rho)$**

($\nu, \rho > 0$.)

$$f(x|\nu, \rho) = \frac{\Gamma((\nu + \rho)/2)\nu^{\rho/2}\rho^{\nu/2}}{\Gamma(\nu/2)\Gamma(\rho/2)}\frac{x^{(\nu-2)/2}}{(\nu + \rho x)^{(\nu+\rho)/2}}\mathbb{I}_{[0,+\infty)}(x).$$

$\mathbb{E}_{\nu,\rho}[X] = \rho/(\rho-2)$ ($\rho > 2$) and $\text{var}_{\nu,\rho}(X) = 2\rho^2(\nu+\rho-2)/[\nu(\rho-4)(\rho-2)^2]$
($\rho > 4$).
The distribution $\mathcal{F}(p, q)$ is also the distribution of $(\mathbf{X} - \theta)^t\Sigma^{-1}(\mathbf{X} - \theta)/p$
when $\mathbf{X} \sim \mathcal{T}_p(q, \theta, \Sigma)$. Moreover, if $X \sim \mathcal{F}(\nu, \rho)$, $\nu X/(\rho + \nu X) \sim$
$\mathcal{B}e(\nu/2, \rho/2)$.

A.6. **Inverse Gamma Distribution, $\mathcal{IG}(\alpha, \beta)$**

($\alpha, \beta > 0$.)

$$f(x|\alpha, \beta) = \frac{\beta^\alpha}{\Gamma(\alpha)}\frac{e^{-\beta/x}}{x^{\alpha+1}}\mathbb{I}_{[0,+\infty[}(x).$$

$\mathbb{E}_{\alpha,\beta}[X] = \beta/(\alpha - 1)$ ($\alpha > 1$) and $\text{var}_{\alpha,\beta}(X) = \beta^2/((\alpha - 1)^2(\alpha - 2))$
($\alpha > 2$).
This distribution is the distribution of X^{-1} when $X \sim \mathcal{G}a(\alpha, \beta)$.

A.7. **Noncentral Chi Squared Distribution, $\chi^2_\nu(\lambda)$**

($\lambda \geq 0$.)

$$f(x|\lambda) = \frac{1}{2}(x/\lambda)^{(p-2)/4}I_{(p-2)/2}(\sqrt{\lambda x})e^{-(\lambda+x)/2}.$$

$\mathbb{E}_\lambda[X] = p + \lambda$ and $\text{var}_\lambda(X) = 3p + 4\lambda$.
This distribution can be derived as the distribution of $X_1^2 + \cdots + X_p^2$ when
$X_i \sim \mathcal{N}(\theta_i, 1)$ and $\theta_1^2 + \ldots + \theta_p^2 = \lambda$.

A.8. **Dirichlet Distribution,** $\mathcal{D}_k(\alpha_1, \ldots, \alpha_k)$

$(\alpha_1, \ldots, \alpha_k > 0$ and $\alpha_0 = \alpha_1 + \cdots + \alpha_k.)$

$$f(x|\alpha_1, \ldots, \alpha_k) = \frac{\Gamma(\alpha_0)}{\Gamma(\alpha_1) \ldots \Gamma(\alpha_k)} x_1^{\alpha_1 - 1} \ldots x_k^{\alpha_k - 1} \mathbb{I}_{\{\sum x_i = 1\}}.$$

$\mathbb{E}_\alpha[X_i] = \alpha_i/\alpha_0$, $\mathrm{var}(X_i) = (\alpha_0 - \alpha_i)\alpha_i/[\alpha_0^2(\alpha_0 + 1)]$ and $\mathrm{cov}(X_i, X_j) = -\alpha_i\alpha_j/[\alpha_0^2(\alpha_0 + 1)]$ $(i \neq j)$.

As a particular case, note that $(X, 1 - X) \sim \mathcal{D}_2(\alpha_1, \alpha_2)$ is equivalent to $X \sim \mathcal{B}e(\alpha_1, \alpha_2)$.

A.9. **Pareto Distribution,** $\mathcal{P}a(\alpha, x_0)$

$(\alpha > 0$ and $x_0 > 0.)$

$$f(x|\alpha, x_0) = \alpha \frac{x_0^\alpha}{x^{\alpha+1}} \mathbb{I}_{[x_0, +\infty[}(x).$$

$\mathbb{E}_{\alpha, x_0}[X] = \alpha x_0/(\alpha - 1)$ $(\alpha > 1)$ and $\mathrm{var}_{\alpha, x_0}(X) = \alpha x_0^2/[(\alpha - 1)^2(\alpha - 2)]$ $(\alpha > 2)$.

A.10. **Binomial Distribution,** $\mathcal{B}(n, p)$

$(0 \leq p \leq 1.)$

$$f(x|p) = \binom{n}{x} p^x (1 - p)^{n-x} \mathbb{I}_{\{0, \ldots, n\}}(x).$$

$\mathbb{E}_p(X) = np$ and $\mathrm{var}(X) = np(1 - p)$.

A.11. **Multinomial Distribution,** $\mathcal{M}_k(n; p_1, \ldots, p_k)$

$(p_i \geq 0$ $(1 \leq i \leq k)$ and $\sum_i p_i = 1.)$

$$f(x_1, \ldots, x_k|p_1, \ldots, p_k) = \binom{n}{x_1 \ \ldots \ x_k} \prod_{i=1}^k p_i^{x_i} \, \mathbb{I}_{\sum x_i = n}.$$

$\mathbb{E}_p(X_i) = np_i$, $\mathrm{var}(X_i) = np_i(1 - p_i)$, and $\mathrm{cov}(X_i, X_j) = -np_ip_j$ $(i \neq j)$.

Note that, if $X \sim \mathcal{M}_k(n; p_1, \ldots, p_k)$, $X_i \sim \mathcal{B}(n, p_i)$, and that the binomial distribution $X \sim \mathcal{B}(n, p)$ corresponds to $(X, n - X) \sim \mathcal{M}_2(n; p, 1 - p)$.

A.12. **Poisson Distribution,** $\mathcal{P}(\lambda)$

$(\lambda > 0.)$

$$f(x|\lambda) = e^{-\lambda} \frac{\lambda^x}{x!} \mathbb{I}_\mathbb{N}(x).$$

$\mathbb{E}_\lambda[X] = \lambda$ and $\mathrm{var}_\lambda(X) = \lambda$.

A.13. **Negative Binomial Distribution,** $\mathcal{N}eg(n, p)$

$(0 \leq p \leq 1.)$

$$f(x|p) = \binom{n + x + 1}{x} p^n (1 - p)^x \mathbb{I}_\mathbb{N}(x).$$

$\mathbb{E}_p[X] = n(1 - p)/p$ and $\mathrm{var}_p(X) = n(1 - p)/p^2$.

A.14. **Hypergeometric Distribution,** $\mathcal{H}yp(N; n; p)$

$(0 \leq p \leq 1,\ n < N$ and $pN \in \mathbb{N}.)$

$$f(x|p) = \frac{\binom{pn}{x}\binom{(1-p)N}{n-x}}{\binom{N}{n}}\mathbb{I}_{\{n-(1-p)N,\dots,pN\}}(x)\mathbb{I}_{\{0,1,\dots,n\}}(x).$$

$\mathbb{E}_{N,n,p}[X] = np$ and $\mathrm{var}_{N,n,p}(X) = (N-n)np(1-p)/(N-1).$

B

Notation

B.1 Mathematical

$\mathbf{h} = (h_1, \ldots, h_n) = \{h_i\}$ boldface signifies a vector

$H = \{h_{ij}\} = \|h_{ij}\|$ uppercase signifies a matrix

$I, \mathbf{1}, J = \mathbf{11}'$ identity matrix, vector of ones, a matrix of ones

$A \prec B$ $(B - A)$ is a positive definite matrix

$|A|$ determinant of the matrix A

$\mathrm{tr}(A)$ trace of the matrice A

a^+ $\max(a, 0)$

$C_n^p, \binom{n}{p}$ binomial coefficient

D_α logistic function

$_1F_1(a; b; z)$ confluent hypergeometric function

F^- generalized inverse of F

$\Gamma(x)$ gamma function $(x > 0)$

$\Psi(x)$ digamma function, $(d/dx)\Gamma(x)$ $(x > 0)$

$\mathbb{I}_A(t)$ indicator function (1 if $t \in A$, 0 otherwise)

$I_\nu(z)$ modified Bessel function $(z > 0)$

$\binom{n}{p_1 \ldots p_n}$ multinomial coefficient

$\nabla f(z)$ gradient of $f(z)$, the vector with coefficients $(\partial/\partial z_i)f(z)$ $(f(z) \in \mathbb{R}$ and $z \in \mathbb{R}^p)$

$\nabla^t f(z)$ divergence of $f(z)$, $\sum(\partial/\partial z_i)f(z)$ $(f(z) \in \mathbb{R}^p$ and $z \in \mathbb{R})$

$\Delta f(z)$ Laplacian of $f(z)$, $\sum(\partial^2/\partial z_i^2)f(z)$

$\|\cdot\|_{TV}$ total variation norm

$|\mathbf{x}| = (\Sigma x_i^2)^{1/2}$ Euclidean norm

$[x]$ or $\lfloor x \rfloor$ greatest integer less than x

$\lceil x \rceil$ smallest integer larger than x

$f(t) \propto g(t)$ the functions f and g are proportional

$\mathrm{supp}(f)$ support of f

$\langle x, y \rangle$ scalar product of x and y in $\mathbb{R}^p$

$x \vee y$ maximum of x and y

$x \wedge y$ minimum of x and y

B.2 Probability

X, Y	random variable (uppercase)
$(\mathcal{X}, \mathcal{P}, \mathcal{B})$	probability triple: sample space, probability distribution, and σ-algebra of sets
β_n	β-mixing coefficient
$\delta_{\theta_0}(\theta)$	Dirac mass at θ_0
$E(\theta)$	energy function of a Gibbs distribution
$\mathcal{E}(\pi)$	entropy of the distribution π
$F(x\|\theta)$	cumulative distribution function of X, conditional on the parameter θ
$f(x\|\theta)$	density of X, conditional on the parameter θ, with respect to Lebesgue or counting measure
$X \sim f(x\|\theta)$	X is distributed with density $f(x\|\theta)$
$\mathbb{E}_\theta[g(X)]$	expectation of $g(x)$ under the distribution $X \sim f(x\|\theta)$
$\mathbb{E}^V[h(V)]$	expectation of $h(v)$ under the distribution of V
$\mathbb{E}^\pi[h(\theta)\|x]$	expectation of $h(\theta)$ under the distribution of θ, conditional on x, $\pi(\theta\|x)$
iid	independent and identically distributed
$\lambda(dx)$	Lebesgue measure, also denoted by $d\lambda(x)$
P_θ	probability distribution, indexed by the parameter θ
$p \star q$	convolution product of the distributions p and q, that is, distribution of the sum of $X \sim p$ and $Y \sim q$
$p^{n\star}$	convolution nth power, that is, distribution of the sum of n iid rv's distributed fr
$\varphi(t)$	density of the Normal distribution $\mathcal{N}(0,1)$
$\Phi(t)$	cumulative distribution function of the Normal distribution
$O(n), o(n)$ or $O_p(n), o_p(n)$	big "Oh", little "oh." As $n \to \infty$, $\frac{O(n)}{n} \to$ constant, $\frac{o(n)}{n} \to 0$, and the subscript p denotes *in probability*

B.3 Distributions

$\mathcal{B}(n, p)$	binomial distribution
$\mathcal{B}e(\alpha, \beta)$	beta distribution
$\mathcal{C}(\theta, \sigma^2)$	Cauchy distribution
$\mathcal{D}_k(\alpha_1, \ldots, \alpha_k)$	Dirichlet distribution
$\mathcal{E}xp(\lambda)$	exponential distribution
$\mathcal{F}(p, q)$	Fisher's F distribution
$\mathcal{G}a(\alpha, \beta)$	gamma distribution
$\mathcal{IG}(\alpha, \beta)$	inverse gamma distribution

$\chi_p^2,\ \chi_p^2(\lambda)$	chi squared distribution,
	noncentral chi squared distribution
	with noncentrality parameter λ
$\mathcal{M}_k(n; p_1, .., p_k)$	multinomial distribution
$\mathcal{N}(\theta, \sigma^2)$	univariate normal distribution
$\mathcal{N}_p(\theta, \Sigma)$	multivariate normal distribution
$\mathcal{N}eg(n, p)$	negative binomial distribution
$\mathcal{P}(\lambda)$	Poisson distribution
$\mathcal{P}a(x_0, \alpha)$	Pareto distribution
$\mathcal{T}_p(\nu, \theta, \Sigma)$	multivariate Student's t distribution
$\mathcal{U}_{[a,b]}$	continuous uniform distribution
$\mathcal{W}e(\alpha, c)$	Weibull distribution
$\mathcal{W}_k(p, \Sigma)$	Wishart distribution

B.4 Markov Chains

α	atom
$AR(p)$	autoregressive process of order p
$ARMA(p, q)$	autoregressive moving average process of order (p, q)
C	small set
$d(\alpha)$	period of the state or atom α
$\dagger$	"dagger," absorbing state
$\Delta V(x)$	drift of V
$\mathbb{E}_\mu[h(X_n)]$	expectation associated with P_μ
$\mathbb{E}_{x_0}[h(X_n)]$	expectation associated with P_{x_0}
η_A	total number of passages in A
$Q(x, A)$	probability that η_A is infinite, starting from x
γ_g^2	variance of $S_N(g)$ for the Central Limit Theorem
K_ϵ	kernel of the resolvant
$MA(q)$	moving average process of order q
$L(x, A)$	probability of return to A starting from x
ν, ν_m	minorizing measure for an atom or small set
$P(x, A)$	transition kernel
$P^m(x, A)$	transition kernel of the chain $(X_{mn})_n$
$P_\mu(\cdot)$	probability distribution of the chain (X_n)
	with initial state $X_0 \sim \mu$
$P_{x_0}(\cdot)$	probability distribution of the chain (X_n)
	with initial state $X_0 = x_0$
π	invariant measure
$S_N(g)$	empirical average of $g(x_i)$ for $1 \le i \le N$
$T_{p,q}$	coupling time for the initial distributions p and q
τ_A	return time to A
$\tau_A(k)$	kth return time to A
$U(x, A)$	average number of passages in A, starting from x

$X_t, X^{(t)}$ generic element of a Markov chain
$\check{X}_n$ augmented or split chain

B.5 Statistics

x, y	realized values (lowercase) of the random variables X and Y (uppercase)
$\mathcal{X}, \mathcal{Y}$	sample space (uppercase script Roman letters)
θ, λ	parameters (lowercase Greek letters)
Θ, Ω	parameter space (uppercase script Greek letters)
$B^\pi(x)$	Bayes factor
$\delta^{JS}(x)$	James–Stein estimator
$\delta^\pi(x)$	Bayes estimator
$\delta^+(x)$	positive-part James–Stein estimator
H_0	null hypothesis
$I(\theta)$	Fisher information
$L(\theta, \delta)$	loss function, loss of estimating θ with δ
$L(\theta|x)$	likelihood function, a function of θ for fixed x, mathematically identical to $f(x|\theta)$
$\ell(\theta|x)$	logarithm of the likelihood function
$L^P(\theta|x)$, $\ell^P(\theta|x)$	profile likelihood
$m(x)$	marginal density
$\pi(\theta)$	generic prior density for θ
$\pi^J(\theta)$	Jeffreys prior density for θ
$\pi(\theta|x)$	generic posterior density θ
$\bar{x}$	sample mean
s^2	sample variance
X^*, Y^*, x^*, y^*	latent or missing variables (data)

B.6 Algorithms

$[A_n]$	symbol of the nth algorithm
B	backward operator
B_T	inter-chain variance after T iterations
W_T	intra-chain variance after T iterations
D_T^i	cumulative sum to i, for T iterations
δ_{rb}, δ^{RB}	Rao–Blackwellized estimators
F	forward operator
$g_i(x_i|x_j, j \neq i)$	conditional density for Gibbs sampling
$\Gamma(\theta, \xi)$	regeneration probability
$K(x, y)$	transition kernel
$\tilde{K}$	transition kernel for a mixture algorithm
$K^\star$	transition kernel for a cycle of algorithms

λ_k	duration of excursion
$q(x\|y)$	transition kernel, typically used for
	an instrumental variable
$\rho(x,y)$	acceptance probability
	for a Metropolis–Hastings algorithm
S_T	empirical average
S_T^C	conditional version of the empirical average
S_T^{MP}	recycled version of the empirical average
S_T^P	importance sampling version of the empirical average
S_T^R	Riemann version of the empirical average
$S_h(\omega)$	spectral density of the function h

References

Aarts, E. and Kors, T. (1989). *Simulated Annealing and Boltzman Machines: A Stochastic Approach to Combinatorial Optimisation and Neural Computing.* John Wiley, New York.

Abramowitz, M. and Stegun, I. (1964). *Handbook of Mathematical Functions.* Dover, New York.

Ackley, D., Hinton, G., and Sejnowski, T. (1985). A learning algorithm for Boltzmann machines. *Cognitive Science*, 9:147–169.

Adams, M. (1987). *William Ockham.* University of Notre Dame Press, Notre Dame, Indiana.

Agresti, A. (1992). A survey of exact inference for contingency tables (with discussion). *Statist. Science*, 7:131–177.

Agresti, A. (1996). *An Introduction to Categorical Data Analysis.* John Wiley, New York.

Ahrens, J. and Dieter, U. (1974). Computer methods for sampling from gamma, beta, Poisson and binomial distributions. *Computing*, 12:223–246.

Akaike, H. (1974). A new look at the statistical model identification. *IEEE Transactions on Automatic Control*, AC-19:716–723.

Albert, J. and Chib, S. (1993). Bayes inference via Gibbs sampling of autoregressive time series subject to Markov mean and variance shifts. *J. Business Economic Statistics*, 1:1–15.

Aldous, D. (1987). On the Markov chain simulation method for uniform combinatorial distributions and simulated annealing. *Pr. Eng. Inform. Sciences*, 1:33–46.

Andrieu, C. and Robert, C. (2001). Controlled Markov chain Monte Carlo methods for optimal sampling. Technical Report 0125, Univ. Paris Dauphine.

Arnold, B. and Press, S. (1989). Compatible conditional distributions. *J. American Statist. Assoc.*, 84:152–156.

Asmussen, S. (1979). *Applied Probability and Queues.* John Wiley, New York.

Asmussen, S., Glynn, P., and Thorisson, H. (1992). Stationarity detection in the initial transient problem. *ACM Trans. Modelling and Computer Simulations*, 2:130–157.

Athreya, K., Doss, H., and Sethuraman, J. (1996). On the convergence of the Markov chain simulation method. *Ann. Statist.*, 24:69–100.

Athreya, K. and Ney, P. (1978). A new approach to the limit theory of recurrent Markov chains. *Trans. Amer. Math. Soc.*, 245:493–501.

Atkinson, A. (1979). The computer generation of Poisson random variables. *Appl. Statist.*, 28:29–35.

Ball, F., Cai, Y., Kadane, J., and O'Hagan, A. (1999). Bayesian inference for ion channel gating mechanisms directly from single channel recordings, using Markov chain Monte Carlo. *Proc. Royal Society London A*, 455:2879–2932.

Barbe, P. and Bertail, P. (1995). *The Weighted Bootstrap*, volume 98 of *Lecture Notes in Statistics*. Springer-Verlag, New York.

Barndorff-Nielsen, O. (1991). Modified signed log-likelihood ratio. *Biometrika*, 78:557–563.

Barndorff-Nielsen, O. and Cox, D. R. (1994). *Inference and Asymptotics*. Chapman and Hall, New York.

Barnett, G., Kohn, R., and Sheather, S. (1996). Bayesian estimation of an autoregressive model using Markov chain Monte Carlo. *J. Econometrics*, 74:237–254.

Barone, P. and Frigessi, A. (1989). Improving stochastic relaxation for Gaussian random fields. *Biometrics*, 47:1473–1487.

Barry, D. and Hartigan, J. (1992). Product partition models for change point problems. *Ann. Statist.*, 20:260–279.

Barry, D. and Hartigan, J. (1993). A Bayesian analysis of change point problems. *J. American Statist. Assoc.*, 88:309–319.

Baum, L. and Petrie, T. (1966). Statistical inference for probabilistic functions of finite state Markov chains. *Ann. Mathemat. Statist.*, 37:1554–1563.

Baum, L., Petrie, T., Soules, G., and Weiss, N. (1970). A maximization technique occurring in the statistical analysis of probabilistic functions of Markov chains. *Ann. Mathemat. Statist.*, 41:164–171.

Bauwens, L. (1984). *Bayesian Full Information of Simultaneous Equations Models Using Integration by Monte Carlo*, volume 232. Springer–Verlag, New York. Lecture Notes in Economics and Mathematical Systems.

Bauwens, L. (1991). The "pathology" of the natural conjugate prior density in the regression model. *Ann. Econom. Statist.*, 23:49–64.

Bauwens, L. and Richard, J. (1985). A 1–1 poly-*t* random variable generator with application to Monte Carlo integration. *J. Econometrics*, 29:19–46.

Bennett, J., Racine-Poon, A., and Wakefield, J. (1996). MCMC for nonlinear hierarchical models. In Gilks, W., Richardson, S., and Spiegelhalter, D., editors, *Markov chain Monte Carlo in Practice*, pages 339–358. Chapman and Hall, New York.

Benveniste, A., Métivier, M., and Priouret, P. (1990). *Adaptive Algorithms and Stochastic Approximations*. Springer-Verlag, New York, Berlin–Heidelberg.

Bergé, P., Pommeau, Y., and Vidal, C. (1984). *Order Within Chaos*. John Wiley, New York.

Berger, J. (1985). *Statistical Decision Theory and Bayesian Analysis*. Springer-Verlag, New York, second edition.

Berger, J. (1990). Robust Bayesian analysis: sensitivity to the prior. *J. Statist. Plann. Inference*, 25:303–328.

Berger, J. (1994). An overview of robust Bayesian analysis (with discussion). *TEST*, 3:5–124.

Berger, J. and Bernardo, J. (1992). On the development of the reference prior method. In Berger, J., Bernardo, J., Dawid, A., and Smith, A., editors, *Bayesian Statistics 4*, pages 35–49. Oxford University Press, London.

Berger, J. and Pericchi, L. (1998). Accurate and stable Bayesian model selection: the median intrinsic Bayes factor. *Sankhya B*, 60:1–18.

Berger, J., Philippe, A., and Robert, C. (1998). Estimation of quadratic functions: reference priors for non-centrality parameters. *Statistica Sinica*, 8(2):359–375.

Berger, J. and Wolpert, R. (1988). *The Likelihood Principle (2nd edition)*, volume 9 of *IMS Lecture Notes — Monograph Series*. IMS, Hayward, second edition.

Bernardo, J. (1979). Reference posterior distributions for Bayesian inference (with discussion). *J. Royal Statist. Soc. Series B*, 41:113–147.

Bernardo, J. and Giròn, F. (1986). A Bayesian approach to cluster analysis. In *Second Catalan International Symposium on Statistics*, Barcelona, Spain.

Bernardo, J. and Giròn, F. (1988). A Bayesian analysis of simple mixture problems. In Bernardo, J., DeGroot, M., Lindley, D., and Smith, A., editors, *Bayesian Statistics 3*, pages 67–78. Oxford University Press, Oxford.

Bernardo, J. and Smith, A. (1994). *Bayesian Theory*. John Wiley, New York.

Berthelsen, K. and Møller, J. (2003). Likelihood and non-parametric Bayesian MCMC inference for spatial point processes based on perfect simulation and path sampling. *Scandinavian J. Statist.*, 30:549–564.

Besag, J. (1974). Spatial interaction and the statistical analysis of lattice systems (with discussion). *J. Royal Statist. Soc. Series B*, 36:192–326.

Besag, J. (1986). On the statistical analysis of dirty pictures. *J. Royal Statist. Soc. Series B*, 48:259–279.

Besag, J. (1989). Towards Bayesian image analysis. *J. Applied Statistics*, 16:395–407.

Besag, J. (1994). Discussion of "Markov chains for exploring posterior distributions". *Ann. Statist.*, 22:1734–1741.

Besag, J. (2000). Markov chain Monte Carlo for statistical inference. Technical Report 9, U. Washington, Center for Statistics and the Social Sciences.

Besag, J. and Clifford, P. (1989). Generalized Monte Carlo significance tests. *Biometrika*, 76:633–642.

Besag, J., Green, E., Higdon, D., and Mengersen, K. (1995). Bayesian computation and stochastic systems (with discussion). *Statistical Science*, 10:3–66.

Best, N., Cowles, M., and Vines, K. (1995). CODA: Convergence diagnosis and output analysis software for Gibbs sampling output, version 0.30. Technical report, MRC Biostatistics Unit, Univ. of Cambridge.

Billingsley, P. (1968). *Convergence of Probability Measures*. John Wiley, New York.

Billingsley, P. (1995). *Probability and Measure*. John Wiley, New York, third edition.

Billio, M. and Monfort, A. (1998). Switching state space models. *J. Statist. Plann. Inference*, 68(1):65–103.

Billio, M., Monfort, A., and Robert, C. (1998). The simulated likelihood ratio method. Technical Report 9821, CREST, INSEE, Paris.

Billio, M., Monfort, A., and Robert, C. (1999). Bayesian estimation of switching ARMA models. *J. Econometrics*, 93:229–255.

Boch, R. and Aitkin, M. (1981). Marginal maximum likelihood estimation of item parameters: Application of an EM algorithm. *Psychometrika*, 46:443–459.

Bollerslev, T., Chou, R., and Kroner, K. (1992). ARCH modeling in finance. a review of the theory and empirical evidence. *J. Econometrics*, 52:5–59.

Booth, J., Hobert, J., and Ohman, P. (1999). On the probable error of the ratio of two gamma means. *Biometrika*, 86:439–452.

Borchers, D., Zucchini, W., and Buckland, S. (2002). *Estimating Animal Abundance: Closed Populations*. Springer-Verlag, New York.

Borovkov, A. and Foss, S. (1992). Stochastically recursive sequences and their generalizations. *Siberian Adv. Math.*, 2(1):16–81.

Bouleau, N. and Lépingle, D. (1994). *Numerical Methods for Stochastic Processes*. John Wiley, New York.

Box, G. and Muller, M. (1958). A note on the generation of random normal variates. *Ann. Mathemat. Statist.*, 29:610–611.

Boyles, R. (1983). On the convergence of the EM algorithm. *J. Royal Statist. Soc. Series B*, 45:47–50.

Bradley, R. (1986). Basic properties of strong mixing conditions. In Ebberlein, E. and Taqqu, M., editors, *Dependence in Probability and Statistics*, pages 165–192. Birkhäuser, Boston.

Breslow, N. and Clayton, D. (1993). Approximate inference in generalized linear mixed models. *J. American Statist. Assoc.*, 88:9–25.

Breyer, L. and Roberts, G. (2000a). Catalytic perfect simulation. Technical report, Department of Statistics, Univ. of Lancaster.

Breyer, L. and Roberts, G. (2000b). Some multistep coupling constructions for Markov chains. Technical report, Department of Statistics, Univ. of Lancaster.

Brockwell, P. and Davis, P. (1996). *Introduction to Time Series and Forecasting*. Springer Texts in Statistics. Springer-Verlag, New York.

Broniatowski, M., Celeux, G., and Diebolt, J. (1984). Reconnaissance de mélanges de densités par un algorithme d'apprentissage probabiliste. In Diday, E., editor, *Data Analysis and Informatics*, volume 3, pages 359–373. North-Holland, Amsterdam.

Brooks, S. (1998a). Markov chain Monte Carlo and its application. *The Statistician*, 47:69–100.

Brooks, S. (1998b). MCMC convergence diagnosis via multivariate bounds on log-concave densities. *Ann. Statist.*, 26:398–433.

Brooks, S. (1998c). Quantitative convergence diagnosis for MCMC via cusums. *Statistics and Computing*, 8(3):267–274.

Brooks, S., Dellaportas, P., and Roberts, G. (1997). A total variation method for diagnosing convergence of MCMC algorithms. *J. Comput. Graph. Statist.*, 6:251–265.

Brooks, S., Fan, Y., and Rosenthal, J. (2002). Perfect forward simulation via simulation tempering. Technical report, Department of Statistics, Univ. of Cambridge.

Brooks, S. and Gelman, A. (1998a). General methods for monitoring convergence of iterative simulations. *J. Comput. Graph. Statist.*, 7(4):434–455.

Brooks, S. and Gelman, A. (1998b). Some issues in monitoring convergence of iterative simulations. Technical report, Univ. of Bristol.

Brooks, S. and Giudici, P. (2000). MCMC convergence assessment via two-way ANOVA. *J. Comput. Graph. Statist.*, 9:266–285.

Brooks, S., Giudici, P., and Philippe, A. (2003a). Nonparametric convergence assessment for MCMC model selection. *J. Comput. Graph. Statist.*, 12(1):1–22.

Brooks, S., Giudici, P., and Roberts, G. (2003b). Efficient construction of reversible jump Markov chain Monte Carlo proposal distributions (with discussion). *J. Royal Statist. Soc. Series B*, 65(1):3–55.

Brooks, S. and Roberts, G. (1998). Assessing convergence of Markov chain Monte Carlo algorithms. *Statistics and Computing*, 8:319–335.

Brooks, S. and Roberts, G. (1999). On quantile estimation and MCMC convergence. *Biometrika*, 86:710–717.

Brown, L. (1971). Admissible estimators, recurrent diffusions, and insoluble boundary-value problems. *Ann. Mathemat. Statist.*, 42:855–903.

Brown, L. (1986). *Foundations of Exponential Families*, volume 6 of *IMS Lecture Notes — Monograph Series*. IMS, Hayward.

Brownie, C., Hines, J., Nichols, J., Pollock, K., and Hestbeck, J. (1993). Capture-recapture studies for multiple strata including non-Markovian transition probabilities. *Biometrics*, 49:1173–1187.

Bucklew, J. (1990). *Large Deviation Techniques in Decision, Simulation and Estimation*. John Wiley, New York.

Cappé, O. (2001). Recursive computation of smoothed functionals of hidden Markovian processes using a particle approximation. *Monte Carlo Methods and Applications*, 7(1-2):81–92.

Cappé, O., Guillin, A., Marin, J., and Robert, C. (2004). Population Monte Carlo. *J. Comput. Graph. Statist.* (to appear).

Cappé, O., Moulines, E., and Rydén, T. (2004). *Hidden Markov Models.* Springer-Verlag, New York.

Cappé, O., Robert, C., and Rydén, T. (2003). Reversible jump, birth-and-death, and more general continuous time MCMC samplers. *J. Royal Statist. Soc. Series B*, 65(3):679–700.

Carlin, B. and Chib, S. (1995). Bayesian model choice through Markov chain Monte Carlo. *J. Roy. Statist. Soc. (Ser. B)*, 57(3):473–484.

Carlin, B., Gelfand, A., and Smith, A. (1992). Hierarchical Bayesian analysis of change point problems. *Applied Statistics (Ser. C)*, 41:389–405.

Carlin, B. and Louis, T. (1996). *Bayes and Empirical Bayes Methods for Data Analysis.* Chapman and Hall, New York.

Carlin, B. and Louis, T. (2001). *Bayes and Empirical Bayes Methods for Data Analysis.* Chapman and Hall, New York, second edition.

Carpenter, J., Clifford, P., and Fernhead, P. (1999). Improved particle filter for nonlinear problems. *IEE Proc.-Radar, Sonar Navig.*, 146:2–4.

Carroll, R. and Lombard, F. (1985). A note on n estimators for the binomial distribution. *J. American Statist. Assoc.*, 80:423–426.

Casella, G. (1986). Stabilizing binomial n estimators. *J. American Statist. Assoc.*, 81:171–175.

Casella, G. (1996). Statistical theory and Monte Carlo algorithms (with discussion). *TEST*, 5:249–344.

Casella, G. and Berger, R. (1994). Estimation with selected binomial information or do you really believe that Dave Winfield is batting .471? *J. American Statist. Assoc.*, 89:1080–1090.

Casella, G. and Berger, R. (2001). *Statistical Inference.* Wadsworth, Belmont, CA.

Casella, G. and George, E. (1992). An introduction to Gibbs sampling. *Ann. Mathemat. Statist.*, 46:167–174.

Casella, G., Lavine, M., and Robert, C. (2001). Explaining the perfect sampler. *The American Statistician*, 55(4):299–305.

Casella, G., Mengersen, K., Robert, C., and Titterington, D. (2002). Perfect slice samplers for mixtures of distributions. *J. Royal Statist. Soc. Series B*, 64(4):777–790.

Casella, G. and Robert, C. (1996). Rao-Blackwellisation of sampling schemes. *Biometrika*, 83(1):81–94.

Casella, G. and Robert, C. (1998). Post-processing accept–reject samples: recycling and rescaling. *J. Comput. Graph. Statist.*, 7(2):139–157.

Casella, G., Robert, C., and Wells, M. (1999). Mixture models, latent variables and partitioned importance sampling. Technical report, CREST.

Castledine, B. (1981). A Bayesian analysis of multiple-recapture sampling for a closed population. *Biometrika*, 67:197–210.

Celeux, G., Chauveau, D., and Diebolt, J. (1996). Stochastic versions of the EM algorithm: An experimental study in the mixture case. *J. Statist. Comput. Simul.*, 55(4):287–314.

Celeux, G. and Clairambault, J. (1992). Estimation de chaînes de Markov cachées: méthodes et problèmes. In *Approches Markoviennes en Signal et Images*, pages 5–19, CNRS, Paris. GDR CNRS Traitement du Signal et Images.

Celeux, G. and Diebolt, J. (1985). The SEM algorithm: a probabilistic teacher algorithm derived from the EM algorithm for the mixture problem. *Comput. Statist. Quater.*, 2:73–82.

Celeux, G. and Diebolt, J. (1990). Une version de type recuit simulé de l'algorithme EM. *Comptes Rendus Acad. Sciences Paris*, 310:119–124.

Celeux, G. and Diebolt, J. (1992). A classification type EM algorithm for the mixture problem. *Stochastics and Stochastics Reports*, 41:119–134.

Celeux, G., Hurn, M., and Robert, C. (2000). Computational and inferential difficulties with mixtures posterior distribution. *J. American Statist. Assoc.*, 95(3):957–979.

Celeux, G., Marin, J., and Robert, C. (2003). Iterated importance sampling in missing data problems. Technical Report 0326, Univ. Paris Dauphine.

Chaitin, G. J. (1982). Algorithmic information theory. *Encyclopedia of Statistical Science*, 1:38–41.

Chaitin, G. J. (1988). Randomness in arithmetic. *Scient. Amer.*, 259:80–85.

Chan, K. and Geyer, C. (1994). Discussion of "Markov chains for exploring posterior distribution". *Ann. Statist.*, 22:1747–1758.

Chauveau, D., Diebolt, J., and Robert, C. (1998). Control by the Central Limit theorem. In Robert, C., editor, *Discretization and MCMC Convergence Assessment*, volume 135, chapter 5, pages 99–126. Springer–Verlag, New York. Lecture Notes in Statistics.

Chen, M. and Schmeiser, B. (1993). Performances of the Gibbs, hit-and-run, and Metropolis samplers. *J. Comput. Graph. Statist.*, 2:251–272.

Chen, M. and Schmeiser, B. (1998). Towards black-box sampling. *J. Comput. Graph. Statist.*, 7:1–22.

Chen, M. and Shao, Q. (1997). On Monte Carlo methods for estimating ratios of normalizing constants. *Ann. Statist.*, 25:1563–1594.

Cheng, B. and Titterington, D. (1994). Neural networks: A review from a statistical point of view (with discussion). *Stat. Science*, 9:2–54.

Cheng, R. (1977). The generation of gamma variables with non-integral shape parameter. *Applied Statistics (Ser. C)*, 26:71–75.

Cheng, R. and Feast, G. (1979). Some simple gamma variate generators. *Appl. Statist.*, 28:290–295.

Chib, S. (1995). Marginal likelihood from the Gibbs output. *J. American Statist. Assoc.*, 90:1313–1321.

Chib, S. (1996). Calculating posterior distributions and modal estimates in Markov mixture models. *J. Econometrics*, 75:79–97.

Chib, S. and Jeliazkov, I. (2001). Marginal likelihood from the Metropolis–Hastings output. *J. American Statist. Assoc.*, 96:270–281.

Chib, S., Nadari, F., and Shephard, N. (2002). Markov chain Monte Carlo methods for stochastic volatility models. *J. Econometrics*, 108:281–316.

Chopin, N. (2002). A sequential particle filter method for static models. *Biometrika*, 89:539–552.

Chopin, N. (2004). Central limit theorem for sequential Monte Carlo methods and its application to Bayesian inference. *Ann. Statist.* (To appear).

Chung, K. (1960). *Markov Processes with Stationary Transition Probabilities.* Springer–Verlag, Heidelberg.

Churchill, G. (1989). Stochastic models for heterogeneous DNA sequences. *Bull. Math. Biol.*, 51:79–94.

Churchill, G. (1995). Accurate restoration of DNA sequences (with discussion). In C. Gatsonis, J.S. Hodges, R. K. and Singpurwalla, N., editors, *Case Studies in Bayesian Statistics*, volume 2, pages 90–148. Springer–Verlag, New York.

Cipra, B. (1987). An introduction to the Ising model. *American Mathematical Monthly*, 94:937–959.

Clapp, T. and Godsill, S. (2001). Improvement strategies for Monte Carlo particle filters. In Doucet, A., deFreitas, N., and Gordon, N., editors, *Sequential MCMC in Practice.* Springer-Verlag, New York.

Clarke, B. and Barron, A. (1990). Information-theoretic asymptotics of Bayes methods. *IEEE Trans. Information Theory*, 36:453–471.

Clarke, B. and Wasserman, L. (1993). Noninformative priors and nuisance parameters. *J. American Statist. Assoc.*, 88:1427–1432.

Cocozza-Thivent, C. and Guédon, Y. (1990). Explicit state occupancy modeling by hidden semi-Markov models: application of Derin's scheme. *Comput. Speech Lang.*, 4:167–192.

Copas, J. B. (1975). On the unimodality of the likelihood for the Cauchy distribution. *Biometrika*, 62:701–704.

Corcoran, J. and Tweedie, R. (2002). Perfect sampling from independent Metropolis–Hastings chains. *J. Statist. Plann. Inference*, 104(2):297–314.

Cowles, M. and Carlin, B. (1996). Markov chain Monte Carlo convergence diagnostics: a comparative study. *J. American Statist. Assoc.*, 91:883–904.

Cowles, M. and Rosenthal, J. (1998). A simulation approach to convergence rates for Markov chain Monte Carlo. *Statistics and Computing*, 8:115–124.

Cox, D. R. and Snell, E. J. (1981). *Applied Statistics.* Chapman and Hall, New York.

Crawford, S., DeGroot, M., Kadane, J., and Small, M. (1992). Modelling lake-chemistry distributions: Approximate Bayesian methods for estimating a finite-mixture model. *Technometrics*, 34:441–453.

Cressie, N. (1993). *Spatial Statistics.* John Wiley, New York.

Crisan, D., Del Moral, P., and Lyons, T. (1999). Discrete filtering using branching and interacting particle systems. *Markov Processes and Related Fields*, 5(3):293–318.

Dalal, S., Fowlkes, E., and Hoadley, B. (1989). Risk analysis of the space shuttle: pre-Challenger prediction of failure. *J. American Statist. Assoc.*, 84:945–957.

Damien, P., Wakefield, J., and Walker, S. (1999). Gibbs sampling for Bayesian non-conjugate and hierarchical models by using auxiliary variables. *J. Royal Statist. Soc. Series B*, 61(2):331–344.

Daniels, H. (1954). Saddlepoint approximations in statistics. *Ann. Mathemat. Statist.*, 25:631–650.

Daniels, H. (1983). Saddlepoint approximations for estimating equations. *Biometrika*, 70:89–96.

Daniels, H. (1987). Tail probability approximations. *International Statistical Review*, 55:37–48.

Davies, R. (1977). Hypothesis testing when a nuisance parameter is present only under the alternative. *Biometrika*, 64:247–254.

Dawid, A. and Lauritzen, S. (1993). Hyper Markov laws in the statistical analysis of decomposable graphical models. *Ann. Statist.*, 21:1272–1317.

Decoux, C. (1997). Estimation de modèles de semi-chaînes de Markov cachées par échantillonnage de Gibbs. *Revue de Statistique Appliquée*, 45:71–88.

Del Moral, P. (2001). On a class of discrete generation interacting particle systems. *Electronic J. Probab.*, 6:1–26.

Del Moral, P. and Doucet, A. (2003). Sequential Monte Carlo samplers. Technical report, Dept. of Engineering, Cambridge Univ.

Del Moral, P. and Miclo, L. (2000). Branching and interacting particle systems approximations of Feynman-Kac formulae with applications to nonlinear filtering. In Azéma, J., Emery, M., Ledoux, M., and Yor, M., editors, *Séminaire de Probabilités XXXIV*, volume 1729 of *Lecture Notes in Mathematics*, pages 1–145. Springer-Verlag, Berlin.

Dellaportas, P. (1995). Random variate transformations in the Gibbs sampler: Issues of efficiency and convergence. *Statis. Comput.*, 5:133–140.

Dellaportas, P. and Forster, J. (1996). Markov chain Monte Carlo model determination for hierarchical and graphical log-linear models. Technical report, Univ. of Southampton.

Dellaportas, P. and Smith, A. (1993). Bayesian inference for generalised linear and proportional hazards models via Gibbs sampling. *Appl. Statist.*, 42:443–459.

Dempster, A., Laird, N., and Rubin, D. (1977). Maximum likelihood from incomplete data via the EM algorithm (with discussion). *J. Royal Statist. Soc. Series B*, 39:1–38.

Denison, D., Mallick, B., and Smith, A. (1998). Automatic Bayesian curve fitting. *J. Royal Statist. Soc. Series B*, 60:333–350.

D'Epifanio, G. (1989). Un approccio all'inferenza basato sulla ricerca di un punto fisso. *Metron*, 47:1–4.

D'Epifanio, G. (1996). Notes on a recursive procedure for point estimate. *TEST*, 5:203–225.

Dette, H. and Studden, W. (1997). *The Theory of Canonical Moments with Applications in Statistics, Probability and Analysis*. John Wiley, New York.

Devroye, L. (1981). The computer generation of Poisson random variables. *Computing*, 26:197–207.

Devroye, L. (1985). *Non-Uniform Random Variate Generation*. Springer-Verlag, New York.

Diaconis, P. and Hanlon, P. (1992). Eigen analysis for some examples of the Metropolis algorithm. *Contemporary Mathematics*, 138:99–117.

Diaconis, P. and Holmes, S. (1994). Gray codes for randomization procedures. *Statistics and Computing*, 4:287–302.

Diaconis, P. and Sturmfels, B. (1998). Algebraic algorithms for sampling from conditionnal distributions. *Ann. Statist.*, 26:363–397.

Diaconis, P. and Ylvisaker, D. (1979). Conjugate priors for exponential families. *Ann. Statist.*, 7:269–281.

DiCiccio, T. J. and Martin, M. A. (1993). Simple modifications for signed roots of likelihood ratio statistics. *J. Royal Statist. Soc. Series B*, 55:305–316.

Diebold, D. and Nerlove, M. (1989). The dynamic of exchange rate volatility: a multivariate latent factor ARCH model. *J. Appl. Econom.*, 4:1–22.

Diebolt, J. and Ip, E. (1996). Stochastic EM: method and application. In Gilks, W., Richardson, S., and Spiegelhalter, D., editors, *Markov chain Monte Carlo in Practice*, pages 259–274. Chapman and Hall, New York.

Diebolt, J. and Robert, C. (1990a). Bayesian estimation of finite mixture distributions, part I: Theoretical aspects. Technical Report 110, LSTA, Univ. Paris VI, Paris.

Diebolt, J. and Robert, C. (1990b). Bayesian estimation of finite mixture distributions, part II: Sampling implementation. Technical Report 111, LSTA, Univ. Paris VI, Paris.

Diebolt, J. and Robert, C. (1990c). Estimation des paramètres d'un mélange par échantillonnage bayésien. *Notes aux Comptes–Rendus de l'Académie des Sciences I*, 311:653–658.

Diebolt, J. and Robert, C. (1993). Discussion of "Bayesian computations via the Gibbs sampler" by A.F.M. Smith and G.O. Roberts. *J. Royal Statist. Soc. Series B*, 55:71–72.

Diebolt, J. and Robert, C. (1994). Estimation of finite mixture distributions by Bayesian sampling. *J. Royal Statist. Soc. Series B*, 56:363–375.

Dimakos, X. K. (2001). A guide to exact simulation. *International Statistical Review*, 69(1):27–48.

Dobson, A. (1983). *An Introduction to Statistical Modelling*. Chapman and Hall, New York.

Doob, J. (1953). *Stochastic Processes*. John Wiley, New York.

Douc, R. (2004). Performances asymptotiques de certains schémas d'échantillonnage. Ecole Polytechnique, Palaiseau, France.

Doucet, A., de Freitas, N., and Gordon, N. (2001). *Sequential Monte Carlo Methods in Practice*. Springer-Verlag, New York.

Doucet, A., Godsill, S., and Andrieu, C. (2000). On sequential Monte Carlo sampling methods for Bayesian filtering. *Statistics and Computing*, 10(3):197–208.

Doucet, A., Godsill, S., and Robert, C. (2002). Marginal maximum a posteriori estimation using Markov chain Monte Carlo. *Statistics and Computing*, 12:77–84.

Doukhan, P., Massart, P., and Rio, E. (1994). The functional central limit theorem for strongly mixing processes. *Annales de l'I.H.P.*, 30:63–82.

Draper, D. (1998). *Bayesian Hierarchical Modeling*. In preparation.

Duflo, M. (1996). Random iterative models. In Karatzas, I. and Yor, M., editors, *Applications of Mathematics*, volume 34. Springer–Verlag, Berlin.

Dupuis, J. (1995). Bayesian estimation of movement probabilities in open populations using hidden Markov chains. *Biometrika*, 82(4):761–772.

Durrett, R. (1991). *Probability: Theory and Examples*. Wadsworth and Brooks Cole, Pacific Grove.

Dykstra, R. and Robertson, T. (1982). An algorithm for isotonic regression for two or more independent variables. *Ann. Statist.*, 10:708–716.

Eaton, M. (1992). A statistical dyptich: Admissible inferences – recurrence of symmetric Markov chains. *Ann. Statist.*, 20:1147–1179.

Eberly, L. E. and Casella, G. (2003). Estimating Bayesian credible intervals. *J. Statist. Plann. Inference*, 112:115–132.

Efron, B. (1979). Bootstrap methods: another look at the jacknife. *Ann. Statist.*, 7:1–26.

Efron, B. (1981). Nonparametric standard errors and confidence intervals (with discussion). *Canadian J. Statist.*, 9:139–172.

Efron, B. (1982). *The Jacknife, the Bootstrap and Other Resampling Plans*, volume 38. SIAM, Philadelphia.

Efron, B. and Tibshirani, R. (1994). *An Introduction to the Bootstrap*. Chapman and Hall, New York.

Enders, W. (1994). *Applied Econometric Time Series*. John Wiley, New York.

Escobar, M. and West, M. (1995). Bayesian prediction and density estimation. *J. American Statist. Assoc.*, 90:577–588.

Ethier, S. and Kurtz, T. (1986). *Markov processes: Characterization and Convergence*. John Wiley, New York.

Evans, G. (1993). *Practical Numerical Integration*. John Wiley, New York.

Evans, M. and Swartz, T. (1995). Methods for approximating integrals in Statistics with special emphasis on Bayesian integration problems. *Statist. Science*, 10:254–272.

Everitt, B. (1984). *An Introduction to Latent Variable Models*. Chapman and Hall, New York.

Fan, J. and Gijbels, I. (1996). *Local Polynomial Modelling and its Applications*. Chapman and Hall, New York.

Fang, K. and Wang, Y. (1994). *Number-Theoretic Methods in Statistics*. Chapman and Hall, New York.

Feller, W. (1970). *An Introduction to Probability Theory and its Applications*, volume 1. John Wiley, New York.

Feller, W. (1971). *An Introduction to Probability Theory and its Applications*, volume 2. John Wiley, New York.

Ferguson, T. (1978). Maximum likelihood estimates of the parameters of the Cauchy distribution for samples of size 3 and 4. *J. American Statist. Assoc.*, 73:211–213.

Fernández, R., Ferrari, P., and Garcia, N. L. (1999). Perfect simulation for interacting point processes, loss networks and Ising models. Technical report, Laboratoire Raphael Salem, Univ. de Rouen.

Ferrenberg, A., Landau, D., and Wong, Y. (1992). Monte Carlo simulations: hidden errors from "good" random number generators. *Physical Review Letters*, 69(23):3382–3385.

Field, C. and Ronchetti, E. (1990). *Small Sample Asymptotics*. IMS Lecture Notes — Monograph Series, Hayward, CA.

Fieller, E. (1954). Some problems in interval estimation. *J. Royal Statist. Soc. Series B*, 16:175–185.

Fill, J. (1998a). An interruptible algorithm for exact sampling via Markov chains. *Ann. Applied Prob.*, 8:131–162.

Fill, J. (1998b). The move-to-front rule: A case study for two perfect sampling algorithms. *Prob. Eng. Info. Sci.*, 8:131–162.

Finch, S., Mendell, N., and Thode, H. (1989). Probabilistic measures of adequacy of a numerical search for a global maximum. *J. American Statist. Assoc.*, 84:1020–1023.

Fishman, G. (1996). *Monte Carlo*. Springer-Verlag, New York.

Fismen, M. (1998). Exact simulation using Markov chains. Technical Report 6/98, Institutt for Matematiske Fag, Oslo. Diploma-thesis.

Fletcher, R. (1980). *Practical Methods of Optimization*, volume 1. John Wiley, New York.

Foss, S. and Tweedie, R. (1998). Perfect simulation and backward coupling. *Stochastic Models*, 14:187–203.

Gamboa, F. and Gassiat, E. (1997). Blind deconvolution of discrete linear systems. *Ann. Statist.*, 24:1964–1981.

Gåsemyr, J. (2002). Markov chain Monte Carlo algorithms with independent proposal distribution and their relation to importance sampling and rejection sampling. Technical Report 2, Department of Statistics, Univ. of Oslo.

Gassiat, E. (1995). Déconvolution aveugle d'un système non-causal. *Matapli*, 45:45–54.

Gauss, C. (1810). *Méthode des Moindres Carrés. Mémoire sur la Combinaison des Observations*. Mallet-Bachelier, Paris. Transl. J. Bertrand.

Gaver, D. and O'Muircheartaigh, I. (1987). Robust empirical Bayes analysis of event rates. *Technometrics*, 29:1–15.

Gelfand, A. and Dey, D. (1994). Bayesian model choice: asymptotics and exact calculations. *J. Roy. Statist. Soc. (Ser. B)*, 56:501–514.

Gelfand, A. and Sahu, S. (1994). On Markov chain Monte Carlo acceleration. *J. Comput. Graph. Statist.*, 3(3):261–276.

Gelfand, A., Sahu, S., and Carlin, B. (1995). Efficient parametrization for normal linear mixed models. *Biometrika*, 82:479–488.

Gelfand, A., Sahu, S., and Carlin, B. (1996). Efficient parametrizations for generalised linear mixed models (with discussion). In Bernardo, J., Berger, J., Dawid, A., and Smith, A., editors, *Bayesian Statistics 5*, pages 165–180. Oxford University Press, Oxford.

Gelfand, A. and Smith, A. (1990). Sampling based approaches to calculating marginal densities. *J. American Statist. Assoc.*, 85:398–409.

Gelman, A. (1996). Inference and monitoring convergence. In Gilks, W., Richardson, S., and Spiegelhalter, D., editors, *Markov chain Monte Carlo in Practice*, pages 131–143. Chapman and Hall, New York.

Gelman, A., Gilks, W., and Roberts, G. (1996). Efficient Metropolis jumping rules. In Berger, J., Bernardo, J., Dawid, A., Lindley, D., and Smith, A., editors, *Bayesian Statistics 5*, pages 599–608. Oxford University Press, Oxford.

Gelman, A. and Meng, X. (1998). Simulating normalizing constants: From importance sampling to bridge sampling to path sampling. *Statist. Science*, 13:163–185.

Gelman, A. and Rubin, D. (1992). Inference from iterative simulation using multiple sequences (with discussion). *Statist. Science*, 7:457–511.

Gelman, A. and Speed, T. (1993). Characterizing a joint probability distribution by conditionals. *J. Royal Statist. Soc. Series B*, 55:185–188.

Geman, S. and Geman, D. (1984). Stochastic relaxation, Gibbs distributions and the Bayesian restoration of images. *IEEE Trans. Pattern Anal. Mach. Intell.*, 6:721–741.

Gentle, J. E. (2002). *Elements of Computational Statistics*. Springer–Verlag, New York.

George, E., Makov, U., and Smith, A. (1993). Conjugate likelihood distributions. *Scand. J. Statist.*, 20:147–156.

George, E. and Robert, C. (1992). Calculating Bayes estimates for capture-recapture models. *Biometrika*, 79(4):677–683.

Geweke, J. (1989). Bayesian inference in econometric models using Monte Carlo integration. *Econometrica*, 57:1317–1340.

Geweke, J. (1991). Efficient simulation from the multivariate normal and student *t*-distributions subject to linear constraints. *Computer Sciences and Statistics: Proc. 23d Symp. Interface.*

Geweke, J. (1992). Evaluating the accuracy of sampling-based approaches to the calculation of posterior moments (with discussion). In Bernardo, J., Berger, J., Dawid, A., and Smith, A., editors, *Bayesian Statistics 4*, pages 169–193. Oxford University Press, Oxford.

Geyer, C. (1992). Practical Monte Carlo Markov chain (with discussion). *Statist. Science*, 7:473–511.

Geyer, C. (1993). Estimating normalizing constants and reweighting mixtures in Markov chain Monte Carlo. Technical Report 568, School of Statistics, Univ. of Minnesota.

Geyer, C. (1994). On the convergence of Monte Carlo maximum likelihood calculations. *J. R. Statist. Soc. B*, 56:261–274.

Geyer, C. (1995). Conditioning in Markov chain Monte Carlo. *J. Comput. Graph. Statis.*, 4:148–154.

Geyer, C. (1996). Estimation and optimization of functions. In Gilks, W., Richardson, S., and Spiegelhalter, D., editors, *Markov chain Monte Carlo in Practice*, pages 241–258. Chapman and Hall, New York.

Geyer, C. and Møller, J. (1994). Simulation procedures and likelihood inference for spatial point processes. *Scand. J. Statist.*, 21:359–373.

Geyer, C. and Thompson, E. (1992). Constrained Monte Carlo maximum likelihood for dependent data (with discussion). *J. Royal Statist. Soc. Series B*, 54:657–699.

Geyer, C. and Thompson, E. (1995). Annealing Markov chain Monte Carlo with applications to ancestral inference. *J. American Statist. Assoc.*, 90:909–920.

Ghosh, M., Carlin, B. P., and Srivastiva, M. S. (1995). Probability matching priors for linear calibration. *TEST*, 4:333–357.

Gilks, W. (1992). Derivative-free adaptive rejection sampling for Gibbs sampling. In Bernardo, J., Berger, J., Dawid, A., and Smith, A., editors, *Bayesian Statistics 4*, pages 641–649. Oxford University Press, Oxford.

Gilks, W. and Berzuini, C. (2001). Following a moving target–Monte Carlo inference for dynamic Bayesian models. *J. Royal Statist. Soc. Series B*, 63(1):127–146.

Gilks, W., Best, N., and Tan, K. (1995). Adaptive rejection Metropolis sampling within Gibbs sampling. *Applied Statist. (Ser. C)*, 44:455–472.

Gilks, W., Clayton, D., Spiegelhalter, D., Best, N., McNeil, A., Sharples, L., and Kirby, A. (1993). Modelling complexity: applications of Gibbs sampling in medicine. *J. Royal Statist. Soc. Series B*, 55:39–52.

Gilks, W. and Roberts, G. (1996). Strategies for improving MCMC. In Gilks, W., Richardson, S., and Spiegelhalter, D., editors, *Markov chain Monte Carlo in Practice*, pages 89–114. Chapman and Hall, New York.

Gilks, W., Roberts, G., and Sahu, S. (1998). Adaptive Markov chain Monte Carlo. *J. American Statist. Assoc.*, 93:1045–1054.

Gilks, W. and Wild, P. (1992). Adaptive rejection sampling for Gibbs sampling. *Appl. Statist.*, 41:337–348.

Giudici, P. and Green, P. (1999). Decomposable graphical Gaussian model determination. *Biometrika*, 86(4):785–801.

Gleick, J. (1987). *Chaos*. Penguin, New York.

Gleser, L. (1989). The gamma distribution as a mixture of exponential distributions. *Am. Statist.*, 43:115–117.

Gleser, L. and Hwang, J. (1987). The non-existence of $100(1-\alpha)\%$ confidence sets of finite expected diameters in errors-in-variable and related models. *Ann. Statist.*, 15:1351–1362.

Godsill, S. J. (2003). Discussion of "Trans-dimensional Markov chain Monte Carlo" by Peter J. Green. In Green, P., Hjort, N., and Richardson, S., editors, *Highly Structured Stochastic Systems*, chapter 6A. Oxford University Press, Oxford.

Goldfeld, S. and Quandt, R. (1973). A Markov model for switching regressions. *J. Econometrics*, 1:3–16.

Gordon, N., Salmond, J., and Smith, A. (1993). A novel approach to non-linear/non-Gaussian Bayesian state estimation. *IEEE Proceedings on Radar and Signal Processing*, 140:107–113.

Gouriéroux, C. (1996). *ARCH Models*. Springer-Verlag, New York.

Gouriéroux, C. and Monfort, A. (1996). *Statistics and Econometric Models*. Cambridge University Press, Cambridge.

Gouriéroux, C., Monfort, A., and Renault, E. (1993). Indirect inference. *J. Applied Econom.*, 8:85–118.

Goutis, C. and Casella, G. (1999). Explaining the saddlepoint approximation. *The American Statistician*, 53.

Goutis, C. and Robert, C. (1998). Model choice in generalized linear models: a Bayesian approach via Kullback–Leibler projections. *Biometrika*, 85:29–37.

Green, P. (1995). Reversible jump MCMC computation and Bayesian model determination. *Biometrika*, 82(4):711–732.

Green, P. and Han, X. (1992). Metropolis methods, Gaussian proposals, and antithetic variables. In Barone, P., Frigessi, A., and Piccioni, M., editors, *Stochastic methods and Algorithms in Image Analysis*, volume 74, pages 142–164. Springer-Verlag, New York, Berlin. Lecture Notes in Statistics.

Green, P. and Murdoch, D. (1999). Exact sampling for Bayesian inference: towards general purpose algorithms. In Berger, J., Bernardo, J., Dawid, A., Lindley, D., and Smith, A., editors, *Bayesian Statistics 6*, pages 302–321. Oxford University Press, Oxford.

Grenander, U. and Miller, M. (1994). Representations of knowledge in complex systems (with discussion). *J. Royal Statist. Soc. Series B*, 56:549–603.

Gruet, M., Philippe, A., and Robert, C. (1999). MCMC control spreadsheets for exponential mixture estimation. *J. Comput. Graph. Statist.*, 8:298–317.

Guihenneuc-Jouyaux, C. and Richardson, S. (1996). Modèle de Markov avec erreurs de mesure : approche bayésienne et application au suivi longitudinal des lymphocytes t4. In *Biométrie et Applications Bayésiennes*. Economica, Paris.

Guihenneuc-Jouyaux, C., Richardson, S., and Lasserre, V. (1998). Convergence assessment in latent variable models: apllication to longitudinal modelling of a marker of HIV progression. In Robert, C., editor, *Discretization and MCMC Convergence Assessment*, volume 135, chapter 7, pages 147–160. Springer–Verlag, New York. Lecture Notes in Statistics.

Guihenneuc-Jouyaux, C. and Robert, C. (1998). Finite Markov chain convergence results and MCMC convergence assessments. *J. American Statist. Assoc.*, 93:1055–1067.

Guillin, A., Marin, J., and Robert, C. (2004). Estimation bayésienne approximative par échantillonnage préférentiel. *Revue Statist. Appliquée.* (to appear).

Gumbel, E. (1940). La direction d'une répartition. *Ann. Univ. Lyon 3A*, 2:39–51.

Gumbel, E. (1958). *Statistics of Extremes.* Columbia University Press, New York.

Haario, H. and Sacksman, E. (1991). Simulated annealing in general state space. *Adv. Appl. Probab.*, 23:866–893.

Haario, H., Saksman, E., and Tamminen, J. (1999). Adaptive proposal distribution for random walk Metropolis algorithm. *Computational Statistics*, 14(3):375–395.

Haario, H., Saksman, E., and Tamminen, J. (2001). An adaptive Metropolis algorithm. *Bernoulli*, 7(2):223–242.

Hàjek, B. (1988). Cooling schedules for optimal annealing. *Math. Operation. Research*, 13:311–329.

Hall, P. (1992). *The Bootstrap and Edgeworth Expansion.* Springer-Verlag, New York.

Hall, P. (1994). On the erratic behavior of estimators of n in the binomial n, p distribution. *J. American Statist. Assoc.*, 89:344–352.

Hammersley, J. (1974). Discussion of Besag's paper. *J. Royal Statist. Soc. Series B*, 36:230–231.

Hammersley, J. and Clifford, M. (1970). Markov fields on finite graphs and lattices. Unpublished manuscript.

Hammersley, J. and Morton, K. (1954). Poor man's Monte Carlo. *J. Royal Statist. Soc. Series B*, 16:23–38.

Handschin, J. and Mayne, D. (1969). Monte Carlo techniques to estimate the conditional expectation in multi-stage nonlinear filtering. *International Journal of Control*, 9:547–559.

Harris, T. (1956). The existence of stationary measures for certain Markov processes. In *Proc. 3rd Berkeley Symp. Math. Statis. Prob.*, volume 2, pages 113–124. University of California Press, Berkeley.

Hastings, W. (1970). Monte Carlo sampling methods using Markov chains and their application. *Biometrika*, 57:97–109.

Heidelberger, P. and Welch, P. (1983). A spectral method for confidence interval generation and run length control in simulations. *Comm. Assoc. Comput. Machinery*, 24:233–245.

Hendry, D. and Richard, J. (1991). Likelihood evaluation for dynamic latent variable models. In Amman, H., Beslsey, D., and Pau, I., editors, *Computational Economics and Econometrics*, Dordrecht. Klüwer.

Hesterberg, T. (1998). Weighted average importance sampling and defensive mixture distributions. *Technometrics*, 37:185–194.

Hills, S. and Smith, A. (1992). Parametrization issues in Bayesian inference. In Bernardo, J., Berger, J., Dawid, A., and Smith, A., editors, *Bayesian Statistics 4*, pages 641–649. Oxford University Press, Oxford.

Hills, S. and Smith, A. (1993). Diagnostic plots for improved parameterization in Bayesian inference. *Biometrika*, 80:61–74.

Hitchcock, D. H. (2003). A history of the Metropolis–Hastings algorithm. *The American Statistician*, 57.

Hjorth, J. (1994). *Computer Intensive Statistical Methods*. Chapman and Hall, New York.

Hobert, J. and Casella, G. (1996). The effect of improper priors on Gibbs sampling in hierarchical linear models. *J. American Statist. Assoc.*, 91:1461–1473.

Hobert, J. and Casella, G. (1998). Functional compatibility, Markov chains, and Gibbs sampling with improper posteriors. *J. Comput. Graph. Statist.*, 7:42–60.

Hobert, J., Jones, G., Presnel, B., and Rosenthal, J. (2002). On the applicability of regenerative simulation in Markov chain Monte Carlo. *Biometrika*, 89(4):731–743.

Hobert, J. and Robert, C. (1999). Eaton's Markov chain, its conjugate partner and p-admissibility. *Ann. Statist.*, 27:361–373.

Hobert, J. and Robert, C. (2004). Moralizing perfect sampling. *Ann. Applied Prob.* (to appear).

Hobert, J., Robert, C., and Goutis, C. (1997). Connectedness conditions for the convergence of the Gibbs sampler. *Statis. Prob. Letters*, 33:235–240.

Hobert, J., Robert, C., and Titterington, D. (1999). On perfect simulation for some mixtures of distributions. *Statistics and Computing*, 9(4):287–298.

Hodgson, M. (1999). A Bayesian restoration of a ion channel signal. *J. Royal Statist. Soc. Series B*, 61(1):95–114.

Hodgson, M. and Green, P. (1999). Bayesian choice among Markov models of ion channels using Markov chain Monte Carlo. *Proc. Royal Society London A*, 455:3425–3448.

Holden, L. (1998). Adaptive chains. Technical report, Norwegian Computing Center.

Holmes, C., Denison, D., Mallick, B., and Smith, A. (2002). *Bayesian methods for nonlinear classification and regression*. John Wiley, New York.

Hougaard, P. (1988). Discussion of the paper by Reid. *Statist. Science*, 3:230–231.

Huber, P. (1972). Robust statistics: A review. *Ann. Mathemat. Statist.*, 47:1041–1067.

Huerta, G. and West, M. (1999). Priors and component structures in autoregressive time series models. *J. Royal Statist. Soc. Series B*, 61(4):881–899.

Hürzeler, M. and Küensch, H. (1998). Monte Carlo approximations for general state space models. *J. Comput. Graph. Statist.*, 7:175–193.

Hwang, C. (1980). Laplace's method revisited: Weak convergence of probability measures. *Ann. Probab.*, 8:1177–1182.

Iba, Y. (2000). Population-based Monte Carlo algorithms. *Trans. Japanese Soc. Artificial Intell.*, 16(2):279–286.

Jacquier, E., Polson, N., and Rossi, P. (1994). Bayesian analysis of stochastic volatility models (with discussion). *J. Business Economic Stat.*, 12:371–417.

Jarrett, R. (1979). A note on the intervals between coal-mining disasters. *Biometrika*, 66:191–193.

Jeffreys, H. (1961). *Theory of Probability (3rd edition)*. Oxford University Press, Oxford.

Jensen, J. (1995). *Saddlepoint Approximations*. Clarendon Press, Oxford.

Jöhnk, M. (1964). Erzeugung von betaverteilten und gammaverteilten Zufall-szahen. *Metrika*, 8:5–15.

Johnson, D. and Hoeting, J. (2003). Autoregressive models for capture-recapture data: A Bayesian approach. *Biometrics*, 59(2):341–350.

Johnson, N. and Kotz, S. (1969–1972). *Distributions in Statistics (4 vols.)*. John Wiley, New York.

Johnson, N., Kotz, S., and Balakrishnan, N. (1994). *Continuous Univariate Distributions (2nd edition)*, volume 1. John Wiley, New York.

Johnson, N., Kotz, S., and Balakrishnan, N. (1995). *Continuous Univariate Distributions (2nd edition)*, volume 2. John Wiley, New York.

Jones, G. and Hobert, J. (2001). Honest exploration of intractable probability distributions via Markov chain Monte Carlo. *Statist. Science*, 16(4):312–334.

Juang, B. (1984). On the hidden Markov model and dynamic time warping for speech recognition – a unified view. *BellTech*, 63:1213–1243.

Juang, B. and Rabiner, L. (1991). Hidden Markov models for speech recognition. *Technometrics*, 33:251–272.

Kass, R. and Raftery, A. (1995). Bayes factors. *J. American Statist. Assoc.*, 90:773–795.

Kemeny, J. and Snell, J. (1960). *Finite Markov Chains*. Van Nostrand, Princeton.

Kendall, W. (1998). Perfect simulation for the area-interaction point process. In Heyde, C. and Accardi, L., editors, *Probability Towards 2000*, pages 218–234. Springer–Verlag, New York.

Kendall, W. and Møller, J. (1999). Perfect Metropolis–Hastings simulation of locally stable point processes. Technical Report R-99-2001, Department of Mathematical Sciences, Aalborg Univ.

Kendall, W. and Møller, J. (2000). Perfect simulation using dominating processes on ordered spaces, with application to locally stable point processes. *Advances in Applied Probability*, 32:844–865.

Kennedy, W. and Gentle, J. (1980). *Statistical Computing*. Marcel Dekker, New York.

Kersting, G. (1987). Some results on the asymptotic behavior of the Robbins-Monro process. *Bull. Int. Statis. Inst.*, 47:327–335.

Kiefer, J. and Wolfowitz, J. (1952). Stochastic estimation of the maximum of a regression function. *Ann. Mathemat. Statist.*, 23:462–466.

Kiiveri, H. and Speed, T. (1982). Structural analysis of multivariate data: A review. In Leinhardt, S., editor, *Sociological Methodology*, pages 209–289. Jossey Bass, San Francisco.

Kim, S., Shephard, N., and Chib, S. (1998). Stochastic volatility: Likelihood inference and comparison with ARCH models. *Rev. Econom. Stud.*, 65:361–393.

Kinderman, A., Monahan, J., and Ramage, J. (1977). Computer methods for sampling from Student's t-distribution. *Math. Comput.*, 31:1009–1018.

Kipnis, C. and Varadhan, S. (1986). Central limit theorem for additive functionals of reversible Markov processes and applications to simple exclusions. *Comm. Math. Phys.*, 104:1–19.

Kirkpatrick, S., Gelatt, C., and Vecchi, M. (1983). Optimization by simulated annealing. *Science*, 220:671–680.

Kitagawa, G. (1996). Monte Carlo filter and smoother for non–Gaussian non–linear state space models. *J. Comput. Graph. Statist.*, 5:1–25.

Knuth, D. (1981). *The Art of Computer Programing. Volume 2: Seminumerical Algorithms (2nd edition)*. Addison-Wesley, Reading, MA.

Kolassa, J. (1994). *Series Approximation Methods in Statistics*. Springer-Verlag, New York.

Kong, A., Liu, J., and Wong, W. (1994). Sequential imputations and Bayesian missing data problems. *J. American Statist. Assoc.*, 89:278–288.

Kong, A., McCullagh, P., Meng, X.-L., Nicolae, D., and Tan, Z. (2003). A theory of statistical models for Monte Carlo integration. *J. Royal Statist. Soc. Series B*, 65(3):585–618. (With discussion.).

Kubokawa, T. and Robert, C. (1994). New perspectives on linear calibration. *J. Mult. Anal.*, 51:178–200.

Kuehl, R. (1994). *Statistical Principles of Research Design and Analysis*. Duxbury, Belmont.

Kullback, S. (1968). *Information Theory and Statistics*. John Wiley, New York.

Künsch, H. (2004). Recursive Monte Carlo filters: algorithms and theoretical analysis. *Ann. Statist.* (to appear).

Lahiri, P., editor (2001). *Model Selection*. Institute of Mathematical Statistics, Lecture Notes-Monograph Series, Hayward, CA.

Laird, N., Lange, N., and Stram, D. (1987). Maximum likelihood computations with repeated measures: Application of the EM algorithm. *J. American Statist. Assoc.*, 82:97–105.

Lange, K. (1999). *Numerical analysis for statisticians*. Springer–Verlag, New York.

Lauritzen, S. (1996). *Graphical Models*. Oxford University Press, Oxford.

Lavielle, M. and Moulines, E. (1997). On a stochastic approximation version of the EM algorithm. *Statist. Comput.*, 7:229–236.

Lavine, M. and West, M. (1992). A Bayesian method for classification and discrimination. *Canad. J. Statist.*, 20:451–461.

Le Cun, Y., Boser, D., Denker, J., Henderson, D., Howard, R., Hubbard, W., and Jackel, L. (1989). Handwritten digit recognition with a backpropagation network. In Touresky, D., editor, *Advances in Neural Information Processing Systems II*, pages 396–404. Morgan Kaufman, San Mateo, CA.

Lee, P. (2004). *Bayesian Statistics: an Introduction*. Oxford University Press, Oxford, third edition.

Legendre, A. (1805). *Nouvelles Méthodes pour la Détermination des Orbites des Comètes*. Courcier, Paris.

Lehmann, E. (1975). *Nonparametrics: Statistical Methods based on Ranks*. McGraw-Hill, New York.

Lehmann, E. (1986). *Testing Statistical Hypotheses*. John Wiley, New York.

Lehmann, E. (1998). *Introduction to Large-Sample Theory*. Springer-Verlag, New York.

Lehmann, E. and Casella, G. (1998). *Theory of Point Estimation (revised edition)*. Springer-Verlag, New York.

Levine, R. (1996). Post-processing random variables. Technical report, Biometrics Unit, Cornell Univ. Ph.D. Thesis.

Liang, F. (2002). Dynamically weighted importance sampling in Monte Carlo computation. *J. American Statist. Assoc.*, 97:807–821.

Lindvall, T. (1992). *Lectures on the Coupling Theory*. John Wiley, New York.

Little, R. and Rubin, D. (1983). On jointly estimating parameters and missing data by maximizing the complete-data likelihood. *The American Statistician*, 37(3):218–220.

Little, R. and Rubin, D. (1987). *Statistical Analysis with Missing Data*. John Wiley, New York.

Liu, C., Liu, J., and Rubin, D. (1992). A variational control variable for assessing the convergence of the Gibbs sampler. In *Proc. Amer. Statist. Assoc.*, pages 74–78. Statistical Computing Section.

Liu, C. and Rubin, D. (1994). The ECME algorithm: a simple extension of EM and ECM with faster monotonous convergence. *Biometrika*, 81:633–648.

Liu, J. (1995). Metropolized Gibbs sampler: An improvement. Technical report, Dept. of Statistics, Stanford Univ., CA.

Liu, J. (1996a). Metropolized independent sampling with comparisons to rejection sampling and importance sampling. *Statistics and Computing*, 6:113–119.

Liu, J. (1996b). Peskun's theorem and a modified discrete-state Gibbs sampler. *Biometrika*, 83:681–682.

Liu, J. (2001). *Monte Carlo Strategies in Scientific Computing*. Springer-Verlag, New York.

Liu, J. and Chen, R. (1995). Blind deconvolution via sequential imputations. *J. American Statist. Assoc.*, 90:567–576.

Liu, J., Liang, F., and Wong, W. (2001). A theory of dynamic weighting in Monte Carlo computation. *J. American Statist. Assoc.*, 96:561–573.

Liu, J. and Sabbati, C. (1999). Simulated sintering: MCMC with spaces of varying dimension. In Berger, J., Bernardo, J., Dawid, A., Lindley, D., and

Smith, A., editors, *Bayesian Statistics 6*, pages 389–413. Oxford University Press, Oxford.

Liu, J., Wong, W., and Kong, A. (1994). Covariance structure of the Gibbs sampler with applications to the comparisons of estimators and sampling schemes. *Biometrika*, 81:27–40.

Liu, J., Wong, W., and Kong, A. (1995). Correlation structure and convergence rate of the Gibbs sampler with various scans. *J. Royal Statist. Soc. Series B*, 57:157–169.

Loh, W. (1996). On latin hypercube sampling. *Ann. Statist.*, 24:2058–2080.

Louis, T. (1982). Finding the observed information matrix when using the EM algorithm. *J. Royal Statist. Soc. Series B*, 44:226–233.

Lugannani, R. and Rice, S. (1980). Saddlepoint approximation for the distribution of the sum of independent random variables. *Adv. Appl. Probab.*, 12:475–490.

MacEachern, S. and Berliner, L. (1994). Subsampling the Gibbs sampler. *The American Statistician*, 48:188–190.

MacEachern, S., Clyde, M., and Liu, J. (1999). Sequential importance sampling for non-parametric Bayes models: the next generation. Technical report, Department of Statistics, Ohio State Univ.

MacLachlan, G. and Basford, K. (1988). *Mixture Models: Inference and Applications to Clustering*. Marcel Dekker, New York.

MacLachlan, G. and Krishnan, T. (1997). *The EM Algorithm and Extensions*. John Wiley, New York.

Madigan, D. and York, J. (1995). Bayesian graphical models for discrete data. *International Statistical Review*, 63:215–232.

Madras, N. and Slade, G. (1993). *The Self-Avoiding Random Walk*. Probability and its Applications. Birkhauser, Boston.

Maguire, B., Pearson, E., and Wynn, A. (1952). The time interval between industrial accidents. *Biometrika*, 38:168–180.

Marin, J., Mengersen, K., and Robert, C. (2004). Bayesian modelling and inference on mixtures of distributions. In Rao, C. and Dey, D., editors, *Handbook of Statistics*, volume 25 (to appear). Springer-Verlag, New York.

Marinari, E. and Parisi, G. (1992). Simulated tempering: A new Monte Carlo scheme. *Europhys. Lett.*, 19:451.

Marsaglia, G. (1964). Generating a variable from the tail of a normal distribution. *Technometrics*, 6:101–102.

Marsaglia, G. (1977). The squeeze method for generating gamma variables. *Computers and Mathematics with Applications*, 3:321–325.

Marsaglia, G. and Zaman, A. (1993). The KISS generator. Technical report, Dept. of Statistics, Univ. of Florida.

Marshall, A. (1965). The use of multi-stage sampling schemes in Monte Carlo computations. In *Symposium on Monte Carlo Methods*. John Wiley, New York.

Matthews, P. (1993). A slowly mixing Markov chain with implications for Gibbs sampling. *Statist. Prob. Letters*, 17:231–236.

McCulloch, C. (1997). Maximum likelihood algorithms for generalized linear mixed models. *J. American Statist. Assoc.*, 92:162–170.

McCulloch, R. and Tsay, R. (1994). Statistical analysis of macroeconomic time series via Markov switching models. *J. Time Series Analysis*, 155:523–539.

McKay, M., Beckman, R., and Conover, W. (1979). A comparison of three methods for selecting values of output variables in the analysis of output from a computer code. *Technometrics*, 21:239–245.

McKeague, I. and Wefelmeyer, W. (2000). Markov chain Monte Carlo and Rao–Blackwellisation. *J. Statist. Plann. Inference*, 85:171–182.

Mead, R. (1988). *The Design of Experiments*. Cambridge University Press, Cambridge.

Mehta, C. R., Patel, N. R., and Senchaudhuri, P. (2000). Efficient Monte Carlo methods for conditional logistic regression. *J. American Statist. Assoc.*, 95:99–108.

Meng, X. and Rubin, D. (1991). Using EM to obtain asymptotic variance-covariance matrices. *J. American Statist. Assoc.*, 86:899–909.

Meng, X. and Rubin, D. (1992). Maximum likelihood estimation via the ECM algorithm: A general framework. *Biometrika*, 80:267–278.

Meng, X. and van Dyk, D. (1997). The EM algorithm—an old folk-song sung to a new tune (with discussion). *J. Royal Statist. Soc. Series B*, 59:511–568.

Meng, X. and Wong, W. (1996). Simulating ratios of normalizing constants via a simple identity: a theoretical exploration. *Statist. Sinica*, 6:831–860.

Mengersen, K. and Robert, C. (1996). Testing for mixtures: A Bayesian entropic approach (with discussion). In Berger, J., Bernardo, J., Dawid, A., Lindley, D., and Smith, A., editors, *Bayesian Statistics 5*, pages 255–276. Oxford University Press, Oxford.

Mengersen, K. and Robert, C. (2003). Iid sampling with self-avoiding particle filters: the pinball sampler. In Bernardo, J., Bayarri, M., Berger, J., Dawid, A., Heckerman, D., Smith, A., and West, M., editors, *Bayesian Statistics 7*. Oxford University Press, Oxford.

Mengersen, K., Robert, C., and Guihenneuc-Jouyaux, C. (1999). MCMC convergence diagnostics: a "reviewww". In Berger, J., Bernardo, J., Dawid, A., Lindley, D., and Smith, A., editors, *Bayesian Statistics 6*, pages 415–440, Oxford. Oxford University Press.

Mengersen, K. and Tweedie, R. (1996). Rates of convergence of the Hastings and Metropolis algorithms. *Ann. Statist.*, 24:101–121.

Metropolis, N., Rosenbluth, A., Rosenbluth, M., Teller, A., and Teller, E. (1953). Equations of state calculations by fast computing machines. *J. Chem. Phys.*, 21:1087–1092.

Metropolis, N. and Ulam, S. (1949). The Monte Carlo method. *J. American Statist. Assoc.*, 44:335–341.

Meyn, S. and Tweedie, R. (1993). *Markov Chains and Stochastic Stability*. Springer-Verlag, New York.

Mira, A. and Geyer, C. (1998). Ordering Monte Carlo Markov chains. Technical report, Univ. of Minnesota.

Mira, A., Møller, J., and Roberts, G. (2001). Perfect slice samplers. *J. Royal Statist. Soc. Series B*, 63:583–606.

Mitra, D., Romeo, F., and Sangiovanni-Vincentelli, A. (1986). Convergence and finite-time behaviour of simulated annealing. *Advances in Applied Probability*, 18:747–771.

Møller, J. (2001). A review of perfect simulation in stochastic geometry. In Basawa, I., Heyde, C., and Taylor, R., editors, *Selected Proceedings of the Symposium on Inference for Stochastic Processes*, volume 37 of *IMS Lecture Notes & Monographs Series*, pages 333–355. IMS.

Møller, J. (2003). *Spatial Statistics and Computational Methods*, volume 173 of *Lecture Notes in Statistics*. Springer-Verlag, New York.

Møller, J. and Nicholls, G. (2004). Perfect simulation for sample-based inference. *Statistics and Computing*.

Møller, J. and Waagepetersen, R. (2003). *Statistical Inference and Simulation for Spatial Point Processes*. Chapman and Hall/CRC, Boca Raton.

Moussouris, J. (1974). Gibbs and Markov random systems with constraints. *J. Statist. Phys.*, 10:11–33.

Müller, P. (1991). A generic approach to posterior integration and Gibbs sampling. Technical report, Purdue Univ., West Lafayette, Indiana.

Müller, P. (1993). Alternatives to the Gibbs sampling scheme. Technical report, Institute of Statistics and Decision Sciences, Duke Univ.

Müller, P. (1999). Simulation based optimal design. In Bernardo, J., Berger, J., Dawid, A., and Smith, A., editors, *Bayesian Statistics 6*, pages 459–474, New York. Springer-Verlag.

Murdoch, D. and Green, P. (1998). Exact sampling for a continuous state. *Scandinavian J. Statist.*, 25(3):483–502.

Murray, G. (1977). Comments on "Maximum likelihood from incomplete data via the EM algorithm". *J. Royal Statist. Soc. Series B*, 39:27–28.

Mykland, P., Tierney, L., and Yu, B. (1995). Regeneration in Markov chain samplers. *J. American Statist. Assoc.*, 90:233–241.

Natarajan, R. and McCulloch, C. (1998). Gibbs sampling with diffuse proper priors: A valid approach to data-driven inference? *J. Comput. Graph. Statist.*, 7:267–277.

Naylor, J. and Smith, A. (1982). Application of a method for the efficient computation of posterior distributions. *Applied Statistics*, 31:214–225.

Neal, R. (1993). Probabilistic inference using Markov chain Monte Carlo methods. Technical report, Dept. of Computer Science, Univ. of Toronto. Ph.D. thesis.

Neal, R. (1995). Suppressing random walks in MCMC using ordered overrelaxation. Technical report, Dept. of Statistics, Univ. of Toronto.

Neal, R. (1997). Markov chain Monte Carlo methods based on "slicing" the density function. Technical report, Univ. of Toronto.

Neal, R. (1999). *Bayesian Learning for Neural Networks*, volume 118. Springer-Verlag, New York. Lecture Notes.

Neal, R. (2003). Slice sampling (with discussion). *Ann. Statist.*, 31:705–767.

Newton, M. and Raftery, A. (1994). Approximate Bayesian inference by the weighted likelihood boostrap (with discussion). *J. Royal Statist. Soc. Series B*, 56:1–48.

Niederreiter, H. (1992). *Random Number Generation and Quasi-Monte Carlo Methods*. SIAM, Philadelphia.

Nobile, A. (1998). A hybrid Markov chain for the Bayesian analysis of the multinomial probit model. *Statistics and Computing*, 8:229–242.

Norris, J. (1997). *Markov Chains*. Cambridge University Press, Cambridge.

Nummelin, E. (1978). A splitting technique for Harris recurrent chains. *Zeit. Warsch. Verw. Gebiete*, 43:309–318.

Nummelin, E. (1984). *General Irreducible Markov Chains and Non-Negative Operators*. Cambridge University Press, Cambridge.

Ó Ruanaidh, J. and Fitzgerald, W. (1996). *Numerical Bayesian Methods Applied to Signal Processing*. Springer-Verlag, New York.

Oakes, D. (1999). Direct calculation of the information matrix via the EM algorithm. *J. Royal Statist. Soc. Series B*, 61:479–482.

Oh, M. (1989). Integration of multimodal functions by Monte Carlo importance sampling, using a mixture as an importance function. Technical report, Dept. of Statistics, Univ. of California, Berkeley.

Oh, M. and Berger, J. (1993). Integration of multimodal functions by Monte Carlo importance sampling. *J. American Statist. Assoc.*, 88:450–456.

Olkin, I., Petkau, A., and Zidek, J. (1981). A comparison of n estimators for the binomial distribution. *J. American Statist. Assoc.*, 76:637–642.

Olver, F. (1974). *Asymptotics and Special Functions*. Academic Press, New York.

Orey, S. (1971). *Limit Theorems for Markov Chain Transition Probabilities*. Van Nostrand, London.

Osborne, C. (1991). Statistical calibration: a review. *International Statistical Review*, 59:309–336.

Pearl, J. (1988). *Probabilistic Reasoning in Intelligent Systems: Networks of Plausible Inference*. Morgan Kaufmann, San Mateo, CA.

Pearson, K. (1915). On certain types of compound frequency distributions in which the individual components can be individually described by binomial series. *Biometrika*, 11:139–144.

Perron, F. (1999). Beyond accept–reject sampling. *Biometrika*, 86(4):803–813.

Peskun, P. (1973). Optimum Monte Carlo sampling using Markov chains. *Biometrika*, 60:607–612.

Peskun, P. (1981). Guidelines for chosing the transition matrix in Monte Carlo methods using Markov chains. *Journal of Computational Physics*, 40:327–344.

Pflug, G. (1994). Distributed stochastic approximation: Does cooperation pay? Technical Report TR-94-07, Institute of Statistics, Univ. of Wien.

Philippe, A. (1997a). Importance sampling and Riemann sums. Technical Report VI, Univ. de Lille. Prepub. IRMA.

Philippe, A. (1997b). Simulation of right and left truncated Gamma distributions by mixtures. *Stat. Comput.*, 7:173–181.

Philippe, A. (1997c). Simulation output by Riemann sums. *J. Statist. Comput. Simul.*, 59(4):295–314.

Philippe, A. and Robert, C. (1998a). Linking discrete and continuous chains. In Robert, C., editor, *Discretization and MCMC Convergence Assessment*, chapter 3, pages 47–66. Springer–Verlag, New York. Lecture Notes in Statistics.

Philippe, A. and Robert, C. (1998b). A note on the confidence properties of the reference prior for the calibration model. *TEST*, 7(1):147–160.

Philippe, A. and Robert, C. (2001). Riemann sums for MCMC estimation and convergence monitoring. *Statistics and Computing*, 11(1):103–115.

Philippe, A. and Robert, C. (2003). Perfect simulation of positive Gaussian distributions. *Statistics and Computing*, 13(2):179–186.

Phillips, D. and Smith, A. (1996). Bayesian model comparison via jump diffusions. In Gilks, W., Richardson, S., and Spiegelhalter, D., editors, *Markov chain Monte Carlo in Practice*, pages 215–240. Chapman and Hall, New York.

Pike, M. (1966). A method of analyis of a certain class of experiments in carcinogenesis. *Biometrics*, 22:142–161.

Pincus, M. (1968). A closed form solution of certain programming problems. *Oper. Research*, 18:1225–1228.

Pitt, M. and Shephard, N. (1999). Filtering via simulation: Auxiliary particle filters. *J. American Statist. Assoc.*, 94(446):590–599.

Preston, C. (1976). Spatial birth-and-death processes. *Bull. Inst. Internat. Statist.*, 46:371–391.

Propp, J. and Wilson, D. (1996). Exact sampling with coupled Markov chains and applications to statistical mechanics. *Random Structures and Algorithms*, 9:223–252.

Qian, W. and Titterington, D. (1990). Parameter estimation for hidden Gibbs chains. *Statis. Prob. Letters*, 10:49–58.

Qian, W. and Titterington, D. (1992). Stochastic relaxations and EM algorithms for Markov random fields. *J. Statist. Comput. Simulation*, 40:55–69.

Raftery, A. (1996). Hypothesis testing and model selection. In Gilks, W., Spiegelhalter, D., and Richardson, S., editors, *Markov Chain Monte Carlo in Practice*, pages 163–188. Chapman and Hall, New York, London.

Raftery, A. and Akman, V. (1986). Bayesian analysis of a Poisson process with a change point. *Biometrika*, 73:85–89.

Raftery, A. and Banfield, J. (1991). Stopping the Gibbs sampler, the use of morphology, and other issues in spatial statistics. *Ann. Inst. Statist. Math.*, 43:32–43.

Raftery, A. and Lewis, S. (1992). How many iterations in the Gibbs sampler? In Bernardo, J., Berger, J., Dawid, A., and Smith, A., editors, *Bayesian Statistics 4*, pages 763–773. Oxford University Press, Oxford.

Raftery, A. and Lewis, S. (1996). Implementing MCMC. In Gilks, W., Richardson, S., and Spiegelhalter, D., editors, *Markov chain Monte Carlo in Practice*, pages 115–130. Chapman and Hall, New York.

Randles, R. and Wolfe, D. (1979). *Introduction to the Theory of Nonparametric Statistics*. John Wiley, New York.

Rao, C. (1973). *Linear Statistical Inference and its Applications*. John Wiley, New York.

Redner, R. and Walker, H. (1984). Mixture densities, maximum likelihood and the EM algorithm. *SIAM Rev.*, 26:195–239.

Reid, N. (1988). Saddlepoint methods and statistical inference (with discussion). *Statist. Science*, 3:213–238.

Reid, N. (1991). *Approximations and asymptotics*, pages 287–334. Chapman and Hall, New York.

Resnick, S. (1994). *Adventures in Stochastic Process*. Birkhaüser, Basel.

Revuz, D. (1984). *Markov Chains (2nd edition)*. North-Holland, Amsterdam.

Richardson, S. and Green, P. (1997). On Bayesian analysis of mixtures with an unknown number of components (with discussion). *J. Royal Statist. Soc. Series B*, 59:731–792.

Ripley, B. (1977). Modelling spatial patterns (with discussion). *J. Royal Statist. Soc. Series B*, 39:172–212.

Ripley, B. (1987). *Stochastic Simulation*. John Wiley, New York.

Ripley, B. (1994). Neural networks and related methods for classification (with discussion). *J. Royal Statist. Soc. Series B*, 56:409–4560.

Ripley, B. (1996). *Pattern Recognition and Neural Networks*. Cambridge University Press, Cambridge.

Ritter, C. and Tanner, M. (1992). Facilitating the Gibbs sampler: The Gibbs stopper and the Griddy-Gibbs sampler. *J. American Statist. Assoc.*, 87:861–868.

Robbins, H. and Monro, S. (1951). A stochastic approximation method. *Ann. Mathemat. Statist.*, 22:400–407.

Robert, C. (1991). Generalized inverse normal distributions. *Statist. Prob. Lett.*, 11:37–41.

Robert, C. (1993). Prior feedback: A Bayesian approach to maximum likelihood estimation. *Comput. Statist.*, 8:279–294.

Robert, C. (1994). Discussion of "Markov chains for exploring posterior distribution". *Ann. Statist.*, 22:1742–1747.

Robert, C. (1995a). Convergence control techniques for MCMC algorithms. *Statis. Science*, 10(3):231–253.

Robert, C. (1995b). Simulation of truncated Normal variables. *Statistics and Computing*, 5:121–125.

Robert, C. (1996a). Inference in mixture models. In Gilks, W., Richardson, S., and Spiegelhalter, D., editors, *Markov Chain Monte Carlo in Practice*, pages 441–464. Chapman and Hall, New York.

Robert, C. (1996b). Intrinsic loss functions. *Theory and Decision*, 40(2):191–214.

Robert, C. (1997). Discussion of Richardson and Green's paper. *J. Royal Statist. Soc. Series B*, 59:758–764.

Robert, C. (1998). *Discretization and MCMC Convergence Assessment*, volume 135. Springer-Verlag, New York. Lecture Notes in Statistics.

Robert, C. (2001). *The Bayesian Choice*. Springer-Verlag, New York, second edition.

Robert, C. (2004). Bayesian computational methods. In Gentle, J., Härdle, W., and Mori, Y., editors, *Handbook of Computational Statistics*. Springer-Verlag, New York.

Robert, C. and Hwang, J. (1996). Maximum likelihood estimation under order constraints. *J. American Statist. Assoc.*, 91:167–173.

Robert, C. and Mengersen, K. (1999). Reparametrization issues in mixture estimation and their bearings on the Gibbs sampler. *Comput. Statis. Data Ana.*, 29:325–343.

Robert, C., Rydén, T., and Titterington, D. (1999). Convergence controls for MCMC algorithms, with applications to hidden Markov chains. *J. Statist. Computat. Simulat.*, 64:327–355.

Robert, C. and Soubiran, C. (1993). Estimation of a mixture model through Bayesian sampling and prior feedback. *TEST*, 2:125–146.

Robert, C. and Titterington, M. (1998). Reparameterisation strategies for hidden Markov models and Bayesian approaches to maximum likelihood estimation. *Statistics and Computing*, 8(2):145–158.

Roberts, G. (1992). Convergence diagnostics of the Gibbs sampler. In Bernardo, J., Berger, J., Dawid, A., and Smith, A., editors, *Bayesian Statistics 4*, pages 775–782. Oxford University Press, Oxford.

Roberts, G. (1994). Methods for estimating L^2 convergence of Markov chain Monte Carlo. In D. Barry, K. C. and Geweke, J., editors, *Bayesian Statistics and Econometrics: Essays in Honor of Arnold Zellner*. John Wiley, New York.

Roberts, G. (1998). Introduction to MCMC. Talk at the TMR Spatial Statistics Network Workshop.

Roberts, G., Gelman, A., and Gilks, W. (1997). Weak convergence and optimal scaling of random walk Metropolis algorithms. *Ann. Applied Prob.*, 7:110–120.

Roberts, G. and Polson, N. (1994). A note on the geometric convergence of the Gibbs sampler. *J. Royal Statist. Soc. Series B*, 56:377–384.

Roberts, G. and Rosenthal, J. (1998). Markov chain Monte Carlo: Some practical implications of theoretical results (with discussion). *Canadian J. Statist.*, 26:5–32.

Roberts, G. and Rosenthal, J. (2003). The polar slice sampler. *Stochastic Models*, 18:257–236.

Roberts, G., Rosenthal, J., and Schwartz, P. (1995). Convergence properties of perturbated Markov chains. Technical report, Statistics Laboratory, Univ. of Cambridge.

Roberts, G. and Sahu, S. (1997). Updating schemes, covariance structure, blocking and parametrisation for the Gibbs sampler. *J. Royal Statist. Soc. Series B*, 59:291–318.

Roberts, G. and Tweedie, R. (1995). Exponential convergence for Langevin diffusions and their discrete approximations. Technical report, Statistics Laboratory, Univ. of Cambridge.

Roberts, G. and Tweedie, R. (1996). Geometric convergence and central limit theorems for multidimensional Hastings and Metropolis algorithms. *Biometrika*, 83:95–110.

Roberts, G. and Tweedie, R. (2004). *Understanding MCMC*. Springer-Verlag, New York.

Robertson, T., Wright, F., and Dykstra, R. (1988). *Order Restricted Statistical Inference*. John Wiley, New York.

Roeder, K. (1992). Density estimation with confidence sets exemplified by superclusters and voids in galaxies. *J. American Statist. Assoc.*, 85:617–624.

Roeder, K. and Wasserman, L. (1997). Practical Bayesian density estimation using mixtures of normals. *J. American Statist. Assoc.*, 92:894–902.

Rosenblatt, M. (1971). *Markov Processes: Structure and Asymptotic Behavior*. Springer-Verlag, New York.

Rosenbluth, M. and Rosenbluth, A. (1955). Monte Carlo calculation of the average extension of molecular chains. *J. Chemical Physics*, 23:356–359.

Rosenthal, J. (1995). Minorization conditions and convergence rates for Markov chain Monte Carlo. *J. American Statist. Assoc.*, 90:558–566.

Rubin, D. (1987). *Multiple Imputation for Nonresponse in Surveys*. John Wiley, New York.

Rubinstein, R. (1981). *Simulation and the Monte Carlo Method*. John Wiley, New York.

Ruelle, D. (1987). *Chaotic Evolution and Strange Attractors*. Cambridge University Press, Cambridge.

Sahu, S. and Zhigljavsky, A. (1998). Adaptation for self regenerative MCMC. Technical report, Univ. of Wales, Cardiff.

Sahu, S. and Zhigljavsky, A. (2003). Self regenerative Markov chain Monte Carlo with adaptation. *Bernoulli*, 9:395–422.

Saxena, K. and Alam, K. (1982). Estimation of the non-centrality parameter of a chi-squared distribution. *Ann. Statist.*, 10:1012–1016.

Scherrer, J. (1997). Monte Carlo estimation of transition probabilities in capture-recapture data. Technical report, Biometrics Unit, Cornell Univ., Ithaca, New York. Masters Thesis.

Schervish, M. (1995). *Theory of Statistics*. Springer-Verlag, New York.

Schervish, M. and Carlin, B. (1992). On the convergence of successive substitution sampling. *J. Comput. Graphical Statist.*, 1:111–127.

Schmeiser, B. (1989). Simulation experiments. Technical report, School of Industrial Engineering, Purdue Univ.

Schmeiser, B. and Shalaby, M. (1980). Acceptance/rejection methods for beta variate generation. *J. American Statist. Assoc.*, 75:673–678.

Schruben, L., Singh, H., and Tierney, L. (1983). Optimal tests for initialization bias in simulation output. *Operation. Research*, 31:1176–1178.

Schukken, Y., Casella, G., and van den Broek, J. (1991). Overdispersion in clinical mastitis data from dairy herds: a negative binomial approach. *Preventive Veterinary Medicine*, 10:239–245.

Searle, S., Casella, G., and McCulloch, C. (1992). *Variance Components.* John Wiley, New York.

Seber, G. (1983). Capture–recapture methods. In Kotz, S. and Johnson, N., editors, *Encyclopedia of Statistical Science.* John Wiley, New York.

Seber, G. (1992). A review of estimating animal abundance (ii). *Workshop on Design of Longitudinal Studies and Analysis of Repeated Measure Data*, 60:129–166.

Seidenfeld, T. and Wasserman, L. (1993). Dilation for sets of probabilities. *Ann. Statist.*, 21:1139–1154.

Serfling, R. (1980). *Approximation Theory of Mathematical Statistics.* John Wiley, New York.

Shephard, N. (1994). Partial non-Gaussian state space. *Biometrika*, 81:115–131.

Shephard, N. and Pitt, M. (1997). Likelihood analysis of non-Gaussian measurement time series. *Biometrika*, 84:653–668.

Silverman, B. (1986). *Density Estimation for Statistics and Data Analysis.* Chapman and Hall, New York.

Smith, A. and Gelfand, A. (1992). Bayesian statistics without tears: A sampling-resampling perspective. *The American Statistician*, 46:84–88.

Smith, A. and Makov, U. (1978). A quasi–Bayes sequential procedure for mixtures. *J. Royal Statist. Soc. Series B*, 40:106–112.

Smith, A. and Roberts, G. (1993). Bayesian computation via the Gibbs sampler and related Markov chain Monte Carlo methods (with discussion). *J. Royal Statist. Soc. Series B*, 55:3–24.

Snedecor, G. W. and Cochran, W. G. (1971). *Statistical Methods.* Iowa State University Press.

Spiegelhalter, D., Best, N., Gilks, W., and Inskip, H. (1996). Hepatitis B: a case study in MCMC methods. In Gilks, W., Richardson, S., and Spiegelhalter, D., editors, *Markov Chain Monte Carlo in Practice*, pages 21–44. Chapman and Hall, New York.

Spiegelhalter, D., Dawid, A., Lauritzen, S., and Cowell, R. (1993). Bayesian analysis in expert systems (with discussion). *Statist. Science*, 8:219–283.

Spiegelhalter, D. and Lauritzen, S. (1990). Sequential updating of conditional probabilities on directed graphical structures. *Networks*, 20:579–605.

Spiegelhalter, D., Thomas, A., Best, N., and Gilks, W. (1995a). BUGS: Bayesian inference using Gibbs sampling. Technical report, Medical Research Council Biostatistics Unit, Institute of Public Health, Cambridge Univ.

Spiegelhalter, D., Thomas, A., Best, N., and Gilks, W. (1995b). BUGS examples. Technical report, MRC Biostatistics Unit, Cambridge Univ.

Spiegelhalter, D., Thomas, A., Best, N., and Gilks, W. (1995c). BUGS examples. Technical report, MRC Biostatistics Unit, Cambridge Univ.

Stavropoulos, P. and Titterington, D. (2001). Improved particle filters and smoothing. In Doucet, A., deFreitas, N., and Gordon, N., editors, *Sequential MCMC in Practice*. Springer-Verlag, New York.

Stein, M. (1987). Large sample properties of simulations using latin hypercube sampling. *Technometrics*, 29:143–151.

Stephens, M. (2000). Bayesian analysis of mixture models with an unknown number of components—an alternative to reversible jump methods. *Ann. Statist.*, 28:40–74.

Stigler, S. (1986). *The History of Statistics*. Belknap, Cambridge.

Stramer, O. and Tweedie, R. (1999a). Langevin-type models I: diffusions with given stationary distributions, and their discretizations. *Methodology and Computing in Applied Probability*, 1:283–306.

Stramer, O. and Tweedie, R. (1999b). Langevin-type models II: Self-targeting candidates for Hastings-Metropolis algorithms. *Methodology and Computing in Applied Probability*, 1:307–328.

Strawderman, R. (1996). Discussion of Casella's article. *TEST*, 5:325–329.

Stuart, A. (1962). Gamma-distributed products of independent random variables. *Biometrika*, 49:564–565.

Stuart, A. and Ord, J. (1987). *Kendall's Advanced Theory of Statistics (5th edition)*, volume 1. Oxford University Press, New York.

Swendson, R. and Wang, J. (1987). Nonuniversal critical dynamics in Monte Carlo simulations. *Physical Review Letters*, 58:86–88.

Swendson, R., Wang, J., and Ferrenberg, A. (1992). New Monte Carlo methods for improved efficiency of computer simulations in statistical mechanics. In Binder, K., editor, *The Monte Carlo Method in Condensed Matter Physics*. Springer–Verlag, Berlin.

Tanner, M. (1996). *Tools for Statistical Inference: Observed Data and Data Augmentation Methods*. Springer-Verlag, New York. (3rd edition).

Tanner, M. and Wong, W. (1987). The calculation of posterior distributions by data augmentation. *J. American Statist. Assoc.*, 82:528–550.

Thisted, R. (1988). *Elements of Statistical Computing: Numerical Computation*. Chapman and Hall, New York.

Tierney, L. (1994). Markov chains for exploring posterior distributions (with discussion). *Ann. Statist.*, 22:1701–1786.

Tierney, L. (1998). A note on Metropolis–Hastings kernels for general state spaces. *Ann. Applied Prob.*, 8(1):1–9.

Tierney, L. and Kadane, J. (1986). Accurate approximations for posterior moments and marginal densities. *J. American Statist. Assoc.*, 81:82–86.

Tierney, L., Kass, R., and Kadane, J. (1989). Fully exponential Laplace approximations to expectations and variances of non-positive functions. *J. American Statist. Assoc.*, 84:710–716.

Tierney, L. and Mira, A. (1998). Some adaptive Monte Carlo methods for Bayesian inference. *Statistics in Medicine*, 18:2507–2515.

Titterington, D. (1996). Mixture distributions. In Kotz, S., Read, C., and Banks, D., editors, *Encyclopedia of Statistical Sciences Update, Volume 1*, pages 399–407. John Wiley, New York.

Titterington, D., Smith, A., and Makov, U. (1985). *Statistical Analysis of Finite Mixture Distributions*. John Wiley, New York.

Tobin, J. (1958). Estimation of relationships for limited dependent variables. *Econometrics*, 26:24–36.

Tukey, J. (1977). *Exploratory Data Analysis*. Addison-Wesley, New York.

Van Dijk, H. and Kloeck, T. (1984). Experiments with some alternatives for simple importance sampling in Monte Carlo integration. In Bernardo, J., DeGroot, M., Lindley, D., and Smith, A., editors, *Bayesian Statistics 4*. North-Holland, Amsterdam.

Van Laarhoven, P. and Aarts, E. (1987). *Simulated Annealing: Theory and Applications*, volume CWI Tract 51. Reidel, Amsterdam.

Verdinelli, I. and Wasserman, L. (1992). Bayesian analysis of outliers problems using the Gibbs sampler. *Statist. Comput.*, 1:105–117.

Vines, S. and Gilks, W. (1994). Reparameterising random interactions for Gibbs sampling. Technical report, MRC Biostatistics Unit, Cambridge Univ.

Vines, S., Gilks, W., and Wild, P. (1995). Fitting multiple random effect models. Technical report, MRC Biostatistics Unit, Cambridge Univ.

von Bortkiewicz, L. (1898). *Das Gesetz der kleinen Zahlen*. Teubner, Leipzig.

Von Neumann, J. (1951). Various techniques used in connection with random digits. *J. Resources of the National Bureau of Standards—Applied Mathematics Series*, 12:36–38.

Wahba, G. (1981). Spline interpolation and smoothing on the sphere. *SIAMSSC*, 2:5–16.

Wakefield, J., Smith, A., Racine-Poon, A., and Gelfand, A. (1994). Bayesian analysis of linear and non-linear population models using the Gibbs sampler. *Applied Statistics (Ser. C)*, 43:201–222.

Walker, S. (1997). Robust Bayesian analysis via scale mixtures of beta distributions. Technical report, Dept. of Mathematics, Imperial College, London.

Wand, M. and Jones, M. (1995). *Kernel Smoothing*. Chapman and Hall, New York.

Warnes, G. (2001). The Normal kernel coupler: An adaptive Markov Chain Monte Carlo method for efficiently sampling from multi-modal distributions. Technical Report 395, Univ. of Washington.

Wasan, M. (1969). *Stochastic Approximation*. Cambridge University Press, Cambridge.

Wei, G. and Tanner, M. (1990a). A Monte Carlo implementation of the EM algorithm and the poor man's data augmentation algorithm. *J. American Statist. Assoc.*, 85:699–704.

Wei, G. and Tanner, M. (1990b). Posterior computations for censored regression data. *J. American Statist. Assoc.*, 85:829–839.

West, M. (1992). Modelling with mixtures. In Berger, J., Bernardo, J., Dawid, A., and Smith, A., editors, *Bayesian Statistics 4*, pages 503–525. Oxford University Press, Oxford.

Whitley, D. (1994). A genetic algorithm tutorial. *Statistics and Computing*, 4:65–85.

Whittaker, J. (1990). *Graphical Models in Applied Multivariate Statistics*. John Wiley, Chichester.

Wilson, D. (2000). How to couple from the past using a read-once source of randomness. *Random Structures and Algorithms*, 16(1):85–113.

Winkler, G. (1995). *Image Analysis, Random Fields and Dynamic Monte Carlo Methods*. Springer–Verlag, New York.

Wolff, U. (1989). Collective Monte Carlo updating for spin systems. *Physical Review Letters*, 62:361.

Wolpert, D. (1992). A rigorous investigation of evidence and Occam factors in Bayesian reasoning. Technical report, Santa Fe Institute.

Wolpert, D. (1993). On the Bayesian "Occam factors" argument for Occam's razor. Technical report, Santa Fe Institute.

Wolter, W. (1986). Some coverage error models for census data. *J. American Statist. Assoc.*, 81:338–346.

Wong, W. and Liang, F. (1997). Dynamic weighting in Monte Carlo and optimization. *Proc. National Academy of Science*, 94:14220–14224.

Wood, A., Booth, J., and Butler, R. (1993). Saddlepoint approximations to the CDF of some statistics with nonnormal limit distributions. *J. American Statist. Assoc.*, 88:680–686.

Wu, C. (1983). On the convergence properties of the EM algorithm. *Ann. Statist.*, 11:95–103.

Yakowitz, S., Krimmel, J., and Szidarovszky, F. (1978). Weighted Monte Carlo integration. *SIAM J. Numer. Anal.*, 15(6):1289–1300.

Yu, B. (1994). Monitoring the convergence of Markov samplers based on estimated L^1 error. Technical Report 9409, Department of Statistics, Univ. of California, Berkeley.

Yu, B. and Mykland, P. (1998). Looking at Markov samplers through cusum path plots: A simple diagnostic idea. *Statistics and Computing*, 8(3):275–286.

Index of Names

Aarts, E., 167
Abramowitz, M., 9, 29, 40, 117, 154
Ackley, D.H., 163
Adams, M., 458
Ahrens, J., 66
Aitkin, M., 415
Akman, V.E., 455
Alam, K., 82
Albert, J.H., 579
Aldous, D., 163, 413
Andrieu, C., 168, 441, 557, 570
Arnold, B.C., 412
Asmussen, S., 462, 490
Athreya, K.B., 216, 219, 263, 490
Atkinson, A., 55

Balakrishnan, N., 581
Banfield, J.D., 500
Barbe, P., 32
Barndorff-Nielsen, O.E., 9, 80, 122
Barnett, G., 440
Barone, P., 382
Barron, A., 32
Barry, D., 455
Basford, K., 367
Baum, L.E., 360, 362
Bauwens, L., 26, 80
Beckman, R.J., 156
Bennett, J.E., 383
Benveniste, A., 202
Bergé, P., 37
Berger, J.O., 6, 12, 14, 22, 27, 28, 31,
        90, 115, 124, 192, 239, 404,
        481, 581

Berger, R., 6, 11, 12, 18, 28, 198, 360,
        362
Berliner, L.M., 463
Bernardo, J.M., 12, 14, 31, 32, 192, 367
Bertail, P., 32
Berthelsen, K.K., 539
Berzuini, C., 548
Besag, J., 295, 314, 318, 326, 344, 372,
        377, 407, 412, 419
Best, N.G., 57, 285, 383, 384, 414–416,
        420, 465, 509
Billingsley, P., 32, 39, 77, 205, 235, 240,
        252, 263, 264, 353
Billio, M., 365, 369, 370, 579
Boch, R.D., 415
Bollerslev, T., 369
Booth, J.G., 122, 283
Borchers, D.L., 58
Borovkov, A.A., 513, 536
von Bortkiewicz, L., 60
Boser, D., 201
Bouleau, N., 202
Box, G.E.P., 43
Boyles, R.A., 178
Bradley, R.C., 264
Breslow, N.E., 415
Brockwell, P.J., 508
Broniatowski, M., 200, 419
Brooks, S.P., 461, 465, 479, 482, 483,
        499, 509, 543
Brown, L.D., 5, 28, 31, 71, 146, 192, 262
Brownie, C., 185
Buckland, S.T., 58

Bucklew, J.A., 37, 119, 120
Butler, R.W., 122

Cappé, O., 437, 439, 448, 550, 551, 562, 566, 569, 573, 579, 580
Carlin, B.P., 18, 82, 396, 397, 411, 451, 455, 461, 465, 509
Carpenter, J., 555
Carroll, R.J., 25
Casella, G., 5, 6, 10–12, 18, 25, 28, 31, 68, 81–83, 86, 95, 104, 105, 120, 130, 131, 133, 134, 151, 198, 296–299, 314, 354, 360, 362, 368, 383, 384, 404, 406, 407, 409, 412, 529, 537, 538
Castledine, B., 348
Celeux, G., 200, 419, 541, 579
Cellier, D., 579
Chaitin, G., 36
Chan, K.S., 223, 257
Chauveau, D., 200, 496, 579
Chen, M.H., 147, 191, 291, 391, 393, 508
Chen, R., 458, 549
Cheng, B., 66, 201
Cheng, R.C.H., 66
Chib, S., 314, 365, 368, 369, 412, 426, 451, 550, 579
Chopin, N., 555, 556, 559
Chou, R.Y., 369
Chung, K.L., 205
Churchill, G.A., 168
Cipra, B.A., 168
Clairambault, J., 579
Clapp, T., 558
Clarke, B., 32
Clayton, D.G., 384, 396, 415
Clifford, M.S., 419
Clifford, P., 344, 377, 555
Cochran, W.G., 414
Cocozza-Thivent, C., 579
Conover, W.J., 156
Copas, J.B., 24
Corcoran, J., 525
Cowell, R.G., 2, 422
Cowles, M.K., 461, 465, 504, 509
Cox, D.R., 9, 80, 193
Crawford, S.L., 400
Cressie, N., 168

Crisan, D., 555

D'Epifanio, G., 171
Damien, P., 326
Daniels, H.E., 120, 122
Davies, R.B., 88
Davis, P.A., 508
Dawid, A.P., 2, 422, 423
Decoux, C., 579
DeGroot, M.H., 400
Del Moral, P., 555, 559, 569, 570
Dellaportas, P., 415, 424, 479, 506
Dempster, A.P., 2, 90, 176, 177, 183, 340
Denison, D.G.T., 437
Denker, J.S., 201
Dette, H., 418, 421
Devroye, L., 41, 43, 44, 56, 66, 77
Dey, D., 115
Diaconis, P., 31, 33, 311, 413
DiCiccio, T.J., 122
Dickey, J.M., 131
Diebold, D., 365
Diebolt, J., 200, 341, 351, 366, 367, 403, 419, 496
Dieter, U., 66
Dimakos, X., 539
Dobson, A.J., 416
Doob, J., 205
Doss, H., 263
Douc, R., 574
Doucet, A., 168, 200, 441, 546, 551, 557, 569, 570
Doukhan, P., 51, 303, 480
Draper, D., 383
Duflo, M., 162, 168, 169, 202, 203
Dupuis, J.A., 184, 363
Durrett, R., 226
Dykstra, R.L., 8, 25, 173

Eaton, M.L., 262, 339
Eberly, L., 384
Eco, U., 458
Efron, 32, 33
Efron, B., 32, 121
Enders, W., 369
Escobar, M.D., 341, 367
Ethier, S.N., 318
Evans, G., 19

Evans, M., 114
Everitt, B.S., 367

Fan, J., 403
Fan, Y., 543
Fang, K.T., 77
Feast, G., 66
Feller, W., 77, 205, 208, 216, 219, 224,
        226, 241, 250
Ferguson, T.S., 24
Fernándes, R., 539
Fernhead, P., 555
Ferrari, P.A., 539
Ferrenberg, A.M., 37, 168, 373
Field, C., 120
Fieller, E.C., 18
Fill, J.A., 512, 532
Finch, S.J., 178
Fishman, G.S., 72, 413
Fitzgerald, W.J., 112, 158, 188, 194, 311
Fletcher, R., 19
Forster, J.J., 424
Foss, S.G., 513, 525, 536
de Freitas, N., 546, 551, 555
Frigessi, A., 382

Gamboa, F., 458
Garcia, N.L., 539
Gassiat, E., 458
Gauss, C.F., 7
Gaver, D.P., 385, 386, 470, 473
Gelatt, C.D., 163
Gelfand, A.E., 68, 113, 115, 130, 296,
        314, 326, 354, 355, 363, 385,
        396, 397, 406, 411, 414, 416,
        419, 455, 466
Gelman, A., 148, 316, 318, 319, 377,
        382, 412, 464, 480, 483, 497,
        499, 509
Geman, D., 167, 168, 314, 419
Geman, S., 167, 168, 314, 419
Gentle, J.E., 19, 158
George, E.I., 58, 71, 314, 348, 404
Geweke, J., 53, 94, 100, 101, 143, 508,
        509, 527
Geyer, C.J., 158, 199, 203, 204, 223,
        244, 257, 312, 396, 403, 464,
        540, 541
Ghosh, M., 18

Gijbels, I., 403
Gilks, W.R., 56–58, 285, 316, 318, 319,
        382–384, 396, 397, 414–416,
        420, 471, 492, 548
Giròn, F.J., 367
Giudici, P., 424, 499
Gleick, J., 37
Gleser, L.J., 18, 46
Glynn, P.W., 462
Godsill, S.J., 200, 441, 557, 558, 570
Goldfeld, S.M., 579
Gordon, N., 546, 551, 552, 555
Gouriéroux, C., 7, 80, 86, 90, 365, 369
Goutis, C., 15, 120, 377, 396
Green, P.J., 244, 295, 296, 326, 407,
        424, 426, 429, 431, 432, 437,
        512, 524, 537
Greenberg, E., 314
Grenander, U., 318
Gruet, M.A., 418
Guédon, Y., 579
Guihenneuc–Jouyaux, C., 385, 461, 503,
        579
Guillin, A., 562, 566, 569, 580
Gumbel, E.J., 23, 71

Haario, H., 168, 302
Hàjek, B., 166
Hall, P., 25, 32, 33, 121
Hammersley, J.M., 155, 344, 377, 577
Han, X.L., 296
Handschin, J., 577
Handscomb, D.C., 155
Hanlon, P., 311
Harris, T., 221
Hartigan, J., 455
Hastings, W.K., 206, 284, 291, 305, 314,
        419, 465, 508
Heidelberger, P., 509
Henderson, D., 201
Hendry, D.F., 549
Hestbeck, J.B., 185
Hesterberg, T., 103
Higdon, D.M., 326, 407
Hills, S.E., 388, 396, 397, 406, 414
Hines, J.E., 185
Hinton, G.E., 163
Hjorth, J.S.U., 32

Hobert, J.P., 263, 283, 377, 396, 406, 407, 412, 466, 492, 537
Holmes, C.C., 437
Holmes, S., 33
Hougaard, P., 122
Howard, R.E., 201
Hubbard, W., 201
Huber, P.J., 81
Huerta, G., 440
Hürzeler, M., 556
Hurn, M.A., 541
Hwang, C.R., 169, 203
Hwang, J.T., 18, 173

Inskip, H., 383
Ip, E.H.S., 200

Jöhnk, M.D., 44
Jackel, L.D., 201
Jacquier, E., 550
Jarrett, R.G., 455
Jeffreys, H., 14, 31, 404
Jeliazkov, I., 451
Jennison, C., 396
Jensen, J.L., 120
Johnson, N. L., 581
Johnson, V.E., 464
Jones, G.L., 466, 492
Jones, M.C., 403
Juang, B.H., 579

Kadane, J.B., 108, 115, 171, 188, 400
Kass, R.E., 15, 108, 115, 171
Kemeny, J.G., 248, 254–257, 311, 500
Kemp, A. W., 581
Kendall, W.S., 512, 520, 523, 539
Kennedy, W.J., 158
Kersting, G., 202
Kiefer, J., 202
Kim, S., 365, 369, 550
Kinderman, A., 65
Kipnis, C., 244
Kirby, A.J., 384
Kirkpatrick, S., 163
Kitagawa, G., 555, 577
Kloeck, T., 95
Knuth, D., 65, 72
Kohn, R., 440
Kolassa, J.E., 120

Kong, A., 130, 349, 355, 361, 376, 402, 458
Kors, T.J., 167
Kotz, S., 581
Krimmel, J.E., 77, 135, 486
Krishnan, T., 182
Kroner, K.F., 369
Kubokawa, T., 18
Kuehl, R.O., 156
Künsch, H., 555, 556, 558
Kullback, S., 192
Kurtz, T.G., 318

Laird, N.M., 2, 90, 176, 177, 182, 183, 340
Landau, D.P., 37
Lange, K., 19
Lange, N., 182
Lauritzen, S.L., 2, 422, 423
Lavielle, M., 200
Lavine, M., 367, 538
Le Cun, Y., 201
Lee, T.M., 12, 416
Legendre, A., 7
Lehmann, E.L., 5, 6, 10, 12, 18, 28, 31, 33, 37, 81, 83, 86, 90, 296, 354, 368, 404, 466
Lépingle, 202
Levine, R., 356
Lewis, S., 403, 464, 500–503, 507, 509
Liang, F., 547, 577
Lindvall, T., 232
Little, R.J.A., 182
Liu, C., 182, 200, 340, 464, 479, 505
Liu, J.S., 126, 130, 279, 311, 349, 355, 361, 376, 382, 394, 395, 402, 411, 458, 464, 479, 505, 549, 577
Loh, W.L., 156
Lombard, F., 25
Louis, T.A., 82, 182
Lugannani, R., 122
Lwin, T., 82
Lyons, T., 555, 559

MacDonald, I. L., 579
MacEachern, S.N., 463
MacLachlan, G., 182, 367
Madigan, D., 422, 424

Madras, N., 577
Maguire, B.A., 455
Makov, U.E., 23, 30, 71, 367
Mallick, B., 437
Marin, J.M., 367, 562, 566, 569, 580
Marinari, E., 540
Maritz, J. S., 82
Marsaglia, G., 37, 38, 53, 54, 72, 75
Marshall, A., 577
Martin, M.A., 122
Massart, P., 51, 303, 480
Matthews, P., 504
Mayne, D., 577
McCulloch, C.E., 82, 313, 407
McCulloch, R., 579
McKay, M.D., 156
McKeague, I.W., 296
McNeil, A.J., 384
Mead, R., 156
Mendell, N.R., 178
Meng, X.L., 148, 182, 200, 340
Mengersen, K.L., 258, 276, 288–290,
        295, 305, 306, 367, 398, 407,
        426, 461, 529, 537
Métivier, M., 202
Metropolis, N., 163, 168, 313, 419
Meyn, S.P., 205, 216, 220, 223, 231, 235,
        238, 241, 248, 259–261
Miclo, L., 559
Miller, M., 318
Mira, A., 312, 326, 329, 526, 537
Mitra, D., 167
Møller, J., 520, 526, 536, 537, 539–541
Monahan, J., 65
Monfort, A., 7, 80, 86, 90, 365, 369, 370,
        579
Monro, S., 201
Morton, K., 577
Moulines, E., 200, 366
Moussouris, J., 344
Muller, B., 43
Müller, P., 316, 393, 541
Murdoch, D.J., 512, 524, 537
Muri, F., 579
Murray, G.D., 194
Mykland, P., 471, 474, 481, 492, 509

Natarajan, R., 407
Naylor, J.C., 21

Neal, R.M., 163, 168, 201, 326, 336, 458,
        540
Nerlove, M., 365
Newton, M.A., 148
Ney, P., 216, 219, 490
Neyman, J., 12
Nicholls, G.K., 540, 541
Nichols, J.D., 185
Niederreiter, H., 76, 77
Nobile, A., 391
Norris, J.R., 205
Norris, M., 258, 318
Nummelin, E., 205, 214, 216

d'Occam, W., 458
Oh, M.S., 22
Ohman, P.A., 283
Olkin, I., 25
O'Muircheartaigh, I.G., 385, 386, 470,
        473
Ord, J.K., 121
Orey, S., 253
Ó Ruanaidh, J.J.K., 112, 158, 188, 194,
        311
Osborne, C., 18

Parisi, G., 540
Pearl, J., 2
Pearson, E.S., 12, 455
Pearson, K., 23
Perron, F., 134
Peskun, P.H., 311, 314, 394, 419, 578
Petkau, A.J., 25
Petrie, T., 360, 362
Pflug, G.C., 202
Philippe, A., 14, 18, 69, 115, 135, 136,
        410, 418, 486, 527, 528
Phillips, D.B., 206, 318, 426
Pike, M.C., 24
Pincus, M., 169
Pitt, M.K., 554, 575
Pollock, K.H., 185
Polson, N.G., 315, 550
Pommeau, Y., 37
Presnel, B., 492
Press, S.J., 412
Priouret, P., 202
Propp, J.G., 512, 513, 515, 518

Qian, W., 182, 419

Quant, R.E., 579

Rabiner, C.R., 579
Racine-Poon, A., 383, 406, 414
Raftery, A.E., 15, 148, 403, 426, 455,
        464, 500–503, 507, 509
Ramage, J., 65
Randles, R., 37
Rao, C.R., 183
Raspe, R.E., 32
Redner, R., 11
Reid, N., 120, 121
Renault, E., 365
Resnick, S.I., 318
Revuz, D., 205
Rice, S., 122
Richard, J.F., 80, 549
Richardson, S., 426, 437, 579
Rio, E., 51, 303, 480
Ripley, B.D., 72–74, 90, 201
Ritter, C., 479
Robbins, H., 82, 201
Robert, C., 2
Robert, C.P., 5, 6, 10, 12, 14, 15, 17,
        18, 28, 29, 31, 32, 53, 58, 68,
        80, 82, 95, 101, 104, 105, 115,
        130, 131, 133, 134, 142, 147,
        151, 169, 172, 173, 192, 200,
        239, 263, 264, 296–299, 341,
        348, 351, 354, 365, 367, 369,
        370, 377, 383, 385, 396, 398,
        399, 403, 404, 410, 418, 422,
        426, 437, 439, 443, 448, 461,
        463, 470, 484, 486, 503, 506,
        527, 529, 537, 538, 541, 562,
        566, 569, 576, 580
Roberts, G.O., 206, 274, 276, 284, 311,
        315, 316, 318, 319, 329, 360,
        362, 382, 396, 407, 461, 465,
        471, 478, 479, 492, 505, 509,
        526, 537
Robertson, T., 8, 25, 173
Roeder, K., 366, 426
Romeo, F., 167
Ronchetti, E., 120
Rosenblatt, M., 264
Rosenbluth, A.W., 163, 168, 313, 419,
        577

Rosenbluth, M.N., 163, 168, 313, 419,
        577
Rosenthal, J.S., 206, 284, 318, 319, 329,
        492, 504, 543
Rossi, P.E., 550
Rubin, D.B., 2, 90, 113, 176, 177, 182,
        183, 340, 464, 479, 480, 483,
        497, 499, 505
Rubinstein, R.Y., 72, 95, 155, 162
Rudin, W., 135
Ruelle, D., 37
Rydén, T., 437, 439, 448, 463, 465, 550,
        551, 579

Sabbati, C., 382
Sacksman, E., 168, 302
Sahu, S.K., 296, 396, 397, 407, 411, 471,
        492
Salmon, D., 552
Sangiovanni-Vincentelli, A., 167
Saxena, M.J., 82
Scherrer, J.A., 185, 197, 363
Schervish, M.J., 108, 171, 188
Schmeiser, B.W., 67, 291, 391, 393, 462
Schruben, L., 509
Schukken, Y.H., 383, 409
Schwartz, P.C., 206, 284
Searle, S.R., 82
Seber, G.A.F., 58
Seidenfeld, T., 252
Sejnowski, T.J., 163
Serfling, R.J., 83
Sethuraman, J., 263
Shalaby, M., 67
Shao, Q.M., 147, 191, 508
Sharples, L.D., 384
Sheather, S., 440
Shephard, N., 365, 369, 550, 554, 575,
        579
Silverman, B.W., 508
Singh, H., 509
Slade, G., 577
Small, M.J., 400
Smith, A.F.M., 12, 14, 21, 23, 30–32,
        68, 71, 113, 130, 192, 206,
        314, 318, 326, 354, 355, 360,
        362, 363, 367, 385, 388, 397,
        406, 414–416, 419, 426, 437,
        455, 466, 552

Snedecor, G.W., 414

Snell, E.J., 193

Snell, J.L., 248, 254–257, 500

Soubiran, C., 169, 365

Soules, G., 360, 362

Speed, T.P., 377, 412

Spiegelhalter, D.J., 2, 383, 384, 414–416, 420, 422

Srivastava, M.S., 18

Stavropoulos, M., 556

Stegun, I., 9, 29, 40, 117, 154

Stein, M.L., 156

Stigler, S., 7

Stram, D., 182

Stramer, O., 276, 313, 320

Strawderman, R.L., 152

Stuart, A., 45, 121

Studden, W.J., 418, 421

Sturmfels, B., 413

Swartz, T., 114

Swendson, R.H., 168, 312, 326, 373

Szidarovszky, F., 77, 135, 486

Tamminen, J., 302

Tan, K.K.C., 57, 285

Tanner, M.A., 182, 183, 295, 314, 340, 347, 364, 385, 394, 419, 479, 486, 507

Teller, A.H., 163, 168, 313, 419

Teller, E., 163, 168, 313, 419

Thisted, R.A., 21, 158, 162

Thode, H.C., 178

Thomas, A., 414–416, 420

Thompson, E.A., 191, 204, 540, 541

Thorisson, H., 462

Tibshirani, R.J., 32, 33

Tierney, L., 108, 115, 171, 188, 223, 244, 264, 286, 291, 304, 312, 326, 329, 389, 390, 464, 471, 474, 492, 509

Titterington, D.M., 23, 172, 182, 201, 367, 399, 419, 437, 463, 465, 529, 537, 556, 576

Tobin, J., 410

Tsay, R., 579

Tukey, J.W., 449

Tweedie, R.L., 205, 206, 216, 220, 223, 231, 235, 238, 241, 248, 258–261, 274, 276, 288–290, 305, 306, 313, 315, 318–320, 513, 525, 536

van den Broek, J., 383, 409

Van Dijk, H.K., 95

van Dyk, D., 182, 200

Van Laarhoven, P.J., 167

Varadhan, S.R., 244

Vecchi, M.P., 163

Verdinelli, I., 341, 367

Vermaak, J., 441

Vidal, C., 37

Vines, K., 396, 397, 465, 509

Von Neumann, J., 36

Waagepetersen, R.P., 526

Wahba, G., 21

Wakefield, J.C., 326, 383

Walker, H., 11

Walker, S., 67, 326

Wand, M.P., 403

Wang, J.S., 37, 168, 312, 326, 373

Wang, Y., 77

Wasan, M.T., 202

Wasserman, L., 32, 252, 341, 366, 367

Wefelmeyer, W., 296

Wei, G.C.G., 183, 486

Weiss, N., 360, 362

Welch, P.D., 509

West, M., 341, 367, 440

Whitley, D., 555

Whittaker, J., 422

Wild, P., 56, 58, 397

Wilson, D.B., 512, 513, 515, 518, 539

Winfield, D., 198

Winkler, G., 162, 165, 166, 202, 312, 314

Wolfe, D., 37

Wolff, U., 37

Wolfowitz, J., 202

Wolpert, D.H., 458

Wolpert, R.L., 6, 12, 239

Wolter, W., 348

Wong, W.H., 130, 148, 295, 314, 340, 347, 349, 355, 361, 376, 394, 402, 419, 458, 547, 577

Wood, A.T.A., 122

Wright, F.T., 8, 25, 173

Wu, C.F.J., 178, 194

Wynn, A.H.A., 455

Yakowitz, S., 77, 135, 486
Ylvisaker, D., 31
York, J., 422, 424
Yu, B., 471, 474, 481, 492, 509

Zaman, A., 38, 72, 75
Zhigljavsky, A.A., 471
Zidek, J.V., 25
Zucchini, W., 58, 579

Index of Subjects

α-mixing, 257, 263, 264
absorbing component, 367
Accept–Reject, 94, 133, 134, 142, 143, 271, 276, 378, 392
  and importance sampling, 103
  and MCMC algorithms, 269
  and Metropolis–Hastings algorithm, 278, 280
  and perfect sampling, 511, 515, 530, 532
  and recycling, 53, 104, 105, 132, 154
  and wrong bounds, 304
  approximative, 286
  bound, 154
  corrected, 304
  criticism, 53
  optimization, 52
  sample distribution, 115
  stopping rule, 104
  tight bound, 51
  with envelope, 53
acceptance probability, 48
acceptance rate, 292
  and simulated annealing, 169
  expected, 278
  optimal, 316, 319
acidity level, 400
actualization, 571
adaptive rejection sampling, 56
adjacency, 439
algorithm

acceleration of, 41, 131, 140, 292, 295, 390
adaptive, 284, 300–302, 545, 559, 563, 569
ARMS, 209, 285–287, 321
ARS, 56–58, 60, 70, 285
Baum–Welch, 571
Boltzman, 314, 578
Box–Muller, 43, 63
CFTP, *see* coupling from the past
comparison, 92, 140, 279
convergence of, 137, 147, 153, 162
domination, 107, 154
ECM, 200, 340
EM, 2, 90, 176, 313, 340, 341
generic, 271
Gibbs, *see* Gibbs sampling
greedy, 20, 124, 181
Green's, 429
hit-and-run, 291, 391, 393
hybrid, 380, 389, 391, 392, 400, 474
initial conditions, 200
Kiss, 38, 73, 75, 76
  C program, 75
Levenberg–Marquardt, 21
MCEM, *see* MCEM
MCMC, *see* MCMC
Metropolis–Hastings, *see* Metropolis–Hastings algorithm
multiscale, 564
Newton–Raphson, 19
  Levenberg–Marquardt, 21

steepest descent, 20
optimization, 51, 140
performances, 136
pool-adjacent-violators, 8, 25
read-once, 539
Robbins–Monro, 201, 202, 268
SAEM, 200
SAME, 200
SEM, 200
simulated annealing, 163, 396
SIR, 113, 552
stability, 137, 138, 140
stochastic, 158
Swendson–Wang, 312
Viterbi, 572
algorithmic complexity, 41
allocation
hybrid, 400
indicator, 189, 190, 367
probability, 368
stable, 401
analysis
conditional, 239
EM, 180
numerical, 1, 19
spectral, 509
anti-monotonicity, 523
aperiodicity, 217, 218, 250, 272
approximation, 284
first-order, 109
Normal, 499
for the binomial distribution, 101
Riemann, 134, 486
saddlepoint, 120
second-order, 109
third-order, 109
asymptotic
normality, 86
relative efficiency, 357
atom, 214, 216, 217, 219, 224, 231, 236
accessible, 214
ergodic, 231, 234
geometrically ergodic, 237
Kendall, 237
attrition, 552
augmentation technique, 490, 493
autoexponential model, 372
uniform ergodicity, 387

auxiliary
distribution, 444
variable, 47, 326, 444
auxiliary particle filter, see particle filter

β-mixing, 264
bandwidth selection, 563
baseball, 197
batch sampling, 403
Baum–Welch formulas, 551, 571
Bayes estimator, 13
admissibility of, 262
linear, 31
Bayes factor, 15
computation, 115, 125, 147
Bayesian
calculation, 50
Bayesian inference, 4, 12–18
and Gibbs sampling, 371, 419
Decision Theory, 83
empirical, 82
estimation, 12–18
integration within, 79, 85
noninformative, 368, 403, 576
pluggin, 82
robust, 90
using Gibbs sampling (BUGS), 420
bijection, see one-to-one
binomial parameter, 25
bits, 74
black box, 104
Boltzman algorithm, 314, 578
Bonferroni's inequality, 517
bootstrap, 32, 79
smooth, 556
bridge estimator, 148
Brownian
bridge, 509
motion, 318, 482
BUGS, 22, 414, 420, 509

C, 75, 420
calibration, 287, 291, 320, 504
algorithmic, 316
linear, 18
step, 317
cancer treatment success, 86
canonical moments, 417, 418, 421

capture–recapture, 58, 184, 196, 348, 362
carcinogen, 24
censoring model, 2, 23, 174
    type of, 2
Cesàro average, 252
chains
    interleaved, 349, 371
    parallel, 407
        versus single, 269
    parallel versus single, 464
    sub, 387, 390
change-point, 445, 454, 455
chaos, 37
Chapman–Kolmogorov equations, 210, 213, 219, 220
character recognition, 201
clique, 423–424
    perfect, 423
CODA, 420, 509
combinatoric explosion, 342, 367
comparison of models, 15
completion, 326, 431
    natural, 374
computational difficulties, 83
computer discretization, 284
computing time, 30, 101
    comparison, 41
    excessive, 38
    expensive, 52
condition
    Doeblin's, 238, 305, 315, 329, 524, 525, 537
    drift, 221, 243, 259–261, 290, 330, 465
        geometric, 260
    Liapounov's, 245
    minorizing, 207, 215, 216, 220, 246
    positivity, 377
    transition kernel regularity, 407
conditional distributions, 343
conditioning, 130, 131, 133, 354
confidence interval, 33
conjugate prior, 30, 341, 367
    and BUGS, 420
    for mixtures, 506
connected
    arcwise, 409
    balls, 381

components, 381
sites, 312
space, 161, 335
support, 271, 275, 299, 377, 379
constrained parameter, 8
contingency table, 86, 413
continuous-time process, 318, 446, 449
control
    binary, 502, 503
    by renewal, 470
    of Markov chain Monte-Carlo algorithms, 474
    of Metropolis–Hastings algorithms, 292
    of simulation methods, 153
control variate, 145, 147, 484
    integration, 145
    optimal constant, 145
convergence, 162
    acceleration, 160
    and duality principle, 351
    and irreducibility, 276, 288
    assessment, 107, 140, 211, 232, 261, 269, 373, 491
        multivariate, 127
        of variability, 124
    conservative criterion for, 153
    control, 486
    of the empirical mean, 292, 462
    ergodic, 315
    geometric, 236, 276, 315, 367, 385, 422
        uniform, 315
    Gibbs sampler, see Gibbs sampling
    graphical evaluation of, 466, 484
    indicator, 495, 502
    monitoring of, 84, 123, 283
    of Markov chain Monte Carlo algorithms, 268, see MCMC
    slow, 96, 292, 481
    speed, 137, 138, 172, 236
    test, 84
    to a maximum, 167
    to the stationary distribution, 462, 464, 479
    of the variance criterion, 498
convolution, 457
Corona Borealis Region, 426, 450
correlation, 140, 142

between components, 502
negative, 143
coupling, 232, 304
automatic, 530
forward, 513
from the past, 513–532
horizontal, 539
fundamental theorem, 515
monotone, 536
multigamma, 525
time, 233
covering, 220
criterion
conservative, 495
convergence, 491
efficiency, 316
unidimensional, 499
cumulant generating function, 8
cumulative sums, 481
curse of dimensionality, 21, 136, 156, 486
cycle, 218
of kernels, 389

DAG, 423
data
augmentation, 176, 183, 199, 357, 524
censored, 2, 178
complete, 175
longitudinal, 454
missing, 162, 175
Data Augmentation, *see* two-stage Gibbs sampling
deconvolution, 457
degeneracy, 552, 556–558, 564
delta method, 126
demarginalization, 174, 176, 326, 373, 408
for recursion formulae, 421
density
instrumental, 47
log-concave, 56–58, 71
multimodal, 295
prior, 12
spectral, 508
target, 47
unbounded, 49
dependence

logistic, 287
Markovian, 239
detailed balance condition, 230, 336, 349, 431, 446, 533
detailed balance property, 272
deviance, 416
diagnostic
autocorrelation, 509
distance, 479
multiple estimates, 489
parallel, 500
Die Hard, 37
diffusion, 206, 318
dilation, 252
dimension
fixed, 439
jump, 430
matching, 431
saturation, 432, 445
variable, 425, 426
dimensionality of a problem, 136
discrete random variable, 44
discretization, 206, 504
of a diffusion, 318
and Laplace approximation, 319
distance
entropy, 191
Kullback–Leibler, 31, 191
to the stationary distribution, 479
unbiased estimator, 478, 505
distribution
arcsine, 37, 62
Beta, 8, 13, 44, 66
beta-binomial, 149, 514
binomial, 25
Burr's, 66
Cauchy, 10, 90
conditional, 403
conjugate, 146
Dirichlet, 151
discrete, 44
double exponential, 52
exponential, 99
Gamma, 52, 106
generalized inverse Gaussian, 71
generalized inverse normal, 29
geometric, 211, 219, 290
Gumbel, 71
heavy tail, 94

improper prior, 404
infinitely divisible, 77
instrumental, 47, 52, 92, 97, 101,
    135, 172, 271, 276, 393
  choice of, 299, 387
invariant, 206
inverse Gaussian, 293
Laplace, 120
limiting, 272
log-concave, 305
log-normal, 92
logarithmic series, 62
logistic, 55, 58
marginal, 14
multinomial, 86, 496
negative binomial, 44
noncentral chi-squared, 9, 25, 46,
    82
nonuniform, 40
Normal, 40, 43, 96, 142, 288
  truncated, 52, 101, 323
Pareto, 10
Poisson, 43, 44, 55
posterior, 12
predictive, 427
prior, 12
proposal, 545
stationary, 223, 224, 272
  (distance to), 478, 479
Student, 10
target, 270
uniform, 35, 40, 98, 143
Weibull, 3, 23, 24, 67, 415
Wishart, 28
witch's hat, 504, 505
divergence, 76, 254, 406
  Kullback–Leibler, *see* information
DNA sequence, 168, 426, 579
Doeblin's condition, *see* condition
domination of a Markov chain, 394, 578
drift condition, 258, 291
  geometric, 260
dynamic
  programming, 572
  weighting, 547

effective sample size, 499
efficiency, 298
  loss of, 463, 479, 500, 509

eigenvalues, 279, 507
  of a covariance matrix, 396
EM algorithm, 176–182
  and simulated annealing, 178
  identity, 176
  introduction, 176
  for mixtures, 181
  monotonicity, 177
  Monte Carlo, 183
  relationship to Gibbs sampling,
      357
  standard error, 186
  steps, 177, 200
empirical
  cdf, 80, 89, 90
  mean, 83, 90, 131, 134, 137
  percentile, 124
empirical Bayes principles, 82
energy function, 163, 168, 203
entropy distance, *see* distance
envelope, 54, 56, 520
  method, 54
  pseudo, 285
equation
  backward, 572
  forward, 573
  forward–backward, 551, 571
ergodicity, **231**, 235, 238
  geometric, 236, 257, 276, 289, 353
  in MCMC algorithms, 268
  of Metropolis–Hastings algorithms,
      272, 274, 292
  sufficient conditions, 230
  uniform, 237, 264, 276
    of Metropolis–Hastings algo-
        rithms, 289
    sufficient conditions, 390
errors in variables, 11
estimation
  Bayesian, *see* Bayesian
  decisional, 86
  nonparametric, 509
  parallel, 124, 153
estimator
  Bayes minimax, 115
  bridge, 148
  empirical Bayes, 81, 82
  James–Stein, 81
    truncated, 142

maximum likelihood, 86, 200, 365
    lack of, 368
    parallel, 140
    Rao–Blackwellized, 151, 346
expert systems, 2, 422
exponential
    family, 5, 8, 10, 13, 28, 118, 146,
        169, 293, 341, 367
    tilting, 121
    variate generation, 39, 42
extra Monte Carlo variation, 554, 555

failure probability, 16
filter
    bootstrap, 552
    particle, 547, 552, 575, 577
filtering, 571
fixed point, 177
forward–backward formulas, 551
function
    analytical properties of, 158
    confluent hypergeometric, 29
    digamma, 7, 9
    dilog, 152, 154
    empirical cumulative distribution,
        32, 124
    evaluation, 53
    logistic, 37
    loss, 80, 81, 141
    modified Bessel, 9, 46
    one-to-one, 431, 445
    potential, 221, 258, 259, 261
    regularity of, 158
    risk, 81
    sigmoid, 201
functional compatibility, 411
fundamental matrix, 254
fundamental theorem of simulation, 47,
    321, 328, 526
funneling property, 537

Gamma variate generator, 45, 106
Gauss, 22, 41
generalized inverse, 39
generalized linear model, 287
    completion, 313
generator
    Central Limit Theorem, 64
    chaotic, 37

congruential, 73, 75
pseudo-random, 22, 35, 39, 284,
    515
    tests for, 36
RANDU, 73
shift register, 73
genetic linkage, 183
geometric
    ergodicity, 366
    rate of convergence, 291, 297
    waiting time, 51
Gibbs
    measure, 169, 203
    random field, 419
Gibbs sampler, see Gibbs sampling
Gibbs sampling, 267, 269, 272, 314, 321,
    371, **371**, 383, 403
    and ECM, 200
    completion, 374
    convergence assessment, 496
    Harris recurrence of, 380
    irreducibility of, 379
    random, 402
      scan, 376
      sweep, 376
    relationship to
      EM algorithm, 357
      slice sampling, 343
    relationship to slice sampling, 337
    reversible, 375
    slow convergence of, 399
    symmetric
      scan, 375
      sweep, 375
    two-stage, 262, 337–360, 374, 378,
      422
      and interleaving, 349
      reversible, 350
    uniform ergodicity of, 385
    validity of, 388
    versus Metropolis–Hastings algo-
      rithms, 270, 360, 373, 378,
      381, 387
gradient
    methods, 19
    stochastic, 162
graph, 2, 422–424
    directed, 422
    directed acyclic, 423, 549

undirected, 422
Gray code, 33
greediness, 20
grouped data, 346, 347

harmonic
    function, 240, 241, 256, 273
        bounded, 241
    mean, 96
Harris positivity, 224, 231, 260
Harris recurrence, 207, 222, 233,
        240–242, 259, 379
    importance of, 222, 223
Hessian matrix, 319
Hewitt–Savage 0–1 Law, 240
hidden
    Markov model, 352, 571, 575, 576,
        579–580
    Poisson, 576
    semi-Markov chains, 579
How long is long enough?, 512
hybrid strategy, 378

identifiability, 89, 181, 398–400
    constraint, 398–400
image processing, 168, 419
implicit equations, 79
importance, 90
importance function, 97
importance sampling, 91, **92**, 203, 268
    accuracy, 94
    and Accept–Reject, 93–96, 102,
        103
    and infinite variance, 103, 488
    and MCMC, 271, 545–547, 557
    and particle filters, 547
    and population Monte Carlo, 562
    bias, 95, 133
    by defensive mixtures, 103, 562,
        563
    difficulties, 102
    dynamic, 577
    efficiency, 96
    for convergence assessment, 484
    identity, 92, 546, 577
    implementation, 102
    improvement, 134
    monitoring, 153
    optimum, 95

principle, 90
    sequential, 549
    variance, 126
IMSL, 22
independence, 462, 463
inference
    asymptotic, 83
    Bayesian, 5, 12–14, 269
        noninformative, 368
    difficulties of, 5
    empirical Bayes, 82
    generalized Bayes, 404
    nonparametric, 5
    statistical, 1, 7, 79, 80
information
    Fisher, 31, 404
    Kullback–Leibler, 31, 191, 552, 560
    prior, 14, 383, 406
    Shannon, 32
informative censoring, 23
initial condition, 293, 380
    dependence on, 292, 379, 462
initial distribution, 211, 274
    and parallel chains, 464
    influence of, 379, 464, 499
initial state, 208
integration, 6, 19, 79, 85, 140
    approximative, 83, 107, 135
    by Riemann sums, 22, **134**, 475
    Monte Carlo, 83, 84
    numerical, 80, 135
        bounds, 22, 83
    problems, 12
    recursive, 172, 365
    weighted Monte Carlo, 135
interleaving, **349**, 462
    and reversibility, 349
    property, 349, 355, 356
intrinsic losses, 80
inversion
    of Gibbs steps, 478
    of the cdf, 39, 40
ion channel, 579
irreducibility, 206, 213, **213**, 215, 218,
        274, 284
Ising model, 37, 168, 188, 326, 373, 408

Jacobian, 431, 432
jump

mode, 295
reversible, 430
stochastic gradient, 163

Kac's representation, *see* mixture
    representation
K_ε-chain, 211, 213, 220
kernel
    estimation, 403
    residual, 229, 493, 524
    reverse, 533
    transition, 206, 208, 209
Kolmogorov–Smirnov test, *see* test
Kuiper test, *see* test
Kullback–Leibler information, *see*
    information

L'Hospital's rule, 330
label switching, 540
Lagrangian, 19
Langevin algorithm, 319
    and geometric ergodicity, 319
    extensions of, 320
Langevin diffusion, 316, 318
Laplace approximation, 107, 115, 188
large deviations, 37, 90, 119
Latin hypercubes, 156
Law of Large Numbers, 83, 125, 239,
    268, 494, 551, 558
    strong, 83, 242
lemma
    Pitman–Koopman, 10, 30
Levenberg–Marquardt algorithm, *see*
    algorithm
Liapounov condition, 245
likelihood, 12
    function, 6, 7
    individual, 326
    integral, 12
    maximum, 82, 83, 172
    multimodal, 11
    profile, 9, 18, 80
    pseudo-, 7
    unbounded, 11
likelihood ratio approximation, 369
linear model, 27
local maxima, 160, 388
log-concavity, 72, 289
longitudinal data, 3, 454

loss
    intrinsic, 80
    posterior, 13
    quadratic, 13, 81
low-discrepancy sequence, 76
lozenge, 522

Manhattan Project, 314
MANIAC, 314
Maple, 22
marginalization, 103, 551
    and Rao–Blackwellization, 354
    Chib's, 451
    Monte Carlo, 114
Markov chain, 99, 101, 102, 205, **208**,
    212, 378
    augmented, 217
    average behavior, 238
    divergent, 406
    domination, 394, 578
    ergodic, 268
    essentials, 206
    Harris positive, *see* Harris
        positivity
    Harris recurrent, *see* Harris
        recurrence
    hidden, 579
    homogeneous, 162, 209, 286, 501
    instrumental, 349
    interleaved, 349
    irreducible, 213
    limit theorem for, 238, 263
    lumpable, 256
    m-skeleton, 207
    nonhomogeneous, 162, 286, 300,
        316, 317
    observed, 239
    positive, 224
    random mapping representation,
        513
    recurrent, 220, 259
    reversible, 244, 261
    semi-, 579
    for simulation, 268
    split, 217
    stability, 219, 223
    strongly aperiodic, 218, 220
    strongly irreducible, 213, 345
    transient, 258

two state, 500

φ-irreducible, 213

weakly mixing, 482

Markov chain Monte Carlo method, *see*
    MCMC algorithm

Markov property

    strong, 212, 247

    weak, 211

mastitis, 383

`Mathematica`, 22, 41, 172

`Matlab`, 22, 41

matrix

    regular, 254

    transition, 208, 211, 217, 224

    tridiagonal, 129

maxima

    global, 167, 169

    local, 162, 163, 202

maximization, 6

maximum likelihood, 5, 7, 10, 83

    constrained, 8, 173

    difficulties of, 10, 12

    estimation, 6, 8

    existence of, 11

    justification, 6

MCEM algorithm, 183, 185, 340

    standard error, 186

MCMC algorithm, 213, 236, 244, 268,
    268

    and Monte Carlo methods, 269

    birth-and-death, 446

    calibration, 317

    convergence of, 459

    heterogeneous, 300, 390

    history, 163, 314

    implementation, 269

    monitoring of, 459, 474, 491

    motivation, 268

measure

    counting, 226

    invariant, 223, 225

    Lebesgue, 214, 225

    maximal irreducibility, 213

method

    Accept–Reject, 47, 51–53

    delta, 126

    gradient, 162

    kernel, 508

    least squares, 7, 173, 201

Markov chain Monte Carlo, *see*
    MCMC algorithm

of moments, 8

monitoring, 406

Monte Carlo, *see* Monte Carlo

nonparametric, 508, 563

numerical, 1, 19, 22, 80, 85, 135,
    157, 158, 497

quasi-Monte Carlo, 21, 75

simulated annealing, 163, 167, 200,
    203

simulation, 19

Metropolis–Hastings algorithm, 165,
    167, 267, 269, 270, **270**, 378,
    403, 481

    and Accept–Reject, 279

    and importance sampling, 271

    autoregressive, 291

    classification of, 276

    drawbacks of, 388

    efficiency of, 321

    ergodicity of, 289

    independent, 276, 279, 296, 391,
        472

    irreducibility of, 273

    random walk, 284, 287, 288

    symmetric, 473

    validity of, 270

Metropolization, 394

mining accidents, 455

missing data, 2, 5, **174**, 340, 346, 374

    simulation of, 200

missing mass, 475

mixing, 489

    condition, 263

    speed, 462

mixture, **4**, 295, 366, 368

    computational difficulties with, 4,
        11

    continuous, 45, 420

    defensive, 103, 562

    dual structure, 351

    exponential, 23, 190, 419

    geometric, 420

    identifiability, 89

    indicator, 341

    of kernels, 389

    negative weight, 493

    nonparametric, 420

Normal, 11, 88, 365, 388, 398, 399, 506
   EM algorithm, 181
   number of components
     test on, 88
Poisson, 23
reparameterization of, 397
representation, 45–46, 77, 229, 375, 471, 524
residual decomposition, 252
for simulation, 45, 67, 77, 103, 131, 373
stabilization by, 103, 113
Student's t, 497
two-stage Gibbs for, 340
unbounded likelihood, 11
with unknown number of components, 426, 437
mode, 160, 482, 499
   global, 172
   local, 10, 22, 172, 201
     attraction towards, 388
     trapping effect of, 390
model
   ANOVA, 416
   AR, 194, 210, 227, 244, 260, 265, 428
   ARCH, 308, 365, 368
   ARMA, 508
   augmented, 183
   autoexponential, 372, 382
   autoregressive, 28
   averaging, 427, 433, 440
   Bernoulli–Laplace, 208, 227
   capture–recapture, see capture–recapture
   censored, 327
   change-point, 454
   choice, 125
   completed, 341
   decomposable, 423
   embedded, 433
   generalized linear, 287
   graphical, 422
   hidden Markov, see hidden Markov, 549
   hierarchical, 383
   Ising, see Ising
   linear calibration, 375

logistic, 145
logit, 15, 168
MA, 4
mixed-effects, 384
multilayer, 201
multinomial, 347, 394
Normal, 404
   hierarchical, 27
   logistic, 58
overparameterized, 295
Potts, 312
probit, 391
   EM algorithm for, 192
random effect, 414
random effects, 397, 406
Rasch, 415
saturation, 432, 444, 450, 451
stochastic volatility, 549
switching AR, 453
tobit, 410
variable dimension, 425–427
modeling, 1, 367
   and reduction, 5
moment generating function, 8
monitoring of Markov chain Monte Carlo algorithms, see MCMC
monotone likelihood ratio (MLR), 72
monotonicity
   of covariance, 354, 361, 463
   for perfect sampling, 518
Monte Carlo, 85, 90, 92, 96, 141, 153
   approximation, 203
   EM, 183, 184
   maximization, 204
   optimization, 158
   population, see population
move
   birth, 437
   death, 437
   deterministic, 432
   merge, 433, 438
   Q,R,M, 578
   split, 433, 438
multimodality, 22
multiple chains, 464

The Name of the Rose, 458
neural network, 201, 579
neuron, 579

Newton–Raphson algorithm, *see* algorithm
nodes, 422
non-stationarity, 466
Normal
 approximation, 85, 283
 variate generation, 40, 43, 69
normality
 asymptotic, 86
 violation, 130
normalization, 485
normalizing constant, 14, 16, 17, 50, 137, 147, 191, 271, 367, 405, 431, 471, 479
 and Rao–Blackwellization, 309
 approximation, 95
 evaluation, 479
Northern Pintail ducks, 59
nuclear pump failures, 386, 470, 474
null recurrence and stability, 242
Nummelin's splitting, 216, 225

Oakes' identity, 186
Occam's Razor, 458
on-line monitoring, 482
optimality, 316
 of an algorithm, 41, 58, 91, 95, 137, 374
 of an estimator, 14, 82
 of the Metropolis–Hastings algorithm, 272
optimization, 19, 52, 79, 293, 294, 313, 315
 exploratory methods, 162
 Monte Carlo, 160
 Newton–Raphson algorithm, 20
 simulated annealing, 166
 stochastic, 268
order statistics, 44, 65
ordering
 natural, 526
 stochastic, 518
Orey's inequality, 253, 304
O-ring, 15
outlier, 384
overdispersion, 383
overfitting, 428
overparameterization, 398

Paharmacokinetics, 384
paradox of trapping states, 368
parallel chains, 462, 479
parameter
 of interest, 9, 18, 25, 31
 location–scale, 398
 natural, 8
 noncentrality, 121
 nuisance, 9, 80, 89
parameterization, 377, 388, 399, 408
 cascade, 398, 419
particle, 545, 547, 552
particle filter, *see* filter
 auxiliary, 554
partition product, 456
partitioning, 155
patch clamp recordings, 579
path, 461
 properties, 479
 simulation, 554
 single, 269
perceptron, 458
perfect simulation, 239, 471, 511, **512**
 and renewal, 474
 first occurrence, 512
performances
 of congruential generators, 73
 of estimators, 81
 of the Gibbs sampler, 367, 388, 396
 of importance sampling, 96
 of integer generators, 72
 of the Langevin diffusion, 320
 of the Metropolis–Hastings algorithm, 292, 295, 317
 of Monte Carlo estimates, 96
 of renewal control, 496
 of simulated annealing, 169
period, 72, 73, 218
Peskun's ordering, 394, 578
Physics, 37
pine tree, 452
Pluralitas non est ponenda sine neccesitate, 458
Poisson variate generation, 55
polar coordinates, 111
population Monte Carlo, 560
positivity, 225, 273, 344, 377
 condition, 345
potential function, 258, 291

power of a test, 12
principle
    duality, 351, 354, 367, 495, 512,
        524
    likelihood, 239
    MCMC, 267
    parsimony, 458
    squeeze, 53, 54
prior
    conjugate, 13, 30, 82
    Dirichlet, 424
    feedback, 169
    hierarchical, 383
    improper, 399, 403, 405
    influence of, 171
    information, 30
    Jeffreys, 28
    noninformative, 31, 383
    pseudo, 452
    reference, 14, 31
        for the calibration model, 18
    robust, 124
probability
    density function (pdf)
        monotone, 72
        unimodal, 72
    integral transform, 39
    Metropolis–Hastings acceptance,
        271
    of acceptance, *see* acceptance
        probability
    of regeneration, 473
problem
    Behrens–Fisher, 12
    Fieller's, 18
procedure
    doubling, 336
    stepping-out, 336
process
    birth-and-death, 446
    branching, 219
    forward recurrence, 253
    jump, 446
    Langevin diffusion, *see* Langevin
    Poisson, 43, 66, 386, 454
    renewal, 232
programming, 41
    parallel, 153
propagation, 571

properties
    mixing, 512
    sample path, 239
pseudo-likelihood, 7
pseudo-prior, 452

quality control, 3
quantile, 500

R, 22, 420, 509
random effects, 406
random mapping, 513
random walk, **206**, 225, 226, 250, 257,
        284, 287, 290, 292, 295, 315,
        317
    with drift, 319
    multiplicative, 249
    on a subgraph, 322
randomness, 36
range of proposal, 295
Rao–Blackwellization, **130**, 133, 296,
        402–403
    and importance, 133
    and monotonicity, 354, 361
    argument, 351
    for continuous time jump processes,
        447
    convergence assessment, 484, 486,
        496
    of densities, 403
    for population Monte Carlo, 564
    implementation, 131
    improvement, 297, 356, 357
    for mixtures, 367
    nonparametric, 295
    parametric, 490
    weight, 298, 310
rate
    acceptance, 286, 292, 293, 295,
        316, 317, 373, 382, 401
    mixing, 464
    regeneration, 471
    renewal, 494
rats carcinoma, 24
raw sequence plot, 483
recurrence, 207, **219**, 222, 225, 259
    and admissibility, 262
    and positivity, 223
    Harris, *see* Harris recurrence

null, 223
    positive, 223, 406
recursion equation, 421
recycling, 354, 485, 499
reference prior, *see* prior
reflecting boundary, 291
regeneration, 124, 246, 302, 471, 473
regime, 549
region
    confidence, 16
    highest posterior density, 16, 18
regression
    isotonic, 25, 172
    linear, 7, 27, 28, 498
    logistic, 15, 146
    Poisson, 59
    qualitative, 15
relativity of convergence assessments,
        465
reliability, 23
    of an estimation, 96
renewal, 261, 470
    and regeneration, 471
    control, 495
    probabilities, 473, 493
    rate, 494
    theory, 215, 261, 470
    time, 217, 229, 491
Renyi representation, 65
reparameterization, 159, 367, 399
resampling, 32
    systematic, 555
    unbiased, 114
resolvant, 211
    chain, 251
reversibility, 229, 244, 429
reversible jump, 540
Riemann integration, 21
Riemann sum
    and Rao–Blackwellization, 410
    control variate, 474, 483
Robbins–Monro, 201–203
    conditions, 202
robustness, 10, 90
running mean plot, 127, 128

saddlepoint, 120, 162
    approximation, 174, 282
sample

independent, 140
    preliminary, 502
    test, 501
    uniform, 141
sampling
    batch, 356
    comb, 574
    Gibbs, *see* Gibbs sampling
    importance, 299, 497
    parallel, 546
    residual, 229, 574
    stratified, 155
sandwiching argument, 520
saturation, *see* model saturation
scale, 163
scaling, 443
seeds, 75, 515
semi-Markov chain, 576
sensitivity, 90, 92
separators, 423
set
    Kendall, 260
    recurrent, 220
    small, 215, 218, 221, 235, 259, 274,
        470, 471, 490, 495
    transient, 249
    uniformly transient, 220, 249
shift
    left, 74
    right, 74
shuttle Challenger, 15, 281
signal processing, 158
Simpson's rule, 21
simulated annealing, 23, 209, 287, 315
    acceleration, 160
    and EM algorithm, 178
    and prior feedback, 171
    and tempering, 540
    principle, 163
simulation, 22, 80, 83, 268, 269, 403
    in parallel, 124, 464
    motivation for, 1, 4, 5, 80
    numerical, 462
    parallel, 464
    path, 556
    philosophical paradoxes of, 36, 37
    recycling of, 295
    Riemann sum, 135–139
    twisted, 119

univariate, 372
versus numerical methods, 19, 21, 22, 40, 85, 157
waste, 269, 295
size
    convergence, 501
    warm-up, 501
slice, 335
slice sampler, 176, 321, **322**, 326
    convergence of, 329
    drift condition, 330
    genesis, 335
    geometric ergodicity of, 332
    polar, 332, 365
    poor performance of, 332
    relationship to Gibbs sampling, 337, 343
    uniform, 364
    uniform ergodicity of, 329
    univariate, 526
small dimension problems, 22
smoothing
    backward, 572
spatial Statistics, 168
spectral analysis, 508
speed
    convergence, 295, 388, 393, 402, 489
        for empirical means, 479
    mixing, 462, 482
splitting, 219
S-Plus, 41, 509
squared error loss, 108
stability, 398, 488
    of a path, 96
stabilizing, 495
start-up table, 73
state, 549
    absorbing, 368
    initial, 209
    period, 217
    recurrent, 219
    transient, 219
    trapping, 368, 399
state-space, 549
    continuous, 210
    discrete, 224, 313, 394
    finite, 168, 206, 227, 284, 352, 495, 512, 579

stationarity, 239, 483
stationarization, 461
stationary distribution, 206, **223**, 512
    and diffusions, 206
    as limiting distribution, 207
statistic
    order, 135, 151
    sufficient, 9, 10, 100, 130
Statistical Science, 33
`Statlib`, 502
stochastic
    approximation, 174, 201
    differential equation, 318
    exploration, 159
    gradient, 287
        equation, 318
    monotonicity, 518
    optimization, 268, 271
    recursive sequence, 514, 536
    restoration, 340
    volatility, 549
stopping rule, 212, 281, 465, 491, 497, 499, 502
stopping time, 516
Student's *t*
    approximation, 283
    posterior, 300
    variate generation, 46, 65
subgraph, 321
subsampling, 462, 500
    and convergence assessment, 463
    and independence, 468
support
    connected, 275
    restricted, 97, 101
    unbounded, 49
sweeping, 397
switching models, 579

tail area, 122, 282
    approximation, 91, 112, 114, 122
tail event, 240
tail probability estimation, 93
temperature, 163, 167
    decrease rate of, 167
tempering, 540–543, 558
    power, 540
termwise Rao–Blackwellized estimator, 151

test
  χ^2, 508
  Die Hard, 37
  Kolmogorov–Smirnov, 37, 76, 466, 509
  Kuiper, 466
  likelihood ratio, 86
  for Markov chains, 500
  nonparametric, 466
  power of, 86, 90
  randomness, 38
  stationarity, 462
  uniform, 36
testing, 86
theorem
  Bayes, 12, 14, 50, 572
  Central Limit, 43, 83, 123, 235–237, 239, 242–244, 246, 261, 264, 376, 481, 491, 492, 559
  Cramér, 119
  Cramér-Wold's, 126
  ergodic, 239–241, 269, 302, 462, 481
  Fubini, 13
  fundamental, simulation, 47
  Glivenko–Cantelli, 32
  Hammersley–Clifford, 343, 344, 408, 485
  Kac's, 224, 229, 471
  Kendall, 236
  Rao–Blackwell, 130, 296
theory
  Decision, 14, 81, 83
  Neyman–Pearson, 12
  renewal, 470
There ain't no free lunch, 558
time
  computation, 486
  interjump, 317
  renewal, 216, 491
  stopping, 211, 212, 216, 459
time series, 4, 508
  for convergence assessment, 508
total variation norm, 207, 231, 236, 253
training sample, 201
trajectory (stability of), 218
transience of a discretized diffusion, 318
transition, 273

Metropolis–Hastings, 270
  pseudo-reversible, 472
transition kernel, 206, 208, 210, 490
  atom, 214
  choice of, 292
  symmetric, 207
traveling salesman problem, 308
tree swallows, 196
turnip greens, 414

unbiased estimator
  density, 486
  L_2 distance, 505
uniform ergodicity, 229, 261, 525
uniform random variable, 39, 47
  generation, 35, 36, 39
unit roots, 453
universality, 58, 275, 295
  of Metropolis–Hastings algorithms, 272

variable
  antithetic, 140, 143
  auxiliary, 3, 340, 374
  latent, 2, 327, 341, 391, 550
variance
  asymptotic, 491, 496
  between- and within-chains, 497
  finite, 94, 95, 97
  infinite, 102
  of a ratio, 125
  reduction, 90, 91, 96, 130–132, 141
    and Accept–Reject, 143
    and antithetic variables, 151
    and control variates, 145, 147
    optimal, 155
variate (control), 145
velocities of galaxies, 426, 439, 450
vertices, 422
virtual observation, 171
volatility, 550

waiting time, 51
Wealth is a mixed blessing, 433, 446, 570

You've only seen where you've been, 464, 470

Springer Texts in Statistics *(continued from page ii)*

Lehmann: Testing Statistical Hypotheses, Second Edition

Lehmann and Casella: Theory of Point Estimation, Second Edition

Lindman: Analysis of Variance in Experimental Design

Lindsey: Applying Generalized Linear Models

Madansky: Prescriptions for Working Statisticians

McPherson: Applying and Interpreting Statistics: A Comprehensive Guide, Second Edition

Mueller: Basic Principles of Structural Equation Modeling: An Introduction to LISREL and EQS

Nguyen and Rogers: Fundamentals of Mathematical Statistics: Volume I: Probability for Statistics

Nguyen and Rogers: Fundamentals of Mathematical Statistics: Volume II: Statistical Inference

Noether: Introduction to Statistics: The Nonparametric Way

Nolan and Speed: Stat Labs: Mathematical Statistics Through Applications

Peters: Counting for Something: Statistical Principles and Personalities

Pfeiffer: Probability for Applications

Pitman: Probability

Rawlings, Pantula and Dickey: Applied Regression Analysis

Robert: The Bayesian Choice: From Decision-Theoretic Foundations to Computational Implementation, Second Edition

Robert and Casella: Monte Carlo Statistical Methods, Second Edition

Rose and Smith: Mathematical Statistics with *Mathematica*

Ruppert: Statistics and Finance: An Introduction

Santner and Duffy: The Statistical Analysis of Discrete Data

Saville and Wood: Statistical Methods: The Geometric Approach

Sen and Srivastava: Regression Analysis: Theory, Methods, and Applications

Shao: Mathematical Statistics, Second Edition

Shorack: Probability for Statisticians

Shumway and Stoffer: Time Series Analysis and Its Applications

Simonoff: Analyzing Categorical Data

Terrell: Mathematical Statistics: A Unified Introduction

Timm: Applied Multivariate Analysis

Toutenburg: Statistical Analysis of Designed Experiments, Second Edition

Wasserman: All of Statistics: A Concise Course in Statistical Inference

Whittle: Probability via Expectation, Fourth Edition

Zacks: Introduction to Reliability Analysis: Probability Models and Statistical Methods

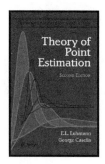